Contents

MW01553343

Contents

Regents Examinations, Answers, and Self-Analysis Charts 83

BARRON'S

Regents Exams and Answers

Chemistry—Physical

Setting Revised Edition

ALBERT S. TARENDASH
Assistant Principal—Supervision (Retired), Department of Chemistry and Physics Stuyvesant High School, New York, New York

KATE GEFELL
Chemistry Teacher, New York State Master Teacher—Chemistry, Southern Tier Region
Ithaca High School, Ithaca, New York

Published by Kaplan, Inc., d/b/a Barron's Educational Series
750 Third Avenue
New York, NY 10017
www.barronseduc.com

ISBN: 978-1-5062-6468-4

10 9 8 7 6 5 4 3 2 1

Kaplan, Inc., d/b/a Barron's Educational Series print books are available at special quantity discounts to use for sales promotions, employee premiums, or educational purposes. For more information or to purchase books, please call the Simon & Schuster special sales department at 866-506-1949.

Preface

A HELPFUL WORD TO THE STUDENT

As you are aware, the purpose of this book is to help you review for the New York State Regents examination in Chemistry—Physical Setting. You can also use it effectively to prepare for classroom, midterm, and final examinations. The book contains a number of special sections and other features designed to aid you in achieving good grades on your examinations. Included are the following:

- *How to Use This Book.* Your journey begins here. This section explains how to use the *entire* book as an effective test-preparation tool. It provides a method of identifying those areas you have mastered and those that will require additional work.
- *Test-Taking Techniques.* In this section, you will learn *how* to take the Regents examination effectively. Included are a list of materials you will need to bring to the examination and directions on how to fill out your answer sheet, how to read the questions, and how to deal with difficult items. In addition, it provides some suggestions on how to approach the extended-response questions found on Part C of the examination.
- *What to Expect on the Chemistry Examination.* This section explains the format, content, and grading of the examination.
- *New York State Physical Setting/Chemistry Core: Topic Outline and Question Index.* The topic outline provides a detailed description of the Regents Physical Setting/Chemistry Core, and the question index keys the Regents examination questions that appear in this book to the topic outline. The index will help you find questions on similar topics and will provide information about the areas that are stressed most heavily on the exam.
- *Glossary of Important Terms.* The glossary will help you understand the technical terms that may appear on the examination.
- *Reference Tables for Chemistry.* A significant part of the Regents examination requires you to be able to find and use the information contained in these reference tables. This section describes each table and the types of information that may be obtained from it.

- *Recent Regents Examinations.* These questions will provide the bulk of your study and review. When you have answered *all* of the questions that appear in this book, you will be able to approach the Regents examination with confidence!
- *Explanation of Answers and Self-Analysis Chart for Each Examination.* The detailed answer explanations will help you to *understand* why certain choices are correct, and others are incorrect. The self-analysis chart will pinpoint your strengths and weaknesses on each practice examination you take.

This book was written to provide you with the basic tools you need to perform well on the Regents Chemistry—Physical Setting examination. If you follow its advice and prepare diligently, you will achieve your goal.

Best wishes for success!

Albert S. Tarendash

How to Use This Book

1. Read the section entitled *Test-Taking Techniques* to learn how to prepare properly for an examination and how to take it with maximum efficiency.
2. Read the section entitled *What to Expect on the Chemistry Examination* to familiarize yourself with the structure and contents of this examination.
3. Read the section entitled *Reference Tables for Chemistry* to familiarize yourself with the contents and use of these tables.
   Note: On occasion, the New York State Education Department may change the content or format of the Regents examination and/or the reference tables. Your classroom teacher is your best source of information about such changes.
4. Read the section entitled *Using the Equations to Solve Chemistry Problems.*
5. Refer to the *Glossary of Important Terms* to learn the meanings of words and terms you do not understand.
6. Take the first Regents examination in this book, answering all of the questions.
7. Check your answers, and then complete the self-analysis chart at the end of the examination to pinpoint your strengths and weaknesses.
8. Read the detailed explanation of *all of the questions,* paying closest attention to the questions you answered incorrectly. Occasionally, the *wrong choices* are explained, and these explanations may help you understand why you chose an incorrect answer.
9. When you have determined your areas of weakness, refer to the *New York State Physical Setting/Chemistry Core: Topic Outline and Question Index* to locate similar questions on other recent examinations. (You can also use this outline and index to determine which areas have been stressed in recent years.)
10. Repeat steps 6–9 for the other examinations, *with the exception of the most recent test.*

11. When you have completed your studying, but no more than 1 or 2 days before the actual examination, take the most recent examination in this book *under strict examination conditions.*

After you have checked your answers to this last examination, you will have a rough idea of how you will perform on the Regents examination you will take.

Test-Taking Techniques

HELPFUL HINTS

The following pages contain 8 tips to help you achieve a good grade on the Chemistry Regents examination.

TIP 1
Be confident and prepared.

SUGGESTIONS

- Review previous tests.
- Use a clock or watch and take the most recent exam at home under examination conditions (i.e., don't have the radio or television on).
- Get a review book. (One useful book is Barron's *Let's Review Regents: Chemistry—Physical Setting.*)
- Talk over the answers to questions on these tests with someone else, such as another student in your class or someone at home.
- Finish all your homework assignments.
- Look over classroom exams that your teacher gave during the term.
- Take class notes carefully.
- Practice good study habits.
- Know that there are answers for every question.
- Be aware that the people who made up the Regents examination want you to succeed.
- Remember that thousands of students over the last few years have taken and passed a Chemistry Regents. You can pass too!
- Complete your study and review at least one day before the examination. Last-minute cramming does not help and may hurt your performance.

- On the night prior to the exam day, lay out all the things you will need.
- Go to bed early; eat wisely.
- Bring the required materials to the examination. This generally means a pen, two sharpened pencils, and a good quality eraser. If your school does not supply a calculator, be certain to bring one to the examination. Some schools require a signed Regents admission card for identification. Good advice: Assume your school will *not* supply you with any materials!
- Once you are in the exam room, arrange things, get comfortable, be relaxed, attend to personal needs (the bathroom).
- Keep your eyes on your own paper; do not let them wander over to anyone else's paper.
- Be polite in making any reasonable demands of the exam-room proctor, such as changing your seat or having window shades raised or lowered.

TIP 2
Read test instructions carefully.

SUGGESTIONS

- Be familiar with the format of the examination.
- Know how the test will be graded.
- If your school supplies an electronic scoring sheet, be certain you are familiar with the additional directions for recording and changing answers.
- If you decide to change an answer, be certain that you erase your original response completely.
- Any stray marks on your answer sheet should be erased completely.
- Be familiar with the directions for Parts B and C questions. Answer each question completely. Explanations must be written as *whole sentences* and substitutions into equations *must* include units. Be certain that your answers are clearly labeled and well organized. Place a box around numerical answers. Be neat!
- Ask for assistance from the exam-room proctor if you do not understand the directions.

TIP 3

Read each question carefully and read each choice before you record your answer.

SUGGESTIONS

- Be sure you understand *what* the question is asking.
- Try to recognize information that is *given* in the question.
- Will a chemistry reference table help you find the answer to the question?
- Some choices may look appealing yet will be incorrect.
- Try to eliminate those choices that are *obviously* incorrect.

TIP 4

Budget your test time (3 hours).

SUGGESTIONS

- Bring a watch or clock to the test.
- The Regents examination is designed to be completed in $1\frac{1}{2}$ to 2 hours.
- If you are absolutely uncertain of the answer to a question, mark your question booklet and move on to the next question.
- If you persist in trying to answer every difficult question *immediately*, you may find yourself rushing or unable to finish the remainder of the examination.
- When you have finished the examination, return to those unanswered questions.
- Good advice: If at all possible, reread the *entire* examination—and your responses—at least one more time. (This will help you eliminate those errors that result from misreading questions.)

TIP 5
Use your reasoning skills.

SUGGESTIONS

- Answer *all* questions.
- Relate (connect) the question to anything that you studied, wrote in your notebook, or heard your teacher say in class.
- Relate (connect) the question to any film, demonstration, or experiment you saw in class, any project you did, or to anything you may have learned from newspapers, magazines, or television.
- Look over the entire test to see whether one part of it can help you answer another part.

TIP 6
Use your reference tables.

SUGGESTIONS

- You should be familiar with the *content* of each table.
- Frequently, the answers to questions can be found from information contained within these tables.
- Some questions refer to specific tables.
- Other questions do not refer to a table, but can be answered by choosing and using the correct table.
- Be especially familiar with the Periodic Table of the Elements, as well as Reference Table *T*, Important Formulas and Equations.

TIP 7
Don't be afraid to guess.

SUGGESTIONS

- Eliminate obvious incorrect choices.
- If still unsure of an answer, make an educated guess.
- There is no penalty for guessing; therefore, answer *all* questions. An omitted answer gets no credit.

TIP 8
Sign the declaration.

SUGGESTIONS

- Be certain that you sign the declaration found on your answer sheet.
- Unless this declaration is signed, your paper cannot be scored.

SUMMARY OF TIPS

1. Be confident and prepared.
2. Read test instructions carefully.
3. Read each question carefully and read each choice before you record your answer.
4. Budget your test time (3 hours).
5. Use your reasoning skills.
6. Use your reference tables.
7. Don't be afraid to guess.
8. Sign the declaration.

HOW TO ANSWER PART C (EXTENDED-RESPONSE) QUESTIONS

An *extended-response question* is an examination question that requires the test taker to do more than choose among several responses or fill in a blank. You may need to perform numerical calculations, draw and interpret graphs, and provide extended written responses to a question or a problem.

Part C of the New York State Regents Examination in Chemistry—Physical Setting contains extended-response questions. This section is designed to provide you with a number of general guidelines for answering them.

SOLVING PROBLEMS INVOLVING NUMERICAL CALCULATIONS

To receive full credit you must:

- Provide the appropriate equation(s).
- Substitute values and units into the equation(s).
- Display the answer, with appropriate units and to the correct number of significant figures.
- If the answer is a vector quantity, include its direction.

You should write as legibly as possible. Teachers are human, and nothing irks them more than trying to decipher a careless, messy scrawl. It is also a good idea to identify your answer clearly, either by placing it in a box or by writing the word "answer" next to it.

A final word: If you provide the correct answer but do not show any work, you will not receive any credit for the problem!

The following is a sample problem and its model solution.

Problem

A 5.00-gram object has a density of 4.00 grams per cubic centimeter. Calculate the volume of this object.

Solution

$$d = \frac{m}{V}$$

Rearranging the equation gives

$$V = \frac{m}{d} = \frac{5.00 \text{ g}}{4.00 \text{ g/cm}^3}$$

$$\boxed{V = 1.25 \text{ cm}^3}$$

GRAPHING EXPERIMENTAL DATA

To receive full credit you must:

- Label both axes with the appropriate variables and units.
- Divide the axes so that the data ranges fill the graph as nearly as possible.
- Plot all data points accurately.
- Draw a best-fit line carefully with a straightedge. The line should pass through the origin *only if the data warrant it.*
- If a part of the question requires that the slope be calculated, calculate the slope *from the line,* not from individual data points.

A graph should have a title, and the *independent variable* is usually drawn along the x-axis.

The following is a sample problem and its model solution.

Problem

A student attempts to estimate absolute zero in the following way: He subjects a sample of gas (at constant pressure and mass) to varying temperatures and measures the gas volume at each of the temperatures. The accompanying table contains his experimental data.

Temperature/°C	Volume/mL
−100	128
−60	148
−40	156
0	204
40	222
80	272
160	310

(a) Using axes that are appropriately labeled and scaled, draw a graph that accurately displays the student's data.

(b) Estimate the student's value for absolute zero by extending the graph to the Celsius temperature at which the volume of the gas is 0 milliliter.

Solution

(a) The first step is to construct an appropriate set of axes if one is not provided. We will assume that you must start from scratch. Since temperature is the independent variable, you need to place it along the x-axis. Also, since absolute zero is $-273.15°C$, you must scale the axes properly. Here is one appropriate set of axes:

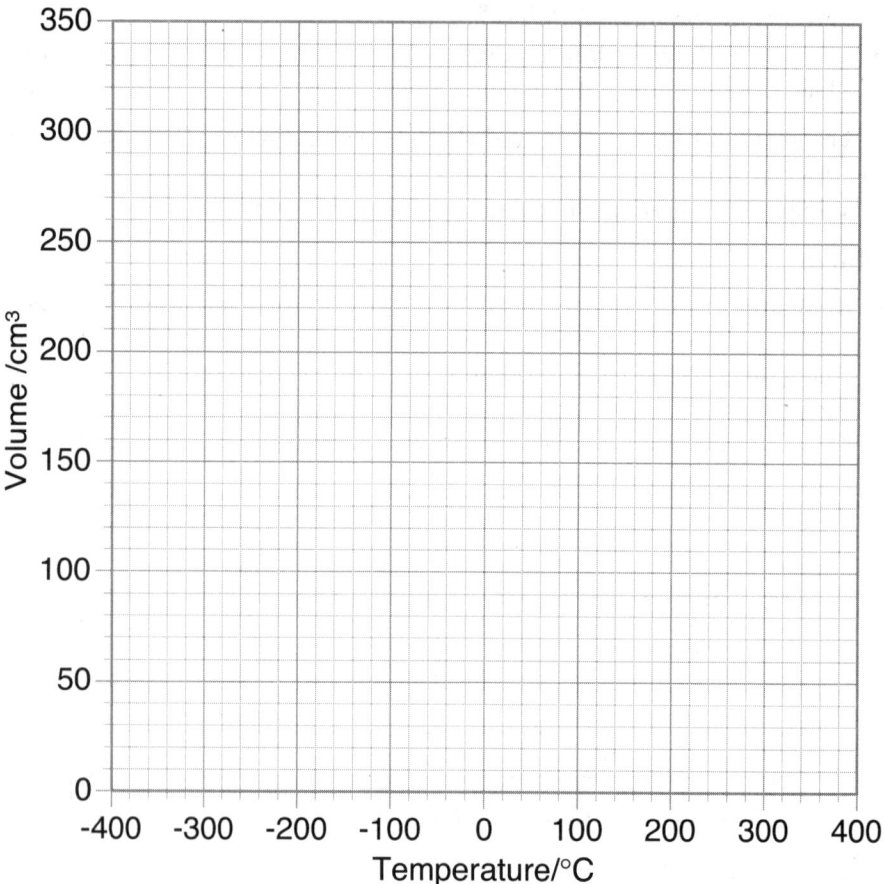

Second, you must plot the data points carefully on the axes as shown below:

Your third task is to draw the graph. You might be tempted to "connect the dots," but then you would miss the significant relationship between volume and temperature. If you examine *all* of the plotted points, you will note that they fall approximately on a *straight line.*

Therefore, the next step is to draw a *best-fit* straight-line graph. This is a graph in which the data points are most closely distributed on both sides of the line. The accompanying graph shows the best-fit straight line.

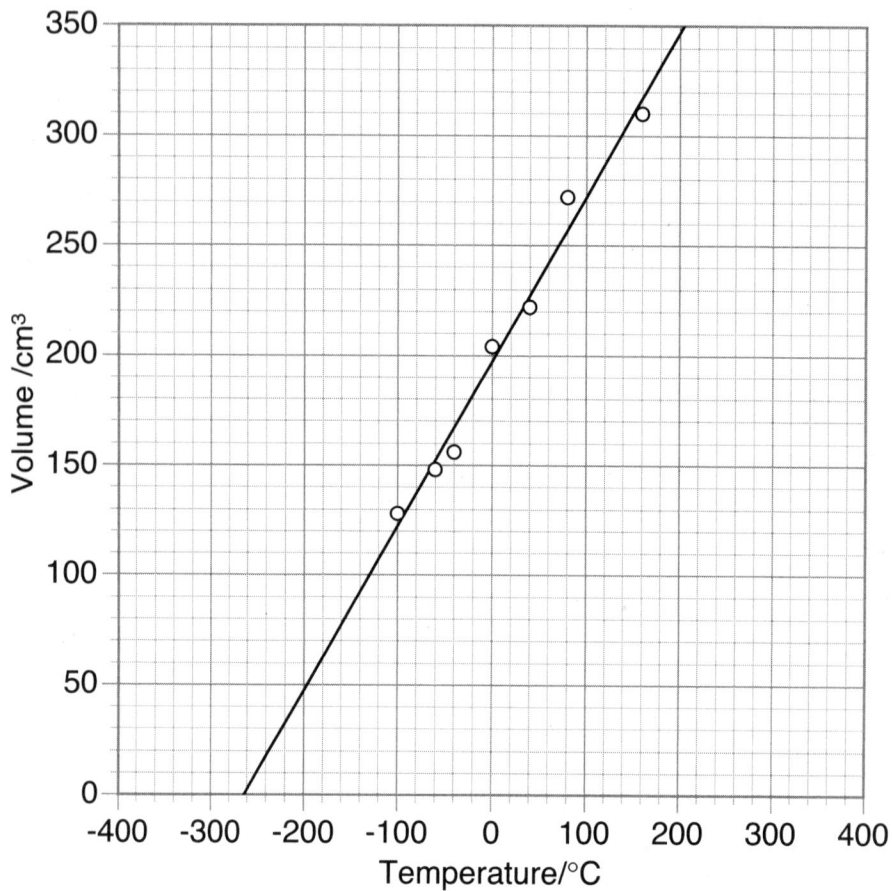

Note that the graph extending beyond the data points (above and below) is a *dashed* line. Such extensions are known as *extrapolated data*, and they are based on the assumption that the gas will continue to behave as it did within the experimental range for which the student collected data.

(You may ask: Why *this* particular line? It seems as though many lines could have been drawn using the plotted data. Actually, there is only *one* best-fit straight line, and it is calculated by using an advanced statistical technique known as *linear regression*. This technique was used to draw the line shown above. At this point, it is sufficient for you to provide an "eyeball" estimate of the best-fit straight line.)

    (b) Your final task is to inspect the graph closely and to estimate absolute zero. The calculated value is −263°C. (This corresponds to an experimental error of 3.7%.)

DRAWING DIAGRAMS

To receive full credit you must:

- Draw your diagrams neatly and label them clearly.
- Bring a straightedge and a protractor with you so that you can draw neat, accurate diagrams.

The accompanying diagram represents a zinc–copper electrochemical cell containing an agar–KCl salt bridge. This is the type of diagram you may be asked to draw as part of an examination question.

Zn(s) → Zn²⁺(aq) + 2e⁻ (oxidation half-reaction)

Cu²⁺(aq) + 2e⁻ → Cu(s) (reduction half-reaction)

WRITING A FREE-RESPONSE ANSWER

To receive full credit you must:

- Use complete, clear sentences that make sense to the reader.
- Use correct chemistry in your explanations.

A sample question and a model answer follow.

Question

Describe in detail the technique used to determine the concentration of a dilute hydrochloric acid solution using a dilute sodium hydroxide solution of known concentration.

Name any equipment or other chemicals that you would use. *You need not describe any mathematical calculations.*

Model Answer

Since this question requires an extended answer containing a series of steps, we decide to use an outline form.

The technique is called *titration*, and it is described in the steps given below.

- Place a known volume of the hydrochloric acid solution in a beaker of suitable size.
- Add a drop or two of an acid–base indicator such as phenolphthalein.
- Pour the sodium hydroxide solution carefully into a burette, using a small funnel.
- Open the stopcock to allow the trapped air to escape. Then close the stopcock and wipe the tip of the burette with a tissue to remove any clinging liquid.
- Record the initial volume in the burette.
- Add the base slowly to the acid solution, with continuous stirring, until the phenolphthalein just changes from colorless to faint pink.
- Record the final volume in the burette.
- Repeat the experiment at least once.
- Rinse the apparatus with water to remove all traces of acid and base.
- Calculate the concentration of the hydrochloric acid solution.

Writing and Balancing Equations in a Chemistry Examination

For all of the suggestions that follow, the reaction that occurs between aqueous solutions of sodium sulfate and barium nitrate is used as an example and is referred to as the given reaction.

- If you are asked to write a *word equation*, be certain to include the correct names of the reactants and products and their phases in the reaction.

For the given reaction, the word equation is

$$\text{barium nitrate(aq) + sodium sulfate(aq)} \rightarrow$$
$$\text{barium sulfate(s) + sodium nitrate(aq)}$$

- If you are asked to write a *balanced* equation, you are usually expected to balance using *smallest whole-number coefficients*.

  For the given reaction, the balanced equation is

  $$Ba(NO_3)_2\,(aq) + Na_2SO_4\,(aq) \rightarrow BaSO_4\,(s) + 2NaNO_3\,(aq)$$

- If you are asked to write an *ionic* equation occurring in aqueous solution, you must reduce everything to its component ions except *insoluble* compounds, such as $BaSO_4(s)$, and (of course!) *covalently* bonded substances, such as $H_2O(\ell)$.

  For the given reaction, the ionic equation is

  $$Ba^{2+}(aq) + 2NO_3^-(aq) + 2Na^+(aq) + SO_4^{2-} \rightarrow$$
  $$BaSO_4(s) + 2Na^+(aq) + 2NO_3^-(aq)$$

- If you are asked to write a *net* ionic equation, you should omit all *spectator ions*, that is, all ions appearing *unchanged on both sides of the equation*.

  For the given reaction, the net ionic equation is

  $$Ba^{2+}(aq) + SO_4^{2-}(aq) \rightarrow BaSO_4(s)$$

What to Expect on the Chemistry Examination

FORMAT OF THE CHEMISTRY EXAMINATION

The chemistry examination will be 3 hours long and will include three parts: A, B, and C. You should be prepared to answer questions in multiple-choice format, as well as answer questions that require a more extended response.

Questions will be content- and skills-based, and you may be required to graph data, complete a data table, label or draw diagrams, design experiments, make calculations, or write short or extended responses.

In addition, you may be required to hypothesize, interpret, analyze, evaluate data, or apply your scientific knowledge and skills to real-world situations.

[In the future, a Part D will be added, which will focus on assessment of laboratory skills. As more information becomes available, the New York State Education Department will inform schools of the development status of the performance test.]

You will be required to answer ALL of the questions on the Physical Setting/Chemistry Regents examination.

Physical Setting/Chemistry Regents Examination Format

PART	ITEM TYPE(S)	DESCRIPTION OF THE ITEMS	APPROXIMATE PERCENT OF TOTAL TEST RAW SCORE
A	Multiple-choice questions	Content-based questions assessing your knowledge and understanding of core material	35
B	Multiple-choice and constructed-response questions	Content- and skills-based questions assessing your ability to apply, analyze, and evaluate material	30
C	Constructed-response and/or extended constructed-response questions	Content-based and application questions assessing your ability to apply knowledge of science concepts and skills to address real-world situations	20

The maximum *raw* score on the examination is 85 points. A teacher's chart will be provided for converting your *raw* score to a *scaled* score that has a maximum of 100 points. A sample conversion table taken from the June 2004 Regents Chemistry—Physical Setting examination is shown below:

Sample Conversion Table

Raw Score	Scaled Score	Raw Score	Scaled Score	Raw Score	Scaled Score	Raw Score	Scaled Score
85	100	63	74	41	58	19	35
84	98	62	73	40	57	18	34
83	97	61	73	39	56	17	33
82	95	60	72	38	55	16	31
81	94	59	71	37	55	15	30
80	93	58	70	36	54	14	28
79	91	57	69	35	53	13	26
78	90	56	69	34	52	12	25
77	89	55	68	33	51	11	23
76	87	54	67	32	50	10	21
75	86	53	66	31	49	9	19
74	85	52	66	30	48	8	17
73	84	51	65	29	47	7	15
72	83	50	64	28	46	6	13
71	82	49	64	27	45	5	11
70	81	48	63	26	44	4	9
69	80	47	62	25	43	3	7
68	79	46	61	24	42	2	5
67	78	45	61	23	41	1	2
66	77	44	60	22	39	0	0
65	76	43	59	21	38		
64	75	42	58	20	37		

The table is used to convert the number of points you actually received on the examination (your "raw" score) to your final score on the examination (your "scaled" score). **Note that this table will change from one examination to another.**

TOPICS COVERED ON THE CHEMISTRY EXAMINATION

All of the questions on the Chemistry examination will test major understandings, skills, and real-world applications drawn from the following 12 subject areas:

M.	Math Skills
R.	Reading Skills
I.	Atomic Concepts
II.	Periodic Table
III.	Moles/Stoichiometry
IV.	Chemical Bonding
V.	Physical Behavior of Matter
VI.	Kinetics/Equilibrium
VII.	Organic Chemistry
VIII.	Oxidation–Reduction
IX.	Acids, Bases, and Salts
X.	Nuclear Chemistry

It is suggested that you read the Topic Outline found on pages 20–38 in order to learn the exact nature of the material that is subject to testing.

New York State Physical Setting/ Chemistry Core

TOPIC OUTLINE

The Topic Outline on pages 20–38 is adapted from Appendix B of the New York State Physical Setting/Chemistry Core. All Regents Chemistry—Physical Setting Examinations are based on this core. The topic outline is divided into 12 sections:

M.	Math Skills
R.	Reading Skills
I.	Atomic Concepts
II.	Periodic Table
III.	Moles/Stoichiometry
IV.	Chemical Bonding
V.	Physical Behavior of Matter
VI.	Kinetics/Equilibrium
VII.	Organic Chemistry
VIII.	Oxidation—Reduction
IX.	Acids, Bases, and Salts
X.	Nuclear Chemistry

Each section contains one or more of the following items:

- The *Major Understandings* that you must have mastered for the examination
- The *Skills* that you need to be able to demonstrate during the examination
- The *Key Points to Remember* that remind you of key chemistry concepts that you should be familiar with to succeed on this examination.

M. MATHEMATICS SKILLS NEEDED FOR CHEMISTRY

M.1 Organize, graph, and analyze data gathered from laboratory activities or other sources.

- Identify independent and dependent variables.
- Create appropriate axes with label and scale.
- Identify graph points clearly.

M.2 Interpret a graph constructed from experimentally determined data.

- Identify direct and inverse relationships.
- Apply data showing trends to predict information.

M.3 Measure and record experimental data and use the data in calculations.

- Choose appropriate measurement scales and use units in recording.
- Show mathematical work stating formula and steps for solution.
- Estimate answers.
- Use appropriate equations and significant digits.
- Identify relationships within variables from data tables.
- Calculate percent error.

M.4 Recognize and convert various scales of measurement.

- Convert between Celsius (°C) and Kelvin (K).
- Convert among kilometers (km), meters (m), centimeters (cm), and millimeters (mm).
- Convert between grams (g) and kilograms (kg).
- Convert between kilopascals (kPa) and atmospheres (atm).

M.5 Employ critical thinking skills in solving problems.

- Apply algebraic or geometric concepts in the solution of mathematical problems.
- Use knowledge of geometric arrangements to predict particle properties or behavior.
- State the assumptions on which a particular mathematical equation is based.
- Evaluate the appropriateness of an answer to a solved problem.

R. READING SKILLS NEEDED FOR CHEMISTRY

R.1 Reading Comprehension: Literal

- Extract an answer from the text when the answer is "right there" in the information provided.
- Read and interpret information provided in the Reference Tables for Chemistry.

R.2 Reading Comprehension: Inferential/Interpretive

- Understand and make connections between prior content knowledge and information from the text when these connections are not explicitly stated.
- Understand and apply new information provided in the text and solve problems using this information.

R.3 Reading Comprehension: Lexical

- Understand and apply key vocabulary in a reading passage.

I. Atomic Concepts

	MAJOR UNDERSTANDINGS	SKILLS The student should be able to:	KEY POINTS TO REMEMBER
I.1	The modern model of the atom has evolved over a long period of time through the work of many scientists.	relate experimental evidence to models of the atom	Examples: 1. J.J. Thomson and cathode ray experiment 2. Ernest Rutherford and gold foil experiment
I.2	Each atom has a nucleus, with an overall positive charge, surrounded by negatively charged electrons.	use models to describe the structure of an atom	
I.3	Subatomic particles contained in the nucleus include protons and neutrons.		
I.4	The proton is positively charged, and the neutron has no charge. The electron is negatively charged.		
I.5	Protons and electrons have equal but opposite charges. The number of protons is equal to the number of electrons in an atom.	determine the number of protons or electrons in an atom or ion when given one of these values	Positive ions form when atoms lose electrons. Negative ions form when atoms gain electrons.
I.6	The mass of each proton and each neutron is approximately equal to one atomic mass unit. An electron is much less massive than a proton or neutron.	calculate the mass of an atom, the number of neutrons, or the number of protons, given the other two values	
I.7	In the wave-mechanical model (electron cloud), the electrons are in orbitals, which are defined as regions of most probable electron location (ground state).	interpret electron configurations listed on the Periodic Table	
I.8	Each electron in an atom has its own distinct amount of energy.		Electron configurations begin with the lowest energy shell near the nucleus and continue in order of increasing energy and distance from the nucleus.
I.9	When an electron in an atom gains a specific amount of energy, the electron is at a higher energy state (excited state).	distinguish between ground state and excited state electron configurations, e.g., 2−8−2 vs. 2−7−3	Excited state configurations contain the same number of electrons as ground state configurations.

I. Atomic Concepts (*Continued*)

	MAJOR UNDERSTANDINGS	SKILLS The student should be able to:	KEY POINTS TO REMEMBER
I.10	When an electron returns from a higher energy state to a lower energy state, a specific amount of energy is emitted. This emitted energy can be used to identify an element.	identify an element by comparing its bright-line spectrum to given spectra	
I.11	The outermost electrons in an atom are called the valence electrons. In general, the number of valence electrons affects the chemical properties of an element.	draw a Lewis electron-dot structure of an atom distinguish between valence and non-valence electrons, given an electron configuration, e.g., 2−8−2	
I.12	Atoms of an element that contain the same number of protons but a different number of neutrons are called isotopes of that element.		
I.13	The average atomic mass of an element is the weighted average of the masses of its naturally occurring isotopes.	given an atomic mass, determine the most abundant isotope calculate the atomic mass of an element, given the masses and ratios of naturally occurring isotopes	

II. Periodic Table

	MAJOR UNDERSTANDINGS	SKILLS The student should be able to:	KEY POINTS TO REMEMBER
II.1	The placement or location of an element on the Periodic Table gives an indication of physical and chemical properties of that element. The elements on the Periodic Table are arranged in order of increasing atomic number.	explain the placement of an unknown element in the Periodic Table based on its properties	
II.2	The number of protons in an atom (atomic number) identifies the element. The sum of the protons and neutrons in an atom (mass number) identifies an Isotope. Common notations that represent isotopes include: ^{14}C, $^{14}_{6}C$, carbon-14, C-14.	interpret and write isotopic notation	Subtract the atomic number from the mass number to find the number of neutrons in an isotope! The term "nuclear charge" refers to the number of protons in an atom's nucleus. For example, a carbon atom with 6 protons would have a nuclear charge of 6+.

II. Periodic Table (*Continued*)

	MAJOR UNDERSTANDINGS	**SKILLS** The student should be able to:	**KEY POINTS TO REMEMBER**
II.3	Elements can be classified by their properties and located on the Periodic Table as metals, nonmetals, metalloids (B, Si, Ge, As, Sb, Te), and noble gases.	classify elements as metals, nonmetals, metalloids, or noble gases by their properties	
II.4	Elements can be differentiated by their physical properties. Physical properties of substances, such as density, conductivity, malleability, solubility, and hardness, differ among elements.	describe the states of the elements at STP suggest a simple separation strategy to separate two or more elements based on physical properties	
II.5	Elements can be differentiated by chemical properties. Chemical properties describe how an element behaves during a chemical reaction.		Metal atoms lose electrons when they bond.
II.6	Some elements exist as two or more forms in the same phase. These forms differ in their molecular or crystal structure and hence in their properties.		Allotropes of carbon include graphite, diamond, and fullerene.
II.7	For Groups 1, 2, and 13–18 on the Periodic Table, elements within the same group have the same number of valence electrons (helium is an exception) and, therefore, similar chemical properties.	determine the group of an element, given the chemical formula of a compound, e.g., XCl or XCl_2	
II.8	The succession of elements within the same group demonstrates characteristic trends: differences in atomic radius, ionic radius, electronegativity, first ionization energy, and metallic/nonmetallic properties.	compare and contrast properties of elements within a group or a period for Groups 1, 2, 13–18 on the Periodic Table	Increasing the distance between the nucleus and an atom's valence electrons causes the ionization energy and the electronegativity to decrease down each group of the Periodic Table.
II.9	The succession of elements across the same period demonstrates characteristic trends: differences in atomic radius, ionic radius, electronegativity, first ionization energy, and metallic/nonmetallic properties.		Increasing nuclear charge causes electronegativity and ionization energy to increase across each period of the Periodic Table.

III. Moles/Stoichiometry

	MAJOR UNDERSTANDINGS	**SKILLS** The student should be able to:	**KEY POINTS TO REMEMBER**
III.1	A compound is a substance composed of two or more different elements that are chemically combined in a fixed proportion. A chemical compound can be broken down by chemical means. A chemical compound can be represented by a specific chemical formula and assigned a name based on the IUPAC system.	determine how many total atoms are in a compound determine how many atoms of each type are in a compound use IUPAC nomenclature rules to name binary ionic, binary molecular, and polyatomic ionic compounds	Compounds, along with elements, are pure substances.
III.2	Types of chemical formulas include empirical, molecular, and structural.	identify examples of empirical, structural, or molecular formulas convert a structural formula into a molecular formula or an empirical formula	
III.3	The empirical formula of a compound is the simplest whole-number ratio of atoms of the elements in a compound. It may be different from the molecular formula, which is the actual ratio of atoms in a molecule of that compound.	determine the molecular formula, given the empirical formula and molecular mass determine the empirical formula from a molecular formula	
III.4	In all chemical reactions there is a conservation of mass, energy, and charge.	interpret balanced chemical equations in terms of conservation of matter and energy	Atoms In = Atoms Out!
III.5	A balanced chemical equation represents conservation of atoms. The coefficients in a balanced chemical equation can be used to determine mole ratios in the reaction.	balance equations, given the formulas for reactants and products interpret balanced chemical equations in terms of conservation of matter and energy create and use models of particles to demonstrate balanced equations calculate simple mole–mole stoichiometry problems, given a balanced equation	Coefficients in a balanced chemical equation represent moles or particles, *not* grams!
III.6	The formula mass of a substance is the sum of the atomic masses of its atoms. The molar mass (gram-formula mass) of a substance equals one mole of that substance.	calculate the formula mass and the gram-formula mass use the molar mass to convert from grams to moles or from moles to grams	Formula masses are reported in atomic mass units (amu). Molar masses are reported in grams/mol.

III. Moles/Stoichiometry (*Continued*)

	MAJOR UNDERSTANDINGS	SKILLS The student should be able to:	KEY POINTS TO REMEMBER
III.7	The percent composition by mass of each element in a compound can be calculated mathematically.	determine the number of moles of a substance, given its mass determine the mass of a given number of moles of a substance use lab data to determine the percent of water in a hydrate	
III.8	Types of chemical reactions include synthesis, decomposition, single replacement, and double replacement.	identify types of chemical reactions	Synthesis reactions create just one product: $A + B \rightarrow AB$ Decomposition reactions have just a single reactant: $AB \rightarrow A + B$ Single replacement reactions have uncombined elements on *both* sides of the reaction: $A + BC \rightarrow AC + B$ Double replacement reactions show "partners switching places": $AB + CD \rightarrow AD + CB$

IV. Chemical Bonding

	MAJOR UNDERSTANDINGS	SKILLS The student should be able to:	KEY POINTS TO REMEMBER
IV.1	Compounds can be differentiated by their chemical and physical properties.	distinguish among ionic, molecular, and metallic substances, given their properties	
IV.2	Two major categories of compounds are ionic and molecular (covalent) compounds.		
IV.3	Chemical bonds are formed when valence electrons are: transferred from one atom to another (ionic); shared between atoms (covalent); mobile within a metal (metallic).	demonstrate bonding concepts using Lewis dot structures representing valence electrons: transferred (ionic bonding); shared (covalent bonding); in a stable octet	Metallic bonds hold samples of pure metals together. Ionic bonds form between metal ions and nonmetal ions. Covalent bonds form between nonmetal or metalloid atoms.

IV. Chemical Bonding (*Continued*)

	MAJOR UNDERSTANDINGS	SKILLS The student should be able to:	KEY POINTS TO REMEMBER
IV.4	In a multiple covalent bond, more than one pair of electrons are shared between two atoms. Unsaturated organic compounds contain at least one double or triple bond.	evaluate the number of electrons shared, or the number of electron pairs shared, in single, double, or triple covalent bonds	
IV.5	Molecular polarity can be determined by the shape and distribution of the charge. Symmetrical (nonpolar) molecules include CO_2, CH_4, and diatomic elements. Asymmetrical (polar) molecules include HCl, NH_3, H_2O.		Molecular symmetry cancels polarity! Draw Lewis electron-dot diagrams to determine molecular shape.
IV.6	When an atom gains one or more electrons, it becomes a negative ion and its radius increases. When an atom loses one or more electrons, it becomes a positive ion and its radius decreases.		
IV.7	When a bond is broken, energy is absorbed. When a bond is formed, energy is released.		
IV.8	Atoms attain a stable valence electron configuration by bonding with other atoms. Noble gases have stable valence electron configurations and tend not to bond.	determine the noble gas configuration an atom will achieve when bonding	Remember the octet rule! Atoms (except hydrogen) tend to bond so as to achieve the stable octet of eight valence electrons.
IV.9	Physical properties of substances can be explained in terms of chemical bonds and intermolecular forces. These properties include conductivity, malleability, solubility, hardness, melting point, and boiling point.	compare and contrast the physical properties of different compounds based on the types of chemical bonds and/or intermolecular forces they have	
IV.10	Electron-dot diagrams (Lewis structures) can represent the valence electron arrangement in elements, compounds, and ions.	demonstrate bonding concepts, using Lewis dot structures representing valence electrons: transferred (ionic bonding); shared (covalent bonding); in a stable octet	Check! Does your drawing show all of the valence electrons? Does each atom (except H) have 8 valence electrons?

IV. Chemical Bonding (*Continued*)

	MAJOR UNDERSTANDINGS	SKILLS The student should be able to:	KEY POINTS TO REMEMBER
IV.11	Electronegativity indicates how strongly an atom of an element attracts electrons in a chemical bond. Electronegativity values are assigned according to arbitrary scales.		
IV.12	The electronegativity difference between two bonded atoms is used to assess the degree of polarity in the bond.	distinguish between nonpolar covalent bonds (two of the same nonmetals) and polar covalent bonds	Bond polarity and ionic character increase with an increasing difference in electronegativity. Ionic character refers to an atom's tendency to transfer electrons when forming a chemical bond. Ionic character, like molecular polarity, is determined by finding the electronegativity difference between the atoms in the bond.
IV.13	Metals tend to react with nonmetals to form ionic compounds. Nonmetals tend to react with other nonmetals to form molecular (covalent) compounds. Ionic compounds containing polyatomic ions have both ionic and covalent bonding.		
IV.14	Ionic compounds will be soluble in water when the attraction between the ions and water molecules is greater than the attractions between ions in the solid compound.	predict solubility/insolubility of ionic compounds using Reference Table *F* identify precipitates in double replacement reactions	

V. Physical Behavior of Matter

	MAJOR UNDERSTANDINGS	SKILLS The student should be able to:	KEY POINTS TO REMEMBER
V.1	Matter is classified as a pure substance or as a mixture of substances.		A pure substance can either be an element or a compound.
V.2	The three phases of matter (solids, liquids, and gases) have different properties.	use a simple particle model to differentiate properties of a solid, a liquid, and a gas	

V. Physical Behavior of Matter (*Continued*)

		MAJOR UNDERSTANDINGS	SKILLS The student should be able to:	KEY POINTS TO REMEMBER
	V.3	A pure substance (element or compound) has a constant composition and has constant properties throughout a given sample and from sample to sample.	use particle models/diagrams to differentiate elements, compounds, and mixtures	
	V.4	Elements are substances that are composed of atoms that have the same atomic number. Elements cannot be broken down by chemical change.		
	V.5	Mixtures are composed of two or more different substances that can be separated by physical means. When different substances are mixed together, a homogeneous or heterogeneous mixture is formed.		
	V.6	The proportions of components in a mixture can be varied. Each component in a mixture retains its original properties.		
	V.7	Differences in properties such as density, particle size, molecular polarity, boiling point and freezing point, and solubility permit physical separation of the components of the mixture.	describe the process and use of filtration, distillation, and chromatography in the separation of a mixture	
	V.8	A solution is a homogeneous mixture of a solute dissolved in a solvent. The solubility of a solute in a given amount of solvent is dependent on the temperature, the pressure, and the chemical natures of the solute and solvent.	interpret and construct solubility curves use solubility curves to distinguish saturated, supersaturated, and unsaturated solutions apply the adage "like dissolves like" to real-world situations	Solid solutes increase in solubility and gas solutes decrease in solubility as the temperature increases.
	V.9	The concentration of a solution may be expressed as: molarity (M), percent by volume, percent by mass, or parts per million (ppm).	describe the preparation of a solution, given the molarity interpret solution concentration data calculate solution concentrations in molarity (M), percent mass, and parts per million (ppm)	

V. Physical Behavior of Matter (*Continued*)

	MAJOR UNDERSTANDINGS	**SKILLS** The student should be able to:	**KEY POINTS TO REMEMBER**
V.10	The addition of a nonvolatile solute to a solvent causes the boiling point of the solvent to increase and the freezing point of the solvent to decrease. The greater the concentration of solute particles, the greater the effect.		Electrolytes have a greater effect on colligative properties than nonelectrolytes do because they dissociate into ions in aqueous solution.
V.11	Energy can exist in different forms, such as chemical, electrical, electromagnetic, heat, mechanical, and nuclear.		
V.12	Heat is a transfer of energy (usually thermal energy) from a body of higher temperature to a body of lower temperature. Thermal energy is associated with the random motion of atoms and molecules.	distinguish between heat energy and temperature in terms of molecular motion and amount of matter qualitatively interpret heating and cooling curves in terms of changes in kinetic and potential energy, heat of vaporization, heat of fusion, and phase changes	
V.13	Temperature is a measure of the average kinetic energy of the particles in a sample of matter. Temperature is not a form of energy.	distinguish between heat energy and temperature in terms of molecular motion and amount of matter explain phase changes in terms of the changes in energy and intermolecular distance convert between Kelvin and Celsius temperature scales using the formula on Reference Table *T*	
V.14	The concept of an ideal gas is a model to explain behavior of gases. A real gas is most like an ideal gas when the real gas is at low pressure and high temperature.		

V. Physical Behavior of Matter (*Continued*)

	MAJOR UNDERSTANDINGS	**SKILLS** The student should be able to:	**KEY POINTS TO REMEMBER**
V.15	Kinetic molecular theory (KMT) for an ideal gas states all gas particles: • are in random, constant, straight-line motion • are separated by great distances relative to their size; the volume of gas particles is considered negligible • have no attractive forces between them • have collisions that may result in a transfer of energy between particles, but the total energy of the system remains constant.		Pressure is caused by collisions of gas particles against the inner walls of a container.
V.16	Kinetic molecular theory describes the relationships of pressure, volume, temperature, velocity, and frequency and force of collisions among gas molecules.	explain the gas laws in terms of KMT solve problems, using the combined gas law	
V.17	Equal volumes of gases at the same temperature and pressure contain an equal number of particles.	convert temperatures in Celsius degrees (°C) to kelvins (K), and kelvins to Celsius degrees	The molar volume of a gas at STP is equal to 22.4 L/mol.
V.18	The concepts of kinetic and potential energy can be used to explain physical processes that include: fusion (melting); solidification (freezing); vaporization (boiling, evaporation), condensation, sublimation, and deposition.	qualitatively interpret heating and cooling curves in terms of changes in kinetic and potential energy, heat of vaporization, heat of fusion, and phase changes calculate the heat involved in a phase or temperature change for a given sample of matter explain phase change in terms of the changes in energy and intermolecular distances	At a substance's melting point, the solid and liquid phases are in equilibrium. At a substance's boiling point, the liquid and gas phases are in equilibrium.
V.19	A physical change results in the rearrangement of existing particles in a substance. A chemical change results in the formation of different substances with changed properties.		

V. Physical Behavior of Matter (*Continued*)

	MAJOR UNDERSTANDINGS	SKILLS The student should be able to:	KEY POINTS TO REMEMBER
V.20	Chemical and physical changes can be exothermic or endothermic.	distinguish between endothermic and exothermic reactions, using energy terms in a reaction equation, ΔH, potential energy diagrams or experimental data	Exothermic reactions have a $-\Delta H$. Endothermic reactions have a $+\Delta H$.
V.21	The structure and arrangement of particles and their interactions determine the physical state of a substance at a given temperature and pressure.	use a simple particle model to differentiate properties of solids, liquids, and gases	
V.22	Intermolecular forces created by the unequal distribution of change result in varying degrees of attraction between molecules. Hydrogen bonding is an example of a strong intermolecular force.	explain vapor pressure, evaporation rate, and phase changes in terms of intermolecular forces use vapor pressure curves (Reference Table *H*) to determine the boiling point of a substance	
V.23	Physical properties of substances can be explained in terms of chemical bonds and intermolecular forces. These properties include conductivity, malleability, solubility, hardness, melting point, and boiling point.	compare the physical properties of substances based upon chemical bonds and intermolecular forces	

VI. Kinetics/Equilibrium

	MAJOR UNDERSTANDINGS	SKILLS The student should be able to:	KEY POINTS TO REMEMBER
VI.1	Collision theory states that a reaction is most likely to occur if reactant particles collide with the proper energy and orientation.	use collision theory to explain how various factors, such as temperature, surface area, and concentration, influence the rate of reaction	The reaction rate depends upon the number of effective collisions between reacting particles.
VI.2	The rate of a chemical reaction depends on several factors: temperature, concentration, nature of reactants, surface area, and the presence of a catalyst.		Reactions between ionic substances are faster than reactions between molecular substances.
VI.3	Some chemical and physical changes can reach equilibrium.	identify examples of physical equilibria as solution equilibrium and phase equilibrium, including the concept that a saturated solution is at equilibrium	

VI. Kinetics/Equilibrium (*Continued*)

	MAJOR UNDERSTANDINGS	SKILLS The student should be able to:	KEY POINTS TO REMEMBER
VI.4	At equilibrium, the rate of the forward reaction equals the rate of the reverse reaction. The measurable quantities of reactants and products remain constant at equilibrium.	describe the concentration of particles and rates of opposing reactions in an equilibrium system	Equilibrium means equal and opposite rates, *not* equal amounts!
VI.5	Le Châtelier's principle can be used to predict the effect of stress (change in pressure, volume, concentration, and temperature) on a system at equilibrium.	qualitatively describe the effect of stress on equilibrium, using Le Châtelier's principle	A "shift to the right" means that product is being produced and reactants are being used up.
VI.6	Energy released or absorbed by a chemical reaction can be represented by a potential energy diagram.	read and interpret potential energy diagrams: PE of reactants and products, activation energy (with or without a catalyst), heat of reaction	
VI.7	Energy released or absorbed by a chemical reaction (heat of reaction) is equal to the difference between the potential energy of the products and the potential energy of the reactants.	use Reference Table I to find the heat of reaction for specified moles produced from a particular reaction	
VI.8	A catalyst provides an alternate reaction pathway that has a lower activation energy than an uncatalyzed reaction.	draw a line (to show the effect of adding a catalyst) on a potential energy diagram	
VI.9	Entropy is a measure of the randomness or disorder of a system. A system with greater disorder has greater entropy.	compare the entropy of phases of matter	
VI.10	Systems in nature tend to undergo changes toward lower energy and higher entropy.		

VII. Organic Chemistry

	MAJOR UNDERSTANDINGS	SKILLS The student should be able to:	KEY POINTS TO REMEMBER
VII.1	Organic compounds contain carbon atoms that bond to one another in chains, rings, and networks to form a variety of structures. Organic compounds can be named using the IUPAC system.	classify an organic compound based on its structural or condensed structural formula	

VII. Organic Chemistry (*Continued*)

	MAJOR UNDERSTANDINGS	SKILLS The student should be able to:	KEY POINTS TO REMEMBER
VII.2	Hydrocarbons are compounds that contain only carbon and hydrogen. Saturated hydrocarbons contain only single carbon–carbon bonds. Unsaturated hydrocarbons contain at least one multiple carbon–carbon bond.	draw structural formulas for alkanes, alkenes, and alkynes containing a maximum of ten carbon atoms use Reference Table Q to classify hydrocarbons as alkanes, alkenes, or alkynes	
VII.3	Organic acids, alcohols, esters, aldehydes, ketones, ethers, halides, amines, amides, and amino acids are types of organic compounds that differ in their structures. Functional groups impart distinctive physical and chemical properties to organic compounds.	classify an organic compound based on its structural or condensed structural formula draw a structural formula with the functional group(s) on a straight chain hydrocarbon backbone, when given the correct IUPAC name for the compound	Use Reference Table R for help in naming and drawing organic compounds.
VII.4	Isomers of organic compounds have the same molecular formula, but different structures and properties.		
VII.5	In a multiple covalent bond, more than one pair of electrons are shared between two atoms. Unsaturated organic compounds contain at least one double or triple bond.		
VII.6	Types of organic reactions include: addition, substitution, polymerization, esterification, fermentation, saponification, and combustion.	identify types of organic reactions determine a missing reactant or product in a balanced equation	

VIII. Oxidation–Reduction

	MAJOR UNDERSTANDINGS	SKILLS The student should be able to:	KEY POINTS TO REMEMBER
VIII.1	An oxidation–reduction (redox) reaction involves transfer of electrons (e^-).	determine a missing reactant or product in a balanced equation	
VIII.2	Reduction is the gain of electrons.		If an element's oxidation number is getting more negative (or less positive), it is reducing (gaining electrons).

VIII. Oxidation–Reduction (*Continued*)

	MAJOR UNDERSTANDINGS	SKILLS The student should be able to:	KEY POINTS TO REMEMBER
VIII.3	A half-reaction can be written to represent reduction.	write and balance half-reactions for oxidation and reduction of free elements and their monatomic ions	Reduction half-reactions show how many electrons are gained.
VIII.4	Oxidation is the loss of electrons.		If an element's oxidation number is getting less negative (or more positive), it is oxidizing (losing electrons).
VIII.5	A half-reaction can be written to represent oxidation.	write and balance half-reactions for the oxidation and reduction of free elements and their monatomic ions	Oxidation half-reactions show how many electrons are lost.
VIII.6	In a redox reaction, the number of electrons lost is equal to the number of electrons gained.		Electrons lost = electrons gained means that charge is conserved.
VIII.7	Oxidation numbers (states) can be assigned to atoms and ions. Changes in oxidation numbers indicate that oxidation and reduction have occurred.	identify which reactions are redox reactions	
VIII.8	An electrochemical cell can be either voltaic or electrolytic. In an electrochemical cell, oxidation occurs at the anode and reduction at the cathode	compare and contrast voltaic and electrolytic cells	
VIII.9	A voltaic cell spontaneously converts chemical energy to electrical energy.	identify and label the parts of a voltaic cell (cathode, anode, salt bridge) and direction of electron flow, given the reaction equation	Remember: Ions flow through the salt bridge. Electrons flow through the wire.
		use an activity series to determine whether a redox reaction is spontaneous	
VIII.10	An electrolytic cell requires electrical energy to produce chemical change. This process is known as electrolysis.	identify and label the parts of an electrolytic cell (anode, cathode) and direction of electron flow, given the reaction equation	

IX. Acids, Bases, and Salts

	MAJOR UNDERSTANDINGS	SKILLS The student should be able to:	KEY POINTS TO REMEMBER
IX.1	Behavior of many acids and bases can be explained by the Arrhenius theory. Arrhenius acids and bases are electrolytes.	given properties, identify substances as Arrhenius acids or Arrhenius bases	
IX.2	An electrolyte is a substance that, when dissolved in water, forms a solution capable of conducting an electric current. The ability of a solution to conduct an electric current depends on the concentration of ions.		The three categories of compounds that are electrolytes are acids, bases, and salts (ionic compounds).
IX.3	Arrhenius acids yield H^+ (hydrogen ion) as the only positive ion in aqueous solution. The hydrogen ion may also be written as H_3O^+, hydronium ion.		
IX.4	Arrhenius bases yield OH^- (hydroxide ion) as the only negative ion in an aqueous solution.		
IX.5	In the process of neutralization, an Arrhenius acid and an Arrhenius base react to form salt and water.	write simple neutralization reactions when given the reactants	$[H^+] = [OH^-]$ at the equivalence point.
IX.6	Titration is a laboratory process in which a volume of solution of known concentration is used to determine the concentration of another solution.	calculate the concentration or volume of a solution, using titration data	
IX.7	There are alternate acid–base theories. One such theory states that an acid is an H+ donor and a base is an H+ acceptor.	identify acids and bases in a reversible reaction using the alternate theory	
IX.8	The acidity and alkalinity of an aqueous solution can be measured by its pH value. The relative level of acidity or alkalinity of a solution can be shown by using indicators.	interpret changes in acid–base indicator color identify solutions as acid, base, or neutral based upon the pH	The pH of pure water is 7. Acids have pH values less than 7, and bases have pH values greater than 7.
IX.9	On the pH scale, each decrease of one unit of pH represents a tenfold increase in hydronium ion concentration.		

X. Nuclear Chemistry

	MAJOR UNDERSTANDINGS	SKILLS The student should be able to:	KEY POINTS TO REMEMBER
X.1	Stability of isotopes is based on the ratio of the neutrons and protons in its nucleus. Although most nuclei are stable, some are unstable and spontaneously decay emitting radiation.		
X.2	Each radioactive isotope has a specific mode and rate of decay (half-life).	calculate the initial amount, the fraction remaining, or the half-life of a radioactive isotope, given two of the three variables use Reference Table N to find the decay mode for a particular radioisotope	
X.3	A change in the nucleus of an atom that converts it from one element to another is called transmutation. This can occur naturally or can be induced by the bombardment of the nucleus of high-energy particles.		Natural transmutation is also called radioactive decay.
X.4	Spontaneous decay can involve the release of alpha particles, beta particles, positrons, and/or gamma radiation from the nucleus of an unstable isotope. These emissions differ in mass, charge, ionizing power, and penetrating power.	determine decay mode and write nuclear equations showing alpha and beta decay	Radioactive decay continues until a stable isotope is reached.
X.5	Nuclear reactions include natural and artificial transmutation, fission, and fusion.	compare and contrast fission and fusion reactions	
X.6	There are benefits and risks associated with fission and fusion reactions.		Fusion generates much more energy than fission. Fission produces far more dangerous radioactive waste than fusion.
X.7	Nuclear reactions can be represented by equations that include symbols that represent atomic nuclei (with the mass number and atomic number), subatomic particles (with mass number and charge), and/or emissions such as gamma radiation.	complete nuclear equations; predict missing particles from nuclear equations	

X. Nuclear Chemistry (*Continued*)

	MAJOR UNDERSTANDINGS	SKILLS The student should be able to:	KEY POINTS TO REMEMBER
X.8	Energy released in a nuclear reaction (fission or fusion) comes from the fractional amount of mass converted into energy. Nuclear changes convert matter into energy.		
X.9	Energy released during nuclear reactions is much greater than the energy released during chemical reactions.		
X.10	There are inherent risks associated with radioactivity and the use of radioactive isotopes. Risks can include biological exposure, long-term storage and disposal, and nuclear accidents.		
X.11	Radioactive isotopes have many beneficial uses. Radioactive isotopes are used in medicine and industrial chemistry, e.g., radioactive dating, tracing chemical and biological processes, industrial measurement, nuclear power, and detection and treatment of diseases.	identify specific uses of some common radioisotopes, such as: I-131 in diagnosing and treating thyroid disorders; C-14 to C-12 ratio in dating living organisms; U-238 to Pb-206 ratio in dating geological formations; Co-60 in treating cancer	

QUESTION INDEX

What follows is an index to the examination questions that are explained in this book. The questions are indexed according to the Sequence numbers given in the Topic Outline found on pages 20–38.

Some questions embrace more than one topic; these questions are marked with a dagger (†).

SEQUENCE	JUNE 2016	AUGUST 2016	JUNE 2017	AUGUST 2017	JUNE 2018	AUGUST 2018	JUNE 2019	AUGUST 2019
M. Math Skills								
M.1		10						
M.2		40			83†			
M.3	69	79	33		69, 80	45	33, 70	
M.4								75
M.5				50†	31†			
R. Reading Skills								
R.1		76	16, 49, 68	55	13†, 28, 39†, 58†, 68	33	18†, 53†	58†, 82
R.2				65	66		60†, 61	54†, 66
R.3								
I. Atomic Concepts								
I.1		48				4		1
I.2			1	1	1	2		
I.3	2						1	
I.4	1						2	
I.5			3	2	2	31	52	2
I.6	29, 84	32	4, 55	3	34, 36	3, 83	3	3
I.7			2			1		4, 52
I.8					9	55	4	5
I.9	72	33	31	31	35	32	32	32
I.10	73, 74	1	5		31†, 32		31	31
I.11		2			4	54	5	
I.12	31	39	56		3	53†	62	33
I.13	32	30	32	32	5			53

SEQUENCE	JUNE 2016	AUGUST 2016	JUNE 2017	AUGUST 2017	JUNE 2018	AUGUST 2018	JUNE 2019	AUGUST 2019
II. Periodic Table								
II.1	3	6				5		
II.2			6, 54	6, 82		53†		51
II.3	33	11, 12	66	61	6	6		7
II.4	12			5	39			
II.5	54			64	7		6	
II.6	4	34		33	8	7	7	
II.7	5	3		62		57	8†, 34	8
II.8		63	51	63		58	8†, 35	
II.9	36	31			11, 51	56		6
III. Moles/Stoichiometry								
III.1	6†, 35	4, 15†	34	34	67	8, 34	9, 66	9, 34
III.2	34		7		74		10	10
III.3	70	51	72	57		35	75	
III.4			8, 35	35		36†	11, 68	35
III.5	52, 60	67	36, 69	69, 72	37, 77	73	36, 69	78
III.6	75	5	74	56	63		53†, 67	68, 74
III.7	37, 59		70	68	55	51	37	67
III.8	38	56	9	75	38	37		
IV. Chemical Bonding								
IV.1						9, 13		
IV.2			52			10		
IV.3	66	36	11, 37	51, 58			77	11, 36
IV.4		7		74†				12
IV.5	39†	8†	38	52	10†, 14	11		
IV.6	68	60, 61	13	8	33	38	73	69
IV.7	41	37	58	12		12		13
IV.8	67	9	57	7			38	70
IV.9				49, 76			39	
IV.10	51	62	59	48		67		71
IV.11	9		60	9		15		14
IV.12	39†	8†, 38	12		41, 75	39	12	
IV.13					54			
IV.14	7		53	11	56		71	

SEQUENCE	JUNE 2016	AUGUST 2016	JUNE 2017	AUGUST 2017	JUNE 2018	AUGUST 2018	JUNE 2019	AUGUST 2019
V. Physical Behavior of Matter								
V.1								
V.2	11, 56	13, 14	21, 41, 67	4		62†	40	19, 54†
V.3	48			36	12			15, 37
V.4		15†		10			13	
V.5				13	40		15	16
V.6			20				14	
V.7	14	53	10	14	15	52		17
V.8	42, 76	52, 70	82	37		59, 60, 61	56	59
V.9		17, 71	81	38	42	40, 70	72	39
V.10	43	73		15		71	57	
V.11			77			14		
V.12		18, 55	14, 62	16		17, 62†, 64		
V.13	19	54	15	18		63	55, 60†	72
V.14	15		17	19		16	58	38
V.15				17	16		16	18
V.16	63, 64, 65	16, 50	18, 50, 71	39	64, 65	42	59	73
V.17	17, 55	59	19	21	43	18		40
V.18	58	41	63, 76	22	17, 60, 62	41	17, 41	56, 57
V.19		49		23†, 40, 41, 67		65		
V.20	71			50†, 66			18†	
V.21					61		19	
V.22	57†	84†			13, 52	19†		58†
V.23	8, 53, 57†	84†	39		10†	19†		55
VI. Kinetics/Equilibrium								
VI.1		19		20	19	43		41
VI.2	13	42, 69		70	72	68	42	
VI.3		20		23†			20	
VI.4	40	21	22	53	71	66	43	20
VI.5	44		42	54		24	44	42
VI.6			43			72, 74		76

SEQUENCE	JUNE 2016	AUGUST 2016	JUNE 2017	AUGUST 2017	JUNE 2018	AUGUST 2018	JUNE 2019	AUGUST 2019
VI.7		22	23	71	44	69	21	43
VI.8	18		24		73		22	77
VI.9	20		75	24				44
VI.10		23			20	20	23	21

VII. Organic Chemistry

SEQUENCE	JUNE 2016	AUGUST 2016	JUNE 2017	AUGUST 2017	JUNE 2018	AUGUST 2018	JUNE 2019	AUGUST 2019
VII.1	21, 47	81	61	73	18	76	24	79
VII.2	45	83	73	59	21	22, 44, 77	25, 46	22, 80, 81
VII.3	23, 77	43, 57, 85	25, 44	42, 60, 77†	53, 70, 76	21, 75	45, 51, 76	23, 45
VII.4			26	77†	22			
VII.5	22			74†				
VII.6	50	82	27	43	45	23	47	

VIII. Oxidation–Reduction

SEQUENCE	JUNE 2016	AUGUST 2016	JUNE 2017	AUGUST 2017	JUNE 2018	AUGUST 2018	JUNE 2019	AUGUST 2019
VIII.1	10				48		74	
VIII.2		24					26	
VIII.3	79	68	79		47		79†	62
VIII.4		58						
VIII.5						46		
VIII.6		35, 75	28	25	79	36†		24
VIII.7	24	74		80	46, 78	25	48	46
VIII.8		25		26		26	79†	25
VIII.9	78, 80	44, 66	78, 80	44	23, 49	47	78, 80†	47, 60, 61
VIII.10	25, 81		29	78	24		80†	

IX. Acids, Bases, and Salts

SEQUENCE	JUNE 2016	AUGUST 2016	JUNE 2017	AUGUST 2017	JUNE 2018	AUGUST 2018	JUNE 2019	AUGUST 2019
IX.1			30†				27	
IX.2	49	26, 72	83	27, 79		48	85	26
IX.3	6†		30†			78		28
IX.4		78	30†	81†	25			
IX.5	26	45	40		26	79	83	48, 64
IX.6	46	80		45	82	80	84	63
IX.7	27	46	45	46	50		49	
IX.8	61	47, 77	84	81†	83†, 84	81	82	27
IX.9	62	27	85	47	81, 85	49	81	65

Glossary of Important Terms

Absolute zero The lowest possible temperature, written as 0 K or $-273°C$.

Accuracy The closeness of a measurement to an accepted value; see also **precision**.

Acid See **Arrhenius acid; Brønsted-Lowry acid**.

Activated complex The intermediate state between reactants and products in a chemical reaction; the peak of the potential energy diagram.

Activation energy The minimum energy needed to initiate a reaction.

Addition polymerization The joining of unsaturated monomers by a series of addition reactions.

Addition reaction The process in which a substance reacts across a double or triple bond in an organic compound.

Alcohol An organic compound containing a hydroxyl (—OH) group.

Aldehyde An organic compound containing a carbonyl group with at least one hydrogen atom attached to the carbonyl carbon.

Alkali metal Any Group 1 element, excluding hydrogen.

Alkaline earth element Any Group 2 element.

Alkane A hydrocarbon containing only single bonds between adjacent carbon atoms.

Alkene A hydrocarbon containing one double bond between two adjacent carbon atoms.

Alkyl group An open-chained hydrocarbon less one hydrogen atom; for example, CH_3 = methyl group, C_2H_5 = ethyl group. Unspecified alkyl groups are designed by the letter R.

Alkyne A hydrocarbon containing one triple bond between two adjacent carbon atoms.

Reprinted with permission from *Let's Review:* Chemistry by Albert S. Tarendash, © 1998 by Barron's Educational Series, Inc.

Allotrope A specific form of an element that can exist in more than one form; graphite and diamond are allotropes of the element carbon.

Alloy A solid metallic solution.

Alpha decay The radioactive process in which an alpha particle is emitted.

Alpha particle (α) A helium-4 nucleus.

Amide An organic compound containing the $CONH_2$ functional group.

Amine A hydrocarbon derivative containing an amino group.

Amino acid An organic compound containing at least one amino group and one carboxyl group.

Amino group An ammonia molecule less one hydrogen atom; —NH_2.

Anhydrous Pertaining to a compound from which the water of crystallization has been removed.

Anode The electrode at which oxidation occurs.

Aqueous Pertaining to a solution in which water is the solvent.

Aromatic hydrocarbon Any ring hydrocarbon whose electronic structure is related to that of benzene.

Arrhenius acid Any substance that releases H^+ ions in water.

Arrhenius base Any substance that releases OH^- ions in water.

Atmospheric pressure 1 standard atmosphere (atm) = 101.3 kilopascals.

Atom The basic unit of an element.

Atomic mass The weighted average of the masses of the isotopes of an element.

Atomic mass unit (u) One-twelfth the mass of a carbon-12 atom.

Atomic number The number of protons in the nucleus of an atom; the atomic number defines the element.

Atomic radius A measure of the size of an atom.

Avogadro's hypothesis Equal volumes of gases, measured at the same temperature and pressure, contain equal numbers of particles.

Avogadro's number (N_A) The number of particles in 1 mole; 6.02×10^{23}.

Battery A commercial voltaic cell.

Benzene C_6H_6; the parent hydrocarbon of all aromatic compounds.

Beta decay The radioactive process in which a beta particle is emitted.

Beta (—) particle (β^-) An electron.

Beta (+) particle (β^+) A positron.

Binary compound A compound containing two elements.

Binding energy The energy released when a nucleus is assembled from its nucleons.

Boiling The transition of liquid to gas; boiling occurs when the vapor pressure of a liquid equals the atmospheric pressure above the liquid.

Boiling point The temperature at which boiling occurs; the temperature at which the liquid and vapor phases of a substance are in equilibrium.

Boiling point elevation The increase in the boiling point of a solvent due to the presence of solute particles.

Bond energy The energy needed to break a chemical bond.

Boyle's law At constant temperature and mass, the pressure of an ideal gas is inversely proportional to its volume; $P_1V_1 = P_2V_2$.

Breeder reactor A fission reactor that generates its own nuclear fuel.

Bright-line spectrum The lines of visible light emitted by elements as electrons fall to lower energy levels.

Brønsted-Lowry acid A substance that can donate H^+ ions.

Brønsted-Lowry base A substance that can accept H^+ ions.

Carbonyl group The functional group characteristic of aldehydes and ketones; $> C=O$.

Carboxyl group The functional group characteristic of organic acids; —COOH.

Catalyst A substance that speeds a chemical reaction by lowering the activation energy of the reaction.

Cathode The electrode at which reduction occurs.

Celsius (C) scale The temperature scale on which the freezing and boiling points of water (at 1 atm) are set at 0 and 100, respectively.

Chain reaction A chemical or nuclear reaction in which one step supplies energy or reactants for the next step.

Charles's law At constant pressure and mass, the volume of an ideal gas is directly proportional to the Kelvin temperature; $\dfrac{V_1}{T_1} = \dfrac{V_2}{T_2}$.

Chemical bond The stabilizing of two atoms by sharing or transferring electrons.

Chemical equation A shorthand listing of reactants, products, and molar quantities in a chemical reaction.

Chemical equilibrium The state in which the rates of the forward and reverse reactions are equal.

Chemical family See **group**.

Coefficient A number in a chemical equation that indicates how many particles of a reactant or product are required or formed in the reaction.

Colligative property A property that depends on the number of particles present rather than the type of particle; see also **boiling point elevation; freezing point depression**.

Combined (ideal) gas law At constant mass, the product of the pressure and volume divided by the Kelvin temperature is a constant; $\dfrac{P_1V_1}{T_1} = \dfrac{P_2V_2}{T_2}$.

Compound A combination of two or more elements with a fixed composition by mass.

Concentrated Pertaining to a solution that contains a relatively large quantity of solute.

Concentration The "strength" of a solution; the quantity of solute relative to the quantity of solvent.

Condensation The change from gas to liquid.

Condensation polymerization The joining of monomers by a series of dehydration reactions.

Control rod The part of a fission reactor that controls the rate of fission by absorbing neutrons.

Coordinate covalent bond A single covalent bond in which the pair of electrons is supplied by one atom.

Covalent bond A chemical bond formed by the sharing of electrons.

Cracking The process of breaking large hydrocarbon molecules into smaller ones in order to increase the yield of compounds such as gasoline.

Crystal A solid whose particles are arranged in a regularly repeating pattern.

Decomposition A reaction in which a compound forms two or more simpler substances.

Density Mass per unit volume;

$$d = \frac{m}{V}.$$

Deposition The direct transition from gas to solid.

Deuterium The isotope of hydrogen with a mass number of 2.

Diatomic molecule A neutral particle consisting of two atoms; Br_2 and CO are diatomic molecules.

Diffusion The movement of one substance through another.

Dihydroxy alcohol An organic compound with two hydroxyl groups.

Dilute (adjective) Pertaining to a solution that contains a relatively small quantity of solute; (verb) to reduce the concentration of a solution by adding solvent.

Dipole An unsymmetrical charge distribution in a neutral molecule.

Dipole–dipole attraction The attractive force between two oppositely charged dipoles of neighboring polar molecules.

Dissociation The separation of an ionic compound in solution into positive and negative ions.

Distillation The simultaneous boiling of a liquid and condensation of its vapor.

Double bond A covalent bond in which two pairs of electrons are shared by two adjacent atoms.

Ductility The property of a substance that allows it to be drawn into a wire; metallic substances possess ductility.

Dynamic equilibrium The state in which the rates of opposing processes are equal; see also **chemical equilibrium; phase equilibrium; solution equilibrium**.

Electrochemical cell—either a *voltaic cell* or an electrolytic cell A device that produces usable electrical energy from a spontaneous redox reaction; see also **battery**.

Electrode A conductor in an electrochemical or electrolytic cell that serves as the site of oxidation or reduction.

Electrolysis A nonspontaneous redox reaction driven by an external source of electricity.

Electrolyte A substance whose aqueous solution conducts electricity.

Electrolytic cell A device for carrying out electrolysis.

Electron The elementary unit of negative charge.

Electron-dot diagram See **Lewis structure**.

Electronegativity The measure of an atom's attraction for a bonded pair of electrons.

Electroplating The use of an electric current to deposit a layer of metal on a negatively charged object.

Element A substance all of whose atoms have the same atomic number.

Empirical formula A formula in which the elements are present in the smallest whole-number ratio; NO_2 is an empirical formula, but C_2H_4 is not.

Endothermic reaction A reaction that absorbs energy; ΔH is positive for an endothermic reaction.

End point The point in a titration that signals that equivalent quantities of reactants have been added.

Energy A quantity related to an object's capacity to do work.

Enthalpy change (ΔH) The heat energy absorbed or released by a system at constant pressure.

Entropy (S) The measure of the randomness or disorder of a system.

Entropy change (ΔS) An increase or decrease in the randomness of a system.

Equilibrium See **dynamic equilibrium**.

Ester The organic product of esterification.

Esterification The reaction of an acid with an alcohol to produce an ester and water.

Ethanoic acid CH_3COOH; acetic acid.

Ethanol CH_3CH_2OH; ethyl (grain) alcohol.

Ethene C_2H_4; ethylene; the parent of the alkene family of hydrocarbons.

Ether An organic compound containing the arrangement R—O—R.

Ethyne C_2H_2; acetylene; the parent of the alkyne family of hydrocarbons.

Evaporation The surface transition of liquid to gas.

Excited state A condition in which one or more electrons in an atom are no longer in the lowest possible energy state.

Exothermic reaction A reaction that releases energy; ΔH is negative for an exothermic reaction.

Fermentation The (anaerobic) oxidation of a sugar such as glucose to produce ethanol and carbon dioxide; the reaction is catalyzed by enzymes.

Filtration A method of separating a liquid from the particles suspended in it.

First ionization energy The quantity of energy needed to remove the most loosely held electron from an isolated neutral atom.

Fission A nuclear reaction in which a heavy nuclide splits to form lighter nuclides and energy.

Fission reactor A device for producing electrical energy by means of a controlled fission reaction.

Formula mass The sum of the masses of the atoms in a formula; units are atomic mass units (u) or grams per mole (g/mol).

Fractional distillation The separation of organic substances based on differences in their boiling points.

Freezing The transition from liquid to solid.

Freezing point The temperature at which freezing occurs.

Freezing point depression (lowering) The decrease in the freezing point of a solvent due to the presence of solute particles.

Fuel rod The part of a nuclear reactor that contains the fissionable material.

Functional group An atom or group of atoms that confers specific properties on an organic molecule.

Fusion A synonym for *melting*; also, a nuclear process in which light nuclides join to form heavier nuclides and produce radiant energy.

Fusion reactor An experimental device for producing a controlled fusion reaction and generating electrical energy from it.

Gas The phase in which matter has neither definite shape nor definite volume.

Gram-atomic mass The molar mass of an element expressed in grams per mole (g/mol).

Gram-formula mass See **molar mass**.

Gram-molecular mass The molar mass of a molecule.

Ground state The electron configuration of an atom in the lowest energy state.

Group The elements within a single vertical column of the Periodic Table.

Half-cell The part of an electrochemical cell in which oxidation or reduction occurs.

Half-life The time needed for a substance to decay to one-half its initial mass.

Half-reaction The oxidation or reduction portion of a redox reaction.

Halogen An element in Group 17 of the Periodic Table; F, Cl, Br, I, At.

Heat energy The energy released or absorbed by a system undergoing a change in temperature, in phase, or in composition.

Heat of fusion (H_f) The heat energy absorbed when a unit mass of solid changes to liquid at its melting point; $H_{f(ice)} = 80$ calories per gram.

Heat of reaction (ΔH) The heat energy absorbed or released as a result of a chemical reaction.

Heat of vaporization (H_v) The heat energy absorbed when a unit mass of liquid changes to gas at its boiling point; $H_{v(water)} = 540$ calories per gram.

Heavy water A molecule of water in which the hydrogen atoms have a mass number of 2; deuterium oxide.

Heterogeneous mixture A nonuniform mixture.

Homogeneous mixture A mixture with a uniform distribution of particles; a solution is one example of a homogeneous mixture.

Homologous series A group of organic compounds with related structures and properties; each successive member of the series differs from the one before it by a specific number of carbon and hydrogen atoms (usually CH_2).

Hydrate A crystalline compound that has water molecules incorporated into its crystal structure; common examples include $CuSO_4 \cdot 5H_2O$ and $Na_2SO_4 \cdot 10H_2O$ [also written as $CuSO_4(H_2O)_5$ and $Na_2SO_4(H_2O)_{10}$].

Hydration The association of water molecules with an ion or another molecule.

Hydride A binary compound of an active metal and hydrogen; the oxidation state of hydrogen is -1.

Hydrocarbon An organic compound composed of carbon and hydrogen.

Hydrogen bond An unusually strong intermolecular attraction that results when hydrogen is bonded to a small, highly electronegative atom such as F, O, or N.

Hydrolysis A reaction in which a water molecule breaks a chemical bond; the reaction between certain salts and water to produce an excess of hydronium or hydroxide ions.

Hydronium ion H_3O^+; the conjugate acid of H_2O; responsible for acidic properties in water solutions.

Hydroxide ion OH^-; the conjugate base of H_2O; responsible for basic properties in water solutions.

Ideal gas A model of a gas in which the particles have no volume, do not attract or repel each other, and collide without loss of energy; real gases approximate ideal gas behavior under conditions of low pressure and high temperature.

Ideal gas law The relationship obeyed by an ideal gas; see **combined (ideal) gas law**.

Indicator A substance that undergoes a color change to signal a change in chemical conditions; acid–base indicators change color over specified pH ranges.

Inert (noble) gas An element in Group 18 of the Periodic Table; Ne, Ar, Kr, Xe, Rn. (He is also associated with Group 18.)

Inorganic compound A compound that is not a hydrocarbon derivative.

Ion A particle in which the numbers of protons and electrons are not equal.

Ion–dipole attraction The attractive force between an ion and the oppositely charged dipole of a neighboring polar molecule.

Ionic bond The electrostatic attraction of positive and negative ions in an ionic compound; an electronegativity difference of 1.7 or greater indicates the presence of an ionic bond.

Ionic compound A substance whose particles consist of positive and negative ions.

Ionization energy The quantity of energy needed to remove an electron from an atom or ion; see also **first ionization energy**.

Isomers Different compounds that have the same molecular formula.

Isotopes Atoms having the same atomic number but different mass numbers; atoms of the same element with differing numbers of neutrons.

IUPAC International Union of Pure and Applied Chemistry; the scientific group responsible for all major policies in chemistry, including the naming of elements and compounds.

Joule (J) The unit of work and energy in the SI (metric) system.

Kelvin (K) A measure of absolute temperature; the Kelvin scale begins at 0 and is related to the Celsius scale by the equation $K = C + 273$; a temperature *difference* of 1 K is equal to a temperature *difference* of 1°C.

Ketone An organic compound containing a carbonyl group with no hydrogen atoms directly attached to the carbonyl carbon.

Kilo- The metric prefix signifying 1,000.

Kilojoule (kJ) 1,000 joules.

Kinetic energy The energy associated with the motion of an object.

Kinetic molecular theory (KMT) The theory that explains the structure and behavior of idealized models of gases, liquids, and solids.

Le Châtelier's principle When a system at equilibrium is subjected to a stress, the system will shift in order to lessen the effects of the stress. Eventually, a new equilibrium point is established.

Lewis structure A shorthand notation for illustrating the ground-state valence electron configuration of an atom.

Liquid The phase in which matter has a definite volume but an indefinite shape; a liquid takes the shape of its container.

Liter (L) A unit of volume in the metric system; 1 liter = 1,000 cubic centimeters; 1 liter = 1 cubic decimeter; 1 liter is approximately equal to 1 quart.

Litmus An acid–base indicator that is red in acidic solutions and blue in basic solutions.

Macromolecule A giant molecule formed by network bonding or by polymerization.

Malleability The property by which a substance is able to be formed into various shapes; metallic substances possess malleability.

Mass number The number of nucleons in a nuclide.

Melting The transition from solid to liquid.

Melting point The temperature at which the vapor pressure of a solid equals the vapor pressure of the liquid; the temperature at which the solid and liquid phases of a substance are in equilibrium; see also **freezing point**.

Meniscus The curved surface of a liquid, caused by the attraction of the particles of the liquid and the container holding the liquid (e.g., water in a graduated cylinder), or by the mutual attraction of the particles of the liquid (e.g., mercury).

Metal A substance composed of atoms with low ionization energies and relatively vacant valence levels; Na, Fe, Ag, and Ba are metallic substances.

Metallic bond The delocalization of the valence electrons among the kernels of the metal atoms; "mobile valence electrons immersed in a sea of positive ions."

Metalloid An element that has both metallic and nonmetallic properties; examples of metalloids include B, Ge, Si, and Te.

Methanal HCHO; formaldehyde; the simplest aldehyde.

Methane CH_4; the parent of the alkane family of hydrocarbons.

Methanoic acid HCOOH; formic acid; the simplest organic acid.

Methanol CH_3OH; methyl (wood) alcohol; the simplest alcohol.

Milli- The metric prefix signifying 1/1,000.

Milliliter (mL) 1/1,000 liter; 1 liter = 1,000 milliliters.

Miscible A solution of liquids that is soluble in all proportions; ethanol and water are a miscible pair of liquids.

Mixture A material consisting of two or more components and having a variable composition.

Moderator A substance used to produce slow neutrons and promote nuclear fission; graphite and heavy water are used as moderators in fission reactors.

Molar mass The mass of any atom, element, ion, or compound expressed in grams per mole (g/mol).

Molarity The concentration of a solution, measured as the number of moles of solute per liter of solution.

Molar volume The volume occupied by 1 mole of an ideal gas; 22.4 liters at STP.

Mole The number of atoms contained in 12 grams of carbon-12; see also **Avogadro's number**.

Molecular formula A chemical formula that lists the number of atoms present but does not show the arrangement of the atoms in space.

Molecular mass The sum of the masses of the atoms in a molecule; units are atomic mass units (u) or grams per mole (g/mol).

Molecule The smallest unit of a nonionic substance; Ar, Cl_2, and NH_3 are molecules.

Monatomic molecule A molecule consisting of one atom; Xe and He are monatomic molecules.

Monomer The basic unit of a polymer; the monomer of a protein is an amino acid; the monomer of starch is glucose.

Network solid A substance formed by a two- or three-dimensional web of covalent bonds to produce a macromolecule; diamond and SiO_2 are network solids.

Neutralization The reaction of equivalent amounts of hydronium and hydroxide ions in aqueous solution; the principal product is water. When the water is evaporated, the spectator ions form a salt; see also **spectator ion**.

Neutron A neutral nuclear particle with a mass comparable to that of a proton.

Noble gas See **inert (noble) gas**.

Nonelectrolyte A substance whose aqueous solution does not conduct electricity; glucose is a nonelectrolyte.

Nonmetal A substance that does not have characteristic metallic properties; C and S are nonmetallic elements.

Nonpolar bond A covalent bond in which the electron pair or pairs are shared equally by both atoms.

Nonpolar molecule A molecule containing only nonpolar bonds, such as N_2, or a molecule with a symmetrical charge distribution, such as CCl_4 and CO_2.

Normal boiling point The boiling temperature of a substance at a pressure of 1 atmosphere.

Nuclear equation A shorthand listing of reactant and product nuclides in a nuclear reaction.

Nucleon A constituent of an atomic nucleus; a proton or a neutron.

Nucleus The portion of the atom that contains more than 99.9 percent of the atom's mass; the nucleus is small, dense, and positively charged.

Ore A native mineral from which a metal or metals can be extracted.

Organic chemistry The study of the hydrocarbons and their derivatives; the chemistry of carbon.

Organic compound A compound that is a hydrocarbon or a hydrocarbon derivative.

Oxidation The loss or apparent loss of electrons in a chemical reaction.

Oxidation number (state) The charge that an atom has or appears to have when certain arbitrary rules are applied; oxidation numbers are useful for identifying the atoms that are oxidized and reduced in a redox reaction.

Oxidizing agent The particle in a redox reaction that causes another particle to be oxidized; as a result, an oxidizing agent is reduced.

Paraffin A common name for a mixture of solid alkanes; another name for paraffin is wax.

Percent composition by mass The number of grams of an element (or group of elements) present in 100 grams of an ion or compound.

Period One of the horizontal rows of the Periodic Table; the period number indicates the valence level of an element.

Periodic law The properties of elements recur at regular intervals and depend on their nuclear charges; "The properties of elements are a periodic function of their atomic numbers."

Peroxide A compound in which oxygen has an oxidation state of -1; H_2O_2 and BaO_2 are peroxides.

Petroleum Crude oil containing a mixture of hydrocarbons.

pH A scale of acidity and basicity based on the hydronium ion concentration in aqueous solution; $pH = -\log[H_3O^+]$. At 298 K, a pH less than 7 indicates an acidic solution; a pH greater than 7, a basic solution; a pH of 7, a neutral solution.

Phase equilibrium The state in which the rates of opposing phase changes (freezing–melting, boiling–condensation, sublimation–deposition) are equal.

Phenolphthalein An acid–base indicator that is colorless in acidic solutions and pink in basic solutions.

Photon A fundamental unit of radiant energy; a quantum of radiation.

Polar bond A covalent bond in which the electron pair or pairs are shared unequally by both atoms; the atom with the larger electronegativity has more of the electron density surrounding it.

Polar molecule A molecule with an unsymmetrical charge distribution, such as H_2O; a dipole.

Polyatomic ion An ion composed of more than one atom; SO_4^{2-} is a polyatomic ion.

Polymer A macromolecule consisting of a chain of simpler units; polyethylene is a polymer of ethene.

Polymerization See **polymer**; **addition polymerization**; **condensation polymerization**.

Positron A positively charged electron; a particle of antimatter.

Potential energy The energy associated with the position of an object; a "stored" form of energy.

Precipitate A deposit formed by the appearance of an excess of solid solute in a saturated solution.

Precision The closeness of a series of measurements to one another; see also **accuracy**.

Pressure The force exerted on an object divided by the surface area of the object;

$$P = \frac{F}{A}.$$

Principal energy level An integer beginning with 1 that describes the approximate distance of an electron from the nucleus of an atom.

Product(s) The substance or substances that are formed in a chemical process; products are on the right side of a chemical equation.

Propanone $(CH_3)_2CO$; acetone; the simplest ketone.

Proton A nuclear particle with a positive charge equal to the negative charge on the electron; a nucleon.

Radiant energy Electromagnetic energy; visible light and X rays are examples of radiant energy.

Radioactive dating The use of radioactive isotopes to measure the age of an object.

Radioactive tracer A radioisotope used to indicate the path of an atom in a chemical reaction.

Radioactivity The spontaneous breakdown of a radioactive nuclide.

Radioisotope A radioactive isotope.

Reactant(s) The substance or substances that react in a chemical process; reactants are on the left side of a chemical equation.

Redox reaction A chemical reaction in which oxidation–reduction takes place.

Reducing agent The particle in a redox reaction that causes another particle to be reduced; as a result, a reducing agent is oxidized.

Reduction The gain or apparent gain of electrons in a chemical reaction.

Salt The spectator-ion product of a neutralization reaction; see also **spectator ion**.

Salt bridge A device for allowing the flow of ions in an electrochemical cell.

Saponification The reaction of an ester with a base to produce an alcohol and the sodium salt of an organic acid; soap is produced by saponifying fats with NaOH.

Saturated hydrocarbon A hydrocarbon containing only single carbon–carbon bonds.

Saturated solution A solution in which the pure solute is in equilibrium with the dissolved solute; a solution that contains the maximum amount of dissolved solute.

Significant digit(s) [figure(s)] The number or numbers that are part of a measurement. If there are two or more, all but the last figure is known; the last figure is the experimenter's best estimate.

Single bond A covalent bond in which one pair of electrons is shared by two adjacent atoms.

Solid The phase in which matter has both definite shape and definite volume.

Solubility The amount of solute needed to produce a saturated solution with a given amount of solvent.

Solute(s) The substance or substances dissolved in a solution.

Solution A homogeneous mixture whose particles are extremely small.

Solution equilibrium The state in which the undissolved and dissolved solutes are in dynamic equilibrium; in a solid–liquid solution, the rate of dissolving equals the rate of crystallization.

Solvent The substance in which the solute is dissolved.

Spectator ion An ion that does not take part in a chemical reaction; in the (acid–base) reaction between NaOH(aq) and HCl(aq), Na^+(aq) and Cl^-(aq) are spectator ions; in the (redox) reaction between Zn(s) and $Cu(NO_3)_2$(aq), NO_3^-(aq) is the spectator ion.

Standard solution A solution whose concentration is accurately known; a standard solution is used for analyzing other substances.

Standard temperature and pressure (STP) 273 K and 1 atmosphere.

Stock system A systematic method of naming chemical compounds in which the *positive* oxidation number is written as a Roman numeral in parentheses after the element. For example, the Stock name of the compound Fe_2O_3 is iron(III) oxide.

Stoichiometry The study of quantitative relationships in substances and reactions; chemical mathematics.

Strong acid In aqueous solution, a substance that ionizes almost completely to hydronium ion.

Strong base In aqueous solution, a substance that ionizes or dissociates almost completely to hydroxide ion.

Structural formula A chemical formula that illustrates the spatial arrangement of each atom.

Sublimation The direct transition from solid to gas; $CO_2(s)$ and $I_2(s)$ sublime at atmospheric pressure.

Substance An element or a compound.

Supersaturated solution A solution that contains more dissolved solute than a saturated solution at the same temperature; supersaturation is an unstable condition.

Temperature A measure of the average kinetic energy of the particles of a substance.

Titration The addition of a known volume of a standard solution in order to determine the concentration of an unknown solution.

Toluene C_7H_8; methylbenzene.

Transition element An element whose atoms contain unfilled d sublevels; an element in Groups 3–11 of the Periodic Table.

Transmutation The conversion of one element to another by a nuclear process.

Triple bond A covalent bond in which three pairs of electrons are shared by two adjacent atoms.

Tritium The radioactive isotope of hydrogen with a mass number of 3.

Unsaturated hydrocarbon A hydrocarbon containing double and/or triple carbon–carbon bonds.

Unsaturated solution A solution that contains less dissolved solute than a saturated solution at the same temperature.

Valence electron An electron in the outermost principle energy level.

Voltaic cell A device that produces usable electrical energy from a spontaneous redox reaction; see also **battery**.

Water of hydration The water molecules that are part of the crystalline structure of certain compounds.

Weak acid In aqueous solution, a substance that is poorly ionized and produces only a small concentration of hydronium ion.

Weak base In aqueous solution, a substance that is poorly ionized or dissociated and produces only a small concentration of hydroxide ion.

Reference Tables for Chemistry

Important Note: This is the 2011 Edition of the Reference Tables, which should be used for all Regents exams.

TABLE A
Standard Temperature and Pressure

Name	Value	Unit
Standard Pressure	101.3 kPa 1 atm	kilopascal atmosphere
Standard Temperature	273 K 0°C	kelvin degree Celsius

TABLE B
Physical Constants for Water

Heat of Fusion	334 J/g
Heat of Vaporization	2260 J/g
Specific Heat Capacity of $H_2O(\ell)$	4.18 J/g•K

TABLE C
Selected Prefixes

Factor	Prefix	Symbol
10^3	kilo-	k
10^{-1}	deci-	d
10^{-2}	centi-	c
10^{-3}	milli-	m
10^{-6}	micro-	μ
10^{-9}	nano-	n
10^{-12}	pico-	p

TABLE D
Selected Units

Symbol	Name	Quantity
m	meter	length
g	gram	mass
Pa	pascal	pressure
K	kelvin	temperature
mol	mole	amount of substance
J	joule	energy, work, quantity of heat
s	second	time
min	minute	time
h	hour	time
d	day	time
y	year	time
L	liter	volume
ppm	parts per million	concentration
M	molarity	solution concentration
u	atomic mass unit	atomic mass

TABLE E
Selected Polyatomic Ions

Formula	Name	Formula	Name
H_3O^+	hydronium	CrO_4^{2-}	chromate
Hg_2^{2+}	mercury(I)	$Cr_2O_7^{2-}$	dichromate
NH_4^+	ammonium	MnO_4^-	permanganate
$C_2H_3O_2^-$ $\}$ CH_3COO^- $\}$ acetate	acetate	NO_2^-	nitrite
		NO_3^-	nitrate
CN^-	cyanide	O_2^{2-}	peroxide
CO_3^{2-}	carbonate	OH^-	hydroxide
HCO_3^-	hydrogen carbonate	PO_4^{3-}	phosphate
$C_2O_4^{2-}$	oxalate	SCN^-	thiocyanate
ClO^-	hypochlorite	SO_3^{2-}	sulfite
ClO_2^-	chlorite	SO_4^{2-}	sulfate
ClO_3^-	chlorate	HSO_4^-	hydrogen sulfate
ClO_4^-	perchlorate	$S_2O_3^{2-}$	thiosulfate

TABLE F
Solubility Guidelines for Aqueous Solutions

Ions That Form _Soluble_ Compounds	Exceptions
Group 1 ions (Li^+, Na^+, etc.)	
ammonium (NH_4^+)	
nitrate (NO_3^-)	
acetate ($C_2H_3O_2^-$ or CH_3COO^-)	
hydrogen carbonate (HCO_3^-)	
chlorate (ClO_3^-)	
halides (Cl^-, Br^-, I^-)	when combined with Ag^+, Pb^{2+}, or Hg_2^{2+}
sulfates (SO_4^{2-})	when combined with Ag^+, Ca^{2+}, Sr^{2+}, Ba^{2+}, or Pb^{2+}

Ions That Form _Insoluble_ Compounds*	Exceptions
carbonate (CO_3^{2-})	when combined with Group 1 ions or ammonium (NH_4^+)
chromate (CrO_4^{2-})	when combined with Group 1 ions, Ca^{2+}, Mg^{2+}, or ammonium (NH_4^+)
phosphate (PO_4^{3-})	when combined with Group 1 ions or ammonium (NH_4^+)
sulfide (S^{2-})	when combined with Group 1 ions or ammonium (NH_4^+)
hydroxide (OH^-)	when combined with Group 1 ions, Ca^{2+}, Ba^{2+}, Sr^{2+}, or ammonium (NH_4^+)

*compounds having very low solubility in H_2O

TABLE G
Solubility Curves at Standard Pressure

TABLE H
Vapor Pressure of Four Liquids

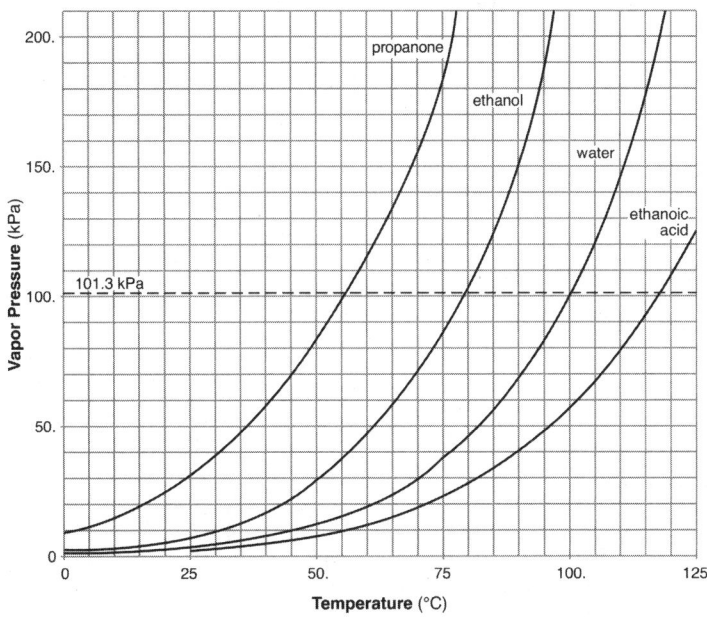

TABLE I
Heats of Reaction at
101.3 kPa and 298 K

Reaction	ΔH (kJ)*
$CH_4(g) + 2O_2(g) \longrightarrow CO_2(g) + 2H_2O(\ell)$	–890.4
$C_3H_8(g) + 5O_2(g) \longrightarrow 3CO_2(g) + 4H_2O(\ell)$	–2219.2
$2C_8H_{18}(\ell) + 25O_2(g) \longrightarrow 16CO_2(g) + 18H_2O(\ell)$	–10943
$2CH_3OH(\ell) + 3O_2(g) \longrightarrow 2CO_2(g) + 4H_2O(\ell)$	–1452
$C_2H_5OH(\ell) + 3O_2(g) \longrightarrow 2CO_2(g) + 3H_2O(\ell)$	–1367
$C_6H_{12}O_6(s) + 6O_2(g) \longrightarrow 6CO_2(g) + 6H_2O(\ell)$	–2804
$2CO(g) + O_2(g) \longrightarrow 2CO_2(g)$	–566.0
$C(s) + O_2(g) \longrightarrow CO_2(g)$	–393.5
$4Al(s) + 3O_2(g) \longrightarrow 2Al_2O_3(s)$	–3351
$N_2(g) + O_2(g) \longrightarrow 2NO(g)$	+182.6
$N_2(g) + 2O_2(g) \longrightarrow 2NO_2(g)$	+66.4
$2H_2(g) + O_2(g) \longrightarrow 2H_2O(g)$	–483.6
$2H_2(g) + O_2(g) \longrightarrow 2H_2O(\ell)$	–571.6
$N_2(g) + 3H_2(g) \longrightarrow 2NH_3(g)$	–91.8
$2C(s) + 3H_2(g) \longrightarrow C_2H_6(g)$	–84.0
$2C(s) + 2H_2(g) \longrightarrow C_2H_4(g)$	+52.4
$2C(s) + H_2(g) \longrightarrow C_2H_2(g)$	+227.4
$H_2(g) + I_2(g) \longrightarrow 2HI(g)$	+53.0
$KNO_3(s) \xrightarrow{H_2O} K^+(aq) + NO_3^-(aq)$	+34.89
$NaOH(s) \xrightarrow{H_2O} Na^+(aq) + OH^-(aq)$	–44.51
$NH_4Cl(s) \xrightarrow{H_2O} NH_4^+(aq) + Cl^-(aq)$	+14.78
$NH_4NO_3(s) \xrightarrow{H_2O} NH_4^+(aq) + NO_3^-(aq)$	+25.69
$NaCl(s) \xrightarrow{H_2O} Na^+(aq) + Cl^-(aq)$	+3.88
$LiBr(s) \xrightarrow{H_2O} Li^+(aq) + Br^-(aq)$	–48.83
$H^+(aq) + OH^-(aq) \longrightarrow H_2O(\ell)$	–55.8

*The ΔH values are based on molar quantities represented in the equations.
A minus sign indicates an exothermic reaction.

TABLE J
Activity Series**

Most Active	Metals	Nonmetals	Most Active
	Li	F_2	
	Rb	Cl_2	
	K	Br_2	
	Cs	I_2	
	Ba		
	Sr		
	Ca		
	Na		
	Mg		
	Al		
	Ti		
	Mn		
	Zn		
	Cr		
	Fe		
	Co		
	Ni		
	Sn		
	Pb		
	H_2		
	Cu		
	Ag		
Least Active	Au		Least Active

**Activity Series is based on the hydrogen standard. H_2 is *not* a metal.

TABLE K
Common Acids

Formula	Name
$HCl(aq)$	hydrochloric acid
$HNO_2(aq)$	nitrous acid
$HNO_3(aq)$	nitric acid
$H_2SO_3(aq)$	sulfurous acid
$H_2SO_4(aq)$	sulfuric acid
$H_3PO_4(aq)$	phosphoric acid
$H_2CO_3(aq)$ or $CO_2(aq)$	carbonic acid
$CH_3COOH(aq)$ or $HC_2H_3O_2(aq)$	ethanoic acid (acetic acid)

TABLE L
Common Bases

Formula	Name
$NaOH(aq)$	sodium hydroxide
$KOH(aq)$	potassium hydroxide
$Ca(OH)_2(aq)$	calcium hydroxide
$NH_3(aq)$	aqueous ammonia

TABLE M
Common Acid–Base Indicators

Indicator	Approximate pH Range for Color Change	Color Change
methyl orange	3.1–4.4	red to yellow
bromthymol blue	6.0–7.6	yellow to blue
phenolphthalein	8–9	colorless to pink
litmus	4.5–8.3	red to blue
bromcresol green	3.8–5.4	yellow to blue
thymol blue	8.0–9.6	yellow to blue

Source: *The Merck Index*, 14th ed., 2006, Merck Publishing Group

TABLE N
Selected Radioisotopes

Nuclide	Half-Life	Decay Mode	Nuclide Name
^{198}Au	2.695 d	β^-	gold-198
^{14}C	5715 y	β^-	carbon-14
^{37}Ca	182 ms	β^+	calcium-37
^{60}Co	5.271 y	β^-	cobalt-60
^{137}Cs	30.2 y	β^-	cesium-137
^{53}Fe	8.51 min	β^+	iron-53
^{220}Fr	27.4 s	α	francium-220
^{3}H	12.31 y	β^-	hydrogen-3
^{131}I	8.021 d	β^-	iodine-131
^{37}K	1.23 s	β^+	potassium-37
^{42}K	12.36 h	β^-	potassium-42
^{85}Kr	10.73 y	β^-	krypton-85
^{16}N	7.13 s	β^-	nitrogen-16
^{19}Ne	17.22 s	β^+	neon-19
^{32}P	14.28 d	β^-	phosphorus-32
^{239}Pu	2.410×10^4 y	α	plutonium-239
^{226}Ra	1599 y	α	radium-226
^{222}Rn	3.823 d	α	radon-222
^{90}Sr	29.1 y	β^-	strontium-90
^{99}Tc	2.13×10^5 y	β^-	technetium-99
^{232}Th	1.40×10^{10} y	α	thorium-232
^{233}U	1.592×10^5 y	α	uranium-233
^{235}U	7.04×10^8 y	α	uranium-235
^{238}U	4.47×10^9 y	α	uranium-238

Source: *CRC Handbook of Chemistry and Physics*, 91st ed., 2010–2011, CRC Press

TABLE O
Symbols Used in Nuclear Chemistry

Name	Notation	Symbol
alpha particle	$^{4}_{2}He$ or $^{4}_{2}\alpha$	α
beta particle	$^{0}_{-1}e$ or $^{0}_{-1}\beta$	β^{-}
gamma radiation	$^{0}_{0}\gamma$	γ
neutron	$^{1}_{0}n$	n
proton	$^{1}_{1}H$ or $^{1}_{1}p$	p
positron	$^{0}_{+1}e$ or $^{0}_{+1}\beta$	β^{+}

TABLE P
Organic Prefixes

Prefix	Number of Carbon Atoms
meth-	1
eth-	2
prop-	3
but-	4
pent-	5
hex-	6
hept-	7
oct-	8
non-	9
dec-	10

TABLE Q
Homologous Series of Hydrocarbons

Name	General Formula	Examples	
		Name	Structural Formula
alkanes	C_nH_{2n+2}	ethane	H—C—C—H (with H atoms)
alkenes	C_nH_{2n}	ethene	C=C (with H atoms)
alkynes	C_nH_{2n-2}	ethyne	H—C≡C—H

Note: n = number of carbon atoms

TABLE R
Organic Functional Groups

Class of Compound	Functional Group	General Formula	Example
halide (halocarbon)	$-F$ (fluoro-) $-Cl$ (chloro-) $-Br$ (bromo-) $-I$ (iodo-)	$R-X$ (X represents any halogen)	$CH_3CHClCH_3$ 2-chloropropane
alcohol	$-OH$	$R-OH$	$CH_3CH_2CH_2OH$ 1-propanol
ether	$-O-$	$R-O-R'$	$CH_3OCH_2CH_3$ methyl ethyl ether
aldehyde	$\overset{O}{\overset{\|}{-C}}-H$	$\overset{O}{\overset{\|}{R-C}}-H$	$CH_3CH_2\overset{O}{\overset{\|}{C}}-H$ propanal
ketone	$\overset{O}{\overset{\|}{-C}}-$	$\overset{O}{\overset{\|}{R-C}}-R'$	$CH_3\overset{O}{\overset{\|}{C}}CH_2CH_2CH_3$ 2-pentanone
organic acid	$\overset{O}{\overset{\|}{-C}}-OH$	$\overset{O}{\overset{\|}{R-C}}-OH$	$CH_3CH_2\overset{O}{\overset{\|}{C}}-OH$ propanoic acid
ester	$\overset{O}{\overset{\|}{-C}}-O-$	$\overset{O}{\overset{\|}{R-C}}-O-R'$	$CH_3CH_2\overset{O}{\overset{\|}{C}}OCH_3$ methyl propanoate
amine	$-\overset{\|}{N}-$	$R-\overset{R'}{\overset{\|}{N}}-R''$	$CH_3CH_2CH_2NH_2$ 1-propanamine
amide	$\overset{O}{\overset{\|}{-C}}-\overset{\|}{N}H$	$R-\overset{O\ \ R'}{\overset{\|\ \ \ \|}{C}}-NH$	$CH_3CH_2\overset{O}{\overset{\|}{C}}-NH_2$ propanamide

Note: R represents a bonded atom or group of atoms.

Periodic Table of the Elements

Period

Group

Period	1
1	1.00794 $+1$ -1 **H** 1 1

Group

	1	2
2	6.941 $+1$ **Li** 3 2-1	9.01218 $+2$ **Be** 4 2-2
3	22.98977 $+1$ **Na** 11 2-8-1	24.305 $+2$ **Mg** 12 2-8-2

Group

	1	2	3	4	5	6	7	8	9
4	39.0983 $+1$ **K** 19 2-8-8-1	40.08 $+2$ **Ca** 20 2-8-8-2	44.9559 $+3$ **Sc** 21 2-8-9-2	47.867 $+2$ $+3$ $+4$ **Ti** 22 2-8-10-2	50.9415 $+2$ $+3$ $+4$ $+5$ **V** 23 2-8-11-2	51.996 $+2$ $+3$ $+6$ **Cr** 24 2-8-13-1	54.9380 $+2$ $+3$ $+4$ $+7$ **Mn** 25 2-8-13-2	55.845 $+2$ $+3$ **Fe** 26 2-8-14-2	58.9332 $+2$ $+3$ **Co** 27 2-8-15-2
5	85.4678 $+1$ **Rb** 37 2-8-18-8-1	87.62 $+2$ **Sr** 38 2-8-18-8-2	88.9059 $+3$ **Y** 39 2-8-18-9-2	91.224 $+4$ **Zr** 40 2-8-18-10-2	92.9064 $+3$ $+5$ **Nb** 41 2-8-18-12-1	95.94 $+6$ **Mo** 42 2-8-18-13-1	(98) **Tc** 43 2-8-18-13-2 $+6$ $+7$	101.07 $+3$ **Ru** 44 2-8-18-15-1	102.906 $+3$ **Rh** 45 2-8-18-16-1
6	132.905 $+1$ **Cs** 55 2-8-18-18-1	137.33 $+2$ **Ba** 56 2-8-18-18-2	138.9055 $+3$ **La** 57 2-8-18-19-2	178.49 $+4$ **Hf** 72 *18-32-10-2	180.948 $+5$ **Ta** 73 -18-32-11-2	183.84 $+6$ **W** 74 -18-32-12-2	186.207 $+4$ **Re** 75 -18-32-13-2 $+6$ $+7$	190.23 $+3$ **Os** 76 -18-32-14-2 $+4$	192.217 $+3$ **Ir** 77 -18-32-15-2 $+4$
7	(223) $+1$ **Fr** 87 -18-32-18-8-1	(226) $+2$ **Ra** 88 -18-32-18-8-2	(227) $+3$ **Ac** 89 -18-32-18-9-2	(261) $+4$ **Rf** 104	(262) **Db** 105	(266) **Sg** 106	(272) **Bh** 107	(277) **Hs** 108	(276) **Mt** 109

140.116 $+3$ $+4$ **Ce** 58	140.908 $+3$ **Pr** 59	144.24 $+3$ **Nd** 60	(145) $+3$ **Pm** 61	150.36 $+2$ $+3$ **Sm** 62
232.038 $+4$ **Th** 90	231.036 $+4$ $+5$ **Pa** 91	238.029 $+3$ $+4$ $+5$ $+6$ **U** 92	(237) $+3$ $+4$ $+5$ $+6$ **Np** 93	(244) $+3$ $+4$ $+5$ $+6$ **Pu** 94

*denotes the presence of (2-8-) for elements 72 and above

**The systematic names and symbols for elements of atomic numbers 113 and above will be used until the approval of trivial names by IUPAC.

Source: *CRC Handbook of Chemistry and Physics*, 91st ed., 2010–2011, CRC Press

KEY

Atomic Mass → 12.011   −4 ← Selected Oxidation States
Symbol → **C**   +2 +4
    Relative atomic masses are based on $^{12}C = 12$ (exact)
Atomic Number → **6**
Electron Configuration → **2-4**

Note: Numbers in parentheses are mass numbers of the most stable or common isotope.

				Group				
			13	14	15	16	17	18

18
4.00260   0
He
2 / 2

13	14	15	16	17	18
10.81 +3 **B** 5 / 2-3	12.011 −4 +2 +4 **C** 6 / 2-4	14.0067 −3 −2 −1 +1 +2 +3 +4 +5 **N** 7 / 2-5	15.9994 −2 **O** 8 / 2-6	18.9984 −1 **F** 9 / 2-7	20.180 0 **Ne** 10 / 2-8
26.98154 +3 **Al** 13 / 2-8-3	28.0855 −4 +2 +4 **Si** 14 / 2-8-4	30.97376 −3 +3 +4 +5 **P** 15 / 2-8-5	32.065 −2 +4 +6 **S** 16 / 2-8-6	35.453 −1 +1 +5 +7 **Cl** 17 / 2-8-7	39.948 0 **Ar** 18 / 2-8-8

10	11	12	13	14	15	16	17	18
58.693 +2 +3 **Ni** 28 / 2-8-16-2	63.546 +1 +2 **Cu** 29 / 2-8-18-1	65.409 +2 **Zn** 30 / 2-8-18-2	69.723 +3 **Ga** 31 / 2-8-18-3	72.64 −4 +4 **Ge** 32 / 2-8-18-4	74.9216 −3 +3 +5 **As** 33 / 2-8-18-5	78.96 −2 +4 +6 **Se** 34 / 2-8-18-6	79.904 −1 +1 +5 **Br** 35 / 2-8-18-7	83.798 0 +2 **Kr** 36 / 2-8-18-8
106.42 +2 +4 **Pd** 46 / 2-8-18-18	107.868 +1 **Ag** 47 / 2-8-18-18-1	112.41 +2 **Cd** 48 / 2-8-18-18-2	114.818 +3 **In** 49 / 2-8-18-18-3	118.71 +2 +4 **Sn** 50 / 2-8-18-18-4	121.760 −3 +3 +5 **Sb** 51 / 2-8-18-18-5	127.60 −2 +4 +6 **Te** 52 / 2-8-18-18-6	126.904 −1 +1 +5 +7 **I** 53 / 2-8-18-18-7	131.29 0 +2 +4 +6 **Xe** 54 / 2-8-18-18-8
195.08 +2 +4 **Pt** 78 / -18-32-17-1	196.967 +1 +3 **Au** 79 / -18-32-18-1	200.59 +1 +2 **Hg** 80 / -18-32-18-2	204.383 +1 +3 **Tl** 81 / -18-32-18-3	207.2 +2 +4 **Pb** 82 / -18-32-18-4	208.980 +3 +5 **Bi** 83 / -18-32-18-5	(209) +2 +4 **Po** 84 / -18-32-18-6	(210) −1 **At** 85 / -18-32-18-7	(222) 0 **Rn** 86 / -18-32-18-8
(281) **Ds** 110	(280) **Rg** 111	(285) **Cn** 112	(284) **Uut** 113**	(289) **Uuq** 114	(288) **Uup** 115	(292) **Uuh** 116	(?) **Uus** 117	(294) **Uuo** 118

151.964 +2 +3 **Eu** 63	157.25 +3 **Gd** 64	158.925 +3 **Tb** 65	162.500 +3 **Dy** 66	164.930 +3 **Ho** 67	167.259 +3 **Er** 68	168.934 +3 **Tm** 69	173.04 +2 +3 **Yb** 70	174.9668 +3 **Lu** 71
(243) +3 +4 +5 +6 **Am** 95	(247) +3 **Cm** 96	(247) +3 +4 **Bk** 97	(251) +3 **Cf** 98	(252) +3 **Es** 99	(257) +3 **Fm** 100	(258) +2 +3 **Md** 101	(259) +2 +3 **No** 102	(262) +3 **Lr** 103

TABLE S
Properties of Selected Elements

Atomic Number	Symbol	Name	First Ionization Energy (kJ/mol)	Electro-negativity	Melting Point (K)	Boiling* Point (K)	Density** (g/cm³)	Atomic Radius (pm)
1	H	hydrogen	1312	2.2	14	20.	0.000082	32
2	He	helium	2372	—	—	4	0.000164	37
3	Li	lithium	520.	1.0	454	1615	0.534	130.
4	Be	beryllium	900.	1.6	1560.	2744	1.85	99
5	B	boron	801	2.0	2348	4273	2.34	84
6	C	carbon	1086	2.6	—	—	—	75
7	N	nitrogen	1402	3.0	63	77	0.001145	71
8	O	oxygen	1314	3.4	54	90.	0.001308	64
9	F	fluorine	1681	4.0	53	85	0.001553	60.
10	Ne	neon	2081	—	24	27	0.000825	62
11	Na	sodium	496	0.9	371	1156	0.97	160.
12	Mg	magnesium	738	1.3	923	1363	1.74	140.
13	Al	aluminum	578	1.6	933	2792	2.70	124
14	Si	silicon	787	1.9	1687	3538	2.3296	114
15	P	phosphorus (white)	1012	2.2	317	554	1.823	109
16	S	sulfur (monoclinic)	1000.	2.6	388	718	2.00	104
17	Cl	chlorine	1251	3.2	172	239	0.002898	100.
18	Ar	argon	1521	—	84	87	0.001633	101
19	K	potassium	419	0.8	337	1032	0.89	200.
20	Ca	calcium	590.	1.0	1115	1757	1.54	174
21	Sc	scandium	633	1.4	1814	3109	2.99	159
22	Ti	titanium	659	1.5	1941	3560.	4.506	148
23	V	vanadium	651	1.6	2183	3680.	6.0	144
24	Cr	chromium	653	1.7	2180.	2944	7.15	130.
25	Mn	manganese	717	1.6	1519	2334	7.3	129
26	Fe	iron	762	1.8	1811	3134	7.87	124
27	Co	cobalt	760.	1.9	1768	3200.	8.86	118
28	Ni	nickel	737	1.9	1728	3186	8.90	117
29	Cu	copper	745	1.9	1358	2835	8.96	122
30	Zn	zinc	906	1.7	693	1180.	7.134	120.
31	Ga	gallium	579	1.8	303	2477	5.91	123
32	Ge	germanium	762	2.0	1211	3106	5.3234	120.
33	As	arsenic (gray)	944	2.2	1090.	—	5.75	120.
34	Se	selenium (gray)	941	2.6	494	958	4.809	118
35	Br	bromine	1140.	3.0	266	332	3.1028	117
36	Kr	krypton	1351	—	116	120.	0.003425	116
37	Rb	rubidium	403	0.8	312	961	1.53	215
38	Sr	strontium	549	1.0	1050.	1655	2.64	190.
39	Y	yttrium	600.	1.2	1795	3618	4.47	176
40	Zr	zirconium	640.	1.3	2128	4682	6.52	164

TABLE S
Properties of Selected Elements (*Continued*)

Atomic Number	Symbol	Name	First Ionization Energy (kJ/mol)	Electro-negativity	Melting Point (K)	Boiling* Point (K)	Density** (g/cm³)	Atomic Radius (pm)
41	Nb	niobium	652	1.6	2750.	5017	8.57	156
42	Mo	molybdenum	684	2.2	2896	4912	10.2	146
43	Tc	technetium	702	2.1	2430.	4538	11	138
44	Ru	ruthenium	710.	2.2	2606	4423	12.1	136
45	Rh	rhodium	720.	2.3	2237	3968	12.4	134
46	Pd	palladium	804	2.2	1828	3236	12.0	130.
47	Ag	silver	731	1.9	1235	2435	10.5	136
48	Cd	cadmium	868	1.7	594	1040.	8.69	140.
49	In	indium	558	1.8	430.	2345	7.31	142
50	Sn	tin (white)	709	2.0	505	2875	7.287	140.
51	Sb	antimony (gray)	831	2.1	904	1860.	6.68	140.
52	Te	tellurium	869	2.1	723	1261	6.232	137
53	I	iodine	1008	2.7	387	457	4.933	136
54	Xe	xenon	1170.	2.6	161	165	0.005366	136
55	Cs	cesium	376	0.8	302	944	1.873	238
56	Ba	barium	503	0.9	1000.	2170.	3.62	206
57	La	lanthanum	538	1.1	1193	3737	6.15	194
			Elements 58–71 have been omitted.					
72	Hf	hafnium	659	1.3	2506	4876	13.3	164
73	Ta	tantalum	728	1.5	3290.	5731	16.4	158
74	W	tungsten	759	1.7	3695	5828	19.3	150.
75	Re	rhenium	756	1.9	3458	5869	20.8	141
76	Os	osmium	814	2.2	3306	5285	22.587	136
77	Ir	iridium	865	2.2	2719	4701	22.562	132
78	Pt	platinum	864	2.2	2041	4098	21.5	130.
79	Au	gold	890.	2.4	1337	3129	19.3	130.
80	Hg	mercury	1007	1.9	234	630.	13.5336	132
81	Tl	thallium	589	1.8	577	1746	11.8	144
82	Pb	lead	716	1.8	600.	2022	11.3	145
83	Bi	bismuth	703	1.9	544	1837	9.79	150.
84	Po	polonium	812	2.0	527	1235	9.20	142
85	At	astatine	—	2.2	575	—	—	148
86	Rn	radon	1037	—	202	211	0.009074	146
87	Fr	francium	393	0.7	300.	—	—	242
88	Ra	radium	509	0.9	969	—	5	211
89	Ac	actinium	499	1.1	1323	3471	10.	201
			Elements 90 and above have been omitted.					

* boiling point at standard pressure
** density of solids and liquids at room temperature and density of gases at 298 K and 101.3 kPa
— no data available
Source: *CRC Handbook for Chemistry and Physics*, 91st ed., 2010–2011, CRC Press

TABLE T
Important Formulas and Equations

Density	$d = \dfrac{m}{V}$	d = density m = mass V = volume
Mole Calculations	number of moles = $\dfrac{\text{given mass}}{\text{gram-formula mass}}$	
Percent Error	% error = $\dfrac{\text{measured value} - \text{accepted value}}{\text{accepted value}} \times 100$	
Percent Composition	% composition by mass = $\dfrac{\text{mass of part}}{\text{mass of whole}} \times 100$	
Concentration	parts per million = $\dfrac{\text{mass of solute}}{\text{mass of solution}} \times 1\,000\,000$	
	molarity = $\dfrac{\text{moles of solute}}{\text{liter of solution}}$	
Combined Gas Law	$\dfrac{P_1 V_1}{T_1} = \dfrac{P_2 V_2}{T_2}$	P = pressure V = volume T = temperature
Titration	$M_A V_A = M_B V_B$	M_A = molarity of H^+ M_B = molarity of OH^- V_A = volume of acid V_B = volume of base
Heat	$q = mC\Delta T$ $q = mH_f$ $q = mH_v$	q = heat H_f = heat of fusion m = mass H_v = heat of vaporization C = specific heat capacity ΔT = change in temperature
Temperature	$K = °C + 273$	K = kelvin $°C$ = degree Celsius

Using the Equations to Solve Chemistry Problems

IMPORTANT FORMULAS/EQUATIONS

In this section, various problems are solved using the formulas and equations found in Reference Table *T*.

1. Density: $d = \dfrac{m}{v}$

Problem

Calculate the mass of a sample of the element niobium (Atomic Number = 41) if the volume of the sample is 13.00 cm^3.

Solution

- Use Reference Table *S* to find the density of niobium (8.570 g/cm^3).
- Rearrange the variables in the equation in order to solve for mass: $m = d \cdot v$.
- Substitute the data in the equation and perform the calculation:

$$m = (8.570 \, \frac{g}{cm^3}) \cdot (13.00 \, cm^3) = \textbf{111.4 g}$$

2. Mole Calculations: number of moles $= \dfrac{\text{given mass}}{\text{gram-formula mass}}$

Problem

Calculate the number of moles in a sample of oxygen gas (O_2, gram-formula mass = 32.0 g/mol) if its mass is 56.0 grams.

Solution

- Substitute the data into the equation:

$$\text{number of moles} = \dfrac{56.0 \text{ g}}{32.0 \dfrac{\text{g}}{\text{mol}}} = \textbf{1.75 mol}$$

---

3. Percent Error: % error $= \dfrac{\text{measured value} - \text{accepted value}}{\text{accepted value}} \times 100$

Problem

A student measured the density of a sample of iron and obtained a value of 8.391 g/cm³. What is the percent error of the student's determination?

Solution

- Use Reference Table S to obtain the density of iron (7.874 g/cm³).
- Substitute the data into the equation:

$$\% \text{ error} = \dfrac{8.391 \dfrac{\text{g}}{\text{cm}^3} - 7.874 \dfrac{\text{g}}{\text{cm}^3}}{7.874 \dfrac{\text{g}}{\text{cm}^3}} \times 100 = \textbf{6.566\%}$$

---

4. Percent Composition: % composition by mass $= \dfrac{\text{mass of part}}{\text{mass of whole}} \times 100$

Problem

Calculate the percent composition by mass of carbon in $C_6H_{12}O_6$.

Solution

- Use the Periodic Table to determine the mass of one mole of the compound (180.2 g) and the mass of the *carbon* in 1 mole of the compound (72.07 g).
- Substitute the data into the equation:

$$\% \text{ composition} = \frac{72.07\,\text{g}}{180.2\,\text{g}} \times 100 = \mathbf{39.99\%}$$

5a. Parts per Million: $\dfrac{\text{grams of solute}}{\text{grams of solution}} \times 1{,}000{,}000$

Problem

The concentration of arsenic in a certain river is 0.267 gram per 2,000 grams of river water. What is the arsenic concentration in parts per million (ppm)?

Solution

- Substitute the data into the equation:

$$\frac{0.267\,\text{g}}{2{,}000\,\text{g}} \times 1{,}000{,}000 = \mathbf{134\ ppm}$$

5b. Molarity: $\text{molarity} = \dfrac{\text{grams of solute}}{\text{liters of solution}}$

Problem

What is the molarity of an aqueous solution of ammonia (gram-formula mass = 17.03 g/mol) if 30.8 grams of solute are dissolved in 2.50 liters of solution?

Solution

- Use the gram-formula mass to convert the mass to moles.
- Substitute the data into the equation:

$$\text{molarity} = \dfrac{(30.8 \text{ g}) \cdot \left(\dfrac{1 \text{ mol}}{17.03 \text{ g}}\right)}{2.50 \text{ L}} = \textbf{0.723 M}$$

---

6. Combined Gas Law: $\dfrac{P_1 V_1}{T_1} = \dfrac{P_2 V_2}{T_2}$

Problem

A 50.0-milliliter sample, initially at STP, has its pressure changed to 0.85 atmosphere and its temperature changed to 330 K. What is the new volume of the gas?

Solution

- Use Reference Table A to obtain the STP values for temperature (273 K) and pressure (1 atm).
- Rearrange the equation in order to solve for V_2: $V_2 = V_2 = \dfrac{P_1 V_1 T_2}{T_1 P_2}$
- Substitute the data into the equation:

$$V_2 = \dfrac{(1 \text{ atm}) \cdot (50.0 \text{ mL}) \cdot (330. \text{ K})}{(273 \text{ K}) \cdot (0.85 \text{ atm})} = \textbf{71.1 mL}$$

---

7. Titration (Acid–Base): $M_A V_A = M_B V_B$

Problem

How many milliliters of 0.25-molar NaOH are needed to neutralize 75 milliliters of 0.17-molar HCl?

Solution

- Rearrange the equation to solve for V_B: $V_B = \dfrac{M_A M_A}{M_B}$.
- Substitute the data into the equation:

$$V_B = \frac{(0.17 \text{ M}) \cdot (75 \text{ mL})}{(0.25 \text{ M})} = \textbf{51 mL (NaOH)}$$

8a. Heat Transferred: $q = mC\Delta T$

Problem

How much heat is released by 120 grams of liquid water as it cools from 340 K to 290 K?

Solution

- Use Reference Table B to obtain the specific heat capacity of liquid water (4.2 J/g·K).
- Substitute the data into the equation:

$$q = (120 \text{ g}) \cdot \left(4.2 \; \frac{\text{J}}{\text{g} \cdot \text{K}}\right) \cdot (290 \text{ K} - 340 \text{ K}) = \textbf{−25,000 J}$$

(The minus sign indicates that heat is *released*.)

8b. Heat of Fusion: $q = mH_f$

Problem

How much heat is absorbed by 200.0 grams of ice as it melts at 273 K?

Solution

- Use Reference Table B to obtain the value for the heat of fusion of ice (333.6 J/g).
- Substitute the data into the equation:

$$q = (200.0 \text{ g}) \cdot \left(333.6 \, \frac{\text{J}}{\text{g}}\right) = \mathbf{66,720 \text{ J}}$$

8c. Heat of Vaporization: $q = mH_v$

Problem

How many grams of steam at 100°C will be condensed to liquid water if 4,000. joules of heat are released?

Solution

- Use Reference Table B to obtain the value for the heat of vaporization of water (2,259 J/g).
- Rearrange the equation to solve for mass: $m = \dfrac{q}{H_v}$
- Substitute the data into the equation:

$$m = \frac{4,000. \text{ J}}{\left(2259 \, \dfrac{\text{J}}{\text{g}}\right)} = \mathbf{1.771 \text{ g}}$$

----

9. Temperature: $K = °C + 273$

Problem

What is the Kelvin temperature equivalent of $34°C$?

Solution

- Substitute the data into the equation:
  $K = 34°C + 273 = \textbf{307 K}$

10a. Radioactive Decay: fraction remaining $= \left(\dfrac{1}{2}\right)^{\frac{t}{T}}$

10b. Radioactive Decay: number of half-life periods $= \dfrac{t}{T}$

Problem

What fraction of a sample of ^{42}K will remain unchanged after 37.2 hours of decay?

Solution

- Use Reference Table N to obtain the half-life of ^{42}K (12.4 h).
- Use equation 10b to determine the number of half-life periods:

$$\text{number of half-life periods} = \frac{37.2\,\text{h}}{12.4\,\text{h}} = 3.00 = \textbf{3}$$

Substitute the number of half-life periods into equation 10a:

$$\text{fraction remaining} = \left(\frac{1}{2}\right)^{3} = \frac{1}{8}$$

That is, one-eighth of the original sample of ^{42}K remains unchanged after 37.2 hours of decay.

Regents Examinations, Answers, and Self-Analysis Charts

Examination
June 2016

Chemistry—Physical Setting

PART A

Answer all questions in this part.

Directions (1–30): For *each* statement or question, write in the answer space the *number* of the word or expression that, of those given, best completes the statement or answers the question. Some questions may require the use of the *2011 Edition Reference Tables for Physical Setting/Chemistry*.

1 Which statement describes the charge of an electron and the charge of a proton?

(1) An electron and a proton both have a charge of +1.
(2) An electron and a proton both have a charge of −1.
(3) An electron has a charge of +1, and a proton has a charge of −1.
(4) An electron has a charge of −1, and a proton has a charge of +1.

1 _____

2 Which subatomic particles are found in the nucleus of an atom of beryllium?

(1) electrons and protons   (3) neutrons and protons
(2) electrons and positrons   (4) neutrons and electrons

2 _____

3 The elements in Period 4 on the Periodic Table are arranged in order of increasing

(1) atomic radius
(2) atomic number
(3) number of valence electrons
(4) number of occupied shells of electrons

3 _____

4 Which phrase describes two forms of solid carbon, diamond and graphite, at STP?

  (1) the same crystal structure and the same properties
  (2) the same crystal structure and different properties
  (3) different crystal structures and the same properties
  (4) different crystal structures and different properties   4 _____

5 Which element has six valence electrons in each of its atoms in the ground state?

  (1) Se   (3) Kr
  (2) As   (4) Ga   5 _____

6 What is the chemical name for $H_2SO_3(aq)$?

  (1) sulfuric acid   (3) hydrosulfuric acid
  (2) sulfurous acid   (4) hydrosulfurous acid   6 _____

7 Which substance is most soluble in water?

  (1) $(NH_4)_3PO_4$ (3) Ag_2SO_4
  (2) $Cu(OH)_2$ (4) $CaCO_3$ 7 _____

8 Which type of bonding is present in a sample of an element that is malleable?

  (1) ionic   (3) nonpolar covalent
  (2) metallic   (4) polar covalent   8 _____

9 Which atom has the greatest attraction for the electrons in a chemical bond?

  (1) hydrogen   (3) silicon
  (2) oxygen   (4) sulfur   9 _____

10 Which type of reaction involves the transfer of electrons?

  (1) alpha decay   (3) neutralization
  (2) double replacement   (4) oxidation-reduction   10 _____

11 A 10.0-gram sample of nitrogen is at STP. Which property will increase when the sample is cooled to 72 K at standard pressure?

(1) mass
(2) volume
(3) density
(4) temperature

11 _____

12 Which element is a gas at STP?

(1) sulfur
(2) xenon
(3) potassium
(4) phosphorus

12 _____

13 A 5.0-gram sample of Fe(s) is to be placed in 100. milliliters of HCl(aq). Which changes will result in the fastest rate of reaction?

(1) increasing the surface area of Fe(s) and increasing the concentration of HCl(aq)
(2) increasing the surface area of Fe(s) and decreasing the concentration of HCl(aq)
(3) decreasing the surface area of Fe(s) and increasing the concentration of HCl(aq)
(4) decreasing the surface area of Fe(s) and decreasing the concentration of HCl(aq)

13 _____

14 Which process is commonly used to separate a mixture of ethanol and water?

(1) distillation
(2) ionization
(3) filtration
(4) titration

14 _____

15 A sample of hydrogen gas will behave most like an ideal gas under the conditions of

(1) low pressure and low temperature
(2) low pressure and high temperature
(3) high pressure and low temperature
(4) high pressure and high temperature

15 _____

16 The collision theory states that a reaction is most likely to occur when the reactant particles collide with the proper

(1) formula masses
(2) molecular masses
(3) density and volume
(4) energy and orientation

16 _____

17 At STP, which sample contains the same number of molecules as 3.0 liters of $H_2(g)$?

(1) 1.5 L of $NH_3(g)$ (3) 3.0 L of $CH_4(g)$

(2) 2.0 L of $CO_2(g)$ (4) 6.0 L of $N_2(g)$ 17 _____

18 The addition of a catalyst to a chemical reaction provides an alternate pathway that

(1) increases the potential energy of reactants

(2) decreases the potential energy of reactants

(3) increases the activation energy

(4) decreases the activation energy   18 _____

19 A sample of water is boiling as heat is added at a constant rate. Which statement describes the potential energy and the average kinetic energy of the water molecules in this sample?

(1) The potential energy decreases and the average kinetic energy remains the same.

(2) The potential energy decreases and the average kinetic energy increases.

(3) The potential energy increases and the average kinetic energy remains the same.

(4) The potential energy increases and the average kinetic energy increases.   19 _____

20 Entropy is a measure of the

(1) acidity of a sample

(2) disorder of a system

(3) concentration of a solution

(4) chemical activity of an element   20 _____

21 Which element has atoms that can bond with each other to form ring, chain, and network structures?

(1) aluminum   (3) carbon

(2) calcium   (4) argon   21 _____

22 What is the number of electrons shared in the multiple carbon-carbon bond in one molecule of 1-pentyne?

(1) 6

(2) 2

(3) 3

(4) 8

22 _____

23 Butanal, butanone, and diethyl ether have different properties because the molecules of each compound differ in their

(1) numbers of carbon atoms

(2) numbers of oxygen atoms

(3) types of functional groups

(4) types of radioactive isotopes

23 _____

24 What occurs when a magnesium atom becomes a magnesium ion?

(1) Electrons are gained and the oxidation number increases.

(2) Electrons are gained and the oxidation number decreases.

(3) Electrons are lost and the oxidation number increases.

(4) Electrons are lost and the oxidation number decreases.

24 _____

25 Energy is required to produce a chemical change during

(1) chromatography

(2) electrolysis

(3) boiling

(4) melting

25 _____

26 The reaction of an Arrhenius acid with an Arrhenius base produces water and

(1) a salt

(2) an ester

(3) an aldehyde

(4) a halocarbon

26 _____

27 One acid-base theory defines an acid as an

(1) H^- acceptor

(2) H^- donor

(3) H^+ acceptor

(4) H^+ donor

27 _____

28 Which phrase describes the decay modes and the half-lives of K-37 and K-42?

(1) the same decay mode but different half-lives

(2) the same decay mode and the same half-life

(3) different decay modes and different half-lives

(4) different decay modes but the same half-life

28 _____

29 Which particle has a mass that is approximately equal to the mass of a proton?

(1) an alpha particle   (3) a neutron
(2) a beta particle   (4) a positron   29 _____

30 Which change occurs during a nuclear fission reaction?

(1) Covalent bonds are converted to ionic bonds.
(2) Isotopes are converted to isomers.
(3) Temperature is converted to mass.
(4) Matter is converted to energy.   30 _____

PART B–1

Answer all questions in this part.

Directions (31–50): For *each* statement or question, write in the answer space the *number* of the word or expression that, of those given, best completes the statement or answers the question. Some questions may require the use of the *2011 Edition Reference Tables for Physical Setting/Chemistry*.

31 Which notations represent hydrogen isotopes?

(1) $_1^1H$ and $_1^2H$

(3) $_2^1H$ and $_3^1H$

(2) $_1^1H$ and $_2^4H$

(4) $_1^2H$ and $_2^7H$

31 _____

32 Naturally occurring gallium is a mixture of isotopes that contains 60.11% of Ga-69 (atomic mass = 68.93 u) and 39.89% of Ga-71 (atomic mass = 70.92 u). Which numerical setup can be used to determine the atomic mass of naturally occurring gallium?

(1) $\dfrac{(68.93\ u + 70.92\ u)}{2}$

(2) $\dfrac{(68.93\ u)(0.6011)}{(70.92\ u)(0.3989)}$

(3) $(68.93\ u)(0.6011) + (70.92\ u)(0.3989)$

(4) $(68.93\ u)(39.89) + (70.92\ u)(60.11)$

32 _____

33 Which list of symbols represents nonmetals, only?

(1) B, Al, Ga

(3) C, Si, Ge

(2) Li, Be, B

(4) P, S, Cl

33 _____

34 In the formula XSO_4, the symbol X could represent the element

(1) Al

(3) Mg

(2) Ar

(4) Na

34 _____

35 What is the chemical formula for lead(IV) oxide?

(1) PbO_2

(3) Pb_2O

(2) PbO_4

(4) Pb_4O

35 _____

36 Which statement describes the general trends in electronegativity and atomic radius as the elements in Period 2 are considered in order from left to right?

(1) Both electronegativity and atomic radius increase.
(2) Both electronegativity and atomic radius decrease.
(3) Electronegativity increases and atomic radius decreases.
(4) Electronegativity decreases and atomic radius increases.   36 _____

37 What is the percent composition by mass of nitrogen in $(NH_4)_2CO_3$ (gram-formula mass = 96.0 g/mol)?

(1) 14.6%   (3) 58.4%
(2) 29.2%   (4) 87.5%   37 _____

38 Given the balanced equation:

$$2KI + F_2 \rightarrow 2KF + I_2$$

Which type of chemical reaction does this equation represent?

(1) synthesis   (3) single replacement
(2) decomposition   (4) double replacement   38 _____

39 Which formula represents a nonpolar molecule containing polar covalent bonds?

H–H   O=C=O   H H H   H   H
                              N   O

(1)   (2)   (3)   (4)   39 _____

40 A reaction reaches equilibrium at 100.°C. The equation and graph representing this reaction are shown below.

$$N_2O_4(g) \rightleftharpoons 2NO_2(g)$$

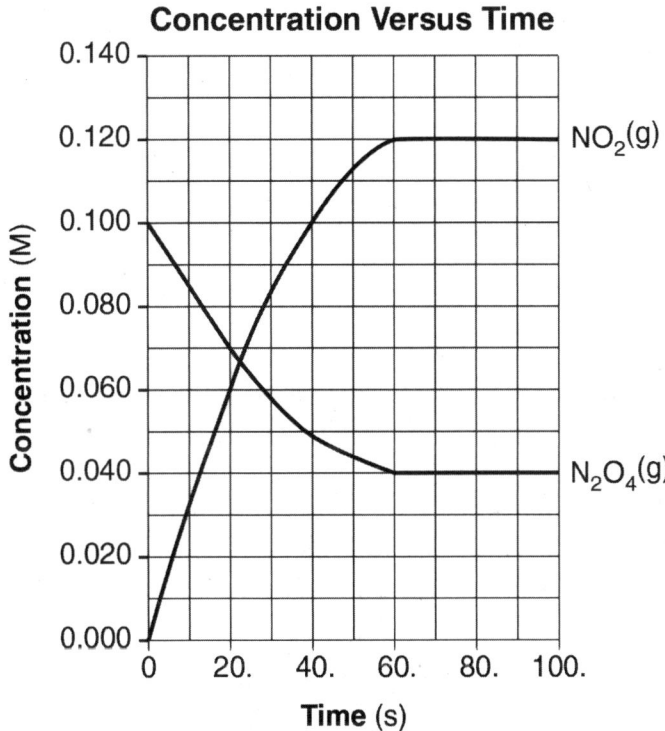

Concentration Versus Time

The graph shows that the reaction is at equilibrium after 60. seconds because the concentrations of both $NO_2(g)$ and $N_2O_4(g)$ are

(1) increasing
(2) decreasing

(3) constant
(4) zero

40 _____

41 Given the balanced equation representing a reaction:

$$2H_2O + energy \rightarrow 2H_2 + O_2$$

Which statement describes the changes in energy and bonding for the reactant?

(1) Energy is absorbed as bonds in H_2O are formed.
(2) Energy is absorbed as bonds in H_2O are broken.
(3) Energy is released as bonds in H_2O are formed.
(4) Energy is released as bonds in H_2O are broken.   41 _____

42 At standard pressure, what is the temperature at which a saturated solution of NH_4Cl has a concentration of 60. g $NH_4Cl/100.$ g H_2O?

(1) 66°C   (3) 22°C
(2) 57°C   (4) 17°C   42 _____

43 Which aqueous solution has the highest boiling point at standard pressure?

(1) 1.0 M KCl(aq)   (3) 2.0 M KCl(aq)
(2) 1.0 M $CaCl_2$(aq)   (4) 2.0 M $CaCl_2$(aq)   43 _____

44 Given the equation representing a system at equilibrium:

$$KNO_3(s) + energy \overset{H_2O}{\rightleftharpoons} K^+(aq) + NO_3^-(aq)$$

Which change causes the equilibrium to shift?

(1) increasing pressure   (3) adding a noble gas
(2) increasing temperature   (4) adding a catalyst   44 _____

45 Which hydrocarbon is saturated?

(1) C_2H_2 (3) C_4H_6
(2) C_3H_4 (4) C_4H_{10} 45 _____

46 Which volume of 0.600 M H_2SO_4(aq) exactly neutralizes 100. milliliters of 0.300 M $Ba(OH)_2$(aq)?

(1) 25.0 mL   (3) 100. mL
(2) 50.0 mL   (4) 200. mL   46 _____

47 Given the formula for an organic compound:

What is the name given to the group in the box?

(1) butyl   (3) methyl

(2) ethyl   (4) propyl   47 _____

48 Given the particle diagram:

Key
○ = atom of an element
● = atom of a different element

Which type of matter is represented by the particle diagram?

(1) an element

(2) a compound

(3) a homogeneous mixture

(4) a heterogeneous mixture   48 _____

49 Which substance is an electrolyte?

(1) O_2 (3) C_3H_8

(2) Xe   (4) KNO_3 49 _____

50 Which type of organic reaction produces both water and carbon dioxide?

(1) addition   (3) esterification

(2) combustion   (4) fermentation   50 _____

PART B–2

Answer all questions in this part.

Directions (51–65): Record your answers on the answer sheet provided. Some questions may require the use of the *2011 Edition Reference Tables for Physical Setting/Chemistry*.

51 Draw a Lewis electron-dot diagram for a chloride ion, Cl^-. [1]

Base your answers to questions 52 and 53 on the information below and on your knowledge of chemistry.

At STP, Cl_2 is a gas and I_2 is a solid. When hydrogen reacts with chlorine, the compound hydrogen chloride is formed. When hydrogen reacts with iodine, the compound hydrogen iodide is formed.

52 Balance the equation *on the answer sheet* for the reaction between hydrogen and chlorine, using the smallest whole-number coefficients. [1]

53 Explain, in terms of intermolecular forces, why iodine is a solid at STP but chlorine is a gas at STP. [1]

Base your answers to questions 54 and 55 on the information below and on your knowledge of chemistry.

Some properties of the element sodium are listed below.

- is a soft, silver-colored metal
- melts at a temperature of 371 K
- oxidizes easily in the presence of air
- forms compounds with nonmetallic elements in nature
- forms sodium chloride in the presence of chlorine gas

54 Identify *one* chemical property of sodium from this list. [1]

55 Convert the melting point of sodium to degrees Celsius. [1]

Base your answers to questions 56 through 58 on the information below and on your knowledge of chemistry.

At standard pressure, water has unusual properties that are due to both its molecular structure and intermolecular forces. For example, although most liquids contract when they freeze, water expands, making ice less dense than liquid water. Water has a much higher boiling point than most other molecular compounds having a similar gram-formula mass.

56 Explain why $H_2O(s)$ floats on $H_2O(\ell)$ when both are at 0°C.   [1]

57 State the type of intermolecular force responsible for the unusual boiling point of $H_2O(\ell)$ at standard pressure.   [1]

58 Determine the total amount of heat, in joules, required to completely vaporize a 50.0-gram sample of $H_2O(\ell)$ at its boiling point at standard pressure.   [1]

Base your answers to questions 59 and 60 on the information below and on your knowledge of chemistry.

At 1023 K and 1 atm, a 3.00-gram sample of $SnO_2(s)$ (gram-formula mass = 151 g/mol) reacts with hydrogen gas to produce tin and water, as shown in the balanced equation below.

$$SnO_2(s) + 2H_2(g) \rightarrow Sn(\ell) + 2H_2O(g)$$

59 Show a numerical setup for calculating the number of moles of $SnO_2(s)$ in the 3.00-gram sample. [1]

60 Determine the number of moles of $Sn(\ell)$ produced when 4.0 moles of $H_2(g)$ is completely consumed. [1]

Base your answers to questions 61 and 62 on the information below and on your knowledge of chemistry.

The incomplete data table below shows the pH value of solutions A and B and the hydrogen ion concentration of solution A.

Hydrogen Ion and pH Data for HCl(aq) Solutions

HCl(aq) Solution	Hydrogen Ion Concentration (M)	pH
A	1.0×10^{-2}	2.0
B	?	5.0

61 State the color of methyl orange in a sample of solution A. [1]

62 Determine the hydrogen ion concentration of solution B. [1]

Base your answers to questions 63 through 65 on the information below and on your knowledge of chemistry.

A sample of helium gas is placed in a rigid cylinder that has a movable piston. The volume of the gas is varied by moving the piston, while the temperature is held constant at 273 K. The volumes and corresponding pressures for three trials are measured and recorded in the data table below. For each of these trials, the product of pressure and volume is also calculated and recorded. For a fourth trial, only the volume is recorded.

**Pressure and Volume Data for
a Sample of Helium Gas at 273 K**

Trial Number	Pressure (atm)	Volume (L)	P x V (L•atm)
1	1.000	0.412	0.412
2	0.750	0.549	0.412
3	0.600	0.687	0.412
4	?	1.373	?

63 State evidence found in the data table that allows the product of pressure and volume for the fourth trial to be predicted.   [1]

64 Determine the pressure of the helium gas in trial 4.   [1]

65 Compare the average distances between the helium atoms in trial 1 to the average distances between the helium atoms in trial 3.   [1]

PART C

Answer all questions in this part.

Directions (66–85): Record your answers on the answer sheet provided. Some questions may require the use of the *2011 Edition Reference Tables for Physical Setting/Chemistry*.

Base your answers to questions 66 through 69 on the information below and on your knowledge of chemistry.

Potassium phosphate, K_3PO_4, is a source of dietary potassium found in a popular cereal. According to the Nutrition-Facts label shown on the boxes of this brand of cereal, the accepted value for a one-cup serving of this cereal is 170. milligrams of potassium. The minimum daily requirement of potassium is 3500 milligrams for an adult human.

66 Identify *two* types of chemical bonding in the source of dietary potassium in this cereal.   [1]

67 Identify the noble gas whose atoms have the same electron configuration as a potassium ion.   [1]

68 Compare the radius of a potassium ion to the radius of a potassium atom.   [1]

69 The mass of potassium in a one-cup serving of this cereal is determined to be 172 mg. Show a numerical setup for calculating the percent error for the mass of potassium in this serving.   [1]

Base your answers to questions 70 and 71 on the information below and on your knowledge of chemistry.

During photosynthesis, plants use carbon dioxide, water, and light energy to produce glucose, $C_6H_{12}O_6$, and oxygen. The reaction for photosynthesis is represented by the balanced equation below.

$$6CO_2 + 6H_2O + \text{light energy} \rightarrow C_6H_{12}O_6 + 6O_2$$

70 Write the empirical formula for glucose. [1]

71 State evidence that indicates photosynthesis is an endothermic reaction. [1]

Base your answers to questions 72 through 74 on the information below and on your knowledge of chemistry.

Fireworks that contain metallic salts such as sodium, strontium, and barium can generate bright colors. A technician investigates what colors are produced by the metallic salts by performing flame tests. During a flame test, a metallic salt is heated in the flame of a gas burner. Each metallic salt emits a characteristic colored light in the flame.

72 Explain why the electron configuration of 2-7-1-1 represents a sodium atom in an excited state.   [1]

73 Explain, in terms of electrons, how a strontium salt emits colored light.   [1]

74 State how bright-line spectra viewed through a spectroscope can be used to identify the metal ions in the salts used in the flame tests.   [1]

Base your answers to questions 75 through 77 on the information below and on your knowledge of chemistry.

The unique odors and flavors of many fruits are primarily due to small quantities of a certain class of organic compounds. The equation below represents the production of one of these compounds.

```
  H  H                    O                      O    H  H
  |  |                    ‖                      ‖    |  |
H−C−C−OH   +   H−C−O−H   ⟶   H−C−O−C−C−H   +   HOH
  |  |                                          |  |
  H  H                                          H  H

 Reactant 1        Reactant 2              Product 1        Product 2
```

75 Show a numerical setup for calculating the gramformula mass for reactant 1. [1]

76 Explain, in terms of molecular polarity, why reactant 2 is soluble in water. [1]

77 State the class of organic compounds to which product 1 belongs. [1]

Base your answers to questions 78 through 81 on the information below and on your knowledge of chemistry.

A student develops the list shown below that includes laboratory equipment and materials for constructing a voltaic cell.

Laboratory Equipment and Materials

- a strip of zinc
- a strip of copper
- a 250-mL beaker containing 150 mL of 0.1 M zinc nitrate
- a 250-mL beaker containing 150 mL of 0.1 M copper(II) nitrate
- wires
- a voltmeter
- a switch
- a salt bridge

78 State the purpose of the salt bridge in the voltaic cell.   [1]

79 Complete and balance the half-reaction equation *on your answer sheet* for the oxidation of the Zn(s) that occurs in the voltaic cell.   [1]

80 Compare the activities of the two metals used by the student for constructing the voltaic cell.   [1]

81 Identify *one* item of laboratory equipment required to build an electrolytic cell that is *not* included in the list.   [1]

Base your answers to questions 82 through 85 on the information below and on your knowledge of chemistry.

In 1896, Antoine H. Becquerel discovered that a uranium compound could expose a photographic plate wrapped in heavy paper in the absence of light. It was shown that the uranium compound was spontaneously releasing particles and high-energy radiation. Further tests showed the emissions from the uranium that exposed the photographic plate were *not* deflected by charged plates.

82 Identify the highly penetrating radioactive emission that exposed the photographic plates.   [1]

83 Complete the nuclear equation *on your answer sheet* for the alpha decay of U-238.   [1]

84 Determine the number of neutrons in an atom of U-233.   [1]

85 Identify the type of nuclear reaction that occurs when an alpha or a beta particle is spontaneously emitted by a radioactive isotope.   [1]

Answer Sheet
June 2016

Chemistry—Physical Setting

PART B–2

51

52 _____ $H_2(g)$ + _____ $Cl_2(g)$ → _____ $HCl(g)$

53 _____

_____

_____

54 _____

_____

55 _____ °C

56 _____

_____

_____

57 _____

58 _____ J

59

60 _____ mol

61 _____

62 _____ M

63 _____

_____

_____

64 _____ atm

65 _____

_____

_____

PART C

66 _____ and _____

67 _____

68 _____

_____

69

70 _____

71 _____

_____

_____

72 _____

_____

_____

73 _____

_____

_____

74 _____

_____

_____

75

76 _____

_____

_____

77 _____

78 _____

_____

79 $Zn(s) \rightarrow$ _____ + _____

80 _____

_____

_____

81 _____

82 _____

83 $^{238}_{92}U \rightarrow {}^{4}_{2}He +$ _____

84 _____

85 _____

Answers
June 2016

Chemistry—Physical Setting

Answer Key

PART A

1. 4	**7.** 1	**13.** 1	**19.** 3	**25.** 2
2. 3	**8.** 2	**14.** 1	**20.** 2	**26.** 1
3. 2	**9.** 2	**15.** 2	**21.** 3	**27.** 4
4. 4	**10.** 4	**16.** 4	**22.** 1	**28.** 3
5. 1	**11.** 3	**17.** 3	**23.** 3	**29.** 3
6. 2	**12.** 2	**18.** 4	**24.** 3	**30.** 4

PART B–1

31. 1	**35.** 1	**39.** 2	**43.** 4	**47.** 2
32. 3	**36.** 3	**40.** 3	**44.** 2	**48.** 2
33. 4	**37.** 2	**41.** 2	**45.** 4	**49.** 4
34. 3	**38.** 3	**42.** 1	**46.** 2	**50.** 2

PART B–2 and **PART C**. *See* **Answers Explained**.

Answers Explained

PART A

1. **4** The charges of all subatomic particles should be committed to memory, but *just to be sure*, you can consult Reference Table O, which shows the symbols, mass numbers, and charges for electrons, neutrons, and protons.

2. **3** The nucleus contains only protons and neutrons. Electrons are found in the electron cloud surrounding the nucleus.

3. **2** The Periodic Table found in the Reference Tables illustrates that the elements are organized in order of increasing atomic number.

WRONG CHOICES EXPLAINED:
(1) Atomic radius data is provided in Reference Table S. The atomic radius decreases in Period 4 of the Periodic Table. This happens because as elements have more protons in their nuclei, the stronger positive charge shrinks the size of the electron cloud.

(3) The number of valence electrons is the last number in the electron configurations provided in the Periodic Table of the Elements. The number of valence electrons does not increase regularly for the transition metals in Period 4.

(4) The number of occupied shells of electrons is 4 for all Period 4 elements. Notice that there are four values present in the electron configurations for Period 4 elements.

4. **4** Diamond and graphite are two *allotropes* of carbon, meaning that they are naturally occurring forms of carbon that have different structural arrangements. Diamond is arranged in a three-dimensional tetrahedral structure, and graphite is arranged in flat sheets, as shown in the illustrations below. Having different structures causes these substances to have different chemical and physical properties.

5. **1** Valence electrons are the electrons in the outer shell, or energy level, of an atom. These are the electrons involved in making chemical bonds with other atoms. The last number in an element's electron configuration is the number of valence electrons for that element. Looking at the Periodic Table of the Elements in the Reference Tables, the electron configuration for selenium (Se) is 2-8-18-6.

WRONG CHOICES EXPLAINED:

(2) The electron configuration for arsenic (As) is 2-8-18-5, so As has 5 valence electrons.

(3) The electron configuration for krypton (Kr) is 2-8-18-8, so Kr has 8 valence electrons.

(4) The electron configuration for gallium (Ga) is 2-8-18-3, so Ga has 3 valence electrons.

6. **2** If a compound's chemical formula begins with hydrogen, there is a good bet that the compound is an acid. One exception is water, H_2O, but even water does at times act as an acid. The chemical name for H_2SO_3 (aq) can be found in Reference Table K.

7. **1** To learn whether or not a compound is soluble in water, consult Reference Table F. The ammonium cation, NH_4^+, is listed in the first column of the table, which means that all compounds containing the ammonium ion are soluble in water, with no exceptions.

WRONG CHOICES EXPLAINED:

(2) The hydroxide ion (OH^-) forms insoluble compounds in water, and Cu^{2+} is not an exception to this insolubility.

(3) The sulfate ion (SO_4^{2-}) forms soluble compounds on water *most of the time*, but Ag^+ is listed as an exception to this solubility (making Ag_2SO_4 insoluble in water).

(4) The carbonate ion (CO_3^{2-}) forms insoluble compounds in water, and Ca^{2-} is not an exception to this insolubility.

8. **2** Malleability, the ability of a material to be hammered or pressed into thin sheets, is a physical property of metallic substances. Metals are also ductile (able to be stretched into wires), shiny (having metallic luster), and highly conductive. These properties all arise from the mobile sea of electrons, the key feature in metallic bonding.

9. **2** This is a "definition question" that requires the use of the Reference Tables. Electronegativity measures an atom's attraction for the electrons in a chemical bond. Electronegativity values can be found in Reference Table *S*. Of all the choices, oxygen's electronegativity of 3.4 is the highest.

10. **4** This is a "definition question!" An oxidation-reduction reaction (nicknamed a "redox" reaction) is defined as a chemical reaction that involves the transfer of electrons.

WRONG CHOICES EXPLAINED:
(1) Alpha decay is a form of radioactive decay, which is a type of nuclear reaction.
(2) A double replacement reaction is a reaction between two ionic compounds in which the cations and anions switch places, forming two new ionic compounds.
(3) A neutralization reaction takes place when an acid reacts with a base to form water and a salt.

11. **3** Reference Table *S* shows that nitrogen boils at a temperature of 77 K. If a sample of nitrogen gas at STP is cooled to 72 K, it will undergo a phase change from gas to liquid. Because particles are closer together in the liquid phase, the liquid phase is always more dense than the gas phase for any substance.

WRONG CHOICES EXPLAINED:
(1) Mass is unaffected by temperature change.
(2) The volume of a gas is directly related to temperature, so cooling a sample of gas will cause the volume to decrease. In addition, the phase change to the liquid state will cause an even more dramatic reduction in volume.
(4) Cooling from 273 K (standard temperature) to 72 K constitutes a decrease in temperature.

12. **2** Xenon is a noble gas, which by definition is a gas at room temperature. Data in Reference Table *S* shows that xenon boils at a temperature of 165 K, which is well below room temperature. All of the other answer choices are solids at room temperature.

13. **1** The rate of a chemical reaction increases when the number of collisions between reactant particles increases. If the concentration of HCl(aq) increases, there will be more HCl particles available to collide with Fe(s). Likewise, if the surface area of the Fe(s) increases by using smaller pieces of Fe(s), there will be more exposed surface available for collisions with HCl.

14. **1** Ethanol is a liquid organic compound that is completely soluble in water. Distillation is used to separate the components of a liquid mixture based on their different boiling points.

15. **2** Real gases behave more like ideal gases when temperature is high and pressure is low. These conditions minimize any possible attractions between gas particles, allowing the gas to adhere more closely to the principles of the kinetic molecular theory of gases.

16. **4** Particles involved in a reaction may collide frequently, but for a reaction to occur the collision must be an "effective" collision. Effective collisions have both sufficient energy and proper orientation.

17. **3** Equal volumes of gas at the same conditions of temperature and pressure contain the same number of particles. This relationship is known as Avogadro's Law. For example, the molar volume of a gas at STP = 22.4 L/mol no matter what type of gas you have.

18. **4** A catalyst lowers the activation energy of a chemical reaction as shown in the energy diagram below.

Since less energy is needed for a reaction to occur, more collisions will be effective and the reaction rate will increase.

19. **3** Temperature is directly related to kinetic energy, which is the energy of motion of particles in any sample of matter. During phase changes, such as the boiling of water, the temperature does not change, as shown in the diagram below.

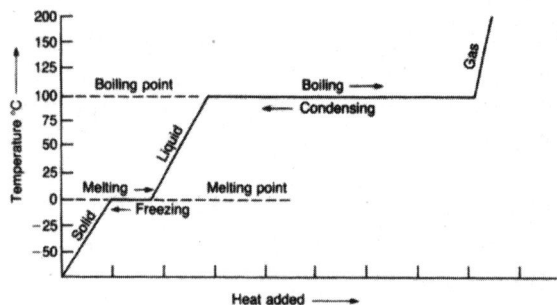

This means that the kinetic energy is constant during boiling. Heat energy added during boiling causes water molecules to overcome their attractions to each other and spread apart, forming water vapor. Increasing the distance between water molecules increases their potential energy.

20. **2** Entropy is defined as the degree of chaos, randomness, or disorder of a system. Remember this definition!

21. **3** Carbon has only four valence electrons, and so atoms of carbon will form four bonds to obtain a stable octet of eight valence electrons. Because carbon forms four bonds, it is capable of forming many different structures, including rings, chains, and networks.

WRONG CHOICES EXPLAINED:
(1), (2) Metals, like aluminum and calcium, form ionic compounds that have lattice structures.
(4) Argon is a very stable, monatomic noble gas that rarely, if ever, forms compounds.

22. **1** Reference Table Q shows that pentyne, a member of the alkyne family of hydrocarbons, contains a triple carbon-carbon bond. A single covalent bond consists of a *shared pair* of electrons. A triple bond would contain *three shared pairs* of electrons, for a total of six electrons shared between these two carbon atoms.

23. **3** Butanal is an aldehyde, butanone is a ketone, and diethyl ether is an ether. These compounds contain different organic functional groups, as shown on Reference Table R.

24. **3** The Periodic Table of the Elements shows that the magnesium ion has a charge of +2. Magnesium must lose two electrons to become the magnesium ion, Mg^{+2}. The oxidation number of simple ions is equal to the charge on the ion, so the magnesium ion has an oxidation number of +2. Since a magnesium atom has an oxidation number of 0, the oxidation number increases when the ion forms.

25. **2** Electrolysis, performed in an electrolytic cell, requires an input of energy from a power source to force electrons to move in a nonspontaneous direction.

WRONG CHOICES EXPLAINED:
(1) Chromatography is a method for separating liquid or gas mixtures. The separation of mixtures is a physical change, not a chemical change.
(3), (4) Boiling and melting are phase changes. Phase changes are reversible physical changes since no new substance forms.

26. **1** An example of an acid-base neutralization reaction is shown below:

$$HCl \ + \ NaOH \ \rightarrow \ H_2O \ + \ NaCl$$
$$(acid) \qquad (base) \qquad (water) \qquad (salt)$$

The products are always water and a salt in neutralization reactions between Arrhenius acids and bases. It might help to think of these reactions as a special type of double replacement reaction in which the ions trade places.

27. **4** The Brønsted-Lowry theory of acids and bases defines acids as proton (H^+) donors and bases as proton (H^+) acceptors. Remember this definition!

28. **3** Consult Reference Table N for information about decay modes and half-lives. K-37 and K-42 differ both in decay modes and half-lives.

29. **3** Consult Reference Table O for information about nuclear particles. Protons and neutrons both have a mass of 1 atomic mass unit (amu).

30. **4** Nuclear fission releases a tremendous amount of energy. This energy results from the conversion of a small amount of matter into energy according to Einstein's famous equation $E = mc^2$.

PART B–1

31. **1** Isotopes are different versions of an element with different numbers of neutrons. All hydrogen atoms must have an atomic number (the bottom number) of 1. Different isotopes of hydrogen will have different mass numbers (the top number) reflecting the different number of neutrons they contain.

WRONG CHOICES EXPLAINED:
   (2), (4) Hydrogen atoms can never have an atomic number of 2.
   (3) The atomic number, which is 1 for all atoms of hydrogen, must be written as the *bottom number* in standard isotopic notation.

32. **3** The average atomic mass of any element is calculated by taking the weighted average of all of its naturally occurring isotopes. The percent abundance of each naturally occurring isotope is multiplied by its atomic mass, resulting in an average that reflects how abundant each isotope is in nature.

33. **4** Nonmetals, with the exception of hydrogen, are located to the right of the staircase dividing line on the Periodic Table.

WRONG CHOICES EXPLAINED:
   (1) Aluminum (Al) and gallium (Ga) are metals (on the left side of the dividing line), and boron (B) is a metalloid. [The six metalloids must be memorized. They are B, Si, As, Te, Sb, and Ge. Look at the positions of these elements on the Periodic Table and visually memorize the physical pattern that they make.]
   (2) Lithium (Li) and beryllium (Be) are metals, and boron (B) is a metalloid.
   (3) Carbon (C) is a nonmetal, but silicon (Si) and germanium (Ge) are metalloids.

34. **3** You can find the charge of the sulfate ion, SO_4^{2-}, on Reference Table *E*. Compounds have no overall charge, so the -2 charge of the sulfate ion must be balanced out by a single ion having a charge of $+2$. Charges for ions of single elements can be found on the Periodic Table provided in your Reference Tables.

35. **1** A Roman numeral is provided in the names of many ionic compounds to indicate the charges on the metal ions in these compounds. The formula for lead(IV) oxide contains the Roman numeral IV, which is equal to 4. Looking at the Periodic Table of the Elements, we confirm that $+4$ is a possible charge for lead. Since all compounds are neutral in charge, and remembering that the oxide ion has a charge of -2 (also found on the Periodic Table), we can put together the following information:

$$Pb^{+4} \quad O^{-2}$$

To finish the chemical formula, two oxide ions are needed to offset the $+4$ charge on lead. The correct formula, PbO_2 is the only answer choice that is balanced in charge.

36. **3** Since electronegativity and atomic radius data for most of the elements is listed on Reference Table *S*, there is never any reason to guess on periodic trends for these properties. Looking at the elements in Period 2 of the Periodic Table, electronegativity *increases* and atomic radius *decreases* as the elements are considered in order from left to right. These trends are caused by increasing nuclear charge, which draws the electron cloud in closer (shrinks that atomic radius) and causes the nucleus of each successive element to be more attractive to electrons in chemical bonds (electronegativity).

37. **2** Using the percent composition formula on Reference Table *T*:

$$\% \text{ composition by mass} = \frac{\text{mass of part}}{\text{mass of whole}} \times 10$$

$\% \text{ composition by mass of nitrogen in } (NH_4)_2CO_3$

$$= \frac{(14.0 \text{ g/mol})(2)}{96.0 \text{ g/mol}} \times 100 = \frac{28.0 \text{ g/mol}}{96.0 \text{ g/mol}} \times 100 \approx 29.2\%$$

38. **3** A single replacement reaction takes place when a more reactive element replaces a less reactive element in a compound. Single replacement reactions can be recognized by the presence of *uncombined elements on each side of the equation*.

WRONG CHOICES EXPLAINED:
(1) Synthesis reactions follow the pattern $A + B \rightarrow AB$. Two reactants combine to form one product.
(2) Decomposition reactions follow the pattern $AB \rightarrow A + B$. One reactant breaks down to form two or more products.
(4) Double replacement reactions follow the pattern $AB + CD \rightarrow AD + CB$. Two compounds appear on both sides of the reaction, and the reactant cations switch places, forming the products.

39. **2** Bonds between two different nonmetal elements with different electronegativities are polar covalent bonds. Although the double bonds between C and O are polar covalent, the symmetry of the CO_2 molecule cancels the polarity of the bonds. The result is an overall nonpolar molecule.

WRONG CHOICES EXPLAINED:

(1) H−H contains a nonpolar covalent bond between two identical nonmetal atoms.

(3), (4) These molecules are not symmetrical. The polarity in their polar covalent bonds does not cancel out, so these will be polar molecules.

40. **3** When chemical equilibrium is established in a reaction system, the rates of the forward and reverse reactions are equal. Since reactant and product molecules are forming at equal rates, the concentrations of all molecules remain constant at equilibrium.

41. **2** Energy is required to break apart chemical bonds, in just the same way as it takes energy to split apart two magnets that are sticking together. The reaction shows water molecules decomposing into their constituent elements. Energy must be absorbed into the water molecules to break apart the bonds.

42. **1** Use Reference Table G to find this answer. Note that the information given, 60. g NH_4Cl in 100. g H_2O, is the y-axis coordinate. Look across to find the temperature that corresponds to this value on the solubility curve for NH_4Cl.

43. **4** Boiling point elevation is one of the colligative properties of solutions. The presence of solute particles causes water to boil at a higher temperature than it normally would, and the more concentrated the solute particles, the higher the boiling temperature becomes. A 2.0 M solution of $CaCl_2$ would yield the highest concentration of ions in solution.

44. **2** Le Châtelier's principle states that when a stress is applied to a system at equilibrium, the system will shift (react) to counteract the stress and restore equilibrium. Increasing the temperature on the reaction system shown would cause a shift to the right to consume the excess heat and bring the temperature back down again, thus restoring equilibrium.

WRONG CHOICES EXPLAINED:

(1), (3) Both refer to increasing the pressure within the reaction system. Pressure is only a stress on equilibrium if one of the reactants or products is a gas. There are no gases in the reactions shown.

(4) Catalysts increase the rate of *both the forward and reverse reactions* in a system at equilibrium, but the position of equilibrium is unaffected.

45. Saturated hydrocarbons contain only single covalent bonds between carbon atoms. Memorize this definition! Reference Table Q shows that alkanes have only single C−C bonds, so the answer will follow the general formula C_nH_{2n+2}.

46. **2** Use the titration formula provided on Reference Table T of your Reference Tables:

$$M_AV_A = M_BV_B$$

in which M_A = the molarity of H^+, V_A = volume of acid, M_B = molarity of OH^-, and V_B = volume of base.

In this case, since H_2SO_4 is a *diprotic* acid, two H^- ions will be released for every molecule that dissociates. We must double the molarity of the H_2SO_4 given to find M_A, the molarity of the H^- ions.

$$M_A = 2(0.600 \text{ M}) = 1.200 \text{ M}$$

Likewise, $Ba(OH)_2$ is a *dihydroxy* base and will release two moles of OH^- ions for every mole of base that dissociates. We must double the molarity of the $Ba(OH)_2$ given to find M_B, the molarity of OH^- ions.

$$M_B = 2(0.300 \text{ M}) = 0.600 \text{ M}$$

Substituting the values we have just determined and the other information given:

$$(1.200 \text{ M})(V_A) = 0.600 \text{ M} (100. \text{ mL})$$

$$V_A = \frac{(0.600 \text{ M})(100. \text{ mL})}{1.200 \text{ M}} = 50.0 \text{ mL}$$

47. **2** The two-carbon branch shown in the box is named following the IUPAC system for naming hydrocarbons. There are two carbons in the branch, so the prefix, found in Reference Table P is "eth-." The suffix for hydrocarbon branches that are not part of the longest chain of carbon atoms is "-yl."

48. **2** A compound consists of two or more elements that are chemically bonded together.

49. **4** The three classes of compounds that are known as electrolytes are acids, bases, and salts (ionic compounds). All of these compounds dissociate into ions in aqueous solution, and the ions formed cause the resulting solution to be capable of conducting electricity. KNO_3 is an ionic compound made from K^+ ions and NO_3^- ions.

50. **2** The general reaction for complete combustion is

$$C_xH_y + O_2 \rightarrow CO_2 + H_2O.$$

WRONG CHOICES EXPLAINED:
(1) Addition reactions typically produce halocarbons.
(3) Esterification reactions produce esters.
(4) Fermentation reactions produce alcohol and carbon dioxide.

PART B–2

[All questions in Part B–2 are worth 1 point.]

51. Lewis electron-dot diagrams for ions are drawn to show the valence electrons in the ion *and* the charge of the ion. Chlorine, as an element, has 7 valence electrons, but the chloride ion contains one additional electron, giving it a charge of -1, as shown in the correct diagram below. The convention is to always draw brackets around the octet of valence electrons for nonmetal ions, showing that any extra electrons have been *transferred completely* to the ion and will not be shared with another atom.

52. The balanced equation (using smallest whole-number coefficients) is

$$\underline{}\,H_2(g) + \underline{}\,Cl_2(g) \rightarrow \underline{2}\,HCl(g)$$

The basic idea when balancing chemical equations can be thought of as "Atoms In = Atoms Out." Since there are two atoms of hydrogen and chlorine on the left of the arrow in the equation above, there must be two atoms of each of these elements on the right side also.

(Note: Credit was granted even if coefficients of "1" were written on the reactant side of the equation.)

53. For iodine to be a solid at STP, its melting temperature must be greater than 0 °C, while chlorine must have a melting point below 0 °C. Iodine has a higher melting point because **it has stronger intermolecular forces between its molecules than chlorine does**.

54. The chemical properties of any substance are observed when the substance undergoes a chemical reaction. From the list of properties given, there are three correct answers: **sodium oxidizes easily in the presence of air, sodium forms compounds with nonmetallic elements in nature, and sodium forms sodium chloride in the presence of chlorine gas**. These properties all involve chemical reactions; the other properties are physical properties of the element sodium and do not involve chemical reactions.

55. The last equation on Reference Table T is the equation used to convert from the Celsius to the Kelvin scale and vice versa.

$$K = °C + 273$$

According to the information given in this problem, sodium melts at 371 K. The formula above can be rearranged to solve for degrees Celsius:

$$°C = K - 273$$

Plugging in:

$$°C = 371 - 273 = \textbf{98°C}$$

56. The answer to this question is stated directly in the information about water provided, making this a "careful reading" question. The second sentence states that **"when [water freezes, it] expands, making ice [$H_2O(s)$] less dense than liquid water [$H_2O(\ell)$]."**

57. **Hydrogen bonding** is an extremely strong type of intermolecular force. Water molecules are strongly attracted to each other because of hydrogen bonding, causing water to have a much higher boiling point than would be expected for such a small molecule. Hydrogen bonding occurs in compounds that have one or more hydrogen atoms *directly bonded to* nitrogen, oxygen, or fluorine atoms, which are all highly electronegative.

58. The last page of the Reference Tables, Reference Table T, lists the formula to find the heat required to vaporize any pure substance in the liquid phase:

$$q = mH_v$$

where q = heat, m = mass, and H_v is the heat of vaporization.
    The mass is given as 50.0 grams, and the heat of vaporization for liquid water, found on Reference Table B, is 2260 J/g. Plug these values into the formula and solve:

$$q = (50.0 \text{ g})(2260 \text{ J/g}) = \mathbf{113,000 \ J}$$

59. Use the Mole Calculations formula found on Reference Table T:

$$\text{number of moles} = \frac{\text{given mass}}{\text{gram-formula mass}}$$

According to the information given in this problem, the sample mass is 3.00 grams, and the gram-formula mass of SnO_2 (s) is 151 g/mol.
    To finish the problem, substitute the known values into the equation:

$$\textbf{number of moles } \mathbf{SnO_2} = \frac{\textbf{3.00 g}}{\textbf{151 g/mol}}$$

(NOTE: This problem only asks for a numerical setup. You are NOT required to solve for the answer. If you DO solve for the actual number of moles and you solve incorrectly, you will LOSE the point for this question even if you have shown the correct numerical setup. If you wish to solve for the number of moles to check that your setup is correct, do so on scrap paper.)

60. All coefficients in chemical reactions represent particles or moles. Mole ratios can be used to find the number of moles of any substance in a chemical equation if the moles of another substance are known. Two methods of setting up this problem are:

$$\frac{4.0 \; \cancel{\text{mol } H_2(g)}}{1} \times \frac{1 \text{ mol } Sn(\ell)}{2 \; \cancel{\text{mol } H_2(g)}} = \textbf{2.0 mol Sn}(\ell)$$

OR

$$\frac{2 \text{ mol } H_2(g)}{1 \text{ mol } Sn(\ell)} = \frac{4.0 \text{ mol } H_2(g)}{x \text{ mol } Sn(\ell)}$$

$$x = \textbf{2.0 mol Sn}(\ell)$$

61. The information presented in Reference Table *M* can be interpreted as follows: below a pH of 3.1, solutions would be red in the presence of methyl orange indicator, and above a pH of 4.4, solutions would appear yellow. Solutions with a pH between 3.1 and 4.4 would appear orange (a blend of the two colors).

This means that, according to Reference Table *M*, methyl orange indicator would appear **red** in a solution that has a pH of 2.0.

62. pH is defined as the *negative of the base 10 exponent* of the hydrogen (hydronium) ion concentration in any aqueous solution. For example, a solution that has a pH of 4 has an H^+ concentration of 1.0×10^{-4} M. Since the pH of solution *B* is 5.0, the molarity of solution *B* is $\textbf{1.0} \times \textbf{10}^{-5}$ **M**.

63. Since **P × V is constant at 0.412 L · atm for the first three trials**, a prediction can be made that the value will remain constant for the fourth trial.

64. Using the assumption that P × V for Trial 4 will be 0.412 L · atm, the pressure of the helium gas in this trial can be calculated.
We know that:

$$P \times V = 0.412 \text{ L} \cdot \text{atm}$$

Plugging in the volume of the gas in Trial 4:

$$P \times 1.373 \text{ L} = 0.412 \text{ L} \cdot \text{atm}$$

To solve, divide both sides by 1.373 L:

$$\frac{(P \times \cancel{1.373\,L})}{\cancel{1.373\,L}} = \frac{0.412\,\cancel{L} \cdot atm}{1.373\,\cancel{L}} = \textbf{0.300 atm (or 0.3 atm or 0.30 atm)}$$

65. When faced with problem-solving in chemistry, it often helps to draw a diagram of the system, or, at the very least, have a good picture in your mind. In this problem, a sample of gas is placed in a rigid cylinder with a movable piston, as shown in the diagram below.

If pressure is put on the piston, the gas sample will be squeezed into a smaller volume. Since the pressure is greater in trial 1, **the average distance between the helium atoms will be smaller in trial 1 than in trial 3**.

PART C

[All questions in Part C are worth 1 point.]

66. Potassium phosphate, K_3PO_4, is an ionic compound that consists of the potassium ion and the polyatomic phosphate ion, which contains the nonmetal elements phosphorus and oxygen. Remember, ionic bonds form when metals bond to nonmetals. However, within the phosphate ion, PO_4^{-3}, the phosphorus is *covalently* bonded to four different oxygen atoms. Covalent bonds form when nonmetal atoms bond to each other. Therefore, there are both **ionic and covalent** bonds present in the compound K_3PO_4.

67. Using the Periodic Table of the Elements provided in the Reference Tables, we learn that potassium makes ions having a charge of +1. A potassium atom must *lose* one electron to become an ion with a +1 charge. Neutral potassium atoms have 19 protons and 19 electrons, but after losing an electron, potassium ions have only 18 electrons, which is the number of electrons found in neutral atoms of the element **argon (Ar)**.

68. A simplified diagram of a potassium atom is shown below:

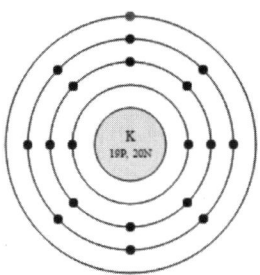

When a potassium atom becomes a potassium ion, K^{+1}, the valence electron in the 4th principal energy level is lost. Since the ion has no electrons occupying the 4th principal energy level, **the radius of the potassium ion is smaller than the radius of the potassium atom**.

69. As in question 59, you are required to write *only* a numerical setup to answer this question. See the note for question 59 concerning numerical setup problems.

From the information provided, the accepted value of potassium in a one-cup serving of cereal is 170. mg. The formula for percent error is found in Reference Table *T*:

$$\% \text{ error} = \frac{\text{measured value} - \text{accepted value}}{\text{accepted value}} \times 100$$

Plug in the measured value of 172 mg given in this problem and the accepted value of 170. mg:

$$\% \textbf{ error} = \frac{\textbf{172 mg} - \textbf{170. mg}}{\textbf{170. mg}} \times \textbf{100}$$

70. Empirical formulas give the smallest whole-number ratio of atoms in a compound, allowing one to see at a glance the proportion of atoms in the compound. An empirical formula reduces the subscripts in a compound formula into "lowest terms." Since the chemical formula of glucose is $C_6H_{12}O_6$, its empirical formula is $\textbf{CH}_2\textbf{O}$.

71. An endothermic reaction absorbs energy from the surroundings. The reading passage for this question states that "plants use carbon dioxide, water, and light energy..." This means that **light energy is absorbed by plants during photosynthesis**. Another way to answer this question is to note that **the energy term is on the left side of the equation** for photosynthesis. If energy is on the left side of the equation, it is consumed, just like the reactants, during the course of the reaction.

72. The Periodic Table in the Reference Tables shows that the ground-state electron configuration of a sodium atom is 2-8-1. A sodium atom would be in an excited state if its configuration were 2-7-1-1 because this configuration shows that **a second shell electron has moved to the fourth shell**. Another acceptable answer is that in the electron configuration 2-7-1-1, **not all of the 11 electrons are in the lowest possible energy levels**.

73. Since the question specifically states that students must answer *in terms of electrons,* you must specifically mention electrons in your response. Electrons within strontium atoms can move to higher energy levels within the atom when heat energy is absorbed from a gas burner during a flame test. **This energy is emitted as photons of light with specific wavelengths and colors as electrons drop back to their lower energy, ground-state positions**.

74. The color of light produced in a flame test can be separated into a bright line spectrum using a spectroscope. Each element produces a unique bright line spectrum. **By comparing the bright line spectrum observed through the spectroscope to spectra for known elements**, the identity of the metal ions in the salt can be established.

75. As in questions 59 and 69, you are required to write *only* a numerical setup to answer this question. See the note for question 59 concerning numerical setup problems.

The gram-formula mass, also known as the molar mass, can be calculated by adding the molar masses of all of the elements in a compound. First, notice that there are two carbon atoms, six hydrogen atoms, and one oxygen atom in Reactant 1. You can find the molar masses of each element in the Periodic Table provided in your Reference Tables.

Algebraically, the gram-formula mass can be expressed as:

$$\left[2 \times (\text{molar mass of carbon})\right] + \left[6 \times (\text{molar mass of hydrogen})\right]$$
$$+ \left[1 \times (\text{molar mass of oxygen})\right]$$

You are asked, however, for a *numerical* setup, so substituting for the molar masses of C, H, and O, the expression becomes:

$$(2 \times 12.0) + (6 \times 1.0) + (1 \times 16.0)$$

76. Water molecules are polar, meaning that water molecules have an end with a partial positive charge and an end with a partial negative charge as shown in the diagram below.

Water molecules will be attracted to other polar molecules since opposite charges will attract. **Reactant 2 will dissolve in water because molecules of both compounds are polar**.

77. Reference Table R is useful for classifying organic compounds. **Product 1 is an ester** because it matches the general formula for esters shown below and in Table R.

78. In a voltaic cell, **a salt bridge allows ions to flow into both compartments to preserve the balance of charge as the cell operates**.

79. Oxidation is defined as the loss of electrons. In this voltaic cell, the solid zinc electrode will oxidize (lose electrons), forming Zn^{+2} ions. This oxidation can be represented by the following half-reaction:

$$Zn(s) \rightarrow Zn^{2+} + 2e^{-}$$

80. Reference Table J lists metals from the most active metals at the top to the least active metals at the bottom. A more active metal will oxidize in the presence of a less active metal. Since zinc is higher than copper on Table J, **zinc (Zn) is more active than copper (Cu)**.

81. This question references the list of equipment needed for a *voltaic* cell, which utilizes a spontaneous redox reaction to generate electricity. *Electrolytic* cells are just the opposite: electricity from an external power source is used to drive a nonspontaneous redox reaction. An **external power source** needs to be added to the list of equipment in order to build an electrolytic cell.

82. The answer to this question is directly tied to two critical pieces of information about uranium emissions contained in the reading passage. First, the uranium compound emitted *high-energy radiation*. Furthermore, the emissions that exposed the photographic plate were *not* deflected by the charged plates. From these clues, we can conclude that **gamma radiation** is emitted by the uranium compound, since gamma radiation has the highest energy on the electromagnetic spectrum and, since it is pure energy, has neither mass nor charge.

83. When a uranium atom emits an alpha particle, it loses two protons and two neutrons from its nucleus, but most of the mass of the nucleus is left behind in the "daughter nuclide." To balance a nuclear equation, the sum of the atomic numbers (bottom) and the mass numbers (top) must be equal on both sides of the arrow. The nuclide that correctly completes the uranium alpha decay equation is

$$^{234}_{90}\text{Th}$$

84. Rewriting U-233 in isotopic notation is helpful for finding the number of neutrons contained in its nucleus. The atomic number of uranium is 92, which means that there are 92 protons contained in the nucleus of every uranium atom.

$$^{233}_{92}\text{U}$$

Subtract the atomic number (the bottom number) from the mass number (the top number) to find that there are **141 neutrons** in an atom of U-233.

85. The spontaneous emission of an alpha or a beta particle happens during the process of **radioactive decay**, which is also called **natural transmutation**.

Mark (✓) the questions you answered correctly. Count the number of checks and follow the formulas given to determine your score on each topic.

Core Area	☐ Questions Answered Correctly

69: (1)

Section M—Math Skills
☐ Number of checks ÷ 1 × 100 = ___ %

1, 2, 29, 31, 32, 72–74, 84: (9)

Section I—Atomic Concepts
☐ Number of checks ÷ 9 × 100 = ___ %

3–5, 12, 33, 36, 54: (7)

Section II—Periodic Table
☐ Number of checks ÷ 7 × 100 = ___ %

6, 34, 35, 37, 38, 52, 59, 60, 70, 75: (10)

Section III—Moles/Stoichiometry
☐ Number of checks ÷ 10 × 100 = ___ %

9, 39, 41, 51, 66–68: (7)

Section IV—Chemical Bonding
☐ Number of checks ÷ 7 × 100 = ___ %

7, 8, 11, 14, 15, 17, 19, 42, 43, 48, 53, 55 58, 63 65, 71, 76: (20)

Section V—Physical Behavior of Matter
☐ Number of checks ÷ 20 × 100 = ___ %

13, 16, 18, 20, 40, 44: (6)

Section VI—Kinetics and Equilibrium
☐ Number of checks ÷ 6 × 100 = ___ %

21–23, 45, 47, 50, 77: (7)

Section VII—Organic Chemistry
☐ Number of checks ÷ 7 × 100 = ___ %

Core Area	☐ Questions Answered Correctly

10, 24, 25, 78–81: (7)

Section VIII—Oxidation–Reduction
☐ Number of checks ÷ 7 × 100 = ___ %

6, 26, 27, 46, 49, 61, 62: (7)

Section IX—Acids, Bases, and Salts
☐ Number of checks ÷ 7 × 100 = ___ %

28, 30, 82, 83, 85: (5)

Section X—Nuclear Chemistry
☐ Number of checks ÷ 5 × 100 = ___ %

Examination
August 2016
Chemistry—Physical Setting

PART A

Answer all questions in this part.

Directions (1–30): For *each* statement or question, write in the answer space the *number* of the word or expression that, of those given, best completes the statement or answers the question. Some questions may require the use of the *2011 Edition Reference Tables for Physical Setting/Chemistry.*

1 Which change occurs when an atom in an excited state returns to the ground state?

(1) Energy is emitted.
(2) Energy is absorbed.
(3) The number of electrons decreases.
(4) The number of electrons increases.

1 _____

2 The valence electrons in an atom of phosphorus in the ground state are all found in

(1) the first shell   (3) the third shell
(2) the second shell   (4) the fourth shell

2 _____

3 Which two elements have the most similar chemical properties?

(1) beryllium and magnesium
(2) hydrogen and helium
(3) phosphorus and sulfur
(4) potassium and strontium

3 _____

4 Which phrase describes a compound that consists of two elements?

   (1) a mixture in which the elements are in a variable proportion
   (2) a mixture in which the elements are in a fixed proportion
   (3) a substance in which the elements are chemically combined in a variable proportion
   (4) a substance in which the elements are chemically combined in a fixed proportion   4 _____

5 The formula mass of a compound is the

   (1) sum of the atomic masses of its atoms
   (2) sum of the atomic numbers of its atoms
   (3) product of the atomic masses of its atoms
   (4) product of the atomic numbers of its atoms   5 _____

6 The arrangement of the elements from left to right in Period 4 on the Periodic Table is based on

   (1) atomic mass
   (2) atomic number
   (3) the number of electron shells
   (4) the number of oxidation states   6 _____

7 Which diatomic molecule is formed when the two atoms share six electrons?

   (1) H_2 (3) O_2
   (2) N_2 (4) F_2 7 _____

8 Which formula represents a polar molecule?

   (1) O_2 (3) NH_3
   (2) CO_2 (4) CH_4 8 _____

9 Which element is *least* likely to undergo a chemical reaction?

   (1) lithium   (3) fluorine
   (2) carbon   (4) neon   9 _____

10 Which element has a melting point higher than the melting point of rhenium?

(1) iridium (3) tantalum

(2) osmium (4) tungsten 10 _____

11 Which property can be defined as the ability of a substance to be hammered into thin sheets?

(1) conductivity (3) melting point

(2) malleability (4) solubility 11 _____

12 Which list of elements consists of a metal, a metalloid, and a noble gas?

(1) aluminum, sulfur, argon

(2) magnesium, sodium, sulfur

(3) sodium, silicon, argon

(4) silicon, phosphorus, chlorine 12 _____

13 Which sample of matter has a crystal structure?

(1) $Hg(\ell)$ (3) $NaCl(s)$

(2) $H_2O(\ell)$ (4) $CH_4(g)$ 13 _____

14 One mole of liquid water and one mole of solid water have *different*

(1) masses (3) empirical formulas

(2) properties (4) gram-formula masses 14 _____

15 Which substance can *not* be broken down by a chemical change?

(1) butanal (3) gold

(2) propene (4) water 15 _____

16 Which statement describes particles of an ideal gas, based on the kinetic molecular theory?

(1) Gas particles are separated by distances smaller than the size of the gas particles.

(2) Gas particles do not transfer energy to each other when they collide.

(3) Gas particles have no attractive forces between them.

(4) Gas particles move in predictable, circular motion. 16 _____

17 Which expression could represent the concentration of a solution?

(1) 3.5 g   (3) 3.5 mL
(2) 3.5 M   (4) 3.5 mol   17 _____

18 Which form of energy is associated with the random motion of the particles in a sample of water?

(1) chemical energy   (3) nuclear energy
(2) electrical energy   (4) thermal energy   18 _____

19 Which change is most likely to occur when a molecule of H_2 and a molecule of I_2 collide with proper orientation and sufficient energy?

(1) a chemical change, because a compound is formed
(2) a chemical change, because an element is formed
(3) a physical change, because a compound is formed
(4) a physical change, because an element is formed   19 _____

20 Which changes can reach dynamic equilibrium?

(1) nuclear changes, only
(2) chemical changes, only
(3) nuclear and physical changes
(4) chemical and physical changes   20 _____

21 What occurs when a reaction reaches equilibrium?

(1) The concentration of the reactants increases.
(2) The concentration of the products increases.
(3) The rate of the forward reaction is equal to the rate of the reverse reaction.
(4) The rate of the forward reaction is slower than the rate of the reverse reaction.   21 _____

22 In terms of potential energy, PE, which expression defines the heat of reaction for a chemical change?

(1) $PE_{products} - PE_{reactants}$ (3) $\dfrac{PE_{products}}{PE_{reactants}}$

(2) $PE_{reactants} - PE_{products}$ (4) $\dfrac{PE_{reactants}}{PE_{products}}$ 22 _____

23 Systems in nature tend to undergo changes that result in

    (1) lower energy and lower entropy
    (2) lower energy and higher entropy
    (3) higher energy and lower entropy
    (4) higher energy and higher entropy   23 _____

24 What occurs when Cr^{3+} ions are reduced to Cr^{2+} ions?

    (1) Electrons are lost and the oxidation number of chromium increases.
    (2) Electrons are lost and the oxidation number of chromium decreases.
    (3) Electrons are gained and the oxidation number of chromium increases.
    (4) Electrons are gained and the oxidation number of chromium decreases.   24 _____

25 Where do reduction and oxidation occur in an electrolytic cell?

    (1) Both occur at the anode.
    (2) Both occur at the cathode.
    (3) Reduction occurs at the anode, and oxidation occurs at the cathode.
    (4) Reduction occurs at the cathode, and oxidation occurs at the anode.   25 _____

26 Which compound is an electrolyte?

    (1) H_2O (3) H_3PO_4
    (2) C_2H_6 (4) CH_3OH 26 _____

27 When the hydronium ion concentration of an aqueous solution is increased by a factor of 10, the pH value of the solution

    (1) decreases by 1   (3) decreases by 10
    (2) increases by 1   (4) increases by 10   27 _____

28 The stability of isotopes is related to the ratio of which particles in the atoms?

    (1) electrons and protons   (3) neutrons and protons
    (2) electrons and positrons   (4) neutrons and positrons   28 _____

29 Which radioisotope has the fastest rate of decay?

(1) ^{14}C (3) ^{53}Fe

(2) ^{37}Ca (4) ^{42}K 29 _____

30 The atomic mass of an element is the weighted average of the atomic masses of

(1) the least abundant isotopes of the element
(2) the naturally occurring isotopes of the element
(3) the artificially produced isotopes of the element
(4) the natural and artificial isotopes of the element   30 _____

PART B–1

Answer all questions in this part.

Directions (31–50): For *each* statement or question, write in the answer space the *number* of the word or expression that, of those given, best completes the statement or answers the question. Some questions may require the use of the *2011 Edition Reference Tables for Physical Setting/Chemistry.*

31 Which list of elements is arranged in order of increasing electronegativity?

(1) Be, Mg, Ca   (3) K, Ca, Sc
(2) F, Cl, Br   (4) Li, Na, K   31 _____

32 The table below gives the masses of two different subatomic particles found in an atom.

Subatomic Particles and Their Masses

Subatomic Particle	Mass (g)
X	1.67×10^{-24}
Z	9.11×10^{-28}

Which of the subatomic particles are each paired with their corresponding name?

(1) X, proton and Z, electron
(2) X, proton and Z, neutron
(3) X, neutron and Z, proton
(4) X, electron and Z, proton   32 _____

33 Which electron configuration represents an excited state for an atom of calcium?

(1) 2-8-7-1   (3) 2-8-7-3
(2) 2-8-7-2   (4) 2-8-8-2   33 _____

34 At STP, graphite and diamond are two solid forms of carbon. Which statement explains why these two forms of carbon differ in hardness?

    (1) Graphite and diamond have different ionic radii.
    (2) Graphite and diamond have different molecular structures.
    (3) Graphite is a metal, but diamond is a nonmetal.
    (4) Graphite is a good conductor of electricity, but diamond is a poor conductor of electricity.   34 _____

35 Which equation shows conservation of charge?

    (1) $Cu + Ag^+ \rightarrow Cu^{2+} + Ag$
    (2) $Mg + Zn^{2+} \rightarrow 2Mg^{2+} + Zn$
    (3) $2F_2 + Br^- \rightarrow 2F^- + Br_2$
    (4) $2I^- + Cl_2 \rightarrow I_2 + 2Cl^-$ 35 _____

36 What occurs when potassium reacts with chlorine to form potassium chloride?

    (1) Electrons are shared and the bonding is ionic.
    (2) Electrons are shared and the bonding is covalent.
    (3) Electrons are transferred and the bonding is ionic.
    (4) Electrons are transferred and the bonding is covalent.   36 _____

37 Given the balanced equation representing a reaction:

$$H_2 + energy \rightarrow H + H$$

What occurs as bonds are broken in one mole of H_2 molecules during this reaction?

    (1) Energy is absorbed and one mole of unbonded hydrogen atoms is produced.
    (2) Energy is absorbed and two moles of unbonded hydrogen atoms are produced.
    (3) Energy is released and one mole of unbonded hydrogen atoms is produced.
    (4) Energy is released and two moles of unbonded hydrogen atoms are produced.   37 _____

38 Which pair of atoms has the most polar bond?

   (1) H−Br   (3) I−Br

   (2) H−Cl   (4) I−Cl   38 _____

39 Which two notations represent isotopes of the same element?

   (1) $^{14}_{7}N$ and $^{18}_{7}N$ (3) $^{14}_{7}N$ and $^{17}_{10}Ne$

   (2) $^{20}_{7}N$ and $^{20}_{10}Ne$ (4) $^{19}_{7}N$ and $^{16}_{10}Ne$ 39 _____

40 The graph below shows the volume and the mass of four different substances at STP.

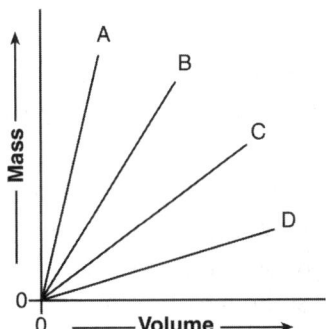

Which of the four substances has the *lowest* density?

   (1) A   (3) C

   (2) B   (4) D   40 _____

41 What is the total amount of heat required to completely melt 347 grams of ice at its melting point?

   (1) 334 J   (3) 116 000 J

   (2) 1450 J   (4) 784 000 J   41 _____

42 As the temperature of a reaction increases, it is expected that the reacting particles collide

   (1) more often and with greater force

   (2) more often and with less force

   (3) less often and with greater force

   (4) less often and with less force   42 _____

43 Given the formula representing a compound:

$$
\begin{array}{ccccccc}
 & H & H & H & O & H & H \\
 & | & | & | & \| & | & | \\
H- & C- & C- & C- & C- & C- & C-H \\
 & | & | & | & & | & | \\
 & H & H & H & & H & H
\end{array}
$$

What is an IUPAC name for this compound?

(1) ethyl propanoate   (3) 3-hexanone

(2) propyl ethanoate   (4) 4-hexanone   43 _____

44 A voltaic cell converts chemical energy to

(1) electrical energy with an external power source

(2) nuclear energy with an external power source

(3) electrical energy without an external power source

(4) nuclear energy without an external power source   44 _____

45 Which acid and base react to form water and sodium sulfate?

(1) sulfuric acid and sodium hydroxide

(2) sulfuric acid and potassium hydroxide

(3) sulfurous acid and sodium hydroxide

(4) sulfurous acid and potassium hydroxide   45 _____

46 Given the equation representing a reaction:

$$H_2CO_3 + NH_3 \rightarrow NH_4^+ + HCO_3^-$$

According to one acid-base theory, the compound NH_3 acts as a base because it

(1) accepts a hydrogen ion

(2) donates a hydrogen ion

(3) accepts a hydroxide ion

(4) donates a hydroxide ion   46 _____

47 Which statement describes characteristics of a 0.01 M KOH(aq) solution?

(1) The solution is acidic with a pH less than 7.
(2) The solution is acidic with a pH greater than 7.
(3) The solution is basic with a pH less than 7.
(4) The solution is basic with a pH greater than 7.   47 ____

48 Four statements about the development of the atomic model are shown below.

A: Electrons have wavelike properties.

B: Atoms have small, negatively charged particles.

C: The center of an atom is a small, dense nucleus.

D: Atoms are hard, indivisible spheres.

Which order of statements represents the historical development of the atomic model?

(1) $C \rightarrow D \rightarrow A \rightarrow B$ (3) $D \rightarrow B \rightarrow A \rightarrow C$
(2) $C \rightarrow D \rightarrow B \rightarrow A$ (4) $D \rightarrow B \rightarrow C \rightarrow A$ 48 ____

49 Five cubes of iron are tested in a laboratory. The tests and the results are shown in the table below.

Iron Tests and the Results

Test	Procedure	Result
1	A cube of Fe is hit with a hammer.	The cube is flattened.
2	A cube of Fe is placed in 3 M HCl(aq).	Bubbles of gas form.
3	A cube of Fe is heated to 1811 K.	The cube melts.
4	A cube of Fe is left in damp air.	The cube rusts.
5	A cube of Fe is placed in water.	The cube sinks.

Which tests demonstrate chemical properties?

(1) 1, 3, and 4   (3) 2 and 4
(2) 1, 3, and 5   (4) 2 and 5   49 ____

50 A rigid cylinder with a movable piston contains a sample of helium gas. The temperature of the gas is held constant as the piston is pulled outward. Which graph represents the relationship between the volume of the gas and the pressure of the gas?

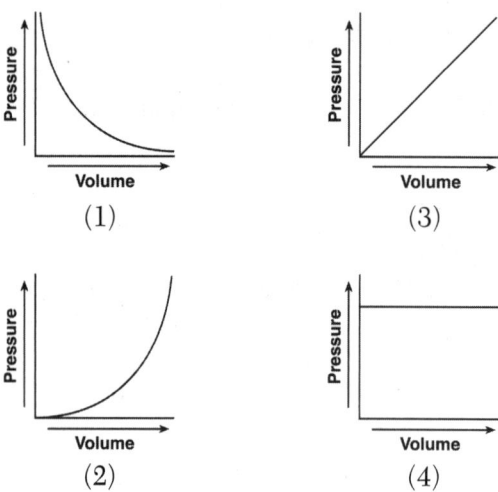

(1)   (3)

(2)   (4)

50 _____

PART B–2

Answer all questions in this part.

Directions (51–65): Record your answers on the answer sheet provided. Some questions may require the use of the *2011 Edition Reference Tables for Physical Setting/Chemistry*.

51 What is the empirical formula for C_6H_{12}? [1]

52 Using Table G, determine the minimum mass of NaCl that must be dissolved in 200. grams of water to produce a saturated solution at 90.°C. [1]

53 State the physical property that makes it possible to separate a solution by distillation. [1]

Base your answers to questions 54 and 55 on the information below and on your knowledge of chemistry.

A beaker contains a liquid sample of a molecular substance. Both the beaker and the liquid are at 194 K. The graph below represents the relationship between temperature and time as the beaker and its contents are cooled for 12 minutes in a refrigerated chamber.

54 State what happens to the average kinetic energy of the molecules in the sample during the first 3 minutes.   [1]

55 Identify the physical change occurring during the time interval, minute 4 to minute 9.   [1]

Base your answers to questions 56 through 59 on the information below and on your knowledge of chemistry.

The equation below represents a reaction between propene and hydrogen bromide.

Cyclopropane, an isomer of propene, has a boiling point of $-33°C$ at standard pressure and is represented by the formula below.

56 Explain why this reaction can be classified as a synthesis reaction.   [1]

57 Identify the class of organic compounds to which the product of this reaction belongs.   [1]

58 Explain, in terms of molecular formulas and structural formulas, why cyclopropane is an isomer of propene.   [1]

59 Convert the boiling point of cyclopropane at standard pressure to kelvins.   [1]

Base your answers to questions 60 through 63 on the information below and on your knowledge of chemistry.

The radius of a lithium atom is 130. picometers, and the radius of a fluorine atom is 60. picometers. The radius of a lithium ion, Li^+, is 59 picometers, and the radius of a fluoride ion, F^-, is 133 picometers.

60 Compare the radius of a fluoride ion to the radius of a fluorine atom.   [1]

61 Explain, in terms of subatomic particles, why the radius of a lithium ion is smaller than the radius of a lithium atom.   [1]

62 In the space *on your answer sheet*, draw a Lewis electron-dot diagram for a fluoride ion.   [1]

63 Describe the general trend in atomic radius as each element in Period 2 is considered in order from left to right.   [1]

Base your answers to questions 64 and 65 on the information below and on your knowledge of chemistry.

Nuclear fission reactions can produce different radioisotopes. One of these radioisotopes is Te-137, which has a half-life of 2.5 seconds. The diagram below represents one of the many nuclear fission reactions.

64 State evidence that this nuclear reaction represents transmutation.   [1]

65 Complete the nuclear equation *on your answer sheet* for the beta decay of Zr-97, by writing an isotopic notation for the missing product.   [1]

PART C

Answer all questions in this part.

Directions (66–85): Record your answers on the answer sheet provided. Some questions may require the use of the *2011 Edition Reference Tables for Physical Setting/Chemistry*.

Base your answers to questions 66 through 69 on the information below and on your knowledge of chemistry.

Stamping an identification number into the steel frame of a bicycle compresses the crystal structure of the metal. If the number is filed off, there are scientific ways to reveal the number.

One method is to apply aqueous copper(II) chloride to the number area. The Cu^{2+} ions react with some iron atoms in the steel frame, producing copper atoms that show the pattern of the number. The ionic equation below represents this reaction.

$$Fe(s) + Cu^{2+}(aq) \rightarrow Fe^{2+}(aq) + Cu(s)$$

Another method is to apply hydrochloric acid to the number area. The acid reacts with the iron, producing bubbles of hydrogen gas. The bubbles form faster where the metal was compressed, so the number becomes visible. The equation below represents this reaction.

$$2HCl(aq) + Fe(s) \rightarrow FeCl_2(aq) + H_2(g)$$

66 Explain why the Fe atoms in the bicycle frame react with the Cu^{2+} ions.   [1]

67 Determine the number of moles of hydrogen gas produced when 0.001 mole of HCl(aq) reacts completely with the iron metal.   [1]

68 Write a balanced half-reaction equation for the reduction of the hydrogen ions to hydrogen gas.   [1]

69 Describe *one* change in the HCl(aq) that will increase the rate at which hydrogen bubbles are produced when the acid is applied to the steel frame.   [1]

Base your answers to questions 70 through 73 on the information below and on your knowledge of chemistry.

In an investigation, aqueous solutions are prepared by completely dissolving a different amount of NaCl(s) in each of four beakers containing 100.00 grams of $H_2O(\ell)$ at room temperature. Each solution is heated and the temperature at which boiling occurred is measured. The data are recorded in the table below.

Boiling Point Data for Four NaCl(aq) Solutions

Beaker Number	Mass of $H_2O(\ell)$ (g)	Mass of NaCl(s) Dissolved (g)	Boiling Point of Solution (°C)
1	100.00	8.76	101.5
2	100.00	17.52	103.1
3	100.00	26.28	104.6
4	100.00	35.04	106.1

70 Identify the solute and the solvent used in this investigation. [1]

71 Show a numerical setup for calculating the percent by mass of NaCl in the solution in beaker 4. [1]

72 Explain, in terms of ions, why the ability to conduct an electric current is greater for the solution in beaker 4 than for the solution in beaker 1. [1]

73 State the relationship between the concentration of ions and the boiling point for these solutions. [1]

Base your answers to questions 74 through 76 on the information below and on your knowledge of chemistry.

One type of voltaic cell, called a mercury battery, uses zinc and mercury(II) oxide to generate an electric current. Mercury batteries were used because of their miniature size, even though mercury is toxic. The overall reaction for a mercury battery is given in the equation below.

$$Zn(s) + HgO(s) \rightarrow ZnO(s) + Hg(\ell)$$

74 Determine the change in the oxidation number of zinc during the operation of the cell.   [1]

75 Compare the number of moles of electrons lost to the number of moles of electrons gained during the reaction.   [1]

76 Using information in the passage, state *one* risk and *one* benefit of using a mercury battery.   [1]

Base your answers to questions 77 through 80 on the information below and on your knowledge of chemistry.

A company produces a colorless vinegar that is 5.0% $HC_2H_3O_2$ in water. Using thymol blue as an indicator, a student titrates a 15.0-milliliter sample of the vinegar with 43.1 milliliters of a 0.30 M NaOH(aq) solution until the acid is neutralized.

77 Based on Table M, what is the color of the indicator in the vinegar solution before any base is added?   [1]

78 Identify the negative ion in the NaOH(aq) used in this titration.   [1]

79 The concentration of the base used in this titration is expressed to what number of significant figures?   [1]

80 Determine the molarity of the $HC_2H_3O_2$ in the vinegar sample, using the titration data.   [1]

Base your answers to questions 81 through 85 on the information below and on your knowledge of chemistry.

In industry, ethanol is primarily produced by two different reactions. One process involves the reaction of glucose in the presence of an enzyme that acts as a catalyst. The equation below represents this reaction.

$$\text{Equation 1: } C_6H_{12}O_6 \xrightarrow{\text{enzyme}} 2CH_3CH_2OH + 2CO_2$$
$$\underset{\text{glucose}}{\phantom{C_6H_{12}O_6}} \qquad \underset{\text{ethanol}}{}$$

In another reaction, ethanol is produced from ethene and water. The equation below represents this reaction in which H_2SO_4 is a catalyst.

$$\text{Equation 2: } CH_2CH_2 + H_2O \xrightarrow{H_2SO_4} CH_3CH_2OH$$

Industrial ethanol can be oxidized using a catalyst to produce ethanal. The equation representing this oxidation is shown below.

$$\text{Equation 3: } CH_3CH_2OH \xrightarrow{\text{catalyst}} CH_3CHO + H_2$$

81 Identify the element that causes the reactant in equation 1 to be classified as an organic compound.   [1]

82 Identify the type of organic reaction represented by equation 1.   [1]

83 Explain why the hydrocarbon in equation 2 is unsaturated.   [1]

84 Explain, in terms of intermolecular forces, why ethanol has a much higher boiling point than ethene, at standard pressure.   [1]

85 Draw a structural formula for the organic product in equation 3.   [1]

Answer Sheet
August 2016

Chemistry—Physical Setting

PART B–2

51 _____

52 _____ g

53 _____

54 _____

_____

_____

55 _____

56 _____

_____

_____

57 _____

58 _____

_____

_____

59 _____ **K**

60 _____

_____

_____

61 _____

_____

_____

62

63 _____

_____

_____

64 _____

_____

_____

65 $^{97}_{40}\text{Zr} \rightarrow \, ^{0}_{-1}\text{e} +$ _____

PART C

66 _____

_____

_____

67 _____ mol

68 _____

69 _____

_____

_____

70 Solute: _____

Solvent: _____

71

72 _____

_____

_____

73 _____

_____

_____

74 From _____ to _____

75 _____

_____

_____

76 Risk: _____

Benefit: _____

77 _____

78 _____

79 _____

80 _____ **M**

81 _____

82 _____

83 _____

_____

_____

84 _____

_____

_____

85

Answers
August 2016

Chemistry—Physical Setting

Answer Key

PART A

1. 1	7. 2	13. 3	19. 1	25. 4
2. 3	8. 3	14. 2	20. 4	26. 3
3. 1	9. 4	15. 3	21. 3	27. 1
4. 4	10. 4	16. 3	22. 1	28. 3
5. 1	11. 2	17. 2	23. 2	29. 2
6. 2	12. 3	18. 4	24. 4	30. 2

PART B–1

31. 3	35. 4	39. 1	43. 3	47. 4
32. 1	36. 3	40. 4	44. 3	48. 4
33. 3	37. 2	41. 3	45. 1	49. 3
34. 2	38. 2	42. 1	46. 1	50. 1

PART B–2 and **PART C**. *See* **Answers Explained**.

Answers Explained

PART A

1. **1** An atom in an excited state contains at least one electron that has previously absorbed energy and has moved to a higher energy orbital within the atom. When the electron returns to the lowest energy position possible, also known as the ground state, the extra energy that was absorbed is emitted as a photon of electromagnetic energy.

2. **3** Valence electrons are the outermost electrons in an atom. The electron configuration for phosphorus, 2-8-5, is shown in the Periodic Table of the Elements in your Chemistry Reference Tables. This configuration shows that three principal energy levels (shells) within atoms of phosphorus contain electrons: 2 electrons in the first shell, 8 electrons in the second shell, and 5 electrons (the valence electrons) in the third shell.

3. **1** Elements in the same vertical group have the same number of valence electrons, which means that they have the same number of electrons available for bonding. Having the same number of valence electrons causes elements in the same group to have similar chemical properties.

4. **4** Compounds are substances composed of two or more elements that are chemically bonded in a fixed proportion. Compounds can be represented by chemical formulas (for example, H_2O), which show the relative proportion of each element in the compound.

WRONG CHOICES EXPLAINED:
(1), (2) A compound is a pure substance, whereas a mixture is made from two or more substances that are blended together but are not chemically bonded.
(3) As a pure substance, a compound is represented by a chemical formula that is always the same and does not vary. For example, water always has the same formula, H_2O.

5. **1** To find the formula mass (molar mass) of a compound, you must add up the atomic masses of all atoms in the compound. The atomic mass for each element is found on the Periodic Table of the Elements in your Reference Tables. For example, the formula mass of the compound carbon dioxide, CO_2, is found by adding the atomic mass of one carbon atom to the atomic mass of two oxygen atoms, as shown below:

$$\text{formula mass of } CO_2 =$$
$$[1 \times (\text{formula mass of C}) + 2 \times (\text{formula mass of oxygen})]$$

Substituting values for the formula masses of carbon and oxygen:

$$\text{formula mass of } CO_2 = [12.0 \text{ amu} + 2(16.0 \text{ amu})] = 44.0 \text{ amu}$$

6. **2** The Periodic Table of the Elements found in the Reference Tables illustrates that the elements are organized in order of increasing atomic number.

WRONG CHOICES EXPLAINED:
(1) The atomic mass *usually* increases from left to right across the Periodic Table of the Elements, but there are exceptions. In Period 4, for example, the atomic mass *decreases* between Co (58.9332 amu) and Ni (58.693 amu).
(3) The number of electron shells is 4 for all Period 4 elements. Notice that there are four values present in all of the electron configurations for Period 4 elements.
(4) The oxidation states for each element are shown in the upper right-hand corner of the element symbol boxes on the Periodic Table of the Elements. Oxidation states vary inconsistently across Period 4.

7. **2** All of the answer choices are covalently-bonded, diatomic molecules. A single covalent bond consists of one shared pair of electrons. For 6 electrons to be shared, there would need to be *three* covalent bonds joining the two atoms. Only nitrogen, N_2, contains a triple covalent bond. It forms this bond because each nitrogen atom has only 5 valence electrons and needs 3 *additional electrons* to form a stable octet of 8 valence electrons.

WRONG CHOICES EXPLAINED:
(1) Each H atom has only 1 valence electron and can form only one covalent bond.
(3) Each O atom has 6 valence electrons and will form two covalent bonds to obtain the stable octet.
(4) Each F atom has 7 valence electrons and will form just one covalent bond to obtain the stable octet.

8. **3** Ammonia (NH_3) molecules are polar, meaning that ammonia molecules have an end with a partial positive charge and an end with a partial negative charge as shown in the diagram below.

$$N^{\delta-}$$

$$H^{\delta+} \quad | \quad H^{\delta+}$$

$$H^{\delta+}$$

These partial charges arise from the polar covalent bonds between the nitrogen and hydrogen atoms. The nitrogen atom is more electronegative than the H atoms, so shared electrons will be attracted more strongly to nitrogen, giving it a partially negative charge.

WRONG CHOICES EXPLAINED:

(1), (2), (4) Each of these compounds contains polar covalent bonds in its molecules, but the overall symmetry of these molecules leads to a symmetrical charge distribution. A good expression to remember is that "symmetry cancels polarity."

9. **4** Neon is a noble gas in Group 18 of the Periodic Table of the Elements. Since elements in this group have 8 valence electrons, they are very stable and do not often form bonds with other elements.

10. **4** According to Reference Table S, the melting point of rhenium (Re) is 3458 K. Tungsten (W) has a melting point of 3695 K.

11. **2** This is a definition question. Malleability is a characteristic property of metals, along with excellent conductivity, ductility (the ability to be drawn into wires), and metallic luster.

12. **3** Metals are located to the left of the staircase dividing line on the Periodic Table of the Elements. Sodium, in Group 1, is a metal. Nonmetals are located to the right of the staircase dividing line on the Periodic Table of the Elements. Argon, in Group 18, is a noble gas. The six metalloids must be memorized. They are B, Si, As, Te, Sb, and Ge. It is easier to remember which elements are metalloids by visually memorizing the physical pattern that they make on the Periodic Table of the Elements.

WRONG CHOICES EXPLAINED:

(1), (2) These answer choices do not include a metalloid.

(4) This answer choice does not include a metal.

13. **3** Solids have a characteristic crystalline lattice structure. Both gases and liquids take the shape of their container because their particles have the ability to move.

14. **2** Water is water, no matter what phase it happens to be in. Properties of liquids and solids are very different, however. For example, ice floats on water, a demonstration of the different densities of these two phases.

WRONG CHOICES EXPLAINED:
(1) One mole of water has a mass of 18.0 grams, no matter what phase it is in.
(3) The empirical formula of water is the same as its molecular formula (H_2O). This formula is the same for all phases of water.
(4) The gram-formula (molar) mass of water is 18.0 g/mol, regardless of its phase.

15. **3** Chemical compounds can be broken down into their constituent elements by chemical change. Elements cannot be broken down by chemical change.

16. **3** One of the principles of the kinetic molecular theory of gases is that gas particles have no attraction for or repulsion to each other. This allows gas particles to have random motion and to expand to fill the space that they occupy. Other principles of the kinetic molecular theory of gases state that an ideal gas consists of particles that:

- Have mass
- Are separated by distances much greater than their size
- Move in random, constant, straight-line motion
- Undergo perfectly elastic collisions in which the total energy of the system remains the same before and after collisions

17. **2** All concentration units give information about how crowded solute particles are in a certain volume of solution. Common solution concentration units include molarity (M), which means moles per liter, percent by volume, percent by mass, and parts per million (ppm).

18. **4** Thermal energy is associated with the random motion of atoms and molecules. Thermal energy is created by the motion of atoms and molecules within an object. The faster the particles move, the more thermal energy the object has and the hotter the object will be. Chemical energy, on the other hand, refers to energy stored in chemical bonds.

19. **1** Proper orientation and sufficient energy are the two requirements for an effective collision. Effective collisions result in chemical reactions (chemical changes) in which chemical bonds are broken and formed. A chemical reaction between H_2 and I_2 would produce the compound HI.

20. **4** Some chemical and physical changes can reach the state of dynamic equilibrium, in which a reversible process occurs in both directions at equal rates. Chemical equilibrium occurs in a closed system when the forward and reverse chemical reactions occur at the same rate. An example of dynamic equilibrium in a physical change is solution equilibrium, when the rates of dissolving and precipitation in a saturated solution are equal.

WRONG CHOICES EXPLAINED:
(1), (3) One reason why nuclear changes can never reach a state of dynamic equilibrium is that the system cannot be closed. Nuclear particles and high-energy electromagnetic radiation that is released during nuclear reactions can travel through the walls of most containers.
(2) Some physical changes can also reach a state of dynamic equilibrium, as in the solution equilibrium described above. Vapor pressure equilibrium is another example of physical equilibrium, in which evaporation and condensation proceed at equal rates in a closed flask containing a pure liquid.

21. **3** When a chemical reaction reaches equilibrium, the rates of the forward and reverse reactions are equal. Since reactant and product molecules are forming at equal rates, the concentrations of all molecules remain constant at equilibrium.

22. **1** The heat of reaction (ΔH) is defined as the potential energy of the products of a chemical reaction minus the potential energy of the reactants. In other words, the heat of reaction is the net change in energy during the reaction. A good analogy for this is elevation change. If you walk up a hill, your elevation increases, which corresponds to a positive change in elevation (elevation at the end of your walk – elevation at the beginning). Likewise, walking down a hill would result in a negative change in elevation.

23. **2** Systems in nature tend to undergo changes that produce lower total energy and greater disorder or randomness (entropy). For example, an avalanche of rock debris down a hillside moves toward a position of lower potential energy and much greater randomness.

24. **4** The chemical process of reduction involves the gain of electrons. This is a definition you must know. A helpful mnemonic device for this definition is OIL RIG, which translates to Oxidation Is Loss and Reduction Is Gain. If the Cr^{3+} ion gains 1 electron, the oxidation number of the ion decreases from 3+ to 2+.

25. **4** The anode is defined as the site of oxidation, and the cathode is defined as the site of reduction in all electrochemical cells, both voltaic and electrolytic. A great mnemonic expression for remembering this definition is the expression "An Ox, Red Cat" (oxidation occurs at the anode and reduction happens at the cathode).

26. **3** Electrolytes are compounds that dissociate into ions when dissolved in water, creating aqueous solutions that are capable of conducting electricity. Acids, bases, and ionic compounds in general are three classes of compounds that have electrolytic properties. H_3PO_4 is an acid that delivers H^+ ions into solution (see Reference Table K), so it is an electrolyte.

27. **1** On the pH scale, each decrease of one unit of pH represents a 10-fold increase in the hydronium ion concentration. This is because the pH scale relates to the base 10 exponent of the hydronium ion concentration.

28. **3** Neutrons and protons reside in the nucleus of an atom. The ratio between the two causes a nucleus to be either stable or unstable.

WRONG CHOICES EXPLAINED:
(1), (2) Electrons are found outside of the nucleus and therefore do not affect its stability.
(2), (4) Positrons are positively charged bits of antimatter. They are the size of an electron but have the opposite charge. Positrons are sometimes emitted by unstable nuclei as they undergo radioactive decay. They have nothing at all to do with the stability of an atom's nucleus.

29. **2** Look at Reference Table N for the half-lives of these radioisotopes. ^{37}Ca, with a half-life of only 182 ms, decays far faster than any of the other isotopes offered as answer choices.

30. **2** The average atomic mass of any element is calculated by taking the weighted average of all of the naturally occurring isotopes of that element. The percent abundance of each naturally occurring isotope is multiplied by its atomic mass, resulting in an average that reflects how abundant each isotope is in nature.

PART B–1

31. **3** Electronegativity data for many elements is listed in Reference Table S. Looking at data for the sets of elements shown as answer choices, only the progression K, Ca, Sc is arranged in order of increasing electronegativity. Electronegativity (attraction for electrons) increases from left to right across the Periodic Table of the Elements because of increasing nuclear charge.

32. **1** Consult Reference Table O for information about nuclear particles. According to this table, protons have a mass of 1 amu, which is actually equal to 1.67×10^{-24} grams. Electrons, which are identical to beta particles, have a mass of 0 according to Reference Table O. Of course, electrons do not really have a mass of 0, but they do have a mass that is far less than the mass of a proton.

33. **3** The ground-state electron configuration for calcium, 2-8-8-2, is found on the Periodic Table of the Elements in the Reference Tables. A calcium atom would be in an excited state if its configuration were 2-8-7-3 because this configuration shows that a third shell electron has moved to the fourth shell.

WRONG CHOICES EXPLAINED:
(1), (2) Calcium atoms have 20 electrons. Neither of these electron configurations has the correct number of electrons.
(4) This is the ground-state electron configuration for calcium. All of the electrons are in the lowest possible energy levels.

34. **2** Diamond and graphite are two *allotropes* of carbon, meaning that they are naturally occurring forms of carbon that have different structural arrangements. Diamond is arranged in a three-dimensional tetrahedral structure, and graphite is arranged in flat sheets, as shown in the illustrations below. Having different structures causes these substances to have different properties.

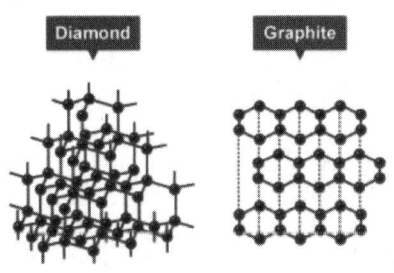

35. **4** In a redox reaction, the number of electrons lost is equal to the number of electrons gained. In the following equation:

$$2I^- + Cl_2 \rightarrow I_2 + 2Cl^-$$

1 electron is lost by each I^- ion as it oxidizes to form I_2, and 1 electron is gained by each Cl atom as it reduces to form Cl^-.

WRONG CHOICES EXPLAINED:
(1) Cu would have to lose 2 electrons to form Cu^{2+}, while Ag^+ would only have to gain 1 electron to form Ag^0.
(2), (3) These equations are not balanced for mass, let alone charge!

36. **3** The key to answering this question is to recognize that electrons are transferred from one atom to another only when an ionic bond forms (electrons are shared in covalent bonds). Ionic bonds form when metals, such as potassium, bond with nonmetals, such as chlorine.

37. **2** Energy is absorbed when bonds break, and energy is released or given off when bonds form. The equation in this question shows a bond breaking as H_2 absorbs energy and breaks apart into two separate atoms.

38. **2** Polarity across a bond is caused by the difference in electronegativity between the two elements involved in the bond. If the electronegativity difference is small, the bond is less polar, whereas if the electronegativity difference is large, the bond will be highly polar with a strongly negative end and a strongly positive end. Look up the electronegativity of the elements in this question on Reference Table *S* to confirm that the H–Cl bond has a higher electronegativity difference than any of the other answer choices.

39. **1** Isotopes are different versions of an element with different numbers of neutrons. All nitrogen atoms must have an atomic number (the bottom number) of 7. Different isotopes of nitrogen will have different mass numbers (the top number), reflecting the number of neutrons they contain.

WRONG CHOICES EXPLAINED:
(2), (3), (4) Nitrogen (N) and neon (Ne) are different elements. Isotopes are different versions of *the same element.*

40. **4** Density equals mass divided by volume. This formula is found on Reference Table T. Since mass is the y-axis coordinate, and volume is the x-axis coordinate, the slope of each line, which is equal to $\dfrac{\Delta y}{\Delta x}$, is equal to the density of the substance. The graph for substance D has the lowest slope and therefore the lowest density.

41. **3** The last page of the Reference Tables, Reference Table T, lists the formula to find the heat required to melt any pure substance in the solid phase:

$$q = mH_f$$

where q = heat, m = mass, and H_f is the heat of fusion. The mass is given as 347 grams, and the heat of fusion for ice (water), found on Reference Table B, is 334 J/g. Plug these values into the formula and solve:

$$q = (347 \text{ g})(334 \text{ J/g}) \approx 116\,000 \text{ J}$$

42. **1** Temperature is directly related to the average kinetic energy of particles in any sample of matter. This relationship is one of the key points in the kinetic molecular theory of matter. Kinetic energy is the energy of motion. As the temperature of a reaction increases, the reacting particles will move faster and will collide more often and with more force.

43. **3** The compound in this question contains the functional group of a ketone, as shown on Reference Table R:

$$
\begin{array}{c}
\text{O} \\
\| \\
-\,\text{C}\,-
\end{array}
$$

From the example for ketones in this table, we know that the suffix for ketones is "-one." There are 6 carbon atoms in the compound in this question, so the name will have the prefix "hex-." Additionally, a number is assigned for the position of the functional group. IUPAC rules for naming organic compounds state that the carbon atoms should be numbered so as to give the functional group the lowest possible numbered position. The functional group is on the third carbon from the right, so putting all of the pieces together, the correct name for this compound is 3-hexanone.

WRONG CHOICES EXPLAINED:
(1), (2) These names are for esters, not ketones.
(4) While the ketone functional group is on the fourth carbon atom from the left, IUPAC rules require us to assign the lowest possible number to the position of the functional group. We can place the ketone group on the third carbon if we number from the right.

44. **3** A voltaic cell spontaneously converts chemical energy to electrical energy. No outside energy source is required.

45. **1** The neutralization reaction between sulfuric acid and sodium hydroxide is

$$H_2SO_4 + 2NaOH \rightarrow Na_2SO_4 + 2H_2O$$

Formulas for both sulfuric acid and sodium hydroxide can be found on Reference Tables K and L, respectively. Sodium sulfate, Na_2SO_4, is a product of this neutralization.

46. **1** The Brønsted–Lowry theory of acids and bases defines a base as a proton (H^+) acceptor. In the equation shown, NH_3 accepts (gains) a H^+ ion to form the ammonium ion, NH_4^+.

47. **4** KOH, potassium hydroxide, is listed on Reference Table L as a common base. Basic solutions have pH values greater than 7, which is the pH of neutral water.

48. **4** The idea of the atom as a hard, indivisible sphere originated in ancient Greece. Electrons were discovered in the late 19th century in J.J. Thomson's famous cathode ray experiment, and, shortly after that, the nucleus was discovered in Ernest Rutherford's gold foil experiment. Most recently, in the quantum model of the atom, electrons have been demonstrated to have properties of both waves and particles, a phenomenon known as wave-particle duality.

49. **3** A chemical property involves how an element or compound will react to form a new substance. Examples of chemical properties include the ability to oxidize (as in the process of rusting) and the ability to react with acid. Physical properties are properties that can be observed without changing the chemical identity of a substance.

50. **1** Pressure and volume are inversely related. This relationship is known as Boyle's Law:

$$P \times V = k$$

where k is a constant. Since the product of pressure and volume is a constant, if one of them increases, the other must decrease. The inverse nature of this relationship is shown in the graph for choice (1).

PART B–2

[All questions in Part B–2 are worth 1 point.]

51. Empirical formulas give the smallest whole-number ratio of atoms in a compound. An empirical formula reduces the subscripts in a compound formula into "lowest terms." The empirical formula for C_6H_{12} is **CH_2**.

52. Reading Table G in the Reference Tables, it can be seen that 40 grams of NaCl dissolved in 100 grams of water produces a saturated solution. To saturate 200 grams of water, **80 grams of NaCl** is the minimum amount that must be dissolved.

(NOTE: Credit was awarded for all answers in the range of **78–82 grams of NaCl.**)

53. Distillation is used to separate the components of a liquid mixture based on their different **boiling points**.

54. Temperature is directly related to kinetic energy, which is the energy of motion of particles in any sample of matter. According to the graph, the temperature of the substance decreases during the first 3 minutes. Due to the direct relationship between temperature and kinetic energy, the **kinetic energy of the molecules in the sample decreases** during the first 3 minutes.

55. The temperature does not change between minutes 4 and 9, which means that the kinetic energy of the sample is constant. Since the sample is still being cooled, energy is leaving the sample in the form of *potential energy*. As potential energy decreases, particles in the liquid sample move closer together, and a phase change to the solid state takes place. This process is known as **freezing**, or **solidification**.

56. Synthesis reactions follow the pattern $A + B \rightarrow AB$. In this case, **two reactants combine to form one product**.

57. Use Reference Table R to classify the organic compound formed in this reaction. Since the compound formed contains a bromine (Br) atom, it is classified as a **halide (halocarbon)**.

58. **Isomers are organic compounds that have the same molecular formula but different structural formulas and properties.** In other words, compounds that are isomers of each other will consist of the same number of atoms of each element, but the atoms will be arranged differently in the different compounds. **Both propene and cyclopropane have the molecular formula, C_3H_6, but one compound has a double bond and the other has a ring structure.**
   (NOTE: Both answers in bold are acceptable.)

59. The last formula on Reference Table T is the formula used to convert from the Celsius to the Kelvin scale and vice versa.

$$K = {}^\circ C + 273$$

According to the information given in this problem, cyclopropane boils at $-33{}^\circ C$.
   Plugging this information into the formula:

$$K = -33{}^\circ C + 273 = \textbf{240. K}$$

60. The answer to this question can be found directly in the information provided. The fluorine atom has a radius of 60. picometers, and the fluoride ion has a radius of 133 picometers. **The radius of the fluoride ion is larger than the radius of the fluorine atom.** The extra electron gained to form the F^{-1} ion is repelled by the 7 electrons already in fluorine's valence shell, and the added repulsion increases the size of the electron cloud.

61. This question states that students must answer "in terms of subatomic particles." In other words, students must include information about protons, neutrons, or electrons in their answer. Electrons are subatomic particles that atoms gain or lose as they form ions, so students must include information about electrons in their response. A simplified diagram of a lithium atom is shown below.

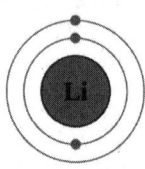

When a lithium atom becomes a lithium ion, Li^+, **the second-shell elec-tron is lost, so the lithium ion only has electrons in the first shell**. Another acceptable answer for this question is simply to say that the **lithium ion has one less electron**.

62. Lewis electron-dot diagrams for ions are drawn to show the valence elec-trons in the ion *and* the charge of the ion. Fluorine, as an element, has 7 valence electrons, but the fluoride ion contains 1 additional electron, giving it a charge of −1, as shown in the correct diagrams below. The common convention is to draw brackets around the octet of valence electrons for nonmetal ions showing that these electrons have been *transferred completely* to the ion and will not be shared with another atom.

$[\overset{..}{\underset{..}{F}}]^-$ *or* $:\overset{..}{\underset{..}{F}}:^-$ are acceptable answers.

63. **The atomic radius decreases as elements are considered in order from left to right across Period 2 of the Periodic Table of the Elements.** This trend is caused by increasing nuclear charge, which draws the electron cloud in closer, thereby shrinking the atomic radius. Since atomic radius data is provided in Reference Table S, you can always look up the values and determine the trend.

64. The term "transmutation" refers to any change in the nucleus of an atom that converts it from one element to another. The diagram shows that **uranium is changing into different elements**.

65. When a zirconium-97 atom emits a beta particle, it loses a particle with the mass and charge of an electron from its nucleus. To balance a nuclear equation, the sum of the atomic numbers (bottom) and the mass numbers (top) must be equal on both sides of the arrow. The nuclide that correctly completes the zirco-nium-97 beta decay equation is

$$^{97}_{41}Nb$$

(Note: Other acceptable forms of this answer are Nb-97 and niobium-97.)

PART C

[All questions in Part C are worth 1 point.]

66. Reference Table *J* lists metals from the most active metals at the top to the least active metals at the bottom. A more active metal will oxidize in the presence of the ion of a less active metal. Since iron is higher than copper on Table *J*, **iron is a more active metal than copper**.

67. All coefficients in chemical reactions represent particles or moles. Mole ratios can be used to find the number of moles of any substance in a chemical equation if the moles of another substance are known. Use the second equation in the information listed, which shows H_2 gas forming from the reaction between Fe and HCl. Two methods of setting up this problem are:

$$\frac{.001 \text{ mol HCL(aq)}}{1} \times \frac{1 \text{ mol } H_2(g)}{2 \text{ mol HCL(aq)}} = \textbf{0.0005 mol } \boldsymbol{H_2(g)}$$

OR

<u>Set up mole ratios:</u>

$$\frac{1 \text{ mol } H_2(g)}{2 \text{ mol HCl(aq)}} = \frac{x \text{ mol } H_2(g)}{0.001 \text{ mol HCl(aq)}}$$

$$\boldsymbol{x = 0.0005 \text{ mol } H_2(g)}$$

68. The second equation shows that the H^+ ions from hydrochloric acid react with solid iron to form hydrogen gas, H_2. Pure elements have an oxidation number of 0, so each H^+ ion that reacts must gain 1 electron to become a hydrogen atom. This reduction (gain of electrons) is shown in the balanced half-reaction below:

$$\boldsymbol{2H^+(aq) + 2e^- \rightarrow H_2{}^0(g)}$$

(NOTE: Showing the oxidation number on H_2 in the equation above is optional.)

69. Hydrogen bubbles will be produced more quickly if the rate of the chemical reaction between hydrochloric acid and iron increases. The rate of a chemical reaction increases when the number of collisions between reactant particles increases. **Increasing the concentration of the HCl(aq)** would cause there to be more HCl particles available to collide with Fe(s). Another approach would be to **increase the temperature of the HCl(aq)**. The HCl molecules would move faster at the higher temperature and would collide more often with Fe(s).

70. The solute in a solution is the substance that dissolves, and the solvent is the substance into which the solute dissolves. **NaCl(s) is the solute, and $H_2O(\ell)$ is the solvent.**

71. Use the percent composition formula on Reference Table T:

$$\% \text{ composition by mass} = \frac{\text{mass of part}}{\text{mass of whole}} \times 100$$

In a solution, the "part" is the mass of the solute, and the "whole" is the mass of the *entire solution*. The solution in beaker 4 weighs a total of 135.04 g, which is the sum of 100.00 g $H_2O(l)$ plus 35.04 g of dissolved NaCl(s). The correct numerical setup for this problem is therefore:

$$\% \textbf{ composition by mass} = \frac{\textbf{35.04 g NaCl(s)}}{\textbf{135.04 g total}} \times \textbf{100}$$

(NOTE: This problem only asks for a numerical setup. You are NOT required to solve for the answer. If you DO solve for the actual percent composition and you solve incorrectly, you will LOSE the point for this question even if you have shown the correct numerical setup. If you wish to solve for the percent composition to check that your setup is correct, do so on scrap paper.)

72. Electrolytes are substances that dissolve in water to form aqueous solutions capable of conducting electricity. The more ions there are in a solution, the more electrically conductive the solution becomes. Since more NaCl(s) was dissolved, **beaker 4 has more ions dissolved in aqueous solution than beaker 1**. The increased ion concentration causes beaker 4 to have a greater ability to conduct electricity.

73. **As the concentration of ions increases, the boiling point of the solution increases.** This relationship is supported by the data in the table provided. The mass of dissolved NaCl(s) increases from beaker 1 through beaker 4, and the boiling point increases as well.

74. The Zn(s) reactant is a pure element, and it has an oxidation number of 0. The compound ZnO(s) that is produced in this reaction has an overall oxidation number of 0. Since oxygen has an oxidation number of -2, the zinc ion in this compound must have an oxidation number of $+2$, making the overall sum of the two oxidation numbers 0. Therefore, zinc's oxidation number changes from **0 to +2** as this cell operates.

75. Redox reactions are reactions that involve the transfer of electrons. The number of electrons lost by the substance that oxidizes must always be equal to

the number of electrons gained by the substance that reduces. In other words, **e$^-$ lost = e$^-$ gained**.

76. The answer to this question is stated directly in the information provided, making this a "careful reading" question. In the second sentence, we learn that **mercury is toxic (the risk), but its use made possible the production of miniature batteries (the benefit)**.

77. The information provided for this question states that vinegar contains acetic acid, $HC_2H_3O_2$. Acetic acid is listed as a common acid on Reference Table K. All acidic solutions have pH values less than 7, which is the pH for pure water. According to Reference Table M, thymol blue, the indicator used in this titration, is **yellow** in all solutions having a pH less than 8.0.

78. NaOH is a strong base that dissociates into ions when it is mixed with water to form an aqueous solution. The two ions produced when NaOH dissociates are Na^+ and **OH$^-$**.

79. The number 0.30 contains **two** significant figures.

80. Using the titration formula on Reference Table T:

$$M_A V_A = M_B V_B$$

in which M_A = the molarity of H^+, V_A = volume of acid, M_B = molarity of OH^-, and V_B = volume of base.
Substituting the values from this problem:

$$M_A (15.0 \text{ mL}) = (0.30 \text{ M})(43.1 \text{ mL})$$

$$M_A = \frac{(0.30 \text{ M})(43.1 \text{ mL})}{(15.0 \text{ mL})} = \textbf{0.86 M}$$

81. All organic compounds contain **carbon** atoms.

82. Equation 1 shows glucose converting into ethanol in the presence of an enzyme catalyst. This is an example of a **fermentation** reaction.

83. Unsaturated hydrocarbons contain at least one carbon–carbon multiple bond. Memorize this definition! The compound CH_2CH_2 has the molecular formula C_2H_4, and, according to Reference Table Q, it must be an alkene since it

follows the general formula C_nH_{2n}. Alkenes have a double carbon–carbon bond, as illustrated in the structural formula for alkenes in Reference Table Q. Since **a double C=C bond is present**, this compound is an unsaturated hydrocarbon.

84. Ethanol has a much higher boiling point than the hydrocarbon ethene because ethanol molecules are attracted to each other by hydrogen bonding. Hydrogen bonding is an extremely strong type of intermolecular attraction. It occurs only in compounds that have one or more hydrogen atoms *directly bonded to* nitrogen, oxygen, or fluorine atoms, which are all highly electronegative. By this definition, it is clear that hydrogen bonding does not exist between ethene molecules. Since **ethanol has stronger intermolecular attractions than ethene**, molecules of ethanol have a harder time separating from each other, so boiling takes place at a higher temperature.

85. Reference Table R shows the structural formulas for major classes of organic compounds. The formula CH_3CHO follows the pattern shown for aldehydes. The aldehyde produced in equation 3 would have the structural formula shown below:

Mark (✓) the questions you answered correctly. Count the number of checks and follow the formulas given to determine your score on each topic.

Core Area	☐ Questions Answered Correctly

10, 40, 79: (3)

Section M—Math Skills
☐ Number of checks ÷ 3 × 100 = ___ %

1, 2, 30, 32, 33, 39, 48: (7)

Section I—Atomic Concepts
☐ Number of checks ÷ 7 × 100 = ___ %

3, 6, 11, 12, 31, 34, 63: (7)

Section II—Periodic Table
☐ Number of checks ÷ 7 × 100 = ___ %

4, 5, 15, 51, 56, 67: (6)

Section III—Moles/Stoichiometry
☐ Number of checks ÷ 6 × 100 = ___ %

7–9, 36–38, 60–62: (9)

Section IV—Chemical Bonding
☐ Number of checks ÷ 9 × 100 = ___ %

13–18, 41, 49, 50, 52–55, 59, 70, 71, 73, 84: (18)

Section V—Physical Behavior of Matter
☐ Number of checks ÷ 18 × 100 = ___ %

19, 20–23, 42, 69: (7)

Section VI—Kinetics and Equilibrium
☐ Number of checks ÷ 7 × 100 = ___ %

43, 57, 81–83, 85: (6)

Section VII—Organic Chemistry
☐ Number of checks ÷ 6 × 100 = ___ %

Core Area	☐ Questions Answered Correctly

24, 25, 35, 44, 58, 66, 68, 74, 75: (9)

Section VIII—Oxidation–Reduction
☐ Number of checks ÷ 9 × 100 = ___ %

26, 27, 45–47, 72, 77, 78, 80: (9)

Section IX—Acids, Bases, and Salts
☐ Number of checks ÷ 9 × 100 = ___ %

28, 29, 64, 65: (4)

Section X—Nuclear Chemistry
☐ Number of checks ÷ 4 × 100 = ___ %

Examination
June 2017
Chemistry—Physical Setting

PART A

Answer all questions in this part.

Directions (1–30): For *each* statement or question, write in the answer space the *number* of the word or expression that, of those given, best completes the statement or answers the question. Some questions may require the use of the *2011 Edition Reference Tables for Physical Setting/Chemistry*.

1 Which statement describes the structure of an atom?

   (1) The nucleus contains positively charged electrons.
   (2) The nucleus contains negatively charged protons.
   (3) The nucleus has a positive charge and is surrounded by negatively charged electrons.
   (4) The nucleus has a negative charge and is surrounded by positively charged electrons.

1 _____

2 Which term is defined as the region in an atom where an electron is most likely to be located?

   (1) nucleus   (3) quanta
   (2) orbital   (4) spectra

2 _____

3 What is the number of electrons in an atom of scandium?

   (1) 21   (3) 45
   (2) 24   (4) 66

3 _____

4 Which particle has the *least* mass?

(1) a proton   (3) a helium atom

(2) an electron   (4) a hydrogen atom   4 _____

5 Which electron transition in an excited atom results in a release of energy?

(1) first shell to the third shell

(2) second shell to the fourth shell

(3) third shell to the fourth shell

(4) fourth shell to the second shell   5 _____

6 On the Periodic Table, the number of protons in an atom of an element is indicated by its

(1) atomic mass

(2) atomic number

(3) selected oxidation states

(4) number of valence electrons   6 _____

7 Which type of formula shows an element symbol for each atom and a line for each bond between atoms?

(1) ionic   (3) empirical

(2) structural   (4) molecular   7 _____

8 What is conserved during all chemical reactions?

(1) charge   (3) vapor pressure

(2) density   (4) melting point   8 _____

9 In which type of reaction can two compounds exchange ions to form two different compounds?

(1) synthesis   (3) single replacement

(2) decomposition   (4) double replacement   9 _____

10 At STP, two 5.0-gram solid samples of different ionic compounds have the same density. These solid samples could be differentiated by their

(1) mass   (3) temperature

(2) volume   (4) solubility in water   10 _____

11 What is the number of electrons shared between the atoms in an
I_2 molecule?

(1) 7   (3) 8

(2) 2   (4) 4   11 _____

12 Which substance has nonpolar covalent bonds?

(1) Cl_2 (3) SiO_2

(2) SO_3 (4) CCl_4 12 _____

13 Compared to a potassium atom, a potassium ion has

(1) a smaller radius   (3) fewer protons

(2) a larger radius   (4) more protons   13 _____

14 Which form of energy is associated with the random motion of
particles in a gas?

(1) chemical   (3) nuclear

(2) electrical   (4) thermal   14 _____

15 The average kinetic energy of water molecules *decreases* when

(1) $H_2O(\ell)$ at 337 K changes to $H_2O(\ell)$ at 300. K

(2) $H_2O(\ell)$ at 373 K changes to $H_2O(g)$ at 373 K

(3) $H_2O(s)$ at 200. K changes to $H_2O(s)$ at 237 K

(4) $H_2O(s)$ at 273 K changes to $H_2O(\ell)$ at 273 K   15 _____

16 The joule is a unit of

(1) concentration   (3) pressure

(2) energy   (4) volume   16 _____

17 Compared to a sample of helium at STP, the same sample of
helium at a higher temperature and a lower pressure

(1) condenses to a liquid

(2) is more soluble in water

(3) forms diatomic molecules

(4) behaves more like an ideal gas   17 _____

18 A sample of a gas is in a sealed, rigid container that maintains a constant volume. Which changes occur between the gas particles when the sample is heated?

(1) The frequency of collisions increases, and the force of collisions decreases.
(2) The frequency of collisions increases, and the force of collisions increases.
(3) The frequency of collisions decreases, and the force of collisions decreases.
(4) The frequency of collisions decreases, and the force of collisions increases.   18 _____

19 At STP, which gaseous sample has the same number of molecules as 3.0 liters of $N_2(g)$?

(1) 6.0 L of $F_2(g)$ (3) 3.0 L of $H_2(g)$
(2) 4.5 L of $N_2(g)$ (4) 1.5 L of $Cl_2(g)$ 19 _____

20 Distillation of crude oil from various parts of the world yields different percentages of hydrocarbons. Which statement explains these different percentages?

(1) Each component in a mixture has a different solubility in water.
(2) Hydrocarbons are organic compounds.
(3) The carbons in hydrocarbons may be bonded in chains or rings.
(4) The proportions of components in a mixture can vary.   20 _____

21 In which 1.0-gram sample are the particles arranged in a crystal structure?

(1) $CaCl_2(s)$ (3) $CH_3OH(\ell)$
(2) $C_2H_6(g)$ (4) $CaI_2(aq)$ 21 _____

22 When a reversible reaction is at equilibrium, the concentration of products and the concentration of reactants must be

(1) decreasing   (3) constant
(2) increasing   (4) equal   22 _____

23 In chemical reactions, the difference between the potential energy of the products and the potential energy of the reactants is equal to the

(1) activation energy   (3) heat of reaction

(2) ionization energy   (4) heat of vaporization   23 _____

24 What occurs when a catalyst is added to a chemical reaction?

(1) an alternate reaction pathway with a lower activation energy

(2) an alternate reaction pathway with a higher activation energy

(3) the same reaction pathway with a lower activation energy

(4) the same reaction pathway with a higher activation energy   24 _____

25 What is the name of the compound with the formula $CH_3CH_2CH_2NH_2$?

(1) 1-propanol   (3) propanal

(2) 1-propanamine   (4) propanamide   25 _____

26 Which compound is an isomer of $C_2H_5OC_2H_5$?

(1) CH_3COOH (3) $C_3H_7COCH_3$

(2) $C_2H_5COOCH_3$ (4) C_4H_9OH 26 _____

27 Ethanoic acid and 1-butanol can react to produce water and a compound classified as an

(1) aldehyde   (3) ester

(2) amide   (4) ether   27 _____

28 During an oxidation-reduction reaction, the number of electrons gained is

(1) equal to the number of electrons lost

(2) equal to the number of protons gained

(3) less than the number of electrons lost

(4) less than the number of protons gained   28 _____

29 Which process requires energy for a nonspontaneous redox reaction to occur?

(1) deposition　　　　　　　　(3) alpha decay

(2) electrolysis　　　　　　　　(4) chromatography　　　29 _____

30 Which pair of compounds represents one Arrhenius acid and one Arrhenius base?

(1) CH_3OH and NaOH　　　　(3) HNO_3 and NaOH

(2) CH_3OH and HCl　　　　　(4) HNO_3 and HCl　　30 _____

PART B–1

Answer all questions in this part.

Directions (31–50): For *each* statement or question, write in the answer space the *number* of the word or expression that, of those given, best completes the statement or answers the question. Some questions may require the use of the *2011 Edition Reference Tables for Physical Setting/Chemistry.*

31 Which electron configuration represents the electrons of an atom of neon in an excited state?

(1) 2-7   (3) 2-7-1
(2) 2-8   (4) 2-8-1   31 _____

32 Some information about the two naturally occurring isotopes of gallium is given in the table below.

**Natural Abundance of
Two Gallium Isotopes**

Isotope	Natural Abundance (%)	Atomic Mass (u)
Ga-69	60.11	68.926
Ga-71	39.89	70.925

Which numerical setup can be used to calculate the atomic mass of gallium?

(1) (0.6011)(68.926 u) + (0.3989)(70.925 u)
(2) (60.11)(68.926 u) + (39.89)(70.925 u)
(3) (0.6011)(70.925 u) + (0.3989)(68.926 u)
(4) (60.11)(70.925 u) + (39.89)(68.926 u)   32 _____

33 A student measures the mass and volume of a sample of copper at room temperature and 101.3 kPa. The mass is 48.9 grams and the volume is 5.00 cubic centimeters. The student calculates the density of the sample. What is the percent error of the student's calculated density?

(1) 7.4%   (3) 9.2%
(2) 8.4%   (4) 10.2%   33 _____

34 What is the chemical formula for sodium sulfate?

   (1) Na_2SO_4 (3) $NaSO_4$

   (2) Na_2SO_3 (4) $NaSO_3$ 34 _____

35 Given the balanced equation representing a reaction:

$$2Na(s) + Cl_2(g) \rightarrow 2NaCl(s) + energy$$

If 46 grams of Na and 71 grams of Cl_2 react completely, what is the total mass of NaCl produced?

   (1) 58.5 g   (3) 163 g

   (2) 117 g   (4) 234 g   35 _____

36 Given the balanced equation representing a reaction:

$$2NO + O_2 \rightarrow 2NO_2 + energy$$

The mole ratio of NO to NO_2 is

   (1) 1 to 1   (3) 3 to 2

   (2) 2 to 1   (4) 5 to 2   36 _____

37 The particle diagram below represents a solid sample of silver.

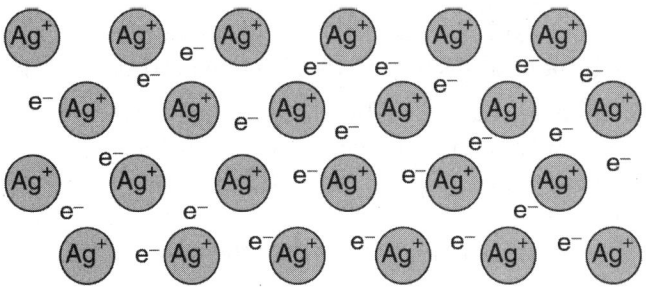

Which type of bonding is present when valence electrons move within the sample?

   (1) metallic bonding   (3) covalent bonding

   (2) hydrogen bonding   (4) ionic bonding   37 _____

38 Given the formula representing a molecule:

$$
\begin{array}{ccccccc}
 & \text{H} & & \text{H} & & \text{H} & \\
 & | & & | & & | & \\
\text{H}-&\text{C}&-&\text{C}&-&\text{C}&-\text{H} \\
 & | & & | & & | & \\
 & \text{H} & & \text{H} & & \text{H} & \\
\end{array}
$$

Which statement explains why the molecule is nonpolar?

(1) Electrons are shared between the carbon atoms and the hydrogen atoms.
(2) Electrons are transferred from the carbon atoms to the hydrogen atoms.
(3) The distribution of charge in the molecule is symmetrical.
(4) The distribution of charge in the molecule is asymmetrical.   38 _____

39 A solid sample of a compound and a liquid sample of the same compound are each tested for electrical conductivity. Which test conclusion indicates that the compound is ionic?

(1) Both the solid and the liquid are good conductors.
(2) Both the solid and the liquid are poor conductors.
(3) The solid is a good conductor, and the liquid is a poor conductor.
(4) The solid is a poor conductor, and the liquid is a good conductor.   39 _____

40 Which statement explains why 10.0 mL of a 0.50 M H_2SO_4(aq) solution exactly neutralizes 5.0 mL of a 2.0 M NaOH(aq) solution?

(1) The moles of H^+(aq) equal the moles of OH^-(aq).
(2) The moles of H_2SO_4(aq) equal the moles of NaOH(aq).
(3) The moles of H_2SO_4(aq) are greater than the moles of NaOH(aq).
(4) The moles of H^+(aq) are greater than the moles of OH^-(aq).   40 _____

41 Which particle diagram represents *one* substance in the gas phase?

41 _____

42 Given the equation representing a chemical reaction at equilibrium in a sealed, rigid container:

$$H_2(g) + I_2(g) + energy \rightleftharpoons 2HI(g)$$

When the concentration of $H_2(g)$ is increased by adding more hydrogen gas to the container at constant temperature, the equilibrium shifts

(1) to the right, and the concentration of HI(g) decreases
(2) to the right, and the concentration of HI(g) increases
(3) to the left, and the concentration of HI(g) decreases
(4) to the left, and the concentration of HI(g) increases

42 _____

43 Which diagram represents the potential energy changes during an exothermic reaction?

Reaction
Coordinate
(1)

Reaction
Coordinate
(3)

Reaction
Coordinate
(2)

Reaction
Coordinate
(4)

43 _____

44 Which compound is classified as an ether?

(1) CH_3CHO

(2) CH_3OCH_3

(3) CH_3COCH_3

(4) CH_3COOCH_3

44 _____

45 Given the equation representing a reversible reaction:

$$HCO_3^-(aq) + H_2O(\ell) \rightleftharpoons H_2CO_3(aq) + OH^-(aq)$$

Which formula represents the H^+ acceptor in the forward reaction?

(1) $HCO_3^-(aq)$

(2) $H_2O(\ell)$

(3) $H_2CO_3(aq)$

(4) $OH^-(aq)$

45 _____

46 What is the mass of an original 5.60-gram sample of iron-53 that remains unchanged after 25.53 minutes?

(1) 0.35 g

(2) 0.70 g

(3) 1.40 g

(4) 2.80 g

46 _____

47 Given the equation representing a nuclear reaction:

$$^1_1H + X \rightarrow ^6_3Li + ^4_2He$$

The particle represented by X is

(1) ^9_4Li (3) $^{10}_5Be$

(2) ^9_4Be (4) $^{10}_6C$ 47 _____

48 Fission and fusion reactions both release energy. However, only fusion reactions

(1) require elements with large atomic numbers
(2) create radioactive products
(3) use radioactive reactants
(4) combine light nuclei   48 _____

49 The chart below shows the crystal shapes and melting points of two forms of solid phosphorus.

Two Forms of Phosphorus

Form of Phosphorus	Crystal Shape	Melting Point (°C)
white	cubic	44
black	orthorhombic	610

Which phrase describes the two forms of phosphorus?

(1) same crystal structure and same properties
(2) same crystal structure and different properties
(3) different crystal structures and different properties
(4) different crystal structures and same properties   49 _____

50 Which graph shows the relationship between pressure and Kelvin temperature for an ideal gas at constant volume?

(1)

(3)

(2)

(4)

50 _____

PART B–2

Answer all questions in this part.

Directions (51–65): Record your answers on the answer sheet provided. Some questions may require the use of the *2011 Edition Reference Tables for Physical Setting/Chemistry*.

Base your answers to questions 51 through 53 on the information below and on your knowledge of chemistry.

The elements in Group 17 are called halogens. The word "halogen" is derived from Greek and means "salt former."

51 State the trend in electronegativity for the halogens as these elements are considered in order of increasing atomic number.   [1]

52 Identify the type of chemical bond that forms when potassium reacts with bromine to form a salt.   [1]

53 Based on Table *F*, identify *one* ion that reacts with iodide ions in an aqueous solution to form an insoluble compound.   [1]

Base your answers to questions 54 through 57 on the information below and on your knowledge of chemistry.

The diagrams below represent four different atomic nuclei.

Four Atomic Nuclei

| Nucleus 1 | Nucleus 2 | Nucleus 3 | Nucleus 4 |

Key
● = proton
○ = neutron

54 Identify the element that has atomic nuclei represented by nucleus 1.   [1]

55 Determine the mass number of the nuclide represented by nucleus 2.   [1]

56 Explain why nucleus 2 and nucleus 4 represent the nuclei of two different isotopes of the same element.   [1]

57 Identify the nucleus above that is found in an atom that has a stable valence electron configuration.   [1]

Base your answers to questions 58 through 60 on the information below and on your knowledge of chemistry.

The equation below represents a chemical reaction at 1 atm and 298 K.

$$2H_2(g) + O_2(g) \rightarrow 2H_2O(g)$$

58 State the change in energy that occurs in order to break the bonds in the hydrogen molecules.   [1]

59 In the space *on your answer sheet*, draw a Lewis electron-dot diagram for a water molecule.   [1]

60 Compare the strength of attraction for electrons by a hydrogen atom to the strength of attraction for electrons by an oxygen atom within a water molecule.   [1]

Base your answers to questions 61 through 63 on the information below and on your knowledge of chemistry.

- A test tube contains a sample of solid stearic acid, an organic acid.
- Both the sample and the test tube have a temperature of 22.0°C.
- The stearic acid melts after the test tube is placed in a beaker with 320. grams of water at 98.0°C.
- The temperature of the liquid stearic acid and water in the beaker reaches 74.0°C.

61 Identify the element in stearic acid that makes it an organic compound. [1]

62 State the direction of heat transfer between the test tube and the water when the test tube was placed in the water. [1]

63 Show a numerical setup for calculating the amount of thermal energy change for the water in the beaker. [1]

Base your answers to questions 64 and 65 on the information below and on your knowledge of chemistry.

A nuclear reaction is represented by the equation below.

$$^{3}_{1}H \rightarrow ^{3}_{2}He + ^{0}_{-1}e$$

64 Identify the decay mode of hydrogen-3. [1]

65 Explain why the equation represents a transmutation. [1]

PART C

Answer all questions in this part.

Directions (66–85): Record your answers on the answer sheet provided. Some questions may require the use of the *2011 Edition Reference Tables for Physical Setting/Chemistry*.

Base your answers to questions 66 through 68 on the information below and on your knowledge of chemistry.

A technician recorded data for two properties of Period 3 elements. The data are shown in the table below.

Two Properties of Period 3 Elements

Element	Na	Mg	Al	Si	P	S	Cl	Ar
Ionic Radius (pm)	95	66	51	41	212	184	181	—
Reaction with Cold Water	reacts vigorously	reacts very slowly	no observable reaction	no observable reaction	no observable reaction	no observable reaction	reacts slowly	no observable reaction

66 Identify the element in this table that is classified as a metalloid. [1]

67 State the phase of chlorine at 281 K and 101.3 kPa. [1]

68 State evidence from the technician's data which indicates that sodium is more active than aluminum. [1]

Base your answers to questions 69 through 71 on the information below and on your knowledge of chemistry.

Ammonia, $NH_3(g)$, can be used as a substitute for fossil fuels in some internal combustion engines. The reaction between ammonia and oxygen in an engine is represented by the unbalanced equation below.

$$NH_3(g) + O_2(g) \rightarrow N_2(g) + H_2O(g) + \text{energy}$$

69 Balance the equation *on the answer sheet* for the reaction of ammonia and oxygen, using the smallest whole-number coefficients. [1]

70 Show a numerical setup for calculating the mass, in grams, of a 4.2-mole sample of O_2. Use 32 g/mol as the gram-formula mass of O_2. [1]

71 Determine the new pressure of a 6.40-L sample of oxygen gas at 300. K and 100. kPa after the gas is compressed to 2.40 L at 900. K. [1]

Base your answers to questions 72 through 76 on the information below and on your knowledge of chemistry.

Fruit growers in Florida protect oranges when the temperature is near freezing by spraying water on them. It is the freezing of the water that protects the oranges from frost damage. When $H_2O(\ell)$ at $0°C$ changes to $H_2O(s)$ at $0°C$, heat energy is released. This energy helps to prevent the temperature inside the orange from dropping below freezing, which could damage the fruit. After harvesting, oranges can be exposed to ethene gas, C_2H_4, to improve their color.

72 Write the empirical formula for ethene. [1]

73 Explain, in terms of bonding, why the hydrocarbon ethene is classified as unsaturated. [1]

74 Determine the gram-formula mass of ethene. [1]

75 Explain, in terms of particle arrangement, why the entropy of the water *decreases* when the water freezes. [1]

76 Determine the quantity of heat released when 2.00 grams of $H_2O(\ell)$ freezes at $0°C$. [1]

Base your answers to questions 77 through 80 on the information below and on your knowledge of chemistry.

A student constructs an electrochemical cell during a laboratory investigation. When the switch is closed, electrons flow through the external circuit. The diagram and ionic equation below represent this cell and the reaction that occurs.

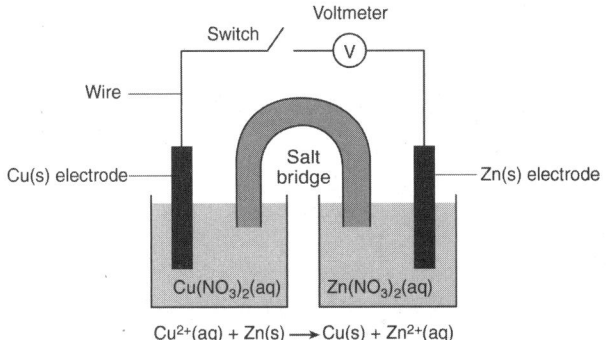

$$Cu^{2+}(aq) + Zn(s) \longrightarrow Cu(s) + Zn^{2+}(aq)$$

77 State the form of energy that is converted to electrical energy in the operating cell.   [1]

78 State, in terms of the Cu(s) electrode and the Zn(s) electrode, the direction of electron flow in the external circuit when the cell operates.   [1]

79 Write a balanced equation for the half-reaction that occurs in the Cu half-cell when the cell operates.   [1]

80 State what happens to the mass of the Cu electrode and the mass of the Zn electrode in the operating cell.   [1]

Base your answers to questions 81 and 82 on the information below and on your knowledge of chemistry.

A solution is made by dissolving 70.0 grams of $KNO_3(s)$ in 100. grams of water at 50.°C and standard pressure.

81 Show a numerical setup for calculating the percent by mass of KNO_3 in the solution. [1]

82 Determine the number of additional grams of KNO_3 that must dissolve to make this solution saturated. [1]

Base your answers to questions 83 through 85 on the information below and on your knowledge of chemistry.

Vinegar is a commercial form of acetic acid, $HC_2H_3O_2(aq)$. One sample of vinegar has a pH value of 2.4.

83 Explain, in terms of particles, why $HC_2H_3O_2(aq)$ can conduct an electric current. [1]

84 State the color of bromthymol blue indicator in a sample of the commercial vinegar. [1]

85 State the pH value of a sample that has ten times *fewer* hydronium ions than an equal volume of a vinegar sample with a pH value of 2.4. [1]

Answer Sheet
June 2017
Chemistry—Physical Setting

PART B–2

51 _____

_____

_____

52 _____

53 _____

54 _____

55 _____

56 _____

_____

_____

57 _____

58 _____

_____

_____

59

60 _____

_____

_____

_____

61 _____

62 _____

_____

_____

63

64 _____

65 _____

PART C

66 _____

67 _____

68 _____

_____

_____

69 ____ $NH_3(g) + O_2(g) \rightarrow$ ____ $N_2(g) +$ ____ $H_2O(g) + $ energy

70

71 _____ **kPa**

72 _____

73 _____

_____

_____

74 _____ **g/mol**

75 _____

_____

_____

76 _____ **J**

77 _____

78 _____

_____

_____

79 _____

80 Cu electrode: _____

Zn electrode: _____

81

82 _____ g

83 _____

_____

_____

84 _____

85 _____

Answers
June 2017

Chemistry—Physical Setting

Answer Key

PART A

1. 3	**7.** 2	**13.** 1	**19.** 3	**25.** 2
2. 2	**8.** 1	**14.** 4	**20.** 4	**26.** 4
3. 1	**9.** 4	**15.** 1	**21.** 1	**27.** 3
4. 2	**10.** 4	**16.** 2	**22.** 3	**28.** 1
5. 4	**11.** 2	**17.** 4	**23.** 3	**29.** 2
6. 2	**12.** 1	**18.** 2	**24.** 1	**30.** 3

PART B–1

31. 3	**35.** 2	**39.** 4	**43.** 4	**47.** 2
32. 1	**36.** 1	**40.** 1	**44.** 2	**48.** 4
33. 3	**37.** 1	**41.** 3	**45.** 1	**49.** 3
34. 1	**38.** 3	**42.** 2	**46.** 2	**50.** 1

PART B–2 and **PART C**. *See* **Answers Explained**.

Answers Explained

PART A

1. **3** The charges of all subatomic particles should be committed to memory, but *just to be sure* you can consult Reference Table O, which shows the symbols, mass numbers, and charges of subatomic particles.

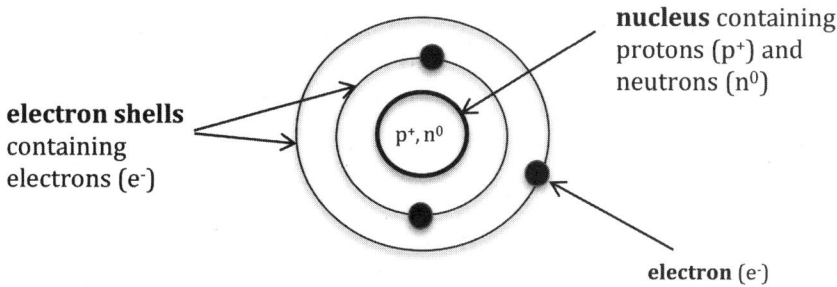

2. **2** In the wave-mechanical, or quantum, model of the atom, an orbital is defined as a three-dimensional region of space in which there is a high probability of locating an electron that has a certain energy.

3. **1** Atoms are neutral in charge, since the number of positively charged protons is equal to the number of negatively charged electrons. Reference Table S confirms that the element scandium has an atomic number of 21. An atomic number is defined as the number of protons in the nucleus of an atom. Since scandium atoms have 21 protons, they must also have 21 electrons.

4. **2** Atoms are built from three subatomic particles: protons, neutrons, and electrons. It stands to reason that all subatomic particles are lighter than the atoms they build. According to Reference Table O, a proton has a mass of approximately 1 amu, while an electron (which is equivalent to a beta particle) has a mass of 0 amu.
   (**NOTE:** Electrons *do* have mass, but when rounded to one significant figure, the mass of an electron is recorded as 0 amu.)

5. **4** Electrons that are located farther from the nucleus have more energy than those that are located closer to the nucleus. Energy is released when electrons move from higher energy shells to lower energy shells within the atom.

(NOTE: The first electron shell, or the first principal energy level, is the shell located closest to the nucleus.)

WRONG CHOICES EXPLAINED:
(1), (2), and (3) all describe electron transitions from a lower energy shell to a higher energy shell. These transitions would require the electron to absorb energy.

6. **2** The atomic number is defined as the number of protons in the nucleus of an atom.

7. **2** Structural formulas are drawn to illustrate the covalent bonding within a molecule. Lines drawn between atoms represent covalent bonds. For example, the structural formula for propene, C_3H_6, is shown below.

WRONG CHOICES EXPLAINED:
(1) Ionic compounds do not contain covalent bonds. Therefore, lines are never drawn between atoms in ionic formulas.
(3) Empirical formulas give the smallest whole-number ratio of atoms in a compound. For example, the empirical formula of propene, C_3H_6, is CH_2.
(4) Molecular formulas are the most common way of representing compounds. These formulas show how many atoms of each type of element are in a compound. In the case of propene, the molecular formula is C_3H_6.

8. **1** Chemical reactions occur when the chemical bonds in compounds are broken and then the atoms form new bonds, creating new compounds or elements. The products of a chemical reaction will have different physical properties than the reactants. Density, vapor pressure, and melting point are physical properties of materials that will change as new materials form. However, all chemical reactions involve the conservation of mass, energy, and charge.

9. **4** Double replacement reactions follow the general pattern AB + CD → AD + CB. AB and CD represent ionic compounds. During the course of the reaction, the positively charged cations (A and C) switch places and form new compounds with the opposite negatively charged anion.

10. **4** Ionic compounds differ widely in solubility, as seen on Reference Table G.

WRONG CHOICES EXPLAINED:
(1) The samples have the same mass, 5.0 grams, as stated in the question.

(2) Since density is the ratio of mass to volume $\left(d = \frac{m}{V} \right)$, if the density is the same and the mass is also the same between the two solids, then the solids must also occupy the same volume.

(3) Two very different solids can be at the same temperature, which indicates the amount of thermal energy in the material. The temperature of a sample does not provide information about the identity of the sample.

11. **2** The Lewis dot diagram for an I_2 molecule is shown below.

Since iodine atoms have seven valence electrons, each atom shares its unpaired electron with the other, giving each atom a stable octet of valence electrons. Two electrons are shared between iodine atoms, forming one covalent bond.

12. **1** Nonpolar covalent bonds are formed when electrons are shared between atoms that have the *same electronegativity*. In the Cl_2 molecule, both atoms have an equal attraction for the electrons in the bond, so partial negative and positive charges do not form.

13. **1** The ground-state electron configuration of a potassium atom is 2-8-8-1. A potassium atom must lose its valence electron to form the potassium ion, K^{+1}. Since it does not have any electrons in the fourth shell, the potassium ion is smaller than the potassium atom. Choices (3) and (4) cannot be correct because potassium atoms and potassium ions both contain 19 protons.

14. **4** Thermal energy is created by the motion of atoms and molecules within an object. The faster the particles move, the more thermal energy the object has and the hotter the object will be.

WRONG CHOICES EXPLAINED:

(1) Chemical energy refers to energy stored in chemical bonds.

(2) Electrical energy is the energy caused by a moving current of electrons.

(3) Nuclear energy is the energy released during nuclear reactions such as fission or fusion.

15. **1** Temperature is directly related to the average kinetic energy of particles in a sample of matter. If the temperature increases, then the average kinetic energy increases. Likewise, if the temperature decreases, then the average kinetic energy decreases. Kinetic energy is constant when the temperature does not change.

16. **2** The joule is a unit of energy in the metric system of measurement. Reference Table *D* provides a handy description of most of the units needed in the study of chemistry.

17. **4** Real gases behave more like ideal gases when temperature is high and pressure is low. These conditions minimize any possible attractions between gas particles, allowing the gas to adhere more closely to the principles of the Kinetic Molecular Theory of Gases.

WRONG CHOICES EXPLAINED:

(1) According to Reference Table S, helium gas condenses to a liquid at 4 K, which is far below standard temperature!

(2) Gases become less soluble at higher temperatures.

(3) Helium is a noble gas. Helium atoms never bond to other atoms.

18. **2** When gas particles are heated, the temperature of the sample increases and the particles move more quickly. This is consistent with the Kinetic Molecular Theory of Gases, which states that temperature is directly related to the average kinetic energy of the gas particles. Remember, kinetic energy is the energy of motion. Imagine that the particles in the container shown below increase in speed. Clearly the particles will strike each other more often, as well as the walls of the container, and since they are moving faster, they will collide with more force.

GAS

19. **3** This question is straightforward if you remember Avogadro's Law, which is shown below as a mathematical expression.

$$V \propto n$$

This law states that the volume of a gas is proportional to the number of particles in a gas sample if the temperature and pressure of the sample are held constant. Simply put, this law means that as the number of particles doubles, the volume doubles. However, if the number of particles is the same, as stated in this question, the volumes must also be the same for the two samples.

20. **4** Mixtures are defined as materials with variable composition containing two or more substances. This distinguishes mixtures from elements and compounds, which are pure substances with definite compositions.

WRONG CHOICES EXPLAINED:
(1), (2), and (3) are all true statements, but none of these statements explains why there are different percentages of hydrocarbons in different samples of crude oil.

21. **1** Most solids are crystalline solids, which mean that the particles are arranged in an orderly, geometric pattern. $CaCl_2$ is the only solid listed among the answer choices.

22. **3** Chemical equilibrium is established when the *rates* of the forward and reverse reactions are equal, NOT the amounts of products and reactants. Imagine cars traveling over a bridge at the same rate in both directions. The number of cars on each side of the bridge would remain constant, but the number of cars on each side of the bridge may not be equal. The condition of chemical equilibrium is illustrated below, showing that the concentration of products and the concentration of reactants become constant at equilibrium.

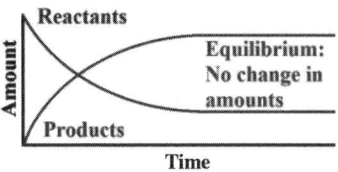

23. **3** The heat of reaction (ΔH) is defined as the potential energy of the products of a chemical reaction minus the potential energy of the reactants. In other words, the heat of reaction is the net change in energy during the reaction. This is a definition that must be memorized. The potential energy diagram below can help you visualize this definition.

24. **1** Many chemical reactions proceed from reactants to products in a series of steps that involve particle collisions rather than as one single collision. A catalyst creates an alternate sequence of steps that leads to a lower activation energy for the complete reaction. Having a lower activation energy overall leads to a faster reaction since more particles will have sufficient kinetic energy to react.

25. **2** Reference Table R lists general structural formulas and examples for some very common organic functional groups. Examine this table and look for nitrogen-containing functional groups. You will see that this compound must be either an amine or an amide. Further examination shows that this compound must be classified as an amine since it has the functional group $-NH_2$. There are three carbons in this amine, so by using the prefixes listed on Reference Table P, you will see that this results in the name 1-propanamine. Incidentally, the compound in this question is the exact same as the example given on Reference Table R. **Use your Reference Tables!**

26. **4** Isomers are organic compounds that have the same numbers and types of atoms but have different structures. In other words, the atoms are attached in a different arrangement. To answer this question, simply choose the answer choice that has 4 carbon atoms, 10 hydrogen atoms, and 1 oxygen atom, just like the original compound written in the question.

27. **3** A reaction between any organic acid and an alcohol is known as an esterification reaction. See the example below.

$$
\underset{\text{carboxylic acid}}{\overset{\overset{\displaystyle O}{\|}}{RC-OH}} \quad + \quad \underset{\text{alcohol}}{R'OH} \quad \underset{\Delta}{\rightarrow} \quad \underset{\text{ester}}{\overset{\overset{\displaystyle O}{\|}}{RC-OR'}} \quad + \quad \underset{\text{water}}{H_2O}
$$

28. **1** Redox reactions involve the transfer of electrons. In these reactions, as in all reactions, charge must be conserved. In practice, this means that the number of electrons lost by the substance that oxidizes must equal the number of electrons gained by the substance that reduces.

29. **2** The process of electrolysis utilizes an electrolytic cell that consumes electrical energy to force a nonspontaneous chemical change to occur. For example, highly reactive alkali metals, such as sodium, can be generated in their pure state using electrolysis to drive an electron back onto alkali metal ions.

WRONG CHOICES EXPLAINED:
(1) Deposition is a phase change from gas to solid directly. Phase changes are always physical changes, not chemical changes.
(3) Alpha decay is a form of radioactive decay, which is a nuclear reaction.
(4) Chromatography is a separation technique for liquid or gaseous mixtures.

30. **3** Reference Tables K and L list common acids and bases. Reference Table K lists HNO_3 (nitric acid) and Reference Table L lists NaOH (sodium hydroxide). Arrhenius acids release H^+ ions in aqueous solution, and Arrhenius bases release OH^- ions in aqueous solution.

WRONG CHOICES EXPLAINED:
(1) Neither of these compounds is an acid. CH_3OH is an organic compound known as an alcohol.
(2) Neither of these compounds is a base. CH_3OH is an organic compound known as an alcohol. HCl is an acid.
(4) Both of these compounds are acids listed on Reference Table K.

PART B–1

31. **3** An atom of neon will always have 10 electrons, regardless of whether it is in the ground state or in the excited state. The Periodic Table in the Reference Tables shows that the ground-state electron configuration of a neon atom is 2-8. A neon atom would be in an excited state if its configuration were 2-7-1 because this configuration shows that *a second-shell electron has moved to the third shell.*

WRONG CHOICES EXPLAINED:
(1), (4) These configurations do not include 10 electrons.
(2) This is the ground-state, lowest energy electron configuration for neon atoms.

32. **1** The atomic mass of an element is found by taking the weighted average of all naturally occurring isotopes of that element. The natural abundance percentages of each isotope are first converted to decimal fractions, and these are then multiplied by the atomic mass of the isotope. Adding the contributions of each isotope together yields the correct atomic mass for that element.

33. **3** The density and percent error formulas can be found on Reference Table *T*. Using the information given in the question, first calculate the density of copper using the student's mass and volume measurements:

$$d = \frac{m}{V}$$

$$d = \frac{48.9 \text{ g}}{5.00 \text{ cm}^3}$$

$$d = 9.78 \text{ g/cm}^3$$

The next step is to calculate the student's percent error for the calculated value shown above by comparing it to the accepted value for the density of copper that is listed on Reference Table *S*.

$$\text{percent error} = \frac{\text{measured value} - \text{accepted value}}{\text{accepted value}} \times 100$$

Plug the calculated value of 9.78 g/cm^3 and the accepted value of 8.96 g/cm^3 into the formula:

$$\text{percent error} = \frac{9.78 \text{ g/cm}^3 - 8.96 \text{ g/cm}^3}{8.96 \text{ g/cm}^3} \times 100$$

$$\text{percent error} = 9.15\% \approx 9.2\%$$

34. **1** Sodium sulfate is an ionic compound made from the sodium ion, Na^{+1}, and the sulfate ion, SO$_4$$^{-2}$. Charges for these two ions can be found on the Periodic Table of the Elements and on Reference Table E, respectively. Since compounds have no overall charge, there must be two sodium ions present in the compound to cancel the -2 charge of the sulfate ion.

WRONG CHOICES EXPLAINED:
(2) This is the formula for sodium sulf*ite*. Look carefully at Reference Table E to notice the difference between the formulas of the sulfite and the sulfate ions.

(3), (4) These formulas are incorrect since the sum of ion charges is not neutral.

35. **2** This answer depends on simple arithmetic involving the Law of Conservation of Mass. Matter is neither created nor destroyed in chemical reactions. The total mass of the products must always equal the mass of the reactants. The mass of both reactants is given in the question. Adding both together will equal the mass of the NaCl produced, as shown below:

$$2\text{Na}(s) + \text{Cl}_2(g) \rightarrow 2\text{NaCl}(s) + \text{energy}$$
$$46 \text{ g} + 71 \text{ g} = 117 \text{ g}$$

36. **1** The coefficients in a balanced chemical equation represent moles or particles. Both NO and NO$_2$ have a coefficient of 2. A ratio of 2:2 expressed in lowest terms is a 1:1 relationship.

37. **1** Metallic bonds hold samples of pure metals together. Metal atoms lose valence electrons to a common "sea of electrons." This mobile pool of delocalized electrons serves as a sort of "glue" for the metal cations created when metal atoms lose their valence electrons.

38. **3** Nonpolar molecules do not have partial positive or negative charges. This is because the charge is distributed symmetrically throughout the molecule. The propane molecule has a symmetrical arrangement of atoms in which a central chain of three carbon atoms is symmetrically surrounded by hydrogen atoms. There is no location on this molecule with greater or lesser electronegativity than any other location, so partial charges do not occur.

39. **4** Ionic solids do not conduct electricity because the ions are locked in place within a crystalline lattice structure. Ions are charged particles and can assist with conduction only if the ions are free to move. Melting the ionic compound creates mobile ions, so the molten compound can conduct electricity.

40. **1** A neutralization reaction involves H^+ ions from an acid reacting with OH^- ions from a base, forming water. These ions must be present in exactly equal amounts for perfect neutralization to occur.

41. **3** A pure substance is represented by just one type of particle throughout the sample. For a pure gas sample, the particles spread out to fill the entire space, just as gases do. The correct answer represents a diatomic gas, like pure O_2 or N_2.

WRONG CHOICES EXPLAINED:
(1) This particle diagram shows a mixture of two different diatomic gases.
(2), (4) These particle diagrams show the regular arrangement of particles in the solid phase.

42. **2** Systems at equilibrium prefer to remain at equilibrium. If additional $H_2(g)$ is added to this system, the reaction "shifts to the right" to use up the extra H_2 that was added. This results in the production of extra $HI(g)$, so when a new equilibrium is established, there will be more $HI(g)$ present than there was initially.

43. **4** Energy is released, or lost, during exothermic chemical reactions, causing these reactions to have a $-\Delta H$ (heat of reaction). The products of these reactions contain *less* stored energy in their bonds than the energy stored in the bonds of the reactants. In order to form the lower energy products, energy must be released.

44. **2** Refer to Reference Table R for all questions that involve organic functional groups. Ethers contain an oxygen atom singly bonded to two carbon atoms in the middle of a hydrocarbon chain.

WRONG CHOICES EXPLAINED:

(1) This compound is an aldehyde, with the oxygen doubly bonded to a carbon at the end of the molecule.

(3) This compound is a ketone. The oxygen atom creates a double bond with a carbon atom in the middle of the molecule.

(4) This compound is an ester, with two oxygens in the middle of the molecule.

45. **1** The Brønsted–Lowry theory of acids and bases defines an acid as a proton (H^+) donor and a base as a proton (H^+) acceptor. $HCO_3^-(aq)$ must accept an H^+ in the forward reaction in order to form the product H_2CO_3.

46. **2** From Reference Table N, the half-life of Fe-53 is 8.51 minutes. In 25.53 minutes, Fe-53 goes through 3 half-lives since 25.53 minutes divided by 8.51 minutes equals 3 half-lives. This question is really asking how much Fe-53 will remain in the sample after 3 half-lives have passed. The mass of a radioisotope reduces by half for each half-life that passes. This problem is solved by taking the original mass and dividing it in half 3 successive times, as shown below:

$$\text{1st half-life:} \quad \frac{5.60 \text{ grams}}{2} = 2.80 \text{ grams}$$

$$\text{2nd half-life:} \quad \frac{2.80 \text{ grams}}{2} = 1.4 \text{ grams}$$

$$\text{3rd half-life:} \quad \frac{1.4 \text{ grams}}{2} = 0.70 \text{ grams}$$

47. **2** To balance a nuclear equation, the sum of the atomic numbers (bottom) and the mass numbers (top) must be equal on both sides of the arrow. A mass number of 9 and an atomic number of 4 correctly balances this nuclear equation. Since the atomic number is 4, the element is beryllium, Be.

$$^1_1H + X \rightarrow \, ^6_3Li + \, ^4_2He$$

$$^9_4Be$$

48. **4** Fusion is a nuclear reaction in which two very lightweight atomic nuclei, usually hydrogen, are combined into a slightly heavier nucleus, usually helium.

WRONG CHOICES EXPLAINED:
(1) Fusion takes place between elements with small atomic numbers.
(2) Both fission and fusion produce radioactive products.
(3) The reactants for nuclear fusion do not need to be radioactive.

49. **3** The chart supplied with this question contains the answer! The two forms of phosphorus have different crystal structures—one is cubic and the other is orthorhombic. The two forms also have different properties, such as the very different melting points shown.

50. **1** Increasing the Kelvin temperature increases the speed of the gas particles in the sample. Faster moving particles will collide more often and with more force against the inside of a container, causing the internal pressure to increase. This is a direct relationship.

PART B–2

[All questions in Part B–2 are worth 1 point.]

51. Electronegativity values can be found on Reference Table S. Simply look up the electronegativity values of the halogens as listed in Group 17 from the top of the group (low atomic number) to the bottom of the group (high atomic number):

Atomic Number	Element	Electronegativity
9	F	4.0
17	Cl	3.2
35	Br	3.0
53	I	2.7
85	At	2.2

Electronegativity decreases as the halogens are considered in order of increasing atomic number.

52. Ionic bonds form when metals bond with nonmetals. Potassium is a metal, and bromine is a nonmetal, so the bond between them would be an **ionic bond**.

53. Three exceptions are listed for the solubility of halide ions (Cl^-, Br^-, and I^-) in Reference Table F: the **silver ion (Ag^+), the lead (II) ion (Pb^{2+}), and the mercury (I) ion (Hg_2^{2+})**. Any one of these is an acceptable answer to this question.

54. The number of protons in an atom (known as the atomic number) identifies the element. There are 9 protons drawn in Nucleus 1, so this must be the element **fluorine (F)**, which has an atomic number of 9.

55. The mass number is defined as the sum of the protons and neutrons in an atom. Each proton and neutron has an approximate mass of 1 amu, so the sum of these particles will be equal to the mass expressed in amu (atomic mass units). There are 8 protons and 10 neutrons drawn in Nucleus 2, so the mass number of this atom would be $8 + 10 = \mathbf{18}$ **(amu)**.
(NOTE: Units were not required to receive full credit on this question.)

56. **Nucleus 2 and Nucleus 4 contain the same number of protons (8) but different numbers of neutrons (10 and 11, respectively).** Atoms of the same element that contain the same number of protons but different numbers of neutrons are called isotopes of that element.

57. The noble gases in Group 18 have a stable valence electron configuration, which is the reason why the noble gases do not tend to form chemical bonds. **Nucleus 3** contains 10 protons, and therefore represents an atom of neon, Ne, which is a noble gas.

58. **Energy must be absorbed** to break the covalent bond in the H_2 molecule, just as energy must be used to break apart two magnets that are stuck together.

59. Lewis electron-dot diagrams are drawn to show how the valence electrons of all atoms are distributed within the compound. These diagrams are used to illustrate the number and types of bonds in a compound. The Lewis electron-dot diagram for water is shown below. Note that in the finished diagrams, the oxygen atom achieves a stable octet of valence electrons by sharing each hydrogen's single valence electron.

$$\mathbf{H - \overset{..}{\underset{..}{O}} - H} \quad \textit{or} \quad \mathbf{H \!:\! \overset{..}{\underset{..}{O}} \!:\! H}$$

60. To answer this question correctly, you need to recall two things: (1) electronegativity is defined as an atom's attraction for an electron in a covalent bond, and (2) values for electronegativity can be found on Reference Table S. Looking at Reference Table S, we find that **oxygen has a much higher electronegativity (3.4) than hydrogen (2.2)**.

61. The element **carbon, C**, is the central atom in organic compounds. Organic chemistry is the study of carbon-containing compounds.

62. Heat always flows from substances at a higher temperature to substances at a lower temperature. In this case, **heat will flow from the water** (at a temperature of 98.0 °C) **to the test tube** (at a temperature of 22.0 °C).

63. "Thermal energy change" is just a fancy way of saying "amount of heat lost or gained." Heat formulas can be found on Reference Table T. Since the temperature of the water in the beaker decreases from 98.0 °C to 74.0 °C, the correct formula is:

$$q = mC\Delta T$$

In this formula, q = the amount of heat in joules (J) lost or gained, m = the mass of the water in grams, C = specific heat capacity of liquid water (as found on Reference Table B), and $\Delta T = T_{final} - T_{initial}$, which equals $74.0 - 98.0 = -24.0°C = -24.0$ K.

Plugging in values from the information given leads to the following expression:

$$q = (320. \text{ g}) (4.18 \text{ J/g} \cdot \text{K}) (-24.0 \text{ K})$$

(NOTE: Credit was given for both a + ΔT value or a – ΔT value.)

64. Nuclear particle symbols are found on Reference Table O. The nuclear equation in this question represents **beta decay** since a **beta particle**, $\beta^- \left(_{-1}^{0}e \right)$ is emitted.

(NOTE: Acceptable answers included the term "beta decay" or either symbol for a beta particle.)

65. "Transmutation" means the conversion of one element into another by a nuclear process. The equation shown in this question represents transmutation because a **H-3 nucleus is converted to a He-3 nucleus**.

PART C

[All questions in Part C are worth 1 point.]

66. The six metalloids must be memorized. They are B, Si, As, Te, Sb, and Ge. These metalloids are easily remembered visually—just look at the positions of these elements on the Periodic Table and memorize the physical pattern that they make. The only metalloid listed in the table for this question is **Si**, or **silicon**.

67. Boiling point information for most of the elements on the Periodic Table can be found on Reference Table S. The asterisk for the "Boiling Point" column heading signifies that the boiling point data listed is for standard pressure, which is equal to 101.3 kPa. We can therefore use the information from Reference Table S to figure out this question. Since the boiling point for chlorine listed on Reference Table S is 239 K, chlorine would be well *above* its boiling point at 281 K and would therefore be a **gas**.

68. Answering this question only requires reading and interpreting the information printed in the table. **Sodium reacts vigorously with cold water whereas aluminum has no observable reaction.** If you struggle to remember which element corresponds to each symbol, remember that most elements and their symbols are listed on Reference Table S.

69. The basic idea when balancing chemical equations can be thought of as "Atoms In = Atoms Out." First, notice that there are two N atoms on the right side of the equation and only one N atom on the left side of the equation. A coefficient of 2 can be written in front of the NH_3 to balance the N atoms. With a coefficient of 2 in front of NH_3, there are 6 H atoms on the reactant side, so a coefficient of 3 is written in front of H_2O on the product side to balance the H atoms. The partially balanced equation is now shown below:

$$\underline{\ \ 2\ \ }\ NH_3(g) + \underline{\ \ \ \ \ }\ O_2(g) \rightarrow \underline{\ \ \ \ \ }\ N_2(g) + \underline{\ \ 3\ \ }\ H_2O(g) + \text{energy}$$

We now run into a classic difficulty when balancing equations that contain diatomic elements, like O_2. Three oxygen atoms are now on the right of this equation, and yet they are only available as pairs on the left side of the equation. In a situation like this, *double all of the coefficients* (except the coefficient for the O_2, which isn't known yet). Doubling an odd number results in an even number. The finished balanced equation becomes:

$$\underline{\ \ 4\ \ }\ \textbf{NH}_3(g) + \underline{\ \ 3\ \ }\ \textbf{O}_2(g) \rightarrow \underline{\ \ 2\ \ }\ \textbf{N}_2(g) + \underline{\ \ 6\ \ }\ \textbf{H}_2O(g) + \textbf{energy}$$

70. The formula that relates moles to mass is found on Reference Table T.

$$\text{number of moles} = \frac{\text{given mass}}{\text{gram-formula mass}}$$

Plug the information given in this question into this formula to write the correct numerical setup:

$$4.2 \text{ mol} = \frac{x \text{ grams}}{32 \text{ g/mol}}$$

71. Use the combined gas law formula found on Reference Table T:

$$\frac{P_1 V_1}{T_1} = \frac{P_2 V_2}{T_2}$$

Substituting information from the problem:

$$\frac{(100. \text{ kPa})(6.40 \text{ L})}{300. \text{ K}} = \frac{(P_2)(2.40 \text{ L})}{900. \text{ K}}$$

Rearranging:

$$P_2 = \frac{(100. \text{ kPa})(6.40 \text{ L})(900. \text{ K})}{(300. \text{ K})(2.40 \text{ L})} = \textbf{800. kPa}$$

72. An empirical formula shows the elements of a compound in their lowest whole number ratio. The molecular formula for ethene, C_2H_4, can be reduced to lowest terms by dividing each subscript by 2. The empirical formula of ethene is **CH_2**.

73. Unsaturated hydrocarbons contain at least one double or triple bond between carbon atoms. The term "unsaturated" implies that other atoms could be added to the compound by breaking the multiple bond between the carbon atoms. The structural formula for ethene is shown below. **Ethene is classified as an unsaturated hydrocarbon because of the double bond between the carbon atoms.**

74. The gram-formula mass, also known as the molar mass, can be calculated by adding the molar masses of all of the elements in a compound. You can find the molar masses of each element on the Periodic Table provided in your Reference Tables.

Algebraically, the gram-formula mass for ethene can be expressed as:

$$[2 \times (\text{molar mass of carbon})] + [4 \times (\text{molar mass of hydrogen})]$$

Plug in the molar masses of each element and solve:

$$(2 \times 12.0) + (4 \times 1.0) = \textbf{28.0 g/mol}$$

75. Entropy is the measure of randomness or disorder in a system. When liquid water freezes into ice, independent, freely moving water molecules become locked into a rigid, geometric, crystalline structure. Entropy decreases during this process because **the molecules in liquid water are arranged more randomly than molecules in ice**.

(NOTE: This problem asks for the explanation to be written *in terms of particle arrangement*. To receive credit, your answer must mention the changing particle arrangement as liquid water freezes.)

76. The last page of the Reference Tables, Reference Table *T*, lists the formula to find the heat of fusion, which is the heat lost or gained by a substance as it freezes or melts, respectively.

$$q = mH_f$$

In this formula, q = heat, m = mass, and H_f = the heat of fusion.

The mass is given as 2.00 grams, and the heat of fusion for liquid water, found on Reference Table *B*, is 334 J/g. Plug these values into the heat of fusion formula and solve:

$$q = (2.00 \text{ g})(334 \text{ J/g}) = \textbf{668 J}$$

77. The electrochemical cell shown, known as a battery, converts **chemical energy** (the potential energy stored in chemical bonds) into electrical energy.

78. To decide the direction of electron flow in this system, you must determine which metal is more active—Zn or Cu. Reference Table J lists metals from the most active metals at the top to the least active metals at the bottom. A more active metal will oxidize (lose electrons) in the presence of a less active metal. Since zinc is higher than copper on Reference Table J, zinc will lose electrons to copper, causing electrons to flow **from the zinc electrode to the copper electrode** in this system.

79. Copper is lower than zinc on Reference Table J, so copper ions will *reduce* in the presence of zinc. Remember that metal *atoms* will never gain electrons to form negative ions—their electronegativity is too low. However, metal *ions* can regain electrons lost previously to re-form the original metal atoms. Recognizing that reduction happens in the Cu half-cell is critical for writing the correct reduction half-reaction for the cell, which is:

$$Cu^{2+} + 2e^- \rightarrow Cu^0$$

[NOTE: Credit was also awarded if the product was written as Cu(s) or even just Cu.]

80. **The mass of the Cu electrode increases, and the mass of the Zn electrode decreases.** This happens because soluble zinc ions form and dissolve as the zinc electrode loses electrons, and insoluble copper atoms form as copper ions in the copper half-cell gain electrons. The copper atoms formed are deposited right onto the copper electrode.

81. Use the percent composition formula found on Reference Table T:

$$\% \text{ composition by mass} = \frac{\text{mass of part}}{\text{mass of whole}} \times 100$$

When applying this formula to a solution, the "part" will always be the dissolved solute, while the "whole" will be the total mass of the solution (solute + solvent). Using the information given in this question, the total mass of the solution described is 70.0 g KNO_3 + 100. g H_2O = 170. g solution.

To write the numerical setup, simply plug the appropriate values into the percent composition formula:

$$\% \text{ composition by mass} = \frac{70.0 \text{ g } KNO_3}{170. \text{ g solution}} \times 100$$

82. Saturated solutions hold the maximum possible dissolved solute at a particular temperature. Reference Table G shows lines of saturation for several solutes. To solve this problem, first read the solubility in a saturated solution of KNO_3 at 50 °C, which is approximately 85. g KNO_3 in 100 g of water.

Therefore 85. g − 70. g = **15 additional grams of KNO_3** must be added to saturate this solution.

(NOTE: Credit was awarded for answers ranging from 12 g of KNO_3 to 16 g of KNO_3.)

83. Electrolytes are compounds that dissociate into ions in aqueous solution. The freely moving charged ions facilitate the flow of electricity through the resulting solution. Acids, bases, and salts (ionic compounds) are three classes of compounds that are electrolytes. Acetic acid solution, $HC_2H_3O_2(aq)$, can conduct an electric current because **it contains mobile, charged ions**.

84. Use Reference Table M to answer this question about an acid–base indicator. Bromthymol blue changes color from yellow to blue in the pH range of 6.0–7.6. Below a pH of 6.0, bromthymol blue is yellow in aqueous solution. Since vinegar has a pH of 2.4, vinegar would be **yellow** in the presence of bromthymol blue indicator.

85. pH is a function of the base 10 exponent of the hydronium ion concentration in an aqueous solution. As an example, a solution with a hydronium concentration of 0.001 M (which equals 10^{-3} M) would have a pH of −(−3), in other words a pH of 3. Likewise, a solution with a hydronium concentration of 0.0001 M (which equals 10^{-4} M) has a pH of 4. Each number on the pH scale differs in hydronium ion concentration by a factor of 10 from the next whole number higher and lower on the scale. In this problem, a sample that has 10 times *fewer* hydronium ions would differ by one whole number from the pH of the original vinegar. Having fewer hydronium ions means that the sample is less acidic than the original vinegar and would have a higher pH value. Putting this information together, the pH of the new sample would be 2.4 + 1.0 = **3.4**.

Mark (✔) the questions you answered correctly. Count the number of checks and follow the formulas given to determine your score on each topic.

Core Area	☐ Questions Answered Correctly

33: (1)

Section M—Math Skills
☐ Number of checks ÷ 1 × 100 = ___ %

16, 49, 68: (3)

Section R—Reading Skills
☐ Number of checks ÷ 3 × 100 = ___ %

1–5, 31, 32, 55, 56: (9)

Section I—Atomic Concepts
☐ Number of checks ÷ 9 × 100 = ___ %

6, 51, 54, 66: (4)

Section II—Periodic Table
☐ Number of checks ÷ 4 × 100 = ___ %

7–9, 34–36, 69, 70, 72, 74: (10)

Section III—Moles/Stoichiometry
☐ Number of checks ÷ 10 × 100 = ___ %

11–13, 37, 38, 52, 53, 57–60: (11)

Section IV—Chemical Bonding
☐ Number of checks ÷ 11 × 100 = ___ %

10, 14, 15, 17–21, 39, 41, 50, 62, 63, 67, 71, 76, 77, 81, 82: (19)

Section V—Physical Behavior of Matter
☐ Number of checks ÷ 19 × 100 = ___ %

22–24, 42, 43, 75: (6)

Section VI—Kinetics and Equilibrium
☐ Number of checks ÷ 6 × 100 = ___ %

25–27, 44, 61, 73: (6)

Section VII—Organic Chemistry
☐ Number of checks ÷ 6 × 100 = ___ %

Core Area	☐ Questions Answered Correctly

28, 29, 78–80: (5)

Section VIII—Oxidation–Reduction
☐ Number of checks ÷ 5 × 100 = ___ %

30, 40, 45, 83–85: (6)

Section IX—Acids, Bases, and Salts
☐ Number of checks ÷ 6 × 100 = ___ %

25–27, 44, 61, 73: (6)

Section X—Nuclear Chemistry
☐ Number of checks ÷ 5 × 100 = ___ %

Examination August 2017
Chemistry—Physical Setting

PART A

Answer all questions in this part.

Directions (1–30): For *each* statement or question, write in the answer space the *number* of the word or expression that, of those given, best completes the statement or answers the question. Some questions may require the use of the *2011 Edition Reference Tables for Physical Setting/Chemistry*.

1 Which phrase describes an Al atom?

  (1) a negatively charged nucleus, surrounded by negatively charged electrons

  (2) a negatively charged nucleus, surrounded by positively charged electrons

  (3) a positively charged nucleus, surrounded by negatively charged electrons

  (4) a positively charged nucleus, surrounded by positively charged electrons   1 _____

2 What is the number of electrons in an atom that has 20 protons and 17 neutrons?

  (1) 37   (3) 3

  (2) 20   (4) 17   2 _____

3 The mass of a proton is approximately equal to the mass of

  (1) an electron   (3) an alpha particle

  (2) a neutron   (4) a beta particle   3 _____

4 When a sample of $CO_2(s)$ becomes $CO_2(g)$, there is a change in

  (1) bond type   (3) molecular polarity

  (2) gram-formula mass   (4) particle arrangement   4 _____

5 Which properties are characteristic of Group 2 elements at STP?

  (1) good electrical conductivity and electronegativities less than 1.7

  (2) good electrical conductivity and electronegativities greater than 1.7

  (3) poor electrical conductivity and electronegativities less than 1.7

  (4) poor electrical conductivity and electronegativities greater than 1.7   5 _____

6 Compared to an atom of C-12, an atom of C-14 has a greater

  (1) number of electrons   (3) atomic number

  (2) number of protons   (4) mass number   6 _____

7 Elements that have atoms with stable valence electron configurations in the ground state are found in

  (1) Group 1   (3) Group 11

  (2) Group 8   (4) Group 18   7 _____

8 A magnesium atom that loses two electrons becomes a

  (1) positive ion with a smaller radius

  (2) negative ion with a smaller radius

  (3) positive ion with a larger radius

  (4) negative ion with a larger radius   8 _____

9 An atom of which element has the strongest attraction for the electrons in a bond?

  (1) aluminum   (3) chlorine

  (2) carbon   (4) lithium   9 _____

10 Which type of matter *cannot* be broken down into simpler substances by a chemical change?

(1) an element   (3) a mixture

(2) a solution   (4) a compound   10 _____

11 According to Table *F*, which substance is most soluble in water?

(1) $AgCl$ (3) Na_2CO_3

(2) $CaCO_3$ (4) $SrSO_4$ 11 _____

12 Given the equation representing a reaction:

$$H + H \rightarrow H_2$$

Which statement describes the energy change in this reaction?

(1) A bond is broken as energy is absorbed.

(2) A bond is broken as energy is released.

(3) A bond is formed as energy is absorbed.

(4) A bond is formed as energy is released.   12 _____

13 Which sample of matter is a mixture?

(1) air   (3) manganese

(2) ammonia   (4) water   13 _____

14 Paper chromatography can separate the components of a mixture of colored dyes because the components have differences in

(1) decay mode   (3) ionization energy

(2) thermal conductivity   (4) molecular polarity   14 _____

15 At standard pressure, the boiling point of an unsaturated $NaNO_3(aq)$ solution increases when

(1) the solution is diluted with water

(2) some of the $NaNO_3(aq)$ solution is removed

(3) the solution is stirred

(4) more $NaNO_3(s)$ is dissolved in the solution   15 _____

16 Which term identifies a form of energy?

(1) combustion   (3) thermal

(2) exothermic   (4) electrolytic   16 _____

17 According to kinetic molecular theory, which statement describes one characteristic of an ideal gas system?

(1) The distance between gas molecules is smaller than the diameter of one gas molecule.

(2) The attractive force between two gas molecules is strong.

(3) The energy of the system decreases as gas molecules collide.

(4) The straight-line motion of the gas molecules is constant and random.   17 _____

18 The temperature of a substance is a measure of the

(1) average kinetic energy of its particles

(2) average potential energy of its particles

(3) ionization energy of its particles

(4) activation energy of its particles   18 _____

19 A real gas behaves most like an ideal gas at

(1) low pressure and high temperature

(2) low pressure and low temperature

(3) high pressure and high temperature

(4) high pressure and low temperature   19 _____

20 A reaction is most likely to occur when the colliding particles have proper orientation and

(1) mass   (3) half-life

(2) volume   (4) energy   20 _____

21 At STP, a 12.0-liter sample of $CH_4(g)$ has the same total number of molecules as

(1) 6.0 L of $H_2(g)$ at STP

(2) 12.0 L of $CO_2(g)$ at STP

(3) 18.0 L of $HCl(g)$ at STP

(4) 24.0 L of $O_2(g)$ at STP   21 _____

22 At standard pressure, during which physical change does the potential energy decrease?

(1) liquid to gas   (3) solid to gas
(2) liquid to solid   (4) solid to liquid   22 _____

23 Which equation represents a chemical equilibrium?

(1) $N_2(\ell) \rightleftharpoons N_2(g)$ (3) $CO_2(s) \rightleftharpoons CO_2(g)$
(2) $2NO_2(g) \rightleftharpoons N_2O_4(g)$ (4) $NH_3(\ell) \rightleftharpoons NH_3(g)$ 23 _____

24 The amount of randomness of the atoms in a system is an indication of the

(1) entropy of the system
(2) polarity of the system
(3) excited state of the atoms
(4) ground state of the atoms   24 _____

25 When a sample of Ca(s) loses 1 mole of electrons in a reaction with a sample of $O_2(g)$, the oxygen

(1) loses 1 mole of electrons
(2) loses 2 moles of electrons
(3) gains 1 mole of electrons
(4) gains 2 moles of electrons   25 _____

26 Which reaction occurs at the anode of an electro-chemical cell?

(1) oxidation   (3) neutralization
(2) reduction   (4) transmutation   26 _____

27 Which substance is an electrolyte?

(1) CCl_4 (3) SiO_2
(2) $C_6H_{12}O_6$ (4) H_2SO_4 27 _____

28 In which process does a heavy nucleus split into two lighter nuclei?

(1) titration   (3) electrolysis
(2) fission   (4) neutralization   28 _____

29 Which process converts mass into energy?

    (1) distillation of ethanol
    (2) filtration of a mixture
    (3) fusion of hydrogen atoms
    (4) ionization of cesium atoms   29 _____

30 Which radioisotope is used to determine the age of once-living organisms?

    (1) carbon-14   (3) iodine-131
    (2) cobalt-60   (4) uranium-238   30 _____

PART B–1

Answer all questions in this part.

Directions (31–50): For *each* statement or question, write in the answer space the *number* of the word or expression that, of those given, best completes the statement or answers the question. Some questions may require the use of the *2011 Edition Reference Tables for Physical Setting/Chemistry*.

31 Which electron configuration represents the electrons in an atom of calcium in an excited state?

(1) 2–8–8   (3) 2–7–8–1

(2) 2–8–8–2   (4) 2–7–8–3   31 _____

32 The table below gives the atomic mass and the abundance of the two naturally occurring isotopes of bromine.

**Naturally Occurring Isotopes
of Bromine**

Isotopes	Atomic Mass (u)	Natural Abundance (%)
Br-79	78.92	50.69
Br-81	80.92	49.31

Which numerical setup can be used to calculate the atomic mass of the element bromine?

(1) $(78.92 \text{ u})(50.69) + (80.92 \text{ u})(49.31)$

(2) $(78.92 \text{ u})(49.31) + (80.92 \text{ u})(50.69)$

(3) $(78.92 \text{ u})(0.5069) + (80.92 \text{ u})(0.4931)$

(4) $(78.92 \text{ u})(0.4931) + (80.92 \text{ u})(0.5069)$ 32 _____

33 Given the formulas of two substances:

$$\ddot{O}=\ddot{O} \qquad \ddot{O}=\ddot{O}\diagdown\ddot{O}:$$

These diagrams represent substances that have

(1) the same molecular structure and the same physical properties
(2) the same molecular structure and different physical properties
(3) different molecular structures and the same physical properties
(4) different molecular structures and different physical properties   33 _____

34 What is the chemical formula of titanium(II) oxide?

(1) TiO   (3) TiO_2
(2) Ti_2O (4) Ti_2O_3 34 _____

35 Which equation shows conservation of mass and energy for a reaction at 101.3 kPa and 298 K?

(1) $2H_2(g) + O_2(g) \rightarrow 2H_2O(g) + 483.6 \text{ kJ}$
(2) $2H_2(g) + O_2(g) \rightarrow 2H_2O(\ell) + 285.8 \text{ kJ}$
(3) $H_2(g) + O_2(g) \rightarrow H_2O(g) + 483.6 \text{ kJ}$
(4) $H_2(g) + O_2(g) \rightarrow H_2O(\ell) + 285.8 \text{ kJ}$ 35 _____

36 Which two particle diagrams each represent a sample of one substance?

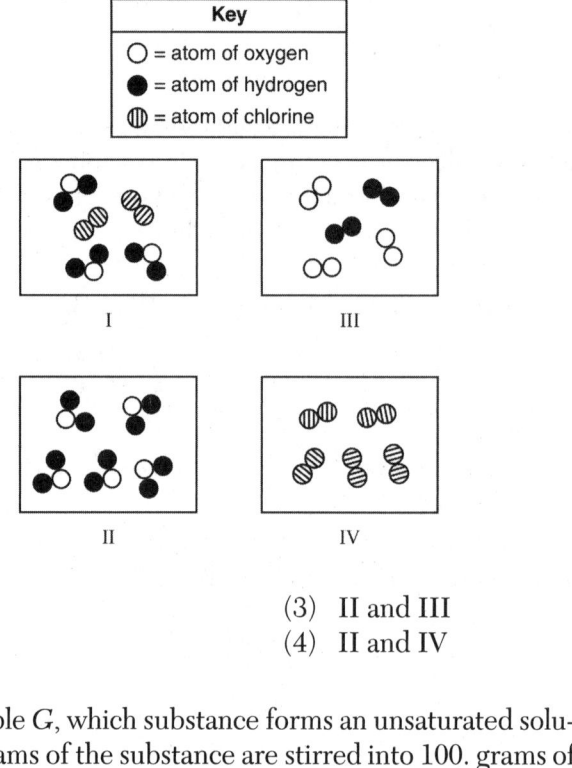

Key
◯ = atom of oxygen
● = atom of hydrogen
◫ = atom of chlorine

I

III

II

IV

(1) I and II
(2) I and III
(3) II and III
(4) II and IV

36 _____

37 According to Table G, which substance forms an unsaturated solution when 80. grams of the substance are stirred into 100. grams of H_2O at 10.°C?

(1) KNO_3
(2) KI
(3) NH_3
(4) NaCl

37 _____

38 What is the concentration of AgCl in an aqueous solution that contains 1.2×10^{-3} gram of AgCl in 800. grams of the solution?

(1) 1.2 ppm
(2) 1.5 ppm
(3) 7.2 ppm
(4) 9.6 ppm

38 _____

39 A sample of gas is in a rigid cylinder with a movable piston. The pressure of the gas is kept constant. If the Kelvin temperature of the gas is doubled, the volume of the gas is

(1) halved
(2) doubled
(3) tripled
(4) unchanged

39 _____

40 What is the amount of heat required to completely melt a 200.-gram sample of $H_2O(s)$ at STP?

(1) 334 J   (3) 66 800 J

(2) 836 J   (4) 452 000 J   40 _____

41 As a 15.1-gram sample of a metal absorbs 48.75 J of heat, its temperature increases 25.0 K. What is the specific heat capacity of the metal?

(1) 0.129 J/g·K   (3) 3.23 J/g·K

(2) 1.95 J/g·K   (4) 7.74 J/g·K   41 _____

42 Given the formula:

$$
\begin{array}{c}
\quad\; H \quad\; H \quad\; O \\
\quad\; | \quad\quad | \quad\quad || \\
H - C - C - C - H \\
\quad\; | \quad\quad | \\
\quad\; H \quad\; H
\end{array}
$$

What is a chemical name of this compound?

(1) propane   (3) propanol

(2) propanal   (4) propanone   42 _____

43 Given the equation representing a reaction:

$$
\begin{array}{c}
\quad\; H \quad\quad\quad\quad\quad\quad\quad\quad H \\
\quad\; | \quad\quad\quad\quad\quad\quad\quad\quad\; | \\
H - C - H \; + \; Cl - Cl \; \rightarrow \; H - C - H \; + \; H - Cl \\
\quad\; | \quad\quad\quad\quad\quad\quad\quad\quad\; | \\
\quad\; H \quad\quad\quad\quad\quad\quad\quad\quad Cl
\end{array}
$$

Which type of reaction is represented by this equation?

(1) addition   (3) polymerization

(2) esterification   (4) substitution   43 _____

44 Atoms of which element react spontaneously with $Mg^{2+}(aq)$?

(1) chromium   (3) iron

(2) barium   (4) zinc   44 _____

45 In a titration, 5.0 mL of a 2.0 M NaOH(aq) solution exactly neutralizes 10.0 mL of an HCl(aq) solution. What is the concentration
of the HCl(aq) solution?

(1) 1.0 M   (3) 10. M

(2) 2.0 M   (4) 20. M   45 _____

46 Given the equation representing a reaction at equilibrium:

$$NH_3(g) + H_2O(\ell) \rightleftharpoons NH_4^+(aq) + OH^-(aq)$$

If an acid is defined as an H^+ donor, what is the acid in the forward
reaction?

(1) $OH^-(aq)$ (3) $NH_3(g)$

(2) $H_2O(\ell)$ (4) $NH_4^+(aq)$ 46 _____

47 Compared to a solution with a pH value of 7, a solution with a
thousand times greater hydronium ion concentration has a pH
value of

(1) 10   (3) 3

(2) 7   (4) 4   47 _____

48 Which Lewis electron-dot diagram represents the bonding in
potassium iodide?

$K^+ \left[:\ddot{\underset{..}{I}}: \right]^-$

(1)

$\left[:\ddot{\underset{..}{K}}: \right]^- I^+$

(2)

$K:\ddot{\underset{..}{I}}:$

(3)

$:\ddot{\underset{..}{K}}:I$

(4)   48 _____

49 The table below shows properties of two compounds at standard pressure.

Selected Properties of Two Compounds

Compound	Melting Point (°C)	Boiling Point (°C)	Electrical Conductivity
1	775	1935	good as a liquid or in an aqueous solution
2	−112.1	46	poor as a liquid

Which statement classifies the two compounds?

(1) Both compounds are ionic.
(2) Both compounds are molecular.
(3) Compound 1 is ionic, and compound 2 is molecular.
(4) Compound 1 is molecular, and compound 2 is ionic.

49 _____

50 Given two balanced equations, each representing a reaction:

Equation 1:

$$^{226}_{88}\text{Ra} \rightarrow\ ^{222}_{86}\text{Rn} + ^{4}_{2}\text{He} + 4.8 \times 10^8 \text{ kJ}$$

Equation 2:

$$C_3H_8 + 5O_2 \rightarrow 3CO_2 + 4H_2O + 2.2 \times 10^3 \text{ kJ}$$

Which statement compares the energy terms in these two equations?

(1) Equation 1 shows 2.2×10^5 times more energy being absorbed.
(2) Equation 2 shows 2.2×10^5 times more energy being absorbed.
(3) Equation 1 shows 2.2×10^5 times more energy being released.
(4) Equation 2 shows 2.2×10^5 times more energy being released.

50 _____

PART B–2

Answer all questions in this part.

Directions (51–65): Record your answers on the answer sheet provided. Some questions may require the use of the *2011 Edition Reference Tables for Physical Setting/Chemistry.*

Base your answers to questions 51 and 52 on the information below and on your knowledge of chemistry.

The formula below represents a molecule of butanamide.

$$H-\underset{\underset{H}{|}}{\overset{\overset{H}{|}}{C}}-\underset{\underset{H}{|}}{\overset{\overset{H}{|}}{C}}-\underset{\underset{H}{|}}{\overset{\overset{H}{|}}{C}}-\overset{\overset{O}{\|}}{C}-N\overset{\diagup H}{\diagdown H}$$

51 State the type of chemical bond between a hydrogen atom and the nitrogen atom in the molecule.   [1]

52 Explain, in terms of charge distribution, why a molecule of butanamide is polar.   [1]

Base your answers to questions 53 and 54 on the information below and on your knowledge of chemistry.

An equilibrium system in a sealed, rigid container is represented by the equation below.

$$CO(g) + H_2O(g) \rightleftharpoons CO_2(g) + H_2(g)$$

53 Compare the rate of the forward reaction to the rate of the reverse reaction at equilibrium. [1]

54 State the effect on the concentrations of $H_2O(g)$ and $CO_2(g)$ when more $H_2(g)$ is added to the system. [1]

Base your answers to questions 55 through 58 on the information below and on your knowledge of chemistry.

The table below contains selected information about chlorine and two compounds containing chlorine. One piece of information is missing for each of the substances in the table.

Chlorine and Two Compounds Containing Chlorine

Name	Formula	Molar Mass (g/mol)	Phase at STP
chlorine	Cl_2	71	?
calcium chloride	$CaCl_2$	?	solid
1,2-dichloroethene	?	97	liquid

55 Identify the phase of the chlorine at STP.   [1]

56 Determine the molar mass for calcium chloride.   [1]

57 The liquid compound has an empirical formula of CHCl. Write the molecular formula for this compound.   [1]

58 Explain, in terms of electrons, why the compound containing calcium and chlorine is classified as an ionic compound.   [1]

Base your answers to questions 59 and 60 on the information below and on your knowledge of chemistry.

The equation below represents the reaction between 2-methylpropene and hydrogen chloride gas.

```
    H   H                    H  Cl  H
    |   |                    |  |   |
H − C − C = C − H  + HCl → H − C − C − C − H
    |   |                    |  |   |
    H   |                    H  |   H
        |                       |
    H − C − H                H − C − H
        |                       |
        H                       H
```

59 Explain, in terms of chemical bonds, why the hydrocarbon is unsaturated. [1]

60 Identify the class of organic compounds to which the product belongs. [1]

Base your answers to questions 61 through 65 on the information below and on your knowledge of chemistry.

Many scientists made observations of the elements that led to the modern Periodic Table. In 1829, Dobereiner found groups of three elements that have similar properties and called each of these groups a triad. Dobereiner noticed a relationship between the atomic masses of the elements in each triad. Triad 1, shown in the table below, consists of sulfur, selenium, and tellurium. The middle element, selenium, has an atomic mass that is close to the sum of the atomic masses of sulfur and tellurium, divided by 2.

For example: $\dfrac{32 \text{ u} + 128 \text{ u}}{2}$ 80. u, which is close to the 79 u value in the table.

The other triads shown in the table below demonstrate the same mathematical relationship.

Dobereiner's Triads

Triad	Triad	Dobereiner's Atomic Masses (u)
1	sulfur selenium tellurium	32 79 128
2	calcium strontium barium	40. 88 137
3	chlorine bromine iodine	35.5 80. 127
4	lithium sodium potassium	7 23 39

61 Identify the triad that contains a metalloid. [1]

62 Explain, in terms of electrons, why the elements in triad 2 have similar chemical properties. [1]

63 State the trend in first ionization energy as the elements in triad 3 are considered in order of increasing atomic number. [1]

64 Compare the volume of a 100.-gram sample of the first element in triad 4 to the volume of a 100.-gram sample of the third element in triad 4 when both samples are at room temperature.   [1]

65 Show a numerical setup that demonstrates Dobereiner's mathematical relationship for triad 2.   [1]

PART C

Answer all questions in this part.

Directions (66–85): Record your answers on the answer sheet provided. Some questions may require the use of the *2011 Edition Reference Tables for Physical Setting/Chemistry*.

Base your answers to questions 66 through 68 on the information below and on your knowledge of chemistry.

Wood is mainly cellulose, a polymer produced by plants. One use of wood is as a fuel in campfires, fireplaces, and wood furnaces. The molecules of cellulose are long chains of repeating units. Each unit of the chain can be represented as $C_6H_{10}O_5$. The balanced equation below represents a reaction that occurs when $C_6H_{10}O_5$ is burned in air.

$$C_6H_{10}O_5 + 6O_2 \rightarrow 6CO_2 + 5H_2O + \text{heat}$$

66 State evidence from the equation that this reaction is exothermic. [1]

67 Explain, in terms of substances in the reaction, why the equation represents a chemical change. [1]

68 Show a numerical setup for calculating the percent composition by mass of carbon in $C_6H_{10}O_5$ (gram-formula mass = 162.1 g/mol). [1]

Base your answers to questions 69 through 72 on the information below and on your knowledge of chemistry.

Millions of tons of ammonia are produced each year for use as fertilizer to increase food production. Most of the hydrogen needed to produce ammonia comes from methane gas reacting with steam. This reaction, which occurs in a container under controlled conditions, is shown below in unbalanced equation 1.
Equation 1:

$$CH_4(g) + H_2O(g) + energy \rightarrow CO(g) + H_2(g)$$

The reaction that produces ammonia is represented by balanced equation 2, shown below. A catalyst can be used to increase the rate of the reaction.
Equation 2:

$$N_2(g) + 3H_2(g) \rightarrow 2NH_3(g) + energy$$

A potential energy diagram for equation 2 is shown below.

69 Balance equation 1 *on your answer sheet*, using the smallest whole-number coefficients. [1]

70 Explain, in terms of collision theory, why an increase in temperature increases the rate of reaction between methane gas and steam. [1]

71 State what is represented by interval A on the potential energy diagram. [1]

72 Determine the number of moles of hydrogen gas required to react completely with 50.0 moles of nitrogen gas in the production of ammonia. [1]

Base your answers to questions 73 through 77 on the information below and on your knowledge of chemistry.

Diethyl ether is used as a laboratory and industrial solvent. The boiling point of diethyl ether at standard pressure is 34.6°C. The equation below represents a reaction that produces diethyl ether.

$$\underset{\text{H}}{\overset{\text{H}}{>}}\text{C=C}\underset{\text{H}}{\overset{\text{H}}{<}} \;+\; \underset{\text{H}}{\overset{\text{H}}{>}}\text{C=C}\underset{\text{H}}{\overset{\text{H}}{<}} \;+\; H_2O \;\xrightarrow{\text{catalyst}}\; \text{H}-\overset{\text{H}}{\underset{\text{H}}{\text{C}}}-\overset{\text{H}}{\underset{\text{H}}{\text{C}}}-\text{O}-\overset{\text{H}}{\underset{\text{H}}{\text{C}}}-\overset{\text{H}}{\underset{\text{H}}{\text{C}}}-\text{H}$$

73 Identify the element in diethyl ether that allows it to be classified as an organic compound.   [1]

74 State the number of electrons shared between the carbon atoms in one molecule of the organic reactant.   [1]

75 State why the reaction is classified as a synthesis reaction.   [1]

76 Explain, in terms of the strength of intermolecular forces, why the boiling point of diethyl ether at standard pressure is *lower* than the boiling point of water at standard pressure.   [1]

77 Draw a structural formula for an isomer of the product that has the same functional group.   [1]

Base your answers to questions 78 through 81 on the information below and on your knowledge of chemistry.

The electrolysis of brine, a concentrated aqueous sodium chloride solution, produces three important industrial chemicals: chlorine gas, hydrogen gas, and sodium hydroxide. The diagram and equation below represent a brine electrolysis cell. Before the battery is connected, the pH value of the brine solution is 7.0.

$$2NaCl(aq) + 2H_2O(\ell) \rightarrow Cl_2(g) + H_2(g) + 2NaOH(aq)$$

78 Explain, in terms of energy, why this cell is an electrolytic cell. [1]

79 Explain, in terms of ions, why the aqueous solution in the cell conducts an electric current. [1]

80 State the oxidation number of oxygen in the aqueous product. [1]

81 Compare the pH value of the solution before the battery is connected to the pH value of the solution after the cell operates for 20 minutes. [1]

Base your answers to questions 82 through 85 on the information below and on your knowledge of chemistry.

The isotope Rn-222 is produced by the decay of uranium in Earth's crust. Some of this isotope leaks into basements of homes in areas where the ground is more porous. An atom of Rn-222 decays to an atom of Pb-206 through a series of steps as shown on the graph below.

Decay Series of Rn-222

Atomic Number

82 Determine the number of neutrons in an atom of Pb-214. [1]

83 Complete the nuclear equation *on your answer sheet* for the decay of Po-218 by writing a notation for the missing product. [1]

84 Determine the fraction of an original sample of Rn-222 that remains unchanged after 7.646 days. [1]

85 Explain, in terms of elements, why the decay of Bi-210 is considered a transmutation. [1]

Answer Sheet
August 2017

Chemistry—Physical Setting

PART B–2

51 _____

52 _____

_____

_____

53 _____

_____

_____

54 $H_2O(g)$: _____

$CO_2(g)$: _____

55 _____

56 _____ g/mol

57 _____

58 _____

_____

_____

59 _____

_____

_____

60 _____

61 _____

62 _____

_____

_____

63 _____

_____

_____

64 _____

_____

_____

65

PART C

66 _____

_____

_____

67 _____

_____

_____

68

69 ___ $CH_4(g)$ + ___ $H_2O(g)$ + energy → ___ $CO(g)$ + ___ $H_2(g)$

70 _____

_____

_____

71 _____

_____

_____

72 _____ **mol**

73 _____

74 _____

75 _____

_____

_____

76 _____

_____

_____

77

78 _____

_____

_____

79 _____

_____

_____

80 _____

81 _____

_____

_____

82 _____

83 $^{218}_{84}\text{PO} \rightarrow\ ^{214}_{82}\text{PB} +$ _____

84 _____

85 _____

_____

_____

Answers
August 2017

Chemistry—Physical Setting

Answer Key

PART A

1. 3	7. 4	13. 1	19. 1	25. 3
2. 2	8. 1	14. 4	20. 4	26. 1
3. 2	9. 3	15. 4	21. 2	27. 4
4. 4	10. 1	16. 3	22. 2	28. 2
5. 1	11. 3	17. 4	23. 2	29. 3
6. 4	12. 4	18. 1	24. 1	30. 1

PART B–1

31. 4	35. 1	39. 2	43. 4	47. 4
32. 3	36. 4	40. 3	44. 2	48. 1
33. 4	37. 2	41. 1	45. 1	49. 3
34. 1	38. 2	42. 2	46. 2	50. 3

PART B–2 and **PART C**. *See* **Answers Explained**.

Answers Explained

PART A

1. **3** The basic structure of an atom is a positively charged nucleus surrounded by a negatively charged electron cloud. The nucleus contains protons, which have a positive charge. Reference Table O provides a handy summary of the names and charges of protons, neutrons, and electrons.

2. **2** Atoms are neutral in charge, since the number of positively charged protons is equal to the number of negatively charged electrons. An atom with 20 protons would have 20 electrons.

3. **2** Reference Table O lists the symbols for all of the particles mentioned in this problem, but you do need to remember that an electron is identical in mass and charge to a beta particle. Particle notations for the neutron and the proton are shown below:

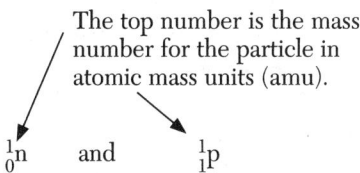

The top number is the mass number for the particle in atomic mass units (amu).

$$^{1}_{0}n \quad \text{and} \quad ^{1}_{1}p$$

Protons and neutrons both have a mass of approximately 1 amu.

WRONG CHOICES EXPLAINED:
(1), (4) An electron has the same mass and charge as a beta particle. Looking at Reference Table O, we can see that the mass of an electron is 0.

In fact, the electron does have a mass of $\dfrac{1}{1840}$ amu, but this is negligible compared with the mass of a proton or a neutron.

(3) Reference Table O shows that the mass of an alpha particle is 4 amu.

4. **4** When a solid becomes a gas, the particles break free of their tightly packed lattice arrangement and become randomly positioned with respect to each other. This represents a dramatic change in particle arrangement! This change in particle arrangement is shown in the diagram below:

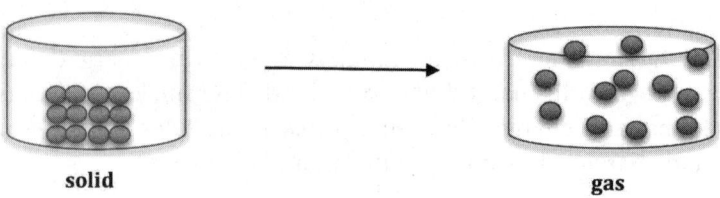

solid   gas

WRONG CHOICES EXPLAINED:

(1) The bonds between carbon and oxygen atoms in CO_2 molecules are non-polar covalent bonds. The bonds within CO_2 molecules do not change when the molecules move farther apart.

(2) The gram-formula mass (also known as the molar mass) of CO_2 is 44.0 g/mol. This is calculated by adding the individual molar masses of the atoms in the compound. Molecules of CO_2 have the same composition in the solid phase as they do in the gas phase, so the molar mass of the molecules does not change.

(3) CO_2 molecules are nonpolar by virtue of their symmetry. Molecules of CO_2 do not change shape as they separate to form a gas.

5. **1** The Group 2 elements are known as the Alkaline Earth Metals. As metals, these elements will all have good electrical conductivity. Reference Table S lists the values for the electronegativity of most elements. The electro-negativity values for Group 2 elements range from a high of 1.6 for beryllium to a low of 0.9 for radium. All of these values are *below* 1.7.

6. **4** In the element notation used in this question, the numbers shown are the mass numbers of the two isotopes. Different isotopes of any element will always have different mass numbers because they have different numbers of neutrons. Remember that the mass number is equal to the number of neutrons in an atom added to the number of protons.

WRONG CHOICES EXPLAINED:

(1) Six electrons are present in every atom of carbon. The number of elec-trons always equals the number of protons in all atoms since atoms have no overall charge.

(2), (3) All atoms of the same element have the same number of protons, which is equal to the element's atomic number. An element is defined by its atomic number. Six protons are present in every atom of carbon since carbon's atomic number is 6.

7. **4** The Group 18 elements are known as the noble gases. Noble gases are considered to be the most stable elements on the Periodic Table because they usually do not react with other elements. These gases rarely form bonds with other elements since they already possess the stable octet of 8 valence electrons.

8. **1** According to the Periodic Table in your Reference Tables, magnesium atoms have a ground-state configuration of 2-8-2. These numbers represent the number of electrons found in the first, second, and third electron shells, respectively. The 2 third-shell electrons are lost when magnesium loses two electrons, leaving the resulting magnesium ion with an electron configuration of 2-8. Since the magnesium ion still has 12 positively charged protons but now only has 10 negatively charged electrons, the magnesium ions have a positive charge. In addition, the ion is smaller than the original atom because the third electron shell is no longer occupied.

9. **3** You must know the definition of electronegativity to answer this question correctly. Electronegativity is a measure of how strongly atoms attract bonding electrons to themselves. Electronegativity values for most elements are listed in Reference Table S. Chlorine's electronegativity of 3.2 is higher than that of any of the other elements given as answer choices for this question. A familiarity with periodic trends can also be helpful when answering this question, namely knowing that electronegativity values increase from left to right across any period of the Periodic Table.

10. **1** Elements are the simplest type of matter. Elements are composed of atoms, which can only be broken down by nuclear change, not by chemical change.

WRONG CHOICES EXPLAINED:
(2), (3) A solution is a homogeneous mixture. All mixtures can be physically separated.
(4) Compounds consist of two or more elements that are chemically bonded together. Compounds can be broken down into their constituent elements by chemical reactions (chemical change).

11. **3** Sodium (Na) is a Group 1 element. The first column in Reference Table *F* shows that all compounds containing sodium ions are soluble.

WRONG CHOICES EXPLAINED:

(1) Silver (Ag) ions are listed in the second column of Reference Table *F* as an exception to the solubility of halide ions, such as the Cl^- ion. Think of the second column as "not soluble," which is what it means to be an "exception to solubility."

(2) Carbonates (compounds containing the CO_3^{2-} ion) are listed in the third column of Reference Table *F* as an example of ions that form *insoluble* compounds in water. The calcium ion is not listed in the last column as an exception.

(4) Most compounds that contain the sulfate ion, SO_4^{2-}, are soluble in water. Notice that sulfates are listed in the first column of Reference Table *F*. However, the strontium ion, Sr^{2+}, is listed in the second column as an exception to the solubility of sulfates. In other words, $SrSO_4$ is *not* soluble.

12. **4** This reaction shows two hydrogen atoms bonding together to form one H_2 molecule. The formation of chemical bonds between elements always causes a *release* of potential energy since elements bond to become more stable.

13. **1** Air is a mixture of several gases, including N_2, O_2, and CO_2.

WRONG CHOICES EXPLAINED:

(2), (4) Ammonia (NH_3) and water (H_2O) are compounds. Compounds are pure substances, not mixtures.

(3) Manganese (Mn) is an element on the Periodic Table (atomic number = 25). Elements are pure substances, not mixtures.

14. **4** Paper chromatography works because the different components, or pigments, in a mixture of colored dyes have different levels of attraction to paper. As the dye mixture advances up the paper, pigment compounds that are more attracted to the paper will absorb first, while compounds with less attraction will move farther up the paper before being absorbed. The attraction between dye compounds and paper is caused by molecular polarity (the attraction of opposite partial charges on different molecules).

15. **4** Dissolved ions raise the boiling point of a solution. This is the colligative property known as boiling point elevation. The more ions that are dissolved, the higher the boiling point. As more $NaNO_3(s)$ is dissolved in the solution, the boiling point will increase.

16. **3** Thermal energy is created by the motion of atoms and molecules within an object.

WRONG CHOICES EXPLAINED:
(1) Combustion is a type of chemical reaction. It is not a form of energy.
(2) The term "exothermic" describes a chemical reaction that releases energy.
(4) "Electrolytic" is a type of electrochemical cell that utilizes electrical energy to produce a chemical change.

17. **4** One of the principles of the kinetic molecular theory of gases is that gas particles move in random, constant, straight-line motion. This allows gas particles to expand to fill the space that they occupy. Other principles of the kinetic molecular theory of gases state that an ideal gas consists of particles that:

- have mass
- have no attractions or repulsions for each other
- are separated by distances much greater than their size
- undergo perfectly elastic collisions in which the total energy of the system remains the same before and after collisions

18. **1** Temperature is defined as a measure of the average kinetic energy of the particles in any substance.

19. **1** Real gases behave more like ideal gases when temperature is high and pressure is low. These conditions minimize any possible attractions between the gas particles, allowing the gas to adhere more closely to the principles of the kinetic molecular theory of gases. (A good way to remember this is to think of summertime, when the temperature is high and the pressure is low.)

20. **4** The basic idea behind collision theory is that particles will only react if they have sufficient energy and proper orientation.

21. **2** Equal volumes of gas at the same conditions of temperature and pressure contain the same number of particles. This relationship is known as Avogadro's Law. For example, the molar volume of a gas at STP equals 22.4 L/mol no matter what type of gas you have.

22. **2** Energy is required to overcome attractions between particles in order to increase spacing between them. Conversely, energy is released when particles that are attracted are able to be closer together. A physical change from the liquid

phase to the solid phase puts particles much closer together, corresponding to a release in potential energy. Consider that to make solid ice, liquid water must be put in the freezer, where energy is taken out of the water.

23. **2** All of the equations shown represent equilibrium systems, but only answer choice (2) shows chemical equilibrium in which forward and reverse chemical reactions are happening at equal rates. The other three answer choices all show changes in phase, and these are physical changes, not chemical changes.

24. **1** Entropy is defined as the measure of randomness, or disorder, in a system.

25. **3** All chemical reactions conserve mass, energy, and charge. If one mole of electrons is lost as Ca(s) oxidizes, then one mole of electrons must be gained by the substance that is reducing, in this case $O_2(g)$.

26. **1** The mnemonic device "An Ox, Red Cat" is helpful for remembering that oxidation (loss of electrons) takes place at the anode, while reduction (gain of electrons) happens at the cathode.

27. **4** Three classes of compounds are electrolytes: acids, bases, and salts. Electrolytes are compounds that dissociate into ions in aqueous solution and, because of these freely moving ions, enable the flow of electricity through the solution. H_2SO_4 is sulfuric acid (check Reference Table K for a list of common acids). It dissociates into H^+ and SO_4^{-2} ions in aqueous solution. All of the other answer choices are nonelectrolytes, meaning that they are covalently bonded and would not dissociate.

28. **2** Fission is the splitting of a large atomic nucleus (usually U-235) into two smaller nuclei. The fission reaction is initiated by the bombardment of the target nucleus with a neutron, and the resulting emission of neutrons as products makes a chain reaction possible. An example of a fission reaction is shown below:

$$^{235}_{92}U + ^{1}_{0}n \rightarrow ^{94}_{38}Sr + ^{140}_{54}Xe + 2^{1}_{0}n$$

WRONG CHOICES EXPLAINED:
(1) Titration is a laboratory process in which the concentration of an unknown solution can be determined by the careful addition of a standard solution with a known concentration. Titration often involves the neutralization of an acid with a base (or vice versa).

(3) Electrolysis is a redox reaction in which electricity is used to force a non-spontaneous chemical change to happen.

(4) Neutralization reactions occur between acids and bases when H^+ ions react with OH^- ions to form water.

29. **3** Nuclear changes produce energy by converting mass into energy. Of the answer choices listed, fusion is the only nuclear reaction.

30. **1** Since living organisms are carbon-based organisms, the radioisotope C-14 is helpful in determining the age of ancient remains from previously living organisms. The process known as radiocarbon dating involves the analysis of the amount of carbon-14 remaining in the sample.

PART B–1

31. **4** An atom of calcium will always have 20 electrons. The Periodic Table in the Reference Tables shows that the ground-state electron configuration of a calcium atom is 2-8-8-2. A calcium atom would be in an excited state if its configuration were 2-7-8-3 because this configuration shows that a second-shell electron has moved to the fourth shell.

WRONG CHOICES EXPLAINED:

(1) There are only 18 electrons shown in this configuration. This is the ground-state configuration for the calcium *ion*, Ca^{+2}, but not for a calcium *atom*.

(2) This is the ground-state configuration for a calcium atom. The question asks for an *excited*-state configuration.

(3) This configuration is short 2 electrons. All calcium atoms have 20 electrons.

32. **3** The atomic mass of an element is found by taking the weighted average of all naturally occurring isotopes of that element. The natural abundance percentages of each isotope are first converted to decimal fractions, and these are then multiplied by the atomic mass of the isotope. Adding the contributions of each isotope together yields the correct atomic mass for that element.

WRONG CHOICES EXPLAINED:

(1) Although the correct percentages are multiplied by each atomic mass, the final sum would have to be divided by 100.

(2) This numerical setup multiplies each isotope's atomic mass by the wrong percentage. Another error is that, if the actual percentage values are used without converting them to decimal fractions, the final step must be to divide by 100, since a percent is literally "parts per hundred."

(4) This numerical setup shows the percentages converted to decimal fractions, but they are then multiplied by the opposite isotope's atomic mass.

33. **4** The substances shown are two different allotropes of the element oxygen. Allotropes are different naturally occurring structures of the same element. Since the molecular structures are different, the molecules would exhibit different physical properties, such as melting point, boiling point, and density.

34. **1** Titanium(II) oxide is an ionic compound made from the titanium(II) ion, Ti^{+2}, and the oxide ion, O^{-2}. Charges for these two ions can be found on the Periodic Table. Since compounds have no overall charge, one titanium(II) ion will cancel the -2 charge of the oxide ion.

WRONG CHOICES EXPLAINED:

(2) This would be the formula for titanium(I) oxide. In this compound, it takes two Ti^{+1} ions to cancel the -2 charge of the oxide ion.

(3) This is the formula for titanium(IV) oxide. One Ti^{+4} ion would cancel the charge of two oxide ions, each having a charge of -2.

(4) This is the formula for titanium(III) oxide. Two Ti^{+3} ions will cancel the charge of three O^{-2} ions.

35. **1** You are not expected to memorize the heat of reaction for chemical reactions, so the only way to know which answer choice is correct is to consult Reference Table *I*, which lists the heat of reaction for some common chemical reactions at 101.3 kPa and 298 K. The reaction in answer choice (1) is identical to one of the equations listed in this table. The ΔH for this reaction is listed as -483.6 kJ. Since the heat of reaction has a negative value, the reaction is exothermic, and the heat term is written as a product in the chemical equation.

WRONG CHOICES EXPLAINED:

(2) This equation is balanced for mass, but the heat of reaction is incorrect. The correct heat of reaction for this reaction is shown on Reference Table *I*.

(3) This equation is not balanced for mass. Two oxygen atoms appear on the reactant side of this equation, but only one oxygen atom appears on the product side.

(4) This equation is neither balanced for mass nor is it balanced for energy.

36. **4** A pure substance is made from just one type of particle. Diagram II shows just one type of molecule, and Diagram IV shows just one type of diatomic element. Diagrams I and III are incorrect because they represent mixtures of more than one substance.

37. **2** The lines on Reference Table G represent lines of solution saturation. A saturated solution is holding all of the solute possible at a given temperature. An unsaturated solution is not fully saturated, meaning that more solute could dissolve. On Reference Table G, look carefully at the intersection between 10°C and 80 grams of solute (in 100 grams of water). This intersection is *above* the saturation curves for KNO_3, NH_3, and NaCl. Adding 80 grams of any of these compounds to 100 grams of water would create *supersaturated* solutions. Only the saturation curve for KI is above the intersection between 10°C and 80 grams of solute. Since 80 grams of KI would not saturate a solution made in 100 grams of water at 10°C, the solution would be *unsaturated*.

38. **2** Look at Reference Table T to find the formula for parts per million:

$$\text{parts per million} = \frac{\text{mass of solute}}{\text{mass of solution}} \times 1{,}000{,}000$$

$$\text{parts per million} = \frac{0.0012 \text{ g}}{800. \text{ g}} \times 1{,}000{,}000 = 1.5 \text{ ppm}$$

39. **2** The Combined Gas Law found on Reference Table T is useful for thinking about this problem. Since the pressure is kept constant, we can reduce the Combined Gas Law to the following expression, which is known as Charles's Law:

$$\frac{V_1}{T_1} = \frac{V_2}{T_2}$$

There is a direct relationship between the volume of a gas and the Kelvin temperature of a gas at constant pressure. If one doubles, the other must also double. For example, in the illustration below, if the Kelvin temperature increases, the gas particles will strike the piston more often and with more force, and the piston will push outward until the pressure inside and outside becomes equal.

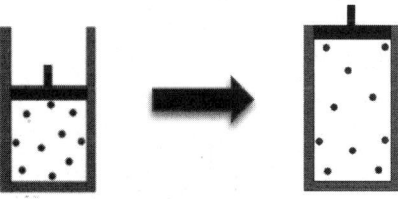

40. **3** The last page of the Reference Tables, Reference Table T, lists the formula to find the heat of fusion, which is the heat lost or gained by a substance as it freezes or melts, respectively.

$$q = mH_f$$

where q = heat, m = mass, and H_f = heat of fusion

The mass is given as 200. grams, and the heat of fusion for liquid water, found on Reference Table B, is 334 J/g. Plug these values into the heat of fusion formula and solve:

$$q = (200.\ g)(334\ J/g) = 66800\ J$$

41. **1** To solve for the specific heat capacity, use the following equation found on Reference Table T:

$$q = mC\Delta T$$

$$\text{where } q = \text{heat, } m = \text{mass,}$$
$$C = \text{specific heat capacity,}$$
$$\text{and } \Delta T = \text{change in temperature}$$

Rearrange this equation to solve for C:

$$C = \frac{q}{m\Delta T}$$

Now just plug in the values given in the problem:

$$C = \frac{48.75\ J}{(15.1\ g)(25.0\ k)} = 0.129\ J/g{\cdot}K$$

42. **2** Refer to Reference Table R for all questions concerning organic functional groups. The structure shown in this question is actually found as an example directly on this table! The compound is definitely an aldehyde since it has the aldehyde functional group shown below:

Since there are three carbon atoms in the molecule, the correct prefix from Reference Table P would be "prop-." To finish the name, use the aldehyde suffix "-al" as shown in the example on Reference Table R. Putting these pieces together, the compound is correctly named propanal.

43. **4** This is an organic reaction known as a substitution reaction because one of the chlorine atoms substitutes for one of the hydrogen atoms on the hydrocarbon.

WRONG CHOICES EXPLAINED:
(1) Addition reactions require a hydrocarbon reactant with a double or triple carbon–carbon bond. The complete diatomic molecule would "add" to the double bond: in other words, the product would contain *both* chlorine atoms, not just one.

(2) Esterification reactions produce esters. The product shown does not contain the ester functional group.

(3) Polymerization reactions create long chains from many simple identical molecules, called monomers. The product in this reaction is a small, single carbon molecule.

44. **2** Reference Table J lists metals from the most active metals at the top to the least active metals at the bottom. Pure metal atoms never gain electrons, but metal ions can gain back the electrons they previously lost in the presence of a more active metal. Mg^{+2} ions will gain electrons back in the presence of barium, which is higher on Reference Table J and is therefore a more active metal.

45. **1** Use the titration formula found on Reference Table T:

$$M_A V_A = M_B V_B$$

$$\text{where } M_A = \text{molarity of } H^+,$$
$$V_A = \text{volume of acid},$$
$$M_B = \text{molarity of } OH^-,$$
$$\text{and } V_B = \text{volume of base}$$

Substitute the values from this problem:

$$\left(M_A\right)(10.0 \text{ mL}) = (2.0 \text{ M})(5.0 \text{ mL})$$

$$M_A = \frac{(2.0 \text{ M})(5.0 \text{ mL})}{10.0 \text{ mL}} = 1.0 \text{ M}$$

46. **2** Compare the reactant formulas to the product formulas. $NH_3(g)$ becomes $NH_4^+(aq)$ by gaining a hydrogen ion (H^+) from $H_2O(\ell)$. In other words, $H_2O(\ell)$ "donates" a hydrogen ion to $NH_3(g)$ to form NH_4^+.

47. **4** pH is a function of the base 10 exponent of the hydronium ion concentration in an aqueous solution. As an example, a solution with a hydronium concentration of 0.001 M $(= 10^{-3}$ M) has a pH of $-(-3)$, in other words a pH of 3. Likewise, a solution with a hydronium concentration of 0.0001 M $(=10^{-4}$ M) has a pH of 4. Each number on the pH scale differs in hydronium ion concentration by a factor of 10 from the next whole number higher and lower on the scale. In this problem, a sample that has one thousand times *greater* hydronium ions would differ by *three* whole numbers from the original pH of 7. Having more hydronium ions means that the sample is more acidic than the original solution and would have a lower pH value. Putting this information together, the pH of the new sample would be $7 - 3 = 4$.

48. **1** Lewis diagrams for ionic compounds show both *ion charges* and *electron transfer*, with all eight valence electrons surrounding the nonmetal, in this case iodine.

WRONG CHOICES EXPLAINED:
(2) The charges and electron positions in this answer choice are incorrect. Metal atoms always lose electrons when they form ionic bonds. Metal ions always have a positive charge. Nonmetal atoms always gain electrons from metal atoms when ionic bonds form. The electrons gained complete the stable octet of valence electrons around the nonmetal atom.

(3), (4) These answer choices show the formation of a shared pair of electrons (covalent bond) between potassium (a metal) and iodine (a nonmetal). Bonds between metal and nonmetal atoms are always ionic bonds, not covalent.

49. **3** Compound 1 has a high melting point and a high boiling point, both of which are characteristic of ionic compounds. In addition, ionic compounds will conduct electricity *only* when the compound is melted or dissolved and the ions are free to move. The extremely low melting point and low boiling point of Compound 2 are characteristic of molecular compounds, which are held together with intermolecular forces. Intermolecular forces are much weaker than ionic bonds, so molecular compounds melt and boil at temperatures that are far lower than those of ionic compounds.

50. **3** Since the heat term is on the product side of both reactions, heat is released in both reactions. This eliminates both answer choices (1) and (2). The heat term in Equation 1 is on the order of 10^5 times larger than the heat term in Equation 2.

PART B–2

[All questions in Part B–2 are worth 1 point.]

51. Since both hydrogen and nitrogen are nonmetals, they will form a covalent bond. Since these elements have different electronegativities, they will form a **polar covalent bond** in which electrons are shared unequally.

(NOTE: **Covalent bond** was also accepted as a correct response to this question.)

52. The oxygen and nitrogen atoms on the end of the butanamide molecule have a far greater electronegativity than the carbon and hydrogen atoms on the other end of the molecule. Electrons will be drawn disproportionately toward the oxygen/nitrogen end of the molecule, giving the overall molecule an **asymmetrical (or unequal) distribution of charge**. Butanamide is a polar molecule because the electrons are not distributed evenly throughout the molecule but are drawn more to one end than the other.

53. When chemical equilibrium is established, **the rate of the forward reaction equals the rate of the reverse reaction**. This is the definition of chemical equilibrium.

54. The addition of $H_2(g)$ is a stress on this equilibrium system, and according to Le Châtelier's principle, systems at equilibrium will "shift" to reduce the stress and restore equilibrium. By "shifting to the left," the extra $H_2(g)$ that was added will be used up. "Shifting to the left" means that the substances on the left of the reaction arrow are produced while the substances on the right of the reaction arrow are consumed. When a new equilibrium is established, **the concentration of $H_2O(g)$ will be higher while the concentration of $CO_2(g)$ will be lower**.

55. Recall that "STP" means "Standard Temperature and Pressure." According to Reference Table A, standard temperature is defined as 273 K. Use Reference Table S to find the melting point and the boiling point information for chlorine. Since chlorine's boiling point is 239 K, this element will be a **gas** at standard temperature.

56. The molar mass of a compound is found by adding together the molar masses of its elements. You can find the molar masses of each element on the Periodic Table provided in your Reference Tables.

Algebraically, the molar mass for calcium chloride, $CaCl_2$, can be expressed as:

$$\left[1 \times (\text{molar mass of calcium})\right] + \left[2 \times (\text{molar mass of chlorine})\right]$$

Plug in the molar masses of each element and solve:

$$(1 \times 40.1) + (2 \times 35.5) = \textbf{111.1 g/mol}$$

(NOTE: All values from **110. g/mol to 111.1 g/mol** were accepted as correct answers to this problem.)

57. An empirical formula is the formula of the actual compound reduced to the smallest whole number ratio. For example, the empirical formula of C_2H_4 is CH_2. To find out what the true formula is for the liquid compound, simply divide the gram-formula mass of the empirical formula into the true gram-formula (molar) mass of the compound to find the "scale-up factor."

The gram-formula mass of the empirical formula $= 12.0 + 1.0 + 35.5 = 48.5$ g/mol for the elements C, H, and Cl, respectively.

Divide this empirical formula mass into the true molar mass of the compound:

$$\frac{97}{48.5} = 2$$

This means that the molecular formula is twice as big as the empirical formula, in other words, $\textbf{C}_2\textbf{H}_2\textbf{Cl}_2$.

58. Since this question specifically asks you to answer "in terms of electrons," the correct answer will state how electrons are involved in the formation of an ionic bond. Ionic bonds form when **electrons are transferred from the metal (calcium) to the nonmetal (chlorine)**. When metals bond with nonmetals, electrons are completely transferred because of the large electronegativity difference between the two elements.

59. Unsaturated hydrocarbons contain at least one double or triple bond between carbon atoms. The term "unsaturated" implies that other atoms could be added to the compound by breaking the multiple bond between the carbon atoms. **The hydrocarbon is unsaturated because it has a double bond between the carbon atoms.**

60. Reference Table R is extremely helpful when classifying organic compounds. The product in the reaction shown is classified as a **halide (or halocarbon)** because of the presence of a halogen (Cl) on the molecule.

61. The metalloids are shown on the diagram of the Periodic Table below. These six metalloids must be committed to memory. **Triad 1** contains the metalloid tellurium (Te).

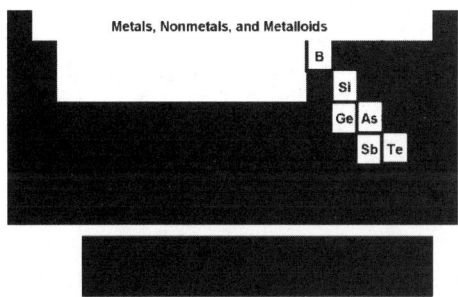

62. The elements in triad 2 are in Group 2 of the Periodic Table. Group 2 elements **each have two valence electrons**, which causes them to have the same chemical properties.

63. Consult Reference Table S for the first ionization energy data for most elements. The elements in triad 3 are all halogen elements in Group 17 of the Periodic Table. They are listed in order of increasing atomic number: Cl (17), Br (35), and I (53). Values for the first ionization energy for these elements are: 1251 kJ/mol, 1140 kJ/mol, and 1008 kJ/mol, respectively. **The first ionization energy decreases as the atomic number increases.** This is because as the atomic number increases, the number of electrons also increases, and the outermost (valence) electrons are located in electron shells that are farther from the nucleus. Less energy is required to remove electrons that are farther from the nucleus.

64. We are comparing equal masses of lithium and potassium, which are the first and third elements in triad 4. Density information from Reference Table S is needed to figure out how the 100. gram samples of these two elements will differ in volume. The element with the higher density will be capable of packing more mass in a smaller volume since the formula for density is $d = \frac{m}{V}$ (see Reference Table T). Lithium has a density of 0.534 g/cm³, while potassium has a density of 0.89 g/cm³. **Since potassium has a higher density, it can pack the same mass into a smaller volume.**

(NOTE: Alternatively, you can solve for the volume occupied by 100. g of each element using the density formula. **You will find that 100. g of Li occupies 187 cm³, while 100. g of K occupies 112 cm³.**)

65. Dobereiner's mathematical relationship for his element triads is stated in the reading passage for this question. He noticed that the atomic mass of the middle element is roughly equal to the *average* of the atomic masses of the first and last element in each triad. Applying this relationship to triad 2 yields the following mathematical expression:

$$\frac{40.\ u + 137\ u}{2}$$

(NOTE: The symbol u stands for atomic mass unit. See Reference Table D.)

PART C

[All questions in Part C are worth 1 point.]

66. An exothermic chemical reaction is a reaction that releases heat to the environment (heat "exits" the system). Since **heat is on the product side of the equation**, the reaction must be exothermic.

67. Chemical changes take place when chemical bonds break and new bonds form, creating different substances than those that were there originally. Since **the products are different substances than the reactants**, a chemical change has taken place.

68. To find the percent composition by mass of carbon in $C_6H_{10}O_5$, use the percent composition formula from Reference Table T:

$$\% \text{ composition by mass} = \frac{\text{mass of part}}{\text{mass of whole}} \times 100$$

In this case, the "part" is the mass of carbon in the compound, and the "whole" is the gram-formula mass of the entire compound.

$$\text{\% composition of carbon by mass} = \frac{6(12.0 \text{ g/mol})}{162.1 \text{ g/mol}} \times 100$$

(NOTE: It is not necessary to solve for the percent composition by mass of carbon in this problem since you are only asked for the numerical setup.)

69. To balance a chemical equation, the number of each type of atom must be equal on both sides of the equation. The unbalanced equation is shown below:

$$\underline{\hspace{1cm}} CH_4(g) + \underline{\hspace{1cm}} H_2O(g) + \text{energy} \rightarrow \underline{\hspace{1cm}} CO(g) + \underline{\hspace{1cm}} H_2(g)$$

As written, there is 1 carbon on each side of the equation, so carbon is already balanced. There is also 1 oxygen on each side of the equation, so oxygen is already balanced. However, there are 6 hydrogen atoms on the left side of the reaction arrow and only 2 hydrogen atoms shown on the right side of the equation, so a coefficient of 3 is needed in front of the hydrogen on the right side of the equation to have 6 H atoms on both sides:

$$\underline{\hspace{1cm}} CH_4(g) + \underline{\hspace{1cm}} H_2O(g) + \text{energy} \rightarrow \underline{\hspace{1cm}} CO(g) + \underline{\text{3}} H_2(g)$$

70. When the temperature of this reaction increases, the particles move more quickly. This is consistent with the kinetic molecular theory of gases, which states that temperature is directly related to the average kinetic energy of gas particles. Faster moving particles will strike each other more often and with more force. Collision theory states that reactant particles must collide with enough force in order to react. By increasing the temperature, **more methane and steam molecules will have sufficient kinetic energy to react**.

71. Interval A represents **the heat of reaction**, or **ΔH**, which is defined as the potential energy of the products (on the right-hand side of the energy "hill") minus the potential energy of the reactants (on the left-hand side of the energy "hill"). Since the products are at a lower potential energy than the reactants, the reaction shown has a $-\Delta H$ and is therefore exothermic.

72. All coefficients in chemical reactions represent particles or moles. Mole ratios can be used to find the number of moles of any substance in a chemical equation if the moles of another substance are known. You could set up this problem as follows:

$$\frac{50.0 \ \cancel{\text{mol N}_2}}{1} \times \frac{3 \ \text{mol H}_2}{1 \ \cancel{\text{mol N}_2}} = \textbf{150.0 mol H}_2$$

You could also set up mole ratios to solve this problem:

$$\frac{50.0 \ \text{mol N}_2}{1} = \frac{x \ \text{mol H}_2}{3}$$
$$x = \textbf{150.0 mol H}_2$$

73. Organic compounds always contain the element **carbon**.

74. There is a double covalent bond present in the reactant molecules. Since single covalent bonds consist of one shared pair of electrons, there are **2 pairs** of electrons in a double covalent bond (in other words, **4 electrons are shared**).

75. Synthesis reactions follow the general pattern A + B → AB. **Two reactants combine to form one product.**

76. Diethyl ether boils at 34.6°C, whereas water boils at 100°C. Diethyl ether does not require as much energy to separate its molecules because **diethyl ether has weaker intermolecular forces than water**.

77. Isomers are molecules with the same chemical formula but different structures. Since the problem asks for the same functional group with a different structure, you will need to draw an ether but change the position of the carbon atoms relative to the oxygen. Two possible isomers with the ether functional group are shown below:

$$
\begin{array}{c}
\quad\ \text{H} \qquad\ \ \text{H}\ \ \text{H}\ \ \text{H} \\
\quad\ | \qquad\quad\ | \ \ \ | \ \ \ | \\
\text{H}-\text{C}-\text{O}-\text{C}-\text{C}-\text{C}-\text{H} \\
\quad\ | \qquad\quad\ | \ \ \ | \ \ \ | \\
\quad\ \text{H} \qquad\ \ \text{H}\ \ \text{H}\ \ \text{H}
\end{array}
$$

$$
\begin{array}{c}
| \quad\ | \qquad\ | \\
-\text{C}-\text{C}-\text{O}-\text{C}- \\
| \quad\ | \qquad\ | \\
\quad -\text{C}- \\
\quad\quad |
\end{array}
$$

78. The presence of the battery is a sure sign that this is an electrolytic cell. **Electrolysis uses electrical energy to force a nonspontaneous chemical change to occur.**

79. All electrolytes dissociate into ions in aqueous solution. **The freely moving ions in this solution allow the conduction of electricity.**

80. The aqueous product in the reaction shown is NaOH (aq). Remember that the sum of the oxidation numbers in any compound must equal zero. The oxidation number of both Na and H in this compound is +1. These oxidation numbers can be found on the Periodic Table in your Reference Tables. The oxidation number of oxygen must be **−2** for the sum of the oxidation numbers to equal zero.

81. The pH value of the brine solution before the battery is connected is 7.0, as stated in the information given in the introductory paragraph for this problem. As the battery operates, the concentration of NaOH increases in the solution, and NaOH is a strong base (see Reference Table L). Since bases have pH values greater than 7, **the pH value will increase as this cell operates**.

82. The notation "Pb-214" is one way of describing a specific isotope of lead. This particular isotope has a mass number of 214, which means that the number of protons plus the number of neutrons is equal to 214. We can represent this mathematical relationship in the expression below:

$$\# \text{ protons} + \# \text{ neutrons} = 214$$

Since lead's atomic number is 82, all lead atoms have 82 protons. Substitute this value into the previous expression to solve for the number of neutrons:

$$82 + \# \text{ neutrons} = 214$$

Rearrange the values:

$$214 - 82 = \# \text{ neutrons}$$

There are **132 neutrons** in Pb-214.

83. To balance a nuclear equation, the sum of the atomic numbers (bottom) and the mass numbers (top) must be equal on both sides of the arrow.

$$^{218}_{84}\text{Po} \rightarrow \,^{214}_{82}\text{Pb} + x$$

$$^4_2\text{He or }^4_2\alpha$$

A mass number of 4 and an atomic number of 2 will correctly balance this nuclear equation. This is an alpha decay equation, and the particle that is emitted is an alpha particle.

84. From Reference Table *N*, the half-life of Rn-222 is 3.823 days. In 7.646 days, Rn-222 goes through two half-lives (7.646 days/3.823 days). One half-life is the time required for a radioisotope to decay to half of its original mass. After one half-life, only one half of the original Rn-222 will remain. After another half-life passes, only $\frac{1}{2}$ of this $\frac{1}{2}$ will remain. $\frac{1}{2} \times \frac{1}{2} = \frac{1}{4}$. The process of radioactive decay is represented below:

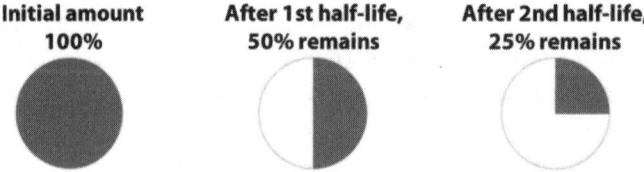

Initial amount **After 1st half-life,** **After 2nd half-life,**
100% **50% remains** **25% remains**

85. Look at the Periodic Table in your Reference Tables for the element bismuth (Bi). Bismuth has an atomic number of 83. Using this number and the mass number of 210, as given in the problem, the Decay Series graph shows that Bi-210 decays into polonium, which has an atomic number of 84. Since **a new element forms**, the decay of Bi-210 is considered a transmutation.

Mark (✔) the questions you answered correctly. Count the number of checks and follow the formulas given to determine your score on each topic.

Core Area	☐ Questions Answered Correctly

50: (1)

Section M—Math Skills
☐ Number of checks ÷ 1 × 100 = ___ %

55, 65: (2)

Section R—Reading Skills
☐ Number of checks ÷ 2 × 100 = ___ %

1, 2, 3, 31, 32: (5)

Section I—Atomic Concepts
☐ Number of checks ÷ 5 × 100 = ___ %

5, 6, 33, 61–64, 82: (8)

Section II—Periodic Table
☐ Number of checks ÷ 8 × 100 = ___ %

34, 35, 56, 57, 68, 69, 72, 75: (8)

Section III—Moles/Stoichiometry
☐ Number of checks ÷ 8 × 100 = ___ %

7–9, 11, 12, 48, 49, 51, 52, 58, 74, 76: (12)

Section IV—Chemical Bonding
☐ Number of checks ÷ 12 × 100 = ___ %

4, 10, 13–19, 21–23, 36–41, 66, 67: (20)

Section V—Physical Behavior of Matter
☐ Number of checks ÷ 20 × 100 = ___ %

20, 24, 53, 54, 70, 71: (6)

Section VI—Kinetics and Equilibrium
☐ Number of checks ÷ 6 × 100 = ___ %

42, 43, 59, 60, 73, 77: (6)

Section VII—Organic Chemistry
☐ Number of checks ÷ 6 × 100 = ___ %

Core Area   ☐ Questions Answered Correctly

25, 26, 44, 78, 80: (5)

Section VIII—Oxidation–Reduction
☐ Number of checks ÷ 5 × 100 = ___ %

27, 45–47, 79, 81: (6)

Section IX—Acids, Bases, and Salts
☐ Number of checks ÷ 6 × 100 = ___ %

28–30, 83–85: (6)

Section X—Nuclear Chemistry
☐ Number of checks ÷ 6 × 100 = ___ %

Examination
June 2018
Chemistry—Physical Setting

PART A

Answer all questions in this part.

Directions (1–30): For *each* statement or question, write in the answer space the *number* of the word or expression that, of those given, best completes the statement or answers the question. Some questions may require the use of the *2011 Edition Reference Tables for Physical Setting/Chemistry*.

1 Which statement describes the charge and location of an electron in an atom?

   (1) An electron has a positive charge and is located outside the nucleus.

   (2) An electron has a positive charge and is located in the nucleus.

   (3) An electron has a negative charge and is located outside the nucleus.

   (4) An electron has a negative charge and is located in the nucleus.   1 _____

2 Which statement explains why a xenon atom is electrically neutral?

   (1) The atom has fewer neutrons than electrons.

   (2) The atom has more protons than electrons.

   (3) The atom has the same number of neutrons and electrons.

   (4) The atom has the same number of protons and electrons.   2 _____

3 If two atoms are isotopes of the same element, the atoms must have

  (1) the same number of protons and the same number of neutrons
  (2) the same number of protons and a different number of neutrons
  (3) a different number of protons and the same number of neutrons
  (4) a different number of protons and a different number of neutrons   3 _____

4 Which electrons in a calcium atom in the ground state have the greatest effect on the chemical properties of calcium?

  (1) the two electrons in the first shell
  (2) the two electrons in the fourth shell
  (3) the eight electrons in the second shell
  (4) the eight electrons in the third shell   4 _____

5 The weighted average of the atomic masses of the naturally occurring isotopes of an element is the

  (1) atomic mass of the element
  (2) atomic number of the element
  (3) mass number of each isotope
  (4) formula mass of each isotope   5 _____

6 Which element is classified as a metalloid?

  (1) Cr   (3) Sc
  (2) Cs   (4) Si   6 _____

7 Which statement describes a chemical property of iron?

  (1) Iron oxidizes.
  (2) Iron is a solid at STP.
  (3) Iron melts.
  (4) Iron is attracted to a magnet.   7 _____

8 Graphite and diamond are two forms of the same element in the solid phase that differ in their

  (1) atomic numbers   (3) electronegativities
  (2) crystal structures   (4) empirical formulas   8 _____

9 Which ion has the largest radius?

(1) Br^- (3) F^-

(2) Cl^- (4) I^- 9 _____

10 Carbon monoxide and carbon dioxide have

(1) the same chemical properties and the same physical properties
(2) the same chemical properties and different physical properties
(3) different chemical properties and the same physical properties
(4) different chemical properties and different physical properties   10 _____

11 Based on Table S, which group on the Periodic Table has the element with the highest electronegativity?

(1) Group 1   (3) Group 17

(2) Group 2   (4) Group 18   11 _____

12 What is represented by the chemical formula $PbCl_2(s)$?

(1) a substance
(2) a solution
(3) a homogeneous mixture
(4) a heterogeneous mixture   12 _____

13 What is the vapor pressure of propanone at 50.°C?

(1) 37 kPa   (3) 83 kPa

(2) 50. kPa   (4) 101 kPa   13 _____

14 Which statement describes the charge distribution and the polarity of a CH_4 molecule?

(1) The charge distribution is symmetrical and the molecule is nonpolar.
(2) The charge distribution is asymmetrical and the molecule is nonpolar.
(3) The charge distribution is symmetrical and the molecule is polar.
(4) The charge distribution is asymmetrical and the molecule is polar.   14 _____

15 In a laboratory investigation, a student separates colored compounds obtained from a mixture of crushed spinach leaves and water by using paper chromatography. The colored compounds separate because of differences in

(1) molecular polarity   (3) boiling point
(2) malleability   (4) electrical conductivity   15 _____

16 Which phrase describes the motion and attractive forces of ideal gas particles?

(1) random straight-line motion and no attractive forces
(2) random straight-line motion and strong attractive forces
(3) random curved-line motion and no attractive forces
(4) random curved-line motion and strong attractive forces   16 _____

17 At which temperature will $Hg(\ell)$ and $Hg(s)$ reach equilibrium in a closed system at 1.0 atmosphere?

(1) 234 K   (3) 373 K
(2) 273 K   (4) 630. K   17 _____

18 A molecule of any organic compound has at least one

(1) ionic bond   (3) oxygen atom
(2) double bond   (4) carbon atom   18 _____

19 A chemical reaction occurs when reactant particles

(1) are separated by great distances
(2) have no attractive forces between them
(3) collide with proper energy and proper orientation
(4) convert chemical energy into nuclear energy   19 _____

20 Systems in nature tend to undergo changes toward

(1) lower energy and lower entropy
(2) lower energy and higher entropy
(3) higher energy and lower entropy
(4) higher energy and higher entropy   20 _____

21 Which formula can represent an alkyne?

  (1) C_2H_4 (3) C_3H_4

  (2) C_2H_6 (4) C_3H_6 21 _____

22 Given the formula representing a compound:

Which formula represents an isomer of this compound?

     (1)   (3)

     (2)   (4)   22 _____

23 Which energy conversion occurs in an operating voltaic cell?

  (1) chemical energy to electrical energy

  (2) chemical energy to nuclear energy

  (3) electrical energy to chemical energy

  (4) electrical energy to nuclear energy   23 _____

24 Which process requires energy to decompose a substance?

  (1) electrolysis   (3) sublimation

  (2) neutralization   (4) synthesis   24 _____

25 The concentration of which ion is increased when LiOH is dissolved in water?

(1) hydroxide ion　　　　　　　(3) hydronium ion
(2) hydrogen ion　　　　　　　(4) halide ion　　　　25 _____

26 Which equation represents neutralization?

(1) $6Li(s) + N_2(g) \rightarrow 2Li_3N(s)$
(2) $2Mg(s) + O_2(g) \rightarrow 2MgO(s)$
(3) $2KOH(aq) + H_2SO_4(aq) \rightarrow K_2SO_4(aq) + 2H_2O(\ell)$
(4) $Pb(NO_3)_2(aq) + K_2CrO_4(aq) \rightarrow 2KNO_3(aq) + PbCrO_4(s)$　　26 _____

27 The stability of an isotope is related to its ratio of

(1) neutrons to positrons　　　(3) electrons to positrons
(2) neutrons to protons　　　　(4) electrons to protons　　27 _____

28 Which particle has the *least* mass?

(1) alpha particle　　　　　　　(3) neutron
(2) beta particle　　　　　　　(4) proton　　　　28 _____

29 The energy released during a nuclear reaction is a result of

(1) breaking chemical bonds
(2) forming chemical bonds
(3) mass being converted to energy
(4) energy being converted to mass　　　　29 _____

30 The use of uranium-238 to determine the age of a geological formation is a beneficial use of

(1) nuclear fusion　　　　　　(3) radioactive isomers
(2) nuclear fission　　　　　　(4) radioactive isotopes　　30 _____

PART B–1

Answer all questions in this part.

Directions (31–50): For *each* statement or question, write in the answer space the *number* of the word or expression that, of those given, best completes the statement or answers the question. Some questions may require the use of the *2011 Edition Reference Tables for Physical Setting/Chemistry.*

Base your answers to questions 31 and 32 on your knowledge of chemistry and the bright-line spectra produced by four elements and the spectrum of a mixture of elements represented in the diagram below.

Bright-Line Spectra

Wavelength (nm)

31 Which elements are present in this mixture?

(1) D and A   (3) X and A

(2) D and Z   (4) X and Z   31 _____

32 Each line in the spectra represents the energy

(1) absorbed as an atom loses an electron

(2) absorbed as an atom gains an electron

(3) released as an electron moves from a lower energy state to a higher energy state

(4) released as an electron moves from a higher energy state to a lower energy state   32 _____

33 The table below shows the number of protons, neutrons, and electrons in four ions.

Four Ions

Ion	Number of Protons	Number of Neutrons	Number of Electrons
A	8	10	10
E	9	10	10
G	11	12	10
J	12	12	10

Which ion has a charge of 2−?

(1) A   (3) G

(2) E   (4) J   33 _____

34 What is the approximate mass of an atom that contains 26 protons, 26 electrons and 19 neutrons?

(1) 26 u   (3) 52 u

(2) 45 u   (4) 71 u   34 _____

35 Which electron configuration represents a potassium atom in an excited state?

(1) 2-7-6   (3) 2-8-8-1

(2) 2-8-5   (4) 2-8-7-2   35 _____

36 What is the total number of neutrons in an atom of K-42?

(1) 19   (3) 23

(2) 20   (4) 42   36 _____

37 Given the equation representing a reaction:

$$2C + 3H_2 \rightarrow C_2H_6$$

What is the number of moles of C that must completely react to produce 2.0 moles of C_2H_6?

(1) 1.0 mol   (3) 3.0 mol

(2) 2.0 mol   (4) 4.0 mol   37 _____

38 Given the equation representing a reaction:

$$Mg(s) + 2HCl(aq) \rightarrow MgCl_2(aq) + H_2(g)$$

Which type of chemical reaction is represented by the equation?

(1) synthesis   (3) single replacement
(2) decomposition   (4) double replacement   38 _____

39 The table below lists properties of selected elements at room temperature.

Properties of Selected Elements at Room Temperature

Element	Density (g/cm³)	Malleability	Conductivity
sodium	0.97	yes	good
gold	19.3	yes	good
iodine	4.933	no	poor
tungsten	19.3	yes	good

Based on this table, which statement describes how two of these elements can be differentiated from each other?

(1) Gold can be differentiated from tungsten based on density.
(2) Gold can be differentiated from sodium based on malleability.
(3) Sodium can be differentiated from tungsten based on conductivity.
(4) Sodium can be differentiated from iodine based on malleability.   39 _____

40 Which particle diagram represents a mixture?

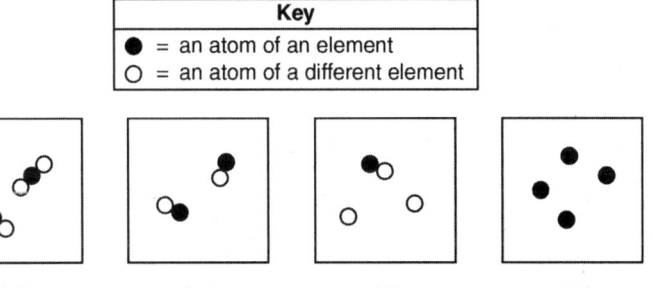

Key
● = an atom of an element
○ = an atom of a different element

(1)   (2)   (3)   (4)   40 _____

41 An atom of which element reacts with an atom of hydrogen to form a bond with the greatest degree of polarity?

(1) carbon   (3) nitrogen
(2) fluorine   (4) oxygen   41 _____

42 What is the concentration of an aqueous solution that contains 1.5 moles of NaCl in 500. milliliters of this solution?

(1) 0.30 M   (3) 3.0 M
(2) 0.75 M   (4) 7.5 M   42 _____

43 The table below shows data for the temperature, pressure, and volume of four gas samples.

Data for Four Gases

Gas Sample	Temperature (K)	Pressure (atm)	Volume (L)
I	600.	2.0	5.0
II	300.	1.0	10.0
III	600.	3.0	5.0
IV	300.	1.0	10.0

Which two gas samples contain the same number of molecules?

(1) I and II   (3) II and III
(2) I and III   (4) II and IV   43 _____

44 Based on Table I, what is the ΔH value for the production of 1.00 mole of $NO_2(g)$ from its elements at 101.3 kPa and 298 K?

(1) +33.2 kJ   (3) +132.8 kJ
(2) −33.2 kJ   (4) −132.8 kJ   44 _____

45 Which equation represents an addition reaction?

(1) $C_3H_8 + Cl_2 \rightarrow C_3H_7Cl + HCl$
(2) $C_3H_6 + Cl_2 \rightarrow C_3H_6Cl_2$
(3) $CaCl_2 + Na_2CO_3 \rightarrow CaCO_3 + 2NaCl$
(4) $CaCO_3 \rightarrow CaO + CO_2$ 45 _____

46 Given the balanced equation representing a reaction:

$$Ni(s) + 2HCl(aq) \rightarrow NiCl_2(aq) + H_2(g)$$

In this reaction, each Ni atom

(1) loses 1 electron   (3) gains 1 electron
(2) loses 2 electrons   (4) gains 2 electrons   46 _____

47 Which equation represents a reduction half-reaction?

(1) $Fe \rightarrow Fe^{3+} + 3e^-$ (3) $Fe^{3+} \rightarrow Fe + 3e^-$
(2) $Fe + 3e^- \rightarrow Fe^{3+}$ (4) $Fe^{3+} + 3e^- \rightarrow Fe$ 47 _____

48 Given the balanced ionic equation representing a reaction:

$$Cu(s) + 2Ag^+(aq) \rightarrow Cu^{2+}(aq) + 2Ag(s)$$

During this reaction, electrons are transferred from

(1) $Cu(s)$ to $Ag^+(aq)$ (3) $Ag(s)$ to $Cu^{2+}(aq)$
(2) $Cu^{2+}(aq)$ to $Ag(s)$ (4) $Ag^+(aq)$ to $Cu(s)$ 48 _____

49 Which metal reacts spontaneously with Sr^{2+} ions?

(1) $Ca(s)$ (3) $Cs(s)$
(2) $Co(s)$ (4) $Cu(s)$ 49 _____

50 Given the balanced equation representing a reaction:

$$HCl + H_2O \rightarrow H_3O^+ + Cl^-$$

The water molecule acts as a base because it

(1) donates an H^+ (3) donates an OH^-
(2) accepts an H^+ (4) accepts an OH^- 50 _____

PART B–2

Answer all questions in this part.

Directions (51–65): Record your answers on the answer sheet provided. Some questions may require the use of the *2011 Edition Reference Tables for Physical Setting/Chemistry*.

51 State the general trend in first ionization energy as the elements in Period 3 are considered from left to right.   [1]

52 Identify a type of strong intermolecular force that exists between water molecules, but does *not* exist between carbon dioxide molecules.   [1]

53 Draw a structural formula for 2-butanol.   [1]

Base your answers to questions 54 through 56 on the information below and on your knowledge of chemistry.

Some compounds of silver are listed with their chemical formulas in the table below.

Silver Compounds

Name	Chemical Formula
silver carbonate	Ag_2CO_3
silver chlorate	$AgClO_3$
silver chloride	$AgCl$
silver sulfate	Ag_2SO_4

54 Explain, in terms of element classification, why silver chloride is an ionic compound.   [1]

55 Show a numerical setup for calculating the percent composition by mass of silver in silver carbonate (gram-formula mass = 276 g/mol).   [1]

56 Identify the silver compound in the table that is most soluble in water.   [1]

Base your answers to questions 57 through 59 on the information below and on your knowledge of chemistry.

When a cobalt-59 atom is bombarded by a subatomic particle, a radioactive cobalt-60 atom is produced. After 21.084 years, 1.20 grams of an original sample of cobalt-60 produced remains unchanged.

57 Complete the nuclear equation by writing a notation for the missing particle.   [1]

58 Based on Table *N*, identify the decay mode of cobalt-60.   [1]

59 Determine the mass of the original sample of cobalt-60 produced.   [1]

Base your answers to questions 60 through 62 on the information below and on your knowledge of chemistry.

A sample of a molecular substance starting as a gas at 206°C and 1 atm is allowed to cool for 16 minutes. This process is represented by the cooling curve below.

Cooling Curve for a Substance

60 Determine the number of minutes that the substance was in the liquid phase, only.   [1]

61 Compare the strength of the intermolecular forces within this substance at 180.°C to the strength of the intermolecular forces within this substance at 120.°C. [1]

62 Describe what happens to the potential energy and the average kinetic energy of the molecules in the sample during interval *DE*. [1]

Base your answers to questions 63 through 65 on the information below and on your knowledge of chemistry.

The diagram below represents a cylinder with a moveable piston containing 16.0 g of $O_2(g)$. At 298 K and 0.500 atm, the $O_2(g)$ has a volume of 24.5 liters.

$O_2(g)$
$P = 0.500$ atm
$V = 24.5$ L
$T = 298$ K

63 Determine the number of moles of $O_2(g)$ in the cylinder. The gram-formula mass of $O_2(g)$ is 32.0 g/mol.   [1]

64 State the changes in *both* pressure and temperature of the gas in the cylinder that would increase the frequency of collisions between the $O_2(g)$ molecules.   [1]

65 Show a numerical setup for calculating the volume of $O_2(g)$ in the cylinder at 265 K and 1.00 atm.   [1]

PART C

Answer all questions in this part.

Directions (66–85): Record your answers on the answer sheet provided. Some questions may require the use of the *2011 Edition Reference Tables for Physical Setting/Chemistry*.

Base your answers to questions 66 through 69 on the information below and on your knowledge of chemistry.

In the late 1800s, Dmitri Mendeleev developed a periodic table of the elements known at that time. Based on the pattern in his periodic table, he was able to predict properties of some elements that had not yet been discovered. Information about two of these elements is shown in the table below.

Some Element Properties Predicted by Mendeleev

Predicted Elements	Property	Predicted Value	Actual Value
eka-aluminum (Ea)	density at STP	5.9 g/cm^3	5.91 g/cm^3
	melting point	low	$30.°C$
	oxide formula	Ea_2O_3	
	approximate molar mass	68 g/mol	
eka-silicon (Es)	density at STP	5.5 g/cm^3	5.3234 g/cm^3
	melting point	high	$938°C$
	oxide formula	EsO_2	
	approximate molar mass	72 g/mol	

66 Identify the phase of Ea at 310. K. [1]

67 Write a chemical formula for the compound formed between Ea and Cl. [1]

68 Identify the element that Mendeleev called eka-silicon, Es. [1]

69 Show a numerical setup for calculating the percent error of Mendeleev's predicted density of Es. [1]

Base your answers to questions 70 through 73 on the information below and your knowledge of chemistry.

Methanol can be manufactured by a reaction that is reversible. In the reaction, carbon monoxide gas and hydrogen gas react using a catalyst. The equation below represents this system at equilibrium.

$$CO(g) + 2H_2(g) \rightleftharpoons CH_3OH(g) + \text{energy}$$

70 State the class of organic compounds to which the product of the forward reaction belongs.   [1]

71 Compare the rate of the forward reaction to the rate of the reverse reaction in this equilibrium system.   [1]

72 Explain, in terms of collision theory, why increasing the concentration of $H_2(g)$ in this system will increase the concentration of $CH_3OH(g)$.   [1]

73 State the effect on the rates of both the forward and reverse reactions if no catalyst is used in the system.   [1]

Base your answers to questions 74 through 76 on the information below and on your knowledge of chemistry.

Fatty acids, a class of compounds found in living things, are organic acids with long hydrocarbon chains. Linoleic acid, an unsaturated fatty acid, is essential for human skin flexibility and smoothness. The formula below represents a molecule of linoleic acid.

```
      H  H  H  H  H  H  H  H  H  H  H  H  H  H  H  H  O
      |  |  |  |  |  |        |  |        |  |  |  |  |  |  |  |  ‖
  H—C—C—C—C—C—C═C—C—C═C—C—C—C—C—C—C—C—O—H
      |  |  |  |  |              |           |  |  |  |  |  |  |
      H  H  H  H  H              H           H  H  H  H  H  H  H
```

74 Write the molecular formula of linoleic acid.   [1]

75 Identify the type of chemical bond between the oxygen atom and the hydrogen atom in the linoleic acid molecule.   [1]

76 On the diagram *on your answer sheet*, circle the organic acid functional group.   [1]

Base your answers to questions 77 through 79 on the information below and on your knowledge of chemistry.

Fuel cells are voltaic cells. In one type of fuel cell, oxygen gas, $O_2(g)$, reacts with hydrogen gas, $H_2(g)$, producing water vapor, $H_2O(g)$, and electrical energy. The unbalanced equation for this redox reaction is shown below.

$$H_2(g) + O_2(g) \rightarrow H_2O(g) + \text{energy}$$

A diagram of the fuel cell is shown below. During operation of the fuel cell, hydrogen gas is pumped into one compartment and oxygen gas is pumped into the other compartment. Each compartment has an inner wall that is a porous carbon electrode through which ions flow. Aqueous potassium hydroxide, $KOH(aq)$, and the porous electrodes serve as the salt bridge.

77 Balance the equation *on your answer sheet* for the reaction in this fuel cell, using the smallest whole-number coefficients. [1]

78 Determine the change in oxidation number for oxygen in this operating fuel cell. [1]

79 State the number of moles of electrons that are gained when 5.0 moles of electrons are lost in this reaction. [1]

Base your answers to questions 80 through 82 on the information below and on your knowledge of chemistry.

In a laboratory investigation, a student compares the concentration and pH value of each of four different solutions of hydrochloric acid, HCl(aq), as shown in the table below.

Data for HCl(aq) Solutions

Solution	Concentration of HCl(aq) (M)	pH Value
W	1.0	0
X	0.10	1
Y	0.010	2
Z	0.0010	3

80 State the number of significant figures used to express the concentration of solution Z.   [1]

81 Determine the concentration of an HCl(aq) solution that has a pH value of 4.   [1]

82 Determine the volume of 0.25 M NaOH(aq) that would exactly neutralize 75.0 milliliters of solution X.   [1]

Base your answers to questions 83 through 85 on the information below and on your knowledge of chemistry.

Carbon dioxide is slightly soluble in seawater. As carbon dioxide levels in the atmosphere increase, more CO_2 dissolves in seawater, making the seawater more acidic because carbonic acid, $H_2CO_3(aq)$, is formed.

Seawater also contains aqueous calcium carbonate, $CaCO_3(aq)$, which is used by some marine organisms to make their hard exoskeletons. As the acidity of the seawater changes, the solubility of $CaCO_3$ also changes, as shown in the graph below.

Solubility of $CaCO_3$ vs. pH of Seawater

83 State the trend in the solubility of $CaCO_3$ as seawater becomes more acidic. [1]

84 State the color of bromcresol green in a sample of seawater in which the $CaCO_3$ solubility is 10^{-2} M. [1]

85 A sample of seawater has a pH of 8. Determine the new pH of the sample if the hydrogen ion concentration is increased by a factor of 100. [1]

Answer Sheet
June 2018

Chemistry—Physical Setting

PART B–2

51 _____

_____

_____

52 _____

53

54 _____

_____

_____

55

56 _____

57 $^{59}_{27}\text{Co} + \underline{\quad} \rightarrow {}^{60}_{27}\text{Co}$

58 _____

59 _____ **g**

60 _____ **min**

61 _____

_____

_____

62 Potential energy: _____

Average kinetic energy: _____

63 _____ **mol**

64 Change in pressure: _____

   Change in temperature: _____

65

PART C

66 _____

67 _____

68 _____

69

70 _____

71 _____

_____

_____

72 _____

_____

_____

73 Rate of forward reaction: _____

Rate of reverse reaction: _____

74 _____

75 _____

76

H-C-C-C-C-C-C=C-C-C=C-C-C-C-C-C-C-C-C-O-H

77 ___ $H_2(g) +$ ___ $O_2(g) \rightarrow$ ___ $H_2O(g) +$ energy

78 From _____ to _____

79 _____ **mol**

80 _____

81 _____ **M**

82 _____ **mL**

83 _____

_____

84 _____

85 _____

Answers
June 2018

Chemistry—Physical Setting

Answer Key

PART A

1. 3	**7.** 1	**13.** 3	**19.** 3	**25.** 1
2. 4	**8.** 2	**14.** 1	**20.** 2	**26.** 3
3. 2	**9.** 4	**15.** 1	**21.** 3	**27.** 2
4. 2	**10.** 4	**16.** 1	**22.** 2	**28.** 2
5. 1	**11.** 3	**17.** 1	**23.** 1	**29.** 3
6. 4	**12.** 1	**18.** 4	**24.** 1	**30.** 4

PART B–1

31. 3	**35.** 4	**39.** 4	**43.** 4	**47.** 4
32. 4	**36.** 3	**40.** 3	**44.** 1	**48.** 1
33. 1	**37.** 4	**41.** 2	**45.** 2	**49.** 3
34. 2	**38.** 3	**42.** 3	**46.** 2	**50.** 2

PART B–2 and **PART C**. *See* **Answers Explained**.

Answers Explained

PART A

1. **3** The basic structure of an atom is a positively charged nucleus surrounded by a negatively charged electron cloud, as shown in the illustration below. Reference Table O provides a summary of the names and charges of protons, neutrons, and electrons (which are identical to beta particles), but you do need to remember where each is located.

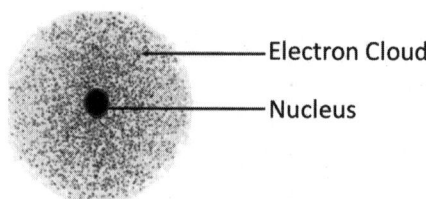

2. **4** Protons are positively charged, and electrons are negatively charged. All atoms have an equal number of protons and electrons, so atoms are electrically neutral (meaning that they have no charge at all).

3. **2** Isotopes are defined as two or more versions of the same element with different numbers of neutrons. For example, C-12 and C-14 are two isotopes of the element carbon. Since isotopes are the same element, they must have the same number of protons since the number of protons defines the element. However, isotopes have different mass numbers since the number of neutrons varies.

4. **2** According to the Periodic Table in your Reference Tables, calcium's electron configuration is 2-8-8-2. Calcium has two electrons in the fourth shell, which means that calcium has two valence electrons. Valence electrons are the outermost electrons in an atom. Valence electrons are involved in forming chemical bonds between atoms and are therefore largely responsible for an atom's chemical properties.

WRONG CHOICES EXPLAINED:

(1), (3), and (4) Calcium's electron configuration is 2-8-8-2. The electrons in the first, second, and third shells are referred to as "core" electrons that do not participate in forming bonds with other atoms.

5. **1** This is a "definition question." The atomic mass of an element is defined as the weighted average of the masses of its naturally occurring isotopes. Memorize this definition!

WRONG CHOICES EXPLAINED:
(2) The atomic number of an element equals the number of protons in the nucleus of any atom of this element.
(3) The mass number of an isotope is found by adding the number of protons to the number of neutrons in the nucleus of any atom of this isotope.
(4) Isotopes do not have formula masses since they consist of just one atom.

6. **4** Metalloids straddle the dividing line between metals and nonmetals. They exhibit properties of both metals and nonmetals. The six metalloid elements are boron (B), silicon (Si), arsenic (As), tellurium (Te), antimony (Sb), and germanium (Ge).

WRONG CHOICES EXPLAINED:
(1) Chromium (Cr) is a transition metal.
(2) Cesium (Cs) is an alkali metal (Group 1).
(3) Scandium (Sc) is a transition metal.

7. **1** Chemical properties describe how a substance will react with other substances. Oxidation is a type of chemical reaction in which a substance loses electrons.

WRONG CHOICES EXPLAINED:
(2), (3), and (4) These choices all describe physical properties that can be observed without changing the original substance (iron).

8. **2** Graphite and diamond are both allotropes of the element carbon. Allotropes are different structural forms of the same element. For example, graphite has a layered structure and diamonds have a three-dimensional network structure.

9. **4** The iodide ion, I^-, has the largest radius because it has the most electrons and therefore has the largest electron cloud.

WRONG CHOICES EXPLAINED:
(1), (2), and (3) The iodide ion contains electrons in the 5th electron shell whereas the other halogen ions mentioned in choices (1), (2), and (3) have no electrons in the 5th electron shell.

10. **4** Carbon monoxide is a linear, polar molecule whereas carbon dioxide is a linear, nonpolar molecule, as shown in the Lewis diagrams below:

$$:C \equiv O:$$
Carbon Monoxide

$$\ddot{O} = C = \ddot{O}$$
Carbon Dioxide

Since their molecular polarities are different, the two compounds will have different chemical and physical properties.

11. **3** When you consider the electronegativity of elements across any period of the Periodic Table, the element in Group 17 always has the highest value. These elements attract electrons more strongly than other elements in the same period. See this for yourself using the data provided on Reference Table *S*.

12. **1** All matter can be classified as either pure substances or mixtures. Pure substances can be either elements or compounds. $PbCl_2(s)$ is a compound.

WRONG CHOICES EXPLAINED:
(2) A solution is a homogeneous mixture, not a pure substance.
(3), (4) Homogeneous and heterogeneous mixtures are not pure substances.

13. **3** Use Reference Table *H* to read the vapor pressure of propanone at 50.°C.

14. **1** The CH_4 molecule is a symmetrical shape known as a tetrahedron. Although each C−H bond is slightly polar, the bonds are symmetrically arranged around the central carbon atom, as shown in the diagram below.

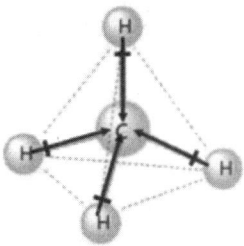

Since the four bonds are symmetrically arranged, electrons will be distributed symmetrically throughout the molecule and the overall molecule will be nonpolar.

15. **1** This question describes the laboratory separation technique known as paper chromatography, which may be used to separate components in some liquid mixtures. This process is illustrated in the diagram below.

Molecules that have stronger polarity will be more attracted to polar solvents and will travel further up the paper than molecules that have weaker polarity.

16. **1** One of the principles of the kinetic molecular theory of gases is that gas particles move in random, constant, straight-line motion. They move in this way because of a second principle of the kinetic molecular theory of gases, namely that an ideal gas consists of particles that have no attractions to or repulsions from each other. Together, these two principles explain why gases expand to fill the space that they occupy.

17. **1** The solid and liquid phases of any pure substance can coexist at equilibrium only at a substance's melting point. According to Reference Table S, the element mercury (Hg) melts at 234 K.

18. **4** All organic compounds contain the element carbon. In fact, a simple definition of "organic chemistry" is the "chemistry of carbon."

19. **3** Collision theory states that there are two requirements for effective collisions (collisions between particles that result in a chemical reaction). These requirements are (1) that the colliding particles have sufficient energy (also known as the activation energy), and (2) that the colliding particles have proper orientation.

20. **2** Systems naturally tend to undergo changes that produce lower total energy and greater disorder or randomness (entropy). For example, when a glass topples off a table, it falls to the ground (a position of lower potential energy) and shatters (the particles in the glass have much greater randomness).

21. **3** Alkynes are one of three homologous series of hydrocarbons listed on Reference Table Q. Alkynes have the general formula C_nH_{2n-2}. C_3H_4 matches this mathematical pattern.

WRONG CHOICES EXPLAINED:
   (1), (4) C_2H_4 and C_3H_6 are alkenes, since they follow the general formula C_nH_{2n}.
   (2) C_2H_6 is an alkane, since it follows the general formula C_nH_{2n+2}.

22. **2** Isomers are organic compounds that have the same chemical formula but different structures. Since the given compound has a chemical formula of C_4H_{10}, its isomer must also have this formula.

WRONG CHOICES EXPLAINED:
   (1) This compound's chemical formula is C_4H_8.
   (3) This compound's chemical formula is C_4H_6.
   (4) This compound's chemical formula is C_4H_{10}, but it is the exact same structure as the given compound and therefore is not an isomer. (Hint: If you can flip a structure over and match it exactly to another compound, the two compounds are exactly the same.)

23. **1** Voltaic cells, also known as batteries, use oxidation–reduction reactions to convert chemical energy to electrical energy.

24. **1** Decomposition is a type of chemical reaction. Energy can be applied to force decomposition to occur in a process known as electrolysis. For example, the electrolysis of water produces oxygen gas and hydrogen gas, as shown in the equation below:

$$2H_2O(\ell) + \text{energy} \rightarrow O_2(g) + 2H_2(g)$$

WRONG CHOICES EXPLAINED:
(2) Neutralization reactions happen when acids react with bases. These reactions are not decomposition reactions.
(3) Sublimation is a phase change directly from the solid phase to the gas phase. Phase changes are not chemical reactions.
(4) Synthesis reactions are the opposite of decomposition reactions.

25. **1** Lithium hydroxide, a base, dissociates in water to form lithium ions (Li^+) and hydroxide ions (OH^-). Remember, you can find the name of the OH^- ion on Reference Table E.

26. **3** A classic neutralization reaction follows the following general pattern:

$$\text{Acid} + \text{Base} \rightarrow \text{Water} + \text{Salt} \left(\text{ionic compound}\right)$$

KOH is a base (see Reference Table L), H_2SO_4 is an acid (see Reference Table K), and water and an ionic compound (K_2SO_4) are produced, so choice (3) is a neutralization reaction. None of the other answer choices contain acids and bases.

WRONG CHOICES EXPLAINED:
(1), (2) These are synthesis reactions.
(4) This is a double replacement reaction.

27. **2** The ratio of neutrons to protons in the nucleus of an atom determines whether the atom will be stable or not. If the ratio gets too large or too small, the nucleus will be unstable and will undergo radioactive decay.

28. **2** Information about all of the particles listed as answer choices for this question can be found on Reference Table *O*. The mass number is the top number on every symbol. Beta particles, which are identical to electrons, have a mass of 0 atomic mass units (amu).

(NOTE: The infinitesimal mass of an electron is much less than 1 amu, so it is considered insignificant.)

29. **3** Nuclear reactions are very different from chemical reactions. While the energy from chemical reactions comes from the formation of chemical bonds, the energy released from nuclear reactions comes from the conversion of a tiny bit of mass into energy according to the equation $E = mc^2$ (*E* is energy, *m* is mass, and *c* is the speed of light).

30. **4** Radioactive dating makes use of the consistent half-life of radio-isotopes to calculate the time elapsed during a certain event. According to Reference Table *N*, U-238 has a half-life of 4.47×10^9 years. By comparing the amount of U-238 remaining in a rock with the amount of its radioactive decay product, an approximate age for the rock can be determined.

WRONG CHOICES EXPLAINED:

(1), (2) Nuclear fusion and nuclear fission are important reactions for energy production.

(3) There is no such thing as a radioactive isomer! If you picked this answer choice, you were confused by the vocabulary. Consult the Glossary of Important Terms section of this book to relearn the difference between an "isomer" and an "isotope."

PART B–1

31. **3** All elements produce unique bright-line spectra when excited electrons in these atoms fall back to ground state. The bright-line spectrum of a mixture of two or more elements will contain all of the wavelength bands of each separate element superimposed together on the same spectrum. Element D can quickly be eliminated since the two wavelength bands at approximately 590 nm are not evident in the mixture. Element Z can also be eliminated since the wavelength band at approximately 450 nm is not found in the mixture. This leaves Elements X and A, and careful inspection shows that all wavelength bands from both elements are also found in the mixture.

32. **4** An atom in the excited state contains at least one electron that has previously absorbed energy and has moved to a higher energy orbital within the atom. When the electron returns to the lowest energy position possible, also known as the ground state, the extra energy that was absorbed is emitted (released) as a photon of electromagnetic energy. The diagram of all wavelengths of electromagnetic energy emitted from an atom is called its bright-line spectrum.

33. **1** Ions have charges because the number of electrons is either greater than or less than the number of protons. An imbalance of electrons and protons occurs when atoms lose or gain electrons during a chemical reaction. Electrons have a negative charge, and protons have a positive charge, so for an ion to have a charge of 2−, there would have to be 2 more electrons than protons.

WRONG CHOICES EXPLAINED:
(2) This ion has a charge of 1−, since the number of electrons is one greater than the number of protons.
(3) This ion has a charge of 1+, since the number of protons is one greater than the number of electrons.
(4) This ion has a charge of 2+, since the number of protons is two greater than the number of electrons.

34. **2** Protons and neutrons both have a mass of approximately 1 amu, while electrons are much tinier and their mass is ignored when calculating the mass of an atom. To find the mass of an atom (the mass number), add the number of protons to the number of neutrons:

$$26 \text{ protons} + 19 \text{ neutrons} = 45 \text{ u}$$

35. **4** The Periodic Table in your Reference Tables shows the ground-state configuration of potassium as 2-8-8-1. A potassium atom would be in an excited state if its configuration were 2-8-7-2 because this configuration shows that a third-shell electron has moved to the fourth shell.

WRONG CHOICES EXPLAINED:
(1), (2) There are only 15 electrons shown in these configurations. Potassium atoms always have 19 electrons. You might have chosen one of these answer choices if you forgot that the element symbol for potassium is K, not P (phosphorus).
(3) This is the ground-state configuration for a potassium atom. The question asked for an *excited*-state configuration.

36. **3** The mass number is defined as the sum of the protons and neutrons in an atom. Each proton and neutron has an approximate mass of 1 amu, so the sum of these particles will be equal to the mass expressed in amu. All potassium atoms have 19 protons (which equals the atomic number of potassium), and the isotope shown has a mass number of 42. The number of neutrons is found by subtracting the number of protons from the mass number:

$$42 - 19 = 23 \text{ neutrons}$$

37. **4** All coefficients in chemical reactions represent particles or moles. Mole ratios can be used to find the number of moles of any substance in a chemical equation if the moles of another substance are known. Two methods of setting up this problem are as follows:

$$\frac{2.0 \text{ mol } C_2H_6}{1} \times \frac{2 \text{ mol } C}{1 \text{ mol } C_2H_6} = 4.0 \text{ mol } C$$

OR

Set up mole ratios:

$$\frac{2.0 \text{ mol } C_2H_6}{1} = \frac{x \text{ mol } C}{2}$$
$$x = 4.0 \text{ mol } C$$

38. **3** Single replacement reactions follow the pattern $A + BC \rightarrow B + AC$. A more reactive element replaces a less reactive element in a compound. Single replacement reactions always have one uncombined element on each side of the equation.

39. **4** Elements can be differentiated, or distinguished, from one another on the basis of having different physical properties. In the data table provided, sodium is malleable whereas iodine is not. Malleability could be used to distinguish sodium from iodine.

40. **3** Mixtures are composed of two or more different substances that can be separated by physical means. Since they can be separated by physical means, the components of a mixture are not chemically bonded to one another.

WRONG CHOICES EXPLAINED:
(1), (2) These are diagrams of two different compounds. Compounds are pure substances, not mixtures.
(4) This is a diagram of an element. Elements are pure substances, not mixtures.

41. **2** The polarity of a bond depends upon the electronegativity difference between the elements in the bond. Electronegativity values for most elements are listed on Reference Table S. Fluorine has the highest electronegativity of all the elements, so an H$-$F bond would have the greatest polarity.

42. **3** Molarity (M) is the concentration of a solution expressed as the number of moles of solute per liter of solution. The formula for calculating molarity is found on Reference Table T:

$$\text{molarity} = \frac{\text{moles of solute}}{\text{liter of solution}}$$

The given volume is in milliliters, and this needs to be converted into liters before plugging it into the molarity formula:

$$\frac{500.\ \text{mL}}{1} \times \frac{1\ \text{L}}{1,000\ \text{mL}} = 0.500\ \text{L}$$

Now, plug this volume into the molarity formula to solve for the molarity:

$$M_{NaCl} = \frac{1.5\ \text{mol NaCl}}{0.500\ \text{L solution}} = 3.0\ \text{M}$$

43. **4** Equal volumes of gas at the same conditions of temperature and pressure contain the same number of particles. This relationship is known as Avogadro's Law. For example, the molar volume of a gas at STP $= 22.4$ L/mol no matter what type of gas you have. Since gas samples II and IV have equal volumes and are at the same temperature and pressure, they will contain the same number of molecules.

44. **1** Reference Table *I* lists the heat of reaction (ΔH) for the production of *two* moles of $NO_2(g)$ as +66.4 kJ. For the production of just one mole of $NO_2(g)$, this value must be divided in two. The sign is still positive, however, because the reaction producing $NO_2(g)$ is endothermic no matter how many moles are produced.

45. **2** Addition reactions are organic reactions in which a diatomic element adds to a double or triple carbon–carbon bond. Addition reactions require an alkene or an alkyne as a reactant, and there is only one product since the multiple bond opens to accept the new atoms. Choice (2) is the only reaction that has an alkene or an alkyne as a reactant, and it is also the only reaction that produces just one product.

WRONG CHOICES EXPLAINED:
(1) This is a different organic reaction called a substitution reaction. A chlorine atom substitutes for a hydrogen atom on the organic compound. The organic reactant is an alkane, and there are two products.
(3) This is a double replacement reaction, which is not a type of organic reaction.
(4) This is a decomposition reaction, which is not a type of organic reaction.

46. **2** As a reactant, nickel is a pure element and has an oxidation number of 0. In the compound $NiCl_2$, each Ni atom has an oxidation number of 2+. Nickel atoms must lose two electrons to go from an oxidation number of 0 to an oxidation number of 2+.

47. **4** Reduction is defined as the gain of electrons. Half-reactions, like all chemical reactions, must be balanced for both mass and charge. The correct answer, $Fe^{3+} + 3e^- \rightarrow Fe$, shows iron(III) ions gaining electrons (they are reducing), and the sum of all charges on both sides of the arrow are equal (conservation of charge).

WRONG CHOICES EXPLAINED:
(1) This is an oxidation half-reaction.
(2) This shows iron atoms gaining electrons, which never happens. Metals are defined as elements that lose electrons when they bond. In addition, this half-reaction does not show conservation of charge (3 negative charges do not equal 3 positive charges).
(3) This is an incorrectly written oxidation half-equation that does not show conservation of charge (3 negative charges do not equal 3 positive charges).

48. **1** $Cu(s)$ loses two electrons to form Cu^{2+}. At the same time, each Ag^+ ion gains one electron to form $Ag(s)$. The transfer of electrons is from $Cu(s)$ to $Ag^+(aq)$.

WRONG CHOICES EXPLAINED:

(2), (3) These answer choices include the products of the reaction. The products are present when the reaction is over. The transfer of electrons is completed in order to form the products.

(4) If electrons were transferred from $Ag^+(aq)$ to $Cu(s)$, the silver ion would become more positive (which is impossible, since silver ions always have a 1+ charge), and the $Cu(s)$ would become negative (which is impossible, since metals can only lose electrons when they bond).

49. **3** Reference Table *J* lists metals from the most active metals at the top to the least active metals at the bottom. Pure metal atoms never gain electrons, but metal ions can gain back the electrons they previously lost in the presence of a more active metal. Sr^{2+} ions will gain electrons back in the presence of $Cs(s)$, which is higher on Reference Table *J* and is therefore a more active metal.

WRONG CHOICES EXPLAINED:

(1), (2), and (4) These are all metals that are lower on Reference Table *J* than Sr. These metals are not as active as strontium and therefore would not be able to force electrons back onto the Sr^{2+} ion.

50. **2** Bases are proton (H^+) acceptors. Compare the reactant formula to the product formula. H_2O must accept a H^+ in order to form the hydronium ion H_3O^+.

PART B–2

[All questions in Part B–2 are worth 1 point.]

51. The **ionization energy increases** as the elements in Period 3 are considered from left to right across the Periodic Table. The increase in the ionization energy is caused by increasing nuclear charge. If you are struggling to remember this trend, you can answer this question using the data listed on Reference Table *S*.

52. Water molecules are attracted to each other by **hydrogen bonding**. Hydrogen bonding is an unusually strong intermolecular force that occurs when hydrogen is bonded to a small, highly electronegative atom such as F, O, or N. Molecules of carbon dioxide do not contain hydrogen and will be attracted by

weaker intermolecular forces. Water's strong hydrogen bonds explain why water is in the liquid state at room temperature, whereas carbon dioxide is a gas at room temperature.

53. Consult Reference Table R for help with questions about organic functional groups. The names of simple organic compounds contain all of the information necessary to figure out their structure. Start with the prefix of the compound 2-butanol. From Reference Table P, we see that 2-butanol contains 4 carbon atoms. Now look at the suffix. From Reference Table R, we find that compounds ending in the suffix −ol are alcohols and contain the functional group −OH. The number in front of the name of this compound gives the position of the functional group on the molecule. Putting all of this information together leads to the following structural formula for 2-butanol:

(NOTE: Other acceptable answers include the following:

However, be careful to make sure to draw the bond on the second carbon specifically to the oxygen in the alcohol functional group. Students who drew imprecise bonds to the functional group did not receive credit.)

54. Silver chloride, AgCl, is classified as an ionic compound because bonds between metals and nonmetals are ionic bonds. **Silver is a metal, and chlorine is a nonmetal.**

(NOTE: Be careful on this type of question. Since you are asked to write your answer *in terms of element classification*, you must write about the kinds of elements included in this compound. Answering this question on the basis of the electronegativity difference between the elements would not be acceptable.)

55. Using the percent composition formula on Reference Table T:

$$\% \text{ composition by mass} = \frac{\text{mass of part}}{\text{mass of whole}} \times 100$$

$$\% \text{ composition by mass of silver in } Ag_2CO_3 = \frac{2(107.9 \text{ g/mol})}{276 \text{ g/mol}} \times 100$$

56. Reference Table F contains information about the solubility of various ionic compounds in water. Only the chlorate ion, ClO_3^-, is listed in the first column of this table, which lists ions that form *soluble* compounds in water. **$AgClO_3$** is therefore the most soluble compound listed in the table of Silver Compounds.

57. To balance a nuclear equation, the sum of the atomic numbers (bottom) and the mass numbers (top) must be equal on both sides of the arrow. Since the atomic numbers are the same on each side of the arrow, the bottom number on the missing particle must be 0. The top number on the missing particle must be 1, since $59 + 1 = 60$. Looking at Reference Table O, the missing particle must be a neutron $\left({}_{0}^{1}\text{n} \right)$.

$$ {}_{27}^{59}\text{Co} + \underline{\quad {}_{0}^{1}\text{n} \quad} \rightarrow {}_{27}^{60}\text{CO} $$

(NOTE: Many students missed this question because they did not notice that the arrow is *after* the particle. Be very careful to pay attention to the position of the arrow.)

58. The decay mode of Co-60, as listed on Reference Table N, is β^-, which is known as a **beta particle**. Remember to look at Reference Table O for the names and symbols of nuclear particles.
(NOTE: Other acceptable answers include ${}_{-1}^{0}\text{e}$, ${}_{-1}^{0}\beta$, or **beta decay**.)

59. From Reference Table N, the half-life of Co-60 is 5.271 years. In 21.084 years, Co-60 undergoes four half-lives (21.084 years/5.271 years = 4 half-lives). The information provided states that after 21.084 years, which is equivalent to four half-lives, only 1.20 grams of the original Co-60 sample remain. Remember, one half-life is the time required for a radioisotope to decay to half of its original mass. However, we need to think about this problem in the opposite way, since we need to work backward from the final mass remaining to calculate the original

mass. Working backward, before each half-life passes, there would be *twice as much* Co-60 present than after the half-life passes. The simplest approach to solving this problem is to take the 1.20 grams remaining after four half-lives pass and *double it four times*, as shown below:

1.20 g × 2 = 2.40 grams present before the 4th half-life
2.40 g × 2 = 4.80 grams present before the 3rd half-life
4.80 g × 2 = 9.6 grams present before the 2nd half-life
9.60 g × 2 = **19.20 grams** present originally, before the 1st half-life

60. The flat regions on the graph, intervals $\overline{BC}$ and $\overline{DE}$, represent times during the cooling period when the substance was changing phase. Since the information provided states that the substance begins as a gas, interval $\overline{BC}$ represents condensation, the phase change from a gas to a liquid. Therefore, during $\overline{CD}$, the substance was a liquid before it froze during $\overline{DE}$. Time period $\overline{CD}$ lasted **five minutes**.

61. Intermolecular forces are inversely proportional to the distance between particles. Particles that are closer together will be subject to far stronger intermolecular forces than particles that are far apart. A helpful analogy would be the strength of attraction between two magnets at different distances apart from each other. When the magnets are close together, they attract each other strongly, yet when they are moved far apart, they appear to have no attraction to each other whatsoever. Of course, the magnets *are* still attracted to each other, but at such a great distance, their attraction is imperceptible. Molecules behave similarly: **at 180°C as a gas, the molecules would experience weaker intermolecular forces than as a liquid at 120°C**, since the distance between the particles is much greater in the gas phase than in the liquid phase.

62. During interval $\overline{DE}$, the temperature does not change. Since the temperature measures the average kinetic energy of the particles in a sample of matter, if the temperature is not changing, then neither is the kinetic energy. However, during interval $\overline{DE}$, the particles in the substance are undergoing a phase change from the liquid state to the solid state. This phase change is exothermic, meaning that energy is released as the freezing process takes place. The energy released is potential energy: as the particles move closer together, the potential energy stored in the sample decreases. Therefore, **the average kinetic energy remains the same, and the potential energy decreases**.

63. Many students were thrown off when answering this question because of all of the information listed for this problem. The important fact is in the introductory information: the cylinder contains 16.0 grams of $O_2(g)$. This problem involves just a simple grams to moles conversion:

$$16.0 \text{ g } O_2(g) \times \frac{1 \text{ mol } O_2(g)}{32.0 \text{ g } O_2(g)} = 0.500 \text{ mol } O_2(g)$$

64. One way to cause $O_2(g)$ molecules to collide more frequently is to push them closer together. $O_2(g)$ molecules would be forced to move closer together if the pressure on the cylinder were to increase. Another way to cause $O_2(g)$ molecules to collide more frequently is to increase their speed by raising the temperature of the $O_2(g)$ molecules. Therefore, the correct two changes are **increasing the pressure and increasing the temperature**.

65. Use the Combined Gas Law equation on Reference Table T to solve this problem:

$$\frac{P_1 V_1}{T_1} = \frac{P_2 V_2}{T_2}$$

Substitute information from the problem:

$$\frac{(0.500 \text{ atm})(24.5 \text{ L})}{298 \text{ K}} = \frac{(1.00 \text{ atm})(V_2)}{265 \text{ K}}$$

You could also rearrange the equation as follows:

$$V_2 = \frac{(0.500 \text{ atm})(24.5 \text{ L})(265 \text{ K})}{(298 \text{ K})(1.00 \text{ atm})}$$

PART C

[All questions in Part C are worth 1 point.]

66. Ea, the predicted element known as eka-aluminum, has a melting point of 30.°C. Convert this melting point to the Kelvin temperature scale using the formula found on Reference Table T:

$$K = °C + 273$$
$$\text{melting point of Ea} = 30°C + 273 = 303 \text{ K}$$

Since the given temperature in the problem is slightly greater than Ea's melting point, Ea would be a **liquid** at this temperature.

67. Ea makes an oxide that has the formula Ea_2O_3. If we assume that Ea is "like aluminum," then its oxide would be an ionic compound (a compound formed when a metal bonds with a nonmetal). In ionic compounds, the negative charges must exactly cancel out the positive charges since the overall compound is neutral in charge. This means that the Ea ion must have a charge of $+3$, since $2 \times +3$ will exactly cancel out oxygen's negative charge of 3×-2.

To determine the formula of the compound that is formed between Ea and Cl, first write charges on both ions:

$$Ea^{+3} \quad Cl^{-1}$$

Remember that ionic compounds must be neutral in overall charge. Three chloride ions are needed to cancel out the $+3$ charge of the Ea ion. Therefore, the chemical formula of the compound formed between Ea and Cl must be **$EaCl_3$**.

(NOTE: $GaCl_3$ was also accepted.)

68. Elements are often identified on the basis of physical properties. Reference Table S contains density and melting point data for many elements. First, use the equation given on Reference Table T to convert the given predicted melting point of 938°C into Kelvin so that it can be compared to the listed values on Reference Table S:

$$K = °C + 273$$
$$\text{melting point of Es} = 938°C + 273 = 1211 \text{ K}$$

Both the actual density of Es given in the problem and the actual melting temperature are equal to the values listed for the element **germanium (Ge)**.

69. The actual value for the density of Es (germanium) listed on the data table for this problem is 5.3234 g/cm^3. The predicted value was 5.5 g/cm^3. Use the percent error formula on Reference Table T to set up the calculation for percent error:

$$\% \text{ error} = \frac{\text{measured value} - \text{accepted value}}{\text{accepted value}} \times 100$$

Plug in the values from the problem:

$$\% \text{ error} = \frac{5.5 \text{ g/cm}^3 - 5.3234 \text{ g/cm}^3}{5.3234 \text{ g/cm}^3} \times 100$$

70. The product of the forward reaction, CH_3OH, is an **alcohol** because it contains the $-OH$ functional group. Reference Table R lists the organic functional groups.

71. Here is another "definition question." A chemical reaction is in a state of equilibrium when **the rate of the forward reaction equals the rate of the reverse reaction**. Memorize this definition!

72. The rate of a chemical reaction increases when the number of effective collisions between reactant particles increases. Effective collisions are collisions that have proper orientation and sufficient energy for a reaction to occur. Just by having more total collisions, there will also be more effective collisions since proper orientation will exist in more collisions. If the concentration of $H_2(g)$ increases, **there will be more effective collisions between CO(g) and H_2(g) molecules, producing more CH_3OH(g)**.

73. Do you remember what catalysts do? Adding a catalyst lowers the activation energy (E_a) for the reaction, making it possible for more collisions to have sufficient energy to react. Logically, removing a catalyst should *raise* the activation energy for *both* the forward reaction and the reverse reaction, making it *harder* for collisions to have sufficient energy to react. This should reduce the rate of reaction in both directions. Therefore, the **rate of the forward reaction will be slower** and the **rate of the reverse reaction will be slower**.

74. The molecular formula of a compound arranges the elements with subscripts, showing how many of each atom is in the compound (i.e., $C_xH_yO_z$). All that is required is to carefully count the number of each type of atom and write this information in the correct format. The answer to this question is $C_{18}H_{32}O_2$.

(NOTE: Many students erroneously wrote a "condensed structural formula" to answer this question (i.e., $CH_3(CH_2)_4 (CH)_2\ldots$, etc.). Knowing the names of the different styles for writing formulas is very important!)

75. Oxygen and hydrogen are two different nonmetals with different electronegativities. Oxygen has a much higher electronegativity than hydrogen, so it will attract the shared electrons in the bond between atoms much more strongly than hydrogen. Since the electrons are shared unequally, the bond between oxygen and hydrogen is **a polar covalent bond**.

76. The introductory information for this question states that the compound being considered, linoleic acid, is an organic acid. Reference Table *R* provides information about organic functional groups. Reference Table *R* shows that the general formula for the organic acid functional group is $-COOH$. Circle this part of the molecular diagram provided on the answer sheet, as shown below. Both diagrams are acceptable answers.

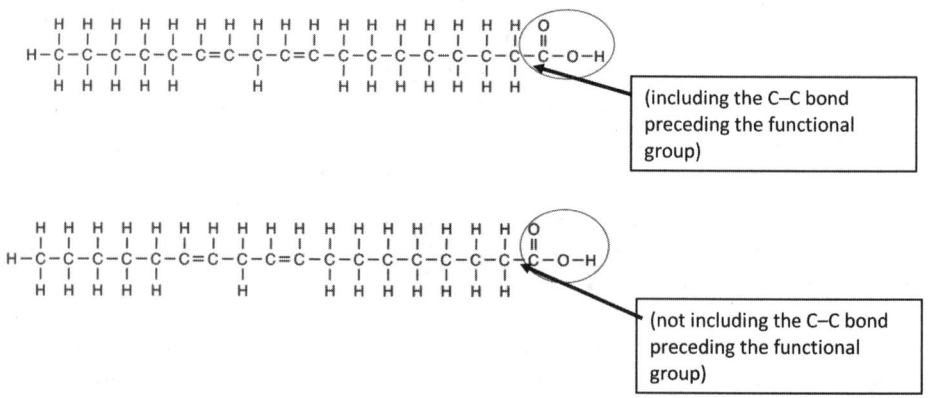

(including the C–C bond preceding the functional group)

(not including the C–C bond preceding the functional group)

(NOTE: The circled functional group *must* contain the hydrogen atom, since acids must contain hydrogen.)

77. To balance a chemical equation, the number of each type of atom must be equal on both sides of the equation. The unbalanced equation is shown below:

$$\underline{\quad} H_2(g) + \underline{\quad} O_2(g) \rightarrow \underline{\quad} H_2O(g) + energy$$

As shown above, there are 2 hydrogen atoms on each side of the equation, so hydrogen is already balanced. However, there are 2 oxygen atoms on the left of the reaction arrow and only 1 oxygen atom on the right side of the equation, so a coefficient of 2 is needed in front of water on the right side of the equation to have 2 O atoms on both sides:

$$\_\_ H_2(g) + \_\_ O_2(g) \rightarrow \underline{2}\,H_2O(g) + \text{energy}$$

With the coefficient of 2 in front of H_2O, hydrogen is no longer balanced. Adding a coefficient of 2 in front of the H_2 yields the final balanced equation:

$$\underline{2}\ \mathbf{H_2(g)} + \_\_ \mathbf{O_2(g)} \rightarrow \underline{2}\ \mathbf{H_2O(g)} + \mathbf{energy}$$

78. The unbalanced redox reaction for the fuel cell, $H_2(g) + O_2(g) \rightarrow H_2O(g) + \text{energy}$, shows that as a reactant, oxygen is a pure and uncombined element. Pure and uncombined elements always have an oxidation number of 0. The reaction produces water, in which oxygen is covalently bonded to hydrogen. Remember that the sum of the oxidation numbers in any compound must equal zero. Oxidation numbers for elements can be found on the Periodic Table in your Reference Tables. The oxidation number of H in this compound is 1+. The oxidation number of oxygen must be 2− for the sum of the oxidation numbers to equal zero. Putting this all together, the oxidation number for oxygen changes from **0** to **2−**.

79. In a redox reaction, the number of electrons lost must equal the number of electrons gained. If 5.0 moles of electrons are lost, then **5.0 moles** of electrons must be gained.

80. The number 0.0010 has **two** significant figures. The zeros preceding the 1 in the thousandths place are referred to as "place-holding zeros" whose only purpose is to push the number over to the correct power of 10. The zero after the 1 in the thousandths place *is* significant, since it does not change the absolute value of the number but only serves to confirm that the number is precise to the ten-thousandths place. With decimals, it can be helpful to rewrite the number in scientific notation to make it easier to observe significant figures:

$$0.0010 = \underline{1.0} \times 10^{-3}$$

two significant digits

81. pH is a function of the base 10 exponent of the hydrogen ion concentration in an aqueous solution. The table provided for this question is very helpful in terms of recognizing the relationship between pH and HCl concentration. Notice that every time the concentration of HCl decreases by a factor of 10, the pH increases by 1. This is because pH is defined as the negative of the exponent of the base 10 concentration of the hydrogen ion. Using the pattern revealed in the data table, simply extend the table by one line to find that if a solution has a pH of 4, the concentration of HCl would be 10^{-4} M, which is **0.00010 M**.

82. Titration is the lab technique used to determine the concentration of an acid or a base by neutralizing it with an acid or a base of known concentration. The titration formula is found on Reference Table T:

$$M_A V_A = M_B V_B$$

$$\begin{aligned} \text{in which } M_A &= \text{molarity of H}^+, \\ V_A &= \text{volume of acid}, \\ M_B &= \text{molarity of OH}^-, \\ \text{and } V_B &= \text{volume of base} \end{aligned}$$

Substitute the values from this problem:

$$(0.10 \text{ M})(75.0 \text{ mL}) = (0.25 \text{ M})(V_B)$$

$$V_B = \frac{(0.10 \text{ M})(75.0 \text{ mL})}{0.25 \text{ M}} = \textbf{30. mL}$$

> HCl concentration in solution X

83. The pH scale was developed to measure how acidic or basic a solution is, with pH values less than 7 indicating that a solution is acidic and pH values greater than 7 indicating that a solution is basic. As seawater becomes more acidic, the pH decreases. The graph shows that **as the pH decreases, the solubility of CaCO$_3$ increases**. Many students answered this question incorrectly because the graph shows descending values from left to right, but the decreasing solubility of CaCO$_3$ occurs as the pH increases (in other words, as it becomes more basic).

84. Use the graph to find that the pH of seawater is approximately 6.5 when the CaCO$_3$ solubility is 10^{-2} M. Reference Table M contains information about common acid–base indicators. The information about bromcresol green on this table states that the approximate pH for color change is 3.8–5.4 and the color

change is yellow to blue. This means that at pH values below 3.8, bromcresol green would be yellow. In the color change range of a pH of 3.8–5.4, the color would be changing from yellow to blue (in other words, green). Above a pH of 5.4, the color of bromcresol green would be **blue**.

85. pH measures the concentration of the hydrogen ion, H^+, in aqueous solution. pH is a logarithmic function, meaning that it is based on the exponent of the base 10 molarity of the hydrogen ion content. Each pH value is 10× different in hydrogen ion molarity than the pH immediately higher and lower than itself. If the hydrogen ion concentration is increased by a factor of 100 ($= 10 \times 10$), the pH of the new solution will differ from the original value by 2. Hydrogen ions make a solution acidic, and since solutions that are more acidic have lower pH values, the new pH will be *lower* than the original. Therefore, $8 - 2 = \mathbf{6}$ is the new pH.

Mark (✓) the questions you answered correctly. Count the number of checks and follow the formulas given to determine your score on each topic.

Core Area	☐ Questions Answered Correctly

69, 80: (2)

Section M—Math Skills
☐ Number of checks ÷ 2 × 100 = ___ %

28, 66, 68: (3)

Section R—Reading Skills
☐ Number of checks ÷ 3 × 100 = ___ %

1–5, 9, 31, 32, 34–36: (11)

Section I—Atomic Concepts
☐ Number of checks ÷ 11 × 100 = ___ %

6–8, 11, 39, 51: (6)

Section II—Periodic Table
☐ Number of checks ÷ 6 × 100 = ___ %

37, 38, 55, 63, 67, 74, 77: (7)

Section III—Moles/Stoichiometry
☐ Number of checks ÷ 7 × 100 = ___ %

14, 33, 41, 54, 56, 75: (6)

Section IV—Chemical Bonding
☐ Number of checks ÷ 6 × 100 = ___ %

10, 12, 13, 15–17, 40, 42, 43, 52, 60–62, 64, 65: (15)

Section V—Physical Behavior of Matter
☐ Number of checks ÷ 15 × 100 = ___ %

19, 20, 44, 71–73: (6)

Section VI—Kinetics and Equilibrium
☐ Number of checks ÷ 6 × 100 = ___ %

18, 21, 22, 45, 53, 70, 76: (7)

Section VII—Organic Chemistry
☐ Number of checks ÷ 7 × 100 = ___ %

Core Area	□ Questions Answered Correctly

23, 24, 46–49, 78, 79: (8)

Section VIII—Oxidation–Reduction
□ Number of checks ÷ 8 × 100 = ___ %

25, 26, 50, 81–85: (8)

Section IX—Acids, Bases, and Salts
□ Number of checks ÷ 8 × 100 = ___ %

27, 29, 30, 57–59: (6)

Section X—Nuclear Chemistry
□ Number of checks ÷ 6 × 100 = ___ %

Examination
August 2018
Chemistry—Physical Setting

PART A

Answer all questions in this part.

Directions (1–30): For *each* statement or question, write in the answer space the *number* of the word or expression that, of those given, best completes the statement or answers the question. Some questions may require the use of the *2011 Edition Reference Tables for Physical Setting/Chemistry*.

1 According to the wave-mechanical model, an orbital is defined as the most probable location of

(1) a proton   (3) a positron

(2) a neutron   (4) an electron   1 _____

2 The part of an atom that has an overall positive charge is called

(1) an electron   (3) the first shell

(2) the nucleus   (4) the valence shell   2 _____

3 Which subatomic particles each have a mass of approximately 1 u?

(1) proton and electron   (3) neutron and electron

(2) proton and neutron   (4) neutron and positron   3 _____

4 The discovery of the electron as a subatomic particle was a result of

(1) collision theory

(2) kinetic molecular theory

(3) the gold-foil experiment

(4) experiments with cathode ray tubes   4 _____

5 The elements on the Periodic Table of the Elements are arranged in order of increasing

(1) atomic mass   (3) atomic number
(2) formula mass   (4) oxidation number   5 _____

6 Which element is classified as a metalloid?

(1) Te   (3) Hg
(2) S   (4) I   6 _____

7 At STP, $O_2(g)$ and $O_3(g)$ are two forms of the same element that have

(1) the same molecular structure and the same properties
(2) the same molecular structure and different properties
(3) different molecular structures and the same properties
(4) different molecular structures and different properties   7 _____

8 Which substance can be broken down by chemical means?

(1) ammonia   (3) antimony
(2) aluminum   (4) argon   8 _____

9 Which statement describes $H_2O(\ell)$ and $H_2O_2(\ell)$?

(1) Both are compounds that have the same properties.
(2) Both are compounds that have different properties.
(3) Both are mixtures that have the same properties.
(4) Both are mixtures that have different properties.   9 _____

10 Which two terms represent major categories of compounds?

(1) ionic and nuclear
(2) ionic and molecular
(3) empirical and nuclear
(4) empirical and molecular   10 _____

11 Which formula represents an asymmetrical molecule?

(1) CH_4 (3) N_2
(2) CO_2 (4) NH_3 11 _____

12 Which statement describes the energy changes that occur as bonds are broken and formed during a chemical reaction?

(1) Energy is absorbed when bonds are both broken and formed.
(2) Energy is released when bonds are both broken and formed.
(3) Energy is absorbed when bonds are broken, and energy is released when bonds are formed.
(4) Energy is released when bonds are broken, and energy is absorbed when bonds are formed.   12 _____

13 A solid sample of copper is an excellent conductor of electric current. Which type of chemical bonds are in the sample?

(1) ionic bonds   (3) nonpolar covalent bonds
(2) metallic bonds   (4) polar covalent bonds   13 _____

14 Which list includes three forms of energy?

(1) thermal, nuclear, electronegativity
(2) thermal, chemical, electromagnetic
(3) temperature, nuclear, electromagnetic
(4) temperature, chemical, electronegativity   14 _____

15 Based on Table S, an atom of which element has the strongest attraction for electrons in a chemical bond?

(1) chlorine   (3) oxygen
(2) nitrogen   (4) selenium   15 _____

16 At which temperature and pressure would a sample of helium behave most like an ideal gas?

(1) 75 K and 500. kPa   (3) 300. K and 50. kPa
(2) 150. K and 500. kPa   (4) 600. K and 50. kPa   16 _____

17 A cube of iron at 20.°C is placed in contact with a cube of copper at 60.°C. Which statement describes the initial flow of heat between the cubes?

(1) Heat flows from the copper cube to the iron cube.
(2) Heat flows from the iron cube to the copper cube.
(3) Heat flows in both directions between the cubes.
(4) Heat does not flow between the cubes.   17 _____

18 Which sample at STP has the same number of atoms as 18 liters of Ne(g) at STP?

(1) 18 moles of Ar(g)　　　　　(3) 18 grams of $H_2O(g)$

(2) 18 liters of Ar(g)　　　　　(4) 18 milliliters of $H_2O(g)$　　18 _____

19 Compared to H_2S, the higher boiling point of H_2O is due to the

(1) greater molecular size of water
(2) stronger hydrogen bonding in water
(3) higher molarity of water
(4) larger gram-formula mass of water　　19 _____

20 In terms of entropy and energy, systems in nature tend to undergo changes toward

(1) lower entropy and lower energy
(2) lower entropy and higher energy
(3) higher entropy and lower energy
(4) higher entropy and higher energy　　20 _____

21 Amines, amides, and amino acids are categories of

(1) isomers　　　　　　　(3) organic compounds

(2) isotopes　　　　　　　(4) inorganic compounds　　21 _____

22 A molecule of which compound has a multiple covalent bond?

(1) CH_4　　　　　　　　(3) C_3H_8

(2) C_2H_4　　　　　　　(4) C_4H_{10}　　22 _____

23 Which type of reaction produces soap?

(1) polymerization　　　　(3) fermentation

(2) combustion　　　　　　(4) saponification　　23 _____

24 For a reaction system at equilibrium, LeChatelier's principle can be used to predict the

(1) activation energy for the system
(2) type of bonds in the reactants
(3) effect of a stress on the system
(4) polarity of the product molecules　　24 _____

25 Which value changes when a Cu atom becomes a Cu^{2+} ion?

  (1) mass number   (3) number of protons

  (2) oxidation number   (4) number of neutrons   25 _____

26 Which reaction occurs at the anode in an electrochemical cell?

  (1) oxidation   (3) combustion

  (2) reduction   (4) substitution   26 _____

27 What evidence indicates that the nuclei of strontium-90 atoms are unstable?

  (1) Strontium-90 electrons are in the excited state.

  (2) Strontium-90 electrons are in the ground state.

  (3) Strontium-90 atoms spontaneously absorb beta particles.

  (4) Strontium-90 atoms spontaneously emit beta particles.   27 _____

28 Which nuclear emission is listed with its notation?

  (1) gamma radiation, $_0^0\gamma$ (3) neutron, $_{-1}^0\beta$

  (2) proton, $_2^4He$ (4) alpha particle, $_1^1H$ 28 _____

29 The energy released by a nuclear fusion reaction is produced when

  (1) energy is converted to mass

  (2) mass is converted to energy

  (3) heat is converted to temperature

  (4) temperature is converted to heat   29 _____

30 Dating once-living organisms is an example of a beneficial use of

  (1) redox reactions   (3) radioactive isotopes

  (2) organic isomers   (4) neutralization reactions   30 _____

PART B–1

Answer all questions in this part.

Directions (31–50): For *each* statement or question, write in the answer space the *number* of the word or expression that, of those given, best completes the statement or answers the question. Some questions may require the use of the *2011 Edition Reference Tables for Physical Setting/Chemistry*.

31 What is the net charge of an ion that has 11 protons, 10 electrons, and 12 neutrons?

(1) 1+   (3) 1–
(2) 2+   (4) 2–   31 _____

32 Which electron configuration represents the electrons of an atom in an excited state?

(1) 2-5   (3) 2-5-1
(2) 2-8-5   (4) 2-6   32 _____

33 Which element is a liquid at 1000. K?

(1) Ag   (3) Ca
(2) Al   (4) Ni   33 _____

34 Which formula represents ammonium nitrate?

(1) NH_4NO_3 (3) $NH_4(NO_3)_2$
(2) NH_4NO_2 (4) $NH_4(NO_2)_2$ 34 _____

35 The empirical formula for butene is

(1) CH_2 (3) C_4H_6
(2) C_2H_4 (4) C_4H_8 35 _____

36 Which equation represents a conservation of charge?

(1) $2Fe^{3+} + Al \rightarrow 2Fe^{2+} + Al^{3+}$
(2) $2Fe^{3+} + 2Al \rightarrow 3Fe^{2+} + 2Al^{3+}$
(3) $3Fe^{3+} + 2Al \rightarrow 2Fe^{2+} + 2Al^{3+}$
(4) $3Fe^{3+} + Al \rightarrow 3Fe^{2+} + Al^{3+}$ 36 _____

37 Given the balanced equation representing a reaction:

$$2H_2 + O_2 \rightarrow 2H_2O + \text{energy}$$

Which type of reaction is represented by this equation?
(1) decomposition   (3) single replacement
(2) double replacement   (4) synthesis   37 _____

38 When a Mg^{2+} ion becomes a Mg atom, the radius increases because the Mg^{2+} ion
(1) gains 2 protons   (3) loses 2 protons
(2) gains 2 electrons   (4) loses 2 electrons   38 _____

39 The *least* polar bond is found in a molecule of
(1) HI   (3) HCl
(2) HF   (4) HBr   39 _____

40 A solution is prepared using 0.125 g of glucose, $C_6H_{12}O_6$, in enough water to make 250. g of total solution. The concentration of this solution, expressed in parts per million, is
(1) 5.00×10^1 ppm   (3) 5.00×10^3 ppm
(2) 5.00×10^2 ppm   (4) 5.00×10^4 ppm   40 _____

41 What is the amount of heat, in joules, required to increase the temperature of a 49.5-gram sample of water from 22°C to 66°C?
(1) 2.2×10^3 J   (3) 9.1×10^3 J
(2) 4.6×10^3 J   (4) 1.4×10^4 J   41 _____

42 A sample of a gas in a rigid cylinder with a movable piston has a volume of 11.2 liters at STP. What is the volume of this gas at 202.6 kPa and 300. K?
(1) 5.10 L   (3) 22.4 L
(2) 6.15 L   (4) 24.6 L   42 _____

43 The equation below represents a reaction between two molecules, X_2 and Z_2. These molecules form an "activated complex," which then forms molecules of the product.

Which diagram represents the most likely orientation of X_2 and Z_2 when the molecules collide with proper energy, producing an activated complex?

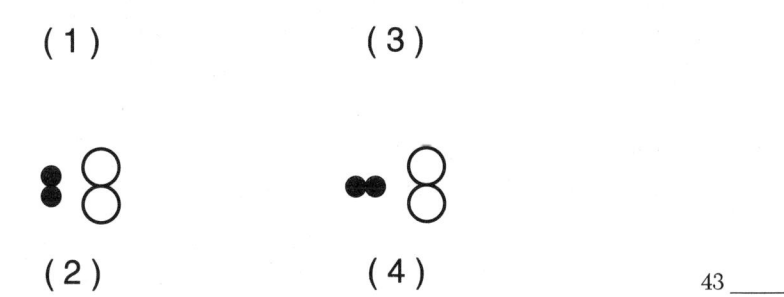

43 _____

44 What is the chemical name for the compound $CH_3CH_2CH_2CH_3$?

  (1) butane   (3) decane
  (2) butene   (4) decene

44 _____

45 In a laboratory activity, the density of a sample of vanadium is determined to be 6.9 g/cm³ at room temperature. What is the percent error for the determined value?

  (1) 0.15%   (3) 13%
  (2) 0.87%   (4) 15%

45 _____

46 Given the equation representing a reaction:

$$Cd + NiO_2 + 2H_2O \rightarrow Cd(OH)_2 + Ni(OH)_2$$

Which half-reaction equation represents the oxidation in the reaction?

(1) $Ni^{4+} + 2e^- \rightarrow Ni^{2+}$ (3) $Cd \rightarrow Cd^{2+} + 2e^-$

(2) $Ni^{4+} \rightarrow Ni^{2+} + 2e^-$ (4) $Cd + 2e^- \rightarrow Cd^{2+}$ 46 _____

47 Which metal reacts spontaneously with $NiCl_2(aq)$?

(1) Au(s)   (3) Sn(s)

(2) Cu(s)   (4) Zn(s)   47 _____

48 Which solution is the best conductor of an electric current?

(1) 0.001 mole of NaCl dissolved in 1000. mL of water

(2) 0.005 mole of NaCl dissolved in 1000. mL of water

(3) 0.1 mole of NaCl dissolved in 1000. mL of water

(4) 0.05 mole of NaCl dissolved in 1000. mL of water   48 _____

49 Compared to a 1.0-liter aqueous solution with a pH of 7.0, a 1.0-liter aqueous solution with a pH of 5.0 contains

(1) 10 times more hydronium ions

(2) 100 times more hydronium ions

(3) 10 times more hydroxide ions

(4) 100 times more hydroxide ions   49 _____

50 Given the equation representing a reaction:

$$^{208}_{82}Pb + ^{70}_{30}Zn \rightarrow ^{277}_{112}Cn + ^{1}_{0}n$$

Which type of reaction is represented by this equation?

(1) neutralization   (3) substitution

(2) polymerization   (4) transmutation   50 _____

PART B–2

Answer all questions in this part.

Directions (51–65): Record your answers on the answer sheet provided. Some questions may require the use of the *2011 Edition Reference Tables for Physical Setting/Chemistry.*

51 Show a numerical setup for calculating the percent composition by mass of oxygen in Al_2O_3 (gram-formula mass = 102 g/mol).   [1]

52 Identify a laboratory process that can be used to separate a liquid mixture of methanol and water, based on the differences in their boiling points.   [1]

Base your answers to questions 53 through 55 on the information below and on your knowledge of chemistry.

The table below shows data for three isotopes of the same element.

Data for Three Isotopes of an Element

Isotopes	Number of Protons	Number of Neutrons	Atomic Mass (u)	Natural Abundance (%)
Atom D	12	12	23.99	78.99
Atom E	12	13	24.99	10.00
Atom G	12	14	25.98	11.01

53 Explain, in terms of subatomic particles, why these three isotopes represent the same element.   [1]

54 State the number of valence electrons in an atom of isotope *D* in the ground state.   [1]

55 Compare the energy of an electron in the first electron shell to the energy of an electron in the second electron shell in an atom of isotope *E*.   [1]

Base your answers to questions 56 through 58 on the information below and on your knowledge of chemistry.

The elements in Group 2 on the Periodic Table can be compared in terms of first ionization energy, electronegativity, and other general properties.

56 Describe the general trend in electronegativity as the metals in Group 2 on the Periodic Table are considered in order of increasing atomic number.   [1]

57 Explain, in terms of electron configuration, why the elements in Group 2 have similar chemical properties.   [1]

58 Explain, in terms of atomic structure, why barium has a lower first ionization energy than magnesium.   [1]

Base your answers to questions 59 through 61 on the information below and on your knowledge of chemistry.

A saturated solution of sulfur dioxide is prepared by dissolving $SO_2(g)$ in 100. g of water at 10.°C and standard pressure.

59 Determine the mass of SO_2 in this solution.   [1]

60 Based on Table G, state the general relationship between solubility and temperature of an aqueous SO_2 solution at standard pressure.   [1]

61 Describe what happens to the solubility of $SO_2(g)$ when the pressure is increased at constant temperature.   [1]

Base your answers to questions 62 through 65 on the information below and on your knowledge of chemistry.

Starting as a solid, a sample of a molecular substance is heated, until the entire sample of the substance is a gas. The graph below represents the relationship between the temperature of the sample and the elapsed time.

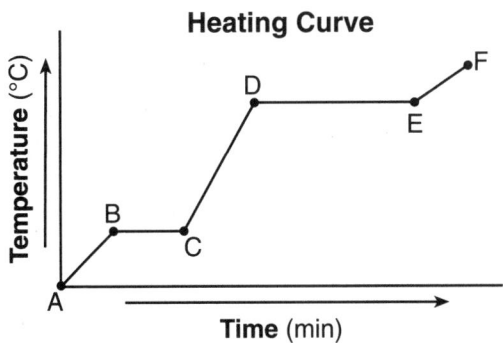

62 Using the key *on your answer sheet*, draw a particle diagram to represent the sample during interval *AB*. Your response must include *at least six* molecules. [1]

63 Compare the average kinetic energy of the molecules of the sample during interval *BC* to the average kinetic energy of the molecules of the sample during interval *DE*. [1]

64 On the graph *on your answer sheet*, mark an **X** on the axis labeled "Temperature (°C)" to indicate the boiling point of the substance. [1]

65 State evidence that indicates the sample undergoes only physical changes during this heating. [1]

PART C

Answer all questions in this part.

Directions (66–85): Record your answers on the answer sheet provided. Some questions may require the use of the *2011 Edition Reference Tables for Physical Setting/Chemistry*.

Base your anwers to questions 66 through 68 on the information below and on your knowledge of chemistry.

"Water gas," a mixture of hydrogen and carbon monoxide, is an industrial fuel and source of commercial hydrogen. Water gas is produced by passing steam over hot carbon obtained from coal. The equation below represents this system at equilibrium:

$$C(s) + H_2O(g) + heat \rightleftharpoons CO(g) + H_2(g)$$

66 State, in terms of the rates of the forward and reverse reactions, what occurs when dynamic equilibrium is reached in this system.   [1]

67 In the space *on your answer sheet*, draw a Lewis electron-dot diagram for a molecule of H_2O.   [1]

68 Explain, in terms of collisions, why increasing the surface area of the hot carbon increases the rate of the forward reaction.   [1]

Base your answers to questions 69 through 71 on the information below and on your knowledge of chemistry.

In a laboratory activity, each of four different masses of $KNO_3(s)$ is placed in a separate test tube that contains 10.0 grams of H_2O at 25°C.

When each sample is first placed in the water, the temperature of the mixture decreases. The mixture in each test tube is then stirred while it is heated in a hot water bath until all of the $KNO_3(s)$ is dissolved. The contents of each test tube are then cooled to the temperature at which KNO_3 crystals first reappear. The procedure is repeated until the recrystallization temperatures for each mixture are consistent, as shown in the table below.

Data Table for the Laboratory Activity

Mixture	Mass of KNO_3 (g)	Mass of H_2O (g)	Temperature of Recrystallization (°C)
1	4.0	10.0	24
2	5.0	10.0	32
3	7.5	10.0	45
4	10.0	10.0	58

69 Based on Table *I*, explain why there is a *decrease* in temperature when the $KNO_3(s)$ was first dissolved in the water. [1]

70 Determine the percent by mass concentration of KNO_3 in mixture 2 after heating. [1]

71 Compare the freezing point of mixture 4 at 1.0 atm to the freezing point of water at 1.0 atm. [1]

Base your answers to questions 72 through 74 on the information below and on your knowledge of chemistry.

The balanced equation below represents the reaction between carbon monoxide and oxygen to produce carbon dioxide.

$$2CO(g) + O_2(g) \rightarrow 2CO_2(g) + energy$$

72 On the potential energy diagram *on your answer sheet*, draw a double-headed arrow ($\updownarrow$) to indicate the interval that represents the heat of reaction. [1]

73 Determine the number of moles of $O_2(g)$ needed to completely react with 8.0 moles of CO(g). [1]

74 On the potential energy diagram *on your answer sheet*, draw a dashed line to show how the potential energy diagram changes when the reaction is catalyzed. [1]

Base your answers to questions 75 through 77 on the information below and on your knowledge of chemistry.

The equation below represents an industrial preparation of diethyl ether.

$$2\left(\begin{array}{c} H\ \ H \\ |\ \ | \\ H-C-C-OH \\ |\ \ | \\ H\ \ H \end{array} \right) \xrightarrow{H_2SO_4} \begin{array}{c} H\ \ H\ \ \ \ \ \ H\ \ H \\ |\ \ |\ \ \ \ \ \ \ |\ \ | \\ H-C-C-O-C-C-H + H_2O \\ |\ \ |\ \ \ \ \ \ \ |\ \ | \\ H\ \ H\ \ \ \ \ \ H\ \ H \end{array}$$

$$\text{Compound A} \qquad\qquad \text{Compound B}$$

75 Write the name of the class of organic compounds to which compound A belongs. [1]

76 Identify the element in compound B that makes it an organic compound. [1]

77 Explain, in terms of elements, why compound B is *not* a hydrocarbon. [1]

Base your answers to questions 78 through 81 on the information below and on your knowledge of chemistry.

A student is to determine the concentration of an NaOH(aq) solution by performing two different titrations. In a first titration, the student titrates 25.0 mL of 0.100 M H_2SO_4(aq) with NaOH(aq) of unknown concentration.

In a second titration, the student titrates 25.0 mL of 0.100 M HCl(aq) with a sample of the NaOH(aq). During this second titration, the volume of the NaOH(aq) added and the corresponding pH value of the reaction mixture is measured. The graph below represents the relationship between pH and the volume of the NaOH(aq) added for this second titration.

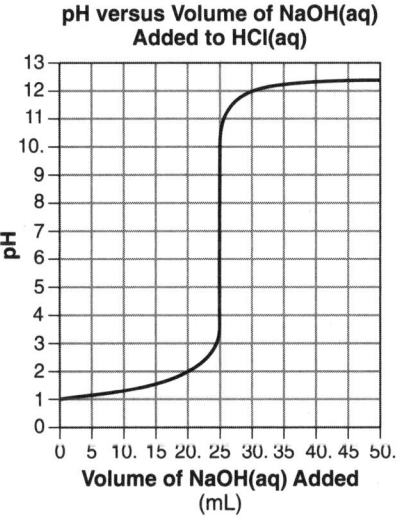

pH versus Volume of NaOH(aq) Added to HCl(aq)

Volume of NaOH(aq) Added (mL)

78 Identify the positive ion present in the H_2SO_4(aq) solution before the titration. [1]

79 Complete the equation *on your answer sheet* for the neutralization that occurs in the first titration by writing a formula of the missing product. [1]

80 Based on the graph, determine the volume of NaOH(aq) used to exactly neutralize the HCl(aq). [1]

81 State the color of phenolphthalein indicator if it were added after the HCl(aq) was titrated with 50. mL of NaOH(aq). [1]

Base your answers to questions 82 through 85 on the information below and on your knowledge of chemistry.

When uranium-235 nuclei are bombarded with neutrons, many different combinations of smaller nuclei can be produced. The production of neodymium-150 and germanium-81 in one of these reactions is represented by the equation below.

$$\ _{0}^{1}n + \ _{92}^{235}U \rightarrow \ _{60}^{150}Nd + \ _{32}^{81}Ge + 5 \ _{0}^{1}n$$

Germanium-81 and uranium-235 have different decay modes. Ge-81 emits beta particles and has a half-life of 7.6 seconds.

82 Explain, in terms of nuclides, why the reaction represented by the nuclear equation is a fission reaction.   [1]

83 State the number of protons and number of neutrons in a neodymium-150 atom.   [1]

84 Complete the equation *on your answer sheet* for the decay of Ge-81 by writing a notation for the missing nuclide.   [1]

85 Determine the time required for a 16.00-gram sample of Ge-81 to decay until only 1.00 gram of the sample remains unchanged.   [1]

Answer Sheet
August 2018

Chemistry—Physical Setting

PART B–2

51

52 _____

53 _____

_____

_____

54 _____

55 _____

_____

_____

56 _____

_____

_____

57 _____

_____

_____

58 _____

_____

_____

59 _____ g

60 _____

_____

_____

61 _____

_____

_____

62

Key
○ = a molecule of the substance

63 _____

_____

_____

64

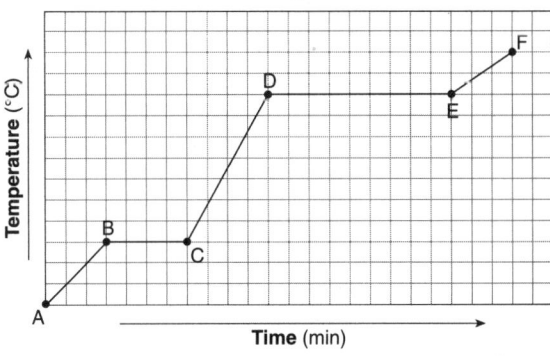

Heating Curve

65 _____

_____

_____

PART C

66 _____

_____

_____

67

68 _____

_____

_____

69 _____

_____

_____

70 _____ %

71 _____

_____

_____

72

73 _____ mol

74

75 _____

76 _____

77 _____

_____

_____

78 _____

79 $2NaOH(aq) + H_2SO_4(aq) \rightarrow 2H_2O(\ell) + $ _____ (aq)

80 _____ **mL**

81 _____

82 _____

_____

_____

83 Protons: _____

Neutrons: _____

84 $^{81}_{32}Ge \rightarrow {}^{0}_{-1}e + $ _____

85 _____ **s**

Answers
August 2018

Chemistry—Physical Setting

Answer Key

PART A

1. 4	7. 4	13. 2	19. 2	25. 2
2. 2	8. 1	14. 2	20. 3	26. 1
3. 2	9. 2	15. 3	21. 3	27. 4
4. 4	10. 2	16. 4	22. 2	28. 1
5. 3	11. 4	17. 1	23. 4	29. 2
6. 1	12. 3	18. 2	24. 3	30. 3

PART B–1

31. 1	35. 1	39. 1	43. 2	47. 4
32. 3	36. 4	40. 2	44. 1	48. 3
33. 2	37. 4	41. 3	45. 4	49. 2
34. 1	38. 2	42. 2	46. 3	50. 4

PART B–2 and **PART C**. *See* **Answers Explained**.

Answers Explained

PART A

1. **4** Electrons in an atom exist in the electron cloud surrounding the nucleus, where the protons and the neutrons are found. According to the wave-mechanical model of the atom, the electron cloud contains many different orbitals, which are three-dimensional regions of space where there is a high probability of finding electrons.

2. **2** Protons have a positive charge and are located in the nucleus of the atom. Refer to Reference Table *O* for help remembering the charges of subatomic particles.

WRONG CHOICES EXPLAINED:
(1) Electrons have a negative charge.
(3), (4) Electrons are found outside of the nucleus in shells or principal energy levels.

3. **2** Protons and neutrons both have a mass of approximately 1 atomic mass unit (u). Refer to Reference Table *O* to verify the masses of subatomic particles. The mass is the upper number associated with each symbol.

4. **4** J. J. Thomson discovered the electron during his famous cathode ray tube experiment. What first appeared to be light in the tube was deflected toward the positive side of an external electric field, proving that the light was caused by negatively charged particles.

5. **3** Each element has a unique atomic number that equals the number of protons in each atom. The Periodic Table is arranged from hydrogen (H, atomic number = 1) to ununoctium (Uuo, atomic number = 118) in order of increasing atomic number.

6. **1** The bolded staircase divider printed on the Periodic Table in your Reference Tables divides the metal elements from the nonmetal elements. Metalloids, elements with properties of both metals and nonmetals, are found next to this dividing line. Six elements are classified as metalloids: B, Si, As, Te, Ge, and Sb. Memorize these elements!

7. **4** O_2 and O_3 are both allotropes of the element oxygen. Allotropes are different physical forms of the same element. As seen in the structures that follow, O_2 is a linear, nonpolar molecule, while O_3 (ozone) is a bent, polar molecule. Their different structures cause these allotropes of oxygen to have different physical and chemical properties.

$$\ddot{\text{O}}=\ddot{\text{O}} \qquad\qquad :\!\overset{..}{\text{O}}.\quad\overset{\overset{..}{\text{O}}}{\diagup}\!\diagdown\quad\overset{..}{\text{O}}:$$

8. **1** Ammonia (NH_3) is a chemical compound that can be broken apart into its elements by a chemical reaction.

WRONG CHOICES EXPLAINED:
(2), (3), and (4) All of these answer choices are elements found on the Periodic Table. Elements are the simplest pure substances and cannot be broken down further by chemical means.

9. **2** H_2O is water, and H_2O_2 is hydrogen peroxide. These are two different compounds. The state symbol (ℓ) represents a liquid that is a pure compound, not a mixture. Since the molecules of these two compounds differ in size, composition, and shape, the intermolecular forces within these substances will be different, resulting in different physical properties.

WRONG CHOICES EXPLAINED:
(1) Different compounds will not have the same properties because their molecules are not the same.
(3), (4) The state symbol for mixtures containing water is (aq), meaning an aqueous solution.

10. **2** Compounds are distinguished by the type of bonding that holds them together. Ionic compounds contain particles held together by ionic bonds, and molecular compounds contain particles held together by covalent bonds.

11. **4** The best way to determine whether or not a molecule is symmetrical is to draw its Lewis structure and evaluate the bonding. The Lewis structure for ammonia is as follows:

$$\text{H}-\overset{..}{\text{N}}-\text{H}$$
$$|$$
$$\text{H}$$

The lone pair of electrons on the central atom pushes away the bonding pairs of electrons, causing this molecule to have an asymmetrical, trigonal pyramidal shape:

WRONG CHOICES EXPLAINED:
  (1) CH_4 molecules are tetrahedral and symmetrical.
  (2) CO_2 molecules are linear and symmetrical.
  (3) N_2 molecules are linear and symmetrical.

12. **3** Energy must be *absorbed* to break the bonds that hold atoms together, just as energy would be needed to pull apart two magnets that are attracted to each other. However, when chemical bonds form, energy is *released* since atoms are more stable once they are bonded together.

13. **2** Copper is a metal element that is held together by metallic bonds. Metals are great conductors of electricity because of the mobile "sea of electrons" that is created when metallic bonds form.

14. **2** Energy can exist in many different forms. The motion of atoms and molecules within a substance creates thermal energy. Chemical energy is stored in chemical bonds, and electromagnetic energy is energy that travels through space in the form of waves.

WRONG CHOICES EXPLAINED:
  (1), (4) Electronegativity is not a form of energy. Electronegativity is the amount of attraction that an atom has for a bonded pair of electrons.
  (3), (4) Temperature is not a form of energy. Temperature is directly related to the kinetic energy of particles in a sample of matter.

15. **3** This question is asking about electronegativity, which is defined as an atom's attraction for the electrons in a chemical bond. Look up electronegativity values for these elements in Reference Table *S*. Oxygen has an electronegativity of 3.4, the second highest value on the table!

16. **4** Real gases behave more like ideal gases when the temperature is high and the pressure is low. These conditions minimize any possible attractions between gas particles, allowing the gas to adhere more closely to the principles of the kinetic molecular theory of gases. (A good way to remember this is to think of summertime, when the temperature is high and the pressure is low.)

17. **1** Heat energy always flows from an object with a higher temperature to an object with a lower temperature. The condition we know as "cold" is merely the absence of heat. Fast-moving particles in the hotter sample collide with the slow-moving particles in the colder sample. Energy is transferred as a result of these collisions.

18. **2** Equal volumes of gas at the same conditions of temperature and pressure contain the same number of particles. This relationship is known as Avogadro's Law. For example, the molar volume of a gas at STP = 22.4 L/mol no matter what type of gas you are working with.

19. **2** Water molecules are attracted to each other by **hydrogen bonding**. Hydrogen bonding is an unusually strong intermolecular force that occurs when hydrogen is bonded to a small, highly electronegative atom, such as F, O, or N. Molecules of H_2S do not contain F, O, or N atoms and will be attracted by weaker intermolecular forces. Water's strong hydrogen bonds explain why water is in the liquid state at room temperature, whereas H_2S is a gas at room temperature.

20. **3** Systems naturally tend to undergo changes that produce lower total energy and greater disorder or randomness (entropy). For example, during a landslide, the debris moves downslope toward a position of lower potential energy and eventually comes to rest in a random, chaotic mess.

21. **3** Amines, amides, and amino acids all contain the element carbon and are different categories of organic compounds. Reference Table R provides a list of organic functional groups. Amines and amides are on this table. Although they are not included on Reference Table R, amino acids are compounds that contain both the amine functional group and the organic acid functional group.

22. **2** Reference Table Q shows the general formulas for the three homologous series of hydrocarbons, known as the alkanes, alkenes, and alkynes. From this table, we see that only alkenes and alkynes contain multiple covalent bonds (double and triple bonds, respectively). C_2H_4 matches the general formula for alkenes, C_nH_{2n}.

WRONG CHOICES EXPLAINED:

(1), (3), and (4) These compounds are alkanes, since they match the general formula for alkanes, C_nH_{2n+2}.

23. **4** In saponification reactions, a fat reacts with a base to produce glycerol and soap. Review the following example:

$$
\begin{array}{l}
CH_2OCOC_{17}H_{35} \\
| \\
CH-OCOC_{17}H_{35} \quad + \quad 3NaOH\,(aq) \\
| \qquad\qquad\qquad\qquad\quad \text{Sodium hydroxide} \\
CH_2OCOC_{17}H_{35} \\
\text{Triester or Triglyceride} \\
\text{(oil or fat)}
\end{array}
\longrightarrow
\begin{array}{l}
CH_2OH \\
| \\
CHOH \quad + \quad 3C_{17}H_{35}COONa \\
| \qquad\qquad\qquad\quad \text{Soap} \\
CH_2OH \\
\text{Glycerol}
\end{array}
$$

(NOTE: You need to be able to recognize seven major types of organic reactions for this exam. These include addition, substitution, esterification, polymerization, fermentation, saponification, and combustion. Memorize these reactions!)

WRONG CHOICES EXPLAINED:

(1) Polymerization reactions create long chains from many simple, identical molecules called monomers.

(2) Combustion reactions occur when fuel (usually a hydrocarbon) reacts with oxygen to produce carbon dioxide and water.

(3) Fermentation reactions involve the conversion of glucose into ethanol in the presence of an enzyme catalyst.

24. **3** Stresses can include changes in concentration, pressure, or temperature. Le Châtelier's principle states that chemical systems that are at equilibrium and experience a stress will respond to restore equilibrium.

25. **2** Cu atoms must lose two electrons to become Cu^{2+} ions. Oxidation numbers are assigned to elements to reflect the loss or gain of electrons.

WRONG CHOICES EXPLAINED:

(1), (3), and (4) The mass number is the sum of protons and neutrons in the nucleus of an atom. Protons and neutrons are not involved in the formation of ions.

26. **1** The mnemonic device "An Ox, Red Cat" is helpful for remembering that oxidation (loss of electrons) takes place at the anode, while reduction (gain of electrons) happens at the cathode.

27. **4** The spontaneous emission of beta particles is a form of radioactive decay. Beta decay occurs when the ratio of neutrons to protons in the nucleus of an atom is too large, causing the nucleus to become unstable.

28. **1** Notations for particles involved in nuclear chemistry are listed in Reference Table O. The only particle that is correctly paired with its notation is gamma radiation.

WRONG CHOICES EXPLAINED:
(2) $^4_2$He is an alpha particle, not a proton.
(3) $^0_{-1}\beta$ is a beta particle, not a neutron.
(4) $^1_1$H is a proton, not an alpha particle.

29. **2** Nuclear fusion produces energy by converting a small amount of mass into energy. The amount of energy produced can be calculated by Albert Einstein's famous equation $E = mc^2$, in which "m" is the mass lost in the fusion reaction and "c" is the speed of light.

30. **3** Radioactive dating makes use of the consistent half-life of radioisotopes to calculate the time elapsed during a certain event.

PART B–1

31. **1** Ions form when electrons enter or leave an atom, creating an imbalance between the number of protons and the number of electrons. Protons have a charge of 1+, and electrons have a charge of 1−. Since the ion described in this question has one more proton than the number of electrons, this ion will have a charge of 1+.

32. **3** The first electron shell (also known as a principal energy level) can only hold 2 electrons. The second electron shell can hold up to 8 electrons. Electrons will always seek the lowest energy positions available within the atom, which are positions near the nucleus. An electron configuration of 2-5-1 represents an electron in an excited state because it shows that an electron occupies the third electron shell before the second electron shell is filled.

33. **2** Reference Table S lists the melting points of the majority of elements on the Periodic Table. An element will be in the liquid phase if it is at a temperature above its melting point and below its boiling point. Aluminum would be a liquid at 1000. K since its melting point is 933 K.

34. **1** Ammonium nitrate is an ionic compound made from two oppositely charged polyatomic ions. The trick is to remember to consult Reference Table E to find the formulas and charges of these ions. The ammonium ion is NH_4^+, and the nitrate ion is NO_3^-. Ionic compounds are neutral in charge, so the formula for ammonium nitrate is NH_4NO_3 because the 1+ charge on the ammonium ion exactly cancels the 1− charge on the nitrate ion.

35. **1** Butene belongs to the homologous series of hydrocarbons known as the alkenes. Alkenes are one of the three homologous series of hydrocarbons listed on Reference Table Q. Alkenes have the general formula C_nH_{2n}. Reference Table P shows that the prefix "but-" means that a hydrocarbon has four carbon atoms, so butene has the molecular formula C_4H_8. However, this question asks for the *empirical formula* of butene, which is a simplified formula that is reduced to the lowest whole-number ratio. The empirical formula for butene, C_4H_8, is CH_2.

WRONG CHOICES EXPLAINED:
(2) C_2H_4 can be further simplified to CH_2.
(3) C_4H_6 is the formula for butyne, not butene.
(4) C_4H_8 is the molecular formula for butene.

36. **4** Conservation of charge means that the number of electrons that are lost through the process of oxidation equals the number of electrons that are gained through the process of reduction. To verify conservation of charge, take the sum of the charges on the reactant side and on the product side of the reaction. The sum of the charges should be the same on both sides. For the reaction displayed in choice (4):

$$3Fe^{3+} + Al \rightarrow 3Fe^{2+} + Al^{3+}$$
$$3(3+) + 0 = 3(2+) + (3+)$$
$$9+ = 9+$$

WRONG CHOICES EXPLAINED:
(1) The sum of the charges on the reactant side is 6+, while the sum of the charges on the product side is 7+.
(2) The sum of the charges on the reactant side is 6+, while the sum of the charges on the product side is 12+.
(3) The sum of the charges on the reactant side is 9+, while the sum of the charges on the product side is 10+.

37. **4** Synthesis reactions follow the general pattern A + B → AB. Two reactants combine to form one product.

WRONG CHOICES EXPLAINED:
(1) Decomposition reactions follow the general pattern AB → A + B. Decomposition reactions are the *opposite* of synthesis reactions.
(2) Double replacement reactions follow the general pattern AB + CD → AD + CB. Ions in two different ionic compounds switch places.
(3) Single replacement reactions follow the general pattern A + BC → B + AC. Uncombined elements appear on *both* sides of the equation.

38. **2** An Mg^{2+} ion must gain 2 electrons to become a neutral Mg atom. The Mg^{2+} ion has an electron configuration of 2-8. The Mg atom has an electron configuration of 2-8-2. Since the Mg atom has electrons in the third electron shell, it will have a larger radius than the Mg^{2+} ion, which does not have any electrons in the third electron shell.

39. **1** The least polar bond will have the lowest electronegativity difference between the atoms involved in the bond. Electronegativity values are found on Reference Table S. The electronegativity of H is 2.2, and the electronegativity of I is 2.7. The electronegativity difference of 0.5 (since $2.7 - 2.2 = 0.5$) in the compound HI is the lowest of all the compounds offered as answer choices for this question.

40. **2** Look at Reference Table T to find the formula for parts per million:

$$\text{parts per million} = \frac{\text{mass of solute}}{\text{mass of solution}} \times 1,000,00$$

$$\text{parts per million} = \frac{0.125\,\text{g}}{250.\,\text{g}} \times 1,000,000 = 500.\,\text{ppm} = 5.00 \times 10^2\,\text{ppm}$$

41. **3** Use the heat equation for temperature changes from Reference Table T:

$$q = mC\Delta T$$

where q = heat, m = mass, C = specific heat capacity, and ΔT = change in temperature

Plugging in the values given in the problem, and consulting Reference Table B for the specific heat capacity of water, results in:
$q = 49.5$ g $(4.18$ J/g·°C$)(66°C - 22°C) = 9104$ J $= 9.1 \times 10^3$ J (when expressed to two significant figures)

(NOTE: The specific heat capacity of water can be written as either 4.18 J/g·°C or as 4.18 J/g·K, since the interval of one Celsius degree equals the interval of one Kelvin degree.)

42. **2** Use the Combined Gas Law equation from Reference Table T to solve this problem:

$$\frac{P_1 V_1}{T_1} = \frac{P_2 V_2}{P_2}$$

Reference Table A provides the values for standard temperature and pressure.

Substituting the values from Reference Table A plus the information from the question yields:

$$\frac{(101.3 \text{ kPa})(11.2 \text{ L})}{273 \text{ K}} = \frac{(202.6 \text{ kPa})(V_2)}{300. \text{ K}}$$

Rearranging this equation leads to:

$$V_2 = \frac{(101.3 \text{ kPa})(11.2 \text{ L})(300. \text{ K})}{(273 \text{ K})(202.6 \text{ kPa})} = 6.15 \text{ L}$$

43. **2** For a chemical reaction to occur, molecules must collide with sufficient energy and proper orientation. Consider the following collision of reactant molecules:

$$\Rightarrow \; \vdots \; 8 \; \Leftarrow$$

This orientation immediately leads to the activated complex, which is a necessary step on the way to forming the products. None of the other orientations shown in the remaining choices immediately lead to the activated complex.

44. **1** The compound shown contains 4 carbon atoms and 10 hydrogen atoms, so its molecular formula is C_4H_{10}. This formula follows the general pattern for the alkane series of hydrocarbons, C_nH_{2n+2}, as shown on Reference Table Q. Reference Table P shows that the prefix for a hydrocarbon with 4 carbon atoms is *but-*. Putting together this prefix with the suffix for the alkane series, *-ane*, leads to the name "butane."

WRONG CHOICES EXPLAINED:
(2) Butene is an alkene with a formula of C_4H_8.
(3), (4) These hydrocarbons contain 10 carbon atoms.

45. **4** The actual value for the density of vanadium, 6.0 g/cm^3, is listed on Reference Table S. Use the percent error formula on Reference Table T to set up the calculation for percent error:

$$\% \text{ error} = \frac{\text{measured value} - \text{accepted value}}{\text{accepted value}} \times 100$$

Plugging in the values from the question:

$$\% \text{ error} = \frac{6.9 \text{ g/cm}^3 - 6.0 \text{ g/cm}^3}{6.0 \text{ g/cm}^3} \times 100 = 15\%$$

46. **3** Oxidation is the loss of electrons. As a reactant, cadmium is a pure element and has an oxidation number of 0. In the compound $Cd(OH)_2$, on the product side of the equation, each Cd atom has an oxidation number of 2+. Cadmium atoms must lose two electrons to go from an oxidation number of 0 to an oxidation number of 2+.

WRONG CHOICES EXPLAINED:
(1) This half-reaction is an example of *reduction*, which is the gain of electrons.
(2) This oxidation half-reaction is not balanced for charge $[4+ \neq (2+) + (2-)]$.
(4) This reduction half-reaction is not balanced for charge $[0 + (2-) \neq (2+)]$.

47. **4** Reference Table J lists the most active metals at the top of the table down to the least active metals at the bottom of the table. Pure metal atoms never gain electrons, but metal ions can gain back the electrons they previously lost in the presence of a more active metal. Ni^{2+} ions will gain electrons back in the presence of zinc (Zn), which is higher on Reference Table J and is therefore a more active metal.

WRONG CHOICES EXPLAINED:
(1), (2), and (3) These metals are found below nickel on Reference Table J, signifying that they have less ability to oxidize than nickel does and consequently they will not be able to push electrons back onto the Ni^{2+} ion.

48. **3** Electrolytes are substances that dissolve in water to form aqueous solutions that are capable of conducting electricity. The more ions there are in a solution, the more electrically conductive the solution becomes.

49. **2** The pH is the negative of the exponent of the base 10 concentration of the hydrogen (or hydronium) ion. Using this definition, a solution with a pH of 7.0 has a hydronium ion concentration of 10^{-7} M, while a solution with a pH of 5.0 has a hydronium ion concentration of 10^{-5} M. The solution with a pH of 5.0 has 100 times more hydronium ions than the solution with a pH of 7.0.

50. **4** The term "transmutation" refers to any change in the nucleus of an atom that converts it from one element to another. The nuclear reaction in this question shows a collision, between two elements with smaller nuclei, that produces an element with a larger nucleus.

WRONG CHOICES EXPLAINED:

(1) Neutralization reactions occur when an acid reacts with a base to form water and a salt.

(2) Polymerization reactions are organic reactions in which small molecules (monomers) combine to make very long chains (polymers).

(3) Substitution reactions are organic reactions in which an atom (usually a halogen) substitutes for a hydrogen in an organic compound.

PART B–2

[All questions in Part B–2 are worth 1 point.]

51. To find the percent composition by mass of oxygen in Al_2O_3, use the percent composition formula found on Reference Table *T*:

$$\% \text{ composition by mass} = \frac{\text{mass of part}}{\text{mass of whole}} \times 100$$

where the "part" is the mass of the oxygen in the compound, and the "whole" is the gram-formula mass of the entire compound

$$\% \text{ composition by mass of oxygen in } Al_2O_3 = \frac{3(16.0\,\text{g/mol})}{102.0\,\text{g/mol}} \times \mathbf{100}$$

(NOTE: It is not necessary to solve for the percent composition by mass of oxygen in this question since you are only asked for the *numerical setup*.)

52. **Distillation** is the laboratory process that is used to separate liquid mixtures based on boiling points. As the liquid mixture begins to boil, the component of the mixture with the lower boiling point will boil away first. By passing the vapors through a condenser, the components of the mixture can be collected separately.

53. All three isotopes have the **same number of protons**. The number of protons is equal to the atomic number of the element. All three isotopes have 12 protons and are different isotopes of the element magnesium.

54. Valence electrons are located in the outermost energy level of an atom. Valence electrons are important because they are the electrons that are involved in chemical bonding. Isotope *D* contains 12 protons, so it represents the element magnesium. Look at the Periodic Table in your Reference Tables to find the electron configuration for magnesium, which is 2-8-**2**. The outermost energy level contains **2** electrons, which are the valence electrons for this element.

55. Electron shells that are closer to the nucleus have lower energy than electron shells that are farther from the nucleus. This is because negatively charged electrons are attracted to the positively charged nucleus. **An electron in the first electron shell of any atom will have less energy than an electron in the second electron shell.**

56. **Electronegativity decreases** as the atomic number increases in all groups on the Periodic Table, including the metals in Group 2. As the atomic number increases going down a group of elements on the Periodic Table, the valence (outermost) electrons are located in electron shells that are farther from the nucleus. The nucleus of an atom has far less force of attraction for new electrons when the new electrons will be added to a more distant electron shell.

57. **The elements in Group 2 on the Periodic Table each have 2 valence electrons.** Elements with the same number of valence electrons will have similar chemical properties. Valence electrons are the electrons involved in bonding, and they have a large influence on the stability of an element.

58. The first ionization energy is the energy that is needed to remove the outermost, or highest energy, electron from an atom. **Barium's valence electrons are in the 6th electron shell, whereas magnesium's valence electrons are in the 3rd electron shell. Less energy is required to remove electrons that are farther from the nucleus.**

59. Refer to Reference Table *G* to answer this question. The table shows that approximately **16 g** of SO_2 would be dissolved in the solution described in this question.
    (NOTE: All values from **15 g to 18 g** were accepted as correct answers to this question.)

60. From Reference Table G it is clear that, **as the temperature increases, the solubility of SO_2 decreases**. SO_2 is a gas, and gases at higher temperatures have sufficient kinetic energy to overcome the intermolecular forces with solvent molecules.

61. **The solubility of SO_2 gas increases at higher pressures** because the increased pressure forces more SO_2 molecules into the solution, as shown in the following diagram:

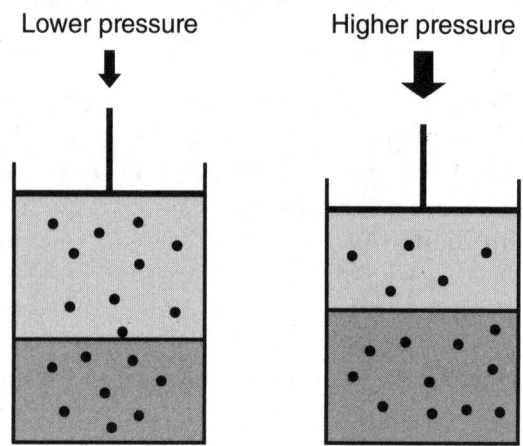

62. The information supplied for this question states that a solid sample is heated continuously until it becomes a gas. The interval AB represents the solid sample before it melts. Particle diagrams of solids need to show high density, fixed size and shape, and regular arrangement of particles, as shown in the following diagram. Make sure to draw *at least six* particles in your diagram, as directed by the question.

63. Temperature is directly related to the average kinetic energy of particles in a sample of matter. **Molecules in the sample have greater average kinetic energy during interval DE** because the sample is at a higher temperature during this period of time.

64. Phase changes occur during the "plateaus," or flat intervals, on a heating curve. The temperature remains constant during phase changes. Boiling happens when particles are transitioning from the liquid phase to the gas phase. This transition occurs during interval *DE* on the heating curve. The directions state that you must mark an X on the axis labeled "Temperature (°C)," so your "X" needs to be drawn directly in line with the boiling plateau, *DE*, as shown in the following diagram:

Heating Curve

65. As heating proceeds along this heating curve, the particles in the sample move faster and spread farther apart, but the molecules in the sample are the very same substance throughout the entire heating curve. This represents a physical change of matter, since **no new substance is formed**. On the other hand, chemical changes involve the formation of new substances with different properties.

PART C

[All questions in Part C are worth 1 point.]

66. The condition known as dynamic equilibrium occurs when **the rate of the forward reaction equals the rate of the reverse reaction**. One real life example of dynamic equilibrium is traffic on a street moving in opposite directions at an equal rate.

67. Lewis electron-dot diagrams are drawn to show the distribution of valence electrons within a molecule. These diagrams help to illustrate the number and types of bonds in a compound. The following illustrations are examples of a Lewis electron-dot diagram for water. Note that, in the finished diagram, the oxygen atom achieves a stable octet of valence electrons by sharing each hydrogen's single valence electron.

$$\text{H}\!-\!\ddot{\text{O}}\!-\!\text{H} \quad \text{OR} \quad \text{H}\!:\!\ddot{\text{O}}\!:\!\text{H}$$

68. In order to react, particles must collide with sufficient energy (the activation energy) and proper orientation. **More effective collisions occur** between water molecules and solid carbon if the carbon is spread out over a larger area.

69. Reference Table I shows that the heat of reaction (ΔH) for dissolving KNO_3 is +34.89 kJ. **This reaction is endothermic, meaning that heat is required for dissolving to take place.** The heat of reaction for dissolving KNO_3 will be supplied by the water, causing the temperature of the water to decrease.

70. The percent by mass (percent composition) formula is given on Reference Table T:

$$\% \text{ composition by mass} = \frac{\text{mass of part}}{\text{mass of whole}} \times 100$$

where the "part" is the mass of the KNO_3 in the solution and the "whole" is the mass of the entire solution

$$\% \text{ composition by mass of } KNO_3 = \frac{5.0 \, \text{grams}}{15.0 \, \text{grams}} \times 100 = \mathbf{33.3\%}$$

(NOTE: The mass of the solution (15.0 grams) equals the mass of KNO_3 in mixture 2 (5.0 grams) added to the mass of the original water (10.0 grams). Also, note that all values from **33%** to **33.3%** were accepted as correct answers to this question.)

71. Dissolved ions lower the freezing point of a solution. This is the colligative property known as freezing point depression. The more ions that are dissolved, the lower the freezing point. **The freezing point of mixture 4 will be lower than the freezing point of water** due to the dissolved KNO_3.

72. The heat of reaction is defined as the difference between the potential energy of the products (present after the reaction) and the potential energy of the reactants (present before the reaction). The heat of reaction is indicated on a potential energy diagram as follows:

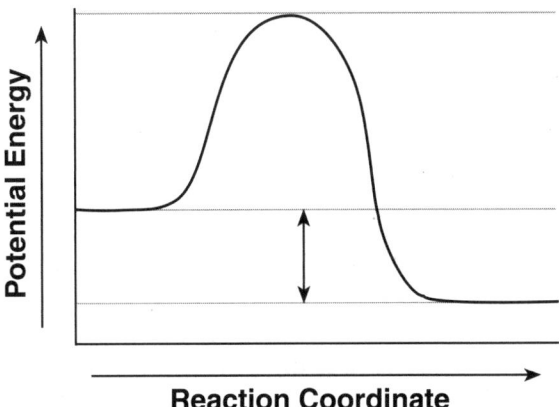

(NOTE: As per the question, the arrow **must** be a double-headed arrow.)

73. All coefficients in chemical reactions represent particles or moles. Mole ratios can be used to find the number of moles of any substance in a cahemical equation if the moles of another substance are known. Two methods of setting up this problem are:

$$\frac{8.0 \, \cancel{mol \, CO}}{1} \times \frac{1 \, mol \, O_2}{2 \, \cancel{mol \, CO}} - \textbf{4.0 mol O}_2$$

You could also set up mole ratios:

$$\frac{8.0 \, mol \, CO}{2} = \frac{x \, mol \, O_2}{1}$$

$$x = \textbf{4.0 mol O}_2$$

74. Catalysts speed up chemical reactions by *lowering* the activation energy (E_a) for the reaction. With a lower E_a, less energy is needed for effective collisions. You can show the lowering of the E_a by drawing the dashed line as follows:

Reaction Coordinate

75. Compound *A* contains the –OH functional group. According to Reference Table *R*, compounds with an –OH functional group are classified as **alcohols**.

76. Organic chemistry is the study of compounds that contain the element **carbon**.

77. Hydrocarbons are organic compounds that contain only two elements: carbon and hydrogen. **Since compound B also contains oxygen**, it cannot be classified as a hydrocarbon.

78. According to the Arrhenius theory of acids and bases, H^+ (the hydrogen ion) is the only positive ion present in acidic solutions.
   (NOTE: Other acceptable answers include H_3O^+, or **the hydronium ion**.)

79. The completed equation should read as follows:

$$2NaOH(aq) + H_2SO_4(aq) \rightarrow 2H_2O(\ell) + \textbf{Na}_2\textbf{SO}_4(aq)$$

The general equation for a neutralization equation is:

$$Acid + Base \rightarrow Water + Salt$$

Note that the salt will be an ionic compound. The Law of Conservation of Mass states that all atoms that enter a chemical reaction must leave the reaction in equal numbers. A neutralization equation is a specific type of double replacement reaction in which the ions switch places and leave with the oppositely charged ion on the other compound. The sodium ion (Na^+) switches places with the hydrogen ion (H^+) to form the ionic compound sodium sulfate (Na_2SO_4).

80. On the pH scale, which ranges from 0 to 14, a pH of 7 is exactly neutral. This is the pH of pure water. The graph shows that a pH of 7 is reached after the addition of **25 mL of NaOH**.
(NOTE: All values from **24 mL to 26 mL** were accepted as correct answers to this question.)

81. The graph shows that after 50. mL of NaOH(aq) were added, the solution reached a pH of approximately 12.5. Reference Table M lists the colors of common acid-base indicators at different pH ranges. Phenolphthalein will be **pink** when the pH is greater than 9.

82. Fission is the splitting of a large atomic nucleus (usually U-235) into two smaller nuclei. The fission reaction is initiated when the target nucleus is struck by a neutron, and the resulting emission of neutrons as products makes a chain reaction possible. The nuclear equation in this question represents a fission reaction because **the $^{235}_{92}U$ nuclide is bombarded by a neutron and splits into two smaller nuclides**.

83. The isotopic notation for neodymium 150, $^{150}_{60}Nd$, shows that Nd 150 has an atomic number of 60 and, therefore, this element has **60 protons**. The neutrons in this isotope of neodymium can be found by subtracting the atomic number (the number of protons) from the mass number of 150, which is the sum of the protons and the neutrons. Thus, this element has **90 neutrons**, since $150 - 60 = 90$ neutrons.

84. The completed equation should read as follows:

$$^{81}_{32}Ge \rightarrow {}^{0}_{-1}e + {}^{81}_{33}As$$

When balancing a nuclear equation, the sum of the atomic numbers (the bottom numbers) and the mass numbers (the top numbers) must be equal on both sides of the arrow.

85. One half-life is the time required for a radioisotope to decay to half of its original mass. To find the total number of half-lives required for Ge-81 to decay from a mass of 16.00 grams to a mass of 1.00 gram, divide the mass in half over a series of steps.

In the first half-life, Ge-81 decays from 16.00 grams to 8.00 grams. (The other 8.00 grams undergo radioactive decay). In the second half-life, the 8.00 grams of Ge-81 remaining decay to 4.00 grams. In the third half-life, the 4.00 grams of Ge-81 remaining decay to 2.00 grams. In the fourth half-life, the 2.00 grams of Ge-81 remaining decay to 1.00 gram. Thus, four half-lives are needed for the radioactive decay of Ge-81 from 16.00 grams to 1.00 gram. The information given for questions 82–85 states that Ge-81 has a half-life of 7.6 seconds. Therefore, the time required for four half-lives is 4×7.6 seconds = **30.4 seconds**.

Mark (✓) the questions you answered correctly. Count the number of checks and follow the formulas given to determine your score on each topic.

Core Area	· □ Questions Answered Correctly
	45: (1)
Section M—Math Skills	
□ Number of checks ÷ 1 × 100 = ___ %	
	33: (1)
Section R—Reading Skills	
□ Number of checks ÷ 1 × 100 = ___ %	
	1–4, 31, 32, 53–55, 83: (10)
Section I—Atomic Concepts	
□ Number of checks ÷ 10 × 100 = ___ %	
	5–7, 56–58: (6)
Section II—Periodic Table	
□ Number of checks ÷ 6 × 100 = ___ %	
	8, 34–37, 51, 73: (7)
Section III—Moles/Stoichiometry	
□ Number of checks ÷ 7 × 100 = ___ %	
	9–13, 15, 38, 39, 67: (9)
Section IV—Chemical Bonding	
□ Number of checks ÷ 9 × 100 = ___ %	
	14, 16–19, 40–42, 52, 59–65, 70, 71: (18)
Section V—Physical Behavior of Matter	
□ Number of checks ÷ 18 × 100 = ___ %	
	20, 24, 43, 66, 68, 69, 72, 74: (8)
Section VI—Kinetics and Equilibrium	
□ Number of checks ÷ 8 × 100 = ___ %	

Core Area	□ Questions Answered Correctly

21–23, 44, 75–77: (7)

Section VII—Organic Chemistry
□ Number of checks ÷ 7 × 100 = ___ %

25, 26, 46, 47: (4)

Section VIII—Oxidation–Reduction
□ Number of checks ÷ 4 × 100 = ___ %

48, 49, 78–81: (6)

Section IX—Acids, Bases, and Salts
□ Number of checks ÷ 6 × 100 = ___ %

27–30, 50, 82, 84, 85: (8)

Section X—Nuclear Chemistry
□ Number of checks ÷ 8 × 100 = ___ %

Examination
June 2019
Chemistry—Physical Setting

PART A

Answer all questions in this part.

Directions (1–30): For *each* statement or question, write in the answer space the *number* of the word or expression that, of those given, best completes the statement or answers the question. Some questions may require the use of the *2011 Edition Reference Tables for Physical Setting/Chemistry.*

1 Which particles are found in the nucleus of an argon atom?

(1) protons and electrons
(2) positrons and neutrons
(3) protons and neutrons
(4) positrons and electrons

1 ＿＿＿

2 The diagram below represents a particle traveling through an electric field.

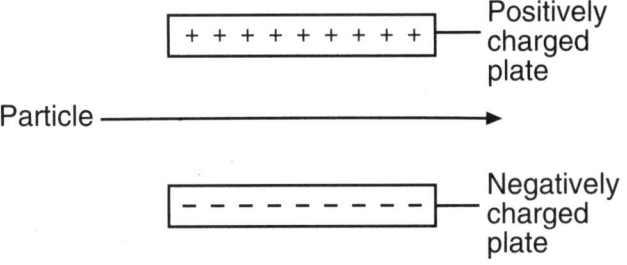

An electric field exists between the two plates.

Which particle remains undeflected when passing through this electric field?

(1) proton   (3) neutron
(2) electron   (4) positron   2 _____

3 The mass of an electron is

(1) equal to the mass of a proton
(2) equal to the mass of a neutron
(3) greater than the mass of a proton
(4) less than the mass of a neutron   3 _____

4 Compared to the energy of an electron in the second shell of an atom of sulfur, the energy of an electron in the

(1) first shell is lower
(2) first shell is the same
(3) third shell is lower
(4) third shell is the same   4 _____

5 In the ground state, an atom of which element has seven valence electrons?

(1) sodium   (3) nitrogen
(2) phosphorus   (4) fluorine   5 _____

6 Which information is sufficient to differentiate a sample of sodium from a sample of silver?

(1) the mass of each sample
(2) the volume of each sample
(3) the reactivity of each sample with water
(4) the phase of each sample at room temperature   6 _____

7 Graphite and diamond are two forms of solid carbon at STP. These forms have

(1) different molecular structures and different properties
(2) different molecular structures and the same properties
(3) the same molecular structures and different properties
(4) the same molecular structures and the same properties   7 _____

8 As the first five elements in Group 14 are considered in order from top to bottom, there are changes in both the

(1) number of valence shell electrons and number of first shell electrons

(2) electronegativity values and number of first shell electrons

(3) number of valence shell electrons and atomic radii

(4) electronegativity values and atomic radii   8 _____

9 Which statement explains why NaBr is classified as a compound?

(1) Na and Br are chemically combined in a fixed proportion.

(2) Na and Br are both nonmetals.

(3) NaBr is a solid at 298 K and standard pressure.

(4) NaBr dissolves in H_2O at 298 K.   9 _____

10 Which two terms represent types of chemical formulas?

(1) fission and fusion

(2) oxidation and reduction

(3) empirical and structural

(4) endothermic and exothermic   10 _____

11 During all chemical reactions, charge, mass, and energy are

(1) condensed   (3) decayed

(2) conserved   (4) decomposed   11 _____

12 The degree of polarity of a covalent bond between two atoms is determined by calculating the difference in their

(1) atomic radii   (3) electronegativities

(2) melting points   (4) ionization energies   12 _____

13 Which substance can *not* be broken down by a chemical change?

(1) ammonia   (3) methane

(2) magnesium   (4) water   13 _____

14 Which statement describes the components of a mixture?

   (1) Each component gains new properties.
   (2) Each component loses its original properties.
   (3) The proportions of components can vary.
   (4) The proportions of components cannot vary.   14 _____

15 Table sugar can be separated from a mixture of table sugar and sand at STP by adding

   (1) sand, stirring, and distilling at 100.°C
   (2) sand, stirring, and filtering
   (3) water, stirring, and distilling at 100.°C
   (4) water, stirring, and filtering   15 _____

16 Which statement describes the particles of an ideal gas, based on the kinetic molecular theory?

   (1) The volume of the particles is considered negligible.
   (2) The force of attraction between the particles is strong.
   (3) The particles are closely packed in a regular, repeating pattern.
   (4) The particles are separated by small distances, relative to their size.   16 _____

17 During which two processes does a substance release energy?

   (1) freezing and condensation
   (2) freezing and melting
   (3) evaporation and condensation
   (4) evaporation and melting   17 _____

18 Based on Table *I*, which compound dissolves in water by an exothermic process?

   (1) NaCl   (3) NH_4Cl
   (2) NaOH   (4) NH_4NO_3 18 _____

19 At STP, which property of a molecular substance is determined by the arrangement of its molecules?

(1) half-life
(2) molar mass
(3) physical state
(4) percent composition   19 _____

20 Equilibrium can be reached by

(1) physical changes, only
(2) nuclear changes, only
(3) both physical changes and chemical changes
(4) both nuclear changes and chemical changes   20 _____

21 Which value is defined as the difference between the potential energy of the products and the potential energy of the reactants during a chemical change?

(1) heat of fusion
(2) heat of reaction
(3) heat of deposition
(4) heat of vaporization   21 _____

22 The effect of a catalyst on a chemical reaction is to provide a new reaction pathway that results in a different

(1) potential energy of the products
(2) heat of reaction
(3) potential energy of the reactants
(4) activation energy   22 _____

23 Chemical systems in nature tend to undergo changes toward

(1) lower energy and lower entropy
(2) lower energy and higher entropy
(3) higher energy and lower entropy
(4) higher energy and higher entropy   23 _____

24 The atoms of which element bond to one another in chains, rings, and networks?

(1) barium   (3) iodine
(2) carbon   (4) mercury   24 _____

25 What is the general formula for the homologous series that includes ethene?

(1) C_nH_{2n} (3) C_nH_{2n-2}

(2) C_nH_{2n-6} (4) C_nH_{2n+2} 25 _____

26 When an F atom becomes an F^- ion, the F atom

(1) gains a proton   (3) gains an electron

(2) loses a proton   (4) loses an electron   26 _____

27 Which substance is an Arrhenius base?

(1) HNO_3 (3) $Ca(OH)_2$

(2) H_2SO_3 (4) CH_3COOH 27 _____

28 In which type of nuclear reaction do two light nuclei combine to produce a heavier nucleus?

(1) positron emission   (3) fission

(2) gamma emission   (4) fusion   28 _____

29 Using equal masses of reactants, which statement describes the relative amounts of energy released during a chemical reaction and a nuclear reaction?

(1) The chemical and nuclear reactions release equal amounts of energy.

(2) The nuclear reaction releases half the amount of energy of the chemical reaction.

(3) The chemical reaction releases more energy than the nuclear reaction.

(4) The nuclear reaction releases more energy than the chemical reaction.   29 _____

30 The ratio of the mass of U-238 to the mass of Pb-206 can be used to

(1) diagnose thyroid disorders

(2) diagnose kidney function

(3) date geological formations

(4) date once-living things   30 _____

PART B–1

Answer all questions in this part.

Directions (31–50): For *each* statement or question, write in the answer space the *number* of the word or expression that, of those given, best completes the statement or answers the question. Some questions may require the use of the *2011 Edition Reference Tables for Physical Setting/Chemistry.*

31 The bright-line spectra of four elements, *G, J, L,* and *M,* and a mixture of *at least two* of these elements is given below.

Bright-Line Spectra

Wavelength (nm)

Which elements are present in this mixture?

(1) *G* and *J*   (3) *M, J,* and *G*
(2) *G* and *L*   (4) *M, J,* and *L*   31 _____

32 Which electron configuration represents an atom of chlorine in an excited state?

(1) 2-8-7-2   (3) 2-8-8
(2) 2-8-7   (4) 2-7-8   32 _____

33 A student measures the mass and volume of a sample of aluminum at room temperature, and calculates the density of Al to be 2.85 grams per cubic centimeter. Based on Table S, what is the percent error for the student's calculated density of Al?

(1) 2.7%   (3) 5.6%
(2) 5.3%   (4) 95%   33 _____

34 Magnesium and calcium have similar chemical properties because their atoms in the ground state have

(1) equal numbers of protons and electrons
(2) equal numbers of protons and neutrons
(3) two electrons in the first shell
(4) two electrons in the outermost shell   34 _____

35 As the elements in Period 2 of the Periodic Table are considered in order from left to right, which property generally *decreases*?

(1) atomic radius   (3) ionization energy
(2) electronegativity   (4) nuclear charge   35 _____

36 Given the balanced equation for the reaction of butane and oxygen:

$$2C_4H_{10} + 13O_2 \rightarrow 8CO_2 + 10H_2O + energy$$

How many moles of carbon dioxide are produced when 5.0 moles of butane react completely?

(1) 5.0 mol   (3) 20. mol
(2) 10. mol   (4) 40. mol   36 _____

37 What is the percent composition by mass of nitrogen in the compound N_2H_4 (gram-formula mass = 32 g/mol)?

(1) 13%   (3) 88%
(2) 44%   (4) 93%   37 _____

38 Which ion in the ground state has the same electron configuration as an atom of neon in the ground state?

(1) Ca^{2+} (3) Li^+
(2) Cl^- (4) O^{2-} 38 _____

39 The molar masses and boiling points at standard pressure for four compounds are given in the table below.

Compound	Molar Mass (g/mol)	Boiling Point (K)
HF	20.01	293
HCl	36.46	188
HBr	80.91	207
HI	127.91	237

Which compound has the strongest intermolecular forces?

(1) HF
(2) HCl
(3) HBr
(4) HI

39 _____

40 Which particle model diagram represents xenon at STP?

Key
O = an atom of xenon

(1)

(3)

(2)

(4)

40 _____

41 What is the amount of heat absorbed when the temperature of 75 grams of water increases from 20.°C to 35°C?

(1) 1100 J
(2) 4700 J

(3) 6300 J
(4) 11 000 J

41 _____

42 Which sample of HCl(aq) reacts at the fastest rate with a 1.0-gram sample of iron filings?

(1) 10. mL of 1 M HCl(aq) at 10.°C
(2) 10. mL of 1 M HCl(aq) at 25°C
(3) 10. mL of 3 M HCl(aq) at 10.°C
(4) 10. mL of 3 M HCl(aq) at 25°C

42 _____

43 Given the equation representing a system at equilibrium:

$$N_2O_4(g) \rightleftharpoons 2NO_2(g)$$

Which statement describes the concentration of the two gases in this system?

(1) The concentration of $N_2O_4(g)$ must be less than the concentration of $NO_2(g)$.
(2) The concentration of $N_2O_4(g)$ must be greater than the concentration of $NO_2(g)$.
(3) The concentration of $N_2O_4(g)$ and the concentration of $NO_2(g)$ must be equal.
(4) The concentration of $N_2O_4(g)$ and the concentration of $NO_2(g)$ must be constant.

43 _____

44 Given the equation representing a system at equilibrium:

$$PCl_5(g) + energy \rightleftharpoons PCl_3(g) + Cl_2(g)$$

Which change will cause the equilibrium to shift to the right?

(1) adding a catalyst
(2) adding more $PCl_3(g)$
(3) increasing the pressure
(4) increasing the temperature

44 _____

45 Given the formula representing a molecule:

$$H-\overset{\overset{\displaystyle H}{|}}{C}-\overset{\overset{\displaystyle H}{|}}{C}-\overset{\overset{\displaystyle H}{|}}{C}-\overset{\overset{\displaystyle H}{|}}{C}-\overset{\overset{\displaystyle H}{|}}{C}-N\overset{\diagup H}{\diagdown H}$$

A chemical name for this compound is

(1) pentanone   (3) 1-pentanamine
(2) 1-pentanol   (4) pentanamide   45 _____

46 Given the formula of a compound:

$$H-C\equiv C-\overset{\overset{\displaystyle H}{|}}{C}-\overset{\overset{\displaystyle H}{|}}{C}-H$$

This compound is classified as an

(1) aldehyde   (3) alkyne
(2) alkene   (4) alcohol   46 _____

47 Which equation represents fermentation?

(1) $C_2H_4 + H_2O \rightarrow CH_3CH_2OH$
(2) $C_2H_4 + HCl \rightarrow CH_3CH_2Cl$
(3) $C_6H_{12}O_6 \rightarrow 2CH_3CH_2OH + 2CO_2$
(4) $2CH_3CHO \rightarrow C_3H_5CHO + H_2O$ 47 _____

48 Given the equation representing a reaction:

$$3CuCl_2(aq) + 2Al(s) \rightarrow 3Cu(s) + 2AlCl_3(aq)$$

The oxidation number of copper changes from

(1) +1 to 0   (3) +2 to +1
(2) +2 to 0   (4) +6 to +3   48 _____

49 Given the equation representing a reversible reaction:

$$CH_3COOH(aq) + H_2O(\ell) \rightleftharpoons$$
$$CH_3COO^-(aq) + H_3O^+(aq)$$

According to one acid-base theory, the two H^+ donors in the equation are

(1) CH_3COOH and H_2O
(2) CH_3COOH and H_3O^+
(3) CH_3COO^- and H_2O
(4) CH_3COO^- and H_3O^+ 49 _____

50 Which nuclear equation represents a spontaneous decay?

(1) $^{222}_{86}Rn \rightarrow ^{218}_{84}Po + ^{4}_{2}He$

(2) $^{27}_{13}Al + ^{4}_{2}He \rightarrow ^{30}_{15}P + ^{1}_{0}n$

(3) $^{235}_{92}U + ^{1}_{0}n \rightarrow ^{139}_{56}Ba + ^{96}_{36}Kr + 3^{1}_{0}n$

(4) $^{7}_{3}Li + ^{1}_{1}H \rightarrow ^{4}_{2}He + ^{4}_{2}He$ 50 _____

PART B–2

Answer all questions in this part.

Directions (51–65): Record your answers on the answer sheet provided. Some questions may require the use of the *2011 Edition Reference Tables for Physical Setting/Chemistry*.

51 Draw a structural formula for methanal.   [1]

Base your answers to questions 52 through 54 on the information below and on your knowledge of chemistry.

The atomic mass and natural abundance of the naturally occurring isotopes of hydrogen are shown in the table below.

Naturally Occurring Isotopes of Hydrogen

Isotope	Common Name of Isotope	Atomic Mass (u)	Natural Abundance (%)
H-1	protium	1.0078	99.9885
H-2	deuterium	2.0141	0.0115
H-3	tritium	3.0160	negligible

The isotope H-2, also called deuterium, is usually represented by the symbol "D." Heavy water forms when deuterium reacts with oxygen, producing molecules of D_2O.

52 Explain, in terms of subatomic particles, why atoms of H-1, H-2, and H-3 are each electrically neutral.   [1]

53 Determine the formula mass of heavy water, D_2O.   [1]

54 Based on Table *N*, identify the decay mode of tritium.   [1]

Base your answers to questions 55 through 57 on the information below and on your knowledge of chemistry.

At 23°C, 85.0 grams of $NaNO_3(s)$ are dissolved in 100. grams of $H_2O(\ell)$.

55 Convert the temperature of the $NaNO_3(s)$ to kelvins.   [1]

56 Based on Table G, determine the additional mass of $NaNO_3(s)$ that must be dissolved to saturate the solution at 23°C.   [1]

57 State what happens to the boiling point and freezing point of the solution when the solution is diluted with an additional 100. grams of $H_2O(\ell)$.   [1]

Base your answers to questions 58 through 61 on the information below and on your knowledge of chemistry.

A 200.-milliliter sample of $CO_2(g)$ is placed in a sealed, rigid cylinder with a movable piston at 296 K and 101.3 kPa.

58 State a change in temperature and a change in pressure of the $CO_2(g)$ that would cause it to behave more like an ideal gas.   [1]

59 Determine the volume of the sample of $CO_2(g)$ if the temperature and pressure are changed to 336 K and 152.0 kPa.   [1]

60 State, in terms of both the frequency and force of collisions, what would result from decreasing the temperature of the original sample of $CO_2(g)$, at constant volume.   [1]

61 Compare the mass of the original 200.-milliliter sample of $CO_2(g)$ to the mass of the $CO_2(g)$ sample when the cylinder is adjusted to a volume of 100. milliliters.   [1]

Base your answers to questions 62 through 65 on the information below and on your knowledge of chemistry.

Cobalt-60 is an artificial isotope of Co-59. The incomplete equation for the decay of cobalt-60, including beta and gamma emissions, is shown below.

$$^{60}_{27}\text{Co} \rightarrow X + {}^{0}_{-1}\text{e} + {}^{0}_{0}\gamma$$

62 Explain, in terms of *both* protons and neutrons, why Co-59 and Co-60 are isotopes of cobalt. [1]

63 Compare the penetrating power of the beta and gamma emissions. [1]

64 Complete the nuclear equation, on your answer sheet, for the decay of cobalt-60 by writing a notation for the missing product. [1]

65 Based on Table N, determine the total time required for an 80.00-gram sample of cobalt-60 to decay until only 10.00 grams of the sample remain unchanged. [1]

PART C

Answer all questions in this part.

Directions (66–85): Record your answers on the answer sheet provided. Some questions may require the use of the *2011 Edition Reference Tables for Physical Setting/Chemistry.*

Base your answers to questions 66 through 69 on the information below and on your knowledge of chemistry.

During a laboratory activity, appropriate safety equipment was used and safety procedures were followed. A laboratory technician heated a sample of solid $KClO_3$ in a crucible to determine the percent composition by mass of oxygen in the compound. The unbalanced equation and the data for the decomposition of solid $KClO_3$ are shown below.

$$KClO_3(s) \rightarrow KCl(s) + O_2(g)$$

Lab Data and Calculated Results

Object or Material	Mass (g)
empty crucible and cover	22.14
empty crucible, cover, and $KClO_3$	24.21
$KClO_3$	2.07
crucible, cover, and KCl after heating	23.41
KCl	?
O_2	0.80

66 Write a chemical name for the compound that decomposed.   [1]

67 Based on the lab data, show a numerical setup to determine the number of moles of O_2 produced. Use 32 g/mol as the gram-formula mass of O_2.   [1]

68 Based on the lab data, determine the mass of KCl produced in the reaction.   [1]

69 Balance the equation on your answer sheet for the decomposition of $KClO_3$, using the smallest whole-number coefficients.   [1]

Base your answers to questions 70 through 73 on the information below and on your knowledge of chemistry.

A bottled water label lists the ions dissolved in the water. The table below lists the mass of some ions dissolved in a 500.-gram sample of the bottled water.

Ions in 500. g of Bottled Water

Ion Formula	Mass (g)
Ca^{2+}	0.040
Mg^{2+}	0.013
Na^+	0.0033
SO_4^{2-}	0.0063
HCO_3^-	0.180

70 State the number of significant figures used to express the mass of hydrogen carbonate ions in the table above.   [1]

71 Based on Table F, write the formula of the ion in the bottled water table that would form the *least* soluble compound when combined with the sulfate ion.   [1]

72 Show a numerical setup for calculating the parts per million of the Na^+ ions in the 500.-gram sample of the bottled water.   [1]

73 Compare the radius of a Mg^{2+} ion to the radius of a Mg atom.   [1]

Base your answers to questions 74 through 77 on the information below and on your knowledge of chemistry.

Ethyl ethanoate is used as a solvent for varnishes and in the manufacture of artificial leather. The formula below represents a molecule of ethyl ethanoate.

$$
\begin{array}{ccccccc}
 & H & O & & H & H \\
 & | & \| & & | & | \\
H- & C- & C- & O- & C- & C- & H \\
 & | & & & | & | \\
 & H & & & H & H
\end{array}
$$

74 Identify the element in ethyl ethanoate that makes it an organic compound.   [1]

75 Write the empirical formula for this compound.   [1]

76 Write the name of the class of organic compounds to which this compound belongs.   [1]

77 Determine the number of electrons shared in the bond between a hydrogen atom and a carbon atom in the molecule.   [1]

Base your answers to questions 78 through 80 on the information below and on your knowledge of chemistry.

An operating voltaic cell has magnesium and silver electrodes. The cell and the ionic equation representing the reaction that occurs in the cell are shown below.

Voltaic Cell

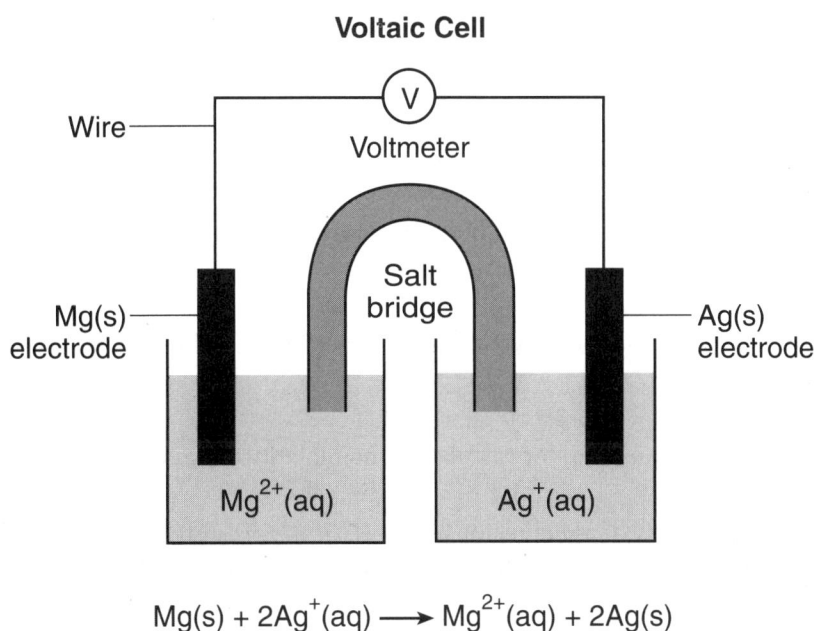

$$Mg(s) + 2Ag^+(aq) \longrightarrow Mg^{2+}(aq) + 2Ag(s)$$

78 State the purpose of the salt bridge in this cell.   [1]

79 Write a balanced equation for the half-reaction that occurs at the magnesium electrode in this cell.   [1]

80 Explain, in terms of electrical energy, how electrolysis reactions differ from voltaic cell reactions.   [1]

Base your answers to questions 81 through 85 on the information below and on your knowledge of chemistry.

In a laboratory investigation, an HCl(aq) solution with a pH value of 2 is used to determine the molarity of a KOH(aq) solution. A 7.5-milliliter sample of the KOH(aq) is exactly neutralized by 15.0 milliliters of the 0.010 M HCl(aq). During this laboratory activity, appropriate safety equipment is used and safety procedures are followed.

81 Determine the pH value of a solution that is ten times less acidic than the HCl(aq) solution. [1]

82 State the color of the indicator bromcresol green if it is added to a sample of the KOH(aq) solution. [1]

83 Complete the equation on your answer sheet by writing the chemical formula for *each* product. [1]

84 Show a numerical setup for calculating the molarity of the KOH solution. [1]

85 Explain, in terms of aqueous ions, why 15.0 mL of a 1.0 M HCl(aq) solution is a better conductor of electricity than 15.0 mL of a 0.010 M HCl(aq) solution. [1]

Answer Sheet
June 2019

Chemistry—Physical Setting

PART B–2

51

52 _____

_____

_____

53 _____ u

54 _____

55 _____ **K**

56 _____ **g**

57 Boiling point: _____

Freezing point: _____

58 Temperature: _____

Pressure: _____

59 _____ **mL**

60 _____

_____

_____

61 _____

_____

_____

62 _____

_____

_____

63 _____

_____

_____

64 $^{60}_{27}\text{Co} \rightarrow$ _____ $+\,^{0}_{-1}\text{e} +\,^{0}_{0}\gamma$

65 _____ y

PART C

66 _____

67

68 _____ g

69 _____ $KClO_3(s) \rightarrow$ _____ $KCl(s) +$ _____ $O_2(g)$

70 _____

71 _____

72

73 _____

_____

_____

74 _____

75 _____

76 _____

77 _____

78 _____

_____

_____

79 _____

80 _____

_____

_____

81 _____

82 _____

83 $HCl(aq) + KOH(aq) \rightarrow$ _____ + _____

84

85

Answers
June 2019

Chemistry—Physical Setting

Answer Key

PART A

1. 3	**7.** 1	**13.** 2	**19.** 3	**25.** 1
2. 3	**8.** 4	**14.** 3	**20.** 3	**26.** 3
3. 4	**9.** 1	**15.** 4	**21.** 2	**27.** 3
4. 1	**10.** 3	**16.** 1	**22.** 4	**28.** 4
5. 4	**11.** 2	**17.** 1	**23.** 2	**29.** 4
6. 3	**12.** 3	**18.** 2	**24.** 2	**30.** 3

PART B–1

31. 2	**35.** 1	**39.** 1	**43.** 4	**47.** 3
32. 4	**36.** 3	**40.** 1	**44.** 4	**48.** 2
33. 3	**37.** 3	**41.** 2	**45.** 3	**49.** 2
34. 4	**38.** 4	**42.** 4	**46.** 3	**50.** 1

PART B–2 and **PART C**. *See* **Answers Explained**.

Answers Explained

PART A

1. **3** Every atom of every element consists of a nucleus surrounded by an electron cloud. The nucleus contains positively charged protons as well as neutrons, which have no charge.

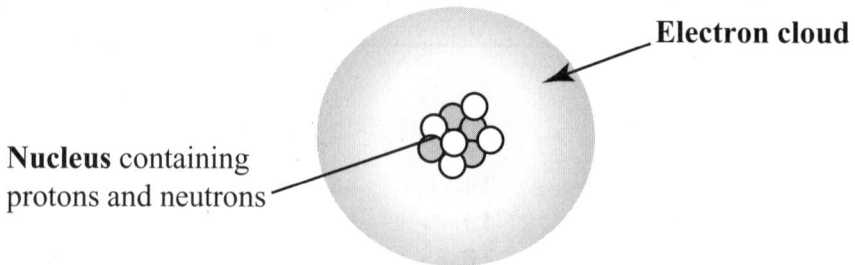

Electron cloud

Nucleus containing
protons and neutrons

2. **3** Neutrons are electrically neutral and would not be attracted to either of the electrically charged plates.

WRONG CHOICES EXPLAINED:
(1) Protons are positively charged and would be attracted to the negatively charged plate, since opposite charges attract each other.
(2) Electrons are negatively charged and would be attracted to the positively charged plate, since opposite charges attract each other.
(4) Positrons are positively charged and would be attracted to the negatively charged plate, since opposite charges attract each other.

3. **4** Neutrons and protons have a mass of approximately 1 amu (atomic mass unit), while the mass of an electron is approximately 0.00054 amu. Electrons are so small compared to protons and neutrons that electron mass is ignored when calculating the atomic mass of an atom.

4. **1** Electron shells closer to the nucleus are lower in energy than electron shells farther from the nucleus. This is because negatively charged electrons are attracted to the positively charged nucleus. An electron in the first electron shell of any atom will have less energy than an electron in the second shell.

5. **4** The outermost electrons in an atom are called the valence electrons. Look at the electron configurations of these elements on the Periodic Table in your Chemistry Reference Tables. The last number in each element's electron configuration represents its valence (outermost) electrons. Fluorine, with an electron configuration of 2-7, has 7 valence electrons.

WRONG CHOICES EXPLAINED:

(1) The electron configuration for sodium (Na) is 2-8-1, meaning that sodium has 1 valence electron.

(2) The electron configuration for phosphorus (P) is 2-8-5, meaning that phosphorus has 5 valence electrons.

(3) The electron configuration for nitrogen (N) is 2-5, meaning that nitrogen has 5 valence electrons.

6. **3** Elements differ in their chemical reactivity because of the number and positioning of their valence electrons. Sodium is an alkali metal and is highly reactive in water, whereas silver is chemically unreactive with pure water.

WRONG CHOICES EXPLAINED:

(1) Mass cannot be used to distinguish between silver and sodium because samples of these metals could have the same mass.

(2) Volume cannot be used to distinguish between silver and sodium because samples of these metals could have the same volume.

(4) Both silver and sodium are solid at room temperature.

7. **1** Graphite and diamond are both allotropes of the element carbon. Allotropes are different structural forms of the same element. For example, graphite has a layered structure, and diamonds have a three-dimensional network structure. Because of their different structures, graphite and diamond will have different properties, such as hardness, conductivity, and melting point.

8. **4** Electronegativity and atomic radii values vary from element to element moving both down a group or across a period on the Periodic Table. Refer to Reference Table *S* to look at these values for C, Si, Ge, Sn, and Pb, and you can confirm that both the atomic radii and electronegativity are different for each element.

WRONG CHOICES EXPLAINED:

(1), (3) Group 14 elements all have four valence electrons. Find the electron configurations for these elements, which are printed on the Periodic Table. The last number in each element's electron configuration is "4," which represents the valence (outermost) electrons.

(2) Only two electrons are needed to fill the first electron shell in any atom. Looking at the electron configurations on the Periodic Table for Group 14 elements, "2" represents the first shell electrons.

9. **1** Compounds are substances formed from two or more different elements that are chemically bonded in fixed proportions. Remember this definition!

WRONG CHOICES EXPLAINED:
(2) Sodium (Na) is a metal found in Group 1 on the Periodic Table.
(3) Sodium bromide (NaBr) is a solid at room temperature and standard pressure, but this is a physical property and is not sufficient evidence to classify NaBr as a chemical compound. Many substances that are not compounds are solids at the conditions stated in this question, such as gold (an element) and sand (a mixture).
(4) The ability to dissolve is a physical property and again is not sufficient evidence to classify a substance as a chemical compound. For example, a mixture of salt and sugar would dissolve, but in a mixture, the components are physically, not chemically, combined.

10. **3** Chemical formulas provide information about the types and proportion of elements in a compound. The empirical formula provides the smallest whole number ratio of elements in a compound, and a structural formula shows how the atoms are arranged in the compound. Consider the example of 2-pentene shown below:

Molecular Formula	Empirical Formula	Structural Formula
C_5H_{10}	CH_2	

$$\begin{array}{ccccc} H & & & H & H \\ | & & & | & | \\ H-C-C=C-C-C-H \\ | & | & | & | & | \\ H & H & H & H & H \end{array}$$

11. **2** Neither charge nor mass nor energy can be created or destroyed during chemical reactions. The charge and mass of the products in a chemical reaction will always equal the charge and mass of the reactants. Energy may change form during a chemical reaction, but the net sum of energy contained in a chemical system and the environment surrounding the system before and after the reaction is the same.

12. **3** Polarity of a covalent bond is caused by the unequal sharing of the electrons in the bond. Electronegativity measures an atom's attraction for a bonded pair of electrons. When atoms with different electronegativities form a covalent bond, the atom with higher electronegativity will have a stronger attraction for the electrons in the bond and will assume a partial negative charge (δ^-), while the atom with lower electronegativity will assume a partial positive charge (δ^+). A larger difference in electronegativity between the atoms in the bond causes stronger partial charges to form, leading to a bond with greater polarity.

13. **2** Magnesium (Mg) is an element found in Group 2 on the Periodic Table. All elements are composed of atoms, and atoms cannot be further broken down during the course of a chemical reaction.

WRONG CHOICES EXPLAINED:
(1), (3), and (4) Ammonia (NH_3), methane (CH_4), and water (H_2O) are all compounds. Compounds are composed of molecules, and molecules can be broken down into their constituent atoms during chemical reactions.

14. **3** Mixtures contain two or more substances (elements or compounds) that are physically mixed together but are not chemically bonded together. The proportions of components in a mixture can vary. For example, consider a solid mixture of salt and sugar. This mixture can contain more salt or less salt, but it is still a mixture of salt and sugar.

WRONG CHOICES EXPLAINED:
(1), (2) Since components of a mixture are not chemically combined, each component retains all of its original properties after mixing.
(4) Proportions of components in mixtures are always variable.

15. **4** Sugar and sand can be separated on the basis of having different solubilities in water. Sugar will dissolve in water, while sand will not dissolve in water. Adding water to a sugar/sand mixture will dissolve the sugar. Passing the water/sugar/sand mixture through a filter will cause sand particles to be trapped by the filter paper, while the dissolved sugar and the water will pass through the filter paper.

16. **1** Gases consist of tiny particles (atoms or molecules) that are very far apart from each other compared with their size. Because their size is so small, the kinetic molecular theory for an ideal gas assumes that the volume of the gas particles themselves is insignificant compared with the volume of the space that they occupy.

WRONG CHOICES EXPLAINED:

(2) Intermolecular forces between gas particles are very weak! This is why gas particles move randomly and separately from each other.

(3) This answer choice describes the spatial arrangement of particles in a solid, not a gas.

(4) This answer choice describes particle arrangement in solid and liquids, not gases. We can see samples of solid and liquid matter because the particles are very close together. Gases are invisible because the tiny particles are spread very far apart.

17. **1** Exothermic phase changes are phase changes in which energy is released. In the process of freezing, the heat of fusion is released, and particles lose potential energy as they assume the closely packed, crystalline arrangement of a solid. In the process of condensation, the transition from the gas to the liquid state, the heat of vaporization is released, and particles lose potential energy as intermolecular forces draw gas particles back together into the much denser liquid phase.

18. **2** Reference Table *I* shows the heat of reaction (ΔH) for several dissolving equations near the bottom of the table. The heat of reaction for dissolving NaOH is –44.51 kJ. The negative sign indicates that this is an exothermic process in which heat is released, as stated in the footnote beginning with an asterisk (*) under the table.

19. **3** The physical state of a sample of matter is determined by the spacing and arrangement of particles within the sample. As shown in the diagrams below, gases contain particles that are very widely spaced and randomly arranged, whereas solids contain particles that are tightly packed and arranged in a regular, repeating geometric pattern.

GAS

SOLID

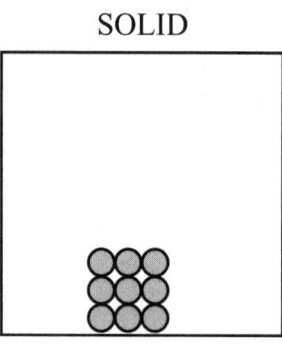

20. **3** A process reaches equilibrium when opposing changes occur at equal rates. An example of a physical change reaching equilibrium is the point at which a solution becomes saturated and the rates of dissolving and precipitation occur at equal rates within the solution. Chemical equilibrium is reached when the rate of the forward reaction equals the rate of the reverse reaction in a reversible reaction. Nuclear changes are not reversible and cannot reach equilibrium.

21. **2** This is a "definition question." The heat of reaction, ΔH, is defined as the potential energy of the products minus the potential energy of the reactants. ΔH can be found on a potential energy diagram, as shown below.

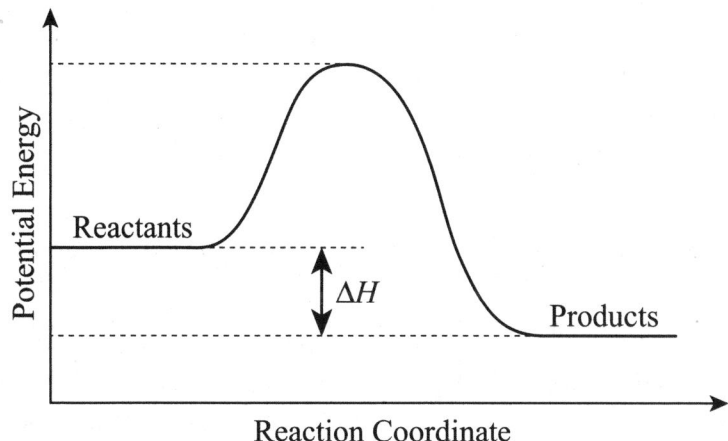

22. **4** Adding a catalyst to a chemical reaction lowers the activation energy (E_a) for the reaction, making it possible for more collisions to have sufficient energy for particles to react.

WRONG CHOICES EXPLAINED:

(1), (2), and (3) None of these are affected by the addition of a catalyst.

23. **2** Systems naturally tend to undergo changes that produce lower total energy and greater disorder or randomness (entropy). For example, when a slab of rock falls from the top of a cliff, it falls to the bottom of the cliff (a position of lower potential energy) and smashes into many pieces (greater randomness).

24. **2** Carbon is a versatile element that forms four bonds to attain a stable octet of eight valence electrons. Because carbon forms four bonds, it is capable of forming many different structures, including chains, rings, and networks, as shown in the diagrams below.

25. **1** Reference Table Q lists the homologous series of hydrocarbons. Ethene, with the suffix "-ene," belongs to the alkene series of hydrocarbons. The general formula for alkenes given in Reference Table Q is C_nH_{2n}.

26. **3** Electrons are the only particles that can move between atoms during chemical reactions. Electrons have a negative charge, so for a neutral fluorine atom to become a negatively charged ion (F^-), it must gain one electron.

27. **3** Arrhenius bases are defined as substances that release hydroxide ions (OH^-) when dissolved in water. $Ca(OH)_2$ contains the hydroxide ion, so it must be the Arrhenius base. Don't forget that common bases are listed on Reference Table L. Arrhenius bases will always contain the OH^- ion.

WRONG CHOICES EXPLAINED:
(1) HNO_3 is nitric acid, as listed on Reference Table K.
(2) H_2SO_3 is sulfurous acid, as listed on Reference Table K.
(4) CH_3COOH is ethanoic acid, as listed on Reference Table K.

28. **4** Nuclear fusion is a nuclear reaction in which two small nuclei fuse together to form a heavier nucleus. This reaction is accompanied by the release of a terrific amount of energy. Hydrogen fusion, for example, is the reaction that powers the sun and the stars.

WRONG CHOICES EXPLAINED:
(1) Positron emission is a type of natural transmutation reaction.
(2) Gamma emission occurs during several types of nuclear reactions, including fusion and fission, and often during alpha decay.
(3) Nuclear fission is the opposite of nuclear fusion. In nuclear fission, a very large nucleus (often U-235) splits into two or more medium-sized nuclei.

29. **4** Nuclear reactions, especially fusion and fission, are dangerous reactions! Consider the destruction caused by the fission bombs that ended World War II. The energy released during nuclear reactions is many orders of magnitude larger than even the most exothermic chemical reactions.

30. **3** U-238 is a radioactive isotope that undergoes radioactive decay in a series of steps, eventually resulting in the formation of Pb-206, a stable isotope of lead. Since the rate of uranium decay into lead is known, finding the ratio of lead to uranium in a rock sample can be used to determine the age of the sample.

PART B–1

31. **2** All elements produce unique bright-line spectra when excited electrons in these atoms fall back to ground state. The bright-line spectrum of a mixture of two or more elements will contain all of the wavelength bands of each separate element superimposed together on the same spectrum. Element J can quickly be eliminated since the two wavelength bands at ~730 nm are not evident in the mixture. If element J is not in the mixture, the answer must be choice (2), and careful inspection shows that all wavelength bands from elements G and L are also found in the mixture.

32. **4** The Periodic Table in the Chemistry Reference Tables shows the ground state configuration of chlorine as 2-8-7. A chlorine atom would be in an excited state if its configuration were 2-7-8 because this configuration shows that a second shell electron has moved to the third shell.

WRONG CHOICES EXPLAINED:
(1) Chlorine atoms have 17 electrons. The electron configuration 2-8-7-2 has a total of 19 electrons and is an excited state electron configuration for a potassium atom.
(2) This is the ground state electron configuration for a chlorine atom. The question asks for an *excited* state configuration.
(3) 2-8-8 is the ground state electron configuration for an atom of argon.

33. **3** The actual value for the density of aluminum, 2.70 g/cm³, is found on Reference Table S. The calculated value of the density of aluminum, as stated in the question, is 2.85 g/cm³. Use the percent error formula, found on Reference Table T, to set up the calculation for percent error:

$$\% \, error = \frac{measured \, value - accepted \, value}{accepted \, value} \times 100$$

Plug in the values from the question:

$$\% \text{ error} = \frac{2.85\,\text{g/cm}^3 - 2.70\,\text{g/cm}^3}{2.70\,\text{g/cm}^3} \times 100$$

$$\% \text{ error} \cong 5.6\%$$

34. **4** Find the elements magnesium and calcium on the Periodic Table in your Reference Tables. Looking at the electron configurations for both elements, we see that each element has 2 electrons in the outermost shell. Electrons in the outermost shell are known as valence electrons. Valence electrons are involved in forming chemical bonds between atoms and are therefore largely responsible for an atom's chemical properties.

WRONG CHOICES EXPLAINED:
(1) All elements, no matter how different their properties are, have atoms with equal numbers of protons and electrons, so this cannot be responsible for the chemical similarity between magnesium and calcium.
(2) Calcium atoms contain 20 protons, whereas magnesium atoms contain 12 protons.
(3) All atoms with two or more electrons in the ground state have two electrons in the first shell. The first shell has the least energy of all electrons shells since it is closest to the nucleus, and electrons in the ground state will occupy the lowest energy positions possible.

35. **1** The atomic radius decreases as elements are considered from left to right across Period 2 of the Periodic Table. This trend is caused by increasing nuclear charge, which draws the electron cloud in closer, thereby shrinking the atomic radius. Since atomic radius data is provided in Reference Table S, you can always look up values and figure out the trend if you forget.

WRONG CHOICES EXPLAINED:
(2) Electronegativity values increase as elements are considered from left to right across Period 2 of the Periodic Table because increasing nuclear charge increases an element's attraction for electrons.
(3) Ionization energy values increase as elements are considered from left to right across Period 2 of the Periodic Table because increasing nuclear charge makes it more difficult to remove electrons from an atom.
(4) The number of protons increases as elements are considered from left to right across Period 2 of the Periodic Table, and nuclear charge is caused by the charge of the protons in the nucleus.

36. **3** All coefficients in chemical reactions represent particles or moles. Mole ratios can be used to find the number of moles of any substance in a chemical equation if the moles of another substance are known. Dimensional analysis or mole ratios can be used to solve this problem:

Using dimensional analysis:

$$\frac{5.0 \, \text{mol} \, C_4H_{10}}{1} \times \frac{8 \, \text{mol} \, CO_2}{2 \, \text{mol} \, C_4H_{10}} = 20. \, \text{mol} \, CO_2$$

Using mole ratios:

$$\frac{5.0 \, \text{mol} \, C_4H_{10}}{2} = \frac{x \, \text{mol} \, CO_2}{8}$$
$$x = 20. \, \text{mol} \, CO_2$$

37. **3** Use the percent composition formula found on Reference Table T:

$$\% \text{ composition by mass} = \frac{\text{mass of part}}{\text{mass of whole}} \times 100$$

$$\% \text{ composition by mass} = \frac{(14.0 \, \text{g/mol})(2)}{32.0 \, \text{g/mol}} \times 100$$

$$\% \text{ composition by mass} \cong 88\%$$

38. **4** The oxygen atom has 8 electrons, but the atom must gain two electrons to form the O^{2-} ion. The ion has 10 electrons, just like all neon atoms.

WRONG CHOICES EXPLAINED:
(1) Ca^{2+} ions have 18 electrons and are isoelectronic with argon.
(2) Cl^- ions have 18 electrons and are isoelectronic with argon.
(3) Li^+ ions have just two electrons and are isoelectronic with helium.

39. **1** Higher boiling points are caused by stronger intermolecular forces between molecules in a liquid sample. Boiling involves a transition from the liquid state to the gas state; as boiling occurs, molecules gain the potential energy that they need to overcome intermolecular attractions and separate from each other to form a gas. When intermolecular forces are strong, substances will heat to a higher temperature in the liquid state before boiling occurs since it is harder for particles to overcome the intermolecular attractions and form a gas.

40. **1** Xenon exists as a monatomic gas at STP.

WRONG CHOICES EXPLAINED:
(2) Xenon is not a diatomic gas.
(3) Xenon is not in the solid phase at STP.
(4) Xenon is not in the liquid phase at STP.

41. **2** Heat formulas are found on Reference Table T. Since the temperature of the water in the beaker increases from 20.°C to 35°C, the correct formula is:

$$q = mC\Delta T$$

where q = the amount of heat in joules (J) lost or gained, m = mass of the water in grams, C = specific heat capacity of water, and ΔT = change in temperature.

In this case, $\Delta T = T_{final} - T_{initial} = 35 - 20. = 15°C$. Plugging in the given values leads to the following expression:

$$q = (75 \text{ g})(4.18 \text{ J/g·K}) (15 \text{ C}) \cong 4700 \text{ J}$$

(NOTE: The specific heat capacity of water (C) is listed on Reference Table B.)

42. **4** Anyone wishing to increase the rate of a chemical reaction needs to increase the number of effective collisions between reactant particles. Increasing the temperature and increasing the concentration of the HCl solution will increase the total number of collisions between reactant particles, and the more total collisions there are, the more effective collisions there will be.

43. **4** The reaction is at equilibrium, so the rate of the forward reaction must equal the rate of the reverse reaction. Since both products and reactants are forming at equal rates, the concentration of each remains constant.

44. **4** Systems at equilibrium prefer to remain at equilibrium. Since this is an endothermic reaction (the heat term is on the reactant side of the equation), heat is absorbed in the forward direction. If this system is stressed by increasing the temperature, the system will "shift to the right" (temporarily favor the forward direction) so that some of the excess heat will be absorbed, which will lower the temperature and restore equilibrium.

WRONG CHOICES EXPLAINED:

(1) Adding a catalyst will cause the reaction rate to increase in *both* the forward and the reverse direction, but equilibrium will not be affected.

(2) Adding more $PCl_3(g)$ will cause a shift to the left, since the reaction will try to use up the excess $PCl_3(g)$ by reacting in the reverse direction.

(3) Increasing the pressure on this system will cause a shift to the left, since there are fewer moles of gas on the reactant side.

45. **3** This organic compound is classified as an amine since the "$-NH_2$" functional group is present (see Reference Table R). The suffix for amines is "–amine."

WRONG CHOICES EXPLAINED:

(1) Pentanone is a ketone.

(2) 1-pentanol is an alcohol.

(4) Pentanamide is an amide. Although amides also contain nitrogen, the compound in this question does not contain oxygen and therefore cannot be an amide.

46. **3** This organic compound is a hydrocarbon, since it contains only carbon and hydrogen. Reference Table Q shows that alkynes contain triple bonds between carbon atoms.

47. **3** Fermentation reactions involve the breakdown of glucose into ethanol and carbon dioxide in the presence of an enzyme catalyst.

WRONG CHOICES EXPLAINED:

(1) Although this reaction forms ethanol, glucose is not a reactant. This is a synthesis reaction.

(2) This is a type of organic reaction known as an addition reaction.

(4) This reaction forms an aldehyde, not ethanol.

(NOTE: You need to be able to recognize seven major types of organic reactions. These include addition, substitution, esterification, polymerization, fermentation, saponification, and combustion. Memorize these reactions!)

48. **2** As a reactant in the compound $CuCl_2$, copper has an oxidation number of $+2$. As a pure element on the product side, copper has an oxidation number of 0.

49. **2** The Brønsted-Lowry theory of acids and bases defines an acid as a proton (H^+) donor and a base as a proton (H^+) acceptor. $CH_3COOH(aq)$ must lose (donate) an H^+ in the forward reaction in order to form the product $CH_3COO^-(aq)$. Likewise, in the reverse reaction, $H_3O^+(aq)$ must lose an H^+ to form $H_2O(\ell)$.

50. **1** Natural transmutation, also known as radioactive decay, involves the spontaneous disintegration of an unstable nucleus. This type of nuclear reaction can always be recognized by the presence of only one substance on the reactant side and the presence of a decay particle (in this case, an alpha particle, ^4_2He) on the product side.

WRONG CHOICES EXPLAINED:
(2) Artificial transmutation reactions take place when a high-speed particle (in this case an alpha particle, ^4_2He) collides with a target nucleus.

(3) Fission reactions occur when a high-speed neutron collides with a U-235 nucleus, forcing it to split into two or more smaller nuclei. The formation of neutrons enables a chain reaction to occur.

(4) Nuclear fusion happens when two lightweight nuclei merge to form heavier nuclei.

(NOTE: You need to be able to recognize the four types of nuclear reactions. These include natural transmutation, artificial transmutation, fission, and fusion. Memorize these reactions!)

PART B–2

[All questions in Part B–2 are worth 1 point.]

51. The suffix "-al" in the compound methanal reveals that this compound is an aldehyde (see Reference Table *R*). The general formula for compounds that contain the aldehyde functional group is shown below:

$$
\begin{array}{c}
\text{O} \\
\|\\
\text{R}-\text{C}-\text{H}
\end{array}
$$

Specifically, however, the compound methanal contains only one carbon atom (refer to Reference Table *P*), so the structural formula of methanal is

$$
\begin{array}{c}
\text{O} \\
\|\\
\text{H}-\text{C}-\text{H}
\end{array}
$$

52. All atoms are electrically neutral because the **number of positively charged protons equals the number of negatively charged electrons**.

(NOTE: This question requires the response to be "in terms of subatomic particles." Acceptable responses must either use the term "subatomic particles" directly or must use the specific names of relevant subatomic particles [protons and electrons].)

53. The formula mass, also known as the molar mass, is calculated by adding the atomic masses of the individual atoms in a compound. The information that precedes the question states that the formula of heavy water is D_2O. The table includes atomic masses for three isotopes of the element hydrogen. Heavy water is composed of two atoms of deuterium (H-2) and one atom of oxygen. Oxygen's atomic mass is found on the Periodic Table in the Reference Tables.

Atomic mass of D_2O = 2(atomic mass of H-2) + 1 (atomic mass of oxygen)

Atomic mass of D_2O = 2(2.0141 u) + 16.0 u

Atomic mass of D_2O = **20.03 u**

(NOTE: The abbreviation "u" stands for "atomic mass unit," or "amu." Acceptable answers included all values from **19.999 u to 20.03 u**, inclusive. Significant figures were not considered when grading this question.)

54. Reference Table N contains information about the decay mode for a number of radioactive isotopes. Consult Reference Table N to find the decay mode of tritium, H-3. Tritium undergoes **beta decay**.
    (NOTE: Other acceptable answers include **beta particle, beta, β^-, $_{1}^{0}e$**, or $_{-1}^{0}\beta$.)

55. The formula to convert temperatures from degrees Celsius to the Kelvin scale is found on the last page of your Reference Tables, Reference Table T:

$$K = {}^{\circ}C + 273$$

Plug in the given Celsius temperature to solve for the temperature on the Kelvin scale:

$$K = 23{}^{\circ}C + 273 = \textbf{296 K}$$

56. Reference Table G shows the solubilities of different solutes in water over a range of temperatures. The lines on the graph represent lines of saturation, meaning that the solution is holding the maximum amount of solute possible.

At 23°C, Reference Table G shows that approximately 90.0 g of $NaNO_3$ will dissolve in 100.0 g of water. The amount of $NaNO_3$ in the initial solution must be subtracted from this calculation to find the additional amount of $NaNO_3$ that would need to be added to reach saturation:

$$90.0 \text{ g} - 85.0 \text{ g} = \textbf{5.0 g of NaNO}_3 \textbf{ must be added}$$

(NOTE: Credit was awarded for any value from 4.0 g to 6.0 g, inclusive.)

57. Boiling point elevation and freezing point depression are two colligative properties of solutions. Dissolved particles raise the boiling point and lower the freezing point of a solution. The more concentrated the solution, the greater the effect on the boiling point and the freezing point. This question is tricky because, instead of adding more solute particles to the solution, *water* is added, leading to a *less* concentrated solution. Having a solution that is less concentrated would result in **a lower boiling point and a higher freezing point**.

58. Real gases behave more like ideal gases when temperature is high and pressure is low. These conditions minimize any possible attractions between gas particles, allowing the gas to adhere more closely to the principles of the Kinetic Molecular Theory of Gases. The original conditions for the temperature and pressure of the gas sample, as described in the information that precedes the question, are 296 K and 101.3 kPa, respectively. This sample will behave more like an ideal gas if **the temperature was higher and the pressure was lower**.
(NOTE: Acceptable responses also included "the temperature increased and the pressure decreased" or "any temperature above 296 K and any pressure lower than 101.3 kPa.")

59. Use the Combined Gas Law Equation on Reference Table T to solve this problem:

$$\frac{P_1 V_1}{T_1} = \frac{P_2 V_2}{T_2}$$

Substitute information from the question:

$$\frac{(101.3 \text{ kPa})(200. \text{ mL})}{296 \text{ K}} = \frac{(152.0 \text{ kPa})(V_2)}{336 \text{ K}}$$

Rearrange the equation:

$$V_2 = \frac{(101.3\,\text{kPa})(200.\,\text{mL})(336\,\text{K})}{(296\,\text{K})(152.0\,\text{kPa})}$$

Solve this equation to find that **V_2 = 151.3 mL**.

(NOTE: Credit was awarded for any value from 151 mL to 151.4 mL, inclusive.)

60. Temperature is directly related to the kinetic energy of particles in a sample of matter. Kinetic energy is the energy of particle motion. Putting these two ideas together leads to the understanding that particles in a sample of matter move faster at higher temperatures and slower at lower temperatures. Picture gas particles in an enclosed space (see the image below). Imagine the gas particles moving slower at a lower temperature. **They would collide with each other less often, and they would also collide with less force.** A good analogy would be bumper cars in a bumper car arena. At lower speeds, the cars collide less often and with considerably less force than they do at higher speeds.

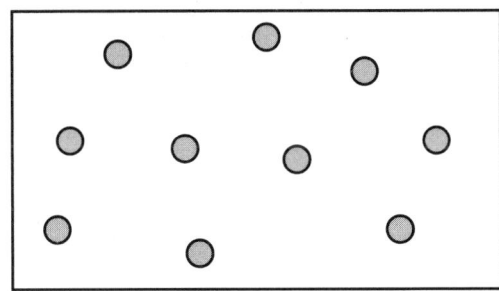

61. The introductory information for this question states that the sample of $CO_2(g)$ is in a *sealed*, rigid container. If the container is sealed, then gas particles cannot enter or leave the container. If the number of gas particles stays constant, then **the mass of gas in the container also stays constant!**

62. Isotopes are different versions of the same element that have different numbers of neutrons. Co-59 and Co-60 are isotopes for two reasons: (1) they have the **same number of protons** (27, the atomic number of cobalt) and (2) they have **different numbers of neutrons**. Co-60 has a mass number of 60. To find the number of neutrons, subtract the number of protons from the mass number: $60 - 27 = 33$ neutrons. Using the same calculation, the number of neutrons in Co-59 is $59 - 27 = 32$ neutrons.

63. **Gamma radiation has greater penetrating power than beta emission.** Gamma radiation is pure energy and has neither mass nor charge. Just as X-rays can penetrate the body to expose the location of a broken bone, gamma radiation can pass right through matter. Beta radiation, on the other hand, has a very small mass and a negative charge. Having a charge causes the beta particles to slow down as they are repelled by electrons and are attracted to nuclei in the matter through which they are traveling.

64. Like regular chemical reactions, nuclear reactions must be balanced for both mass and charge. To balance a nuclear equation, the sum of the atomic numbers (bottom) and the mass numbers (top) must be equal on both sides of the arrow. Consider the reaction below:

$$^{60}_{27}\text{Co} \rightarrow \underline{\hspace{1cm}} + ^{0}_{-1}\text{e} + ^{0}_{0}\gamma$$

Since the beta particle and the gamma radiation both have a mass number of 0, the missing product must have a mass number of 60. To solve for the atomic number of the missing product, ask yourself, "What number, when added to –1, will equal 27?" The atomic number of the missing product must be 28. Look on the Periodic Table to find that the missing product is an atom of **nickel-60**.

(NOTE: $^{60}_{28}$ **Ni**, **Ni-60**, and ^{60}**Ni** are all acceptable responses as well.)

65. One half-life is the time required for a radioisotope to decay to half of its original mass. To find the total number of half-lives required for Co-60 to decay from an original mass of 80.00 grams to a mass of 10.00 grams, divide the mass in half over a series of steps.

In the first half-life, Co-60 decays from 80.00 grams to 40.00 grams. (The other 40.00 grams undergo radioactive decay.)

In the second half-life, the 40.00 grams of Co-60 that remain decay to 20.00 grams.

In the third half-life, the 20.00 grams of Co-60 that remain decay to 10.00 grams.

Three half-lives are needed for the radioactive decay of Co-60 from 80.00 grams to 10.00 grams. Reference Table N shows that Co-60 has a half-life of 5.271 years. The total time required for three half-lives is 3×5.271 years = **15.813 years**.

PART C

[All questions in Part C are worth 1 point.]

66. $KClO_3$ is an ionic compound that is formed when the K^+ ion bonds to the ClO_3^- ion. To name this compound, look up the name of ClO_3^- in Reference Table E: chlorate. Put this together with the name of the cation to find the compound name: **potassium chlorate**.

67. The data table shows that 0.80 g of $O_2(g)$ are produced. To convert this mass into moles of $O_2(g)$, use the following formula found on Reference Table T:

$$\text{number of moles} = \frac{\text{given mass}}{\text{gram-formula mass}}$$

Plug in the numbers:

$$\textbf{number of moles} = \frac{\textbf{0.80 g O}_2}{\textbf{32 g/mol O}_2}$$

(NOTE: This question asks ONLY for the numerical setup to determine the number of moles of $O_2(g)$ produced. The numerical setup is all that you should record on your answer sheet. If you would like to check your work by solving for the number of moles, do this on scrap paper.)

68. While stoichiometry can be used to solve this problem, there is a much easier and faster method, using the Law of Conservation of Mass. The unbalanced equation is:

$$KClO_3(s) \rightarrow KCl(s) + O_2(g)$$

We know the starting mass of $KClO_3(s)$ (2.07 g), and we know the mass of $O_2(g)$ produced (0.80 g). Since matter is neither created nor destroyed in a chemical reaction, the total mass of the products must equal the mass of the original reactant. Subtract the mass of $O_2(g)$ produced from the mass of $KClO_3(s)$ used to find the mass of $KCl(s)$ produced:

$$2.07 \text{ g} - 0.80 \text{ g} = \textbf{1.27 g of KCl(s)}$$

69. To balance a chemical equation, the number of each type of atom must be equal on both sides of the equation. The unbalanced equation is shown below:

$$\_\_\_\_ KClO_3(s) \rightarrow \_\_\_\_ KCl(s) + \_\_\_\_ O_2(g)$$

As written, there is 1 potassium atom and 1 chlorine atom on each side of the equation, so these two elements are already balanced. However, there are 3 oxygen atoms on the left of the reaction arrow and only 2 oxygen atoms on the right side of the equation, so the equation as it stands is not balanced for oxygen. This odd/even situation often occurs when balancing equations that contain diatomic elements, like O_2. The best way to handle this situation is to *double all of the coefficients* (except the coefficient for the O_2, which isn't known yet). Doubling an odd number results in an even number. The finished balanced equation becomes:

$$\underline{\ 2\ }\ KClO_3(s) \rightarrow \underline{\ 2\ }\ KCl(s) + \underline{\ 3\ }\ O_2(g)$$

70. If you have any doubt about which of the ions listed are the hydrogen carbonate ions, consult Reference Table E. The table for this question shows that 0.180 g of HCO_3^- ions are present in a 500. g sample of bottled water. The number 0.180 contains **three significant figures**. The zero in the thousandths place does not change the absolute value of the number, but it does show the precision of the measurement.

71. The formula for the sulfate ion, SO_4^{2-}, can be found in Reference Table E. Reference Table F contains information about the solubility of different ions in aqueous solutions. SO_4^{2-} is listed as soluble under most conditions *except* when combined with the five cations listed in the second column of that table. If any of these five ions is present, an insoluble compound will form. Of the ions listed in the table for this question, only **Ca^{2+}** will form an insoluble compound in the presence of SO_4^{2-} in the bottled water.

72. Look at Reference Table T to find the formula for parts per million:

$$\text{parts per million} = \frac{\text{mass of solute}}{\text{mass of solution}} \times 1,000,000$$

Next, plug in the information for this question:

$$\textbf{parts per million} = \frac{\textbf{0.0033 g}}{\textbf{500. g}} \times \textbf{1,000,000}$$

(NOTE: This question asks ONLY for the numerical setup for the calculation of the concentration of Na^+ ions in parts per million. The numerical setup is all that you should record on your answer sheet. If you would like to check your work by solving for the number of ppm, do this on scrap paper.)

73. An Mg^{2+} ion must gain two electrons to become a neutral Mg atom. The Mg^{2+} ion has an electron configuration of 2-8. The Mg atom has an electron configuration of 2-8-2. **The Mg atom has a larger radius than that of the Mg^{2+} ion** since the Mg atom has electrons in the third electron shell, whereas the Mg^{2+} ion has no electrons in the third electron shell.

74. Organic chemistry, by definition, is the chemistry of **carbon**-based compounds. Carbon atoms form four covalent bonds and can easily bond with other carbon atoms to form chains, rings, and networks.

75. An empirical formula shows the elements of a compound in their lowest whole number ratio. Use the structural diagram given for this question, and count the atoms to find the molecular formula, $C_4H_8O_2$. Next, reduce this molecular formula to lowest terms by dividing each subscript by the common factor of 2. Thus, the empirical formula is **C_2H_4O**.
(NOTE: The order of elements in the answer may vary.)

76. The functional group for the organic compound in this question is marked in the diagram below:

$$
\begin{array}{ccccc}
H & O & & H & H \\
| & || & & | & | \\
H-C & -C-O & -C & -C-H \\
| & & & | & | \\
H & & & H & H \\
\end{array}
$$

Look at the second column in Reference Table *R*, which shows functional groups for several classes of organic compounds. This compound is an **ester**.

77. All of the covalent bonds between hydrogen atoms and carbon atoms in the molecule are single covalent bonds, represented by C–H. There are **two electrons shared** in every single covalent bond.

78. The salt bridge **allows ions to flow into each compartment in the voltaic cell in order to maintain electrical neutrality** in the system. A buildup of positive or negative charges in either compartment would cause electrons to stop flowing.

79. Magnesium is higher than silver on Reference Table J, so magnesium will *oxidize* in the presence of silver. Oxidation is defined as the "loss of electrons." Magnesium will lose two electrons to form Mg^{2+} ions as shown in the following half-cell equation:

$$Mg(s) \rightarrow Mg^{2+}(aq) + 2e^-$$

80. **Voltaic cells produce electrical energy, while electrolytic cells use electrical energy.** Voltaic cells are commonly known as batteries. In batteries, a spontaneous redox reaction generates electricity, whereas in electrolytic cells, electricity from an external power source is used to drive a nonspontaneous redox reaction.

81. pH measures the concentration of hydrogen ion, H^+, in aqueous solution. pH is a logarithmic function, meaning that it is based on the exponent of the base 10 molarity of the hydrogen ion content. Each pH value is 10 times different in hydrogen ion molarity than the pH immediately higher and lower than itself. Hydrogen ions make a solution acidic, and solutions that are more acidic have lower pH values. The problem states that the original HCl solution had a pH of 2. The pH of a solution that is ten times *less* acidic will be one number *higher* than the original pH—in other words, a pH of **3**.

82. KOH(aq) is a base that is listed on Reference Table L. All basic solutions have pH values greater than 7. The KOH solution will be **blue** in the presence of the bromcresol green indicator, since Reference Table M shows that bromcresol green is blue in all solutions with a pH greater than 5.4.

83. The reaction between HCl(aq) and KOH(aq) is a neutralization reaction. Neutralization reactions are really double replacement reactions between acids and bases. The word equation for neutralization reactions between Arrhenius acids and bases is:

$$ACID + BASE \rightarrow WATER + SALT$$

Water is always a product of classic neutralization reactions, and the salt that forms completes the double replacement reaction (ions switching places):

$$HCl(aq) + KOH(aq) \rightarrow \mathbf{H_2O(\ell) + KCl(aq)}$$

84. Titration is the lab technique that is used to determine the concentration of an acid or a base by neutralizing it with an acid or a base of known concentration. The titration formula is found on Reference Table T:

$$M_A V_A = M_B V_B$$

in which M_A = molarity of H^+, V_A = volume of acid,

M_B = molarity of OH^-, and V_B = volume of base

Substitute the values from this question:

$$(0.010\,\text{M})(15.0\,\text{mL}) = (M_B)(7.5\,\text{mL})$$

$$M_B = \frac{(0.010\,\text{M})(15.0\,\text{mL})}{7.5\,\text{mL}}$$

(NOTE: Once again, this question asks ONLY for the numerical setup for calculating the molarity of the KOH solution. The numerical setup is all that you should record on your answer sheet. If you would like to check your work by solving for the molarity of the KOH solution, do this on scrap paper.)

85. HCl is a strong electrolyte because it dissociates completely into freely moving ions in an aqueous solution. Freely moving dissociated ions allow charge to flow through the solution. **The 1.0 M HCl solution has a greater concentration of mobile ions**, making it possible for more electrical current to flow through the solution.

Mark (✓) the questions you answered correctly. Count the number of checks and follow the formulas given to determine your score on each topic.

Core Area   ☐ Questions Answered Correctly

33, 70: (2)

Section M—Math Skills
☐ Number of checks ÷ 2 × 100 = ___ %

61: (1)

Section R—Reading Skills
☐ Number of checks ÷ 1 × 100 = ___ %

1–5, 31, 32, 52, 62: (9)

Section I—Atomic Concepts
☐ Number of checks ÷ 9 × 100 = ___ %

6–8, 34, 35: (5)

Section II—Periodic Table
☐ Number of checks ÷ 5 × 100 = ___ %

9–11, 36, 37, 53, 66–69, 75: (11)

Section III—Moles/Stoichiometry
☐ Number of checks ÷ 11 × 100 = ___ %

12, 38, 39, 71, 73, 77: (6)

Section IV—Chemical Bonding
☐ Number of checks ÷ 6 × 100 = ___ %

13–19, 40, 41, 55–60, 72: (16)

Section V—Physical Behavior of Matter
☐ Number of checks ÷ 16 × 100 = ___ %

20–23, 42–44: (7)

Section VI—Kinetics and Equilibrium
☐ Number of checks ÷ 7 × 100 = ___ %

Core Area	☐ Questions Answered Correctly

24, 25, 45–47, 51, 76: (7)

Section VII—Organic Chemistry
☐ Number of checks ÷ 7 × 100 = ___ %

26, 48, 74, 78–80: (6)

Section VIII—Oxidation-Reduction
☐ Number of checks ÷ 6 × 100 = ___ %

27, 49, 81–85: (7)

Section IX—Acids, Bases, and Salts
☐ Number of checks ÷ 7 × 100 = ___ %

28–30, 50, 54, 63–65: (8)

Section X—Nuclear Chemistry
☐ Number of checks ÷ 8 × 100 = ___ %

Examination August 2019
Chemistry—Physical Setting

PART A

Answer all questions in this part.

Directions (1–30): For *each* statement or question, write in the answer space the *number* of the word or expression that, of those given, best completes the statement or answers the question. Some questions may require the use of the *2011 Edition Reference Tables for Physical Setting/Chemistry.*

1 Which statement describes the earliest model of the atom?

   (1) An atom is an indivisible hard sphere.
   (2) An atom has a small, dense nucleus.
   (3) Electrons are negative particles in an atom.
   (4) Electrons in an atom have wave-like properties.   1 _____

2 In all atoms of bismuth, the number of electrons must equal the

   (1) number of protons
   (2) number of neutrons
   (3) sum of the number of neutrons and protons
   (4) difference between the number of neutrons and protons   2 _____

3 Which symbol represents a particle that has a mass approximately equal to the mass of a neutron?

   (1) α (3) β^-
   (2) β^+ (4) p   3 _____

4 An orbital is a region in an atom where there is a high probability
of finding

(1) an alpha particle   (3) a neutron
(2) an electron   (4) a positron   4 _____

5 Which electron shell in an atom of calcium in the ground state has
an electron with the greatest amount of energy?

(1) 1   (3) 3
(2) 2   (4) 4   5 _____

6 As the elements in Period 2 are considered in order from lithium
to fluorine, there is an increase in the

(1) atomic radius
(2) electronegativity
(3) number of electron shells
(4) number of electrons in the first shell   6 _____

7 Which element is classified as a metalloid?

(1) boron   (3) sulfur
(2) potassium   (4) xenon   7 _____

8 Strontium and barium have similar chemical properties because
atoms of these elements have the same number of

(1) protons   (3) electron shells
(2) neutrons   (4) valence electrons   8 _____

9 Which term represents the fixed proportion of elements in a
compound?

(1) atomic mass   (3) chemical formula
(2) molar mass   (4) density formula   9 _____

10 Which two terms represent types of chemical formulas?

(1) mechanical and structural
(2) mechanical and thermal
(3) molecular and structural
(4) molecular and thermal   10 _____

11 Which element has metallic bonds at room temperature?

(1) bromine   (3) krypton
(2) cesium   (4) sulfur   11 _____

12 What is the number of electrons shared between the atoms in a molecule of nitrogen, N_2?

(1) 8   (3) 3
(2) 2   (4) 6   12 _____

13 Given the equation representing a reaction:

$$H + H \rightarrow H_2$$

What occurs during this reaction?

(1) A bond is broken and energy is absorbed.
(2) A bond is broken and energy is released.
(3) A bond is formed and energy is absorbed.
(4) A bond is formed and energy is released.   13 _____

14 An atom of which element has the strongest attraction for electrons in a chemical bond?

(1) chlorine   (3) phosphorus
(2) carbon   (4) sulfur   14 _____

15 At STP, a 50.-gram sample of $H_2O(\ell)$ and a 100.-gram sample of $H_2O(\ell)$ have

(1) the same chemical properties
(2) the same volume
(3) different temperatures
(4) different empirical formulas   15 _____

16 Which statement describes a mixture of sand and water at room temperature?

(1) It is heterogeneous, and its components are in the same phase.
(2) It is heterogeneous, and its components are in different phases.
(3) It is homogeneous, and its components are in the same phase.
(4) It is homogeneous, and its components are in different phases. 16 _____

17 Distillation is a process used to separate a mixture of liquids based on different

(1) boiling points   (3) freezing points
(2) densities   (4) solubilities   17 _____

18 According to the kinetic molecular theory, which statement describes the particles in a sample of an ideal gas?

(1) The particles are constantly moving in circular paths.
(2) The particles collide, decreasing the total energy of the system.
(3) The particles have attractive forces between them.
(4) The particles are considered to have negligible volume.   18 _____

19 Which sample of matter has the greatest distance between molecules at STP?

(1) $N_2(g)$ (3) $C_6H_{14}(\ell)$
(2) $NH_3(aq)$ (4) $C_6H_{12}O_6(s)$ 19 _____

20 For a chemical system at equilibrium, the concentrations of both the reactants and the products must

(1) decrease   (3) be constant
(2) increase   (4) be equal   20 _____

21 In terms of disorder and energy, systems in nature have a tendency to undergo changes toward

(1) less disorder and lower energy
(2) less disorder and higher energy
(3) greater disorder and lower energy
(4) greater disorder and higher energy   21 _____

22 The only two elements in alkenes and alkynes are

(1) carbon and nitrogen
(2) carbon and hydrogen
(3) oxygen and nitrogen
(4) oxygen and hydrogen   22 _____

23 Which functional group contains a nitrogen atom and an oxygen atom?

(1) ester   (3) amide

(2) ether   (4) amine   23 _____

24 When a sample of Mg(s) reacts completely with O_2(g), the Mg(s) loses 5.0 moles of electrons. How many moles of electrons are gained by the O_2(g)?

(1) 1.0 mol   (3) 5.0 mol

(2) 2.5 mol   (4) 10.0 mol   24 _____

25 Which statement describes the reactions in an electrochemical cell?

(1) Oxidation occurs at the anode, and reduction occurs at the cathode.

(2) Oxidation occurs at the cathode, and reduction occurs at the anode.

(3) Oxidation and reduction both occur at the cathode.

(4) Oxidation and reduction both occur at the anode.   25 _____

26 A 0.050 M aqueous solution of which compound is the best conductor of electric current?

(1) C_3H_7OH (3) $MgSO_4$

(2) $C_6H_{12}O_6$ (4) K_2SO_4 26 _____

27 What is the color of bromcresol green indicator in a solution with a pH value of 2.0?

(1) blue   (3) red

(2) green   (4) yellow   27 _____

28 Which formula can represent hydrogen ions in an aqueous solution?

(1) OH^-(aq)   (3) H_3O^+(aq)

(2) Hg_2^{2+}(aq)   (4) NH_4^+(aq)   28 _____

29 In which reaction is an atom of one element converted into an atom of another element?

(1) combustion
(2) fermentation
(3) oxidation-reduction
(4) transmutation   29 _____

30 In which type of nuclear reaction do nuclei combine to form a nucleus with a greater mass?

(1) alpha decay   (3) fusion
(2) beta decay   (4) fission   30 _____

PART B–1

Answer all questions in this part.

Directions (31–50): For *each* statement or question, write in the answer space the *number* of the word or expression that, of those given, best completes the statement or answers the question. Some questions may require the use of the *2011 Edition Reference Tables for Physical Setting/Chemistry.*

31 The bright-line spectra produced by four elements are represented in the diagram below.

Bright-Line Spectra of Four Elements

Given the bright-line spectrum of a mixture formed from two of these elements:

Which elements are present in this mixture?

(1) *A* and *X*   (3) *D* and *X*

(2) *A* and *Z*   (4) *D* and *Z*   31 _____

32 Which electron configuration represents the electrons in an atom of sulfur in an excited state?

(1) 2 – 8 – 6   (3) 2 – 8 – 7

(2) 2 – 7 – 7   (4) 2 – 7 – 8   32 _____

33 Which notations represent atoms that have the same number of protons but a different number of neutrons?

(1) H-3 and He-3   (3) Cl-35 and Cl-37

(2) S-32 and S-32   (4) Ga-70 and Ge-73   33 _____

34 What is the chemical name of the compound NH_4SCN?

(1) ammonium thiocyanate

(2) ammonium cyanide

(3) nitrogen hydrogen cyanide

(4) nitrogen hydrogen sulfate   34 _____

35 Which equation represents a conservation of atoms?

(1) $2Fe + 2O_2 \rightarrow Fe_2O_3$

(2) $2Fe + 3O_2 \rightarrow Fe_2O_3$

(3) $4Fe + 2O_2 \rightarrow 2Fe_2O_3$

(4) $4Fe + 3O_2 \rightarrow 2Fe_2O_3$ 35 _____

36 Which compound has covalent bonds?

(1) H_2O (3) Na_2O

(2) Li_2O (4) K_2O 36 _____

37 Which particle diagram represents a sample of oxygen gas at STP?

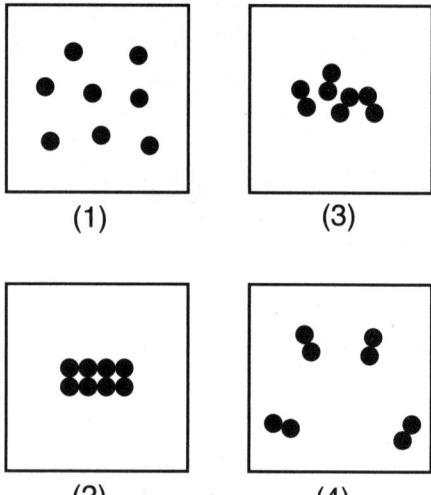

       37 _____

38 At which temperature and pressure will a sample of neon gas behave most like an ideal gas?

(1) 300. K and 2.0 atm　　　　(3) 500. K and 2.0 atm
(2) 300. K and 4.0 atm　　　　(4) 500. K and 4.0 atm

       38 _____

39 What is the molarity of 2.0 liters of an aqueous solution that contains 0.50 mole of potassium iodide, KI?

(1) 1.0 M　　　　　　　　　(3) 0.25 M
(2) 2.0 M　　　　　　　　　(4) 0.50 M

       39 _____

40 The volumes of four samples of gaseous compounds at 298 K and 101.3 kPa are shown in the table below.

Sample	Compounds	Volume (L)
1	$NH_3(g)$	44.0
2	$CO_2(g)$	33.0
3	$HF(g)$	44.0
4	$CH_4(g)$	22.0

Which two samples contain the same number of molecules?

(1) 1 and 2   (3) 2 and 3
(2) 1 and 3   (4) 2 and 4   40 _____

41 Hydrochloric acid reacts faster with powdered zinc than with an equal mass of zinc strips because the greater surface area of the powdered zinc

(1) decreases the frequency of particle collisions
(2) decreases the activation energy of the reaction
(3) increases the frequency of particle collisions
(4) increases the activation energy of the reaction   41 _____

42 Given the equation representing a system at equilibrium in a sealed, rigid container:

$$2HI(g) \rightleftharpoons H_2(g) + I_2(g) + energy$$

Increasing the temperature of the system causes the concentration of

(1) HI to increase
(2) H_2 to increase
(3) HI to remain constant
(4) H_2 to remain constant   42 _____

43 Based on Table *I*, which equation represents a reaction with the greatest difference between the potential energy of the products and the potential energy of the reactants?

(1) $4Al(s) + 3O_2(g) \rightarrow 2Al_2O_3(s)$
(2) $2H_2(g) + O_2(g) \rightarrow 2H_2O(\ell)$
(3) $C_3H_8(g) + 5O_2(g) \rightarrow 3CO_2(g) + 4H_2O(\ell)$
(4) $C_6H_{12}O_6(s) + 6O_2(g) \rightarrow 6CO_2(g) + 6H_2O(\ell)$ 43 _____

44 Which phase change results in an increase in entropy?

(1) $I_2(g) \rightarrow I_2(s)$ (3) $Br_2(\ell) \rightarrow Br_2(g)$
(2) $CH_4(g) \rightarrow CH_4(\ell)$ (4) $H_2O(\ell) \rightarrow H_2O(s)$ 44 _____

45 Given the formula for a compound:

$$
\begin{array}{ccccccc}
H & H & H & O & & H & \\
| & | & | & \| & & | & \\
H-C-&C-&C-&C-&O-&C-&H \\
| & | & | & & & | & \\
H & H & H & & & H &
\end{array}
$$

What is the name of this compound?

(1) methyl butanoate   (3) pentanone
(2) methyl butyl ether   (4) pentanoic acid   45 _____

46 Given the equation representing a reaction:

$$2Ca(s) + O_2(g) \rightarrow 2CaO(s)$$

During this reaction, each element changes in
(1) atomic number
(2) oxidation number
(3) number of protons per atom
(4) number of neutrons per atom   46 _____

47 Which equation represents a spontaneous reaction?

(1) $Ca + Ba^{2+} \rightarrow Ca^{2+} + Ba$
(2) $Co + Zn^{2+} \rightarrow Co^{2+} + Zn$
(3) $Fe + Mg^{2+} \rightarrow Fe^{2+} + Mg$
(4) $Mn + Ni^{2+} \rightarrow Mn^{2+} + Ni$ 47 _____

48 Which equation represents a neutralization reaction?

(1) $6HClO \rightarrow 4HCl + 2HClO_3$
(2) $CH_4 + 2O_2 \rightarrow CO_2 + 2H_2O$
(3) $Ca(OH)_2 + H_2SO_4 \rightarrow CaSO_4 + 2H_2O$
(4) $Ba(OH)_2 + Cu(NO_3)_2 \rightarrow Ba(NO_3)_2 + Cu(OH)_2$ 48 _____

49 Which radioisotope requires long-term storage as the method of disposal, to protect living things from radiation exposure over time?

(1) Pu-239 (3) Fe-53
(2) Fr-220 (4) P-32 49 _____

50 Given the equation representing a reaction:

$$^{235}_{92}U + ^{1}_{0}n \longrightarrow ^{140}_{56}Ba + ^{93}_{36}Kr + 3^{1}_{0}n + \text{energy}$$

$\underbrace{\qquad}$ total mass equals 236.053 u $\qquad$ $\underbrace{\qquad}$ total mass equals 235.868 u

Which statement explains the energy term in this reaction?

(1) Mass is gained due to the conversion of mass to energy.
(2) Mass is gained due to the conversion of energy to mass.
(3) Mass is lost due to the conversion of mass to energy.
(4) Mass is lost due to the conversion of energy to mass. 50 _____

PART B–2

Answer all questions in this part.

Directions (51–65): Record your answers on the answer sheet provided. Some questions may require the use of the *2011 Edition Reference Tables for Physical Setting/Chemistry*.

Base your answers to questions 51 through 53 on the information below and on your knowledge of chemistry.

The only naturally occurring isotopes of nitrogen are N-14 and N-15.

51 State the number of protons in an atom of N-15.   [1]

52 State the number of electrons in each shell of a N-14 atom in the ground state.   [1]

53 Based on the atomic mass of the element nitrogen on the Periodic Table, compare the relative abundances of the naturally occurring isotopes of nitrogen.   [1]

Base your answers to questions 54 through 56 on the information below and on your knowledge of chemistry.

The melting points and boiling points of five substances at standard pressure are listed on the table below.

Melting Points and Boiling Points of Five Substances

Substance	Melting Point (K)	Boiling Point (K)
HCl	159	188
NO	109	121
F_2	53	85
Br_2	266	332
I_2	387	457

54 Identify the substance in this table that is a liquid at STP. [1]

55 State, in terms of the strength of intermolecular forces, why I_2 has a higher boiling point than F_2. [1]

56 State what happens to the potential energy of a sample of $NO(\ell)$ at 121 K as it changes to $NO(g)$ at constant temperature and standard pressure. [1]

Base your answers to questions 57 through 59 on the information below and on your knowledge of chemistry.

A 100.-gram sample of liquid water is heated from 20.0°C to 50.0°C. Enough $KClO_3(s)$ is dissolved in the sample of water at 50.0°C to form a saturated solution.

57 Using the information on Table B, determine the amount of heat absorbed by the water when the water is heated from 20.0°C to 50.0°C. [1]

58 Based on Table H, determine the vapor pressure of the water sample at its final temperature. [1]

59 Based on Table G, determine the mass of $KClO_3(s)$ that must dissolve to make a saturated solution in 100. g of H_2O at 50.0°C. [1]

Base your answers to questions 60 through 62 on the information below and on your knowledge of chemistry.

The diagram and ionic equation below represent an operating voltaic cell.

Voltaic Cell

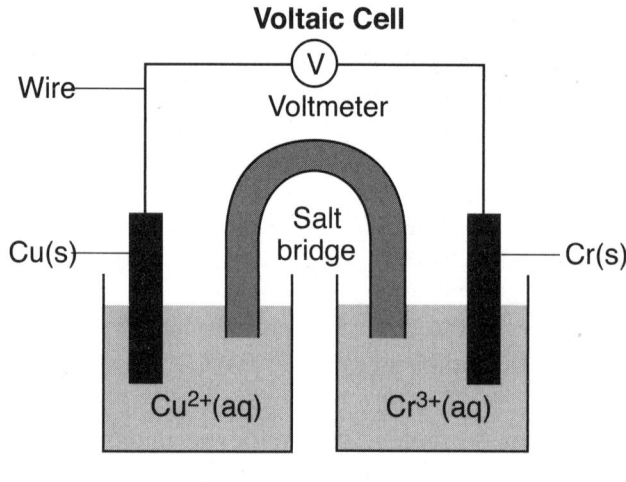

$$3Cu^{2+}(aq) + 2Cr(s) \longrightarrow 3Cu(s) + 2Cr^{3+}(aq)$$

60 Identify the subatomic particles that flow through the wires as the cell operates. [1]

61 State the purpose of the salt bridge in completing the circuit in this cell. [1]

62 Write a balanced equation for the half-reaction that occurs in the copper half-cell when the cell operates. [1]

Base your answers to questions 63 through 65 on the information below and on your knowledge of chemistry.

A NaOH(aq) solution with a pH value of 13 is used to determine the molarity of a HCl(aq) solution. A 10.0-mL sample of the HCl(aq) is exactly neutralized by 16.0 mL of 0.100 M NaOH(aq). During this laboratory activity, appropriate safety equipment was used and safety procedures were followed.

63 Determine the molarity of the HCl(aq) sample, using the titration data.   [1]

64 Compare the hydronium ion concentration to the hydroxide ion concentration when the HCl(aq) solution is exactly neutralized by the NaOH(aq) solution.   [1]

65 Determine the pH value of a solution that has a H^+(aq) ion concentration 10 times greater than the original NaOH(aq) solution.   [1]

PART C

Answer all questions in this part.

Directions (66–85): Record your answers on the answer sheet provided. Some questions may require the use of the *2011 Edition Reference Tables for Physical Setting/Chemistry*.

Base your answers to questions 66 through 68 on the information below and on your knowledge of chemistry.

A hydrate is a compound that has water molecules within its crystal structure. Magnesium sulfate heptahydrate, $MgSO_4 \cdot 7H_2O$, is a hydrated form of magnesium sulfate. The hydrated compound has 7 moles of H_2O for each mole of $MgSO_4$. When 5.06 grams of $MgSO_4 \cdot 7H_2O$ are heated to at least 300.°C in a crucible by using a laboratory burner, the water molecules are released. The sample was heated repeatedly, until the remaining $MgSO_4$ had a constant mass of 2.47 grams. During this laboratory activity, appropriate safety equipment was used and safety procedures were followed.

66 Explain why the sample in the crucible was heated repeatedly until the sample had a constant mass. [1]

67 Using the lab data, show a numerical setup for calculating the percent composition by mass of water in the hydrated compound. [1]

68 Determine the gram-formula mass of the magnesium sulfate heptahydrate. [1]

Base your answers to questions 69 through 71 on the information below and on your knowledge of chemistry.

Solid sodium chloride, also known as table salt, can be obtained by the solar evaporation of seawater and from underground mining. Liquid sodium chloride can be decomposed by electrolysis to produce liquid sodium and chlorine gas, as represented by the equation below.

$$2NaCl(\ell) \rightarrow 2Na(\ell) + Cl_2(g)$$

69 State, in terms of electrons, why the radius of a Na^+ ion in the table salt is smaller than the radius of a Na atom. [1]

70 Identify the noble gas that has atoms with the same number of electrons as a chloride ion in table salt.   [1]

71 In the space *on your answer sheet*, draw a Lewis electron-dot diagram of a Cl_2 molecule.   [1]

Base your answers to questions 72 through 75 on the information below and on your knowledge of chemistry.

The enclosed cabin of a submarine has a volume of 2.4×10^5 liters, a temperature of 312 K, and a pressure of 116 kPa. As people in the cabin breathe, carbon dioxide gas, $CO_2(g)$, can build up to unsafe levels. Air in the cabin becomes unsafe to breathe when the mass of $CO_2(g)$ in this cabin exceeds 2156 grams.

72 State what happens to the average kinetic energy of the gas molecules if the cabin temperature *decreases*.   [1]

73 Show a numerical setup for calculating the pressure in the submarine cabin if the cabin temperature changes to 293 K.   [1]

74 Determine the number of moles of $CO_2(g)$ in the submarine cabin at which the air becomes unsafe to breathe. The gram-formula mass of CO_2 is 44.0 g/mol.   [1]

75 Convert the original air pressure in the cabin of the submarine to atmospheres.   [1]

Base your answers to questions 76 through 78 on the information below and on your knowledge of chemistry.

Automobile catalytic converters use a platinum catalyst to reduce air pollution by changing emissions such as carbon monoxide, $CO(g)$, into carbon dioxide, $CO_2(g)$. The uncatalyzed reaction is represented by the balanced equation below.

$$2CO(g) + O_2(g) \rightarrow 2CO_2(g) + heat$$

76 On the labeled axes *on your answer sheet*, draw a potential energy diagram for the reaction represented by this equation. [1]

77 Compare the activation energy of the catalyzed reaction to the activation energy of the uncatalyzed reaction. [1]

78 Determine the number of moles of $O_2(g)$ required to completely react with 28 moles of $CO(g)$ during this reaction. [1]

_____

Base your answers to questions 79 through 81 on the information below and on your knowledge of chemistry.

The solvent 2-chloropropane can be made when chemists react propene with hydrogen chloride, as shown in the equation below.

```
    H   H                              H   H   H
    |   |                              |   |   |
H − C − C = C − H  +  HCl  ⟶   H − C − C − C − H
    |       |                          |   |   |
    H       H                          H   Cl  H
```

79 Identify the element in propene that is in all organic compounds. [1]

80 Explain, in terms of chemical bonds, why the hydrocarbon reactant is classified as unsaturated. [1]

81 Write the general formula for the homologous series to which propene belongs. [1]

Base your answers to questions 82 through 85 on the information below and on your knowledge of chemistry.

Radioactive emissions can be detected by a Geiger counter. When radioactive emissions enter the Geiger counter probe, which contains a noble gas such as argon or helium, some of the atoms are ionized. The ionized gas allows for a brief electric current. The current causes the speaker to make a clicking sound. To make sure that the Geiger counter is measuring radiation properly, the device is tested using the radioisotope Cs-137.

To detect gamma radiation, an aluminum shield can be placed over the probe window, to keep alpha and beta radiation from entering the probe. A diagram that represents the Geiger counter is shown below.

82 Compare the first ionization energy of argon to the first ionization energy of helium.   [1]

83 State evidence from the passage that gamma radiation has greater penetrating power than alpha or beta radiation.   [1]

84 Determine the time required for a sample of cesium-137 to decay until only $\frac{1}{8}$ of the original sample remains unchanged.   [1]

85 Complete the nuclear equation *on your answer sheet* for the decay of Cs-137 by writing a notation for the missing product.   [1]

Answer Sheet
August 2019

Chemistry—Physical Setting

PART B–2

51 _____

52 First shell: _____

   Second shell: _____

53 _____

   _____

   _____

54 _____

55 _____

   _____

   _____

56 _____

   _____

   _____

57 _____ **J**

58 _____ **kPa**

59 _____ **g**

60 _____

61 _____

_____

_____

62 _____

63 _____ **M**

64 _____

_____

_____

65 _____

PART C

66 _____

_____

_____

67

68 _____ g/mol

69 _____

_____

_____

70 _____

71

72 _____

_____

_____

73

74 _____ mol

75 _____ atm

76

77 _____

_____

_____

78 _____ mol

79 _____

80 _____

_____

_____

81 _____

82 _____

_____

_____

83 _____

_____

_____

84 _____ y

85 $^{137}_{55}\text{Cs} \rightarrow {}^{0}_{-1}\text{e} +$ _____

Answers
August 2019

Chemistry—Physical Setting

Answer Key

PART A

1. 1	**7.** 1	**13.** 4	**19.** 1	**25.** 1
2. 1	**8.** 4	**14.** 1	**20.** 3	**26.** 4
3. 4	**9.** 3	**15.** 1	**21.** 3	**27.** 4
4. 2	**10.** 3	**16.** 2	**22.** 2	**28.** 3
5. 4	**11.** 2	**17.** 1	**23.** 3	**29.** 4
6. 2	**12.** 4	**18.** 4	**24.** 3	**30.** 3

PART B–1

31. 2	**35.** 4	**39.** 3	**43.** 1	**47.** 4
32. 2	**36.** 1	**40.** 2	**44.** 3	**48.** 3
33. 3	**37.** 4	**41.** 3	**45.** 1	**49.** 1
34. 1	**38.** 3	**42.** 1	**46.** 2	**50.** 3

PART B–2 and **PART C**. *See* **Answers Explained**.

Answers Explained

PART A

1. **1** The idea that atoms were the smallest particles of matter began with the ancient Greeks in the 5th century B.C.E. Credit is given to Democritus and Leucippus. These philosophers imagined atoms to be tiny, hard, indivisible spheres of matter.

WRONG CHOICES EXPLAINED:
(2) Ernest Rutherford developed the nuclear model of the atom in 1911 as a result of his famous gold foil experiment.

(3) J. J. Thompson discovered the presence of negatively charged electrons in atoms in 1897 as a result of his famous cathode ray tube experiment.

(4) Wave-particle duality is a cornerstone premise in the quantum model of the atom. Louis de Broglie, a French physicist, first proposed this concept in 1923.

2. **1** All atoms are electrically neutral. The number of negatively charged electrons must equal the number of positively charged protons.

3. **4** Protons and neutrons each have a mass of 1 amu (atomic mass unit). Reference Table *O* contains information about nuclear particles. Check to verify that protons and neutrons have the same mass (the upper number on each symbol).

4. **2** This is a definition question. Know this definition! The modern quantum model of the atom places electrons in orbitals of varying energies around the nucleus.

5. **4** Electron shells closer to the nucleus are lower in energy than electron shells farther from the nucleus. This is because negatively charged electrons are attracted to the positively charged nucleus.

6. **2** Electronegativity is a measure of an atom's attraction for electrons and specifically for a bonded pair of electrons. Electronegativity increases across any period of the Periodic Table of the Elements because of the increasing nuclear charge of each successive element. Use Reference Table *S* for any questions involving periodic trends! Using the electronegativity data for elements in Period 2, you can see that the values increase as elements are considered from left to right across the period.

WRONG CHOICES EXPLAINED:

(1) Look at the atomic radius data shown on Reference Table S. The atomic radius decreases as elements in Period 2 are considered from lithium to fluorine. As nuclear charge (the number of protons in the nucleus) increases, the atomic radius decreases.

(3) Look at the electron configurations on the Periodic Table of the Elements. The electrons in each element in Period 2 occupy the first and second electron shell.

(4) There are 2 electrons in the first electron shell of every atom in the ground state except for hydrogen, which has only 1 electron.

7. **1** Six elements are classified as metalloids, and these must be memorized! Metalloids straddle the dividing line between metals and nonmetals. They exhibit properties of both metals and nonmetals. The six metalloid elements are boron (B), silicon (Si), arsenic (As), tellurium (Te), antimony (Sb), and germanium (Ge).

WRONG CHOICES EXPLAINED:

(2) Potassium (K) is an alkali metal (Group 1).
(3) Sulfur (S) is a nonmetal.
(4) Xenon (Xe) is a noble gas.

8. **4** Strontium and barium are both in Group 2 on the Periodic Table of the Elements, and they both have 2 valence electrons. Valence electrons are involved in forming chemical bonds between atoms and are therefore largely responsible for an atom's chemical properties.

9. **3** The chemical formula for a compound, for example H_2O, describes the exact proportion of elements in the compound. In H_2O, there are 2 hydrogen atoms for every oxygen atom.

10. **3** Chemical formulas provide information about the types and proportion of elements in a compound. The molecular formula provides the number of atoms of each element present in 1 molecule of a specific compound. The structural formula shows the arrangement of atoms in the compound. Consider the example of butane shown below:

Molecular Formula   Structural Formula

C_4H_{10}

11. **2** Elements classified as metals are held together by metallic bonds. All metals are found on the left of the staircase dividing line on the Periodic Table of the Elements. Cesium (Cs) is an alkali metal in Group 1.

WRONG CHOICES EXPLAINED:
(1) Bromine (Br) is a nonmetal in Group 17.
(3) Krypton (Kr) is a noble gas in Group 18.
(4) Sulfur (S) is a nonmetal in Group 16.

12. **4** Draw a Lewis structure to show how many covalent bonds are shared between nitrogen atoms in a nitrogen diatomic molecule. The Lewis structure for nitrogen, N_2, is shown below:

$$:N\equiv N:$$

Since the nitrogen atoms form 3 covalent bonds and since 2 electrons are shared in each covalent bond, there are 6 electrons shared between nitrogen atoms.

13. **4** The reaction shows two independent hydrogen atoms reacting to form a stable diatomic hydrogen molecule. The formation of chemical bonds releases energy since atoms are more stable once they are bonded together.

14. **1** To answer this question correctly, you need to recall two things: (1) electronegativity is defined as an atom's attraction for electrons in a covalent bond, and (2) values for electronegativity can be found on Reference Table S. Looking at Reference Table S, we find that chlorine's electronegativity is greater than that of carbon, phosphorus, or sulfur.

15. **1** Chemical properties involve the chemical reactivity of a substance, and this depends on the bonding and chemical composition within the substance. Physical changes to a substance, like having more or less mass, do not affect chemical properties.

WRONG CHOICES EXPLAINED:
(2) The 100.-gram sample of $H_2O(\ell)$ has a larger volume than a 50.-gram sample of $H_2O(\ell)$.
(3) Standard temperature is 0°C, and both samples are at STP.
(4) Both samples are made of H_2O. So, both samples have the same chemical formula and also the same empirical formula.

16. **2** Sand does not dissolve in water. When the two are mixed together, the sand will settle to the bottom since it is more dense than water. Since a sand/water mixture has a layer of solid sand and a layer of liquid water, it is a heterogeneous mixture in two different phases.

17. **1** Distillation is the laboratory process used to separate liquid mixtures based on boiling point. As the liquid mixture begins to boil, the component of the mixture with the lower boiling point will boil away first. By passing the vapors through a condenser, the components of the mixture can be collected separately.

18. **4** One of the principles of the kinetic molecular theory of gases is that since gas particles are so small, the volume occupied by the particles themselves is considered to be negligible compared with the volume of their container.

WRONG CHOICES EXPLAINED:
(1) The kinetic molecular theory states that gas particles move in random, constant, straight-line motion.
(2) The kinetic molecular theory states that gas particles undergo perfectly elastic collisions in which the total energy of the system remains the same before and after collisions.
(3) The kinetic molecular theory states that gas particles have no attractions or repulsions for each other.

19. **1** The gas phase is far less dense than any of the other phases. Because gas particles are extremely tiny and because the distances between gas particles are very large relative to their size, we cannot see gases. We can see liquids, solids, and solutions because their particles are much closer together.

20. **3** At equilibrium, the rate of the forward reaction must equal the rate of the reverse reaction. Since both products and reactants are forming at equal rates, the concentration of each remains constant. A common misconception is that equilibrium involves equal amounts of reactants and products, but this is not the case! Equilibrium is about equal rates, not equal amounts.

21. **3** Systems naturally tend to undergo changes that produce lower total energy and greater disorder or randomness (entropy). For example, when leaves fall from a tree, they fall to the ground (a position of lower potential energy) and scatter in many directions (greater randomness).

22. **2** Alkenes and alkynes are two homologous series of hydrocarbons (see Reference Table Q). All hydrocarbons contain just two elements: carbon and hydrogen.

23. **3** Reference Table R lists the general formulas of common organic functional groups. Only amides contain both nitrogen and oxygen.

24. **3** Electrons are tiny charged particles of matter. The moles of electrons lost by Mg(s) must be equal to the number of moles of electrons gained by the O_2(g). This conservation of charge is present during all chemical reactions.

25. **1** The anode is defined as the electrode where oxidation takes place, and the cathode is defined as the electrode where reduction takes place. The mnemonic device "An Ox, Red Cat" can help you remember these definitions!

26. **4** K_2SO_4 and $MgSO_4$ are both electrolytes, meaning that they dissociate completely into freely moving ions in aqueous solution. Freely moving dissociated ions allow charge to flow through the solution. However, K_2SO_4 will dissociate into *more* ions when it dissociates, and the more ions in solution, the greater the solution's capacity to conduct electricity.

WRONG CHOICES EXPLAINED:

(1), (2) These compounds are molecular compounds. They do not dissociate into ions in aqueous solution, and therefore they are nonelectrolytes.

(3) $MgSO_4$ dissociates into fewer ions than does K_2SO_4 when placed into an aqueous solution. Therefore, $MgSO_4$ is less effective as an electrolyte than K_2SO_4.

27. **4** Reference Table M lists the colors of common acid-base indicators at different pH levels. Bromcresol green is yellow at all pH values less than 3.8.

WRONG CHOICES EXPLAINED:

(1) Bromcresol green is blue at pH values greater than 5.4.
(2) Bromcresol green is green at pH values between 3.8 and 5.4.
(3) Bromcresol green is never red.

28. **3** H_3O^+(aq) is the hydronium ion. Hydronium ions form when hydrogen ions (H^+) are released into water. The H^+ protonates the water molecules, resulting in hydronium ions.

29. **4** Both natural and artificial transmutation reactions are nuclear reactions that convert atoms of one element into atoms of another element. Natural transmutation, also known as radioactive decay, involves the spontaneous disintegration of an unstable nucleus. Artificial transmutation reactions take place when a high-speed particle collides with a target nucleus.

30. **3** Nuclear fusion happens when two lighter weight nuclei merge to form heavier nuclei. An equation representing a nuclear fusion reaction is shown below:

$$\frac{2}{1}H + \frac{3}{1}H \rightarrow \frac{4}{2}He + \frac{1}{0}n$$

WRONG CHOICES EXPLAINED:
(1), (2) Both alpha decay and beta decay are types of natural transmutation reactions.
(4) Fission reactions occur when a high-speed neutron collides with a U-235 nucleus, forcing the nucleus to split into two or more smaller nuclei. The formation of neutrons enables a chain reaction to occur.
(NOTE: You need to be able to recognize the four types of nuclear reactions. These include natural transmutation, artificial transmutation, fission, and fusion. Memorize these reactions!)

PART B–1

31. **2** All elements produce unique bright-line spectra when excited electrons in these atoms fall back to the ground state. The bright-line spectrum of a mixture of two or more elements contains all of the wavelength bands of each separate element superimposed together on the same spectrum. Element X can quickly be eliminated since the two wavelength bands at ~740 nm are not evident in the mixture. Likewise, element D can be eliminated because the two wavelength bands at ~590 nm are not present in the mixture. Careful inspection shows that all wavelength bands from elements A and Z are found in the mixture.

32. **2** The Periodic Table of the Elements in the Chemistry Reference Tables shows the ground state configuration of sulfur as 2–8–6. A sulfur atom would be in an excited state if its configuration were 2–7–7 because this configuration shows that a second shell electron has moved to the third shell.

WRONG CHOICES EXPLAINED:
(1) This is the ground state electron configuration for a sulfur atom. The question asks for an *excited* state configuration.
(3) 2–8–7 is the ground state configuration for an atom of chlorine.
(4) Sulfur atoms have 16 electrons. The electron configuration 2–7–8 has a total of 17 electrons and is an excited state electron configuration for a chlorine atom.

33. **3** Atoms with the same number of protons and different numbers of neutrons are isotopes of one another. Cl-35 and Cl-37 are both atoms of chlorine. So, they both have an atomic number of 17, and therefore each has 17 protons. The mass numbers differ because Cl-37 has 2 more neutrons than does Cl-35.

WRONG CHOICES EXPLAINED:
(1) Hydrogen (H) and helium (He) are different elements and have different numbers of protons.
(2) S-32 is identical to S-32. These atoms have the same number of protons and the same number of neutrons.
(4) Gallium (Ga) and germanium (Ge) are different elements and have different numbers of protons.

34. **1** NH_4SCN is an ionic compound formed from two polyatomic ions listed on Reference Table *E*. NH_4^+ is the ammonium ion, and SCN^- is the thiocyanate ion.

35. **4** Conservation of atoms means that the number of atoms of each type of element must be the same before and after the reaction. In answer choice (4), there are 4 iron (Fe) atoms and 6 oxygen (O) atoms on each side of the equation.

36. **1** Covalent bonds form between nonmetal atoms, such as between hydrogen (H) and oxygen (O). When metal atoms bond with nonmetal atoms, ionic bonds form.

37. **4** Oxygen is a diatomic gas at STP.

WRONG CHOICES EXPLAINED:
(1) This model shows a monatomic gas, like one of the noble gases.
(2) This model shows the regular arrangement and tight packing of a solid.
(3) This model does not show the large spaces between particles that exist in samples of gas.

38. **3** Real gases behave more like ideal gases when temperature is high and pressure is low. These conditions minimize any possible attractions between gas particles, allowing the gas to adhere more closely to the principles of the kinetic molecular theory of gases. A good way to remember this is to think of summertime, when the temperature is high and the pressure is low!

39. **3** The molarity formula is found on Reference Table T:

$$\text{molarity} = \frac{\text{moles of solute}}{\text{liter of solution}}$$

Plug the given information into this equation to solve for the molarity:

$$\text{molarity} = \frac{0.50 \text{ mol}}{2.0 \text{ L}} = 0.25 \text{ M}$$

40. **2** Equal volumes of gas at the same conditions of temperature and pressure contain the same number of particles. This relationship is known as Avogadro's law. For example, the molar volume of a gas at STP equals 22.4 L/mol no matter what type of gas you have. Since gas samples 1 and 3 have equal volumes and are at the same temperature and pressure, they contain the same number of particles.

41. **3** Reactions occur only when reactant particles collide. The powdered zinc exposes more zinc atoms to collisions with HCl molecules.

42. **1** Le Châtelier's principle states that chemical systems at equilibrium that experience a stress will respond to restore equilibrium. If temperature is increased in this equilibrium system, the reaction will respond by favoring the reverse reaction to use up the extra energy and bring the temperature back down again, restoring equilibrium. This "shift to the left" causes more HI(g) to be produced.

43. **1** The heat of reaction, ΔH, is defined as the difference between the potential energy of the products in a chemical reaction and the potential energy of the reactants. Know this definition! Reference Table I lists the heats of reaction for various chemical reactions. Use this table to find the largest value of the heat of reaction for the equations shown. Since you are only looking for the difference between the potential energy of the products and the potential energy of the reactants, the sign on the ΔH does not matter. The difference is greatest for $4\text{Al}(s) + 3\text{O}_2(g) \rightarrow 2\text{Al}_2\text{O}_3(s)$, with a ΔH of -3351 kJ.

44. **3** Entropy is defined as the degree of chaos, randomness, or disorder within a system or a substance. Increasing entropy is one important factor that leads to spontaneous chemical or physical change. Phase changes that lead to a less orderly arrangement of atoms increase the entropy of a sample of matter. Particles in a gas are far less ordered than particles in a liquid.

WRONG CHOICES EXPLAINED:

(1) A gas to solid change results in a much more ordered arrangement of atoms. This is the opposite of entropy!

(2) A gas to liquid change increases the order (and decreases the entropy) in a sample of matter. Liquid particles have far greater attractions to one another than do gas particles, and their motion is less random than the motion of gas particles.

(4) A liquid to solid change results in a much more ordered arrangement of atoms. This is the opposite of entropy!

45. **1** To name the compound in this problem (and shown below), classify the compound according to which organic functional group it contains. Organic functional groups are listed in Reference Table R:

$$
\begin{array}{c}
\text{H} \quad \text{H} \quad \text{H} \quad \boxed{\text{O}} \quad \quad \text{H} \\
| \quad\quad | \quad\quad | \quad\quad || \quad\quad\quad | \\
\text{H}-\text{C}-\text{C}-\text{C}-\boxed{\text{C}-\text{O}}-\text{C}-\text{H} \\
| \quad\quad | \quad\quad | \quad\quad\quad\quad | \\
\text{H} \quad \text{H} \quad \text{H} \quad\quad\quad\quad \text{H}
\end{array}
$$

This compound contains the ester functional group, and the name will end with the suffix –*oate*. There is only one answer choice with the correct suffix: choice (1).

46. **2** This synthesis does not affect the nuclei of the atoms involved. However, before the reaction, Ca(s) has an oxidation number of 0. After the reaction, the Ca^{2+} ion has an oxidation number of +2. Oxygen also changes in oxidation number, from 0 before the reaction to −2 after the reaction.

47. **4** Reference Table J organizes metals from the most active metals at the top to the least active metals at the bottom. Metal atoms oxidize when they react, and they can give electrons to the ions of less reactive metals. For a spontaneous reaction, a pure metal must be paired with the ion of a metal that is lower on Reference Table J.

WRONG CHOICES EXPLAINED:
(1), (2), (3) The pure metals in these reactions are located below the ions with which they are paired. This means that the pure metals are less reactive than the ions, and consequently, the pure metals will not be able to oxidize (no reaction will happen).

48. **3** The reaction between $Ca(OH)_2$ and H_2SO_4 is a neutralization reaction. Neutralization reactions are really double replacement reactions between acids and bases. The word equation for neutralization reactions between Arrhenius acids and bases is the following:

$$ACID + BASE \rightarrow WATER + SALT$$

$Ca(OH)_2$ is a base listed on Reference Table L, and H_2SO_4 is an acid listed on Reference Table K.

WRONG CHOICES EXPLAINED:
(1) This reaction is an example of a decomposition reaction.
(2) This reaction is an example of a combustion reaction.
(4) This reaction is an example of a double replacement reaction between two ionic compounds.

49. **1** Reference Table N shows that Pu-239 has a half-life of 2.410×10^4 years. With a half-life this long, this radioisotope would continue to emit radiation for thousands of years. The other radioisotopes have half-lives on the order of seconds, minutes, or days and would not remain radioactive for very long.

50. **3** Nuclear fission is depicted in the equation for this question. Nuclear fission produces energy by converting a small amount of mass into energy. The amount of energy produced can be calculated by Einstein's famous equation $E = mc^2$, in which m is the mass lost in the fission reaction and c is the speed of light.

PART B–2

[All questions in Part B–2 are worth 1 point.]

51. All nitrogen atoms have **7** protons. Nitrogen's atomic number is 7, which equals the number of protons in the nucleus of every nitrogen atom.

52. The Periodic Table of the Elements shows that the electron configuration for the element nitrogen is 2–5. This means that **2 electrons are in the first electron shell and 5 electrons are in the second electron shell**.

53. The atomic mass of an element is the weighted average of the masses of the naturally occurring isotopes of that element. The atomic mass of nitrogen as shown on the Periodic Table of the Elements is 14.0067 amu. Since this weighted average is much closer to 14 than to 15, there must be a **far greater amount of N-14 than N-15**.

54. **Bromine, Br$_2$**, is a liquid because standard temperature, 273 K, lies between bromine's melting point and boiling point.

55. **Iodine, I$_2$, has stronger intermolecular forces than fluorine, F$_2$.** Stronger intermolecular forces between molecules in a liquid sample cause higher boiling points. More energy is required to allow molecules to overcome the forces of attraction between them and separate completely to form a gas.

56. **Potential energy increases** as NO changes phase from a liquid to a gas at 121 K. Potential energy is stored energy. Particles gain potential energy as they move farther apart to form a gas.

57. Heat formulas are found on Reference Table T. Since the temperature of the water increases from 20.0°C to 50.0°C, use the following formula:

$$q = mC\Delta T$$

In the formula above, q is the amount of heat in joules (J) lost or gained, m is the mass of the water in grams, and ΔT is the change in temperature in Kelvin. To determine the change in temperature, remember that 1 degree Celsius equals 1 Kelvin and calculate the following:

$$\Delta T = T_{final} - T_{initial}$$
$$= 50.0°C - 20.0°C$$
$$= 30.0°C$$
$$= 30.0 \text{ K}$$

Using the value for the specific heat capacity shown on Reference Table B and plugging in the given values leads to the following expression:

$$q = (100. \text{ g})(4.18 \text{ J/g·K})(30.0 \text{ K}) \cong \textbf{12,500 J}$$

(NOTE: All values from **12,500 J to 13,000 J** were accepted as correct answers to this problem.)

58. The vapor pressure of water at 50.0°C is **12 kPa**. Read this value directly from Reference Table H.

(NOTE: All values from **11 kPa to 13 kPa** were accepted as correct answers to this problem.)

59. **22 g** of $KClO_3$ must be added to 100. g of water at 50.0°C to make a saturated solution. Read this value directly from Reference Table G.

(NOTE: All values from **20 g to 23 g** were accepted as correct answers to this problem.)

60. **Electrons** are lost as oxidation occurs at the anode. These electrons travel through the wire to the cathode, where they are gained during the process of reduction.

61. **The salt bridge allows ions to flow into each compartment** in the voltaic cell **to maintain electrical neutrality** in the system. A buildup of positive or negative charges in either compartment would cause electrons to stop flowing.

62. Copper is below chromium on Reference Table J, so copper ions reduce in the presence of chromium. Reduction is defined as the "gain of electrons." Copper ions gain 2 electrons to form neutral copper atoms, Cu^0, as shown in the following reduction half-cell equation:

$$Cu^{2+}(aq) + 2e^- \rightarrow Cu^0(s)$$

(NOTE: Including state symbols is not required to earn credit.)

63. Titration is the lab technique used to determine the concentration of an acid or base by neutralizing it with an acid or base of known concentration. The titration formula is found on Reference Table T:

$$M_A V_A = M_B V_B$$

In this formula, M_A is the molarity of H^+, V_A is the volume of acid, M_B is the molarity of OH^-, and V_B is the volume of base.

Substitute the values given in the problem and rearrange to solve:

$$M_{(A)}(10.0 \text{ mL}) = (0.100 \text{ M})(16.0 \text{ mL})$$

$$M_A = \frac{(0.100 \text{ M})(16.0 \text{ mL})}{10.0 \text{ mL}}$$

$$= \mathbf{0.160 \text{ M}}$$

64. At the point of neutralization, **the concentration of hydronium ions is exactly equal to the concentration of hydroxide ions**. The net ionic equation for neutralization is $H^+(aq) + OH^-(aq) \rightarrow H_2O(\ell)$. Since H^+ and OH^- are present in a 1:1 molar relationship in this equation, these ions must be present in equal amounts for neutralization to occur. Remember that $H^+(aq)$ is equivalent to a hydronium ion $H_3O^+(aq)$.

65. pH measures the concentration of hydrogen ion, H^+, in aqueous solution. pH is a logarithmic function, meaning that it is based on the exponent of the base 10 molarity of the hydrogen ion content. Each pH value is $10\times$ different in hydrogen ion molarity than the pH immediately higher and lower than itself. Hydrogen ions make a solution acidic, and solutions that are more acidic have lower pH values. The problem states that the original NaOH solution has a pH of 13. The pH of a solution that has 10 times *more* $H^+(aq)$ ion is one number *lower* than the original pH—in other words, **12**.

PART C

[All questions in Part C are worth 1 point.]

66. The reading passage for this question states that when the magnesium sulfate heptahydrate ($MgSO_4 \cdot 7H_2O$) is heated to a high temperature in a crucible, water molecules are released from the compound. **The sample was heated repeatedly to ensure that all of the water in the sample was removed.**

67. The formula for percent composition is found on Reference Table *T*. In this problem, you are specifically asked to solve for the percent composition by mass of water in the hydrate:

$$\% \text{ composition by mass of water} = \frac{\text{mass of water}}{\text{total mass of hydrate}} \times 100$$

The mass of water can be found by subtracting the dry solid mass left in the crucible after heating from the initial mass of the hydrate that was placed into the crucible. The difference between these masses is the mass of the water that was driven off during the heating process:

$$5.06 \text{ g hydrate} - 2.47 \text{ g anhydrous residue} = 2.59 \text{ g water}$$

Plugging this value into the equation above:

$$\% \text{ composition by mass of water} = \frac{2.59 \text{ g water}}{5.06 \text{ g hydrate}} \times 100$$

(NOTE: This question asks ONLY for the numerical setup for calculating the mass percent of water in the hydrate. The numerical setup is all that you should record on your answer sheet. If you would like to check your work by solving for mass percent of water, do it on scrap paper.)

68. Calculate the gram-formula mass, also known as the molar mass, by adding the atomic masses of each element in the compound. The molar mass of each element is found on the Periodic Table of the Elements:

Calculating Gram-Formula Mass of $MgSO_4 \cdot 7H_2O$			
Element	**Atomic Mass**	**Number of Atoms**	**Total Mass of Element**
Mg	24.3 g/mol	1	24.3 g/mol
S	32.1 g/mol	1	32.1 g/mol
O	16.0 g/mol	4	64.0 g/mol
H	1.0 g/mol	14	14.0 g/mol
O	16.0 g/mol	7	112.0 g/mol
		Gram-Formula Mass	**246.4 g/mol**

(NOTE: Acceptable values range from **245.989 g/mol to 247 g/mol**.)

69. The electron configuration of the element sodium is 2–8–1. When the Na^+ ion forms, the outmost electron is lost, so the sodium ion has an electron configuration of 2–8. The radius of the Na^+ ion is smaller than the radius of a Na atom

because **the sodium atom has an electron in the third electron shell whereas the sodium ion only has electrons in the first and second electron shells**.

70. The chloride ion, Cl^-, forms when an atom of chlorine gains 1 electron, giving the ion a total of 18 electrons (which is 1 more than an atom of chlorine). Atoms of the noble gas **argon** also contain 18 electrons.

71. Lewis electron-dot diagrams show the distribution of valence electrons within a molecule. Lewis electron-dot diagrams help to illustrate the number and types of bonds in a compound. Each Cl atom has 7 valence electrons. By sharing one pair of electrons in a single covalent bond, each chlorine atom achieves a stable octet of valence electrons:

$$:\ddot{C}l-\ddot{C}l:$$

72. Temperature is directly related to the kinetic energy of particles in a sample of matter. Kinetic energy is the energy of particle motion. Putting these two ideas together leads to the understanding that particles in a sample of matter move faster at higher temperatures and slower at lower temperatures. If the cabin temperature decreases, **the average kinetic energy of the gas molecules would also decrease**.

73. Use the combined gas law equation on Reference Table T to solve this problem:

$$\frac{P_1 V_1}{T_1} = \frac{P_2 V_2}{T_2}$$

Since the volume of the submarine cabin does not change, $V_1 = V_2$, and we can simplify the formula to the following:

$$\frac{P_1}{T_1} = \frac{P_2}{T_2}$$

Plug the given information into the equation:

$$\frac{116 \text{ kPa}}{312 \text{ K}} = \frac{P_2}{293 \text{ K}}$$

(NOTE: This question asks ONLY for the numerical setup for calculating the pressure in the submarine cabin. The numerical setup is all that you should record on your answer sheet. If you would like to check your work by solving for the pressure, do it on scrap paper.)

74. The introductory paragraph to this question states, "Air in the cabin becomes unsafe to breathe when the mass of $CO_2(g)$ in this cabin exceeds 2156 grams." To convert this mass into moles of $CO_2(g)$, use the mole calculation formula found on Reference Table T:

$$\text{number of moles} = \frac{\text{given mass}}{\text{gram-formula mass}}$$

Plugging in the numbers:

$$\text{number of moles} = \frac{2156 \text{ g } CO_2}{44.0 \text{ g/mol } CO_2} = \textbf{49.0 mol } CO_2$$

(NOTE: All values from **49 mol to 50 mol** of CO_2 were accepted as correct answers to this problem.)

75. The relationship between kPa and atm is given on Reference Table A: $101.3 \text{ kPa} = 1 \text{ atm}$. The conversion from the given pressure in kPa to pressure in atm can be accomplished using either dimensional analysis or ratios.

Using dimensional analysis:

$$116 \text{ \cancel{kPa}} \times \frac{1 \text{ atm}}{101.3 \text{ \cancel{kPa}}} = 1.15 \text{ atm}$$

Using ratios:

$$\frac{1 \text{ atm}}{101.3 \text{ kPa}} = \frac{x \text{ atm}}{116 \text{ kPa}}$$

$$x = \textbf{1.15 atm}$$

(NOTE: All values from **1.14 atm to 1.16 atm** were accepted as correct answers to this problem.)

76. The equation given for the conversion of CO(g) to CO$_2$(g) shows the heat term on the product side of the reaction. Since heat is produced, this is an exothermic reaction. Some of the potential energy of the reactant is lost during this reaction as a lower energy product is produced, as shown in the potential energy diagram below:

77. Adding a catalyst to a chemical reaction lowers the activation energy (E_a) for the reaction, making it possible for more collisions to have sufficient energy for particles to react. **The catalyzed reaction has a lower activation energy than the uncatalyzed reaction.**

78. All coefficients in chemical reactions represent particles or moles. Mole ratios can be used to find the number of moles of any substance in a chemical equation if the moles of another substance are known. Solve this stoichiometry problem using either dimensional analysis or ratios.

Using dimensional analysis:

$$\frac{28 \, \cancel{\text{mol CO(g)}}}{1} \times \frac{1 \, \text{mol O}_2(g)}{2 \, \cancel{\text{mol CO(g)}}} = 14 \, \text{mol O}_2(g)$$

Using mole ratios:

$$\frac{28 \, \text{mol CO(g)}}{2} = \frac{x \, \text{mol O}_2(g)}{1}$$

$$x = \textbf{14 mol O}_2\textbf{(g)}$$

79. Organic chemistry, by definition, is the chemistry of **carbon**-based compounds. Carbon atoms form 4 covalent bonds and can easily bond with other carbon atoms to form chains, rings, and networks.

80. Unsaturated hydrocarbons contain at least one double or triple bond between carbon atoms. The term "unsaturated" implies that other atoms could be added to the compound by breaking the double or triple bond between carbon atoms. **The hydrocarbon reactant is classified as an unsaturated hydrocarbon because of the double bond between the carbon atoms.**

81. Propene belongs to the alkene family of hydrocarbons. The general formula for alkenes, shown on Reference Table Q, is C_nH_{2n}.

82. First ionization energy data is listed on Reference Table S. **Argon has a first ionization energy of 1521 kJ/mol, while helium has a first ionization energy of 2372 kJ/mol.**

83. The passage states that an aluminum shield can be placed over the probe window, preventing alpha and beta radiation from entering the probe. The passage says that gamma radiation can still be measured when the aluminum shield is over the probe window. Therefore, **gamma radiation has greater penetrating power than either alpha or beta radiation because gamma radiation can pass through aluminum, but alpha and beta radiation cannot pass through aluminum**.

84. From Reference Table N, the half-life of Cs-137 is 30.2 years. One half-life is the time required for a radioisotope to decay to half of its original mass. After one half-life, only half of the original Cs-137 remains. After another half-life passes, only half of this half remains. This continues for each half-life:

$$\frac{1}{2} \times \frac{1}{2} \times \frac{1}{2} = \frac{1}{8}$$

The process of radioactive decay is represented below:

Initial Amount 100%	After one half-life, $\frac{1}{2}$ remains	After 2nd half-life, $\frac{1}{4}$ remains	After 3rd half-life, $\frac{1}{8}$ remains

Three half-lives must pass for Cs-137 to decay until only $\frac{1}{8}$ of the original sample remains unchanged. Since each half-life is 30.2 years long, calculate the total time elapsed:

$$3 \times 30.2 \text{ years} = \textbf{90.6 years}$$

85. To balance a nuclear equation, the sum of the atomic numbers (bottom) and the mass numbers (top) must be equal on both sides of the arrow. The missing product must have a mass number of 137 since the mass of a beta particle is 0. The atomic number of the missing product must be 56 since $56 - 1 = 55$. The missing product is an atom of the element barium, Ba, since barium has an atomic number of 56. The missing product is therefore $^{137}_{56}\textbf{Ba}$.

(NOTE: **Ba-137**, **barium-137**, and ^{137}**Ba** are all acceptable responses.)

Mark ($\checkmark$) the questions you answered correctly. Count the number of checks and follow the formulas given to determine your score on each topic.

Core Area	☐ Questions Answered Correctly

75: (1)

Section M—Math Skills
☐ Number of checks ÷ 1 × 100 = ___ %

66, 82: (2)

Section R—Reading Skills
☐ Number of checks ÷ 2 × 100 = ___ %

1–5, 31–33, 52, 53: (10)

Section I—Atomic Concepts
☐ Number of checks ÷ 10 × 100 = ___ %

6–8, 51: (4)

Section II—Periodic Table
☐ Number of checks ÷ 4 × 100 = ___ %

9, 10, 34, 35, 67, 68, 74, 78: (8)

Section III—Moles/Stoichiometry
☐ Number of checks ÷ 8 × 100 = ___ %

11–14, 36, 69–71: (8)

Section IV—Chemical Bonding
☐ Number of checks ÷ 8 × 100 = ___ %

15–19, 37–40, 54–59, 72, 73: (17)

Section V—Physical Behavior of Matter
☐ Number of checks ÷ 17 × 100 = ___ %

20, 21, 41–44, 76, 77: (8)

Section VI—Kinetics and Equilibrium
☐ Number of checks ÷ 8 × 100 = ___ %

Core Area	☐ Questions Answered Correctly

22, 23, 45, 79–81: (6)

Section VII—Organic Chemistry
☐ Number of checks ÷ 6 × 100 = ___ %

24, 25, 46, 47, 60–62: (7)

Section VIII—Oxidation–Reduction
☐ Number of checks ÷ 7 × 100 = ___ %

26–28, 48, 63–65: (7)

Section IX—Acids, Bases, and Salts
☐ Number of checks ÷ 7 × 100 = ___ %

29, 30, 49, 50, 83–85: (7)

Section X—Nuclear Chemistry
☐ Number of checks ÷ 7 × 100 = ___ %

NOTES

NOTES

NOTES

NOTES

NOTES

NOTES

Let's Review Regents:

Chemistry—
Physical Setting
Revised Edition

Albert S. Tarendash, M.S.
Assistant Principal—Supervision (Retired)
Department of Chemistry and Physics
Stuyvesant High School
New York, New York

Chemistry/Physics Faculty (Retired)
The Frisch School
Paramus, New Jersey

Published by Kaplan, Inc., d/b/a Barron's Educational Series
750 Third Avenue
New York, NY 10017
www.barronseduc.com

ISBN: 978-1-5062-6469-1

10 9 8 7 6 5 4 3 2 1

Kaplan, Inc., d/b/a Barron's Educational Series print books are available at special quantity discounts to use for sales promotions, employee premiums, or educational purposes. For more information or to purchase books, please call the Simon & Schuster special sales department at 866-506-1949.

TABLE OF CONTENTS

PREFACE

To the Student:

This book has been written to help you understand and review high school chemistry. Chemistry is not a particularly easy subject. No book—no matter how well written—can give you *instant* insight. Nevertheless, if you read this book carefully and do all of the problems and review questions, you will have a pretty decent understanding of what chemistry is all about.

I have designed this book to be your "chemistry companion." It is more detailed than most review books, but it is probably less detailed than your hardcover text. The book is divided into 16 chapters. Each chapter begins with an overview (*Key Ideas*) and a summary of what you should learn in the chapter (*Key Objectives*). Most of the 16 chapters are then divided into two sections. Section I contains basic material that follows the New York State Regents Core syllabus very closely. Section II contains additional material that is *not* in the Regents Core but may well be part of your chemistry course (and is likely to appear on other standardized tests such as the SAT Subject Test in Chemistry).

This book provides a large number of problems with detailed solutions. Many of the problems are usually followed by a "Try It Yourself" exercise and its answer. You are encouraged to work out each and every exercise.

Each chapter ends with multiple-choice questions and, usually, one or more constructed-response and free-response questions. These are drawn from both Sections I and II. The answers appear in Appendix 3. Appendix 1 contains the official New York State Regents reference tables that are used throughout the book, and Appendix 2 contains additional (non-Regents) tables that are used for Section II topics. Appendix 4 provides some information on how to answer various types of constructed-response questions.

If you have any comments about this book or if you find errors (for which I apologize in advance), please e-mail me at:

ast-lrc@optimum.net

I wish you much success.

A. S. Tarendash

To the Teacher:

This book was written according to the following philosophy: A worthwhile review book should provide clear, careful, and detailed explanations that will, in effect, "hold the student's hand" while he or she is studying the subject.

Features of This Book

- Most of the 16 chapters are divided into two sections. Section I contains basic material that follows the New York State Regents Core syllabus. Section II contains additional material that goes beyond the New York State Core and may well appear on such standardized tests such as the SAT Subject Test in Chemistry.
- The additional reference tables have been revised to reflect recent discoveries, such as elements 117 and 118, and the naming of elements 113–118.
- Thanks to reader correspondence, errors have been corrected extensively throughout the book.
- The end-of-chapter questions have been revised so that Section I questions contain more recent Regents examination material.

You may find that my ordering of the material is a bit unorthodox. Then again, we all have our own visions as to how a high school chemistry course should be sequenced. In the final analysis, you may not feel that the way I subdivided the material is appropriate for your teaching style or course. If you are more comfortable using your own order or the order shown in your textbook, you should certainly continue with that approach.

Rather than padding each section of the text with multiple-choice questions, I spent considerable time in working out sample problems. The text heavily emphasizes the factor label method (dimensional analysis) because I believe that this one area is where students experience considerable difficulties. In this vein, I have also included an additional appendix (Appendix 4) on how to answer constructed-response and free-response items. This appendix is designed to train your students to compose answers that are logical and orderly and that will not drive you crazy when you grade them!

I reserved the end of each chapter for the multiple-choice and constructed-response questions, including free-response questions. I have *not* included the type of questions found on the SAT Subject Test. If you are preparing your class to take the SAT Subject Test, you should obtain a book published by The College Board that contains a variety of *authentic*, released SAT Subject Test examinations in all subjects, including chemistry. You will find the chemistry section of that book particularly helpful. You should make every effort to include at least a few SAT Subject Test-type questions on your class examinations throughout the year.

A final thought: *Let's Review Regents: Chemistry—Physical Setting* was never meant to be one of those thin "Regents prep books." It was always designed to be a review of a *complete* introductory one-year course in high school chemistry as taught in most of the country. To New York teachers: Remember, the New York State Education Department took great pains to indicate that "[the] Core is *not* a syllabus" [*Chemistry Core* p. 3.] but a document to assist in planning a year-long chemistry course. In effect, the Core is merely a blueprint for questions that students might be asked on the Regents Chemistry examination. I firmly believe that no New York State high school chemistry teacher should be using the Core as his or her *sole* curriculum guide. In this vein, many New York teachers have recognized that omitting certain topics makes other basic concepts mysterious to their students. (Can one really teach Lewis electron-dot diagrams in a meaningful way if atomic sublevels and orbitals have been omitted?)

I hope that this book provides some new insights and that it will help you to ensure that you approach your lesson plans and lessons in a spirit of inquiry and problem solving. I also hope that *Let's Review Regents: Chemistry— Physical Setting* will be successful for both you and your students.

If you have any comments about this book or if you find errors (for which I apologize in advance), please e-mail me at:

ast-lrc@optimum.net

Acknowledgments

No book is ever the work of one person. Once again, I wish to acknowledge and thank:

- The editors at Barron's for their guidance, patience, understanding, and gentle prodding;
- My wife, Bea, whose love, support, and help (particularly with the end-of-chapter questions) made the preparation of this edition considerably easier;
- My children, Franci, David, Jeff, and Janet, and my grandchildren, Brittany, Alexa, Danielle, Raina, and Jonathan, just for being near;
- All of my students—past and present—that have inspired me over the last 50 years and have made teaching a happy and satisfying endeavor.

ALBERT S. TARENDASH
Nanuet, New York

Chapter One
INTRODUCTION TO CHEMISTRY

KEY IDEAS

This chapter introduces some of the basic concepts in chemistry that serve as a foundation for the chapters that follow. In particular, measurement and problem solving are studied.

KEY OBJECTIVES
At the conclusion of this chapter you will be able to:
- Define the terms *chemistry*, *matter*, *pure substance*, and *mixture*.
- Distinguish among elements, compounds, and mixtures.
- Distinguish between physical and chemical properties.
- Indicate how mixtures are separated.
- Distinguish between physical and chemical changes.
- List the various forms of energy.
- List the common metric units used in chemistry.
- Name the most commonly used metric prefixes and their numerical equivalents.
- Express numbers in scientific notation.
- Perform simple operations on numbers expressed in scientific notation.
- Define the terms *volume* and *density* as they apply to chemistry.
- Describe how measurement readings are taken and estimated.
- Define and apply the terms *accuracy*, *precision*, and *significant digits* (*figures*).
- Apply the rules for adding (subtracting) and multiplying (dividing) measurements.
- Define the term *percent error*, and calculate the percent error of a measurement.
- Understand the factor-label method (FLM), and use it to solve problems.
- Use equations and graphs to solve problems.

Reminder: If you have not already done so, read Preface: To the Student (page xi) at this time.

SECTION I—BASIC (REGENTS-LEVEL) MATERIAL

NYS REGENTS CONCEPTS AND SKILLS

By the time you have finished Section I, you should have mastered the concepts and skills listed below. The Regents chemistry examination will test your knowledge of these items and your ability to apply them.

Concepts are the *basic ideas* that form the body of the Regents chemistry course (what you need to know!).

Skills are the *activities* that demonstrate your mastery of these concepts (how you show that you know them!).

Following each concept or skill is a page reference (given in parentheses) to this chapter.

1.1 Concept:
Matter is classified as a pure substance or as a mixture of substances. (Page 4)

1.2 Concept:
The structure and arrangement of particles determine the phase of a substance.
The three phases of matter (solids, liquids, and gases) have different properties. (Page 4)

Skill:
Use a simple particle model/diagram to differentiate among the properties of a solid, a liquid, and a gas. (Pages 4–5)

1.3 Concept:
A pure substance (element or compound) has a constant composition and constant properties throughout a given sample, and from sample to sample. (Pages 4–5)

Skill:
Use a simple particle model/diagram to differentiate among elements, compounds, and mixtures. (Pages 5–6)

1.4 Concept:
Elements cannot be broken down by chemical change. (Pages 4–5)

1.5 Concept:
Mixtures are composed of two or more different substances that can be separated by physical means. When different substances are mixed together, a homogeneous or heterogeneous mixture is formed. (Pages 5–6)

1.6 Concept:
The proportions of components in a mixture can be varied. Each component in a mixture retains its original properties. (Page 5)

1.7 Concept:
A physical change results in the rearrangement of existing particles in a substance. A chemical change results in the formation of different substances with changed properties. (Page 5)

1.8 Concept:
Energy can exist in different forms, such as chemical, electrical, electromagnetic, thermal, mechanical, and nuclear. (Pages 12–13)

1.9 Concept:
Algebraic and geometric representations can be used to describe and compare data.

Skills:
- Organize, graph, and analyze scientific data. (Pages 27–31)
- Recognize and convert various scales of measurement. (Pages 23–25)

1.10 Concept:
Algebraic and geometric representations can be used to recognize patterns and relationships in mathematics.

Skills:
- Identify direct and inverse relationships. (Pages 28–29)
- Use data trends (such as those shown in graphs) to predict information. (Pages 28–29, 30–31)

1.11 Concept:
Algebraic and geometric concepts can be used to solve problems.

Skills:
- State what assumptions are made when a mathematical equation is used. (Pages 25–27)
- Evaluate whether an answer to a numerical problem is reasonable. (Pages 25–27)

1.12 Concept:
Differences in properties such as density, particle size, molecular polarity, boiling point and freezing point, and solubility permit physical separation of the components of a mixture. (Pages 6–12)

Skill:
Describe the processes and uses of filtration, distillation, and chromatography in the separation of a mixture. (Pages 6–12)

1.1 CHEMISTRY IS . . . ?

Chemistry is the science that focuses on the structure, composition, and properties of matter. In our study, we will also learn about the changes that matter undergoes and the energy that accompanies these changes.

1.2 MATTER AND ENERGY

Our universe is composed of matter and energy. **Matter** is anything that has mass and volume (that is, anything that has *density,* a concept we will develop in Section 1.6).

Matter commonly exists in three phases: *solid*, *liquid*, and *gas*. Generally, solids are composed of particles that are tightly packed and have a regular arrangement. The particles present in gases have no regular arrangement and no appreciable packing. In liquids, the arrangement and packing of the particles are somewhere between those in solids and gases. As a consequence, a solid has a definite shape and volume; a liquid has a definite volume but takes the shape of its container; a gas has neither a definite shape nor a definite volume: its shape and volume are those of its container.

We can use *particle diagrams* to visualize solids, liquids, and gases and to distinguish among them, as shown below.

| Gas | Liquid | Solid |

Substances

For convenience, we divide matter into two classes: *pure substances* and *mixtures*.

A pure **substance** is any variety of matter that is homogeneous (uniform) and has a fixed composition by mass. **Elements** are classified as substances because they contain atoms of a single type and cannot be decomposed further. Although **compounds** contain more than one type of element, they are also classified as substances because their compositions are fixed. For example, in every sample of water, the ratio, by mass, of oxygen to hydrogen is eight to one. Because this ratio cannot be changed, we usually say that the elements in a compound are **chemically combined**. The accompanying particle diagrams represent two different elements and a compound.

This represents
an **element**.

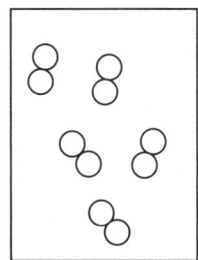

This also represents
an **element**.

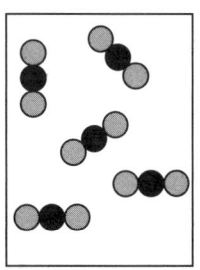

This represents
a **compound**.

Every substance has a unique set of *properties* that allow it to be distinguished from other substances. These properties can be grouped into two categories: physical and chemical.

Physical properties can be measured without changing the identity and composition of a substance, and include color, odor, density, melting point, and boiling point. For example, when the melting point of a substance is measured, its phase changes from solid to liquid. The appearance of the substance changes, but not its composition. Changes that accompany the measurement of physical properties are called **physical changes**. For example, all phase changes are physical changes.

Chemical properties are properties that lead to changes in the identity and composition of a substance. For example, combustibility is a chemical property. When hydrogen gas is burned in air, the hydrogen combines with the oxygen in the air and changes into water, a substance whose composition is entirely different from that of hydrogen. Changes that accompany the measurement of chemical properties are called **chemical changes**; they are also called **chemical reactions**. For example, when a battery provides electrical energy, the changes that occur *inside* it are chemical changes because entirely new substances are produced.

Mixtures

A **mixture** is composed of two or more distinct substances, but the compositions of mixtures may be varied. A solution of sugar and water is a mixture because the ratio of sugar to water can be changed. A solution is an example

5

of a **homogeneous mixture** because each component (sugar and water in this case) is uniformly dispersed throughout the solution. An ice cream soda, on the other hand, is not homogeneous because each component (ice cream, syrup, soda, whipped cream) is not uniformly dispersed. This is an example of a **heterogeneous mixture**. The accompanying particle diagram represents a heterogeneous mixture.

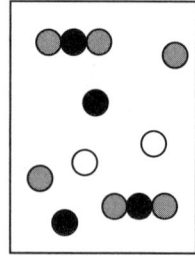

This represents a
heterogeneous mixture.

Separation of Mixtures

One way of distinguishing between mixtures and substances involves the ability to separate mixtures by "physical" methods. For example, it is possible to separate a sand-water mixture by allowing the water to evaporate. When a physical method is used to separate a mixture, the chemical properties of the mixture's components are *not* changed. We discuss these separation techniques below.

Filtration

Suppose we have a sand-salt mixture, and we want to separate the sand from the salt. The technique known as **filtration** will accomplish the task of recovering the sand from the mixture. First, we add water to the mixture, a procedure that dissolves the salt, but not the sand. Second, we fit a funnel with a cone of **filter paper**, and carefully pour the well-stirred mixture into the funnel, as shown in the accompanying diagram.

Filtering Diagram

Filter paper in funnel

Filtrate

The salt-water solution that passes through the filter paper into the beaker below is known as the **filtrate**. The sand remains on the filter paper. To remove the last traces of salt from the moist sand, we wash the sand several times with pure water. We then allow the sand to dry, either in air or in a **heating oven**.

Evaporation

The next step is to recover the salt from the salt-water solution. Drying would work but would take a very long time. In situations such as this, we use the technique known as **evaporation**. We pour the solution carefully into an **evaporating dish**, and cover the dish with a **watch glass**. Then we heat the entire assembly *gently over a low flame* until the water boils and is completely evaporated. The watch glass prevents the salt from spattering during the heating process. The accompanying diagram illustrates this technique.

Evaporation-1 Diagram

In the event that *strong heating* is required, a **crucible** is used. Crucibles are frequently made of porcelain and are supported by a **pipestem triangle** while being heated. The accompanying diagram illustrates the use of a crucible.

Evaporation-2 Diagram

Crystallization

Yet another technique we can use to separate a dissolved solid from its solvent is known as **crystallization**. In this technique, we prepare a concentrated solution at a relatively high temperature and allow the solution to cool slowly. As it cools solid crystals form at the bottom of the vessel containing the solution. At times, the solution becomes *supersaturated* and the crystals do *not* form immediately. In these cases, we can initiate crystallization by adding a single "seed crystal" of the solid to the solution.

Distillation

Suppose we have an aqueous solution containing dissolved solids, but it is the *solvent* that we want to recover. In this instance, we use the technique known as **distillation**. We place the solution in a **boiling flask** and heat it to boiling. As the solvent vapor passes out of the flask, it passes through a **condenser**, a tube that is surrounded by an outer tube containing cold water. The cold water *condenses* the vapor in the tube, and the solvent is collected in a second flask. The collected solvent is known as the **distillate**. The accompanying diagram illustrates one way of distilling a solution.

Distillation Diagram

There are a number of other types of distillation. In Chapter 11, we will learn that the process of *fractional distillation* is used to separate petroleum into simpler mixtures of *liquid* hydrocarbons, and that this process is dependent on the *differences in the boiling points* of the hydrocarbon fractions.

Chromatography

Another technique used by chemists to separate mixtures is **chromatography**. The word *chromatography* comes from the Greek words *chroma* (color) and *graphein* (to write). The term was coined by the Russian botanist Mikhail Tsvett, who used the technique in 1906 to separate a mixture of plant pigments. In chromatography, a spot of a mixture is placed on a **stationary phase**, such as paper. The stationary phase is then placed in a sealed container of solvent, which is known as the **moving phase**. As the solvent moves along the stationary phase, it carries the mixture with it. If different components of the mixture have different degrees of attraction for the stationary phase, they will travel at different speeds along it and will be *separated* from one another.

The accompanying diagrams illustrate how one technique, known as *ascending paper chromatography*, is used to separate the components of a mixture.

Distinguishing Types of Matter

Chromatography paper (stationary phase)

Spot of mixture to be separated

Solvent front

Solvent (moving phase)

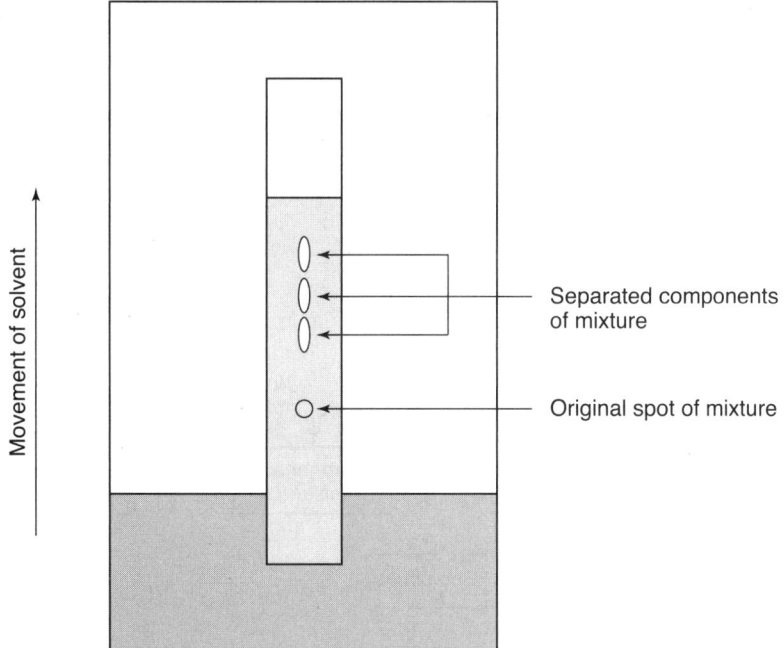

Movement of solvent

Separated components of mixture

Original spot of mixture

An element *cannot* be separated, and a compound can be separated only by more drastic "chemical" methods. These methods will change the chemical properties of the components that are part of the compound. For example, when the compound water is separated by using an electric current, this familiar liquid is replaced by two gaseous elements, hydrogen and oxygen. The chemical properties of these elements are very different than those of water.

The accompanying diagram presents a scheme for distinguishing among the various types of matter.

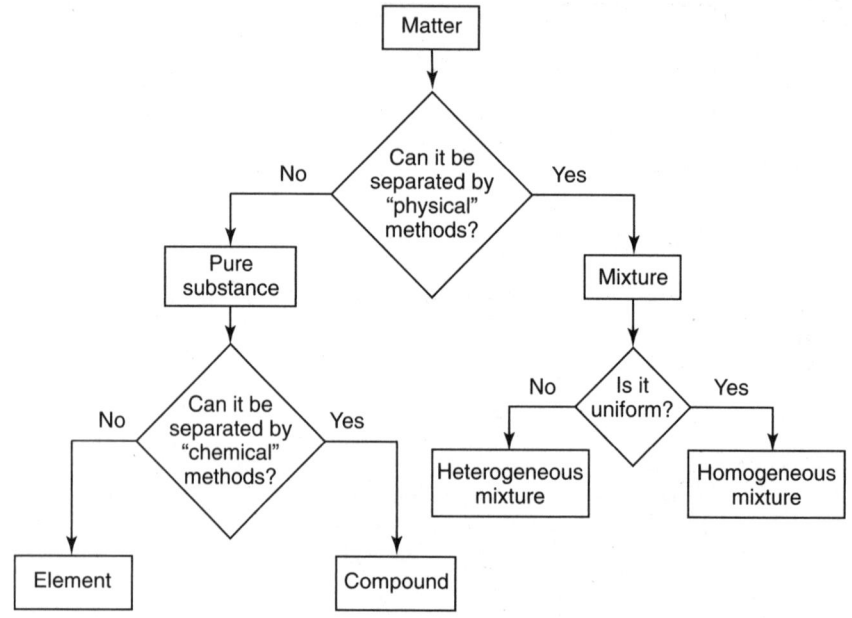

Energy

It is not easy to define **energy** precisely, but we are all familiar with at least some of its forms: thermal (heat), chemical, electrical, electromagnetic radiation (light), nuclear, and mechanical energy. Mechanical energy is divided into **kinetic energy** (the energy associated with motion) and **potential energy** (the energy associated with position). As students of chemistry, we will be particularly interested in heat and in mechanical and electrical energy.

All forms of energy are associated with the concept of mechanical work, which associates a force with a change of position. For example, we use a vacuum cleaner to transform electrical energy into the mechanical work needed to lift and remove dirt from a carpet.

The concept of energy is of great importance to chemists. For example, chemical reactions may absorb energy (*endothermic* reactions) or release it (*exothermic* reactions). Many reactions involve the conversion of one form of energy to another. Yet, as a result of countless experiments, chemists have arrived at the conclusion that energy is never destroyed in a chemical change; rather, it is *conserved*. The role of energy in chemical reactions will be treated more fully in Chapter 5.

Although work and energy may seem very different, they are closely related; in fact, the same unit is used to measure both quantities. In the SI (as shown in the next paragraph) this unit is the joule (J); its multiple is the kilojoule (kJ).

An older method of measuring energy involves calculating how much heat energy is absorbed or released by a substance. This older unit of energy is the calorie (cal); its multiple is the kilocalorie (kcal). Originally, the calorie was defined as the amount of heat energy needed to change the temperature of 1 gram of liquid water by 1 Celsius degree. The calorie is now defined in terms of the joule: 1 calorie = 4.2 joules.

1.3 MEASUREMENT AND THE METRIC SYSTEM

The basis of all science lies in the ability to measure quantities. For example, we can easily measure the length or mass of this book if we have some standardized system of measurement. We use the Système International (SI) and its derivatives because scientists all over the world express measurements in metric units. The SI has established seven fundamental quantities upon which all measurement is based: (1) length, (2) mass, (3) temperature, (4) time, (5) number of particles, (6) electric current, and (7) luminous intensity. For each of these quantities there is a unit of measure based on a standard that can be duplicated easily and does not vary appreciably.

Length

The unit of length is the meter (m), which is approximately 39 inches. In chemistry, we also use various subdivisions of the meter: picometer (pm, 10^{-12} meter), nanometer (nm, 10^{-9} meter), millimeter (mm, 0.001 meter), centimeter (cm, 0.01 meter), and decimeter (dm, 0.1 meter). The prefixes *pico-*, *nano-*, *centi-*, and so on, are known as **metric prefixes**. A list of these prefixes appears in Section 1.4.

Mass

The unit of mass is the kilogram (kg), which has an approximate weight (on Earth) of 2.2 pounds. (We note that mass and weight are not the same quantity. *Mass* is the amount of matter an object contains, while *weight* is the force with which gravity attracts matter.) In chemistry we also use the gram (g; 1000 grams equals 1 kilogram) as a unit of mass.

Temperature

Temperature measures the "hotness" of an object. The SI unit of temperature is the kelvin (K). The fixed points on the Kelvin scale of temperature are the zero point (0 K, known as "absolute zero") and 273.16 K, which is known as the "triple point" of water. (More will be said about the Kelvin scale of temperature in Chapter 6.)

In chemistry, we also use the Celsius temperature scale (°C). Originally the fixed points on this scale were set at the normal freezing and boiling points of water (0°C and 100°C, respectively).

We can convert between the Celsius and Kelvin scales using the simplified equation that appears in Reference Table T in Appendix 1:

$$K = °C + 273$$

Time

The unit of time is the second (s). In chemistry, we also use the familiar minute (min), hour (h), day (d), and year (y) as units of time.

Number of Particles

The unit of number of particles is the mole (mol). The mole is a very large number— 6.02×10^{23}. In chemistry, we frequently need to know the number of particles (such as atoms or molecules) in a sample of matter in order to predict its behavior.

Electric Current

The unit of electric current is the ampere (A). Electric current measures the flow of electric charge. In the SI system of measurement, electric charge is a *derived* quantity that is based on electric current and time. The SI unit of electric charge is the coulomb (C).

Luminous Intensity

The unit of luminous intensity is the candela (cd). As its name implies, luminous intensity measures the brightness of light. We will not need to use this unit in our study of chemistry.

1.4 METRIC PREFIXES

Very often, an SI unit is not convenient for expressing measurements. For example, the meter is not useful for describing the diameter of a red blood cell or the distance between Earth and the Sun. There are two ways to deal with this dilemma: we can use scientific notation (which is introduced in the next section), or we can use metric prefixes to create multiples and subdivisions of any unit of measure. The accompanying table lists some metric prefixes along with their symbols and values.

Factor	Prefix	Symbol	Factor	Prefix	Symbol
10^{15}	peta-	P	0.1	deci-	d
10^{12}	tera-	T	0.01	centi-	c
10^9	giga-	G	0.001	milli-	m
10^6	mega-	M	10^{-6}	micro-	μ
1000	kilo-	k	10^{-9}	nano-	n
100	hecto-	h	10^{-12}	pico-	p
10	deka-	da	10^{-15}	femto-	f

For example, we use the term picogram (pg) to designate 10^{-12} gram (one-trillionth of a gram). This is a very small quantity of matter! An abbreviated list of metric prefixes appears in Reference Table C in Appendix 1.

1.5 SCIENTIFIC NOTATION

Once we have established a system of measurement, we need to be able to express small and large numbers easily. Scientific notation accomplishes this purpose. In scientific notation, a number is expressed as a power of 10 and takes the following form:

$$M \times 10^n$$

M is called the *mantissa*, and n is the *exponent*. The mantissa is a real number that is greater than or equal to 1 and is less than 10 ($1 \le M < 10$). The exponent is an integer that can be positive, negative, or zero. For example, the number 2300 is written in scientific notation as 2.3×10^3 (*not* as 23×10^2 or 0.23×10^4). The number 0.0000578 is written as 5.78×10^{-5}.

To write a number in scientific notation, we move the decimal place until the mantissa is a number between 1 and 10. If we move the decimal place to the left, the exponent is a positive number; if we move it to the right, the exponent is a negative number.

To *multiply* two numbers expressed in scientific notation, we *multiply the mantissas and add the exponents*. The final result must always be expressed in proper scientific notation. Here are two examples:

$$(2.0 \times 10^5)(3.0 \times 10^{-2}) = 6.0 \times 10^3$$
$$(4.0 \times 10^4)(5.0 \times 10^3) = (20. \times 10^7) = 2.0 \times 10^8$$

Division is accomplished by *dividing the mantissas and subtracting the exponents.*

To *add* (*subtract*) two numbers expressed in scientific notation, *both numbers must have the same exponent. The mantissas are then added* (*subtracted*). The following examples illustrate the application of this rule:

$$(2.0 \times 10^3) + (3.0 \times 10^2) = (2.0 \times 10^3) + (0.30 \times 10^3)$$
$$= 2.3 \times 10^3$$
$$(5.0 \times 10^{-5}) - (2.0 \times 10^{-5}) = 3.0 \times 10^{-5}$$

Every calculator is different. Become familiar with how to input and manipulate scientific notation on your calculator. Doing so will help you work with calculations involving scientific notation.

1.6   VOLUME AND DENSITY

In addition to the fundamental metric quantities, there are many derived quantities that are combinations of the fundamental ones. For example, the speed of an object is the ratio of the distance (length) it travels to a given amount of time. In chemistry, two derived quantities are of immediate interest to us: volume and density.

Volume

The **volume** of an object is the amount of three-dimensional space the object occupies. The SI unit of volume is the cubic meter (m^3), which is too large for practical use in chemistry. We will use the cubic centimeter (cm^3), which is approximately equal to one-thousandth of a quart, and the liter (L), which is approximately equal to 1 quart. One liter is *exactly* equal to 1000 cubic centimeters.

We often designate one-thousandth of a liter as a milliliter (mL). It follows that 1 cubic centimeter equals 1 milliliter (1 cm^3 = 1 mL).

Density

Density is the ratio of mass to volume:

$$\text{Density} = \frac{\text{Mass}}{\text{Volume}}$$

Symbolically, this is written as follows:

$$d = \frac{m}{V}$$

Density measures the "compactness" of a substance. A substance such as lead has a large density because a relatively small volume of lead has a relatively large mass. A substance such as Styrofoam (the material used to make coffee cups) has a small density because a relatively large volume of Styrofoam has a relatively small mass.

Density is a property that depends only on the nature of a substance, not on the size of any particular sample of the substance. For example, a solid gold coin and a solid gold brick have very different masses and volumes, but they have the same density because both are pure gold.

In the SI system, the unit of density is the kilogram per cubic meter (kg/m^3). In chemistry, the densities of solids and liquids are commonly reported in grams per cubic centimeter (g/cm^3), while the densities of gases are given in grams per cubic decimeter (g/dm^3) or, equivalently, in grams per liter (g/L). We will solve some problems involving density in Section 1.8.

1.7 REPORTING MEASURED QUANTITIES

Every scientific discipline, including chemistry, is concerned with making measurements. Since no instrument is perfect, a degree of uncertainty is associated with every measurement. In a well-designed experiment, however, the uncertainty of each measurement is reduced to the smallest possible value.

Accuracy refers to how well a measurement agrees with an accepted value. For example, if the accepted density of a material is 1220 grams per cubic decimeter and a student's measurement is 1235 grams per cubic decimeter, the difference (15 g/dm^3) is an indication of the accuracy of the measurement. The smaller the difference, the more accurate is the measurement.

Precision describes how well a measuring device can reproduce a measurement. The limit of precision depends on the design and construction of the device. No matter how carefully we measure, we can never obtain a result more precise than the limit built into our measuring device. A good general rule is that the limit of precision of a measuring device is equal to plus or minus one-half of its smallest division. For example, in the following diagram:

0.1   0.2   0.3   0.4   0.5   0.6   0.7   0.8   0.9

←———————————— 1 meter ————————————→

the smallest division of the meter stick is 0.1 meter, and therefore the limit of its precision is ±0.05 meter. When we read any measurement using this meter stick, we must attach this limit to the measurement, for example, 0.27 ± 0.05 meter.

Significant digits (also called **significant figures**) are the digits that are part of any valid measurement. The number of significant digits is a direct result of the number of divisions the measuring device contains. The following diagram shows a meter stick with no divisions. How should the measurement indicated by the arrow be reported?

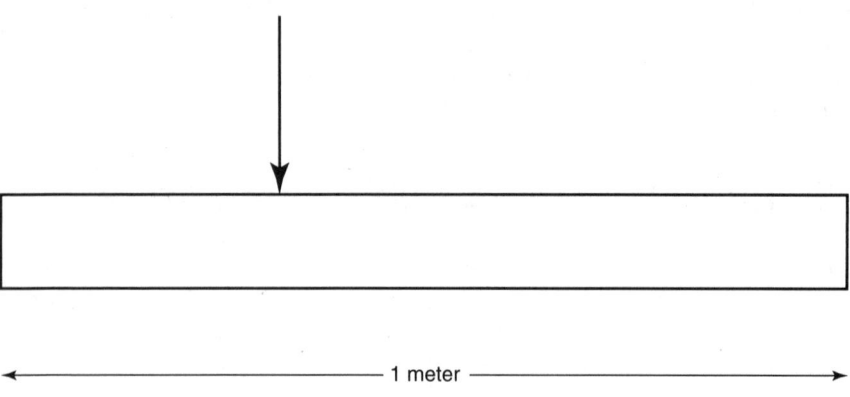

←———————————— 1 meter ————————————→

Since there are no divisions, all we know is that the measurement is some-where between 0 and 1 meter. The best we can do is to make an educated guess based on the position of the arrow. To obtain the best educated guess, we divide the meter stick mentally into ten divisions and report the mea-surement as 0.3 meter. This meter stick allows us to measure length to one significant digit.

Suppose we now use a meter stick that has been divided into tenths, as shown on the next page, and repeat the measurement.

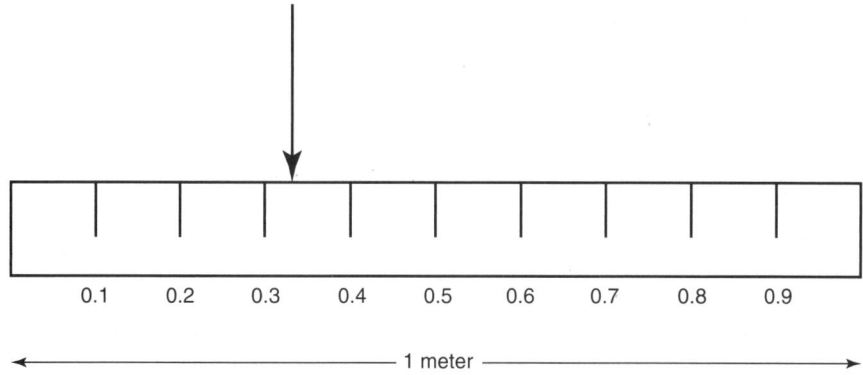

With this device we can report the measurement with less uncertainty because we know that the indicated length lies between 0.3 and 0.4 meter. Allowing ourselves one educated guess, we will report the length as 0.33 meter. This measurement has two significant digits. The more significant digits a measurement has, the more confidence we have in our ability to reproduce the measurement because only the last digit is in doubt.

Measurements that contain zeros can be particularly troublesome. For example, we say that the average distance between Earth and the Moon is 238,000 miles. Do we really know this number to six significant digits? If so, we would have measured the distance to the nearest mile. Actually, this measurement contains only three significant figures. The distance is being reported to the nearest thousand miles. The zeros simply indicate how large the measurement is.

To avoid confusion, a number of rules have been established for determining how many significant digits a measurement has.

Rules for Determining the Number of Significant Digits in a Measurement

1. All nonzero numbers are significant. The measurement 2.735 grams has four significant digits.
2. Zeros located between nonzero numbers are also significant. The measurements 1.0285 liters and 202.03 torr each have five significant digits.
3. For numbers greater than or equal to 1, zeros located at the end of the measurement are significant only if a decimal point is present. The measurement 60 grams has one significant digit. In this case, the zero indicates the size of the number, not its significance. The measurements 60. grams and 60.000 grams, however, have two and five significant digits, respectively.

4. For numbers less than 1, leading zeros are not significant; they merely indicate the size of the number or numbers that follow. Thus, the measurements 0.00**2** kilogram, 0.0**20** kilogram, and 0.000**200** kilogram have one, two, and three significant digits, respectively; the significant digits are indicated in bold type.

Using Significant Digits in Calculations

Significant digits are particularly important in calculations involving measured quantities, and it is crucial that the result of a calculation does not imply a greater precision than any of the individual measurements. Calculators routinely give us answers with ten digits. It is incorrect to believe, however, that the results of most of our calculations have this many significant digits.

Multiplication and Division of Measurements

When two measurements are multiplied (or divided), the answer should contain as many significant digits as the less precise measurement. For example, if the measurement 2.3 centimeters (two significant digits) is multiplied by 7.45 centimeters (three significant digits), the answer must contain only two significant digits:

$$(2.3 \text{ cm}) \cdot (7.45 \text{ cm}) = 17.135 \text{ cm}^2 = 17 \text{ cm}^2$$

(Note that the units are also multiplied.)

TRY IT YOURSELF
Divide the following measurements, and express the result to the correct number of significant figures: 6.443 grams/8.91 liters.

ANSWER
0.723 g/L

Addition and Subtraction of Measurements

When two measurements are added (or subtracted), the answer should contain as many decimal places as the measurement with the smaller number of decimal places. For example, when 8.11 liters (two decimal places) and 2.476 liters (three decimal places) are added, the answer must be taken only to the second decimal place:

$$8.11 \text{ L} + 2.476 \text{ L} = 10.586 \text{ L} = 10.59 \text{ L}$$

TRY IT YOURSELF

Add the following measurements, and express the result to the correct number of significant figures: 246.213 milliliters, 79.91 milliliters, 8786.268 milliliters.

ANSWER

9112.39 mL

Note: If *counted* numbers (such as six atoms) or *defined* numbers (such as 273.16 K) are used in any calculations, they are treated as though they have an infinite number of significant digits or decimal places.

Rounding a Number

Suppose we had a measurement with five significant figures. If we wished to express it to three significant figures, we would need to round this measurement.

We *round* a measurement by looking one digit beyond the precision we need. If the digit is less than 5, we do not change the value of the preceding digit. If it is 5 or greater, we raise the preceding digit by 1.

For example, if our measurement was 127.36 grams and we wanted to round it to three significant figures, we would first examine the fourth digit. Since 3 is less than 5, we would leave the third digit, 7, unchanged, and the rounded measurement would be 127 grams expressed to three significant figures.

If, however, we wished to round the measurement 127.36 grams to four significant figures, we would need to examine the fifth digit. Since 6 is greater than 5, we would raise the preceding digit, 3, by 1. Expressed to four significant figures, this measurement would be 127.4 grams.

TRY IT YOURSELF

Round each of the following measurements to three significant digits:
(a) 903.04 L   (b) 298.86 K   (c) 0.002259 mol

ANSWERS

(a) 903 L   (b) 299 K   (c) 0.00226 mol

Order of Magnitude

There are times when we are interested in the relative size of a measurement rather than its actual value. The *order of magnitude* of a measurement is the power of 10 closest to its value. For example, the order of magnitude of 1284 grams (1.284×10^3) is 10^3, while the order of magnitude of 8756 grams (8.756×10^3) is 10^4. Orders of magnitude are very useful for comparing quantities, such as mass or distance, and for estimating the answers to problems involving complex calculations.

TRY IT YOURSELF
What is the order of magnitude of Avogadro's number (6.02×10^{23})?

ANSWER
10^{24}

Percent Error of a Measurement

The **percent error** of a measurement indicates how closely the measurement agrees with an accepted value of the same quantity. Its definition is as follows:

$$\text{Percent Error} = \frac{\text{Measured Value} - \text{Accepted Value}}{\text{Accepted Value}} \times 100$$

The *measured value* is the value that is experimentally determined, and the *accepted value* is the value that is generally accepted as the most probable value of the measurement. Generally, we are interested only in the *magnitude* of the percent error, not in its algebraic sign.

PROBLEM
A student measures the density of an object to be 5600 grams per liter. The accepted density of the object is 6400 grams per cubic liter. What is the percent error of this measurement?

SOLUTION

$$\text{Percent Error} = \frac{\text{Measured Value} - \text{Accepted Value}}{\text{Accepted Value}} \times 100$$

$$= \frac{5600 \text{ g/L} - 6400 \text{ g/L}}{6400 \text{ g/L}} \times 100$$

$$= \frac{-800 \text{ g/L}}{6400 \text{ g/L}} \times 100$$

$$= 12\%$$

Note that the answer is reported as a positive value even though the numerator of the fraction in this example is negative.

TRY IT YOURSELF
The accepted value for the volume of a gas is 22.4 liters. A student reports her measurement as 23.8 liters. What is the percent error of her measurement?

ANSWER
6.25%

1.8 SOLVING PROBLEMS

Using the Factor-Label Method (FLM) in Chemistry

We now introduce a general technique for solving problems known as the *factor–label method* (FLM). Suppose we want to solve this problem: How many inches equal 40. feet? We know that, by definition, 12 inches equals 1 foot. FLM allows us to use this relation to form two fractions called *conversion factors:*

$$\frac{12 \text{ inches}}{1 \text{ foot}} \quad \text{and} \quad \frac{1 \text{ foot}}{12 \text{ inches}}$$

These fractions are what we will use to solve the problem. First we write a tentative solution as follows:

$$x \text{ inches} = 40. \text{ feet}$$

Because we need to convert feet to inches, we must cancel the unit "feet" on the right and substitute "inches" in its place. This can be done by using the first conversion factor as follows:

$$x \text{ inches} = 40. \; \cancel{\text{feet}} \cdot \frac{12 \text{ inches}}{1 \; \cancel{\text{foot}}} = 480 \text{ inches}$$

Note that units can be multiplied and divided just as numbers are. In the example above, the "feet" units cancel, leaving "inches" in the numerator—which is exactly what we needed to do. All that remains is the arithmetic (40. · 12), and our answer is 480 inches.

Now let us solve a slightly more complicated problem.

PROBLEM
How many hours are there in 5 years?

SOLUTION
First, we need a plan of attack—a "road map" that gives us direction:

$$\text{YEARS} \rightarrow ??? \rightarrow \text{HOURS}$$

The symbol "???" must represent some unit that connects years to hours and is easily related to both these units. The unit "days" will do very nicely since there are 365 days in 1 year and 24 hours in 1 day. We set up our problem for solution:

$$x \text{ hours} = 5 \text{ years} \cdot \frac{365 \text{ days}}{1 \text{ year}} \cdot \frac{24 \text{ hours}}{1 \text{ day}} = 43{,}800 \text{ hours}$$

(40,000 hours to 1 significant figure)

Our next problem involves units with exponents.

PROBLEM

How many cubic centimeters are there in 1 cubic decimeter?

SOLUTION

If we examine the metric prefix table in Section 1.4, we can conclude that 1 meter equals 100 centimeters and also equals 10 decimeters. Let us set up the solution to the problem as though the exponent 3 were not present. Our "road map" for this problem is:

DECIMETERS → METERS → CENTIMETERS

Using FLM, and standard abbreviations for the units, we can write:

$$x \text{ cm} = 1 \text{ dm} \cdot \left(\frac{1 \text{ m}}{10 \text{ dm}} \right) \cdot \left(\frac{100 \text{ cm}}{1 \text{ m}} \right)$$

Adding the exponent 3 is easy: we simply add it to each term in the equation:

$$x \text{ cm}^3 = 1 \text{ dm}^3 \cdot \left(\frac{1 \text{ m}}{10 \text{ dm}} \right)^3 \cdot \left(\frac{100 \text{ cm}}{1 \text{ m}} \right)^3 = 1000 \text{ cm}^3$$

Our final problem involves units that appear in the numerator and denominator of a fraction.

PROBLEM

A substance called sulfuric acid flows out of a pipe at the rate of 2500 grams per second (2500 g/s). What is the flow rate, in kilograms per hour (kg/h), of the sulfuric acid?

SOLUTION

We will need two "road maps" to solve this problem:

GRAMS → KILOGRAMS

SECONDS → MINUTES → HOURS

We know that there are 60 seconds in 1 minute, 60 minutes in 1 hour, and 1000 grams in 1 kilogram. The FLM solution is:

$$x \ \frac{kg}{h} = 2500 \ \frac{g}{s} \cdot \left(\frac{1 \ kg}{1000 \ g}\right) \cdot \left(\frac{60 \ s}{1 \ min}\right) \cdot \left(\frac{60 \ min}{1 \ h}\right)$$

$$= 9.0 \times 10^3 \ \frac{kg}{h}$$

TRY IT YOURSELF

The density of water is very nearly 1 gram per cubic centimeter (g/cm^3). What is the density of water in kilograms per cubic meter (kg/m^3)? (*Hint:* Use the techniques of the last two problems solved above.)

ANSWER

$$\text{Density} = 1\frac{g}{cm^3} = 1000 \ \frac{kg}{m^3}$$

As you can see, FLM is a very powerful tool for solving a wide variety of chemistry problems. Whenever you can find simple connections between the units in a problem, you should use FLM to solve that problem. We will use FLM as a problem-solving tool throughout this book.

Using Mathematical Equations in Chemistry

In our study of chemistry, we are frequently required to solve problems in which the variables are related by an equation. We can solve such problems quite easily if we follow five simple steps:

1. Prepare a list of the variables and their values. (Remember to include units!)
2. Write the equation that relates the variables of the problem.
3. Rewrite the equation in order to isolate the unknown variable.
4. Substitute the known values and their units into the rewritten equation.
5. Perform the arithmetic operations indicated by the equation to arrive at a solution.

The following problem shows how this technique is employed.

PROBLEM
The density of a substance is 1200 grams per liter (g/L). Calculate the mass of a sample of this substance if its volume is 0.50 liter (L).

SOLUTION
(The numbers correspond to the steps given on the previous page.)

1. List of values:

$$d = 1200 \text{ g/L}$$
$$V = 0.50 \text{ L}$$
$$m = ???$$

2. Relevant equation:

$$d = \frac{m}{V}$$

3. Rewritten equation:

$$m = d \cdot V$$

4. Substitution of values:

$$m = 1200\frac{g}{\cancel{L}} \cdot 0.50 \, \cancel{L}$$

5. Performance of arithmetic operations:

$$m = 6.0 \times 10^2 \text{ g}$$

TRY IT YOURSELF
The density of a substance is 8.0 grams per liter (g/L). What is the volume of a sample of this substance if its mass is 24 grams?

ANSWER
$V = 3.0$ L

PROBLEM
What is the boiling point of mercury (630 K) on the Celsius temperature scale?

SOLUTION

$$K = °C + 273$$
$$°C = K - 273$$
$$= 630 - 273$$
$$= 357°C$$

PROBLEM
What is the normal boiling point of water (100°C) on the Kelvin temperature scale?

SOLUTION

$$K = °C + 273$$
$$= 100 + 273$$
$$= 373 \text{ K}$$

TRY IT YOURSELF
(a) Convert 155 K to the Celsius temperature scale.
(b) Convert 37°C to the Kelvin temperature scale.

ANSWERS
(a) −118°C   (b) 310 K

Using Graphs in Chemistry

Sometimes data are presented as a table of values, and we are asked to interpret the data and draw conclusions from them. In such cases a graph of the data values may provide information about values that do not appear in the original table. In addition, the slope and the x- and y-intercepts of the graph may provide useful information. The following problem illustrates how a graph can be used to obtain additional information.

PROBLEM
A student measures the masses and volumes of several samples of the same substance and obtains the results shown in the table.

Volume/L	Mass/g
6.80	20.0
13.5	40.0
20.9	60.0
28.8	80.0
34.2	100.0

(a) Draw a graph of the data provided in the table.
(b) What is the mass of a sample of the substance if its volume is 25.0 liters?
(c) What is the density of the substance?

SOLUTIONS
(a) We begin by graphing the data points. We plot volume on the x-axis and mass on the y-axis because this arrangement will be useful in answering part (c) of the problem. Note that the axes are "scaled" so that the plotted points fill nearly the entire graph, allowing any trend in the data to be observed.

After we plot the data points, we can see that they form almost a straight line. Instead of playing "connect the dots," we draw a *best-fit* straight line (that is, a line that is most closely associated with its data points) as shown.

27

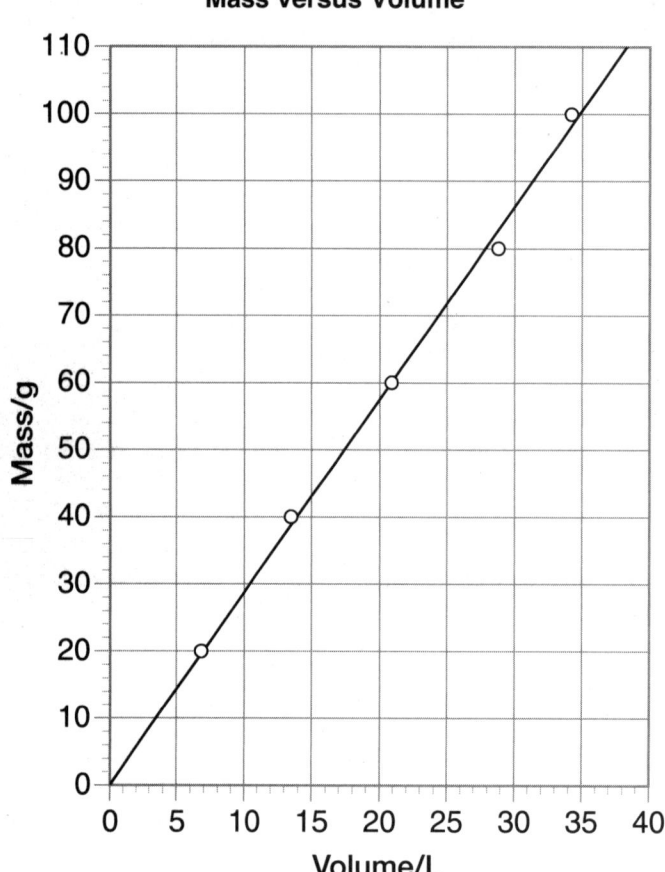

In this example, we draw our best-fit straight line so that it passes through the origin, because a mass of 0.0 g corresponds to a volume of 0.0 L.

(b) By inspecting the graph, we can see that a volume of 25.0 L corresponds to a mass of 71.8 g.

(c) Since every data point does not fit on the line, we cannot use individual data points to calculate the density of the substance. The slope of the line, however, will give us the density because it measures the ratio of mass to volume for the entire set of data. The diagram on the graph shows the quantities Δm and ΔV, and the slope of the line is calculated from the ratio $\dfrac{\Delta m}{\Delta V}$.

Mass versus Volume

A careful calculation of the slope of the line yields a value of 2.87 g/L for the density of the substance.

TRY IT YOURSELF
Use the graph drawn above to calculate the volume of a sample whose mass is 90.0 grams.

ANSWER
According to the graph, a mass of 90.0 g corresponds to a volume of 31.4 L.

In other situations, we may be presented with data and asked to connect all of the data points, as shown in the following problem.

PROBLEM

The following table lists the atomic numbers (numbers used to identify elements) and melting points of several elements:

Atomic Number	Melting Point /K
3	454
19	337
37	312
55	302

(a) Plot the data points on a graph in which the atomic numbers lie along the x-axis and the melting points lie along the y-axis. Your graph should be scaled so that a trend is clearly observed.

(b) Connect each adjacent pair of data points with a straight line. Describe the trend associated with the plotted data points.

(c) Estimate the melting point of the element whose atomic number is 11. (To see how close to the accepted value your graph came, refer to Reference Table S in Appendix 1.)

SOLUTIONS

(a) The completed graph is shown below:

Notice that the y-axis doesn't start at zero since you are plotting melting points that lie between 300 K and 500 K.

(b) The trend is fairly obvious: as the atomic number increases, the melting point decreases. This trend is known as an *inverse relationship*.

(c) According to the graph, the element whose atomic number is 11 has a melting point of approximately 395 K. According to Reference Table S, the accepted value for the melting point is 371 K. (This corresponds to a 6.5% error.)

END-OF-CHAPTER QUESTIONS

1. In which pair are the members classified as substances?
   (1) mixtures and solutions   (2) compounds and solutions
   (3) elements and mixtures   (4) compounds and elements

2. Which can *not* be decomposed by a chemical change?
   (1) a compound   (2) a heterogeneous mixture
   (3) a homogeneous mixture   (4) an element

3. An example of a heterogeneous mixture is
   (1) an ice cream soda   (2) a sugar solution
   (3) table salt   (4) carbon dioxide

4. Which statement describes a characteristic of all compounds?
   (1) Compounds contain one element, only.
   (2) Compounds contain two elements, only.
   (3) Compounds can be decomposed by chemical means.
   (4) Compounds can be decomposed by physical means.

5. Which is a characteristic of all mixtures?
   (1) They are homogeneous.
   (2) They are heterogeneous.
   (3) Their composition is generally fixed.
   (4) Their composition generally can be varied.

6. Which is the equivalent of 750. joules?
   (1) 0.750 kJ   (2) 7.50 kJ   (3) 75.0 kJ   (4) 750. kJ

7. Which temperatures originally represented the fixed points on the Celsius temperature scale?
   (1) 32° and 100°   (2) 32° and 212°
   (3) 0° and 212°   (4) 0° and 100°

8. Expressed in proper scientific notation, the number 0.00213 is
(1) 0.213×10^2 (2) 2.13×10^{-2}
(3) 2.13×10^{-3} (4) 213×10^0

9. When the numbers 1.2×10^2 and 3.4×10^1 are added, the result, expressed in proper scientific notation, is
(1) 1.5×10^2 (2) 4.6×10^2 (3) 1.5×10^1 (4) 4.6×10^1

10. When the numbers 4.2×10^2 and 8.4×10^1 are multiplied, the exponent of the result, expressed in proper scientific notation, is
(1) 1   (2) 2   (3) 3   (4) 4

Base your answers to questions 11 and 12 on the following table, which represents measurements made on four rectangular blocks in a chemistry laboratory.

Block	Mass (g)	Length (cm)	Width (cm)	Height (cm)
A	72.	4.0	3.0	2.0
B	60.	5.0	2.0	1.0
C	30.	10.	1.0	1.0
D	6.0	4.0	2.0	1.5

11. Which block has the smallest density?
(1) *A*   (2) *B*   (3) *C*   (4) *D*

12. Which two blocks may be made of the same material?
(1) *A* and *B*   (2) *A* and *C*   (3) *B* and *C*   (4) *B* and *D*

13. A cube has a volume of 8.00 cubic centimeters and a mass of 21.6 grams. The density of the cube is best expressed as
(1) 2.7 g/cm^3 (2) 2.70 g/cm^3
(3) 0.37 g/cm^3 (4) 0.370 g/cm^3

14. In a laboratory exercise to determine the density of a substance, a student found the mass of the substance to be 6.00 grams and the volume to be 2.0 milliliters. Expressed to the correct number of significant figures, the density of the substance is
(1) 3.000 g/mL   (2) 3.00 g/mL   (3) 3.0 g/mL   (4) 3 g/mL

15. Which measurement contains a total of three significant figures?
(1) 0.012 g   (2) 0.125 g   (3) 1205 g   (4) 12.050 g

16. The diagram below shows a section of a 100-milliliter graduated cylinder:

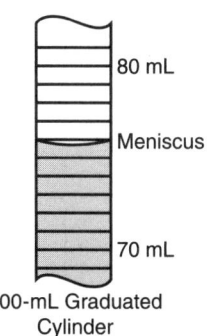

80 mL

Meniscus

70 mL

100-mL Graduated
Cylinder

When the bottom of the meniscus is read to the correct number of significant figures, the volume of water in the cylinder will be recorded as
(1) 75.7 mL   (2) 75.70 mL   (3) 84.3 mL   (4) 84.30 mL

17. A student determined the melting point of a substance to be 328 K. If the accepted value is 323 K, the percent error in her determination was
(1) 0.152%   (2) 1.52%   (3) 1.55%   (4) 5.00%

18. When the rules for significant figures are used, the sum of 0.027 gram and 0.0023 gram should be expressed as
(1) 0.029 g   (2) 0.0293 g   (3) 0.03 g   (4) 0.030 g

19. Which measurement contains three significant figures?
(1) 0.05 g   (2) 0.050 g   (3) 0.056 g   (4) 0.0563 g

20. The graph below represents an experiment in which the decay of a radioactive substance is measured.

Mass versus Time

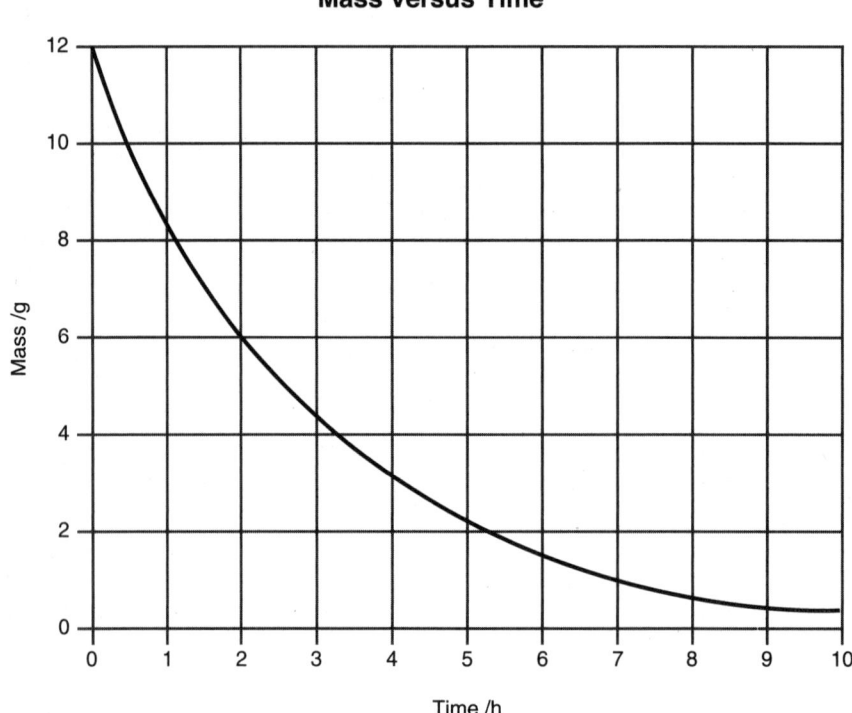

Time /h

In how many hours will this substance have decayed to one-half its initial value?
(1) 1   (2) 2   (3) 3   (4) 6

21. In an experiment, a student found that the percent of oxygen in a sample was 42.3%. If the accepted value is 39.3%, the magnitude of the experimental percent error is
(1) $42.3/39.3 \times 100\%$ (2) $3.0/42.3 \times 100\%$
(3) $39.3/42.3 \times 100\%$ (4) $3.0/39.3 \times 100\%$

22. According to an accepted chemistry reference, the energy needed to vaporize a substance is 5400 joules per gram. A student determined in the laboratory that this energy was 6200 joules per gram. The student's results had a percent error of
(1) 13   (2) 15   (3) 80.   (4) 87

23. Expressed to the correct number of significant figures, what is the sum of (3.04 g + 4.134 g + 6.1 g)?

(1) 13 g   (2) 13.3 g   (3) 13.27 g   (4) 13.274 g

24. In the calculation 41.06 centimeters × 10.2 centimeters, how many significant figures should the product of the two values contain?

(1) 5   (2) 6   (3) 3   (4) 4

25. In an experiment, the density of a gas was determined to be 2.47 grams per liter. Compared to the accepted value of 2.43 grams per liter, the percent error for this determination was

(1) 0.400   (2) 1.65   (3) 24.7   (4) 98.4

26. The graph below shows the volume and the mass of four different substances at STP.

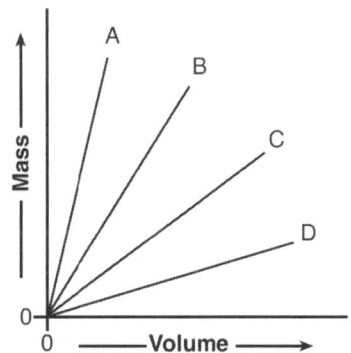

Which of the four substances has the *lowest* density?

(1) A   (2) B   (3) C   (4) D

27. Five cubes of iron are tested in a laboratory. The tests and results are shown in the table below.

Iron Tests and the Results

Test	Procedure	Result
1	A cube of Fe is hit with a hammer.	The cube is flattened.
2	A cube of Fe is placed into 3 M HCl(aq).	Bubbles of gas form.
3	A cube of Fe is heated to 1811 K.	The cubes melts.
4	A cube of Fe is left in damp air.	The cube rusts.
5	A cube of Fe is placed into water.	The cube sinks.

Which tests demonstrate chemical properties?

(1) 1, 3, and 4   (2) 1, 3, and 5   (3) 2 and 4   (4) 2 and 5

28. The ratio of chromium to iron to carbon varies among the different types of stainless steel. Therefore, stainless steel is classified as
    (1) a compound
    (2) an element
    (3) a mixture
    (4) substance

Constructed-Response Questions

1. Express the following numbers in scientific notation:
   (a) 63,477   (b) 0.000230

2. Express the following numbers in standard decimal form:
   (a) 6.54×10^5 (b) 5.55×10^{-3}

3. In a certain foreign country, length is measured in "piddles" and "daddles." One piddle is defined to be exactly 3.33 daddles. How many piddles are there in 7.00 daddles?

4. The compound ethanol has a density of 0.84 gram per milliliter. What is the mass of 25.0 milliliters of ethanol?

5. Solve the following relationship for f: $ab = \dfrac{cd}{ef}$.

6. The table below represents the volumes and corresponding masses of seven samples of a substance.

Volume/cm³	Mass/g
1.33	8.22
2.41	11.5
3.72	18.7
4.98	25.8
6.81	34.6
8.13	43.8
10.3	54.9

   (a) Plot the data points on the grid on page 37 and draw the best-fit line through the points.

(b) Determine the density of the substance by measuring the slope ($\Delta m / \Delta V$) of the best-fit line.

Mass versus Volume

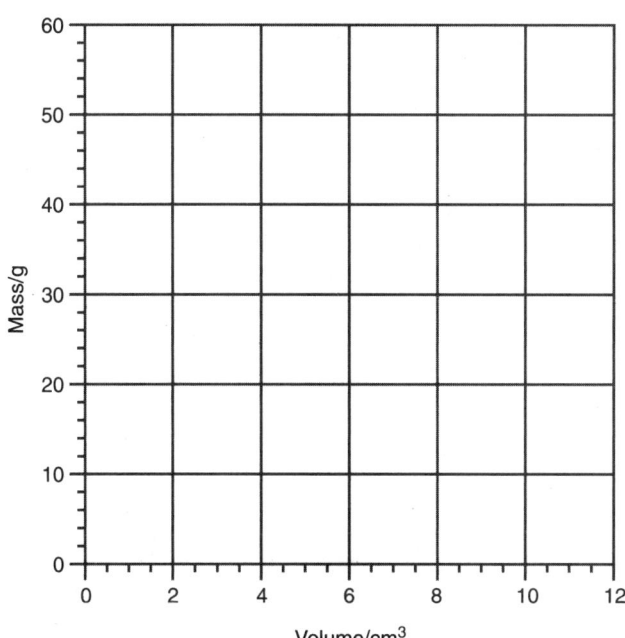

7. How many significant figures are there in each of the following measurements?
   (a) 2.450 mL
   (b) 0.10007 mol
   (c) 0.00023 kg
   (d) 7.01 J
   (e) 50.00 torr

8. Express each of the following operations to the correct number of significant figures:
   (a) 21.2 m + 9.7654 m + 312.22 m
   (b) 0.91765 kJ − 0.9012 kJ
   (c) 6.72 cm × 0.32 cm
   (d) 12.316 g ÷ 2.3 L

Base your answers to questions 9 through 11 on the particle diagrams below, which show atoms and/or molecules in three different samples of matter.

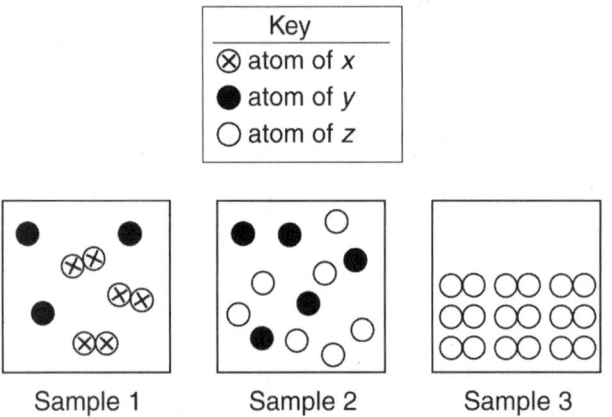

Sample 1   Sample 2   Sample 3

9. Which sample represents a pure substance?

10. When two atoms of *y* react with one atom of *z*, a compound forms. Using the number of atoms shown in sample 2, what is the maximum number of molecules of this compound that can be formed?

11. Explain why ⊗⊗ does not represent a compound.

The answers to these questions are found in Appendix 3.

Chapter Two

ATOMS, MOLECULES, AND IONS

KEY IDEAS

This chapter traces the development of the early atomic models of matter from Dalton to Thomson and Rutherford. The nuclear model is then studied with particular emphasis on the placement of protons, neutrons, and electrons within atoms. Molecules, ions, and the Periodic Table of the Elements are also introduced in this chapter.

KEY OBJECTIVES

At the conclusion of this chapter you will be able to:

- Compare the Dalton, Thomson, and Rutherford models of the atom.
- Describe the placement of protons, neutrons, and electrons according to the nuclear model of the atom.
- Define and apply the terms *atomic number*, *mass number*, and *isotope*.
- Calculate the atomic mass of an element, given the mass and abundance of each of its naturally occurring isotopes.
- Define the terms *group* and *period* as they relate to the Periodic Table of the Elements.
- Define and apply the terms *molecule*, *ion*, *monoatomic*, and *polyatomic*.

SECTION I—BASIC (REGENTS-LEVEL) MATERIAL

NYS REGENTS CONCEPTS AND SKILLS

Note: By the time you have finished Section I, you should have mastered the concepts and skills listed below. The Regents chemistry examination will test your knowledge of these items and your ability to apply them.

Concepts are the *basic ideas* that form the body of the Regents chemistry course (what you need to know!).

Skills are the *activities* that demonstrate your mastery of these concepts (how you show that you know them!).

Following each concept or skill is a page reference (given in parentheses) to this chapter.

2.1 Concept:
The modern model of the atom has evolved over a long period of time through the work of many scientists. (Pages 42–44)

Skill:
Relate experimental evidence to models of the atom. (Pages 42–44)

2.2 Concept:
Each atom has a nucleus, with an overall positive charge, surrounded by one or more negatively charged electrons. (Page 44)

Skill:
Use models to describe atoms. (Page 47)

2.3 Concept:
Subatomic particles contained in the nucleus include protons and neutrons. (Pages 44–45)

2.4 Concept:
The proton is positively charged, and the neutron has no charge. The electron is negatively charged. (Page 45)

2.5 Concept:
Protons and electrons have equal but opposite charges. The number of protons equals the number of electrons in an atom. (Page 45)

Skill:
Determine the number of protons or electrons in an atom or ion. (Pages 46–47)

2.6 Concept:
The mass of a proton is nearly equal to the mass of a neutron. An electron is much less massive than a proton or a neutron. (Page 46)

2.7 Concept:
Atoms of an element that contain the same number of protons but different numbers of neutrons are called isotopes of that element. (Page 46)

2.8 Concept:
The number of protons in an atom (atomic number) identifies the element. The sum of the protons and neutrons in an atom (the mass number) identifies an isotope. Common notations that represent isotopes include ^{6}C, C, carbon-14, C-14. (Page 46)

Skills:
- Interpret and write isotopic notation. (Page 46)
- Given any two of the following values: number of protons, number of neutrons, mass number, calculate the third value. (Pages 46–47)

2.9 Concept:
When an atom gains one or more electrons, it becomes a negative ion. When an atom loses one or more electrons, it becomes a positive ion. (Page 49)

2.1 INTRODUCTION TO THE ATOMIC MODEL OF MATTER

If we wish to explain the *properties* of matter, for example, how substances combine with one another, we will need a detailed description of how matter is constructed. Because of very strong experimental evidence, we accept the *atomic model* as a reliable basis on which to explain the properties of matter.

The term *model* needs further explanation. Scientific models are very similar to road maps, because they answer questions and make predictions. For example, if we wanted to travel from New York to Chicago, a map of the United States would indicate the appropriate interstate highways we ought to take. In addition, we would be able to predict through which major towns and cities we would pass on our journey. But our map would *not* tell us the condition of the highways or provide a street-by-street depiction of Chicago. For this information, we would need more detailed maps. However, no matter how detailed our maps became, they could never provide answers to *all* of our questions because, as models, they only *represent* the landscape; they are not the landscape itself.

A scientific model, such as the atomic model, also gives us only an approximate picture of reality. Nevertheless, we use such models because they provide some answers for us and they allow us to make predictions as to how a part of the universe will behave in a given situation. As we gather more information, we refine and revise our models so that they become even more useful to us. The atomic model of matter has a long history, and with each revision it has provided more scientific information.

2.2 DEVELOPMENT OF THE EARLY MODELS OF THE ATOM

The Dalton Model

In the period 1803–1807, an English school teacher named John Dalton proposed a model of atomic structure based on experimental work. This model can be summarized as follows:

- Each element is composed of indivisible particles called *atoms*.
- In an *element*, all of the atoms are identical; atoms of different elements have different properties, including mass.
- In a chemical reaction, atoms are not created, destroyed, or changed into other types of atoms; the atoms are simply rearranged.
- *Compounds* are formed when atoms of more than one element combine.
- Samples of a given compound *always* have the same relative numbers of atoms. (For example, all samples of water contain two hydrogen atoms for every oxygen atom present.)

According to Dalton's model, atoms are the basic building blocks of matter. The model explains why elements and compounds have fixed compositions by mass. One serious shortcoming of this model, however, is its failure to provide clues as to the *internal structure* of atoms.

Cathode Rays and Electrons

By 1850, experimental data suggested that atoms are composed of smaller **subatomic particles**. In one such experiment, high-voltage electricity was passed across an evacuated glass tube containing metal electrodes, as shown in the accompanying diagram. The electricity produced a stream of radiation that had a negative electric charge and flowed from the negative electrode (the *cathode*) to the positive electrode (the *anode*). This radiation became known as *cathode rays*, and the glass tube is called a *cathode-ray tube*.

Cathode Ray Tube

42

Although cathode rays could not be seen, they were made visible by beaming them at certain *fluorescent* materials. In 1897, the British physicist J. J. Thomson examined cathode rays produced from many different sources and concluded that the rays consisted of streams of identical, negatively charged *particles*, which he called *electrons*. Thomson's experiments enabled him to measure the ratio of the electron's charge to its mass without knowing the value of either the charge or the mass. (This is analogous to our knowing that there are four quarters in every dollar without knowing the number of quarters or dollars we have in our pockets.)

In 1909, the American physicist Robert Millikan determined the charge on an electron by measuring a series of charges on irradiated oil droplets. He discovered that all of the measured charges were multiples of a single number. That number, which is the charge on an electron, is also known as an *elementary charge*. In SI units, the currently accepted value for an elementary charge is 1.6022×10^{-19} coulomb. As a result of Thomson's and Millikan's work, it was possible to calculate the mass of the electron. In SI units, the currently accepted value is 9.1094×10^{-31} kilogram.

The Thomson Model

Thomson's work with cathode rays enabled him to propose a model of the atom that had internal structure. In the Thomson model, an atom consists of a (positively charged) jellylike mass with (negative) electrons scattered throughout it—much as raisins are spread throughout a plum pudding. (You don't know what a plum pudding is? Ask your teacher or any other older person!) This model did not contradict the Dalton model; it merely extended it by providing the atom with some structure.

The Rutherford Model

In 1910, the New Zealand physicist Ernest Rutherford, working at the University of Manchester in England, and two of his students (Hans Geiger and Ernest Marsden) bombarded thin gold foils with the nuclei of helium atoms and observed how these particles were scattered by the foils. (These positively charged particles were named *alpha particles;* they consist of two protons and two neutrons each.) Rutherford and his coworkers observed that most of the alpha particles passed through the foil with little or no deflection. However, a small number of particles were deflected at large angles. This is shown in the accompanying diagram.

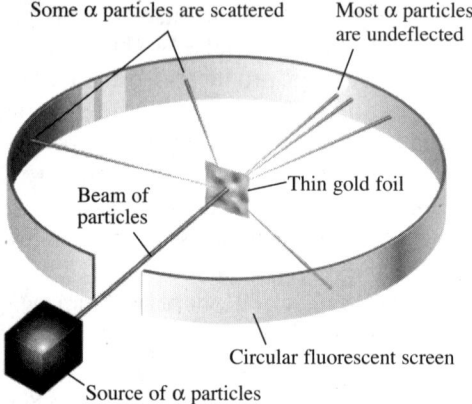

Some α particles are scattered

Most α particles are undeflected

Beam of particles

Thin gold foil

Circular fluorescent screen

Source of α particles

This work showed that the Thomson model of the atom was not consistent with the experimental evidence. Consequently, in 1911, Rutherford proposed a new model with the following characteristics:

- Most of the (volume of the) atom is empty space.
- Most of the mass of the atom is concentrated in a dense, positively charged nucleus.
- Electrons are present in the space surrounding the nucleus.

Rutherford developed the *nuclear model* of the atom, and we credit him with the discovery of the atomic nucleus. The nucleus is so important that we will devote all of Chapter 7 to it.

2.3 THE CURRENT VIEW OF ATOMIC STRUCTURE

We present a simplified view of the atom that is based on the experimental work carried out during the first three decades of the twentieth century. The structure of the atom is based on its principal subatomic particles. An atom consists of a *nucleus* that is positively charged and contains most of the mass of the atom. In the nucleus, we find the *nucleons*: the positively charged *protons* and the *neutrons*, which have no charge. Outside the nucleus are the negatively charged *electrons*. The properties of these particles are summarized in the accompanying table.

| Particle | Location | CHARGE | | MASS | | |
		Charge/C*	Relative Charge†	Mass/kg	Mass/u‡	Relative Mass§
Proton	Nucleus	$+1.6022 \times 10^{-19}$	$1+$	1.6726×10^{-27}	1.0073	1
Neutron	Nucleus	0	0	1.6749×10^{-27}	1.0087	≈ 1
Electron	Outside nucleus	-1.6022×10^{-19}	$1-$	9.1094×10^{-31}	0.00054858	≈ 0.0005

*The symbol C stands for coulomb, the SI unit of electric charge.
†A **relative charge** is one whose magnitude is compared with the magnitude of an elementary charge, which is assigned a value of 1. In this book, we will adopt the convention that the magnitude of a relative charge is written before its sign.
‡The symbol u stands for atomic mass unit. It will be defined in Chapter 4.
§A **relative mass** is based on a comparison with the mass of a proton, which is assigned a value of 1.

2.4 IDENTIFYING ELEMENTS: NAMES, SYMBOLS, AND ATOMIC NUMBERS

Every element (as well as its atoms) is associated with three unique identifiers: its *name*, *symbol*, and *atomic number*. Reference Table S in Appendix 1 lists most of the elements by atomic number, symbol, and name.

The **name** of an element may be based on a person (e.g., *einsteinium*), a place (e.g., *francium*), or a characteristic of the element itself (e.g., *chlorine* from the Greek word *chloros*, meaning green-yellow).

Each element with a permanent name has a unique **symbol** that consists of one or two letters. The first letter is *always* capitalized; the second letter (if there is one) is *never* capitalized. Newly discovered elements are assigned temporary names and three-letter symbols that are Latin translations of their atomic numbers. (We will learn more about this in Chapter 9.)

The **atomic number** of an element is the number of protons that is contained in the nucleus of each of its atoms. For example, sodium (Na) has the atomic number 11. Each atom of sodium contains 11 protons in its nucleus. Since an atom is electrically neutral, *the number of protons in its nucleus must be equal to the number of electrons surrounding its nucleus*. For example, an atom of sodium also contains 11 electrons surrounding its nucleus.

The **Periodic Table of the Elements** in Appendix 1 lists the elements by atomic numbers and symbols. The Periodic Table is arranged in such a way that the chemical and physical properties of elements display a repeating pattern. The similarity of these properties is most pronounced within a *vertical column* called a **group**. The groups are numbered left to right, from 1 to 18. Elements in Groups 1, 2, and 13–18 are known as the *representative* elements.

Some groups have been given special names, as shown in the accompanying table.

Group Number	Group Name	Elements in Group
1	Alkali metals	Li, Na, K, Rb, Cs, Fr
2	Alkaline earth metals	Be, Mg, Ca, Sr, Ba, Ra
17	Halogens	F, Cl, Br, I, At, Ts
18	Noble gases	He, Ne, Ar, Kr, Xe, Rn, Og

Each *horizontal row* is called a **period**. The periods are numbered from top to bottom, from 1 to 7. As one proceeds from left to right across a period, the properties of the elements change from metallic to nonmetallic. We shall examine periodic properties more fully in Chapter 9.

2.5 NEUTRONS, ISOTOPES, AND MASS NUMBERS

Neutrons add mass to an atom but do not change the atom's identity as an element. For example, an atom containing 6 protons, 6 electrons, and 7 neutrons is classified as a carbon atom—as is an atom with 6 protons, 6 electrons, and 8 neutrons. These two atoms are **isotopes** of the element carbon; isotopes contain the same number of protons but have different numbers of neutrons. We may refer to an isotope by its name and its **mass number**, that is, the sum of the number of protons and neutrons in the nucleus of the atom. The first isotope described above has a mass number of 13, and we may refer to it as carbon-13 or C-13. The second isotope, with a mass number of 14, is referred to as carbon-14 or C-14. There are two other ways of designating a particular isotope. In the first method, the mass number is placed as a *superscript* to the *left* of the symbol for the element, for example, ^{13}C or ^{14}C. The second method gives also the atomic number, which is added as a *subscript* to the *left* of the symbol, for example, $^{13}_{6}C$ or $^{14}_{6}C$.

PROBLEM
How many protons, neutrons, and electrons are contained in an atom of uranium-238? The atomic number of uranium (U) is 92.

SOLUTION
From its atomic number, we see that a neutral atom of uranium-238 has 92 protons and 92 electrons. We also know that it has 238 protons and neutrons. If we subtract the atomic number from the mass number, the answer will be the number of neutrons:

$$238 \text{ [protons + neutrons]} - 92 \text{ protons} = 146 \text{ neutrons}$$

A neutral atom of uranium-238 has 92 protons, 146 neutrons, and 92 electrons.

The preceding problem establishes an important relationship:

We can use a simple model to represent an atom, that is, the contents of the nucleus and the electrons that surround it. We will use the symbols p^+, e^-, and n for protons, electrons, and neutrons, respectively. The model for the sample problem is shown below:

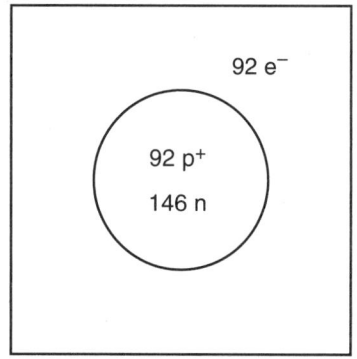

92 e^-

92 p^+
146 n

Model of U-238 ATOM

TRY IT YOURSELF
How many protons, neutrons, and electrons are in a neutral atom of $^{63}_{29}\text{Cu}$? Draw a model of this atom.

ANSWER
29 protons, 34 neutrons, and 29 electrons

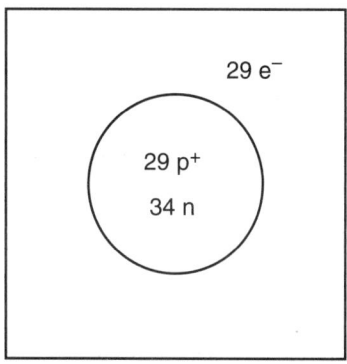

29 e^-

29 p^+
34 n

Model of Cu-63 ATOM

2.6 MOLECULES

A **molecule** is the smallest identifiable sample of a substance. In nature, the *noble gases* (the elements of group 18) exist as isolated atoms and are called **monatomic molecules**.

Some substances exist as molecules in which two or more atoms are linked together. Molecules containing two or more atoms are called **polyatomic molecules.** Polyatomic molecules containing exactly two atoms are called **diatomic molecules**.

Every molecule can be represented by a **molecular formula**, in which the symbol of the element is succeeded by a *subscript* that indicates the number of atoms present. We indicate the presence of *one atom* by writing the symbol *without* the subscript 1. The accompanying list gives the molecular formulas of some common molecules.

MONATOMIC MOLECULES
Elements:
He, Ne, Ar, Kr, Xe, Rn

DIATOMIC MOLECULES
Elements:
$H_2, O_2, N_2, F_2, Cl_2, Br_2, I_2$

Compounds:
CO (carbon monoxide: 1 carbon atom, 1 oxygen atom)
HCl (hydrogen chloride)

POLYATOMIC MOLECULES
Elements:
O_3, P_4, S_8, C_{60}

Compounds:
H_2O (water: 2 hydrogen atoms, 1 oxygen atom)
CO_2 (carbon dioxide), NH_3 (ammonia), H_2O_2 (hydrogen peroxide), CH_4 (methane)

Note that a molecular formula indicates only the *number* of atoms of each element present in the molecule; it does *not* indicate *how* the atoms are linked together.

2.7 IONS

Simple Ions

When atoms combine chemically, the *nuclei* of the atoms cannot change: otherwise, the elements themselves would be changed. In many cases, however, atoms can lose or gain *electrons* when they combine. The loss or gain of electrons by an atom results in the formation of an electrically charged particle called an **ion**. The ion has a charge because the numbers of protons and electrons are no longer equal. If there are more protons than electrons, the ion is *positively* charged; if there are fewer protons than electrons, the ion is *negatively* charged. The following diagrams show how positive and negative ions are formed from atoms:

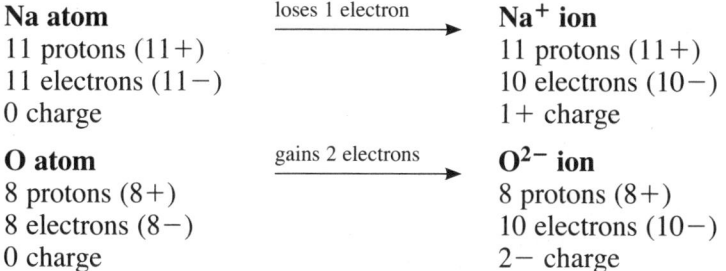

Note that the charge of an ion is written as a *superscript* to the right of the symbol, and because it is a relative charge, the magnitude precedes the sign (as in O^{2-}). We indicate charges of $1+$ and $1-$ by writing only the *sign* of the charge, and omit the magnitude (as in Na^+). If no charge or sign appears to the right of an atomic symbol, it is understood that we are dealing with an *atom* (e.g., Na).

The Periodic Table of the Elements in Appendix 1 can be used to find the charges of simple ions. The table below relates the group numbers of the representative elements to their ionic charges:

Group	1	2	13	14	15	16	17
Charge	1+	2+	3+	4±	3–	2–	1–

In Chapter 3 we will learn how to use the Periodic Table to find the charges of other ions.

Polyatomic Ions

A **polyatomic ion** is an electrically charged particle that consists of two or more atoms linked together in much the same way as in a neutral molecule. This linkage serves to make the polyatomic ion behave as a unit rather than as separate atoms or simple ions. The charge on a polyatomic ion does not belong to any individual atom; it belongs to the ion as a whole. For example, the polyatomic *hydroxide ion* (OH^-) has a 1− charge that belongs to the entire OH unit, rather than to the oxygen or the hydrogen atom. Reference Table E in Appendix 1 includes the names and charges of selected polyatomic ions.

END-OF-CHAPTER QUESTIONS

1. Rutherford's famous experiment using alpha particles to bombard a thin sheet of gold foil indicated that most of the volume of the atoms in the foil is taken up by
   (1) electrons   (2) protons   (3) neutrons   (4) empty space

2. Compared to the entire atom, the nucleus of the atom is
   (1) smaller and contains most of the atom's mass
   (2) smaller and contains little of the atom's mass
   (3) larger and contains most of the atom's mass
   (4) larger and contains little of the atom's mass

3. Which of the following particles has the *least* mass?
   (1) an electron   (2) a proton   (3) a deuteron   (4) a neutron

4. What is the total number of protons and neutrons in an atom of ^{79}Se?
   (1) 34   (2) 45   (3) 79   (4) 113

5. What particles are found in the nucleus of an atom?
   (1) protons and electrons   (2) protons and neutrons
   (3) neutrons and electrons   (4) neutrons and positrons

6. The atoms in a sample of an element must contain nuclei with the same number of
   (1) electrons   (2) protons   (3) neutrons   (4) nucleons

7. The atomic number of an atom is always equal to the total number of
   (1) neutrons in the nucleus
   (2) protons in the nucleus
   (3) neutrons plus protons in the atom
   (4) protons plus electrons in the atom

8. What is the total number of electrons in an atom with an atomic number of 13 and a mass number of 27?
(1) 13   (2) 14   (3) 27   (4) 40

9. All atoms of potassium must have the same
(1) atomic mass   (2) atomic weight
(3) mass number   (4) atomic number

10. Compared to an atom of calcium-40, an atom of potassium-39 contains fewer
(1) protons   (2) neutrons
(3) occupied sublevels   (4) occupied principal energy levels

11. The number of protons in the nucleus of ^{32}P is
(1) 15   (2) 17   (3) 32   (4) 47

12. The number of protons in an atom of ^{36}Cl is
(1) 17   (2) 18   (3) 35   (4) 36

13. The nucleus of an atom of $^{127}_{53}I$ contains
(1) 53 neutrons and 127 protons
(2) 53 protons and 127 neutrons
(3) 53 protons and 74 neutrons
(4) 53 protons and 74 electrons

14. When an atom becomes a 2^+ ion, it
(1) gains 2 electrons   (2) gains 2 protons
(3) loses 2 electrons   (4) loses 2 protons

15. How many protons, neutrons, and electrons are contained in a $^{34}S^{2-}$ ion?
(1) 17 protons, 17 neutrons, 16 electrons
(2) 18 protons, 16 neutrons, 20 electrons
(3) 16 protons, 18 neutrons, 18 electrons
(4) 16 protons, 34 neutrons, 16 electrons

16. Which particle has a negative charge and a mass that is approximately 1/1836 the mass of a proton?
(1) a neutron   (2) an alpha particle
(3) an electron   (4) a positron

17. The element found in Group 13 and in Period 2 of the Periodic Table is
(1) Be   (2) Mg   (3) B   (4) Al

18. Which element is in Group 2 and Period 7?
(1) Mg   (2) Mn   (3) Ra   (4) Rn

19. In which group of the Periodic Table is oxygen located?
(1) 1   (2) 2   (3) 16   (4) 17

20. Which of the following elements is classified as an alkali metal?
(1) magnesium   (2) potassium   (3) argon   (4) phosphorus

21. Iodine belongs to the family of elements known as
(1) halogens   (2) alkaline earth elements
(3) noble gases   (4) transition elements

22. All of the following gases exist as diatomic molecules *except*
(1) hydrogen   (2) fluorine   (3) nitrogen   (4) argon

23. When a magnesium atom loses 2 electrons, it forms an ion whose charge is
(1) 1−   (2) 2−   (3) 1+   (4) 2+

24. The name of the polyatomic ion whose formula is HCO_3^- is
(1) hydrogen sulfate   (2) sulfate
(3) hydrogen carbonate   (4) carbonate

25. Which statement best describes electrons?
(1) They are positive subatomic particles and are found in the nucleus.
(2) They are positive subatomic particles and are found surrounding the nucleus.
(3) They are negative subatomic particles and are found in the nucleus.
(4) They are negative subatomic particles and are found surrounding the nucleus.

26. An atom of carbon-12 and an atom of carbon-14 differ in
(1) atomic number
(2) mass number
(3) nuclear charge
(4) number of electrons

27. Hydrogen has three isotopes with mass numbers of 1, 2, and 3 and has an average atomic mass of 1.00794 amu. This information indicates that
(1) equal numbers of each isotope are present
(2) more isotopes have an atomic mass of 2 or 3 than of 1
(3) more isotopes have an atomic mass of 1 than of only 2 or 3
(4) isotopes have an atomic mass of only 1

28. The atomic number of an atom is always equal to the number of its
   (1) protons, only
   (2) neutrons, only
   (3) protons plus neutrons
   (4) protons plus electrons

29. Which subatomic particle has no charge?
   (1) alpha particle
   (2) beta particle
   (3) neutron
   (4) electron

30. When the electrons of an excited atom return to a lower energy state, the energy emitted can result in the production of
   (1) alpha particles
   (2) isotopes
   (3) protons
   (4) spectra

31. Which phrase describes the charge and mass of the neutron?
   (1) a charge of +1 and no mass
   (2) a charge of +1 and an approximate mass of 1 u
   (3) no charge and no mass
   (4) no charge and an approximate mass of 1 u

32. What is the number of electrons in a potassium atom?
   (1) 18   (2) 19   (3) 20   (4) 39

33. Which substance *cannot* be broken down by a chemical change?
   (1) ethane   (2) propanone   (3) silicon   (4) water

34. What is the number of electrons in an Al^{3+} ion?
   (1) 10   (2) 13   (3) 3   (4) 16

Constructed-Response Questions

1. Fill in the gaps in the following table:

Symbol	^{39}K				$^{31}P^{3-}$			
Protons		23				82		28
Neutrons		28		136		124	45	31
Electrons						80	36	
Atomic no.			79	86				
Mass no.			197					
Net charge	0	0	0	0			2−	2+

2. Draw a model for $^{31}P^{3-}$.

The answers to these questions are found in Appendix 3.

FORMULAS, EQUATIONS, AND CHEMICAL REACTIONS

Chapter Three

KEY IDEAS

This chapter introduces the language of chemistry: formulas and equations. A chemical formula provides information about the atomic composition of a substance, while an equation allows us to describe a chemical reaction using the names or formulas of the reactants and products involved in the process.

KEY OBJECTIVES
At the conclusion of this chapter you will be able to:
• Distinguish between ionic compounds and molecular compounds.
• Define the following terms: *ionic formula*, *molecular formula*, *empirical formula*, *structural formula*, and *binary compound*.
• Write the formulas for ionic compounds containing simple and/or polyatomic ions and name these compounds using the IUPAC system.
• Define the term *oxidation number* (*state*) and assign oxidation numbers to elements, simple ions, and the elements contained in compounds and polyatomic ions.
• Write the formulas for molecular compounds and name these compounds using the IUPAC system.
• Write chemical equations using names and chemical formulas.
• Balance chemical equations.
• Classify a chemical reaction as *synthesis*, *decomposition*, *single replacement*, or *double replacement*.

SECTION I—BASIC (REGENTS-LEVEL) MATERIAL

NYS REGENTS CONCEPTS AND SKILLS

Note: By the time you have finished this chapter, you should have mastered the concepts and skills listed on page 56. The Regents chemistry examination will test your knowledge of these items and your ability to apply them.

Concepts are the *basic ideas* that form the body of the Regents chemistry course (what you need to know!).

Skills are the *activities* that demonstrate your mastery of these concepts (how you show that you know them!).

Following each concept or skill is a page reference (given in parentheses) to this chapter.

3.1 Concept:
A chemical compound can be represented by a specific chemical formula and assigned a name based on the IUPAC system.
(Pages 57, 59–64)

3.2 Concept:
Types of chemical formulas include empirical, molecular, and structural. (Pages 57–58)

3.3 Concept:
In all chemical reactions there is a conservation of mass, energy, and charge. (Page 66)

Skills:
• Interpret balanced chemical equations in terms of conservation of matter, charge, and energy. (Pages 65–67)
• Balance a chemical equation, given the formulas of reactants and products. (Pages 65–67)
• Balance a chemical equation using smallest-whole-number coefficients. (Pages 65–67)
• Use a particle diagram to represent a balanced chemical equation. (Pages 68–69)

3.4 Concept:
Types of chemical reactions include synthesis, decomposition, single replacement, and double replacement. (Pages 69–72)

Skill:
Identify types of chemical reactions. (Pages 69–72)

3.1 CHEMICAL FORMULAS

A **chemical formula** is an important part of the language of chemistry because it tells us something about the composition of an element or a compound. A formula can be as simple as He (helium) or as complicated as $C_{27}H_{46}O$ (cholesterol).

Elemental Formulas

The formula for an atom of an element is its symbol, such as Ca, Fe, or Kr. In cases where the element exists as a diatomic or polyatomic molecule, we use molecular formulas such as F_2, P_4, and S_8.

Ionic Formulas

Ionic compounds such as NaCl and $Al(NO_3)_3$ consist of a combination of positive and negative ions. The positive ions consist of simple *metallic* ions (such as Na^+) or *polyatomic* ions (such as NH_4^+). The negative ions consist of simple *nonmetallic* ions (such as S^{2-}) or polyatomic ions (such as ClO_3^-). The compound CaF_2 contains *exactly two elements* (calcium and fluorine) and is known as a **binary compound**. The ions in ionic compounds do *not* exist as individual molecules.

An **ionic formula** is the chemical formula for an ionic compound. The formula tells us the relative numbers of ions present in the compound. In NaCl, the ratio of Na^+ ions to Cl^- ions is 1 to 1; in $Al(NO_3)_3$, the ratio of Al^{3+} ions to NO_3^- ions is 1 to 3.

Molecular Formulas

Molecular compounds such as NH_3 and H_2O usually consist of combinations of *nonmetallic* elements in which the bonds between the atoms are formed by the *sharing* of electrons. As the name implies, molecular compounds exist as individual molecules.

The formula NH_3 is a **molecular formula** because NH_3 is a molecular compound, that is, NH_3 exists in the form of individual molecules. A molecular formula tells us how many atoms of each element are present in one molecule of the compound. For example, one molecule of C_2H_6 (ethane) contains two carbon and six hydrogen atoms. Note that the compounds NH_3 and C_2H_6 are also binary compounds (why?).

Empirical Formulas

In an **empirical formula** the elements appear in smallest whole-number ratios. For example, the molecular formula H_2O is also an empirical formula because the ratio of hydrogen atoms to oxygen atoms (2 to 1) cannot be reduced further. In contrast, the molecular formula N_2O_4 has NO_2 as its empirical formula. Note that an empirical formula need not be related in any special way to its parent formula: it may represent another compound, or it may not even exist as a compound.

PROBLEM
What is the empirical formula of $C_6H_{12}O_6$?

SOLUTION
Divide the formula by 6 to obtain CH_2O.

TRY IT YOURSELF
What are the empirical formulas of NH_3 and C_6H_6?

ANSWER
NH_3 and CH

Structural Formulas

A **structural formula** shows how the atoms in a molecule are bonded to one another. (It is usual to represent chemical bonds by lines and unshared pairs of electrons by pairs of dots.) A structural formula may also show the shape of the molecule. For example, the water molecule may be represented as shown in the accompanying diagram:

$$\overset{\cdot\cdot\,\,\cdot\cdot}{\underset{H \qquad\qquad H}{O}}$$

The diagram shows that each hydrogen atom is bonded to the oxygen atom, that the molecule is "V-shaped," and that the oxygen atom contains two unshared pairs of electrons.

Particle diagrams can also be used to represent structural formulas, as shown in the accompanying particle diagram of CO_2:

The particle diagram shows that each oxygen atom is bonded to the carbon atom, and that the molecule is linear.

3.2 WRITING AND NAMING CHEMICAL FORMULAS

Writing Formulas for Ionic Compounds

Since all compounds are electrically neutral, the total electric charge on an ionic compound must be 0. When we write the formula of an ionic compound, we must be certain that the *total* charge on the positive ions equals the *total* charge on the negative ions.

In the compound that contains K^+ ions and F^- ions, this is easily accomplished if the ion ratio is 1 to 1: the formula is KF. In the compound containing Li^+ ions and O^{2-} ions, the ion ratio must be 2 to 1: the formula is Li_2O.

PROBLEM

Write the formula of the compound that contains Al^{3+} ions and S^{2-} ions.

SOLUTION

We need a common total charge for both positive and negative ions. If we multiply 3 and 2 (the ionic charges), we obtain 6, a number that is evenly divisible by 3 and 2. Two Al^{3+} ions will have a total charge of 6+, and three S^{2-} ions will have a total charge of 6–.

The formula for the compound is Al_2S_3.

TRY IT YOURSELF

The element lead forms *two* ions: Pb^{2+} and Pb^{4+}. Write the formulas for the compounds that contain these ions combined with the O^{2-} ion.

ANSWER

PbO and PbO_2

Note that, in general, when we write a formula the positive ion appears first, followed by the negative ion. (There are a small number of exceptions to this rule, but we will deal with them as necessary.)

Formulas for ionic compounds that contain polyatomic ions follow the same rules as those for simple ions. When more than one polyatomic ion is needed in a formula, it is enclosed in parentheses, as in $Fe(NO_3)_2$. This notation tell us that the ratio of Fe^{2+} ions to NO_3^- ions is 1 to 2.

PROBLEM

Write the formula for the compound that contains Zn^{2+} ions and PO_4^{3-} ions.

SOLUTION

The ratio of Zn^{2+} ions to PO_4^{3-} ions must be 3 to 2 in order for the compound to have a total charge of 0. The formula is $Zn_3(PO_4)_2$.

TRY IT YOURSELF

Write the formula for the compound that contains NH_4^+ ions and SO_4^{2-} ions.

ANSWER

$(NH_4)_2SO_4$

We can also work backwards: given the formula, we can determine the charges on the ions present in the compound. In Chapter 2 we related the charge on an ion of a representative element to its group number (page 49). Moreover, the charges on selected polyatomic ions are given in Reference Table E in Appendix 1.

PROBLEM

Given the ionic formula CuI_2, what are the charges on the ions?

SOLUTION

Refer to the Periodic Table of the Elements in Appendix 1. The representative element I is located in Group 17, its ion has a 1– charge. Since there are two I^- ions in the formula, the total negative charge is 2–. To balance this charge, a total positive charge of 2+ must be present on the single Cu ion, that is, Cu^{2+}.

TRY IT YOURSELF

Determine the ionic charges in the compound Ag_2S.

ANSWER

Ag^+ and S^{2-}

Naming Binary Ionic Compounds

We name a binary ionic compound by combining the names of the two elements that make up the compound. The ion that has a positive charge is placed first; the ion with the negative charge is second; and the suffix *-ide* is added at the end of the name. Thus, NaCl is named sodium chloride, CaO is calcium oxide, and AlF_3 is aluminum fluoride.

A list of some elements with negative oxidation numbers and their name endings is given below:

H^-	hydride	C^{4-}	carbide	N^{3-}	nitride	O^{2-}	oxide
F^-	fluoride	P^{3-}	phosphide	S^{2-}	sulfide	Cl^-	chloride
Br^-	bromide	I^-	iodide				

Many metallic elements can have more than one positive charge. (Refer to the Periodic Table of the Elements in Appendix 1, and examine "Selected Oxidation States" in the upper right corner of each element.)

A metallic element such as iron (Fe) can form more than one compound. For example, iron can exist as the Fe^{2+} or Fe^{3+} ion and, therefore, can form two oxides: FeO and Fe_2O_3. How should we name these compounds? Clearly, both are entitled to be called iron oxide! Since it is iron whose charge varies, we provide Fe with a *Roman numeral in parentheses*, which corresponds to its positive charge. Therefore, FeO is given the name iron(II) oxide, and Fe_2O_3 is called iron(III) oxide. If a metallic ion can have only *one charge* (such as Na^+ or Mg^{2+}), no Roman numeral is used in the name. This method of naming compounds is known as the *IUPAC system,* where IUPAC stands for the International Union of Pure and Applied Chemistry.

TRY IT YOURSELF
Write the formulas for the following compounds:
(a) copper(II) sulfide   (b) iron(III) bromide
(c) tin(IV) oxide   (d) aluminum nitride

ANSWERS
(a) CuS   (b) $FeBr_3$ (c) SnO_2 (d) AlN

TRY IT YOURSELF
Name the following compounds:
(a) MgF_2 (b) KH   (c) NiI_3 (d) V_2O_5

(Use the Periodic Table and Reference Table S in Appendix 1.)

ANSWERS
(a) magnesium fluoride   (b) potassium hydride
(c) nickel(III) iodide   (d) vanadium(V) oxide

Naming Compounds that Contain Polyatomic Ions

Ionic compounds that contain polyatomic ions are named in the same fashion as binary ionic compounds. The trick is to recognize that a polyatomic ion is actually present in the formula. At this point, you should review the names and formulas of the polyatomic ions contained in Reference Table E in Appendix 1.

PROBLEM
Name the following compounds:
(a) K_2SO_4 (b) NH_4NO_3 (c) $Fe(C_2H_3O_2)_3$

SOLUTIONS
(a) Note that two K^+ ions are needed to balance the SO_4^{2-} ion for charge. The name of this compound is potassium sulfate.

(b) This compound contains two polyatomic ions: NH^+ and NO_3^-. The name of this compound is ammonium nitrate.
(c) This compound contains the Fe^{3+} ion and the $C_2H_3O_2^-$ ion. The name of this compound is iron(III) acetate.

TRY IT YOURSELF
Name the following compounds:
(a) Na_3PO_4
(b) $Hg(OH)_2$
(c) $(NH_4)_2SO_3$

ANSWERS
(a) sodium phosphate
(b) mercury(II) hydroxide
(c) ammonium sulfite

Writing Formulas for Molecular Compounds

Atoms in molecular compounds do not have *real* charges (as ions do in ionic compounds). Nevertheless, atoms are electrical in nature, and they can behave as though they possessed charges, even in molecular compounds. The type of charge found in molecular compounds is known as an *apparent charge*. To be able to work with both real and apparent charges, the concept of **oxidation numbers** (also known as **oxidation states**) was developed.

An oxidation number can be positive, negative, or zero. When we write oxidation numbers, we place the sign of the number *before* the number itself, for example, +2. (This distinguishes oxidation numbers from actual charges, in which the sign is placed *after* the number, as in 2+.) The following list gives six simplified rules for assigning oxidation numbers to elements.

1. The oxidation number of every atom in a *free element* is 0. By "free element" we mean an element that is not combined with another element. Therefore, Na, He, O_2, and P_4 are free elements, and each of the atoms that comprise them has an oxidation number of 0.
2. In a molecular or ionic compound, the *algebraic sum* of the oxidation numbers must be 0 since all compounds are electrically neutral.
3. The oxidation number of a simple ion is its charge. For example, the oxidation number of Cl^- is −1, and the oxidation number of Al^{3+} is +3.
4. In a polyatomic ion, the *algebraic sum* of the individual oxidation numbers must equal the *charge* on the ion.
5. Hydrogen in combination usually has an oxidation number of +1. An exception is the group of compounds known as *hydrides* (such as sodium hydride, NaH), in which the hydride ion has a charge of 1^-.

6. Oxygen in combination usually has an oxidation number of –2. Exceptions include *peroxides* (such as hydrogen peroxide, H_2O_2), in which the oxidation number of oxygen is –1, and oxygen–fluorine compounds, in which the oxidation number of oxygen has a *positive* value.

The Periodic Table of the Elements in Appendix 1 shows the common oxidation states of the elements. These values appear in the upper right side of each element's box.

We consider *binary* molecular compounds first. The rules are simple: (1) Elements with positive oxidation numbers generally precede elements with negative oxidation numbers, and (2) the algebraic sum of the numbers must be 0.

For example, sulfur can combine with oxygen to form *two* compounds. Since O must have an oxidation number of –2, S must have a *positive* oxidation number. Referring to the Periodic Table, we see that the positive oxidation numbers of S are +4 and +6. Therefore, the two possible formulas are SO_2 and SO_3.

TRY IT YOURSELF
Write the formulas for the binary compounds corresponding to the following combinations of elements:
(a) N and O   (b) P and Cl   (c) H and S   (d) H and N

ANSWERS
(a) There are *five* possibilities: N_2O, NO, N_2O_3, NO_2, N_2O_5.
(b) There are *two* possibilities: PCl_3, PCl_5.
(c) H_2S
(d) NH_3 (This is relatively straightforward, but we write the formula placing nitrogen as the first element. The name of this compound is *ammonia*.)

We can also work backwards: Given the formula of a molecular compound we can determine the oxidation number of each element in the compound. We can also do this for a polyatomic ion, since the atoms within the ion behave pretty much as those in a molecular compound: they *share* electrons.

PROBLEM
In the compound Al_2O_3, aluminum has an oxidation number of +3. Show that the sum of the oxidation numbers is 0.

SOLUTION
Since there are two Al atoms, the oxidation number total for Al is $2 \cdot (+3) = +6$. The oxidation number total for O is $3 \cdot (-2) = -6$. The sum of these two oxidation number totals is 0: $(+6) + (-6) = 0$.

PROBLEM

Find the oxidation number of each of the following elements:

(a) S in H_2SO_3 (b) Cr in $Na_2Cr_2O_7$

(c) Fe and Cl in $FeCl_3$ (d) P in PO_4^{3-}

(e) O in OF_2

SOLUTIONS

(a) The oxidation number total for H is $2 \cdot (+1) = +2$. The oxidation number total for O is $3 \cdot (-2) = -6$. These numbers add to -4.

     If the compound is to be neutral, S must have an oxidation number of $+4$.

(b) The oxidation number total for Na is $2 \cdot (+1) = +2$. The oxidation number total for O is $7 \cdot (-2) = -14$. These numbers add to -12.

     If the compound is to be neutral, the two Cr atoms must have an oxidation number total of $+12$. Therefore, each Cr atom has an oxidation number of $+6$.

(c) We must rely on the fact that Cl will be negative in this compound. According to the Periodic Table, the negative oxidation number of Cl is -1. In this compound, then, the oxidation number total for Cl is $3 \cdot (-1) = -3$.

     For the compound to be neutral, the oxidation number of Fe must be $+3$.

(d) The oxidation number total for O is $4 \cdot (-2) = -8$. In this polyatomic ion the sum of the oxidation number totals must equal the charge on the ion, that is, -3.

     Therefore, P must have an oxidation number of $+5$.

(e) According to the Periodic Table, F has an oxidation number of -1. The oxidation number total for F is $2 \cdot (-1) = -2$.

     If the compound is to be neutral, O must have an oxidation number of $+2$. (Note that this oxidation number is *not* shown on the Periodic Table.)

TRY IT YOURSELF

Find the oxidation number of each of the following elements:

(a) Cl in ClO_4^- (b) N in N_2O_5 (c) C in $C_6H_{12}O_6$

ANSWERS

(a) $+7$.   (b) $+5$.

(c) If you followed all of the rules, you calculated an oxidation number for C equal to 0! This is entirely possible: according to our rules, some *combined* elements may also have oxidation numbers of 0.

Naming Molecular Compounds

The IUPAC system for naming molecular compounds uses *prefixes* that indicate how many atoms of an element are present. The following prefixes are used:

Prefix	Number of Atoms
mono	1
di	2
tri	3
tetra	4
penta	5
hexa	6

The prefix "mono" is never used for the first element. For example, SO_2, CO, and N_2O_3, are named, respectively, sulfur dioxide, carbon monoxide, and dinitrogen trioxide.

TRY IT YOURSELF

Name the following molecular compounds:
(a) PCl_5
(b) SO_3
(c) SF_6
(d) H_2O

ANSWERS

(a) phosphorus pentachloride
(b) sulfur trioxide
(c) sulfur hexafluoride
(d) dihydrogen monoxide—but we will continue to use its common name: *water*!

3.3 CHEMICAL EQUATIONS

An *equation* provides a "recipe" for carrying out a chemical reaction. Consider the following description of a chemical reaction:

> Methane (CH_4) gas reacts with oxygen gas to produce
> carbon dioxide gas, liquid water, and heat.

This is quite a mouthful! Let us develop a chemical equation for this reaction, step by step.

Developing an Equation

First, we write a *word equation:*

methane(g) + oxygen(g) → carbon dioxide(g) + water(ℓ) + heat

Note that an arrow separates what we react (the *reactants*) from what we produce (the *products*), and that the *phases* of the substances are included in parentheses: "s" for solid, "ℓ" for liquid, "g" for gas, and "aq" for a substance dissolved in water.

Next, we replace the name of each substance by its formula:

$$CH_4(g) + O_2(g) \rightarrow CO_2(g) + H_2O(\ell) + \text{heat}$$

And that's about it. We have changed our original, wordy statement into a shortened form that tells what happens when methane gas reacts with oxygen. This equation is still incomplete. We will "balance" it in Section 3.4.

TRY IT YOURSELF
Write word and formula equations for the following process:

Solid zinc reacts with a water solution of hydrogen sulfate to produce hydrogen gas, a water solution of zinc sulfate, and heat.

ANSWER
zinc(s) + hydrogen sulfate(aq) → hydrogen(g) + zinc sulfate(aq) + heat
$$Zn(s) + H_2SO_4(aq) \rightarrow H_2(g) + ZnSO_4(aq) + \text{heat}$$

3.4   BALANCING A CHEMICAL EQUATION

All chemical reactions must obey three *conservation laws:* conservation of matter (mass), conservation of energy, and conservation of electric charge. This means that the total mass, energy content, and electric charge of the reactants must equal those of the products.

If we inspect the equation written in Section 3.3 for the reaction between CH_4 and O_2 carefully, we find something seriously wrong with it. On the left side of the equation we have four hydrogen atoms and two oxygen atoms. On the right side we have two hydrogen atoms and three oxygen atoms. This is intolerable—our equation violates the law of conservation of matter!

If, however, we inspect the equation written in answer to "Try It Yourself," we find that the numbers of zinc, hydrogen, sulfur, and oxygen atoms on each side of the equation are equal. This equation is said to be *balanced,* while the other one is unbalanced. To correct this serious defect, we must *balance* the equation for the reaction between CH_4 and O_2.

To balance an equation, we use a *coefficient:* a number that is placed before a substance to indicate how many units of the substance we need. For example, the notation $5NH_3$ means that five molecules of NH_3 are required.

We now proceed to balance the faulty equation by inspection. (First, we will drop the heat term since it will not play a direct part in what we do.) We have four hydrogen atoms on the left, so we must produce four on the right. Since hydrogen appears only in the H_2O, it follows that we need $2H_2O$ to account for all of the hydrogen:

$$CH_4(g) + O_2(g) \rightarrow CO_2(g) + 2H_2O(\ell)$$
(The equation is balanced for hydrogen.)

If we count the oxygen atoms on the *right* in our new equation, we see that we have four—two from the CO_2 and two from the $2H_2O$. (Remember: the coefficient applies to the *entire* H_2O, not only the hydrogen.) We can easily produce four oxygen atoms on the *left* side by using $2O_2$:

$$CH_4(g) + 2O_2(g) \rightarrow CO_2(g) + 2H_2O(\ell)$$
(The equation is completely balanced.)

Now inspect the three equations given below. Are they balanced?

$$2CH_4(g) + 4O_2(g) \rightarrow 2CO_2(g) + 4H_2O(\ell)$$

$$6CH_4(g) + 12O_2(g) \rightarrow 6CO_2(g) + 12H_2O(\ell)$$

$$\frac{1}{2}CH_4(g) + O_2(g) \rightarrow \frac{1}{2}CO_2(g) + H_2O(\ell)$$

Clearly, the first two equations are balanced because we count the same number of atoms of each kind on the left and right sides. Note that the coefficients are *proportional* to the coefficients in the equation we just balanced.

Although the last equation looks strange (it has *fractional* coefficients), it is balanced because its coefficients are also in proportion to those in our original equation. (In Chapter 5 we will see how such equations are used.)

Our original (balanced) equation:

$$CH_4(g) + 2O_2(g) \rightarrow CO_2(g) + 2H_2O(\ell)$$

is called an equation balanced with *smallest whole-number coefficients*. For the present, we will use this form.

TRY IT YOURSELF

Write an equation for each of the following reactions, and balance it using smallest whole-number coefficients:
(a) nitrogen(g) + hydrogen(g) → ammonia(g)
(b) hydrogen(g) + oxygen(g) → water(ℓ)
(c) copper(II) oxide(s) + carbon(s) → copper(s) + carbon dioxide(g)
(d) magnesium(s) + hydrogen sulfate(aq) → hydrogen(g) + magnesium sulfate(aq)
(c) ammonia(g) + oxygen(g) → nitrogen monoxide(g) + water(ℓ)

ANSWERS
(a) $N_2(g) + 3H_2(g) \rightarrow 2NH_3(g)$
(b) $2H_2(g) + O_2(g) \rightarrow 2H_2O(\ell)$
(c) $2CuO(s) + C(s) \rightarrow 2Cu(s) + CO_2(g)$
(d) $Mg(s) + H_2SO_4(aq) \rightarrow H_2(g) + MgSO_4(aq)$
(e) $4NH_3(g) + 5O_2(g) \rightarrow 4NO(g) + 6H_2O(\ell)$

Particle Diagrams and Balanced Chemical Equations

We have used particle diagrams to represent atoms and molecules. We can also use particle diagrams to represent balanced equations. For example, consider the balanced equation $2H_2 + O_2 \rightarrow 2H_2O$. The following diagram specifies how each molecule in the equation is represented:

$\bullet\bullet$ $= H_2$

$\circ\circ$ $= O_2$

$\bullet\circ\bullet$ $= H_2O$

Now let us use these diagrams to form the balanced equation:

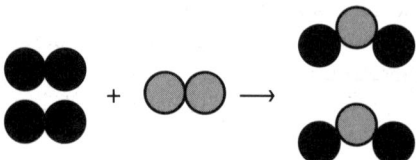

We can see that the equation is balanced because we can count the number of each type of atoms on each side of the equation.

TRY IT YOURSELF

(a) First, balance the chemical equation $C + O_2 \rightarrow CO$ using smallest whole-number coefficients. (b) Then use the following particle diagrams to represent the balanced equation:

⬤ = C

◯◯ = O_2

⬤◯ = CO

ANSWERS

(a) $2C + O_2 \rightarrow 2CO$ (b)

3.5 CLASSIFYING CHEMICAL REACTIONS

As noted in Section 3.4, an equation describes a specific chemical reaction. Since there are so many chemical reactions, chemists find it useful to place them in various categories based on their similarities. Throughout this book, we will be examining these categories. We begin, in this section, with the four simplest types of chemical reactions.

Direct Combination (Synthesis) Reactions

A *direct combination* (or *synthesis*) reaction involves the combination of *two or more reactants* to produce *one product*. The reactants can be elements or compounds. In general, we can write a direct combination reaction in the form

$$X + Y \rightarrow XY$$

As an example, the production of iron(II) chloride from iron and chlorine is a direct combination reaction:

$$Fe(s) + Cl_2(g) \rightarrow FeCl_2(s)$$

69

Another example involves the production of carbon dioxide from the combination of carbon monoxide and oxygen:

$$2CO(g) + O_2(g) \rightarrow 2CO_2(g)$$

TRY IT YOURSELF

Write a balanced equation for the direct combination of sodium(s) and bromine(ℓ) to produce a solid product.

ANSWER

$2Na(s) + Br_2(\ell) \rightarrow 2NaBr(s)$

Decomposition Reactions

A *decomposition* reaction involves the breakdown of a *single reactant* into *two or more products*. It is the opposite of a direct combination reaction. In general, we can write a decomposition reaction in the form

$$AB \rightarrow A + B$$

As an example, hydrogen peroxide decomposes into water and oxygen:

$$2H_2O_2 \, (\ell) \rightarrow 2H_2O(\ell) + O_2(g)$$

TRY IT YOURSELF

Write a balanced equation for the decomposition of ammonia gas (NH_3) into its elements.

ANSWER

$2NH_3(g) \rightarrow N_2(g) + 3H_2(g)$

Single-Replacement Reactions

In a *single-replacement* reaction, an *uncombined element* replaces another element that is part of a compound. As a result, the replaced element becomes uncombined. In general, we can write a single-replacement reaction in the form

$$E + FG \rightarrow EG + F$$

As an example, the element zinc replaces the element copper from an aqueous solution of the compound copper(II) sulfate:

$$Zn(s) + CuSO_4(aq) \rightarrow ZnSO_4(aq) + Cu(s)$$

Single-replacement reactions occur only under certain conditions. For example, the replacement of zinc by copper (the reverse of the reaction shown on the previous page) will *not* occur. Since zinc replaces copper from solution, zinc is said to be a *more active metal* than copper. In other words, zinc will form a *positive* ion (in solution) more easily than will copper.

We can use Reference Table J in Appendix 1 to predict which single-replacement reactions of metals will occur. (Note that, even though H_2 is not a metallic element, it is included in the table because it forms H^+ ions; in fact, the table is based on H_2 as a standard.) The *higher* a metallic element appears in the table, the more active it is; and a metal will always replace one that is *lower* in the table.

PROBLEM
(a) Will magnesium (Mg) replace lead (Pb) [as $Pb(NO_3)_2$] in aqueous solution? If so, write a balanced equation for the reaction.
(b) Will silver (Ag) replace sodium (Na) [as NaCl] in aqueous solution? If so, write a balanced equation for the reaction.

SOLUTIONS
(a) Yes, since Mg is higher than Pb in Reference Table J in Appendix 1. The equation is

$$Mg + Pb(NO_3)_2 \rightarrow Pb + Mg(NO_3)_2$$

(b) No, since Ag is lower in the table than Na.

We can also compare the activities of *nonmetals* (i.e., their abilities to form *negative* ions in solution) by using Reference Table J. Consider the reaction between liquid bromine (Br_2) and aqueous potassium iodide (KI):

$$Br_2(\ell) + 2KI(aq) \rightarrow I_2(s) + 2KBr$$

Since Br_2 is higher than I_2 in Reference Table J, Br_2 is a more active non-metal than I_2; that is, Br_2 replaces I_2 because Br_2 forms a negative ion more easily than does I_2.

PROBLEM
Write the single-replacement reaction that occurs between gaseous Cl_2 and NaBr in aqueous solution.

SOLUTION
Since Cl_2 is higher than Br_2 on reference Table J, a single-replacement reaction will occur:

$$Cl_2(g) + 2NaBr(aq) \rightarrow Br_2(\ell) + 2NaCl$$

Double-Replacement Reactions

In a *double-replacement* reaction, *two elements in different compounds* replace each other. In general, we can write a double-replacement reaction in the form

$$PQ + RS \rightarrow PS + RQ$$

As an example, when aqueous solutions of barium nitrate and sodium sulfate are mixed, barium and sodium replace each other:

$$Ba(NO_3)_2(aq) + Na_2SO_4(aq) \rightarrow BaSO_4(s) + 2NaNO_3(aq)$$

Double-replacement reactions also occur only under certain conditions: at least one of the products must form a solid (known as a **precipitate**), a gas, or a molecular compound. In the example above, the reaction occurs because a precipitate of $BaSO_4$ forms. We can use Reference Table F in Appendix 1 to predict when a precipitate will form. All of the ionic compounds that are classified as **insoluble** will form precipitates; **soluble** compounds will remain in solution.

PROBLEM

A solution of $AgNO_3$ is mixed with a solution of NaCl. Does a double-replacement reaction occur?

SOLUTION

First, we write the equation as though the reaction does occur:

$$AgNO_3 (\ ) + NaCl(\ ) \rightarrow AgCl(\ ) + NaNO_3(\ )$$

Next, we refer to Table F. According to the table:

- All compounds containing Na^+ or NO_3^- ions are *soluble*: they remain in solution.
- Compounds containing Ag^+ and Cl^- or Br^- or I^- are *insoluble*: they form precipitates.

Since AgCl is an insoluble product, the reaction will occur.

Finally, we complete the equation:

$$AgNO_3 (aq) + NaCl(aq) \rightarrow AgCl(s) + NaNO_3(aq)$$

TRY IT YOURSELF

(a) A solution of sodium sulfide is mixed with a solution of zinc nitrate. Does a reaction occur? Use Reference Table F in Appendix 1.

(b) A solution of potassium chloride is mixed with a solution of sodium iodide. Does a reaction occur? Use Reference Table F in Appendix 1.

ANSWERS

(a) Yes. The balanced equation is:

$$Na_2S(aq) + Zn(NO_3)_2(aq) \rightarrow ZnS(s) + 2NaNO_3(aq)$$

(b) No. The products (potassium iodide and sodium chloride) are soluble, as are the reactants (potassium chloride and sodium iodide).

SECTION II—ADDITIONAL MATERIAL

3.1A OTHER WAYS OF NAMING IONIC COMPOUNDS

There are other, older systems for naming ionic compounds. For metallic elements that can form exactly *two* positive ions, the suffixes *-ous* and *-ic* are added to distinguish between the lower and higher charge. The first suffix (*-ous*) is used with the element's *smaller* charge, and the second (*-ic*) with its *larger* charge. These suffixes are usually (but not always) attached to the *Latin* names of the elements. Some examples are shown below:

$FeCl_2$ ferr*ous* chloride $FeCl_3$ ferr*ic* chloride
Cu_2O cupr*ous* oxide CuO cupr*ic* oxide
Hg_2Br_2 mercur*ous* bromide $HgBr_2$ mercur*ic* bromide

We will *not* employ this system in naming compounds; rather, we will use the IUPAC system throughout this book.

END-OF-CHAPTER QUESTIONS

1. Which represents a substance dissolved in water?
   (1) HCl(aq) (2) HCl(ℓ) (3) HCl(g) (4) HCl(s)

2. Which may represent a crystalline material?
   (1) K_2SO_4(s) (2) $Br_2(\ell)$ (3) NaCl(aq) (4) CO_2(g)

3. Which represents a compound?
   (1) Ca (2) Cr (3) CO (4) Co

4. Which is the formula of a binary compound?
   (1) KOH (2) $NaClO_3$ (3) Al_2S_3 (4) $Bi(NO_3)_3$

5. In a sample of solid $Ba(NO_3)_2$, the ratio of barium ions to nitrate ions is
(1) 1:1   (2) 1:2   (3) 1:3   (4) 1:6

6. A chemical formula is an expression used to represent
(1) mixtures, only
(2) elements, only
(3) compounds, only
(4) compounds and elements

7. In which compound does chlorine have the highest oxidation number?
(1) $KClO$ (2) $KClO_2$ (3) $KClO_3$ (4) $KClO_4$

8. What is the oxidation number of sulfur in H_2SO_4?
(1) 0   (2) -2 (3) $+6$ (4) $+4$

9. If element X forms the oxides XO and X_2O_3, the oxidation numbers of element X are
(1) $+1$ and $+2$ (2) $+2$ and $+3$
(3) $+1$ and $+3$ (4) $+2$ and $+4$

10. For the compound $(NH_4)_2SO_4$, the oxidation numbers of all the atoms must add to
(1) 1   (2) 0   (3) 3   (4) 11

11. What is the oxidation number of sulfur in $Na_2S_2O_7$?
(1) -2 (2) $+2$ (3) $+6$ (4) $+4$

12. Sulfur exhibits a negative oxidation state in
(1) H_2S (2) H_2SO_3 (3) S_8 (4) SO_3

13. In which compound does hydrogen have an oxidation number of -1?
(1) NH_3 (2) KH (3) HCl (4) H_2O

14. The formula for iron(III) oxide is
(1) FeO_3 (2) Fe_2O_3 (3) Fe_3O (4) Fe_3O_2

15. Which is the formula for potassium hydride?
(1) KH (2) KH_2 (3) KOH (4) $K(OH)_2$

16. Which is the formula for dinitrogen monoxide?
(1) NO (2) N_2O (3) NO_2 (4) N_2O_4

17. Which formula represents mercury(I) chloride?
(1) Hg_2Cl (2) $HgCl_2$ (3) Hg_2Cl_2 (4) Hg_2Cl_4

18. Which is the formula for titanium(III) oxide?
    (1) Ti_2O_3 (2) TiO (3) Ti_3O_2 (4) Ti_2O_4

19. Which is the formula for sulfur dioxide?
    (1) SO (2) SO_2 (3) SO_3 (4) SO_4

20. Which is the correct formula for carbon monoxide?
    (1) CO (2) CO_2 (3) C_2O (4) C_2O_3

21. What is the correct name for the compound with the formula $CrPO_4$?
    (1) chromium(II) phosphate   (2) chromium(III) phosphate
    (3) chromium(II) phosphide   (4) chromium(III) phosphide

22. In an equation, which symbol indicates a mixture?
    (1) $NH_3(s)$ (2) $NH_3(\ell)$ (3) $NH_3(aq)$ (4) $NH_3(g)$

23. Given this unbalanced equation:

$$Ca(OH)_2 + (NH_4)_2SO_4 \rightarrow CaSO_4 + NH_3 + H_2O$$

What is the *sum* of the coefficients when the equation is completely balanced using smallest whole-number coefficients?
    (1) 5   (2) 7   (3) 9   (4) 11

24. When the equation

$$\_Na(s) + \_H_2O(\ell) \rightarrow \_NaOH(aq) + \_H_2(g)$$

is correctly balanced using smallest whole-number coefficients, the coefficient of the water is
    (1) 1   (2) 2   (3) 3   (4) 4

25. Given this unbalanced equation:

$$(NH_4)_3PO_4 + Ba(NO_3)_2 \rightarrow Ba_3(PO_4)_2 + NH_4NO_3$$

What is the coefficient of NH_4NO_3 when the equation is correctly balanced using smallest whole-number coefficients?
    (1) 6   (2) 2   (3) 3   (4) 4

26. When the equation

$$\_NH_3 + \_O_2 \rightarrow \_HNO_3 + \_H_2O$$

is correctly balanced using smallest whole-number coefficients, the coefficient of O_2 is
    (1) 1   (2) 2   (3) 3   (4) 4

27. When the equation

$$\_SiO_2 + \_C \rightarrow \_SiC + \_CO$$

is correctly balanced using smallest whole-number coefficients, the sum of all the coefficients is
(1) 5   (2) 7   (3) 8   (4) 9

28. When the equation

$$\_Al_2(SO_4)_3 + \_ZnCl_2 \rightarrow \_AlCl_3 + \_ZnSO_4$$

is correctly balanced using smallest whole-number coefficients, the sum of the coefficients is
(1) 9   (2) 8   (3) 5   (4) 4

29. When the equation

$$\_H_2 + \_Fe_3O_4 \rightarrow \_Fe + \_H_2O$$

is correctly balanced using smallest whole-number coefficients, the coefficient of H_2 is
(1) 1   (2) 2   (3) 3   (4) 4

30. When the equation

$$\_C_2H_4 + \_O_2 \rightarrow \_CO_2 + \_H_2O$$

is correctly balanced using smallest whole-number coefficients, what is the coefficient of O_2?
(1) 1   (2) 2   (3) 3   (4) 4

31. Given this unbalanced equation:

$$\_AlPO_4 + \_Ca(OH)_2 \rightarrow \_Al(OH)_3 + \_Ca_3(PO_4)_2$$

When the equation is correctly balanced using smallest whole-number coefficients, the sum of the coefficients is
(1) 7   (2) 8   (3) 3   (4) 4

32. Which equation illustrates the conservation of mass?
(1) $H_2 + Cl_2 \rightarrow HCl$ (2) $H_2 + Cl_2 \rightarrow 2HCl$
(3) $H_2 + O_2 \rightarrow H_2O$ (4) $H_2 + O_2 \rightarrow 2H_2O$

33. When the equation

$$C_2H_6 + O_2 \rightarrow CO_2 + H_2O$$

is correctly balanced, the coefficient of O_2 is
(1) 7   (2) 10   (3) 3   (4) 4

34. When the equation

$$Cu + H_2SO_4 \rightarrow CuSO_4 + H_2O + SO_2$$

is correctly balanced, what is the coefficient of $CuSO_4$?
(1) 1   (2) 2   (3) 3   (4) 4

35. Given this balanced equation:

$$2Na + 2H_2O \rightarrow 2X + H_2$$

What is the correct formula for the product represented by the letter X?
(1) NaO   (2) Na_2O (3) NaOH   (4) Na_2OH

36. The products, X and Y, of this balanced equation:

$$CS_2 + 3O_2 \rightarrow X + 2Y$$

could be
(1) CO_2 and SO_3 (2) CO and SO
(3) CO and SO_2 (4) CO_2 and SO_2

Base your answers to questions 37–40 on the four types of reactions given below. Match each reaction with its appropriate reaction type, (1), (2), (3), or (4). A choice may be used more than once or not at all.

(1) synthesis
(2) decomposition
(3) single replacement
(4) double replacement

37. $CaCO_3 \rightarrow CaO + O_2$

38. $Cl_2 + 2KI \rightarrow I_2 + 2KCl$

39. $MgSO_4 + Ba(NO_3)_2 \rightarrow BaSO_4 + Mg(NO_3)_2$

40. $Zn + 2AgNO_3 \rightarrow 2Ag + Zn(NO_3)_2$

41. Given the balanced equation representing a reaction:

$$K_2CO_3(aq) + BaCl_2(aq) \rightarrow 2KCl(aq) + BaCO_3(s)$$

Which type of reaction is represented by this equation?
(1) synthesis
(2) decomposition
(3) single replacement
(4) double replacement

Constructed-Response Questions

1. Below is a grid of positive and negative ions. In the space where a positive and a negative ion intersect, write the correct formula for the compound formed by those two ions. A number of correct formulas have already been inserted to assist you.

	bromide Br^-	carbonate CO_3^{2-}	chlorate ClO_3^-	chloride Cl^-	chromate CrO_4^{2-}	nitrate NO_3^-	phosphate PO_4^{3-}	sulfate SO_4^{2-}	sulfide S^{2-}
aluminum Al^{3+}					$Al_2(CrO_4)_3$				
ammonium NH_4^+					$(NH_4)_2CrO_4$				
barium Ba^{2+}					$BaCrO_4$				
calcium Ca^{2+}					$CaCrO_4$				
copper (II) Cu^{2+}					$CuCrO_4$				
iron (II) Fe^{2+}					$FeCrO_4$				
iron (III) Fe^{3+}					$Fe_2(CrO_4)_3$				
lead (II) Pb^{2+}	$PbBr_2$	$PbCO_3$	$Pb(ClO_3)_2$	$PbCl_2$	$PbCrO_4$	$Pb(NO_3)_2$	$Pb_3(PO_4)_2$	$PbSO_4$	PbS
lead (IV) Pb^{4+}					$Pb(CrO_4)_2$				
magnesium Mg^{2+}					$MgCrO_4$				
mercury (I) Hg_2^{2+}					Hg_2CrO_4				
mercury (II) Hg^{2+}					$HgCrO_4$				
potassium K^+					K_2CrO_4				
silver Ag^+					Ag_2CrO_4				
sodium Na^+					Na_2CrO_4				
zinc Zn^{2+}					$ZnCrO_4$				

2. Determine the oxidation numbers of the elements in SO_3^{2-}.

3. Determine the oxidation numbers of the elements in $KMnO_4$.

4. Balance each of the following equations using smallest whole-number coefficients.
   (a) $Zn(OH)_2 + H_3PO_4 \rightarrow H_2O + Zn_3(PO_4)_2$
   (b) $CO_2 + H_2O \rightarrow C_6H_{12}O_6 + O_2$

5. Name each of the following compounds using the IUPAC system:
   (a) NaI   (b) CS_2 (c) $FeCl_3$ (d) P_4O_{10} (e) Cl_2O_7

6. Write a balanced equation for a synthesis reaction with magnesium and Br_2 as reactants.

7. Write a balanced equation for the decomposition of FeO.

8. Write a balanced equation for the single-replacement reaction that occurs between zinc and aluminum. Assume that NO_3^- is the negative ion.

9. Write a balanced equation for the double-replacement reaction between Na_3PO_4 and $Pb(NO_3)_2$. Use Reference Table F in Appendix 1 to determine which compound is the precipitate.

Base your answers to questions 10 through 12 on the balanced chemical equation below.

$$2H_2O \rightarrow 2H_2 + O_2$$

10. What type of reaction does this equation represent?

11. How does the balanced chemical equation show the law of conservation of mass?

12. What is the total number of molecules of O_2 produced when 8 molecules of H_2O are completely consumed?

The answers to these questions are found in Appendix 3.

Chapter Four

CHEMICAL CALCULATIONS

KEY IDEAS

This chapter focuses on problem solving in chemistry. The concept of the mole is stressed throughout the chapter. A variety of problems involving single substances and equations is explored.

KEY OBJECTIVES

At the conclusion of this chapter you will be able to:

- Define the terms *atomic mass unit*, *isotopic mass*, and *average atomic mass*.
- Calculate the atomic mass of an element, given the masses of its naturally occurring isotopes and the abundances of these isotopes.
- Calculate the formula mass of a substance.
- Define the term *mole* in relation to number of particles and the mass of a substance.
- Define the term *molar mass* (also called *gram-formula mass*).
- Calculate the molar masses of various types of substances.
- Solve mole-mass problems.
- Solve percent composition problems.
- Calculate the molecular formula of a substance, given its empirical formula and molar mass.
- Calculate the empirical formula of a substance from its percent composition by mass.
- Solve mole and mass problems involving chemical equations.
- Solve problems involving percent yields and limiting reactants.

SECTION I—BASIC (REGENTS-LEVEL) MATERIAL

NYS REGENTS CONCEPTS AND SKILLS

Note: By the time you have finished Section I, this chapter, you should have mastered the concepts and skills listed below. The Regents chemistry examination will test your knowledge of these items and your ability to apply them.

Concepts are the *basic ideas* that form the body of the Regents chemistry course (what you need to know!).

Skills are the *activities* that demonstrate your mastery of these concepts (how you show that you know them!).

Following each concept or skill is a page reference (given in parentheses) to this chapter.

4.1 Concept:
The mass of each proton and of each neutron is approximately equal to one atomic mass unit. An electron is much less massive than a proton or a neutron. The mass of an atom is very nearly equal to its mass number. (Pages 83–84)

4.2 Concept:
The average atomic mass of an element is the weighted average of the masses of its naturally occurring isotopes. (Pages 83–84)

Skills:
• Given the atomic mass of an element, determine the mass number of the element's most abundant isotope. (Pages 83–84)
• Calculate the atomic mass of an element, given the masses and ratios of the element's naturally occurring isotopes. (Pages 83–85)

4.3 Concept:
The empirical formula of a compound is the simplest whole number ratio of atoms of the elements in a compound. It may be different from the molecular formula, which is the actual ratio of atoms in a molecule of that compound. (Pages 88–89)

Skills:
• Determine the molecular formula of a compound, given the empirical formula and the molecular mass. (Pages 88–89)
• Determine the empirical formula from the molecular formula. (Pages 88–89)

4.4 Concept:

The formula mass of a substance is the sum of the atomic masses of its atoms. The molar mass (gram formula mass) of a substance is the mass of 1 mole of that substance. (Pages 85–86)

Skills:

- Calculate the formula mass and the molar mass (gram-formula mass) of a substance. (Pages 87–88)
- Determine the number of moles of a substance, given its mass. (Pages 85–86)
- Determine the mass of a given number of moles of a substance. (Pages 85–86)

4.5 Concept:

The percent composition by mass of each element in a compound can be calculated mathematically. (Pages 87–88)

Skill:

Determine the percent composition(s) of one or more elements in a compound. (Pages 87–88)

4.6 Concept:

A balanced chemical equation represents conservation of atoms. The coefficients in a balanced chemical equation can be used to determine mole ratios in the reaction. (Pages 90–91)

Skill:

Solve simple mole-mole stoichiometry problems, given a balanced equation. (Pages 90–91)

4.1 INTRODUCTION

As its name implies, this chapter is devoted to solving various numerical problems in chemistry. The ability to solve numerical problems is essential to chemists because it provides the means to translate theory into practice. We will solve many of the problems in this chapter by using the factor-label method (FLM) introduced in Chapter 1. You may wish to review the method at this time.

4.2 AVERAGE ATOMIC MASS

How do scientists measure the masses of atoms and their components? They use an instrument called a *mass spectrometer*. First, though, a standard mass must be established and, it must be given a value. Chemists have agreed that a neutral atom of the isotope carbon-12 is the appropriate standard. Its mass has been set at exactly 12 atomic mass units (12 u or 12 amu). It follows that 1 atomic mass unit is 1/12 of the mass of a neutral carbon-12 atom. All other atomic masses are measured relative to this value. With this system, the masses of a proton and a neutron are nearly equal to 1 atomic mass unit, while the electron has a mass close to 0.0005 atomic mass unit.

We are now able to report the mass of any atom or subatomic particle. But how should we report the masses of the atoms in a sample of an element? There are problems here because elements may contain atoms with different mass numbers (i.e., isotopes). The solution is to measure the mass of each individual isotope in the element and then calculate a *weighted average* based on the *abundance* of each isotope in the sample of the element.

This weighted average is known as the **average atomic mass** of the element. For example, the average atomic mass of the element carbon is 12.0111 atomic mass units, indicating that a sample of carbon contains isotopes whose mass numbers are other than 12. We can think of the quantity 12.0111 atomic mass units as the mass of an *average* atom the element carbon.

If we examine The Periodic Table of the Elements in Appendix 1, we note that the average atomic masses of most elements are nearly whole numbers. If an average atomic mass is rounded to a whole number, it usually represents the mass number of the most abundant isotope of the element.

PROBLEM

What is the most abundant isotope of the element oxygen (average atomic mass = 15.9994 u)?

SOLUTION

The average atomic mass is very nearly equal to 16 atomic mass units, so we conclude that oxygen-16 is the most abundant isotope of the element oxygen.

TRY IT YOURSELF

What is the most abundant isotope of the element gold (average atomic mass = 196.9665 u)?

ANSWER

Gold-197

PROBLEM

An element consists of two isotopes. Isotope *A* has an abundance of 75.00 percent, and its mass is 14.000 atomic mass units. Isotope *B* has an abundance of 25.00 percent, and its mass is 15.000 atomic mass units. What is the average atomic mass of the element?

SOLUTION

The term *percentage abundance* indicates how many atoms of an isotope are contained in 100 atoms of the element. In this example, 100 atoms of the element would contain 75 atoms of isotope *A* and 25 atoms of isotope *B*.

To calculate the average atomic mass of the element, we multiply the mass of each isotope by the *decimal equivalent* of its percentage abundance and then add the results:

$$\text{Average atomic mass} = (14.000 \text{ u} \cdot 0.7500) + (15.000 \text{ u} \cdot 0.2500)$$
$$= 14.25 \text{ u}$$

TRY IT YOURSELF

The element boron occurs in nature as two isotopes. Boron-10 has a mass of 10.0130 atomic mass units, and its abundance is 19.90 percent; boron-11 has a mass of 11.0093 atomic mass units, and its abundance is 80.10 percent. From these data, calculate the average atomic mass of the element boron.

ANSWER

10.81 u

From this point forward, we will replace the term *average atomic mass* with the simpler term **atomic mass**.

4.3   THE FORMULA MASS OF A SUBSTANCE

The **formula mass** is the sum of the masses of all of the atoms in a given formula. We recall that atomic masses are measured in atomic mass units (u). The formula mass of the molecule O_2 equals 32 atomic mass units since there are two oxygen atoms in the formula and each oxygen atom has a mass of 16.00 atomic mass units.

Suppose we want to find the formula mass of the compound $CuSO_4$. We need to do a little planning: first, we obtain the atomic masses from the Periodic Table of Elements in Appendix 1 of this book, and then we set up a table to keep things organized:

Element	Atomic Mass	×	Number of Atoms in Formula	=	Mass of Element
Cu	63.55 u	×	1	=	63.55 u
S	32.07 u	×	1	=	32.07 u
O	16.00 u	×	4	=	64.00 u
			Formula mass (total)	=	159.62 u

TRY IT YOURSELF

Calculate the formula mass of NH_3.

ANSWER

17.03 u

4.4   THE MOLE CONCEPT AND MOLAR MASS

The *mole* is one of the most useful and important concepts in chemistry. Mole (abbreviated as *mol*) is a word with two faces: it represents a number and a mass. In Chapter 6, we will see that the mole has a third face: the volume of a gas under certain specified conditions.

One mole refers to *Avogadro's number* ($N_A = 6.02 \times 10^{23}$) of particles of anything. For example, 1 mole of silver refers to 6.02×10^{23} *atoms* of Ag, while 1 mole of CO_2 refers to 6.02×10^{23} *molecules* of CO_2. And 1 mole of slices of pizza? That's correct, 6.02×10^{23} slices!

One mole of particles has a special mass associated with it: *the formula mass of the substance expressed in grams*. For example, 1 mole of H_2O has a mass of 18.00 grams (because its formula mass is 18.00 u). In this context, we refer to 1 mole as the **molar mass** (also known as the **gram-formula mass**), whose units are grams per mole (g/mol). In this book, we will use these two terms interchangeably and we will use the symbol $\mathcal{M}$ to represent them. If the formula happens to be that of an element, such as calcium, the molar mass will refer to the atomic mass of the element; if the formula represents a molecule or an ionic compound, the molar mass will refer to the formula mass of the substance.

PROBLEM

Calculate the molar mass of each of the following:
(a) Ne   (b) Cl_2 (c) SO_3 (d) KBr

SOLUTIONS
(a)   We use the atomic mass of Ne: $\mathcal{M} = 20.18$ g/mol.
(b)   We calculate the molar mass of Cl_2: $\mathcal{M} = 70.91$ g/mol.
(c)   We calculate the molar mass of SO_3: $\mathcal{M} = 80.07$ g/mol.
(d)   We calculate the molar mass of KBr: $\mathcal{M} = 118.0$ g/mol.

TRY IT YOURSELF

Calculate the molar mass of $CaCO_3$.

ANSWER

$\mathcal{M} = 100.1$ g/mol

4.5 PROBLEMS INVOLVING A SINGLE SUBSTANCE

We can solve a variety of problems involving a single substance because any substance can be represented by a chemical formula, and any formula can be associated with the mole concept. For example, the formula H_2SO_4 can be interpreted as follows: 1 mole of H_2SO_4 contains 2 moles of H atoms, 1 mole of S atoms, and 4 moles of O atoms. Similarly, 1 mole of $CaCl_2$ contains 1 mole of Ca^{2+} ions and 2 moles of Cl^- ions.

Mole-Mass Conversions

Reference Table T in Appendix 1 gives us a formula for calculating among mass, molar mass, and number of moles:

$$\text{number of moles} = \frac{\text{given mass (g)}}{\text{molar mass (g/mol)}}$$

In symbols, we can represent this formula as:

$$n = \frac{m}{\mathcal{M}}$$

where n is the number of moles, m is the mass of the substance in grams, and $\mathcal{M}$ is the molar mass in grams per mole.

PROBLEM
(a) Calculate the number of moles in 250. grams of $CaCO_3$ ($\mathcal{M}$ = 100. g/mol).
(b) Calculate the mass of 6.50 moles of KBr ($\mathcal{M}$ = 118 g/mol).

SOLUTIONS

(a) $n = \dfrac{m}{\mathcal{M}} = \dfrac{250.\ \text{g}}{100.\ \text{g/mol}} = 2.50\ \text{mol}$

(b) $n = \dfrac{m}{\mathcal{M}}$

$m = n \cdot \mathcal{M} = (6.50\ \text{mol}) \cdot (118\ \text{g/mol}) = 767\ \text{g}$

TRY IT YOURSELF
Calculate the molar mass of a substance if 0.25 mole of the substance has a mass of 45 grams.

ANSWER
$\mathcal{M} = 180$ g/mol

Mole-mass conversion problems can also be solved by using the factor-label method (FLM). When FLM is used, the molar mass becomes the conversion factor. For example, the molar mass of NH_3 is 17.03 grams per mole.

Therefore, we can form two possible fractions:

$$\frac{17.03 \text{ g NH}_3}{1 \text{ mol NH}_3} \quad \text{or} \quad \frac{1 \text{ mol NH}_3}{17.03 \text{ g NH}_3}$$

Either of these two fractions can be used in solving a mole-mass problem.

PROBLEM
How many moles of molecules are contained in 67.25 grams of NH_3 ($\mathcal{M} = 17.03$ grams per mole)?

SOLUTION
The appropriate conversion factor is $\dfrac{1 \text{ mol NH}_3}{17.03 \text{ g NH}_3}$. Our problem is then set up for solution:

$$x \text{ mol NH}_3 = 67.25\text{g NH}_3 \cdot \frac{1 \text{ mol NH}_3}{17.03 \text{ g NH}_3} = 3.949 \text{ mol NH}_3$$

TRY IT YOURSELF
Calculate the mass of 3.00 moles of nitrogen gas (N_2) using the FLM.

ANSWER
84.04 g N_2

Empirical Formulas, Molecular Formulas, and Molar Mass

In Chapter 3, we learned how to calculate the empirical formula of a substance, given its molecular formula. Now, we learn how to calculate the molecular formula of a substance, given its empirical formula and its molar mass. The following steps show how this is accomplished:

1. Calculate the molar mass of the *empirical formula*.
2. *Divide* the molar mass of the substance by the molar mass of the empirical formula.
3. *Multiply* the empirical formula by the number obtained in Step 2.

PROBLEM

The empirical formula of a molecular substance is CH_2, and the molar mass of the substance is 70.05 grams per mole. Determine the molecular formula of the substance.

SOLUTION

1. The molar mass of the empirical formula CH_2 is 14.01 g/mol.

2. $\dfrac{70.05 \text{ g/mol}}{14.01 \text{ g/mol}} = 5.000 = 5$

3. $5 \cdot (CH_2) = C_5H_{10}$

TRY IT YOURSELF

The empirical formula of a substance is NO_2, and the molar mass of the substance is 92.01 g/mol. Determine the molecular formula of the substance.

ANSWER

N_2O_4

Percent Composition by Mass

The term *percent* is always associated with a number meaning "parts per hundred." We calculate all percentages in a very simple way: we divide the quantity under consideration by the total quantity involved, and multiply the result by 100. For example, if 7 persons in 20 wear eyeglasses, then the percentage of persons wearing eyeglasses is calculated from this relationship: $(7/20) \cdot 100 = 35$ percent. Reference Table T in Appendix 1 provides a general formula for percent composition.

$$\% \text{ composition by mass} = \frac{\text{mass of part}}{\text{mass of whole}} \times 100$$

Similarly, we can calculate the percent composition of an element in a compound by dividing the mass of the element by the molar mass of the substance and multiplying the result by 100.

PROBLEM

Calculate the percentage of oxygen by mass in $CuSO_4$.

SOLUTION

We construct the accompanying table to calculate the molar mass of $CuSO_4$.

Element	Atomic Mass	×	Number of Atoms in Formula	=	Mass of Element
Cu	63.55 g/mol	×	1	=	63.55 g/mol
S	32.07 g/mol	×	1	=	32.07 g/mol
O	16.00 g/mol	×	4	=	64.00 g/mol
			Molar mass (total)	=	159.62 g/mol

We note that the mass of oxygen in the formula is 64.00 g/mol, while the molar mass is 159.62 g/mol. The percent composition of oxygen in $CuSO_4$ is found from this relationship:

$$\frac{64.00 \text{ g/mol}}{159.62 \text{ g/mol}} \cdot 100 = 40.10\%$$

TRY IT YOURSELF

Calculate the percentage of carbon by mass in the sugar glucose ($C_6H_{12}O_6$).

ANSWER

40.00%

4.6 PROBLEMS INVOLVING CHEMICAL EQUATIONS

Let us consider this balanced chemical equation:

$$N_2 + 3H_2 \rightarrow 2NH_3$$

Previously, we interpreted the coefficients (1, 3, and 2) as numbers of molecules. Since moles and numbers of molecules are closely related, these coefficients can also be read as numbers of *moles* of molecules; then the equation can be read as follows:

1 *mole* of N_2 molecules and 3 *moles* of H_2 molecules react to yield 2 *moles* of NH_3 molecules.

Since we can convert among moles, mass, and numbers of particles, it is possible for us to solve a wide variety of problems by applying the factor-label method to chemical equations.

Mole-Mole Problems

Any two substances in a balanced chemical equation (reactants, products, or both) can be connected by means of their respective coefficients, since the coefficients stand for the numbers of moles involved in the chemical reaction. For example, let us consider the following problem:

PROBLEM
In the equation $N_2 + 3H_2 \rightarrow 2NH_3$, how many moles of N_2 are needed to produce 5.0 moles of NH_3?

SOLUTION
From the equation we know that 1 mol of N_2 will produce 2 mol of NH_3. We can now solve the problem by using FLM. Our solution map is as follows:

$$5.0 \text{ mol NH}_3 \xrightarrow[\text{coefficients}]{\text{equation}} \text{mol N}_2$$

We need to use the appropriate coefficients of the equation because a mole-mole conversion is involved. The solution is as follows:

$$\text{mol N}_2 = 5.0 \text{ mol NH}_3 \cdot \frac{1 \text{ mol N}_2}{2 \text{ mol NH}_3} = 2.5 \text{ mol N}_2$$

Note that the equation coefficients form the conversion factor in the solution to the problem.

TRY IT YOURSELF
In the equation $2NO + O_2 \rightarrow 2NO_2$, how many moles of O_2 are needed to produce 3.50 moles of NO_2?

ANSWER
1.75 mol

SECTION II—ADDITIONAL MATERIAL

4.1A CONVERTING BETWEEN MOLES AND NUMBERS OF PARTICLES

Since 1 mole is equal to Avogadro's number of particles ($N_A = 6.02 \times 10^{23}$), we can use this number as the conversion factor.

PROBLEM
How many particles are in 2.00 moles of SO_2?

SOLUTION

$$N \text{ molecules SO}_2 = 2.00 \text{ mol SO}_2 \cdot \frac{6.02 \times 10^{23} \text{ molecules SO}_2}{1 \text{ mol SO}_2} = 1.20 \times 10^{24} \text{ molecules SO}_2$$

TRY IT YOURSELF

How many moles of molecules are equivalent to 9.03×10^{23} molecules of H_2O?

ANSWER

1.50 mol H_2O

We can also solve more complicated problems involving moles, mass, and number of particles by the factor-label method. In this case, *both* the molar mass and Avogadro's number are used as conversion factors.

PROBLEM

What is the mass of 2.40×10^{24} molecules of O_2?

SOLUTION

We need to convert the number of molecules to moles using Avogadro's number, and then convert the number of moles to mass using the molar mass of oxygen. The factor-label method allows us to solve this problem in *one step*:

$$(2.40 \times 10^{24} \text{ molecules O}_2) \cdot \left(\frac{1 \text{ mol O}_2}{6.02 \times 10^{23} \text{ molecules O}_2} \right) \cdot \left(\frac{32.0 \text{ g O}_2}{1 \text{ mol O}_2} \right) = 128 \text{ g O}_2$$

TRY IT YOURSELF

What is the number of atoms contained in 12.01 grams of helium (He)?

ANSWER

1.81×10^{24} atoms of He

4.2A EMPIRICAL FORMULA FROM PERCENT COMPOSITION

Recall that the **empirical formula** of a compound is a formula reduced to simplest numbers by division. For example, the empirical formula for N_2O_4 is NO_2, and the empirical formula for C_6H_6 is CH. The empirical formula for H_2O, however, remains H_2O because it cannot be reduced further.

Since we calculate percent compositions from chemical formulas, we are able to reverse the process and determine formulas from percent composition. However, when we use this process, we are able to determine only *empirical* formulas. The technique we use relies on our connection of a formula with the mole concept.

PROBLEM

Calculate the empirical formula of a substance that contains 50.00 percent sulfur and 50.00 percent oxygen by mass.

SOLUTION

If we had 100.0 g of this compound, we would know from the percent composition given in the problem that the compound contains 50.00 g of S and 50.00 g of O, but formulas indicate numbers of *moles,* not masses in grams. If, however, we divide the given masses by the atomic masses of the two elements, we transform mass into moles as follows:

$$S_{\frac{50.00\,\cancel{g}}{32.0\frac{\cancel{g}}{mol}}}\ O_{\frac{50.00\,\cancel{g}}{16.0\frac{\cancel{g}}{mol}}} = S_{1.56\,mol}O_{3.125\,mol}$$

As strange as this formula looks, it is accurate because it shows the *relative number of moles* of each element. Now, our task is to make the formula *look* like a formula, that is, to convert it into an expression with whole numbers. Usually we accomplish this by dividing each element by the *smallest* number of moles, as illustrated below:

$$S_{\frac{1.56}{1.56}}O_{\frac{3.125}{1.56}} = S_{1.00}O_{2.00} = SO_2$$

TRY IT YOURSELF

A hydrocarbon consists of 80.00 percent carbon and 20.00 percent hydrogen by mass.

(a) Express the formula of the compound in terms of the relative numbers of moles of carbon and hydrogen.

(b) Determine the empirical formula of the hydrocarbon.

ANSWERS

(a) $C_{6.661}H_{19.84} = C_{1.000}H_{2.979}$ (b) CH_3

4.3A MOLE-MASS PROBLEMS

PROBLEM

In the equation $4Al + 3O_2 \rightarrow 2Al_2O_3$, how many grams of aluminum will combine with 1.50 moles of oxygen?

SOLUTION

From the equation we know that 4 mol of Al will combine with 3 mol of O_2. We can solve the problem by constructing a "solution map" and using FLM. Here is our solution map:

$$1.50 \text{ mol } O_2 \xrightarrow[\text{coefficients}]{\text{equation}} \text{mol Al} \longrightarrow \text{g Al}$$

The solution is as follows:

$$\text{g Al} = 1.50 \text{ mol } O_2 \left(\frac{4 \text{ mol Al}}{3 \text{ mol } O_2} \right) \cdot \left(\frac{27.0 \text{ g Al}}{1 \text{ mol Al}} \right) = 54.0 \text{ g Al}$$

Notice that the equation coefficients and the molar mass of Al form the conversion factors in the solution to the problem.

TRY IT YOURSELF

In the equation given in the problem above, how many moles of oxygen are needed to produce 51.0 grams of Al_2O_3?

ANSWER

0.750 mol

4.4A   MASS-MASS PROBLEMS

PROBLEM

In the equation $CH_4 + 2O_2 \rightarrow CO_2 + 2H_2O$, how many grams of CO_2 are formed when 8.0 grams of CH_4 reacts with an excess of O_2?

SOLUTION

Again, the problem is solved by using FLM. Since equations are given in terms of *moles*, not grams, we must convert to and from moles to arrive at an answer. Here is our solution map:

$$8.0 \text{ g } CH_4 \longrightarrow \text{mol } CH_4 \xrightarrow[\text{coefficients}]{\text{equation}} \text{mol } CO_2 \longrightarrow \text{g } CO_2$$

The problem is solved as follows:

$$\text{g } CO_2 = 8.0 \text{ g } CH_4 \cdot \frac{1 \text{ mol } CH_4}{16.0 \text{ g } CH_4} \cdot \frac{1 \text{ mol } CO_2}{1 \text{ mol } CH_4} \cdot \frac{44.0 \text{ g } CO_2}{1 \text{ mol } CO_2} = 22 \text{ g } CO_2$$

The term *excess* in the problem statement means that we have more than enough O_2 to react with the CH_4. Therefore, the mass of CO_2 that will be produced depends solely on the quantity of CH_4 present (i.e., 8.0 g). For this reason, CH_4 is called a *limiting reactant*. At the end of the reaction, all of the CH_4 will have been consumed and some of the O_2 will remain unreacted. We will explore limiting reactant problems in more detail later in this section (see pages 96–98).

TRY IT YOURSELF
In the equation $2H_2O_2 \rightarrow 2H_2O + O_2$, how many grams of O_2 will be formed from the decomposition of 17.0 grams of H_2O_2?

ANSWER
8.00 g

4.5A PERCENT YIELD

In the real world, reactions do not always go as planned and the yield produced by a reaction may be *less* than the yield predicted by a pencil-and-paper solution (e.g., a mass-mass problem). Usually, the result of a real-world reaction is reported in terms of *percent yield,* given by the following expression:

$$\text{percent yield} = \frac{\text{actual yield}}{\text{predicted yield}} \times 100$$

PROBLEM
Now, suppose that something goes wrong in the reaction in the "Try It Yourself" immediately preceding, and we produce only 6.00 grams of O_2. What is the percent yield of this reaction?

SOLUTION

$$\% \text{ yield} = \frac{6.00 \text{ g}}{8.00 \text{ g}} \times 100 = 75.0\%$$

TRY IT YOURSELF

In the equation $CH_4 + 2O_2 \rightarrow CO_2 + 2H_2O$:

(a) How many grams of H_2O are expected when 8.00 grams of CH_4 reacts with an excess of O_2?

(b) If only 17.0 grams of H_2O are produced, what is the percent yield of this reaction?

ANSWERS

(a) 18.0 g   (b) 94.4%

4.6A   LIMITING REACTANTS

When we discussed mass-mass problems earlier in this chapter, we introduced the concept of a *limiting reactant,* that is, a reactant that is *not* present in excess and is entirely consumed in a reaction. We now focus on a technique that will allow us to determine *which reactant* is present in a limiting quantity.

Let us consider this equation:

$$N_2 + 3H_2 \rightarrow 2NH_3$$

Suppose 2.0 moles of N_2 and 2.0 moles of H_2 are brought together to react. Which of the reactants (N_2 or H_2) is present in a limiting quantity?

We begin by calculating the *mole ratio* of the reactants (N_2 to H_2) *as they appear in the problem:*

$$\text{mole ratio (problem)} \quad \frac{2.0 \text{ mol } N_2}{2.0 \text{ mol } H_2} = 1.0 \quad \frac{\text{mol } N_2}{\text{mol } H_2}$$

Now we calculate the *mole ratio* of the reactants *as they appear in the equation.* To do this, we use the equation coefficients:

$$\text{mole ratio (equation)} \quad \frac{1 \text{ mol } N_2}{3 \text{ mol } H_2} = 0.33 \quad \frac{\text{mol } N_2}{\text{mol } H_2}$$

We now *compare* the two mole ratios. If the mole ratio (problem) is *larger* than the mole ratio (equation), then the *reactant in the numerator is present in excess*; if the mole ratio (problem) is *smaller* than the mole ratio (equation), then the *reactant in the denominator is present in excess.* The reactant that is *not* present in excess is the *limiting reactant.*

In the example shown above, the mole ratio given in the *problem* is larger than the mole ratio given in the *equation, so* the reactant in the *numerator* (N_2) is present in excess. *Therefore, H_2 is the limiting reactant* and all of it will be consumed in the reaction.

The next question we need to ask is: What quantity of N_2 will actually react? We solve this by the factor-label method, using the limiting reactant (H_2) and the equation coefficients:

$$\text{mol } N_2 \text{ reacting} = 2.0 \text{ mol } H_2 \cdot \frac{1 \text{ mol } N_2}{3 \text{ mol } H_2} = 0.67 \text{ mol } N_2$$

Therefore, we conclude that all 2.0 moles of H_2 but only 0.67 mole of N_2 will react according to the equation given above.

PROBLEM

Consider the reaction $C_3H_8 + 5O_2 \rightarrow 3CO_2 + 4H_2O$, in which 1.5 moles of C_3H_8 and 15 moles of O_2 are brought together.

(a) Calculate the mole ratio C_3H_8/O_2 according to the problem and the equation.

(b) Determine which reactant is present in excess and which is present in limiting quantity.

(c) Calculate the quantity of the excess reactant that reacts with the limiting reactant.

SOLUTIONS

(a) mole ratio (problem) $= \dfrac{1.5 \text{ mol } C_3H_8}{15 \text{ mol } O_2} = 0.10 \, \dfrac{\text{mol } C_3H_8}{\text{mol } O_2}$

mole ratio (equation) $= \dfrac{1 \text{ mol } C_3H_8}{5 \text{ mol } O_2} = 0.20 \, \dfrac{\text{mol } C_3H_8}{\text{mol } O_2}$

(b) Since the mole ratio of C_3H_8 to O_2 is smaller in the problem than in the equation, it follows that O_2 is present in excess and that C_3H_8 is the limiting reactant.

(c) mol O_2 reacting $= 1.5 \text{ mol } C_3H_8 \cdot \dfrac{5 \text{ mol } O_2}{1 \text{ mol } C_3H_8} = 7.5 \text{ mol } O_2$

As a final example, let us tackle a mass-mass problem that involves limiting reactants.

PROBLEM

How many grams of water can be made according to this equation:

$$2H_2 + O_2 \rightarrow 2H_2O$$

if 100.0 grams of H_2 and 160.0 grams of O_2 are brought together to react?

SOLUTION

Warning! This is not a simple mass-mass problem! We need first to deter-mine the limiting reactant by converting the masses to moles (using the molar masses of H_2 and O_2) and then to calculate the H_2/O_2 mole ratios:

$$\text{moles } H_2 \text{ in problem} = 100. \text{ g } H_2 \cdot \frac{1 \text{ mol } H_2}{2.00 \text{ g } H_2} = 50.0 \text{ mol } H_2$$

$$\text{moles } O_2 \text{ in problem} = 160. \text{ g } O_2 \cdot \frac{1 \text{ mol } O_2}{32.0 \text{ g } O_2} = 5.00 \text{ mol } O_2$$

$$\text{mole ratio (problem)} = \frac{50.0 \text{ mol } H_2}{5.00 \text{ mol } O_2} = 10.0 \frac{\text{mol } H_2}{\text{mol } O_2}$$

$$\text{mole ratio (equation)} = \frac{2 \text{ mol } H_2}{1 \text{ mol } O_2} = 2.00 \frac{\text{mol } H_2}{\text{mol } O_2}$$

Since the mole ratio in the problem is *larger* than the mole ratio in the equa-tion, we conclude that H_2 is present in excess and that O_2 is the limiting reactant. Then, since O_2 is the limiting reactant, we know that all of it will be consumed, and this fact will determine how much water is produced.

We now proceed as we did in the problems just given, using the solution map:

$$\text{mol } O_2 \rightarrow \text{mol } H_2O \rightarrow \text{g } H_2O$$

the equation coefficients, and the molar mass of water:

$$5.00 \text{ mol } O_2 \cdot \frac{2 \text{ mol } H_2O}{1 \text{ mol } O_2} \cdot \frac{18.0 \text{ g } H_2O}{1 \text{ mol } H_2O} = 180. \text{ g } H_2O$$

END-OF-CHAPTER QUESTIONS

Some questions have the symbol "§2" in front of the question number. This symbol means that the question is based on Section II material.

1. The standard of atomic mass is the isotope
   (1) ^{16}S, which is equal to 16.000 u
   (2) ^{32}S, which is equal to 32.000 u
   (3) ^{12}C, which is equal to 12.000 u
   (4) ^{14}N, which is equal to 14.000 u

2. Element X has three naturally occurring isotopes. The table below lists the mass numbers and percent abundances of these isotopes.

Mass Number	Percent Abundance
10	10.0
11	20.0
12	70.0

The average atomic mass of element X is closest to
(1) 11.0 u   (2) 11.6 u   (3) 12.0 u   (4) 12.4 u

3. What is the total mass of oxygen in 1.00 mole of $Al_2(CrO_4)_3$?
(1) 192 g   (2) 112 g   (3) 64.0 g   (4) 48.0 g

4. How many moles of $N_2(g)$ molecules will contain exactly 4.0 moles of nitrogen atoms?
(1) 1.0   (2) 2.0   (3) 3.0   (4) 4.0

5. The gram formula mass of NH_4Cl is closest to
(1) 22.4 g/mol   (2) 28.0 g/mol   (3) 53.5 g/mol   (4) 95.5 g/mol

6. Which is an empirical formula?
(1) P_2O_5 (2) P_4O_6 (3) C_2H_4 (4) C_3H_6

7. Which molecular formula is also an empirical formula?
(1) H_2O_2 (2) H_2O (3) C_2H_6 (4) C_6H_6

8. What mass contains 6.0×10^{23} atoms?
(1) 6.0 g of C   (2) 16 g of S
(3) 3.0 g of He   (4) 28 g of Si

9. How many moles of *hydrogen atoms* are in 1 mole of $C_6H_{12}O_6$ molecules?
(1) $24(6.0 \times 10^{23})$ (2) $12(6.0 \times 10^{23})$
(3) 24   (4) 12

10. What is the molar mass of $CuSO_4 \cdot 5H_2O$?
(1) 160. g/mol   (2) 178 g/mol   (3) 186 g/mol   (4) 250. g/mol

11. What is the molecular formula of a compound whose empirical formula is CH_4 and whose molar mass is 16 grams per mole?

(1) CH_4 (2) C_2H_4 (3) C_4H_8 (4) C_8H_{18}

12. The percent by mass of hydrogen in NH_3 is closest to

(1) $\frac{17}{1} \times 100$ (2) $\frac{17}{3} \times 100$ (3) $\frac{1}{17} \times 100$ (4) $\frac{3}{17} \times 100$

13. What is the approximate percent by mass of sulfur in H_2SO_4? [formula mass = 98]

(1) 16   (2) 33   (3) 65   (4) 98

14. A hydrated salt is a solid that includes water molecules within its crystal structure. A student heated a 9.10-gram sample of a hydrated salt to a constant mass of 5.41 grams. What percent by mass of water did the salt contain?

(1) 3.69 %   (2) 16.8%   (3) 40.5%   (4) 59.5%

§2 **15.** A compound consists of 85 percent silver and 15 percent fluorine by mass. What is its empirical formula?

(1) AgF (2) AgF_2 (3) Ag_2F (4) Ag_6F

16. The percent by mass of carbon in CO_2 is equal to

(1) $44/12 \times 100$ (2) $12/44 \times 100$
(3) $28/12 \times 100$ (4) $12/28 \times 100$

§2 **17.** A sample of a compound contains 24 grams of carbon and 64 grams of oxygen. What is the empirical formula of this compound?

(1) CO (2) CO_2 (3) C_2O_2 (4) C_2O_4

18. The percent by mass of oxygen in $Ca(OH)_2$ (molar mass = 74 g/mol) is closest to

(1) 16   (2) 22   (3) 43   (4) 74

19. What is the molecular formula of a compound that has a molar mass of 92 grams per mole and an empirical formula of NO_2?

(1) NO_2 (2) N_2O_4 (3) N_3O (4) N_4O_8

§2 **20.** At STP, 32 grams of O_2 will occupy the same volume as

(1) 64 g of H_2 (2) 32 g of SO
(3) 8.0 g of CH_4 (4) 4.0 g of He

21. The percent by mass of oxygen in MgO (molar mass = 40 g/mol) is closest to
(1) 16   (2) 24   (3) 40   (4) 60

§2 **22.** What is the empirical formula of an ion whose composition by mass is 57.14. percent sulfur and 42.86. percent oxygen?
(1) SO_2^{2-} (2) SO_3^{2-} (3) $S_2O_3^{2-}$ (4) SO_4^{2-}

23. A compound has the empirical formula NO_2. Its molecular formula could be
(1) NO_2 (2) N_2O (3) N_4O_2 (4) N_4O_4

24. Which represents the greatest mass of chlorine?
(1) 1 mole of Cl   (2) 1 atom of Cl
(3) 1 gram of Cl   (4) 1 molecule of Cl

25. What is the percent by mass of hydrogen in CH_3COOH (molar mass = 60 g/mol)?
(1) 1.7   (2) 5.0   (3) 6.7   (4) 7.1

26. An example of an empirical formula is
(1) C_2H_2 (2) H_2O_2 (3) C_2Cl_2 (4) $CaCl_2$

27. A 254-gram sample of I_2 contains approximately the same number of molecules as
(1) 14 g of N_2 (2) 2.0 g of H_2
(3) 36 g of H_2O (4) 40 g of Ne

28. Given this reaction

$$CH_4(g) + 2O_2(g) \rightarrow CO_2(g) + 2H_2O(g)$$

How many moles of oxygen are needed for the complete combustion of 3.0 moles of $CH_4(g)$?
(1) 6.0   (2) 2.0   (3) 3.0   (4) 4.0

29. Given this equation:

$$6CO_2 + 6H_2O \rightarrow C_6H_{12}O_6 + 6O_2$$

What is the total number of moles of water needed to make 2.5 moles of $C_6H_{12}O_6$?
(1) 2.5   (2) 6.0   (3) 12   (4) 15

30. Given this equation:

$$2C_2H_2(g) + 5O_2(g) \rightarrow 4CO_2(g) + 2H_2O(g)$$

How many moles of oxygen are required to react with 1.0 mole of C_2H_2?

(1) 2.5   (2) 2.0   (3) 5.0   (4) 10.

§2 **31.** Given this reaction:

$$S + O_2 \rightarrow SO_2$$

What is the total number of grams of oxygen needed to react completely with 2.0 moles of sulfur?

(1) 20   (2) 32   (3) 64   (4) 128

32. In the reaction:

$$Zn + 2HCl \rightarrow ZnCl_2 + H_2$$

how many moles of hydrogen will be formed when 4 moles of HCl is consumed?

(1) 6   (2) 2   (3) 8   (4) 4

§2 **33.** Given this reaction:

$$2H_2 + O_2 \rightarrow 2H_2O$$

The total number of grams of O_2 needed to produce 54 grams of water is

(1) 36   (2) 48   (3) 61   (4) 75

34. Given this reaction:

$$2Al + 3H_2SO_4 \rightarrow 3H_2 + Al_2(SO_4)_3$$

The total number of moles of H_2SO_4 needed to react completely with 5.0 moles of aluminum is

(1) 2.5   (2) 5.0   (3) 7.5   (4) 9.0

§2 **35.** According to the reaction $H_2 + Cl_2 \rightarrow 2HCl$, the production of 2.0 moles of HCl would require 70. grams of Cl_2 and

(1) 1.0 g of H_2 (2) 2.0 g of H_2
(3) 3.0 g of H_2 (4) 4.0 g of H_2

§2 **36.** Given this reaction:

$$Cu + 4HNO_3 \rightarrow Cu(NO_3)_2 + 2H_2O + 2NO_2$$

What is the total mass of H_2O produced when 32 grams of copper is completely consumed?
(1) 9.0 g   (2) 18 g   (3) 36 g   (4) 72 g

§2 **37.** Given this reaction:

$$2C_2H_6 + 7O_2 \rightarrow 4CO_2 + 6H_2O$$

What is the total number of CO_2 molecules produced when 1 mole of C_2H_6 is consumed?
(1) 6.02×10^{23} (2) $2(6.02 \times 10^{23})$
(3) $3(6.02 \times 10^{23})$ (4) $4(6.02 \times 10^{23})$

§2 **38.** Given this reaction:

$$4Al + 3O_2 \rightarrow 2Al_2O_3$$

How many moles of Al_2O_3 will be formed when 27 grams of aluminum reacts completely with O_2?

(1) 1.0   (2) 2.0   (3) 0.50   (4) 4.0

39. The percent, by mass, of water in $BaCl_2 \cdot 2H_2O$ (molar mass = 243 g/mol) is equal to
(1) $18/243 \times 100$ (2) $36/243 \times 100$
(3) $243/18 \times 100$ (4) $243/36 \times 100$

40. What species contains the greatest percent by mass of hydrogen?
(1) OH   (2) H_2O (3) H_3O^+ (4) H_2O_2

41. In which list are the elements arranged in order of increasing atomic mass?
(1) Cl, K, Ar   (2) Fe, Co, Ni
(3) Te, I, Xe   (4) Ne, F, Na

42. What is the percent by mass of oxygen in H_2SO_4? [formula mass = 98]
(1) 16%   (2) 33%   (3) 65%   (4) 98%

43. A hydrate is a compound that includes water molecules within its crystal structure. During an experiment to determine the percent by mass of water in a hydrated crystal, a student found the mass of the hydrated crystal to be 4.10 grams. After heating to constant mass, the mass was 3.70 grams. What is the percent by mass of water in this crystal?
(1) 90%   (2) 11%   (3) 9.8%   (4) 0.40%

44. Given the equation:

$$2C_2H_2(g) + 5O_2(g) \rightarrow 4CO_2(g) + 2H_2O(g)$$

How many moles of oxygen are required to react completely with 1.0 mole of C_2H_2?
(1) 2.5   (2) 2.0   (3) 5.0   (4) 10

45. The atomic mass of an element is calculated using the
(1) atomic number and the ratios of its naturally occurring isotopes
(2) atomic number and the half-lives of each of its isotopes
(3) masses and the ratios of its naturally occurring isotopes
(4) masses and the half-lives of each of its isotopes

46. Given the reaction:

$$PbCl_2(aq) + Na_2CrO_4(aq) \rightarrow PbCrO_4(s) + 2NaCl(aq)$$

What is the total number of moles of NaCl formed when 2 moles of Na_2CrO_4 react completely?
(1) 1 mole   (2) 2 moles
(3) 3 moles   (4) 4 moles

47. In which compound is the percent by mass of oxygen greatest?
(1) BeO   (2) MgO
(3) CaO   (4) SrO

48. The formula mass of a compound is the
(1) sum of the atomic masses of its atoms
(2) sum of the atomic numbers of its atoms
(3) product of the atomic masses of its atoms
(4) product of the atomic numbers of its atoms

49. The atomic mass of an element is the weighted average of the atomic masses of
(1) the least abundant isotopes of the element
(2) the naturally occurring isotopes of the element
(3) the artificially produced isotopes of the element
(4) the natural and artificial isotopes of the element

Constructed-Response Questions

1. The grid below is based on the mole concept developed in this chapter and is partially filled in. Complete the grid by using the data given in the grid.

Substance	Molar Mass ($\mathcal{M}$) (g/mol)	Mass of Substance (m) (g)	Moles of Substance (n) (mol)	Number of Particles of Substance (N)
O_3		24		
NH_3		170		
F_2		38		
CO_2			0.10	
NO_2			0.20	
Ne				1.5×10^{23}
N_2O				1.2×10^{24}
***		8.5		3.0×10^{23}

*** One of the seven substances listed above.

2. Calculate the molar mass of each of the following:
   (a) H_2SO_3 (b) $C_{21}H_{30}O_2$ (the active ingredient in marijuana)

3. In a 64-gram sample of SO_2:
   (a) How many moles are present?
   (b) How many molecules are present?

4. Calculate the mass of 3.33 moles of C_2H_5OH ($\mathcal{M}$ = 46.1 g/mol).

§2 5. Calculate the empirical formula of a compound containing 22.7 percent sodium, 21.6 percent boron, and 55.7 percent oxygen by mass.

§2 6. What is the molecular formula of a binary compound of sulfur and chlorine if it contains 47.5 percent sulfur by mass and its molar mass is 135.1 grams per mole?

§2 7. The net process of photosynthesis may be represented by this balanced equation:

$$6CO_2 + 6H_2O \rightarrow C_6H_{12}O_6 + 6O_2$$

   (a) How many grams of $C_6H_{12}O_6$ are produced when 0.131 mole of CO_2 is consumed?
   (b) How many grams of O_2 can be obtained when 65.0 grams of CO_2 is consumed?

§2 **8.** Given the equation
$$2H_2 + O_2 \rightarrow 2H_2O$$
how many grams of water can be produced from the reaction of 77.7 grams of oxygen with 15.0 grams of hydrogen?

§2 **9.** Given this equation:
$$N_2 + 3H_2 \rightarrow 2NH_3$$
if 27.7 grams of ammonia is produced by the reaction of 18.4 grams of hydrogen with an excess of nitrogen, what is the percent yield of the reaction?

10. (a) Calculate the molar mass of $Mg(OH)_2$
(b) How many moles of $Mg(OH)_2$ are present in an 8.40-gram sample?

11. The table below gives information about two isotopes of element X.

Isotope	Mass	Relative Abundance
X-10	10.01	19.91%
X-11	11.01	80.09%

Calculate the average atomic mass of element X. (Show a correct numerical setup, and express your answer to the correct number of significant figures.)

Base your answers to questions 12 and 13 on the information below. Gypsum is a mineral that is used in the construction industry to make drywall (sheetrock). The chemical formula for this hydrated compound is $CaSO_4 \cdot 2H_2O$. A hydrated compound contains water molecules within its crystalline structure. Gypsum contains 2 moles of water for each mole of calcium sulfate.

12. Calculate the gram formula mass of $CaSO_4 \cdot 2H_2O$.

13. Calculate the percent composition by mass of water in this compound.

Base your answers to questions 14–16 on the information below and on your knowledge of chemistry. The two naturally occurring isotopes of antimony are Sb-121 and Sb-123. The table below shows the atomic mass and the percent natural abundance for these isotopes.

Naturally Occurring Isotopes of Antimony

Isotope	Atomic Mass (u)	Natural Abundance (%)
Sb-121	120.90	57
Sb-123	122.90	43

Antimony and sulfur are both found in the mineral stibnite, Sb_2S_3. To obtain antimony, stibnite is roasted (heated in air), producing oxides of antimony and sulfur. The following unbalanced equation represents one of the reactions that occurs during the roasting:

$$Sb_2S_3(s) + O_2(g) \rightarrow Sb_2O_3(s) + SO_2(g)$$

14. Determine the percent composition by mass of antimony in stibnite (molar mass = 340. g/mol).

15. Show a correct numerical setup for calculating the atomic mass of antimony.

16. Balance the equation given above using smallest whole-number coefficients.

The answers to these questions are found in Appendix 3.

Chapter Five

ENERGY AND CHEMICAL REACTIONS

KEY IDEAS

This chapter focuses on the role energy plays in chemical reactions and the factors that determine whether a chemical process will occur under a given set of conditions.

KEY OBJECTIVES
At the conclusion of this chapter you will be able to:
- Define the terms *system* and *surroundings* as they relate to chemical processes.
- Define the terms *internal energy* and *heat*.
- Distinguish between *heat* and *temperature*.
- Distinguish between exothermic and endothermic reactions.
- Define the term *specific heat*, and use specific heats to solve calorimetry problems.
- Relate the first law of thermodynamics to the law of conservation of energy.
- Define the term *heat of reaction*, and solve problems involving heats of reaction.
- Define the terms *standard heat of formation* and *formation reaction*, and use the appropriate reference tables to solve problems related to the standard heat of formation.
- Interpret a potential energy diagram.
- Define the term *activation energy*.
- Define the term *spontaneous reaction*, and name and describe the factors that drive spontaneous reactions.
- Define the term *entropy*, and predict whether a given reaction leads to an increase or a decrease in entropy.

SECTION I—BASIC (REGENTS-LEVEL) MATERIAL

NYS REGENTS CONCEPTS AND SKILLS

Note: By the time you have finished Section I, you should have mastered the concepts and skills listed on the following pages. The Regents chemistry examination will test your knowledge of these items and your ability to apply them.

Concepts are the *basic ideas* that form the body of the Regents chemistry course (what you need to know!).

Skills are the *activities* that demonstrate your mastery of these concepts (how you show that you know them!).

Following each concept or skill is a page reference (given in parentheses) to this chapter.

5.1 Concepts:
- Heat is a transfer of energy (usually thermal energy) from a body of higher temperature to a body of lower temperature. (Page 112)
- Thermal energy is the energy associated with the random motion of atoms and molecules. (Page 112)

Skills:
- Distinguish between heat energy and temperature. (Page 112)
- Calculate the heat involved in a temperature change. (Pages 112–113)

5.2 Concepts:
- Chemical and physical changes can be exothermic or endothermic. (Page 112)
- Energy released or absorbed during a chemical reaction (the heat of reaction) is equal to the difference between the potential energy of the products and the potential energy of the reactants. (Page 113)

Skill:
Distinguish between endothermic and exothermic reactions in terms of:
(a) a reaction equation (Page 114);
(b) ΔH (Pages 113–114);
(c) experimental data (Page 114).

5.3 Concept:
Energy released or absorbed by a chemical reaction can be represented by a potential energy diagram. (Pages 114–116)

Skill:
Read and interpret potential energy diagrams in terms of:
(a) potential energies of reactants and products (Page 114);
(b) activation energy (Pages 115–116);
(c) heat of reaction (ΔH) (Pages 115–116).

5.4 Concept:
Entropy is a measure of the randomness or disorder of a system. A system with greater disorder has greater entropy. (Pages 117–118)

Skill:
Compare the entropies of solids, liquids, and gases.
(Pages 117–118)

5.5 Concept:
Systems in nature tend to undergo changes toward lower energy and higher entropy. (Pages 117–118)

5.1 ENERGY AND ITS MEASUREMENT

In Chapter 1, we related the concept of *energy* to the concept of *mechanical work*. In this chapter, we will study more closely the concept of energy and its role in chemistry. First, we need to define some basic terms that will be used throughout the chapter.

The part of the universe that a chemist chooses to study is called a **system**. The rest of the universe is known as the **surroundings**. For example, suppose you decide to study the melting of ice in a beaker, as shown in the accompanying diagram.

Surroundings

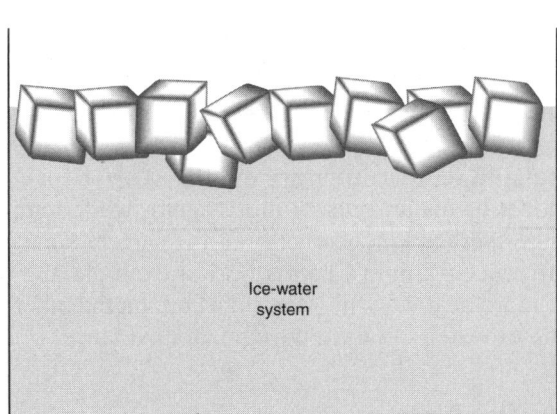

Ice-water
system

The ice and water constitute the system; the beaker, the air, and everything else constitute the surroundings.

In chemical processes, a system can *interact* with its surroundings in a number of ways, as shown in the following examples:

- A beaker of liquid absorbs heat from a burner; in this example, *energy* is transferred between the system (the beaker of liquid) and its surroundings.
- A cylinder gas is compressed by a piston; in this example, *work* is done *on* the system (the gas) by its surroundings.

- Water is poured into a solution; in this example, *mass* has been transferred to the system (the solution) from its surroundings.

As we see from the examples given above, energy transfers involve transfers of heat and/or work. These transfers affect the *total* energy within a system, known as the **internal energy** (*E*), which has two main components: **thermal energy**, the energy associated with random molecular motions, and **chemical energy**, the energy associated with chemical bonds and attractions between the particles of a system.

Since energy and work are so closely related, the same unit is used to measure both quantities. In the SI system, this unit is the **joule** (J); its multiple is the **kilojoule** (kJ). In this book, we will use the joule and the kilojoule *exclusively*.

Heat

Heat is the energy transferred between a system and its surroundings as a result of a *temperature difference*. In the first example given on page 111, heat passes from the burner to the beaker of liquid because the temperature of the burner is higher than the temperature of the beaker and liquid. Heat is represented by the symbol *q*. When energy is transferred *into* a system as a result of a temperature difference, heat is reported as a *positive* number, and we say that the process is **endothermic**; when energy is transferred *out of* a system as a result of a temperature difference, heat is reported as a *negative* number, and we say that the process is **exothermic**. For simplicity, we will state that in endothermic processes *heat is absorbed*, and in exothermic processes *heat is released*.

Note that **heat and temperature are not the same**. Heat energy is related to the total amount of matter present in a system, while temperature is not: it is related to the average kinetic energy of the system's particles. (We will discuss this difference again in Chapter 6.) For example, the water in a lake can transfer far more heat than the water in a cup; on the other hand, both the lake and the cup of water can have the same temperature.

Calorimetry Problems

To measure the heat absorbed or released by a substance that is not undergoing a phase change, we need to know three things about the substance: its identity (what it is!), its mass, and the change in its temperature. These three quantities can be combined into a simple equation that will allow us to solve many problems involving heat. In mathematical form, the equation is as follows:

$$q = c_p \cdot m \cdot \Delta T_C$$

This equation (in a slightly different form) is found on Reference Table T in Appendix 1 of this book. In this equation, where q represents the amount of heat, in joules; m is the mass of the substance, in grams; and ΔT_C is the temperature change of the substance, in Celsius degrees (C°). (*Note*: The unit "C°" stands for temperature *changes*, while the unit "°C" represents temperature *readings*.) The symbol c_p stands for the **specific heat capacity** (or more simply, the **specific heat**) of the substance. The substance is identified by the value of its specific heat. The smaller the specific heat of a substance, the more readily the substance will change temperature as it absorbs or releases heat.

Reference Table B in Appendix 1 of this book lists the specific heat of water as 4.2 joules per gram · Kelvin. *Note that a Kelvin (K) and a Celsius degree (C°) are completely equivalent.*

PROBLEM
How much heat is needed to raise the temperature of 20.0 grams of liquid water from 5.0°C to 20.0°C?

SOLUTION
$q = c_p \cdot m \cdot \Delta T_C$
$q = ???$ (This is the quantity we are asked to determine.)
$c_p = 4.2 \text{ J/g} \cdot \text{K}$ (See Reference Table B in Appendix 1.)
$m = 20.0$ g
$\Delta T_C = T_{final} - T_{initial} = 20.0°C - 5.0°C = 15.0C°$

We now solve the problem by direct substitution of the values into the equation:

$$q = c_p \cdot m \cdot \Delta T_C$$

$$= \frac{4.2 \text{ J}}{\text{g} \cdot \text{K}} \cdot 20.0 \text{ g} \cdot 15.0 \text{ C°}$$

$$= 1260 \text{ J } (1300 \text{ J to 2 significant figures.})$$

5.2 HEAT OF REACTION

When applied to a chemical reaction, ΔH is known as the **heat of reaction**. It is the difference in enthalpy (potential energy) between the products of the reaction and the reactants:

$$\Delta H = H_{products} - H_{reactants}$$

In an *endothermic* reaction, the enthalpy of the products is *larger* than the enthalpy of the reactants and the sign of ΔH is *positive*. In an *exothermic* reaction, the enthalpy of the products is *smaller* than the enthalpy of the reactants and the sign of ΔH is *negative*.

We indicate ΔH for an exothermic reaction in either of two ways:

$$2CO(g) + O_2(g) \rightarrow 2CO_2(g) \qquad \Delta H = -566.0 \text{ kJ}$$

or

$$2CO(g) + O_2(g) \rightarrow 2CO_2(g) + 566.0 \text{ kJ}$$

The two forms mean exactly the same thing: 2 moles of CO gas combine with 1 mole of O_2 gas to produce 2 moles of CO_2 gas and release 566.0 kilojoules of heat.

To illustrate an endothermic reaction, we need only write the reactions given above in reverse:

$$2CO_2(g) \rightarrow 2CO(g) + O_2(g) \qquad \Delta H = +566.0 \text{ kJ}$$

or

$$566.0 \text{ kJ} + 2CO_2(g) \rightarrow 2CO(g) + O_2(g)$$

This example also illustrates the important principle that a reaction that is exothermic in one direction is endothermic in the opposite direction. Reference Table I in Appendix 1 lists selected heats of reaction.

5.3   POTENTIAL ENERGY DIAGRAMS

A **potential energy diagram** is used to illustrate the progress of a chemical reaction and to provide qualitative information about energy changes within a reaction. We need to define one more term: The **activation energy** (or **energy of activation**), represented by the symbol E_a, is the minimum energy needed by the reactants in order for a reaction to occur.

Let us consider the *exothermic* reaction

$$\text{Reactants} \rightarrow \text{Products} + \text{Heat}$$

and the accompanying potential energy diagram for this reaction. We note that the potential energy of the system is plotted along the y-axis, and a quantity called the *reaction coordinate* (which monitors the progress of the reaction) along the x-axis.

Reaction Profile of an Exothermic Reaction

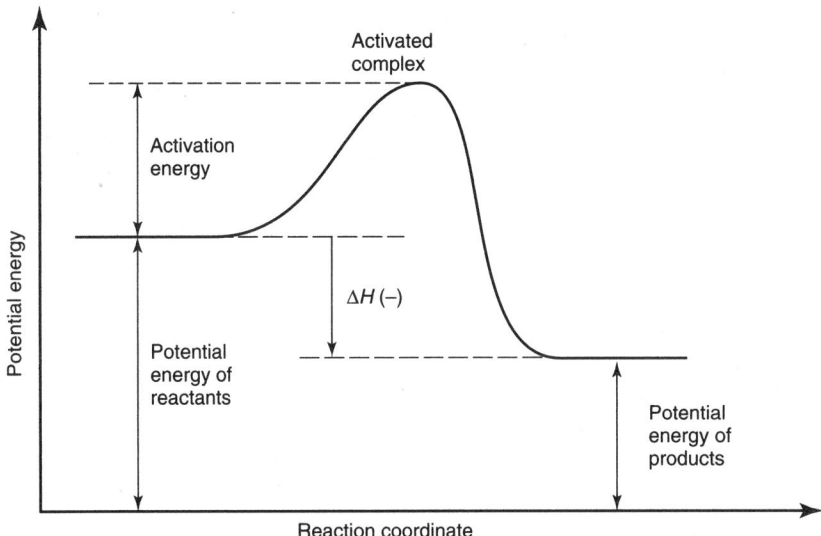

Let us trace along the curve in the diagram and describe the events that occur during the course of this reaction:

- At the beginning, the reactants contain a specific amount of energy (potential energy of reactants).
- Additional energy is *absorbed* by the reactants and serves to break certain bonds and initiate the reaction (activation energy).
- As the reactants absorb energy, they are transformed into an intermediate known as the *activated complex.*
- As the activated complex is converted into products, new bonds are formed and energy is released.
- The products also contain a specific amount of energy (potential energy of products).
- Since *more* energy is released than absorbed, the products are at a *lower* energy state than the reactants and the overall reaction is exothermic.
- The *difference* between the energy of the products and the energy of the reactants is the heat of reaction, ΔH.

Many exothermic reactions are *self-sustaining*; that is, they continue to occur because the heat they liberate provides the activation energy for the reactants. A burning match is an example of a self-sustaining reaction.

Now let us consider the *endothermic* reaction

$$\text{Heat} + \text{Reactants} \rightarrow \text{Products}$$

Reaction Profile of an Endothermic Reaction

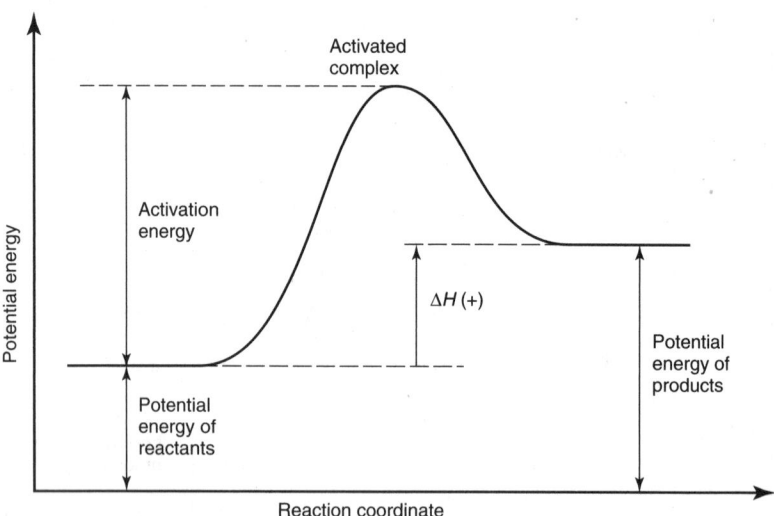

and the accompanying potential energy diagram for this reaction. The sequence of events in an endothermic reaction is *exactly the same* as the sequence of events in an exothermic reaction. At the end of the reaction, however, *less* energy is released than absorbed and the products are at a *higher* energy state than the reactants.

5.4 SPONTANEOUS REACTIONS

When we think of the word *spontaneous* in everyday language, we tend to think of an event happening by itself, without outside interference. In chemistry, nothing happens by itself!

The term **spontaneous reaction** refers to a reaction that can occur under a given set of conditions without the application of external work. During the course of a spontaneous reaction, heat will be absorbed from, or released to, the surroundings without any outside assistance. For example, at +25°C and 1 atmosphere, ice is able to melt but water is *not* able to freeze.

$H_2O(s) \rightarrow H_2O(\ell)$ is a spontaneous reaction at +25°C and 1 atmosphere.
$H_2O(\ell) \rightarrow H_2O(s)$ is *not* a spontaneous reaction at +25°C and 1 atmosphere.

At –25°C and 1 atmosphere, the opposite is true:

$H_2O(s) \rightarrow H_2O(\ell)$ is *not* a spontaneous reaction at –25°C and 1 atmosphere.
$H_2O(\ell) \rightarrow H_2O(s)$ is a spontaneous reaction at –25°C and 1 atmosphere.

We note that, when a reaction is spontaneous in one direction, it is not spontaneous in the reverse direction. The fact that a reaction is not spontaneous does *not* mean that it can never occur. If external work is applied to a system, the reaction can be *forced* to take place. For example, water will not decompose into hydrogen and oxygen spontaneously. The application of a direct electric current, however, will force this decomposition.

All spontaneous reactions are able to contribute a portion of their energy to perform *useful work*. For example, certain reactions can be used to produce electric energy that will drive a mechanical device, such as a CD player. The portion of energy that is available for useful work is known as the *Gibbs free energy* and is symbolized by the letter *G*. As a spontaneous reaction proceeds, its capacity to do further work *decreases* (i.e., its Gibbs free energy is lowered). After a time, the system is no longer able to convert any energy to work and the reaction ceases to be spontaneous. At this point, we say that the system is in a state of *equilibrium*. A practical example is the operation of a battery: as it functions, its capacity decreases until it "dies" and can function no more.

What factors enable a reaction to occur spontaneously under a given set of conditions? In nature, two fundamental tendencies—the *energy factor* and the *disorder factor*—govern every system.

The Energy Factor

We know from experience that systems tend to change from higher to lower energy states. Water drops from famous Niagara Falls from a higher to a lower potential energy state under the influence of gravity. Many chemical reactions release energy, with the result that the products have less energy than the reactants.

From a chemical point of view, there is a strong tendency for a reaction to occur spontaneously when it is *exothermic*: that is, when ΔH is *negative*.

The Disorder Factor: Entropy

We also know from experience that systems tend to reach a state of higher disorder: smoke spontaneously diffuses through the air, sugar dissolves uniformly in water, and your bedroom becomes chaotic after only a few days of neglect!

Disorder is measured by a quantity called **entropy**, and it is represented by the symbol *S*. When a system achieves greater disorder, its entropy increases.

Events Leading to Higher Disorder

Chemically, a system can reach a state of higher disorder by means of a number of events or circumstances:

117

- The temperature of the system increases, leading to an increase in the random motion of the particles present.
- There is a phase change: gases have the most disorder; solids, the least.
- The products of a chemical reaction are simpler in structure than the reactants.
- There are more products than reactants in a chemical reaction.
- A substance is placed in solution.

As an example, let us consider the reaction

$$2KClO_3(s) \rightarrow 2KCl(s) + 3O_2(g)$$

This reaction leads to an increase in disorder because the products are simpler in structure, there are more of them, and one product is a gas.

From a chemical point of view, there is a strong tendency for a reaction to occur spontaneously when its entropy *increases*: that is, when ΔS is *positive*.

PROBLEM

Predict whether each of the following changes is accompanied by an increase or a decrease in disorder:
(a) $NH_4HS(s) \rightarrow NH_3(g) + H_2S(g)$
(b) $2H_2(g) + O_2(g) \rightarrow 2H_2O(g)$
(c) $C_6H_{12}O_6(s) \rightarrow C_6H_{12}O_6(aq)$
(d) $He(g) [300\,K] \rightarrow He(g) [200\,K]$

SOLUTIONS

(a) Increase: The solid is converted into two gases.
(b) Decrease: The number of particles is decreased from 3 mol to 2 mol.
(c) Increase: The solid is dissolved in water.
(d) Decrease: The temperature is lowered from 300 K to 200 K.

SECTION II—ADDITIONAL MATERIAL

5.1A ADDITIONAL CALORIMETRY PROBLEMS

PROBLEM

How much heat is released by 200.0 grams of solid aluminum as it cools from 200.0°C to 150.0°C?

SOLUTION

Use Reference Table W-1 in Appendix 2, which contains a more complete list of specific heats.

$$q = c_p \cdot m \cdot \Delta T_C$$

$q = $??? (This is the quantity we are asked to determine.)

$c_p = 0.897$ J/g $\cdot$ C° (aluminum)

$m = 200.0$ g

$\Delta T_C = T_{final} - T_{initial} = 150.0°C - 200.0°C = -50.0 C°$

$$q = c_p \cdot m \cdot \Delta T_C$$

$$= \left(0.897 \frac{J}{\cancel{g} \cdot \cancel{C°}} \right) \cdot (200.0 \ \cancel{g}) \cdot (-50.0 \cancel{C°}) = -8970 \text{ J}$$

That is, 8970 J of heat is released.

TRY IT YOURSELF

How much heat is absorbed by 5.00 grams of copper when it is heated from 0.0°C to 200.0°C?

ANSWER

385 J

5.2A TRANSFER OF ENERGY AND EQUILIBRIUM TEMPERATURE

What happens if two objects with different temperatures are brought into contact? Experience tells us that heat is always transferred from the hotter object to the colder one until they both reach the same final temperature. When this occurs, we say that the objects have reached *thermal equilibrium*, and the final temperature is known as the *equilibrium temperature*. This will always occur; it does not matter what the objects are composed of or how massive they are. (The compositions of the objects and their masses will determine the value of the final temperature, however.)

Problems Involving Transfer of Energy

Problems involving heat transfer also utilize the relationship $q = c_p \cdot m \cdot \Delta T_C$. However, this relationship must be used twice—once for the object that releases ("loses") heat, and once for the object that absorbs ("gains") it. Because heat is assumed to be conserved in these problems, the heat "lost" by the hotter object is set equal to the heat energy "gained" by the colder object. The equation looks like this:

$$|q|_{lost} = |q|_{gained}$$
$$q = c_p \cdot m \cdot |\Delta T_C|_{lost} = c_p \cdot m \cdot |\Delta T_C|_{gained}$$

The vertical bars (| |) mean that we are interested only in the (absolute) value of the quantity, not in whether it is positive or negative. The quantity $|\Delta T_C|$ is evaluated by subtracting the smaller temperature from the larger one.

PROBLEM
A 10.0-gram block of copper at 60.0°C is placed in contact with an identical 10.0-gram block of copper at 20.0°C. What is the equilibrium temperature of the two blocks?

SOLUTION
We could guess at the answer to this (40.0°C—the midpoint temperature between 20.0°C and 60.0°C), and we would be correct! But let us see why this is so. The two blocks have equal masses (10.0 g) and are composed of the same substance (copper). The change in heat energy should therefore affect the temperature of each block equally (but in opposite directions). If both blocks must reach the same final temperature, the midpoint temperature of 40.0°C is the only temperature that satisfies these requirements. Now let us *prove* that this is the case by solving the pair of equations given above:

$$|q|_{lost} = |q|_{gained}$$
$$q = c_p \cdot m \cdot |\Delta T_C|_{lost} = c_p \cdot m \cdot |\Delta T_C|_{gained}$$

Hotter Object (loses heat)   Colder Object (gains heat)

$c_p = 0.385$ J/g · C° $c_p = 0.385$ J/g · C°

$m = 10.0$ g $m = 10.0$ g

$|\Delta T_C|_{lost} = 60.0°C - T_{final}$ $|\Delta T_C|_{gained} = T_{final} - 20.0°C$

Substitution into the equations yields:

$$(0.385 \text{ J/g} \cdot C°) \cdot (10.0 \text{ g}) \cdot (60.0°C - T_{final}) = (0.385 \text{ J/g} \cdot C°) \cdot (10.0 \text{ g}) \cdot (T_{final} - 20.0°C)$$

Simplifying yields: $2T_{final}$ $= 80.0°C$

And our answer is T_{final} $= 40.0°C$

TRY IT YOURSELF
Calculate the equilibrium temperature when a 5.00-gram sample of lead at 20.0°C is placed in contact with a 10.0-gram block of silver at 40.0°C. (Use Reference Table W-1.)

ANSWER

35.7°C

Note that the equilibrium temperature is closer to the initial temperature of the silver (40.0°C) than to that of the lead (20.0°C). This is due to the larger mass of the silver block and to the higher specific heat of silver.

5.3A THE ROLE OF ENERGY IN CHEMICAL REACTIONS

If we are to understand anything about how chemical reactions occur, we must first recognize the fundamental role played by energy.

The First Law of Thermodynamics

The role of energy in a chemical reaction is embodied in a fundamental rule that relates *internal energy*, *heat*, and *work*. This rule is known as the **first law of thermodynamics** (or, more briefly, as the *first law*). The first law is represented by this equation:

$$\Delta E = q + w$$

This equation tells us that the *change* in the *internal energy* of a system is equal to the sum total of the *heat absorbed* by the system and the *work done on* the system. In this situation, the heat and work are expressed as *positive* numbers; if heat is *released*, or work is *done by* the system, these two quantities will be *negative* numbers.

The first law is another way of expressing the *law of conservation of energy*: If a *system* gains (loses) a given amount of energy, then the energy lost (gained) by the *surroundings* must be *exactly* the same amount; that is, *the energy of the universe is always constant*. We can express this idea by the equation

$$\Delta E_{system} + \Delta E_{surroundings} = \Delta E_{universe} = 0$$

Throughout history, many attempts have be made to circumvent the first law. Numerous patent applications have been made for machines that promise to deliver more energy than the amount with which they were supplied. As of this date, *not one patent has been granted* for a machine of this type. We also hear of "foolproof" reducing diets that promise weight loss without calorie-counting or exercise. According to the first law, this is utter nonsense! If the quantity of food energy ingested is greater than the quantity of energy expended through exercise, the excess energy is stored as—you guessed it—extra pounds.

How does a chemist *measure* ΔE? When a reaction occurs in a *sealed* container so that the *volume is constant*, the heat transferred under these conditions (q_v) is equal to the internal energy change:

$$\Delta E = q_v$$

If heat is absorbed by the system at constant volume, the internal energy will *increase* ($\Delta E > 0$); if heat is released by the system at constant volume, the internal energy will *decrease* ($\Delta E < 0$).

Under laboratory conditions, however, most reactions take place in *open* containers at nearly *constant pressure*. In this case, the heat transferred (q_p) is equal to a change in a quantity that is closely related to the internal energy and is known as **enthalpy** (H). (Sometimes, the term *heat content* is used as a synonym for enthalpy, but this is not really accurate, since it implies that heat is a *substance*, rather than a *transfer of energy*.) The equation relating an enthalpy change to the heat transferred is

$$\Delta H = q_p$$

If heat is absorbed by the system at constant pressure, the enthalpy will *increase* ($\Delta H > 0$); if heat is released by the system at constant pressure, the enthalpy will *decrease* ($\Delta H < 0$). In this book, we will assume that all chemical reactions occur under conditions of *constant pressure*.

5.4A   ADDITIONAL ASPECTS OF HEATS OF REACTION

When a heat of reaction is measured under *standard thermodynamic conditions* (pressure = 100.00 kPa; all reactants and products are present in their pure states), we use the special symbol $\Delta H°$. Generally a reference temperature of 298.15 K (25°C) is also used. Standard heats of reaction are usually measured in kilojoules.

PROBLEM
How much heat is released when 0.5000 mole of $CH_4(g)$ is reacted with O_2 if $\Delta H° = -890.4$ kilojoule per mole of CH_4 reacted?

SOLUTION
Since 1.000 mol of $CH_4(g)$ *releases* 890.4 kJ of heat, 0.5000 mol of $CH_4(g)$ will release *half* this amount, or 445.2 kJ.

TRY IT YOURSELF

How much heat is absorbed by 0.300 mole of $KNO_3(s)$ when it dissolves in H_2O if $\Delta H° = +34.89$ kilojoules per mole of KNO_3 dissolved?

ANSWER

10.5 kJ

Hess's Law

In the nineteenth century, the Swiss chemist G.H. Hess found that, when two or more chemical reactions are combined ("added"), the heat of reaction of the *combined* reaction is simply the *sum* of the *individual* heats of reaction. The example below illustrates this technique:

$$N_2(g) + O_2(g) \rightarrow 2NO(g) \qquad \Delta H_1 = +180 \text{ kJ}$$

$$2NO(g) + O_2(g) \rightarrow 2NO_2(g) \qquad \Delta H_2 = -112 \text{ kJ}$$

$$N_2(g) + 2O_2(g) \rightarrow 2NO_2(g) \qquad \Delta H_{rxn} = \Delta H_1 + \Delta H_2 = +180 \text{ kJ} - 112 \text{ kJ}$$
$$= +68 \text{ kJ}$$

The letters "rxn" form a widely used abbreviation for the word *reaction*. Notice that $NO(g)$ does not appear in the combined reaction because this compound was present *in equal amounts on opposite sides of the individual* reactions. Hess's law is really an application of the first law of thermodynamics, which in itself is a statement of the law of conservation of energy applied to chemical reactions.

PROBLEM

Given the following reactions:

$$2SO_2(g) + O_2(g) \rightarrow 2SO_3(g) \qquad \Delta H°_1 = -196 \text{ kJ}$$
$$2S(s) + 3O_2(g) \rightarrow 2SO_3(g) \qquad \Delta H°_2 = -790 \text{ kJ}$$

Calculate $\Delta H°_{rxn}$ for this reaction: $S(s) + O_2(g) \rightarrow SO_2(g)$.

SOLUTION

The compound $SO_3(g)$ does *not* appear in the combined reaction. Therefore, one of the individual reactions will have to be reversed, along with the *sign* of its $\Delta H°$. We will reverse the *first* reaction because $SO_2(g)$ needs to appear on the *right side*, and $S(s)$ on the *left side*, of the combined reaction. This step gives us:

$$2SO_3(g) \rightarrow 2SO_2(g) + O_2(g) \qquad -\Delta H°_1 = +196 \text{ kJ}$$
$$2S(s) + 3O_2(g) \rightarrow 2SO_3(g) \qquad \Delta H°_2 = -790 \text{ kJ}$$

The combined reaction is:

$$2S(s) + 2O_2(g) \rightarrow 2SO_2(g) \qquad \Delta H°_{rxn} = (+196 \text{ kJ} -790 \text{ kJ}) = -594 \text{ kJ}$$

If we compare the combined reaction with the reaction asked for in the problem statement, we find that all of the coefficients are *doubled*. Therefore, all we need do is multiply the entire equation — and its $\Delta H°$ — by 1/2 and our answer is:

$$S(s) + O_2(g) \rightarrow SO_2(g) \qquad \Delta H°_{rxn} = -297 \text{ kJ}$$

TRY IT YOURSELF

Given the following reactions:

$$H_2S(g) + \frac{3}{2}O_2(g) \rightarrow H_2O(\ell) + SO_2(g) \qquad \Delta H°_1 = -563 \text{ kJ}$$

$$CS_2(\ell) + 3O_2(g) \rightarrow CO_2(g) + 2SO_2(g) \qquad \Delta H°_2 = -1075 \text{ kJ}$$

Calculate $\Delta H°_{rxn}$ for this reaction: $CS_2(\ell) + 2H_2O(\ell) \rightarrow 2H_2S(g) + CO_2(g)$

(*Hint:* Get rid of the fraction first!)

ANSWER
+51 kJ

Standard Heat of Formation of a Compound

The **standard heat of formation** of a compound is a special type of heat of reaction. It is defined as the heat released or absorbed when *1 mole* of a compound is formed from its *elements* under standard conditions.

The symbol for standard heat of formation is $\Delta H°_f$, where the subscript f stands for the word *formation*. The elements that form the compound are in their *stable states* (i.e., oxygen would be a gas; iron, a solid; etc.) at 298.15 K. The standard heat of formation of an *element* is *assigned* a value of 0 kilojoule per mole.

Reference Table W-4 in Appendix 2 lists the standard heats of formation of various compounds. In the table, standard heat of formation is measured in *kilojoules per mole of product formed*. Let us examine what these values mean. According to the table, $\Delta H°_f$ for ammonia [$NH_3(g)$] is –45.9 kilojoule per mole. This means that, when *1 mole* of $NH_3(g)$ is formed from its elements (nitrogen and hydrogen) under standard conditions, 45.9 kilojoules of heat is released.

We can write a **formation reaction** to illustrate this event:

$$\frac{1}{2}N_2(g) + \frac{3}{2}H_2(g) \rightarrow NH_3(g) + 45.9 \text{ kJ}$$

We note that the reaction is written so that *1 mole of product* is formed. Therefore, we require that 0.5 mole of $N_2(g)$ react with 1.5 moles of $H_2(g)$.

PROBLEM
(a) Write the formation reaction for KCl(s).
(b) How much heat is released when 2.00 moles of KCl(s) is formed from its elements under standard conditions?
(c) Suppose we could *decompose* 1.00 mole of KCl(s) into its elements under standard conditions. What would $\Delta H°$ for this reaction be?

SOLUTIONS
(a) Under standard conditions, potassium exists as a solid, and chlorine as diatomic gas molecules. Therefore, the formation reaction is

$$K(s) + \frac{1}{2} Cl_2(g) \rightarrow KCl(s)$$

(b) From Reference Table W-4 we see that $\Delta H°_f$ for KCl(s) is –436.5 kJ/mol, as the formation of 2.00 mol of KCl will release 873 kJ.
(c) Decomposition is the *opposite* of formation. Therefore, the decomposition of 1.00 mol of KCl(s) would require the *absorption* of 436.5 kJ. (In other words, $\Delta H° = +436.5$ kJ/mol KCl(s) decomposed.)

PROBLEM
Can an element have a $\Delta H°_f$ value different from 0?

SOLUTION
Yes, if the element is not in its stable state. For example, note $\Delta H°_f$ for white and gray tin in Reference Table W-4.

Heats of Reaction from Standard Heats of Formation

It is possible to use the standard heats of formation of individual substances to calculate the standard enthalpy change ($\Delta H°$) of a reaction involving these substances. For example, we will consider this reaction:

$$4NH_3(g) + 5O_2(g) \rightarrow 4NO(g) + 6H_2O(g)$$

Reference Table W-4 gives us the standard heats of formation for the three compounds in the equation; oxygen is an element and so has a standard heat of formation of 0. These ΔH_f values are given in the second column of the accompanying table.

Substance	ΔH°_{f} / (kJ/mol)	× Equation Coefficient / (mol)	= Product of 2nd and 3rd Columns / (kJ)
$NH_3(g)$	−45.9	4	−183.6
$O_2(g)$	0.0	5	0.0
$NO(g)$	+91.3	4	+365.2
$H_2O(g)$	−241.8	6	−1450.8

In the third column we have listed the coefficients of the four substances as they appear in the equation. The last column shows the *products* of the second and third columns.

We calculate the heat of reaction by *subtracting* the sum of the values for the *reactants* (NH_3 and O_2) of the reaction from the sum of the values for the *products* (NO and H_2O), as given in the last column of the table:

$$\Delta H^{\circ}_{rxn} = (-1450.8 \text{ kJ} + 365.2 \text{ kJ}) - (-183.6 \text{ kJ} + 0.0 \text{ kJ}) = -902.0 \text{ kJ}$$

Why does this method work? First, we assume that a reaction takes place by breaking down the reactants into their elements and then reassembling those elements as products:

$$\text{Reactants} \rightarrow \text{Elements} \rightarrow \text{Products}$$

The reaction does not *really* take place in this way, but *from an energy point of view* this assumption, which follows from the first law of thermodynamics, is perfectly logical. We make this assumption when we *subtract* the reactant values from the product values. Second, the coefficients of the equation tell us how many moles of each substance are involved in the reaction. Since a standard heat of formation is based on *1 mole* of substance, we need to *multiply* by the coefficient for each substance in the equation.

If you MUST have a mathematical relationship for this technique, here it is:

$$\Delta H^{\circ}_{rxn} = \left(\Sigma n \cdot \Delta H^{\circ}_{f}\right)_{\text{products}} - \left(\Sigma m \cdot \Delta H^{\circ}_{f}\right)_{\text{reactants}}$$

Don't let the fancy symbols throw you! The "Σ" symbol means that you add the individual values of the products and reactants separately *before* you subtract. The symbols m and n represent the coefficients associated with all of the substances in the reaction.

TRY IT YOURSELF

Calculate ΔH° for this reaction:

$$C_2H_4(g) + H_2(g) \rightarrow C_2H_6(g)$$

using the values given in Reference Table W-4.

ANSWER
−136.4 kJ (Did you forget that hydrogen is an *element* in its stable state?)

5.5A THE SECOND LAW OF THERMODYNAMICS

The role of entropy in spontaneous processes is governed by a rule known as the **second law of thermodynamics** (or, more briefly, as the *second law*). This rule states that in any spontaneous process the total entropy of the system and its surroundings, that is, the entropy of the *universe*, must always *increase*. We can state the second law in the form of an equation:

$$\Delta S_{system} + \Delta S_{surroundings} = \Delta S_{universe} > 0$$

Unlike energy, entropy is *not conserved* in a spontaneous process. One consequence of the second law is the assurance that heat will always flow spontaneously from a hotter object to a colder one, an assumption we made in solving calorimetry problems in Section 5.1.

Another consequence of the second law is that it allows some spontaneous processes to occur, even when a system experiences a *decrease* in entropy. For example, when water freezes to ice at −10°C, the entropy of the *system* decreases as the solid is formed. However, the entropy of the surroundings *increases to an even greater extent*, ensuring that there will be an increase in the entropy of the universe.

During the course of a chemical reaction, (1) energy may be absorbed or released and (2) the entropy of the system may increase or decrease. If we focus on the reaction itself and *ignore* the surroundings, we know that a release of energy and an increase in entropy favor a spontaneous reaction, while an absorption of energy and a decrease in entropy do not. Given these possible changes in energy and entropy, we need to raise the following question: How can we *predict* when a reaction will be spontaneous? During the nineteenth century, J. Willard Gibbs, a professor of physics at Yale University, developed a relationship that allows us to make this prediction.

The Gibbs Free-Energy Change

The *Gibbs free-energy change*, denoted by the symbol ΔG, enables us to predict whether a reaction will be spontaneous under a given set of conditions. The free-energy change is defined as follows:

$$\Delta G = \Delta H - T\Delta S$$

Note that ΔG takes both ΔH and ΔS into account. Moreover, the absolute temperature plays an important part and is associated with the disorder factor. As the temperature rises, the disorder factor becomes more important.

When the equation is applied, a *negative* ΔG indicates that the reaction is spontaneous; if ΔG is *positive*, the reaction is *not* spontaneous. Let us see how the equation can be used in practice.

PROBLEM

The reaction $N_2O_4(g) \rightarrow 2NO_2(g)$ is *endothermic*. At low temperatures it is not spontaneous, yet it is spontaneous at higher temperatures. How does the free-energy equation explain this fact?

SOLUTION

This reaction leads to an *increase* in disorder. (Why?) Therefore, the disorder factor favors a spontaneous reaction, but the energy factor does not.

At low temperatures, the disorder factor $(T\Delta S)$ is small and the unfavorable energy factor is more important. At higher temperatures, however, $T\Delta S$ becomes the more important factor and the reaction occurs spontaneously.

We can summarize these results in the accompanying table, which relates the *algebraic signs* of the variables to the possibility of a spontaneous reaction. Note that absolute temperature is *not* included in this table because its sign is always *positive*. As noted on page 123, "rxn" is an abbreviation for the word *reaction*.

Rxn	ΔH	ΔS	ΔG	Result	Example
A	$(-)$	$(+)$	$(-)$	Rxn is spontaneous at *all* temperatures.	$2NO_2(g) \rightarrow 2N_2(g) + O_2(g)$
B	$(-)$	$(-)$	???	Rxn is spontaneous at *low* temperatures.	$H_2O(\ell) \rightarrow H_2O(s)$
C	$(+)$	$(+)$	???	Rxn is spontaneous at *high* temperatures.	$2NH_3(g) \rightarrow N_2(g) + 3H_2(g)$
D	$(+)$	$(-)$	$(+)$	Rxn is *not* spontaneous at *any* temperature.	$3O_2(g) \rightarrow 2O_3(g)$

Remember: A reaction that is not spontaneous can still be made to occur. In reaction D, for example, $O_3(g)$ (ozone gas) is produced by the action of an electrical discharge, such as lightning, on the $O_2(g)$ present in air.

END-OF-CHAPTER QUESTIONS

Some questions have the symbol "§2" in front of the question number. This symbol means that the question is based on Section II material.

1. What is the maximum number of grams of liquid water at 10°C that can be heated to 30°C by the addition of 84 joules of heat?
    (1) 1.0   (2) 2.0   (3) 20.   (4) 30.

§2 2. How many joules of heat are released when 50 grams of solid magnesium is cooled from 70°C to 60°C?
    (1) 10   (2) 50   (3) 500   (4) 1000

§2 3. If 4 grams of solid aluminum at 1°C absorbs 8 joules of heat, the temperature of the aluminum will change by
    (1) 1 C°   (2) 2 C°   (3) 3 C°   (4) 4 C°

§2 4. The temperature of 50 grams of liquid toluene ($C_6H_5CH_3$) was raised to 50°C by the addition of 900 joules of heat. What was the initial temperature of the toluene?
    (1) 0°C   (2) 20°C   (3) 30°C   (4) 40°C

§2 5. The temperature of solid silver increases from 30.00°C to 40.00°C by the addition of 117.5 joules of heat. What is the mass of the silver?
    (1) 1.00 g   (2) 5.00 g   (3) 10.0 g   (4) 50.0 g

6. The temperature of 100.00 grams of water changes from 16.00°C to 20.00°C. What is the total number of joules of heat energy absorbed by the water?
    (1) 250.0   (2) 400.0   (3) 1000.   (4) 1680.

7. In a reversible reaction, the difference between the activation energy of the forward reaction and the activation energy of the reverse reaction is equal to the
    (1) activated complex   (2) heat of reaction
    (3) potential energy of reactants   (4) potential energy of products

8. A piece of Mg(s) ribbon held in a Bunsen burner flame begins to burn according to the equation

$$2Mg(s) + O_2(g) \rightarrow 2MgO(s)$$

The reaction begins because the reactants
(1) are activated by heat from the Bunsen burner flame
(2) are activated by heat from the burning magnesium
(3) underwent an increase in entropy
(4) underwent a decrease in entropy

9.

In the diagram above, the reaction

$$A(g) + B(g) \rightarrow C(g) + D(g)$$

has a forward activation energy of 20 kilojoules. What is the activation energy for the reverse reaction?
(1) 10 kJ   (2) 20 kJ   (3) 30 kJ   (4) 50 kJ

§2 10. What is the standard heat of formation, in kilojoules per mole of liquid water at 298.15 K?
(1) −285.8   (2) −241.8   (3) +241.8   (4) +285.8

11. Given the reaction

$$H_2(g) + \frac{1}{2}O_2(g) \rightarrow H_2O(g) + 243 \text{ kJ}$$

If the activation energy for the forward reaction is 168 kilojoules, the activation energy for the reverse reaction, in kilojoules, will be
(1) 163   (2) 168   (3) 243   (4) 411

12. Assume that the potential energy of the products in a chemical reaction is 60 kilojoules. This reaction would be exothermic if the potential energy of the reactants was
(1) 50 kJ   (2) 20 kJ   (3) 30 kJ   (4) 80 kJ

Base your answers to questions 13–15 on the diagram below.

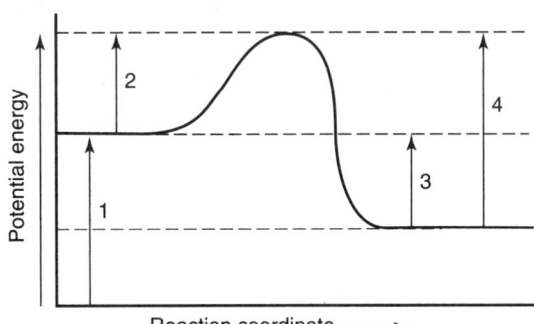

Reaction coordinate ⟶

13. The activation energy for the reverse reaction is represented by
(1) 1   (2) 2   (3) 3   (4) 4

14. The heat of reaction (ΔH) is represented by
(1) 1   (2) 2   (3) 3   (4) 4

15. The potential energy of the activated complex is represented by
(1) 1 + 2   (2) 2 + 3   (3) 3 + 4   (4) 4 + 1

§2 **16.** According to Reference Table W-4, how many kilojoules of heat are released when 0.5000 mole of MgO(s) is formed from its elements?
(1) 150.5   (2) 300.8   (3) 601.6   (4) 1204

§2 **17.** According to Reference Table W-4, the formation of 1 mole of which of the following substances *absorbs* the greatest amount of energy?
(1) C_2H_2 (2) C_2H_4 (3) $CuSO_4$ (4) $BaSO_4$

18. Which potential energy diagram represents an exothermic reaction?

(1)

(2)

(3)

(4)

19. When a substance was dissolved in water, the temperature of the water decreased. This process is described as
(1) endothermic, with the release of energy
(2) endothermic, with the absorption of energy
(3) exothermic, with the release of energy
(4) exothermic, with the absorption of energy

20. In a chemical reaction, the difference between the potential energy of the products and the potential energy of the reactants is called the
(1) activation energy   (2) activated complex
(3) kinetic energy   (4) heat of reaction

21. For this reaction:

$$A + B \rightarrow C + \text{heat,}$$

the potential energy of the products, as compared with the potential energy of the reactants, is
(1) less and the reaction is exothermic
(2) less and the reaction is endothermic
(3) greater and the reaction is exothermic
(4) greater and the reaction is endothermic

§2 **22.** According to Reference Table W-4, which compound has a higher potential energy than the elements from which it is formed?
(1) aluminum oxide(s)   (2) water(ℓ)
(3) nitrogen(II) oxide(g)   (4) carbon dioxide(g)

§2 **23.** The quantity of heat liberated when 2.00 moles of aluminum oxide is formed under standard conditions is closest to
(1) 419 kJ   (2) 838 kJ   (3) 1680 kJ   (4) 3350 kJ

24. Given the reaction

$$A + B \rightarrow AB + 50 \text{ kJ}$$

If an activation energy of 5 kilojoules is required for the forward reaction, the activation energy of the reverse reaction is
(1) 5 kJ   (2) 45 kJ
(3) 50 kJ   (4) 55 kJ

25. Which of the following best describes exothermic chemical reactions?
(1) They never release heat.
(2) They always release heat.
(3) They are always decomposition reactions.
(4) They are always synthesis reactions.

26. Which type of reaction does this equation represent?

$$N_2(g) + 3H_2(g) \rightarrow 2NH_3(g) + \text{heat}$$

(1) exothermic, with an increase in entropy
(2) exothermic, with a decrease in entropy
(3) endothermic, with an increase in entropy
(4) endothermic, with a decrease in entropy

§2 **27.** A chemical reaction will always occur spontaneously if the reaction has a
(1) negative ΔG (2) positive ΔG
(3) negative ΔH (4) positive ΔH

28. Systems in nature tend to undergo changes toward
(1) lower energy and lower entropy
(2) lower energy and higher entropy
(3) higher energy and lower entropy
(4) higher energy and higher entropy

29. A 1-gram sample of a substance has the greatest entropy when it is in the
(1) solid state   (2) liquid state
(3) crystalline state   (4) gaseous state

§2 **30.** Which change represents an increase in the entropy of a system?
(1) $C_6H_{12}O_6(s) \rightarrow C_6H_{12}O_6(aq)$
(2) $H_2O(\ell) \rightarrow H_2O(s)$
(3) $CO_2(g) \rightarrow CO_2(s)$
(4) $C_2H_5OH(g) \rightarrow C_2H_5OH(\ell)$

31. Which change represents an increase in entropy?
(1) $I_2(s) \rightarrow I_2(g)$ (2) $I_2(g) \rightarrow I_2(\ell)$
(3) $H_2O(g) \rightarrow H_2O(\ell)$ (4) $H_2O(\ell) \rightarrow H_2O(s)$

§2 **32.** A reaction that has a positive ΔG must be
(1) exothermic   (2) endothermic
(3) spontaneous   (4) nonspontaneous

§2 **33.** A chemical reaction releases heat, and the products are in a more disordered state than the reactants. The ΔG for this reaction
(1) must be negative
(2) must be positive
(3) could be negative or positive
(4) could be zero

34. A sample of $H_2O(\ell)$ at 20°C is in equilibrium with its vapor in a sealed container. When the temperature increases to 25°C, the entropy of the system will
(1) decrease   (2) increase   (3) remain the same

35. As 1 gram of $H_2O(\ell)$ changes to 1 gram of $H_2O(s)$, the entropy of the system
(1) decreases   (2) increases   (3) remains the same

Base your answers to questions 36 and 37 on the table below, which represents the energy and entropy changes in four chemical reactions, *A–D*.

Reaction	ΔH	ΔS
A	+	+
B	+	−
C	−	+
D	−	−

§2 **36.** Which reaction *must* be spontaneous?
(1) *A*   (2) *B*   (3) *C*   (4) *D*

§2 **37.** Which reaction *must* have a free-energy change (ΔG) that is positive?
(1) *A*   (2) *B*   (3) *C*   (4) *D*

38. Given the reaction:

$$S(s) + O_2(g) \rightarrow SO_2(g) + energy$$

Which diagram best represents the potential energy changes for this reaction?

Reaction coordinate
(1)

Reaction coordinate
(2)

Reaction coordinate
(3)

Reaction coordinate
(4)

39. Which phase change represents a *decrease* in entropy?
(1) solid to liquid   (2) gas to liquid
(3) liquid to gas   (4) solid to gas

40. Which change is exothermic?
(1) freezing of water   (2) melting of iron
(3) vaporization of ethanol   (4) sublimation of iodine

41. Which statement correctly describes an endothermic reaction?
   (1) The products have a higher potential energy than the reactants, and the ΔH is negative.
   (2) The products have a higher potential energy than the reactants, and the ΔH is positive.
   (3) The products have a lower potential energy than the reactants, and the ΔH is negative.
   (4) The products have a lower potential energy than the reactants, and the ΔH is positive.

42. Systems in nature tend to undergo changes that result in
   (1) lower energy and lower entropy
   (2) lower energy and higher entropy
   (3) higher energy and lower entropy
   (4) higher energy and higher entropy

43. Given the balanced equation representing a reaction:

$$H_2 \rightarrow H + H$$

What occurs during this reaction?
   (1) Energy is absorbed as bonds are formed.
   (2) Energy is absorbed as bonds are broken.
   (3) Energy is released as bonds are formed.
   (4) Energy is released as bonds are broken.

44. Given the potential energy diagram for a reaction:

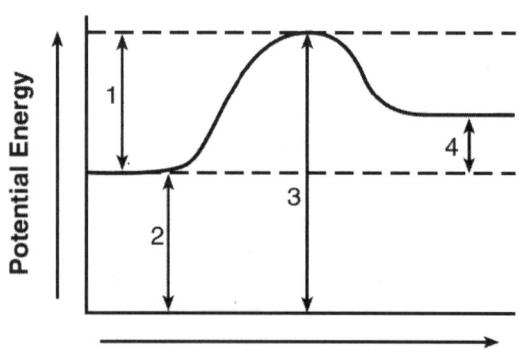

Reaction Coordinate

Which intervals are affected by the addition of a catalyst?
   (1) 1 and 2   (2) 1 and 3   (3) 2 and 4   (4) 3 and 4

Constructed-Response Questions

§2 **1.** The standard heat of formation of C_2H_5OH is -277.6 kilojoule per mole. How much heat will be released when 1.000 gram of this compound is formed from its elements under standard conditions?

§2 **2.** Use the data given below to calculate $\Delta H°$ for this reaction:

$$CaO(s) + CO_2(g) \rightarrow CaCO_3(s)$$

Compound	$\Delta H_f°$(kJ/mol)
CaO (s)	-635.1
CO_2 (g)	-393.5
$CaCO_3$ (s)	-1207.6

§2 **3.** Given the following information:

$2Cu_2O(s) + O_2(g) \rightarrow 4CuO(s)$ $\Delta H° = -288.0$ kJ
$CuO(s) + Cu(s) \rightarrow Cu_2O(s)$ $\Delta H° = -11.3$ kJ

Calculate the standard heat of formation of $Cu_2O(s)$.

Base your answers to questions 4 through 6 on the information and potential energy diagram below.

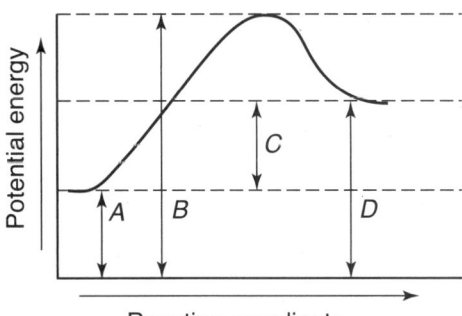

Chemical cold packs are often used to reduce swelling after an athletic injury. The diagram represents the potential energy changes when a cold pack is activated.

4. Which lettered interval on the diagram represents the potential energy of the products?

5. Which lettered interval on the diagram represents the heat of reaction?

6. Identify a reactant listed in Reference Table I that could be mixed with water for use in a chemical cold pack.

Base your answers to questions 7–9 on the information below and on your knowledge of chemistry.

A student made a copper bracelet by hammering a small copper bar into the desired shape. The bracelet had a mass of 30.1 grams and was at a temperature of 21°C in the classroom. After the student wore the bracelet, the bracelet reached a temperature of 33°C. Later, the student removed the bracelet and placed it onto a desk at home, where it cooled from 33°C to 19°C. The specific heat capacity of copper is 0.385 J/g · K.

7. Explain, in terms of heat flow, the change in temperature when the student wore the bracelet.

8. Determine the number of moles of copper in the bracelet.

9. Show a numerical setup for calculating the amount of heat released by the bracelet as it cooled on the desk.

The answers to these questions are found in Appendix 3.

Chapter Six

THE PHASES OF MATTER

KEY IDEAS

This chapter focuses on the gaseous, liquid, and solid phases of matter, as well as changes in phase. In particular, the ideal gas laws and the kinetic-molecular theory are introduced.

KEY OBJECTIVES

At the conclusion of this chapter you will be able to:

- Distinguish among the solid, liquid, and gaseous phases of matter.
- Define the terms *standard atmosphere* and *standard state pressure*.
- Define the term *standard temperature and pressure* (STP).
- Describe the measurement of gas pressure using closed-tube and open-tube manometers.
- Describe and apply the various gas laws to numerical problems.
- Convert between the Celsius and Kelvin temperature scales.
- Calculate the density of a gas at STP.
- Calculate the molar mass of a gas, given its density at STP.
- Solve stoichiometry problems involving gas volumes.
- State the hypotheses of the kinetic-molecular theory (KMT) of gas behavior.
- Define the term *ideal gas*, and the terms *volume*, *pressure*, and *temperature* in relation to the KMT.
- List the conditions under which real gases exhibit most nearly ideal and least ideal behavior.
- Relate vapor pressure to the boiling point of a liquid.
- Solve problems involving gases collected over water.
- Define the terms associated with phase changes.
- Interpret the parts of a heating curve and a cooling curve.
- Solve problems involving changes of phase.
- Interpret phase diagrams.
- Define the terms *triple point*, *critical point*, *critical temperature*, and *critical pressure*.

SECTION I—BASIC (REGENTS-LEVEL) MATERIAL

NYS REGENTS CONCEPTS AND SKILLS

Note: By the time you have finished Section I, you should have mastered the concepts and skills listed below. The Regents chemistry examination will test your knowledge of these items and your ability to apply them.

Concepts are the *basic ideas* that form the body of the Regents chemistry course (what you need to know!).

Skills are the *activities* that demonstrate your mastery of these concepts (how you show that you know them!).

Following each concept or skill is a page reference (given in parentheses) to this chapter.

6.1   Concept:
Temperature is a measurement of the average kinetic energy of the particles in a gas. Temperature is not a form of energy. (Page 154)

6.2   Concept:
The concept of an ideal gas is a model to explain the behavior of gases. A real gas is most like an ideal gas when the real gas is at low pressure and high temperature. (Pages 154–156)

6.3   Concepts:
The kinetic molecular theory (KMT) for an ideal gas states that all gas particles:
(a) are in random, constant, straightline motion;
(b) are separated by great distances relative to their size; the volume of the gas particles is considered negligible;
(c) have no attractive forces between them;
(d) have collisions that may result in the transfer of energy between gas particles, but the total energy of the system remains constant. (Page 154)

6.4   Concept:
The kinetic molecular theory describes the relationships of pressure, volume, temperature, velocity, and frequency and force of collisions among gas molecules. (Page 154)

Skills:
• Explain the gas laws in terms of KMT. (Pages 154–155)
• Solve problems using the combined gas law. (Pages 151–153)

6.5 Concept:
Equal volumes of different gases at the same temperature and pressure contain equal numbers of particles. (Pages 152–153)

Skill:
Convert temperatures between the Celsius and Kelvin scales. (Pages 147–148)

6.6 Concept:
The concepts of kinetic and potential energy can be used to explain physical processes that include fusion (melting), solidification (freezing), vaporization (boiling, evaporation), condensation, sublimation, and deposition. (Pages 158–161)

Skills:
• Interpret heating and cooling curves in terms of changes in kinetic and potential energy, heat of fusion, heat of vaporization, and phase changes. (Pages 159–161)
• Calculate the heat involved in a phase change. (Page 161)

6.1 INTRODUCTION

Matter commonly exists in the solid, liquid, or gaseous phase. The phase of a substance is usually recognized by the characteristics of the substance's shape and volume. Gases have neither definite shape nor definite volume, while solids have both definite shape and definite volume. Liquids do not have definite shape, but they do have definite volume. What phase a particular sample of matter is in depends on the nature of the sample, its temperature, and the pressure exerted on it.

6.2 GASES

As we begin our study of gases, we assume that the sample of gas under discussion behaves *ideally*. While no gas is truly ideal, many samples of gases do exhibit ideal behavior under appropriate conditions. (We will indicate exactly what we mean by the term *ideal behavior* in Section 6.4.) We make this assumption about ideal behavior because, when we do, the "laws" that govern gas behavior become very simple.

To describe the behavior of a sample of an ideal gas, we need to know only four characteristics of the sample: the pressure it exerts, its volume, its temperature, and the number of particles (or numbers of *moles* of particles) it contains. We have already learned how volume, temperature, and number of moles are measured.

Pressure is defined as the force per unit area of surface. Mathematically, this means that, when we apply a force on a surface, we calculate the pressure by *dividing* the force by the area of the surface (pressure = force/area). Pressure gives us a means of describing how a force is distributed over an entire surface. In the English system, the unit of pressure is the *pound per square inch* (lb/in.^{2}). Under normal atmospheric conditions, air exerts a pressure of approximately 14.7 pounds per square inch.

In the SI metric system, the units of pressure are the *pascal* (Pa) and its multiple, the *kilopascal* (kPa). The pascal is a very small unit, and under normal atmospheric conditions air exerts a pressure of approximately 1.01×10^5 pascals (101 kPa).

Currently, there are *two standards* of pressure in the scientific world. The **standard atmosphere** (or *atmosphere*, atm) is defined as *exactly* 101.325 kilopascals; **standard state pressure** is defined as *exactly* 100.000 . . . kilopascals. While the values of standard atmosphere and standard state pressure are not equal, they are very close. Generally, chemists use the *standard atmosphere* in applications involving gas behavior, and the *standard state pressure* in applications involving thermodynamics (Chapter 5).

An older metric unit of pressure, the *millimeter of mercury* (mmHg, or *torr*), is associated with the measurement of air pressure using an instrument known as a *Torricelli barometer*. This instrument consists of a long tube filled with mercury and inverted in a dish containing mercury. After inversion, the mercury in the tube falls to a specific height, and this height (measured in *millimeters*) is a measure of the pressure exerted by the air. One standard atmosphere of pressure supports a column of mercury very nearly equal to 760 millimeters (approximately 30 inches).

The accompanying chart shows the pressure relationships that we will use in this chapter.

1 standard atmosphere (atm) =	**1.01325 × 10^5 Pa [exact]** **101.325 kPa [exact]** **760 mmHg [approximate]** **760 torr [approximate]**

In the study of gas behavior, it is often necessary to refer *jointly* to a standard of pressure and temperature. The term **standard temperature and pressure (STP)** is defined as the ice point of water (273.15 K or 0°C) and 1 standard atmosphere (101.325 kPa or 760 mmHg).

6.3 THE GAS LAWS

Boyle's Law

To investigate the behavior of a gas, we need to change one of its char-acteristics (pressure, for example) and observe how another characteristic (volume, for example) will change in response. We must be certain, though, that the other two variables (temperature and number of particles, in this case) are not allowed to vary during the investigation. **Boyle's law** is the result of measuring how the volume of a gas varies with pressure (at constant temperature and number of particles). The accompanying table illustrates the results of one such experiment carried out on an ideal gas at 25°C.

Pressure (P) / Pa	Volume (V) / mL	$P \times V$ / (Pa·mL)
500	400	200,000
1000	200	200,000
2000	100	200,000
4000	50	200,000
5000	(Solved in the problem on page 145)	

According to the first four pairs of data given in the table, the volume varies *inversely* with the pressure. In other words, an *increase* in pressure is accom-panied by a *decrease* in volume (and vice versa) at constant temperature and number of particles. Moreover, the product of pressure and volume is constant under these conditions. This relationship is known as *Boyle's law*:

> **pressure · volume = constant**
>
> **(at constant temperature and number of particles)**

If we graph the data given in the table, we produce the curve shown on the next page, known as a *rectangular hyperbola*. This curve is characteristic of inverse relationships.

Volume versus Pressure

PROBLEM
According to the graph shown above, what will the volume of the gas be when the pressure is 3000 pascals?

SOLUTION
We locate the point on the curve that corresponds to a pressure of 3000 Pa, and then locate the corresponding value on the volume axis. The volume is approximately 70 mL.

TRY IT YOURSELF
According to the graph, what pressure corresponds to a volume of 150 milliliters?

ANSWER
Approximately 1300 Pa.

Each point on the graph obeys the Boyle's law equation:

$$P \cdot V = \text{constant}$$

If we choose *any* two points, whose respective values for pressure and volume are $P_1 \cdot V_1$ and $P_2 \cdot V_2$, it follows that

$$P_1 \cdot V_1 = P_2 \cdot V_2$$

This is another way of writing Boyle's law. This formula is very useful for solving mathematical problems involving pressure and volume. Note that we can use *any* units of pressure and volume in solving a Boyle's law problem as long as we remain consistent within the problem.

PROBLEM
According to the table on page 143, what will be the volume of the gas when the pressure is 5000 pascals?

SOLUTION
Boyle's law indicates that we can use any other set of values in the table for solving our problem. We will use the second set.

$$P_1 \cdot V_1 = P_2 \cdot V_2$$
$$P_1 = 1000 \text{ Pa}$$
$$V_1 = 200 \text{ mL}$$
$$P_2 = 5000 \text{ Pa}$$
$$V_2 = ???$$
$$1000 \text{ Pa} \cdot 200 \text{ mL} = 5000 \text{ Pa} \cdot V_2$$

$$V_2 = \frac{1000 \text{ P\!\!\!/a} \cdot 200 \text{ mL}}{5000 \text{ P\!\!\!/a}}$$

$$= 40 \text{ mL}$$

PROBLEM
The volume occupied by a gas at STP is 250 liters. At what pressure (in *atmospheres*) will the gas occupy 1500 liters, if the temperature and number of particles remain constant?

SOLUTION
We apply Boyle's law and remember that standard atmospheric pressure equals 1 atm.

$$P_1 \cdot V_1 = P_2 \cdot V_2$$
$$P_1 = 1 \text{ atm}$$
$$V_1 = 250 \text{ L}$$
$$P_2 = ???$$
$$V_2 = 1500 \text{ L}$$
$$1 \text{ atm} \cdot 250 \text{ L} = P_2 \cdot 1500 \text{ L}$$

$$P_2 = \frac{1 \text{ atm} \cdot 250 \text{ L\!\!\!/}}{1500 \text{ L\!\!\!/}}$$

$$= 0.17 \text{ atm}$$

TRY IT YOURSELF
The volume of an ideal gas is 32. liters at a pressure of 10. kilopascals. What will the volume be at a pressure of 2.0 kilopascals, temperature and number of particles remaining constant?

ANSWER
160 L

Charles's Law and the Kelvin Temperature Scale

To investigate the relationship between the temperature and the volume of an ideally behaving gas, we must hold the pressure and the number of particles constant. We begin at 0°C and measure the volumes of two samples of an ideally behaving gas. We then vary the temperature of each sample and measure the respective volumes of the samples. The accompanying table provides the data for these investigations.

	Experiment 1	Experiment 2
Temperature/°C	Volume/mL	Volume/mL
−100	173	346
−50	223	446
−20	253	506
−10	263	526
0	273	546
+10	283	566
+20	293	586
+50	323	646
+100	373	746

From the table, we can draw two conclusions:

1. The volume of the gas increases (decreases) with increasing (decreasing) temperature.
2. The change in volume with change in temperature is regular: In experiment 1 the volume changes by 1 milliliter for each 1° change (1 mL/C°); in experiment 2 the volume changes by 2 milliliters for each 1° change (2 mL/C°).

A graph of these results follows.

146

Volume versus Temperature

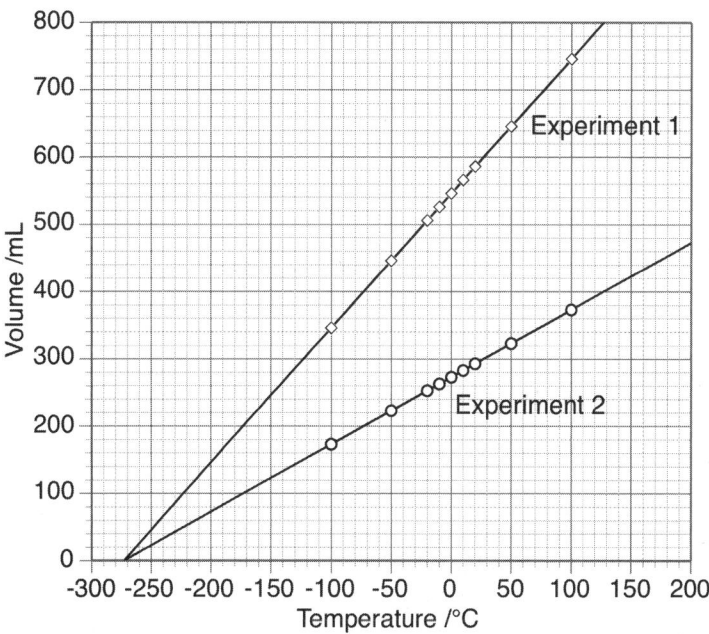

When we inspect the graph, we note that the two experiments produce two straight lines that converge to a point. This point represents a volume of 0 milliliter. By inspecting the table above, we can deduce that this value will be reached—in both experiments—at approximately −273°C. What is truly amazing is that *every* similar experiment will yield a straight-line graph that drops to zero volume at approximately −273°C! Obviously, there is something very special about this temperature. Moreover, it is the lowest temperature possible because no gas, ideally behaving or not, can have a negative volume. The precise value of this lowest temperature is –273.15°C.

Scientists found it very useful to define a new temperature scale in which this lowest possible temperature (−273.15°C) was given a value of 0 and was referred to as **absolute zero**. Today, this temperature scale is called the **Kelvin** or **absolute** scale, and it is related to the Celsius scale by this equation:

$$T_C = T_K - 273.15$$

A simplified version of this equation is given in Reference Table T in Appendix 1:

$$K = °C + 273$$

We can now assume that the value of absolute zero is −273°C. With this relationship, −273°C is assigned the value 0 K, while 0°C equals 273 K and 100°C equals 373 K. Note that there are *no* negative Kelvin temperatures.

PROBLEM
What Kelvin temperature corresponds to 35°C?

SOLUTION
$$K = °C + 273$$
$$= 35°C + 273$$
$$= 308 \text{ K}$$

TRY IT YOURSELF
What Celsius temperature corresponds to 260 K?

ANSWER
−13°C

If we adopt the Kelvin scale as our scale of choice, the relationship between the volume of a gas and its temperature is simplified, as shown in this modification of the preceding graph:

Volume versus Temperature

148

The simple relationship that can be drawn from this graph is that the volume of an ideally behaving gas is *directly proportional* to the Kelvin temperature:

$$\frac{\text{volume}}{\text{temperature}} = \text{constant}$$

(at constant pressure and number of particles)

We can write this relationship, which is known as **Charles's law**, in its equivalent form:

$$\frac{V_1}{T_1} = \frac{V_2}{T_2}$$

where the temperature is *always* specified in the Kelvin scale.

PROBLEM

The volume of an ideally behaving gas is 300 liters at 227°C. What volume will the gas occupy at 27°C, pressure and number of particles remaining constant?

SOLUTION

$$\frac{V_1}{T_1} = \frac{V_2}{T_2}$$

$$V_1 - 300 \text{ L}$$

$$T_1 = 227°C + 273 = 500 \text{ K}$$

$$V_2 = ???$$

$$T_2 = 27°C + 273 = 300 \text{K}$$

$$\frac{300 \text{ L}}{500 \text{ K}} = \frac{V_2}{300 \text{ K}}$$

$$V_2 = \frac{300 \text{ L} \cdot 300 \text{ K}}{500 \text{ K}}$$

$$= 180 \text{ L}$$

TRY IT YOURSELF

The volume of an ideally behaving gas is 50 cubic centimeters at 250 K. At what temperature will the volume of the gas be 200 cubic centimeters, pressure and number of particles remaining constant?

ANSWER

1000 K

Relating Pressure and Temperature (Gay-Lussac's Law)

At constant volume and number of particles, the relationship between the pressure and the Kelvin temperature of an ideally behaving gas is similar to that given by Charles's law:

$$\frac{\text{pressure}}{\text{temperature}} = \text{constant}$$

(at constant volume and number of particles)

Or, in equivalent form:

$$\frac{P_1}{T_1} = \frac{P_2}{T_2}$$

This relationship is known as **Gay-Lussac's law**.

PROBLEM

The pressure exerted by an ideally behaving gas is 7.00 kilopascals at 200. K. What pressure does the gas exert at 500. K, volume and number of particles remaining constant?

SOLUTION

$$\frac{P_1}{T_1} = \frac{P_2}{T_2}$$

$$P_1 = 7.00 \text{ kPa}$$

$$T_1 = 200. \text{ K}$$

$$P_2 = ???$$

$$T_2 = 500 \text{ K}$$

$$\frac{7.00 \text{ kPa}}{200. \text{ K}} = \frac{P_2}{500 \text{ K}}$$

$$P_2 = \frac{7.00 \text{ kPa} \cdot 500 \; \cancel{K}}{200. \; \cancel{K}}$$

$$= 17.5 \text{ kPa}$$

TRY IT YOURSELF
The pressure exerted by an ideally behaving gas is 50,000 pascals at a temperature of 327°C. At what temperature, in degrees Celsius, will the pressure exerted by the gas be 25,000 pascals, volume and number of particles remaining constant? (*Hint:* Remember that Gay-Lussac's law requires *Kelvin* temperature.)

ANSWER
27°C

The Combined Gas Law

It is possible to combine the three laws for ideally behaving gases into a single relationship:

$$\frac{\textbf{pressure} \cdot \textbf{volume}}{\textbf{temperature}} = \textbf{constant}$$
(at constant number of particles)

or, in equivalent form:

$$\frac{P_1 \cdot V_1}{T_1} = \frac{P_2 \cdot V_2}{T_2}$$

This **combined gas law** replaces the three individual laws—Boyle's, Charles's, and Gay-Lussac's—for ideally behaving gases. For example, if the pressure is held constant in a given problem, then $P_1 = P_2$ and the pressure term can be eliminated from the equation, leaving

$$\frac{V_1}{T_1} = \frac{V_2}{T_2} \qquad \text{(Charles's law)}$$

The additional benefit we obtain from using the combined law lies in the fact that it enables us to solve problems in which pressure, volume, and temperature may be changing.

PROBLEM
An ideally behaving gas occupies 500 milliliters at STP. What volume does it occupy at 546 K and 0.5 atmosphere, number of particles remaining constant?

SOLUTION

$$\frac{P_1 \cdot V_1}{T_1} = \frac{P_2 \cdot V_2}{T_2}$$

$$P_1 = 1 \text{ atm}$$
$$V_1 = 500 \text{ mL}$$
$$T_1 = 273 \text{ K}$$
$$P_2 = 0.5 \text{ atm}$$
$$V_2 = ???$$
$$T_2 = 546 \text{ K}$$

$$\frac{1 \text{ atm} \cdot 500 \text{ mL}}{273 \text{ K}} = \frac{0.5 \text{ atm} \cdot V_2}{546 \text{ K}}$$

$$V_2 = \frac{1 \text{ atm} \cdot 500 \text{ mL} \cdot 546 \text{ K}}{273 \text{ K} \cdot 0.5 \text{ atm}}$$

$$= 2000 \text{ mL}$$

TRY IT YOURSELF
An ideally behaving gas occupies 12 cubic decimeters at 0.5 atmosphere and 300 K. At what temperature will this gas occupy 6 cubic decimeters at 0.25 atmosphere, number of particles remaining constant?

ANSWER
75 K

Relating Volume and Moles (Avogadro's Law)

We now ask this question: How does the number of gas particles in a container affect the volume of a gas when the pressure and temperature are held constant? Imagine that we have a container whose volume is 5.0 liters and that contains n moles of ideally behaving gas particles at STP. Now suppose we have a second container that is identical to the first in every respect. If we join the containers, as shown in the diagram, it is obvious that we have a volume of 10 liters and $2n$ moles of gas particles.

Container 1	Container 2
5.0 liters n moles at STP	5.0 liters n moles at STP

Joined Containers

10. liters $2n$ moles at STP

Amadeo Avogadro first proposed an hypothesis about gases in 1811: *At the same temperature and pressure, equal volumes of any (ideally behaving) gas contain equal numbers of particles.* Therefore, each of the three containers in the accompanying diagram holds the same number, n, of particles, even though there is a different gas in each container.

Gas: CO	Gas: H_2	Gas: Ar
Pressure: 10. kPa	Pressure: 10. kPa	Pressure: 10. kPa
Volume: 5.0 L	Volume: 5.0 L	Volume: 5.0 L
Temperature: 800 K	Temperature: 800 K	Temperature: 800 K
Number of particles: n	Number of particles: n	Number of particles: n

Avogadro's law states also that the volume of an ideally behaving gas is *directly proportional* to the number of gas particles present *if the temperature and pressure are held constant.* This relationship, known as *Avogadro's law*, is written as follows:

$$\frac{\text{volume}}{\text{number of particles}} = \text{constant}$$

(at constant temperature and pressure)

or, in equivalent form:

$$\frac{V_1}{n_1} = \frac{V_2}{n_2}$$

This basic law of nature follows from the observation that all the gas laws we have studied do *not* depend on the *identity* of the gas. The only requirement is that the sample of gas behaves ideally.

6.4 THE KINETIC-MOLECULAR THEORY (KMT) OF GAS BEHAVIOR

So far, we have described gas behavior in terms of *experimental laws*; that is, laws that can be investigated in the laboratory. We have assumed that all of the gases behaved ideally, but we have never actually defined the term *ideal gas*.

The KMT Model of an Ideal Gas

Now we take a different approach. We propose a model (the **kinetic-molecular theory** or **KMT model**) of an *ideal* gas, and we manipulate the model according to the basic laws of (shudder!) mathematics and physics. Then we look at the results. (We will not do all of the fancy math; we will be content to accept the conclusions.)

An **ideal gas** is a collection of particles that:

* Have mass but negligible volume.
* Move randomly in straight lines.
* Are *not* subject to any attractive or repulsive forces (except during collisions with each other or with the walls of their container).
* Collide in a *perfectly elastic* fashion; that is, no energy is lost during a collision.

The KMT model of gas behavior enables us to draw three conclusions:

1. The **volume** occupied by an ideal gas is essentially the volume of its container.
2. The **pressure** exerted by an ideal gas is related to the number of collisions that the particles make with the walls of the container in a given amount of time.
3. The **absolute (Kelvin) temperature** of an ideal gas is directly proportional to the *average* kinetic energy of the gas particles. (In other words, if the absolute temperature is doubled, the average kinetic energy of the particles is also doubled.)

Astounding as it seems, *all* of the gas laws we have developed in this chapter can be derived as a direct result of the KMT model. Therefore, we can conclude that the KMT is a well-constructed model for explaining the behavior of ideal gases.

PROBLEM

Use the KMT to explain why reducing the volume of a gas at constant temperature increases the pressure that the gas exerts on the walls of the container.

SOLUTION

Refer to the accompanying diagram.

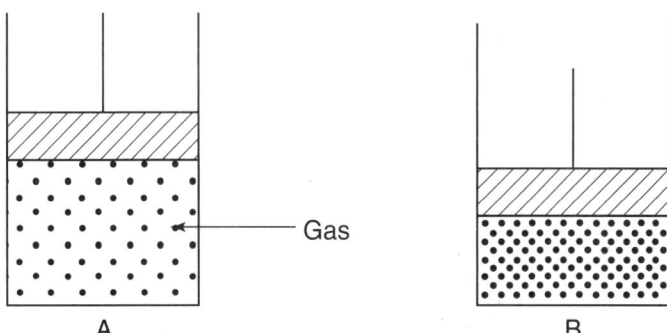

In the cylinder with the reduced volume (cylinder *B*), the molecules of the gas have less space in which to move. As a result, the molecules collide more frequently with the walls of the container. Since pressure depends on the frequency of these collisions, the pressure exerted by the gas increases.

PROBLEM

Use the KMT to explain why lowering the temperature of a gas at constant volume decreases the pressure that the gas exerts on the walls of the container.

SOLUTION

Refer to the accompanying diagram.

As the temperature of the gas is lowered, the average kinetic energy of the gas molecules decreases. Since the molecules are, on the average, traveling more slowly, they collide less frequently with the walls of the container and the pressure exerted on the walls decreases.

The Behavior of Real Gases

Real gases, such as oxygen, hydrogen, helium, and carbon dioxide, may or may not exhibit ideal behavior, depending on the conditions of the environment. For a real gas to behave ideally, its particles must be relatively far apart and be moving with relatively high speeds. Under these conditions, the attractive forces among the particles will be almost nonexistent, and the volume of the particles themselves will be negligible in relation to the volume of the container. What conditions will bring this behavior about? *High temperature and low pressure enable a real gas to exhibit most nearly ideal behavior.*

Conversely, *low temperature and high pressure have the opposite effect: the real gas fails to exhibit ideal behavior* because either the particles attract each other significantly or the volume of the particles is no longer negligible in relation to the volume of the container. As a consequence, gases under these conditions do not obey the gas laws completely and chemists have had to devise other relationships to correct for this nonideal behavior.

6.5   LIQUIDS

Liquids exist because particles of matter do *not* exhibit ideal behavior: they have a measurable volume and they exert forces on one another. When the temperature of a gas is sufficiently low and the pressure is sufficiently high, the gas becomes a liquid. The motion of the liquid particles becomes more restricted (the particles cannot move freely through space). As a result, the volume of the liquid becomes definite, but the liquid continues to take the shape of its container.

Evaporation and Condensation

According to the KMT, as applied to liquids, the absolute temperature of a liquid is related to the *average* kinetic energy of the particles of the liquid. However, some of the particles have enough kinetic energy to overcome the attractive forces of their neighboring particles, and they are able to escape from the surface of the liquid. We call this process **evaporation** because the particles that escape enter the *vapor* or gas phase above the liquid. The reverse process, called **condensation**, also occurs: particles of vapor with lower kinetic energies are recaptured by the liquid.

The rate of evaporation depends on three factors:

1. The nature of the liquid (ethanol evaporates more quickly than water).
2. The temperature of the liquid (hot water evaporates more quickly than cold water).
3. The surface area of the liquid (a puddle of water evaporates more quickly than water in a bottle).

Vapor Pressure

If we place a liquid in a sealed container, the vapor produced by the evaporating liquid will exert pressure on the surface of the liquid. Eventually, this pressure will seem to prevent the liquid from evaporating further. Actually, the *rate* of evaporation becomes equal to the *rate* of condensation: a condition we call **dynamic equilibrium**. The pressure that exists between the vapor and the liquid is known as the **equilibrium vapor pressure** (or simply the **vapor pressure**). Vapor pressure depends on the nature of the liquid and its temperature.

In the accompanying graph of vapor pressure as a function of temperature for five liquids, note that the vapor pressure rises slowly at low temperatures and climbs more rapidly at higher temperatures. At any given temperature, acetone has a higher vapor pressure than ethanol. As a result, the evaporation rate for acetone is greater than the rate for ethanol. A similar graph is shown in Reference Table H in Appendix 1.

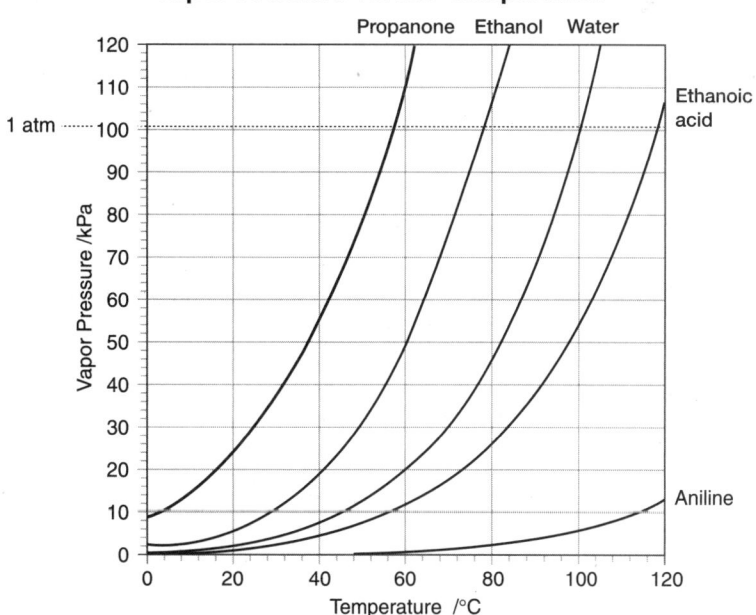

Vapor Pressure versus Temperature

157

If a liquid is heated in an open container, its vapor pressure will rise until it equals the pressure of the atmosphere above the liquid. At this point the liquid will begin to boil, and the temperature at which this occurs is known as the *boiling point* of the liquid. A liquid can have any number of boiling points, depending on the value of the pressure above the liquid.

When the atmospheric pressure is 1 atmosphere, the boiling temperature is known as the **normal boiling point** of the liquid. If we refer to the previous graph, we can see that propanone has the lowest normal boiling point (56°C) because it reaches a vapor pressure of 760 kilopascals at a lower temperature than any of the other liquids. According to the graph, the normal boiling points of ethanol, water, and ethanoic acid are 78°C, 100°C, and 118°C, respectively. The normal boiling point of aniline (184°C) is *not* shown on the graph because its vapor pressure rises too slowly as temperature is increased.

6.6 SOLIDS

As the temperature of a liquid is lowered, the forces of attraction between the particles become stronger and the particles begin to arrange themselves in an orderly fashion. The motion of the particles becomes severely restricted (largely vibrations), and the substance is said to be in the solid phase. All true solids have crystalline structure, and we say that the particles occupy positions within the crystal *lattice* of the substance. Solids and their crystal systems are studied in more advanced courses in chemistry.

6.7 CHANGE OF PHASE

We now take a closer look at how substances change phase. A change of phase is dependent on the nature of the substance and on the pressure and temperature of its environment.

Terms Associated with Phase Changes

The various terms associated with phase changes and the accompanying energy changes are shown in the diagram on page 159.

Reverse phase changes—melting and freezing, for example—occur at the same temperature. Therefore, the melting and freezing points of a substance refer to the same temperature. If heat energy is being added to the system, melting occurs; if heat energy is being removed, freezing occurs. Otherwise, the phases are said to coexist in *dynamic equilibrium*.

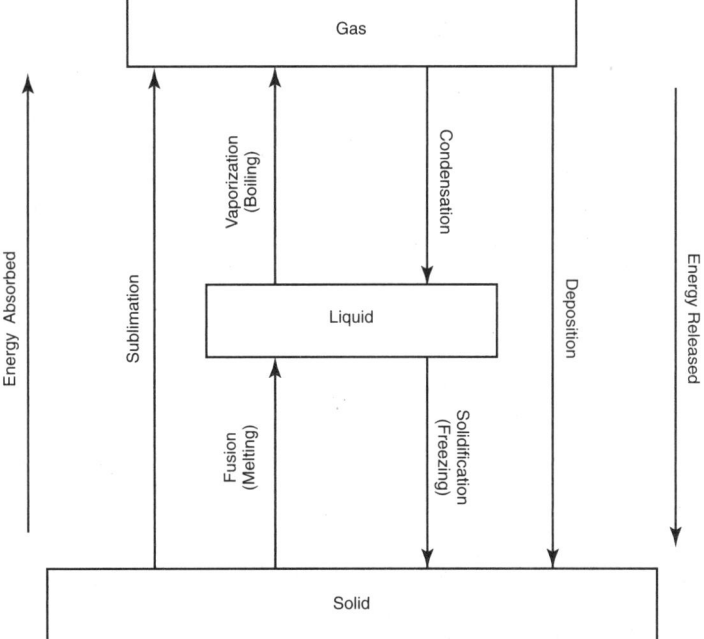

Sublimation is the most peculiar phase change because no liquid phase is encountered; the solid passes directly into the gas phase. Substances that sublime have relatively high vapor pressures (compared to the atmospheric pressure), as well as relatively weak forces of attraction between the particles of the solid. Dry ice [$CO_2(s)$] sublimes readily at room temperature and pressure. *Deposition*, the reverse of sublimation, is demonstrated when a carbon dioxide fire extinguisher is released. The rapid cooling of the carbon dioxide gas causes crystals of dry ice to form.

Heating and Cooling Curves

Suppose we take a solid that does not sublime, gradually add heat energy to it, and note how the temperature of the system changes as time progresses. If we graphed the results, we would produce a curve similar to the one shown that follows the next paragraph.

As heat energy is added, the temperature of the solid rises, indicating that the heat energy is being used to increase the average kinetic energy of the particles. After a while, however, the temperature remains constant, even though heat energy is still being added to the system. At this point the phase change of melting is occurring. According to the heating curve, the melting point of this solid is 100°C. The energy is being stored as potential energy. When melting has been completed, the temperature of the liquid begins to rise again as the average kinetic energy of the particles is increasing.

159

Temperature versus Heat Energy Added

When the boiling point is reached, added heat energy does not raise the temperature (and the average kinetic energy of the particles) but is converted to potential energy instead. According to the heating curve, the boiling point of this liquid is 140°C. At the completion of boiling, the temperature of the gas again rises with the addition of heat energy.

If we remove heat energy from a gas, we obtain the cooling curve given below. Note that it is the reverse of the heating curve above.

Temperature versus Heat Energy Added

Inspection of the cooling curve confirms that condensation takes place at the boiling point and freezing occurs at the melting point. The only change is that heat energy is being removed from the system.

During melting, a specific amount of heat energy was added to the solid in order to convert it to a liquid. This energy, known as the *heat of fusion* (H_f), is sometimes called a *latent* (hidden) heat because its addition is not accompanied by a temperature change. Heat of fusion is measured in joules per gram (J/g). According to Reference Table B in Appendix 1 of this book, the heat of fusion of ice at 0°C is 333.6 joules per gram. This means that 333.6 joules of heat energy is needed to convert each gram of ice into liquid water. Conversely, during freezing, 333.6 joules of heat energy is released for each gram of water that is converted into ice.

Similarly, during boiling, a specific amount of heat energy, known as the *heat of vaporization* (H_v), is needed to convert a liquid into a gas. The heat of vaporization of water at 100°C is 2259 joules per gram. Therefore, as each gram of water is converted into steam, 2259 joules of heat energy is absorbed by the system. As each gram of steam condenses to liquid water, 2259 joules of heat energy is released.

PROBLEM

How much heat energy is needed to melt 6.00 grams of solid water (ice) at its melting point?

SOLUTION

We find the heat of fusion of water in Reference Table B and use this value and the factor-label method to solve the problem.

$$\Delta H = 6.00 \; \cancel{g} \cdot \frac{333.6 \text{ J}}{1 \; \cancel{g}}$$

$$= 2.00 \times 10^3 \text{ J}$$

(Note the use of the symbol ΔH to represent the quantity of the heat absorbed or released.)

SECTION II—ADDITIONAL MATERIAL

6.1A MEASURING GAS PRESSURE IN THE LABORATORY

In laboratories, the pressure of a gas enclosed in a container is usually measured by a device called a **manometer**, which comes in two forms: closed-tube and open-tube. The operation of a manometer is similar to that of a barometer. A bent tube is filled with a liquid, such as mercury, and the tube is connected to the container.

Closed-tube manometers are usually used when the pressure of the enclosed gas is significantly less than atmospheric pressure. The accompanying diagram illustrates a closed-tube manometer.

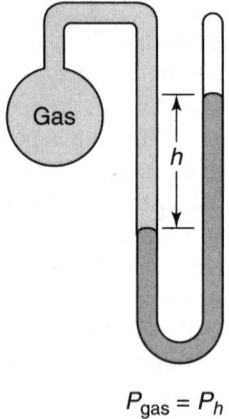

$$P_{gas} = P_h$$

The pressure of the enclosed gas is simply pressure represented by the difference in the heights (h) of the two liquid levels:

$$P_{gas} = P_h$$

Open-tube manometers are used when the pressure of the enclosed gas is at or near atmospheric pressure. The accompanying diagram shows three open-tube manometer arrangements, (a), (b), and (c):

$$P_{gas} = P_{atm}$$

(a)

$$P_{gas} = P_{atm} + P_h$$

(b)

$$P_{gas} = P_{atm} - P_h$$

(c)

When the pressure of the enclosed gas is *equal* to the atmospheric pressure, both liquid levels are *equal*, as in arrangement (a):

$$P_{gas} = P_{atm}$$

When the gas pressure is *greater* than the atmospheric pressure, the liquid level nearest the enclosed gas is *lower* than the level in the open tube, as in arrangement (b):

$$P_{gas} = P_{atm} + P_h$$

When the gas pressure is *less* than the atmospheric pressure, the liquid level nearest the enclosed gas is *higher* than the level in the open tube, as in arrangement (c):

$$P_{gas} = P_{atm} - P_h$$

6.2A   THE IDEAL (UNIVERSAL) GAS LAW

We are now ready to combine all of the preceding gas laws into a single, universal relationship that can be written as follows:

$$\frac{\text{pressure} \cdot \text{volume}}{\text{number of particles} \cdot \text{temperature}} = \text{constant}$$

Mathematically, the **ideal gas law** can be expressed as:

$$\frac{P \cdot V}{n \cdot T} = R$$

or, in its more usual form:

$$P \cdot V = n \cdot R \cdot T$$

where R is a constant (called the **molar gas constant**) that depends on the units of pressure and volume. If the pressure is measured in atmospheres and the volume in liters, then R has the value 0.0821 liter · atmospheres per mole · kelvin (0.0821 L · atm/mol · K). In the SI system, where pressure is measured in pascals and volume in cubic meters, R has the value 8.31 joules per mole · kelvin (8.31 J/mol · K).

PROBLEM
How many moles of an ideally behaving gas occupy 400 liters at 0.821 atmosphere and 200 K?

SOLUTION

$$P \cdot V = n \cdot R \cdot T$$
$$P = 0.821 \text{ atm}$$
$$V = 400 \text{ L}$$
$$n = ???$$
$$R = 0.0821 \text{ L} \cdot \text{atm/mol} \cdot \text{K}$$
$$T = 200 \text{ K}$$
$$0.821 \text{ atm} \cdot 400 \text{ L} = n \cdot 0.0821 \text{ L} \cdot \text{atm/mol} \cdot \text{K} \cdot 200 \text{ K}$$
$$n = \frac{0.821 \text{ atm} \cdot 400 \text{ L}}{0.0821 \text{ L} \cdot \text{atm} / \text{mol} \cdot \text{K} \cdot 200 \text{ K}}$$
$$= 20 \text{ moles of gas}$$

PROBLEM

What volume will an ideally behaving gas containing 1.00 mole of particles occupy at STP?

SOLUTION

$$P \cdot V = n \cdot R \cdot T$$
$$P = 1.00 \text{ atm}$$
$$V = ???$$
$$n = 1.00 \text{ mol}$$
$$R = 0.0821 \text{ L} \cdot \text{atm/mol} \cdot \text{K}$$
$$T = 273 \text{ K}$$
$$1.00 \text{ atm} \cdot V = 1.00 \text{ mol} \cdot 0.0821 \text{ L} \cdot \text{atm/mol} \cdot \text{K} \cdot 273 \text{ K}$$
$$V = \frac{1.00 \text{ mol} \cdot 0.0821 \text{ L} \cdot \text{atm} / \text{mol} \cdot \text{K} \cdot 273 \text{ K}}{1.00 \text{ atm}}$$
$$V = 22.4 \text{ L}$$

This volume, 22.4 liters, appears so often in chemistry that it is given a special name — the **molar volume** *at STP* (denoted as V_m); its units are liters per mole (L/mol).

PROBLEM

What is the volume occupied by 3.00 moles of nitrogen gas (N_2) at STP?

SOLUTION

$$3.00 \text{ mol } N_2 \left(\frac{22.4 \text{ L } N_2}{1 \text{ mol } N_2} \right) = 67.2 \text{ L}$$

TRY IT YOURSELF
What volume is occupied by 8.00 grams of helium gas (He) at STP?

ANSWER
44.8 L

6.3A   THE DENSITY OF AN IDEAL GAS AT STP

Since the mole is associated with a mass (molar mass) and a gas volume (molar volume) at STP, we can calculate the gas *density* for any substance that is assumed to behave as an ideal gas at STP. We recall that density is calculated by dividing the mass of a substance by its volume. Therefore, we can calculate this special density from the following relationship:

$$d_{STP} = \frac{M}{V_m}$$

where V_m is 22.4 liters per mole at STP.

PROBLEM
Calculate the density of CO gas at STP.

SOLUTION
The molar mass of CO is 28.01 g/mol. Therefore:

$$d_{STP} = \frac{28.01 \, \frac{g}{mol}}{22.4 \, \frac{L}{mol}} = 1.25 \frac{g}{L} \, (\text{at STP})$$

TRY IT YOURSELF
Calculate the density of O_2 gas at STP.

ANSWER
1.43 g/L

Molar Mass from Gas Density

Since the molar mass of a gaseous substance can be used to calculate the density of the substance at STP, we can reverse the process; that is, we can calculate the molar mass from the gas density at STP. This is accomplished by multiplying the density by V_m (22.4 L/mol at STP), as given by the following relationship:

$$M = d_{STP} \cdot V_m$$

PROBLEM
The density of a gas is 1.96 grams per liter at STP. Calculate its molar mass.

SOLUTION

$$\frac{1.96 \text{ g}}{\cancel{L}} \cdot \frac{22.4 \text{ } \cancel{L}}{\text{mol}} = 43.9 \text{ g/mol}$$

TRY IT YOURSELF
The density of a gas is 2.59 grams per liter at STP. Calculate its molar mass.

ANSWER
58.0 g/mol

Determination of Molecular Formula from Gas Density

If the empirical formula and the density of a gaseous substance at STP are known, the molecular formula of the substance is easily calculated.

PROBLEM
The empirical formula of a gaseous substance is CH, and its density at STP is 3.48 grams per liter. Calculate the molecular formula of this substance.

SOLUTION
The molar mass of the substance is calculated from this relationship:

$$d_{STP} \cdot 22.4 \frac{L}{\text{mol}} = M$$

$$M = \frac{3.48 \text{ g}}{\cancel{L}} \cdot \frac{22.4 \text{ } \cancel{L}}{\text{mol}} = 78.0 \text{ g/mol}$$

Since the empirical formula is CH, the *empirical* molar mass is 13.0 g/mol.

The molecular formula *must* be a multiple of the empirical formula. We can find this multiple by dividing the two molar masses:

$$\frac{78.0 \text{ g/mol}}{13.0 \text{ g/mol}} = 6.00$$

Therefore, the molecular formula is *six times* the empirical formula:

$$\text{Molecular formula} = 6 \cdot CH = C_6H_6$$

6.4A GASES AND CHEMICAL REACTIONS

Mass-Volume Gas Problems

PROBLEM
In the equation $2CO(g) + O_2(g) \rightarrow 2CO_2(g)$, how many liters of $CO_2(g)$ at STP are produced by the reaction of 64.0 grams of $O_2(g)$?

SOLUTION
This type of problem is also solved by using FLM. We need the connections among mass, volume, and moles. Here is the solution map:

$$64.0 \text{ g } O_2 \xrightarrow{\hspace{1cm}} \text{mol } O_2 \xrightarrow[\text{coefficients}]{\text{equation}} \text{mol } CO_2 \xrightarrow{\hspace{1cm}} \text{L } CO_2$$

The solution is as follows:

$$\text{L } CO_2 = 64.0 \text{ g } O_2 \cdot \frac{1 \text{ mol } O_2}{32.0 \text{ g } O_2} \cdot \frac{2 \text{ mol } CO_2}{1 \text{ mol } O_2} \cdot \frac{22.4 \text{ L } CO_2}{1 \text{ mol } CO_2} = 89.6 \text{ L } CO_2$$

TRY IT YOURSELF
In the equation $2CO(g) + O_2(g) \rightarrow 2CO_2(g)$, how many liters of $O_2(g)$ at STP are able to react with 28.0 grams of $CO(g)$?

ANSWER
11.2 L

Volume-Volume Gas Problems

PROBLEM
In the equation $4NH_3(g) + 5O_2(g) \rightarrow 4NO(g) + 6H_2O(l)$, how many liters of $NH_3(g)$ at STP are needed to react with 200. liters of $O_2(g)$ at STP?

SOLUTION

To solve this problem, we need to convert between volume and numbers of moles. Our solution map is as follows:

$$200.\ \text{L O}_2 \xrightarrow{} \text{mol O}_2 \xrightarrow[\text{coefficients}]{\text{equation}} \text{mol NH}_3 \xrightarrow{} \text{L NH}_3$$

The solution is given below:

$$\text{L NH}_3 = 200.\ \cancel{\text{L O}_2} \cdot \frac{1\ \cancel{\text{mol O}_2}}{22.4\ \cancel{\text{L}}} \cdot \frac{4\ \cancel{\text{mol NH}_3}}{5\ \cancel{\text{mol O}_2}} \cdot \frac{22.4\ \text{L NH}_3}{1\ \cancel{\text{mol NH}_3}} = 160.\ \text{L NH}_3$$

Volume-volume problems are particularly easy to solve because the factor "22.4" *always* cancels in the calculations. (Why?)

TRY IT YOURSELF

In the equation $4NH_3(g) + 5O_2(g) \rightarrow 4NO(g) + 6H_2O(\ell)$, how many liters of $NH_3(g)$ at STP are needed to produce 1000 liters of $NO(g)$ at STP?

ANSWER

1000 L

6.5A DALTON'S LAW OF PARTIAL PRESSURES

Imagine that we have a container with a *mixture* of three ideally behaving gases, *A*, *B*, and *C*. We know that the gases would exert a combined pressure on the walls of the container (P_{total}). What we would like to know is how this combined pressure is related to the individual pressures (P_A, P_B, and P_C) of the three gases in the container. These individual pressures are called *partial pressures*. A partial pressure is the pressure that a single gas would exert *if it were the only gas in the container*.

Dalton discovered that a simple relationship exists between the total pressure of a gas mixture and the partial pressures of its components. **Dalton's law of partial pressures** states that the *sum* of the partial pressures equals the total pressure exerted on the container. For our example, this relationship is as follows:

$$P_{total} = P_A + P_B + P_C$$

The total pressure of the three gas components in the mixture described on the previous page is 55.0 kilopascals. If P_A = 20.0 kilopascals and P_B = 7.5 kilopascals, what is the partial pressure of gas C (P_C)?

SOLUTION

$$P_{total} = P_A + P_B + P_C$$
$$P_C = P_{total} - P_A - P_B$$
$$= 55.0 \text{ kPa} - 20.0 \text{ kPa} - 7.5 \text{ kPa}$$
$$= 27.5 \text{ kPa}$$

TRY IT YOURSELF
Four gases in a tank have respective partial pressures of 0.20 atmosphere, 0.35 atmosphere, 0.15 atmosphere, and 0.60 atmosphere. What is the total gas pressure inside the tank?

ANSWER
1.30 atm

The simple relationship expressed in Dalton's law is a direct result of the fact that each gas in a mixture occupies the *entire* volume of the container. For example, if we sample the air in various parts of a room, we expect to find nitrogen, oxygen, and other gases in every sample we take. Since the volume and temperature are the same for each of the components in a gas mixture, it follows that the partial pressure of each component is directly related to the *number of moles* of that component present in the mixture. There is a simple relationship among the partial pressure of any component in a gas mixture, the total pressure of the mixture, and the number of moles of gas present:

$$P_i = \left(\frac{n_i}{n_{total}} \right) \cdot P_{total}$$

where P_i and n_i represent the partial pressure and number of moles, respectively, of any component in the mixture, and P_{total} and n_{total} represent the pressure and number of moles, respectively, of the *entire mixture*.

PROBLEM
A tank contains 1.5 moles of gas A, 2.0 moles of gas B, and 2.5 moles of gas C. If the total pressure of the gas mixture is 0.80 atmosphere, what is the partial pressure of gas A?

SOLUTION

$$P_A = \left(\frac{n_A}{n_{total}}\right) \cdot P_{total}$$

$$= \left(\frac{1.5 \text{ mol}}{1.5 \text{ mol} + 2.0 \text{ mol} + 2.5 \text{ mol}}\right) \cdot 0.80 \text{ atm}$$

$$= \left(\frac{1.5 \text{ mol}}{6.0 \text{ mol}}\right) \cdot 0.80 \text{ atm}$$

$$= 0.20 \text{ atm}$$

TRY IT YOURSELF

Calculate the partial pressure of gas C in the problem given on page 169.

ANSWER

0.33 atm

6.6A GRAHAM'S LAW OF EFFUSION (DIFFUSION)

Effusion is the process by which gases escape through small openings or *pores* in materials. We have all had the experience of buying a balloon filled with helium and discovering that, one day later, the helium has escaped and the balloon has lost its firmness. This is an example of a gas effusing.

When gases are brought into contact, they mix with one another. This process is known as *diffusion*.

In 1832 the Scottish chemist Thomas Graham investigated the effusion and diffusion of gases and discovered that both processes obey the same simple law known as **Graham's law**: At a given temperature and pressure, the greater the density of a gas, the *slower* the gas effuses (and diffuses). Helium and hydrogen, the least dense gases, effuse (and diffuse) most rapidly, while gases with greater densities, such as oxygen and sulfur dioxide, effuse (and diffuse) more slowly.

Graham's law relates the relative rates of effusion (and diffusion) of gases to their molar masses: the greater the molar mass of a gas, the *slower* it effuses (and diffuses). Gases with small molar masses effuse (and diffuse) most rapidly, while gases with large molar masses effuse (and diffuse) more slowly.

Mathematically, *Graham's law* is expressed as a comparison between two ideally behaving gases, A and B, at a specified temperature and pressure. Their rates of effusion (and diffusion) are represented by R_A and R_B, and their molar masses by M_A and M_B:

$$\frac{R_A}{R_B} = \sqrt{\frac{\mathcal{M}_B}{\mathcal{M}_A}}$$

PROBLEM

Calculate the rate of effusion (and diffusion) of hydrogen gas ($\mathcal{M}$ = 2.0 g/mol) relative to that of oxygen gas ($\mathcal{M}$ = 32.0 g/mol) at the same temperature and pressure.

SOLUTION

$$\frac{R_{H_2}}{R_{O_2}} = \sqrt{\frac{\mathcal{M}_{O_2}}{\mathcal{M}_{H_2}}}$$

$$= \sqrt{\frac{32\ 0\ \text{g/mol}}{2.0\ \text{g/mol}}} = \sqrt{\frac{16.0}{1.0}} = \frac{4.0}{1.0}$$

This result means that, under the same conditions of temperature and pressure, hydrogen gas will effuse (and diffuse) four times faster than oxygen gas.

TRY IT YOURSELF

Compare the rate of effusion (and diffusion) of O_2 gas ($\mathcal{M}$ = 32.0 g/mol) relative to that of SO_2 gas ($\mathcal{M}$ = 64.0 g/mol) at STP.

ANSWER

O_2 will effuse (and diffuse) 1.41 times faster than SO_2.

6.7A GASES COLLECTED OVER WATER

When a gas, such as hydrogen, is collected over water, there are actually *two* gases present in the container: hydrogen and water vapor. According to Dalton's law of partial pressures, both gases occupy the entire volume of the container but each gas exerts its own partial pressure on the container. The pressure of the water vapor may be determined by referring to Table X in Appendix 2. The pressure of the hydrogen is found by *subtracting* the vapor pressure of the water from the total pressure of the gas mixture.

PROBLEM

Hydrogen is collected in a 250-milliliter container over water at 30°C. The total pressure is 101.246 kilopascals.
(a) What is the pressure of the "dry" hydrogen gas?
(b) What is the volume occupied by the water vapor?

SOLUTIONS

(a) According to Table X in Appendix 2, the vapor pressure of water at 30°C is 4.246 kPa. According to Dalton's law:

$$P_{H_2} + P_{H_2O} = P_{total}$$
$$P_{H_2} = P_{total} - P_{H_2O}$$
$$= 101.246 \text{ kPa} - 4.246 \text{ kPa}$$
$$= 97.000 \text{ kPa}$$

(b) The volume of the water vapor is 250 mL (as is the volume of the hydrogen gas).

TRY IT YOURSELF

When hydrogen gas is collected over water at 20°C, it is found that the partial pressure of the hydrogen is 96.349 kilopascals. At what (total) pressure was the gas collected?

ANSWER

98.688 kPa

PROBLEM

Five hundred milliliters of a gas is collected over water at 298 K (25°C). The atmospheric pressure is 101.169 kilopascals. What would the volume of the "dry" gas be at STP?

SOLUTION

This problem involves Dalton's law and the combined gas law. When the gas is collected, the *total* gas pressure in the container becomes equal to the atmospheric pressure (101.169 kPa). The volume of the "dry" gas in the original container is (as always) 500 mL.

The partial pressure of the dry gas is 98.000 kPa (101.169 kPa − 3.169 kPa). We now apply the combined gas law:

$$\frac{P_1 \cdot V_1}{T_1} = \frac{P_2 \cdot V_2}{T_2}$$

$$\frac{98.000 \text{ kPa} \cdot 500 \text{ mL}}{298 \text{ K}} = \frac{101.325 \text{ kPa} \cdot V_2}{273 \text{ K}}$$

$$V_2 = \frac{98.000 \text{ kPa} \cdot 500 \text{ mL} \cdot 273 \text{ K}}{298 \text{ K} \cdot 101.325 \text{ kPa}}$$

$$= 443 \text{ mL}$$

TRY IT YOURSELF

A 940.0-milliliter sample of gas is collected over water at 40.0°C. The total pressure of the gas is 85.592 kilopascals. Calculate the volume of the "dry" gas at 25.0°C and 72.661 kilopascals.

ANSWER

963.3 mL

6.8A ADDITIONAL FUSION AND VAPORIZATION PROBLEMS

Reference Table W-2 in Appendix 2 lists the heats of fusion and vaporization of selected substances. (For convenience, the melting points and boiling points of these substances are also included.)

When we inspect the table, we see that the heat of vaporization of a substance is always considerably larger than the corresponding heat of fusion. We may conclude that more energy is needed to convert a liquid into a gas than to convert a solid into a liquid. This fact can be explained by noting that a gas is a collection of particles that have little or no attractions among them. The large energy absorbed during the boiling process is used to free the particles of a liquid from their mutual attractions.

TRY IT YOURSELF

How much heat energy is absorbed when 20.0 grams of solid ethanol is converted to liquid at its melting point?

ANSWER

2080 J

PROBLEM

How many grams of oxygen gas will release 639 joules of heat energy as the gas condenses at its boiling point?

SOLUTION

The heat of vaporization of oxygen, as obtained from Reference Table W-2, is 213 J/g. Also:

$$x\,g = \frac{639\ \cancel{J}}{213\ \cancel{J}}$$

$$x = 3\ g$$

$$639\ J = x\,g \cdot \frac{213\ J}{1\ g}$$

$$x = 3.00\ g$$

TRY IT YOURSELF

How much heat energy is released by 10.0 grams of liquid copper as it solidifies at its melting point?

ANSWER
$\Delta H = 1340$ J

6.9A   PHASE DIAGRAMS

Every phase change (solid $\rightleftarrows$ liquid, liquid $\rightleftarrows$ gas, solid $\rightleftarrows$ gas) occurs at a specific combination of temperature and pressure. For example, the phase change between ice and liquid water occurs at 0°C when the pressure is 1 atmosphere. The *direction* of the phase change depends on whether heat energy is being added or removed from the system. In our example, liquid water will freeze if heat energy is being *removed* from the ice-water system.

If *no* heat energy is added or removed, the two phases will remain in a state of *dynamic equilibrium* at that temperature and pressure. In our example, the rate of melting ice will be exactly equal to the rate of freezing water and the quantities of ice and water will remain constant over time.

We can summarize all of these relationships graphically by using a *phase diagram*. A general phase diagram follows.

Pressure versus Temperature

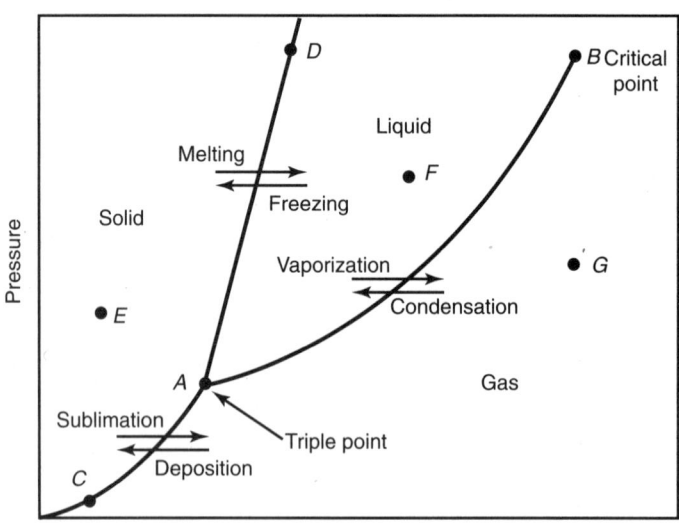

Each line represents the pressure-temperature combination at which two phases are in dynamic equilibrium. For example, line *CA*, the sublimation-deposition line, represents the conditions under which the solid and gas phases can be in dynamic equilibrium. At any combination of temperature and pressure that does *not* lie on one of the lines (points *E*, *F*, and *G*, for example) *only one phase* can exist.

The phase diagram includes two points of special interest:

1. Point *A* is known as the **triple point**. It is the only point at which all three phases (solid, liquid, and gas) can coexist in dynamic equilibrium.
2. Point *B* is known as the **critical point**. The temperature and pressure at the critical point are known, respectively, as the **critical temperature** and **critical pressure**. Above the critical temperature, only the gas phase can exist, no matter how high the pressure, because the kinetic energies of the gas particles are too high to allow the formation of a liquid.

The accompanying table lists the critical temperatures and pressures for four substances. We note that the critical pressures are generally quite large.

Substance	Critical Temperature/°C	Critical Pressure/atm
Ammonia (NH_3)	132.6	111.5
Carbon dioxide (CO_2)	31.3	73.0
Nitrogen (N_2)	−146.9	33.5
Water (H_2O)	374.6	217.7

The accompanying phase diagrams for (a) water and (b) carbon dioxide are *not* drawn to scale.

Pressure versus Temperature

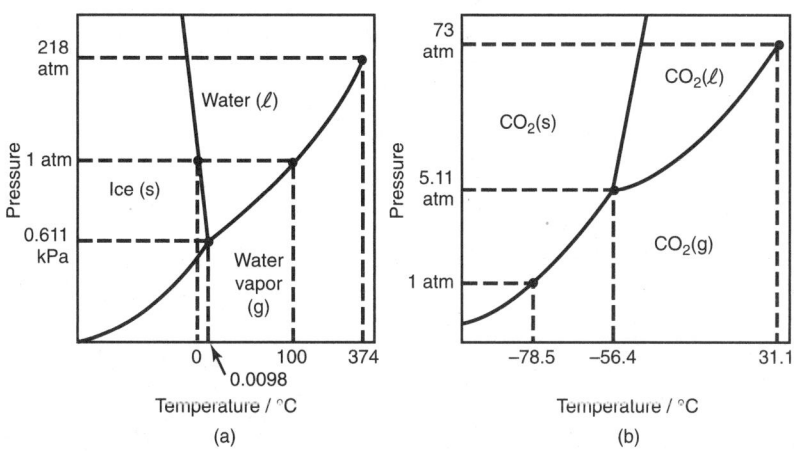

(a)   (b)

In the phase diagram for water, we note that the solid-liquid line tilts toward the left. This is a direct result of the fact that liquid water is more dense than ice at temperatures close to 0°C. It is also the reason why ice melts when the pressure is increased sufficiently. The triple point of water (0.611 kPa, 0.0098°C) provides the basis for defining the Kelvin temperature scale. *This point is assigned an exact value of 273.16 K.*

The phase diagram for carbon dioxide has the same appearance as the general phase diagram shown on page 174 because solid carbon dioxide is more dense than the liquid phase. We also note that, at a constant pressure of 1 atmosphere, carbon dioxide passes directly from the solid to the gas phase as the temperature is increased; the liquid phase does not even come into existence until the pressure reaches 5.11 atmospheres.

END-OF-CHAPTER QUESTIONS

Some questions have the symbol "§2" in front of the question number. This symbol means that the question is based on Section II material.

§2 **1.** A 1-liter flask contains two gases at a total pressure of 3.0 atmospheres. If the partial pressure of one of the gases is 0.5 atmosphere, the partial pressure of the other gas must be
(1) 1.0 atm   (2) 2.5 atm   (3) 1.5 atm   (4) 0.50 atm

2. At constant temperature, which line in the graph best shows the relationship between the volume of an ideal gas and its pressure?

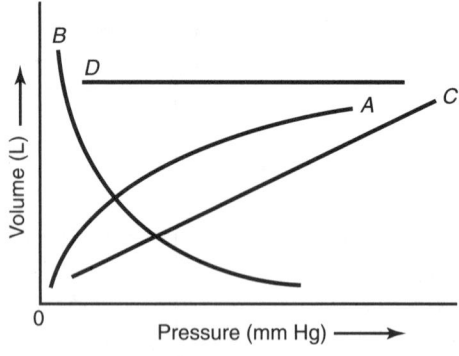

(1) *A*   (2) *B*   (3) *C*   (4) *D*

§2 **3.** A container that is filled with 1.00 mole of oxygen and 2.00 moles of hydrogen has a total pressure of 75.0 kilopascals. What is the partial pressure of the oxygen?
(1) 10.0 kPa   (2) 20.0 kPa
(3) 25.0 kPa   (4) 50.0 kPa

4. At STP, 44.8 liters of CO_2 contains the same number of molecules as
(1) 1.00 mole of He   (2) 2.00 moles of Ne
(3) 0.500 mole of H_2 (4) 4.00 moles of N_2

5. The pressure of 200. milliliters of a gas at constant temperature is changed from 380. torr to 760. torr. The new volume of the gas is
(1) 100. mL   (2) 200. mL
(3) 400. mL   (4) 800. mL

§2 **6.** At STP, the volume occupied by 32 grams of a gas is 11.2 liters. The molar mass of this gas is closest to
(1) 8.0 g/mol   (2) 16 g/mol
(3) 32 g/mol   (4) 64 g/mol

§2 **7.** In a laboratory experiment, hydrogen gas is collected over water. If the atmospheric pressure is 101.645 kPa and the room temperature is 22°C, the partial pressure of the hydrogen is
(1) 104.294 kPa   (2) 103.605 kPa
(3) 101.645 kPa   (4) 99.000 kPa

§2 **8.** If 2.5 moles of a gas occupy 30. liters at a certain temperature and pressure, what volume will 7.5 moles of the gas occupy at the same temperature and pressure?
(1) 10. L   (2) 30. L   (3) 90. L   (4) 180 L

§2 Base your answers to questions 9–11 on the accompanying graph and on the following information:

The graph represents an experiment in which the volumes of four cylinders (A, B, C, and D) of the same ideal gas were measured at different temperatures. Each cylinder contained 1.0 mole of this gas. During the experiment, the pressure of each cylinder was not allowed to change.

Volume versus Temperature

§2 **9.** In which cylinder was the pressure of the gas least?
 (1) *A* (2) *B* (3) *C* (4) *D*

§2 **10.** At 546 K, in which cylinder would the gas occupy a volume of 400 milliliters?
 (1) *A* (2) *B* (3) *C* (4) *D*

§2 **11.** For each 1° decrease in temperature, the volume occupied by the gas in cylinder *A* decreased by

(1) $\dfrac{1}{273}$ mL (2) $\dfrac{100}{273}$ mL (3) $\dfrac{273}{100}$ mL (4) $\dfrac{273}{1}$ mL

12. At which temperature does a water sample have the highest average kinetic energy?
(1) 0°C   (2) 100°C   (3) 0 K   (4) 100 K

13. Which gas has properties that are most similar to those of an ideal gas?
(1) N_2 (2) O_2 (3) He   (4) Xe

14. When the average kinetic energy of a gaseous system is increased, the average molecular velocity of the system
(1) increases and the molecular mass increases
(2) decreases and the molecular mass increases
(3) increases and the molecular mass remains the same
(4) decreases and the molecular mass remains the same

15. Under which conditions does a real gas behave most like an ideal gas?
(1) at high temperatures and low pressures
(2) at high temperatures and high pressures
(3) at low temperatures and low pressures
(4) at low temperatures and high pressures

16. Which temperature represents absolute zero?
(1) 0 K   (2) 0°C   (3) 273 K   (4) 273°C

17. Which of the following substances is made up of particles with the highest average kinetic energy?
(1) Fe(s) at 35°C   (2) $Br_2(\ell)$ at 20°C
(3) $H_2O(\ell)$ at 30°C   (4) $CO_2(g)$ at 25°C

18. When a sample of a gas is heated at constant pressure, the average kinetic energy of its molecules
(1) decreases, and the volume of the gas increases
(2) decreases, and the volume of the gas decreases
(3) increases, and the volume of the gas increases
(4) increases, and the volume of the gas decreases

§2 **19.** Which gas has, under high pressure and low temperature, behavior closest to that of an ideal gas?
(1) $H_2(g)$ (2) $O_2(g)$ (3) $NH_3(g)$ (4) $CO_2(g)$

20. Compared to the average kinetic energy of 1 mole of water at 0°C, the average kinetic energy of 1 mole of water at 298 K is
(1) the same, and the number of molecules is the same
(2) the same, but the number of molecules is greater
(3) greater, and the number of molecules is greater
(4) greater, but the number of molecules is the same

21. When the vapor pressure of a liquid in an open container equals the atmospheric pressure, the liquid will
(1) freeze   (2) crystallize   (3) melt   (4) boil

22. As the temperature of a liquid increases, its vapor pressure
(1) decreases   (2) increases   (3) remains the same

23. Which sample of water has the greatest vapor pressure?
(1) 100 mL at 20°C   (2) 200 mL at 25°C
(3) 20 mL at 30°C   (4) 40 mL at 35°C

§2 24. The temperature of a sample of water is changed from 60°C to 70°C. According to Reference Table X, the change in the vapor pressure of the water is
(1) 11.25 kPa   (2) 19.93 kPa   (3) 31.18 kPa   (4) 51.11. kPa

§2 25. Water will boil at a temperature of 40°C when the pressure on its surface is
(1) 1.938 kPa   (2) 2.339 kPa   (3) 7.381 kPa   (4) 101.3 kPa

26. If 1.00 mole of $H_2(g)$ at STP is compared to 1.00 mole of He(g) at STP, the volumes of the gases will be found to be
(1) equal, and their masses unequal
(2) equal, and their masses equal
(3) unequal, and their masses unequal
(4) unequal, and their masses equal

27. The graph below shows the equilibrium vapor-pressure curves of liquids *A, B, C,* and *D*.

Pressure versus Temperature

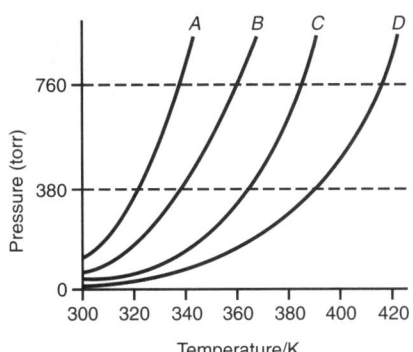

Which liquid has the lowest normal boiling point?
(1) *A*   (2) *B*   (3) *C*   (4) *D*

28. Which sample contains particles arranged in a regular geometric pattern?
(1) $CO_2(\ell)$ (2) $CO_2(s)$ (3) $CO_2(g)$ (4) $CO_2(aq)$

29. The particles in a typical crystalline solid are arranged
(1) randomly and relatively far apart
(2) randomly and relatively close together
(3) regularly and relatively far apart
(4) regularly and relatively close together

30. The accompanying graph represents changes of state for an unknown substance. What is the boiling temperature of the substance?

Temperature versus Time

(1) 0°C   (2) 20°C   (3) 70°C   (4) 40°C

31. What is the approximate number of joules of heat needed to change 150 grams of ice to liquid water at 0°C?
(1) 50,000   (2) 90,000   (3) 130,000   (4) 230,000

32. Which change of phase represents fusion?
(1) gas to liquid   (2) gas to solid
(3) solid to liquid   (4) liquid to gas

33. The density of a gas is 2.0 grams per liter at STP. Its molar mass is approximately
(1) 67 g/mol   (2) 45 g/mol
(3) 22 g/mol   (4) 8.0 g/mol

34. Which equation represents sublimation?
(1) $NH_3(g) \rightarrow NH_3(\ell)$ (2) $H_2O(\ell) \rightarrow H_2O(g)$
(3) $HCl(g) \rightarrow HCl(aq)$ (4) $CO_2(s) \rightarrow CO_2(g)$

§2 **35.** According to Reference Table W-2 in Appendix 2, which of the following solids requires the most heat energy to change it to a liquid at its melting point?
(1) aluminum   (2) copper   (3) lead   (4) water

§2 **36.** The temperature above which a gas can no longer be liquefied is known as the
(1) fusion point   (2) vaporization point
(3) critical point   (4) triple point

37. At STP, 32 grams of O_2 will occupy the same volume as
(1) 64 g of H_2 (2) 32 g of SO_2
(3) 8.0 g of CH_4 (4) 4.0 g of He

38. Which change of phase is exothermic?
(1) $H_2O(s) \rightarrow H_2O(g)$ (2) $CO_2(s) \rightarrow CO_2(\ell)$
(3) $H_2S(g) \rightarrow H_2S(\ell)$ (4) $NH_3(\ell) \rightarrow NH_3(g)$

39. Which term represents the change of a substance from the liquid phase to the gaseous phase?
(1) condensation   (2) vaporization
(3) sublimation   (4) fusion

40. The number of joules per gram required to melt ice at its melting point is called
(1) sublimation   (2) vapor pressure
(3) heat of vaporization (4) heat of fusion

§2 **41.** Which gas is most dense at STP?
(1) $CO(g)$ (2) $NO(g)$ (3) $N_2(g)$ (4) $O_2(g)$

§2 **42.** Given the reaction

$$2C_2H_6(g) + 7O_2(g) \rightarrow 4CO_2(g) + 6H_2O(g)$$

At STP, what is the total volume of $CO_2(g)$ formed when 6.0 liters of $C_2H_6(g)$ is completely oxidized?
(1) 24 L   (2) 12 L   (3) 6.0 L   (4) 4.0 L

§2 **43.** Given this reaction at STP:

$$N_2(g) + 3H_2(g) \rightarrow 2NH_3(g)$$

What is the total number of liters of NH_3 formed when 20 liters of N_2 reacts completely?
(1) 10   (2) 20   (3) 30   (4) 40

§2 **44.** Given this balanced equation:

$$3Fe + 4H_2O \rightarrow Fe_3O_4 + 4H_2$$

What is the total number of liters of H_2 produced at STP when 36.0 grams of H_2O is consumed?
(1) 22.4   (2) 33.6   (3) 44.8   (4) 89.6

45. According to Reference Table H in Appendix 1, what is the vapor pressure of propanone at 45°C?
(1) 22 kPa   (2) 33 kPa   (3) 70 kPa   (4) 98 kPa

46. The freezing point of bromine is
(1) 539°C   (2) 7°C   (3) –539°C   (4) –7°C

47. The graph below represents the heating curve of a substance that starts as a solid below its freezing point.

What is the freezing point of this substance?
(1) 30°C   (2) 55°C   (3) 90°C   (4) 120°C

48. An increase in the average kinetic energy of a sample of copper atoms occurs with an increase in
(1) concentration
(2) temperature
(3) pressure
(4) volume

49. Which statement describes particles of an ideal gas, based on the kinetic-molecular theory?
(1) Gas particles are separated by distances smaller than the size of the gas particles.
(2) Gas particles do not transfer energy to each other when they collide.
(3) Gas particles have no attractive forces between them.
(4) Gas particles move in predictable, circular movements.

50. Compared to a 1.0-gram sample of chlorine gas at standard pressure, a 1.0-gram sample of solid aluminum at standard pressure has
(1) a lower melting point
(2) a higher boiling point
(3) a lower density
(4) a greater volume

51. At STP, a 1-liter sample of Ne(g) and a 1-liter sample of Kr(g) have the same
(1) mass
(2) density
(3) number of atoms
(4) number of electrons

52. A rigid cylinder with a movable piston contains a sample of gas. At 300. K, this sample has a pressure of 240. kilopascals and a volume of 70.0 milliliters. What is the volume of this sample when the pressure is changed to 150. K and the pressure is changed to 160. kilopascals?

(1) 35.0 mL   (2) 52.5 mL   (3) 70.0 mL   (4) 105 mL

Constructed-Response Questions

1. A sample of gas in a container at 30.0°C exerts a pressure of 700. torr. If the temperature of the container is raised to 200.°C at constant volume, what pressure will the gas exert?

2. A sample of gas at 30.00°C and standard pressure occupies a volume of 250.0 milliliters. Calculate the temperature, in degrees Celsius, at which the gas will occupy a volume of 1000. milliliters at standard pressure.

§2 **3.** A sample of oxygen gas is collected over water at 27°C. If the gas is collected under a pressure of 101.567 kPa, what is the partial pressure of the oxygen gas in the collection vessel?

§2 **4.** A 750.-milliliter sample of hydrogen gas is collected over water at 30°C. If the gas is collected under a pressure of 98.659 kPa, what volume will the *dry* hydrogen gas occupy at STP?

§2 **5.** Calculate the relative rates of effusion of N_2(g) and CO_2(g).

§2 **6.** Calculate the density of CH_4 gas at 303 K and a pressure of 0.658 atmosphere. (*Hint:* Assume that you have 1.00 mole of CH_4, and use $PV = nRT$ to calculate the volume of the gas.)

§2 **7.** Given the equation

$$Ni (s) + 4CO (g) \rightarrow Ni(CO)_4 (g):$$

  (a) How many liters of CO(g) at STP are needed to react with 15 grams of Ni(s)?
  (b) How many liters of Ni(CO)$_4$ (g) at STP will be produced by the reaction of 1000. liters of CO(g) at STP with excess Ni(s)?

Base your answers to questions 8 and 9 on the diagram below, which shows a piston confining a gas in a cylinder.

8. Using the set of axes below, sketch the general relationship between the pressure and the volume of an ideal gas at constant temperature.

9. The volume of gas in the cylinder is 6.2 milliliters. The pressure of the gas is 1.4 atmospheres. The piston is then pushed in until the volume of gas is 3.1 milliliters while the temperature remains constant. Calculate the pressure of the gas, in atmospheres, after the change in volume.

Base your answers to questions 10 and 11 on the information below and on your knowledge of chemistry. A beaker contains a liquid sample of a molecular substance. Both the beaker and the liquid are at 194 K. The graph below represents the relationship between temperature and time as the beaker and its contents are cooled for 12 minutes in a refrigerated chamber.

Cooling Curve for a Substance

10. State what happens to the average kinetic energy of the molecules of the sample during the first 3 minutes.

11. Identify the physical change occurring during the time interval minute 4 to minute 9.

Base your answers to questions 12 and 13 on the information below and on your knowledge of chemistry. The diagram below represents a cylinder with a movable piston. The cylinder contains 1.0 liter of oxygen gas at STP. The movable piston is pushed downward at constant temperature until the volume of the $O_2(g)$ is 0.50 liter.

Movable piston

$O_2(g)$

12. Determine the new pressure of the $O_2(g)$ in the cylinder in atmospheres.

13. State the effect on the frequency of gas molecule collisions when the movable piston is pushed farther downward into the cylinder.

Base your answers to questions 14–16 on the graph below, which shows the vapor pressure curves for liquids *A* and *B*.

Pressure versus Temperature

14. What is the vapor pressure of liquid A at 70°C? Your answer must include correct units.

15. At what temperature does liquid *B* have the same vapor pressure as liquid *A* at 70°C? Your answer must include correct units.

16. Which liquid will evaporate more rapidly? Explain your answer in terms of intermolecular forces.

The answers to these questions are found in Appendix 3.

| Chapter Seven | **NUCLEAR CHEMISTRY** |

KEY IDEAS

This chapter focuses on the structure and reactivity of the atomic nucleus. Applications of radioactivity are discussed, as is energy production from nuclear fission and fusion. The chapter concludes with a brief discussion of fundamental interactions and particles.

KEY OBJECTIVES

At the conclusion of this chapter you will be able to:

- List the common nuclear particles, and write their symbols.
- Balance nuclear equations.
- Define the terms *nuclide, natural radioactivity, transmutation,* and *radioactive decay.*
- Describe the various types of radioactive decay.
- Explain how radioactive uranium–238 decays to the stable nuclide lead–206.
- Indicate how radioactive emanations can be separated and detected.
- Define the term *half-life,* and solve simple half-life problems.
- Describe the uses of radioactive isotopes.
- Describe how nuclear reactions can be induced, and indicate the role of accelerators in this process.
- Define the term *nuclear fission,* and describe how energy can be obtained from fission.
- List the parts of a fission reactor, and describe their functions.
- Indicate the safety procedures used in the disposal of radioactive wastes.
- Define the term *nuclear fusion,* and explain the role of this process in energy production.
- Describe the four fundamental interactions (forces) present in the universe.
- Describe the role that quarks play in the composition of protons, neutrons, and other nuclear particles.

SECTION I—BASIC (REGENTS-LEVEL) MATERIAL

NYS REGENTS CONCEPTS AND SKILLS

Note: By the time you have finished Section I, you should have mastered the concepts and skills listed below. The Regents chemistry examination will test your knowledge of these items and your ability to apply them.

Concepts are the *basic ideas* that form the body of the Regents chemistry course (what you need to know!).

Skills are the *activities* that demonstrate your mastery of these concepts (how you show that you know them!).

Following each concept or skill is a page reference (given in parentheses) to this chapter.

7.1 Concept:
- The stability of an isotope is based on the ratio of neutrons and protons in its nucleus. (Page 193)
- Although most nuclei are stable, some are unstable and decay spontaneously, emitting radiation. (Page 193)

7.2 Concept:
Each radioactive isotope has a specific mode and rate of decay (half-life). (Pages 193–197)

Skill:
Given the appropriate data or radioactive decay curve, calculate the:
    (a) initial amount,
    (b) fraction remaining,
    (c) half-life
of a radioactive isotope. (Pages 195–196)

7.3 Concepts:
- A change in the nucleus of an atom that converts the atom from one element to another is called transmutation. (Page 193)
- Transmutation can occur naturally or can be induced by bombardment of the nucleus by high-energy particles. (Pages 193, 197)

7.4 Concepts:
- Spontaneous decay can involve the release of:
    (a) alpha particles,
    (b) beta particles,
    (c) positrons,
    (d) gamma radiation

from the nucleus of an unstable isotope. (Pages 193–194)
- These emissions differ in mass, charge, ionizing power, and penetrating power. (Pages 193–194)

Skills:
- Determine the decay mode of a nuclear reaction. (Pages 193–194)
- Write nuclear equations that represent alpha and beta decay. (Pages 193–194)

7.5 Concept:
Nuclear reactions include:
   (a) natural and artificial transmutation (Pages 193, 197);
   (b) fission (Page 198);
   (c) fusion (Pages 205–206).

Skill:
Compare and contrast fission and fusion reactions. (Pages 198, 205–206)

7.6 Concept:
Both benefits and risks are associated with fission and fusion reactions. (Pages 198, 205–206)

7.7 Concept:
Nuclear reactions can be represented by equations that include:
   (a) symbols that represent atomic nuclei (with mass numbers and atomic numbers) (Page 192);
   (b) subatomic particles (with mass numbers and charges) (Pages 192–193);
   (c) emissions such as alpha, beta, and gamma radiation. (Pages 193–194, 199–200)

Skills:
- Complete nuclear equations. (Pages 199–200)
- Predict missing particles from nuclear equations. (Pages 199–200)

7.8 Concept:
The energy released in a nuclear reaction (fission or fusion) comes from the fractional amount of mass converted into energy. (Page 198)

7.9 Concept:
The energy released during nuclear reactions is much greater than the energy released during chemical reactions. (Page 198)

7.10 Concept:
There are inherent risks associated with radioactivity and the use of radioactive isotopes. Risks can include:

(a) biological exposure (Page 205);

(b) longterm storage and disposal of isotopes (Page 205);

(c) nuclear accidents (Page 205).

7.11 Concept:

Radioactive isotopes have many beneficial uses. Radioactive isotopes are used in research, medicine, and industry for such applications as:

(a) radioactive dating (Pages 196–197);

(b) tracing chemical and biological processes (Pages 196–197);

(c) industrial measurement (Pages 196–197);

(d) nuclear power (Page 198);

(e) detection and treatment of disease (Pages 196–197).

Skill:

Identify specific uses of common radioactive isotopes such as:

(a) I-131 in diagnosing and treating thyroid disorders (Pages 196–197);

(b) Co-60 in treating cancer (Pages 196–197);

(c) the C-14 to C-12 ratio to date organic material (Pages 196–197);

(d) the U-238 to Pb-206 ratio to date geological formations (Pages 196–197).

7.1 NUCLEAR PARTICLES

In Chapter 2 we discussed and defined the important terms *nucleon, atomic number, mass number,* and *isotope*. In addition, we used a symbol such as $^{14}_{6}C$ to describe a particular atomic nucleus (or **nuclide**). You should review this material now.

As we study nuclear chemistry, we will see that certain nuclear particles appear frequently in nuclear processes. For this reason, the special names and symbols are listed in Reference Table O in Appendix 1.

7.2 NUCLEAR EQUATIONS

A nuclear reaction is a change that occurs within or among atomic nuclei and is represented by a *nuclear equation* such as the following:

$$^{15}_{7}N + {}^{1}_{1}H \rightarrow {}^{12}_{6}C + {}^{4}_{2}He$$

If we examine this equation carefully, we note that the sum of the atomic numbers on the left side $(7 + 1)$ equals the sum of the atomic numbers on the right side $(6 + 2)$. This equality demonstrates the fact that electric charge must be conserved in a nuclear reaction. Similarly, the sum of the mass numbers on the left side of the equation (15 + 1) equals the sum of the mass numbers on the right side (12 + 4).

This nuclear equation is considered to be *balanced* because both charge and mass number are conserved. All of the nuclear equations that we write will be balanced equations.

7.3 NATURAL RADIOACTIVITY AND RADIOACTIVE DECAY

At the end of the nineteenth century, the French scientist Henri Becquerel discovered the phenomenon of **natural radioactivity**, that is, the spontaneous disintegration of an atomic nucleus. When the nucleus disintegrates, the process is accompanied by the emission of subatomic particles and/or photons such as X-rays. During this process, one element may change to another; this change is known as **transmutation**.

A large number of elements have radioactive isotopes, such as hydrogen-3 (tritium) and carbon-14, in addition to their stable isotopes, such as hydrogen-1 and carbon-12. However, *elements whose atomic numbers are greater than 83 have no known stable isotopes*.

The earliest radioactive emanations were identified as positively charged alpha particles, negatively charged beta particles, and uncharged gamma radiation that is similar to X-rays but far more energetic. (These are listed in Reference Table O in Appendix 1.) Each of these emanations is different in mass (alpha particles being the most massive) and penetrating power (gamma radiation is the most penetrating; alpha particles are the least).

Why is a particular nuclide stable or radioactive? The answer is related, in part, to the *ratio of neutrons to protons* within the nucleus of the nuclide. Radioactive nuclei break down naturally by a series of processes, known collectively as **radioactive decay** or **natural transmutation**, in order to become more stable; radioactive decay is an attempt to "correct" the ratio of neutrons to protons. However, stability may not occur immediately; a series of decay reactions may be required before a stable nucleus is finally produced.

Alpha Decay

The following nuclear reaction:

$$^{238}_{92}\text{U} \rightarrow {}^{234}_{90}\text{Th} + {}^{4}_{2}\text{He}$$

is an example of *alpha decay*. The uranium-238 nucleus (the *parent nucleus*) breaks down to produce a thorium-234 nucleus (the *daughter nucleus*) and a helium-4 nucleus (an *alpha particle*).

Whenever alpha decay occurs, the atomic number of the daughter nucleus (as compared with that of its parent) is *decreased* by 2 and its mass number is *decreased* by 4. Many heavier radioactive nuclei, especially those with atomic numbers greater than 83, undergo alpha decay as a way of reducing the numbers of protons and neutrons present.

PROBLEM

The nuclide $^{220}_{87}$ Fr undergoes alpha decay. Write a balanced equation that illustrates this process.

SOLUTION

$^{220}_{87}$ Fr $\rightarrow$ $^{216}_{85}$ At $+$ $^{4}_{2}$ He

TRY IT YOURSELF

When a certain radioactive nuclide undergoes alpha decay, $^{218}_{84}$ Po is formed as a daughter nucleus. Identify the parent nuclide.

ANSWER

$^{222}_{86}$ Rn

Beta (−) Decay

Certain nuclei undergo radioactive decay and produce an *electron* [a *beta (−) particle*] in the reaction. In *beta (−) decay,* the atomic number of the daughter nucleus is *increased* by 1 while the mass number remains *unchanged*. Beta (−) decay serves to *lower* the neutron to proton ratio. The following equation illustrates the process of beta (−) decay:

$$^{234}_{90} \text{Th} \rightarrow \,^{234}_{91} \text{Pa} + \,^{0}_{-1}\text{e}$$

PROBLEM

The nuclide $^{32}_{15}$ P undergoes beta (−) decay. Write a balanced nuclear equation that illustrates this process.

SOLUTION

$$^{32}_{15}\text{P} \rightarrow \,^{32}_{16}\text{S} + \,^{0}_{-1}\text{e}$$

TRY IT YOURSELF

Write a balanced nuclear equation that shows how the daughter nucleus $^{14}_{7}$N is produced by beta (−) decay.

ANSWER

$$^{14}_{6}\text{C} \rightarrow \,^{14}_{7}\text{N} + \,^{0}_{-1}\text{e}$$

7.4 HALF-LIFE

Another method of measuring radioactivity involves the use of a quantity called *half-life*. The **half-life** is the time required for a substance to decay to one-half of its initial value. (Half-lives are not limited to radioactive substances; they are used also to measure the persistence of medicines in the human body and of pesticides in soil.)

As an example, consider the radioactive isotope (*radioisotope*, for short) iodine-131, whose half-life is approximately 8 days. If we have a 16-milligram sample of this isotope, its decay over a period of time occurs as follows:

$$16 \text{ mg} \xrightarrow{8d} 8 \text{ mg} \xrightarrow{8d} 4 \text{ mg} \xrightarrow{8d} 2 \text{ mg} \xrightarrow{8d} 1 \text{ mg} \ldots$$

After 32 days of decay, 1 milligram of iodine-131 remains unchanged. The other 15 milligrams have not simply disappeared; they have been transformed into other substances. Moreover, the isotope will *never* decay to 0 milligram; it will simply come closer and closer to this value.

The half-life is constant: it cannot be altered by changes in pressure, temperature, or chemical combination. It depends solely on the nature of the radioactive nucleus. Half-lives range from fractions of a second (fermium-244, half-life = 3.3×10^{-3} s) to *billions* of years (uranium-238, half-life = 4.51×10^9 y). Reference Table N in Appendix 1 gives the half-lives of a number of radioactive isotopes.

The accompanying graph shows how a 16-milligram sample of iodine-131 decays over a period of time. The shape of this graph is characteristic of *all* radioactive decay processes.

Mass of Iodine 131 versus Time

PROBLEM

According to the graph shown on the previous page, approximately how much of the radioisotope iodine-131 remains after 20 days of decay?

SOLUTION

Inspection of the graph shows that approximately 3 mg of the radioisotope remains.

TRY IT YOURSELF

According to the graph, how long must the radioisotope iodine-131 decay so that 12 milligrams remain unchanged?

ANSWER

Approximately 3 d

PROBLEM

According to Reference Table N in Appendix 1, how much of a 100.-microgram (μg) sample of nitrogen-16 will remain after 28.8 seconds of decay?

SOLUTION

According to Reference Table N, the half-life of nitrogen-16 is 7.2 s. Since 28.8 s represents four half-lives (28.8/7.2), the decay is as follows:

$$100. \,\mu g \xrightarrow{7.2\ s} 50.0 \,\mu g \xrightarrow{7.2\ s} 25.0 \,\mu g \xrightarrow{7.2\ s} 12.5 \,\mu g \xrightarrow{7.2\ s} 6.25 \,\mu g$$

After 28.8 s, 6.25 μg of nitrogen-16 will remain.

7.5 USES OF RADIOISOTOPES

Radioactive isotopes can be used in a variety of applications that depend on the properties of the particular isotope. Because radioisotopes (such as carbon-14) are chemically similar to their stable counterparts (such as carbon-12), they can be used as *tracers* to follow the course of a reaction. This application is particularly useful in determining how certain biochemical reactions occur.

Radioisotopes are used also in a variety of medical applications that depend on the isotope's radioactivity and short half-life. These applications include medical diagnoses, such as the use of iodine-131 in uncovering thyroid disorders and technetium-99m in performing bone scans. A short half-life is necessary to ensure rapid decay and elimination of the radioisotope from the body. Also, radioisotopes such as cobalt-60 and iodine-131 are used in treating certain cancerous tumors that are considered inoperable.

Certain radioisotopes have been used to determine the age of a sample of material. For example, carbon-14 has a half-life close to 5,700 years. This radioisotope is produced in the upper atmosphere and is incorporated into carbon dioxide (CO_2) along with nonradioactive carbon-12. Plants, such as trees, take in the CO_2 and incorporate it into the body of the plant (wood, in the case of a tree). As long as the tree is alive, the ratio of carbon-14 to carbon-12 remains constant. When the tree dies or is cut down, however, the carbon-14 decays and is not replaced. Therefore, the ratio of carbon-14 to carbon-12 becomes smaller as time progresses. By comparing the ratio of carbon-14 to carbon-12 in a wooden bowl to the ratio of these isotopes in the atmosphere, scientists can determine the age of the bowl with considerable accuracy. To determine the ages of geological formations, radioactive isotopes with much longer half-lives, such as uranium-238 and lead-206, are needed for the ratios.

7.6 INDUCED NUCLEAR REACTIONS

All of the nuclear reactions we have studied so far have been *natural transmutations*. We now turn our attention to nuclear changes that have been produced artificially, or *induced*. To induce a nuclear reaction, a target nucleus is bombarded with a nuclear particle.

Artificial Transmutation

The first induced nuclear reactions used alpha particles (because of their large masses) as the bombarding particles. In 1919, Rutherford bombarded nitrogen-14 nuclei with alpha particles to produce a reaction in which the nitrogen-14 was artificially changed or *transmuted* into oxygen-17:

$$^{14}_{7}N + ^{4}_{2}He \rightarrow ^{17}_{8}O + ^{1}_{1}H$$

This process is known as **artifical transmutation**.

In another artificial transmutation, English physicist Sir James Chadwick bombarded beryllium-9 and identified a stream of uncharged particles that we now call *neutrons*:

$$^{9}_{4}Be + ^{4}_{2}He \rightarrow ^{1}_{0}n + ^{12}_{6}C$$

Two years later, in 1934, French physicists Frédéric Joliot-Curie and Irène Joliot-Curie bombarded aluminum-27 and produced the first artificially radioactive isotope, phosphorus-30:

$$^{27}_{13}Al + ^{4}_{2}He \rightarrow ^{1}_{0}n + ^{30}_{15}P$$

The phosphorus-30 undergoes positron decay (see page 200) and forms silicon-30.

Nuclear Energy

In the bombardment reactions we have studied so far, only minor changes to the target nuclei occurred. We will now discuss processes that cause major nuclear changes and produce a great deal of energy. Originally, applications of these reactions were used in warfare. More recently, however, the reactions have been applied to generate electrical energy.

The origin of this *nuclear energy* is the conversion of a small amount of matter into radiant energy. If the masses of the reactants and the products of an energy-producing nuclear reaction are measured very precisely, the products are found to have *less* mass than the reactants. This *mass defect* has been converted into an equivalent amount of energy. This conversion is a consequence of Einstein's famous formula ($E = mc^2$), which states that mass and energy can be transformed into one another. The two principal reactions used as sources of energy are nuclear *fission* and nuclear *fusion*.

Nuclear Fission

In the 1930s, Enrico Fermi, an Italian-American physicist, suggested that neutrons be used in bombardment reactions because they are uncharged particles and therefore would not be repelled by the target nuclei. In 1938, German physicists Otto Hahn and Fritz Strassman bombarded uranium atoms with neutrons and discovered that some of the uranium atoms split into two roughly equal fragments. This reaction, known as **nuclear fission**, also produced more neutrons and a large amount of energy.

There are many types of fission reactions; one type is shown below:

$$^{235}_{92}U + ^{1}_{0}n \rightarrow \left[^{236}_{92}U\right] \rightarrow ^{141}_{56}Ba + ^{92}_{36}Kr + 3^{1}_{0}n + \text{energy}$$

The enormous amount of energy released is due to the fact that the combined mass of the reactants (uranium-235 and the neutron) is considerably greater than the combined mass of the products (barium-141, krypton-92, and the 3 neutrons). In addition, the 3 neutrons that are released can be used to cause other uranium-235 nuclei to fission, producing, in turn, 9 neutrons, then 27 neutrons, ... If the fission process is allowed to continue in this manner, an *uncontrolled chain reaction* will occur and the resulting explosion will release a great deal of *destructive* energy. This is the stuff of which fission bombs are made.

SECTION II—ADDITIONAL MATERIAL

7.1A THE URANIUM-238 DECAY SERIES

As we noted previously, radioactive decay occurs in order to produce stable nuclei. Sometimes, this process requires more than one step. The unstable nuclide $^{238}_{92}$U decays to the stable nuclide $^{206}_{82}$Pb in a series of steps involving both alpha and beta $(-)$ decay.

The accompanying graph indicates how the decay of uranium-238 occurs. Atomic numbers are plotted along the x-axis, and mass numbers along the y-axis. Each diagonal line represents an alpha decay; each horizontal line, a beta $(-)$ decay. The daughter nucleus at each step (atomic number and mass number) is indicated by a dot (•).

PROBLEM
How does the nuclide $^{214}_{82}$Pb decay?

SOLUTION
We locate the dot corresponding to $^{214}_{82}$Pb on the graph given above and then follow the (horizontal) arrow to the next dot, which corresponds to $^{214}_{83}$Bi. Therefore, $^{214}_{82}$Pb decays to $^{214}_{83}$Bi by beta $(-)$ decay.

TRY IT YOURSELF
How many alpha and beta decays are there in the uranium-238 decay series?

ANSWER
According to the graph, the series has eight alpha decays and six beta $(-)$ decays.

Positron (beta plus) Decay

The name *positron* is a combination of the word parts *posi*tive elec*tron*. The positron is the *antiparticle* (i.e., the *antimatter* counterpart) of the electron. The following equation is an example of *positron decay*:

$$^{19}_{10}\text{Ne} \rightarrow {}^{19}_{9}\text{F} + {}^{0}_{+1}\text{e}$$

In positron decay, the atomic number of the daughter nucleus is *decreased* by 1 while the mass number remains *unchanged*. Positron decay *increases* the neutron to proton ratio.

Electron Capture

This type of reaction occurs when a nucleus captures one of the inner electrons of an atom. The following equation is an example of *electron capture:*

$$^{7}_{4}\text{Be} + {}^{0}_{-1}\text{e} \rightarrow {}^{7}_{3}\text{Li}$$

Electron capture also serves to *increase* the neutron to proton ratio.

7.2A ISOMERIC TRANSITION

The nucleus of an atom contains energy levels. Occasionally, a nucleus will enter an excited state, known as a *metastable* or *mesomeric* state. We symbolize a metastable nucleus by adding *m* to its mass number (e.g., 99m). Eventually, as the metastable nucleus returns to its normal state, a high-energy gamma photon is emitted. The following equation is an example of *isomeric transition*:

$$^{99\text{m}}_{43}\text{Tc} \rightarrow {}^{99}_{43}\text{Tc} + \gamma$$

The nuclides ^{99}Tc and $^{99\text{m}}\text{Tc}$ are known as *nuclear isomers*.

Reference Table N in Appendix 1 of this book lists the modes of decay for selected radioactive isotopes.

7.3A DETECTION AND MEASUREMENT OF RADIOACTIVITY

Charged nuclear particles may be separated by electric and magnetic fields. For example, in an electric field (produced by oppositely charged parallel plates) alpha particles are deflected toward the negative plate, and beta $(-)$ particles toward the positive plate. Neutrons and gamma ray photons, being uncharged, are not deflected.

Radioactivity may be detected by a number of devices such as photographic film, scintillation counters, electroscopes, bubble and cloud chambers, and (perhaps the best known) the Geiger counter. The tube in the Geiger counter contains a gas that is ionized (electrically charged) by incoming radioactive particles. Each charged gas particle creates a pulse of electric current that produces an audible "click" and/or is counted electronically. The rate of the recorded counts or the frequency of the audible clicks is a measure of the degree of radioactivity present. The details of the other detection devices are described in standard chemistry texts.

Rate of Radioactive Decay

The radioactivity of a substance may be *measured* by the number of disintegrations taking place per unit of time. A substance has an activity of 1 becquerel (Bq) if it experiences one disintegration per second. An older unit, the curie (Ci), was originally based on the activity of 1 gram of radium. One curie equals 3.7×10^{10} becquerels.

7.4A SOLVING RADIOACTIVE DECAY PROBLEMS

Radioactive decay problems are easily solved if the half-life of the radioactive material is given. We begin by considering a simple problem.

PROBLEM
Supposed that 256 mg of iodine-131 (half-life = 8.0 days) decays for 40 days.
(a) After 40. days, what fraction of the I-131 remains?
(b) What mass of I-131 is present after 40. days of decay?

SOLUTION

(a) We can calculate the number of half-life periods that 40. days represents. if t is the decay period and $t_{0.5}$ is the nuclide's half-life:

$$\text{number of half-life periods} = \frac{t}{t_{0.5}}$$

$$\frac{t}{t_{0.5}} = \frac{40.\,\text{d}}{8.0\,\text{d}} = 5.0$$

For each half-life period, the amount of I-131 is reduced by one-half. Therefore, in 5.0 half-life periods (for 40. days), the fraction of I-131 that remains is as follows:

$$1 \xrightarrow{1} \frac{1}{2} \xrightarrow{2} \frac{1}{4} \xrightarrow{3} \frac{1}{8} \xrightarrow{4} \frac{1}{16} \xrightarrow{5} \frac{1}{32}(0.03125)$$

After 5.0 half-life periods, only $\dfrac{1}{32}$ or 0.03125 of the original amount of I-131 remains.

(b) We can calculate the amount remaining by multiplying the original amount of I-131 by the fraction remaining after 5.0 half-life periods:

$$\text{amount remaining} = 256\,\text{mg} \cdot 0.03125 = 8.0\,\text{g}$$

We can also use the radioactive decay formula. If A_0 = the original amount of a radioactive nuclide, t = the amount of time that the nuclide decays, and $t_{0.5}$ = the nuclide's half-life:

$$\text{fraction remaining} = \frac{1}{2}^{\frac{t}{t_{0.5}}}$$

$$\text{amount remaining} = A_0 \cdot \frac{1}{2}^{\frac{t}{t_{0.5}}}$$

We can easily figure out the answer to the previous problem using a graphing calculator such as the TI-83 or TI-84. To calculate the fraction remaining in part (a), key in the following strokes in order:

| 0 | . | 5 | 2nd | ^ | (| 4 | 0 | ÷ | 8 |) | ENTER |

To calculate the amount remaining in part (b), key in the following strokes in order:

| 2 | 5 | 6 | × | 2nd | [ANS] | ENTER |

Using these formulas provides a real bonus — one does not have to use "easy" numbers. For example, consider the following problem.

PROBLEM
The half-life of Kr-85 is 10.73 years. How much of a 412.2-gram sample will remain after 39.22 years of decay?

SOLUTION
Use a calculator:

$$\text{amount remaining} = 412.2 \text{ g} \cdot 0.5^{\left(\frac{39.22 \text{ y}}{10.73 \text{ y}}\right)} = \mathbf{32.72 \text{ g}}$$

TRY IT YOURSELF
Use a calculator:

The half-life of P-32 is 14.28 days. How much of a 0.6768-gram sample will remain after 128.2 days of decay?

ANSWER
0.001343 grams of P-32 remains.

7.5A PARTICLE ACCELERATORS

A *particle accelerator* is a device that uses electric and magnetic fields to provide a (charged) bombarding nuclear particle with sufficient energy to induce a desired nuclear reaction. As an analogy, consider a bullet fired at a wall at a speed of 10 miles per hour. At this slow speed, the energy of the bullet will have hardly any effect on the wall. If the bullet were fired at 1000 miles per hour, however, its effect on the wall would be devastating!

Examples of modern particle accelerators include the *Van de Graaff accelerator*, the *linear accelerator*, the *cyclotron*, the *synchrotron*, and the *large hadron collider* (*LHC*). Detailed descriptions of these devices, which can supply bombarding particles with a range of energies, can be found in standard chemistry and physics texts.

7.6A FISSION REACTORS

A number of physicists recognized that in a *controlled* chain reaction, the fission could provide a continuous source of useful energy. This is the principle behind the *fission reactor*. Fission reactors are used in the production of electrical energy, as research tools, and as a means of producing other radioactive isotopes. The accompanying diagram illustrates a typical fission reactor used to produce electricity.

The primary system of a fission reactor has the following components:

- The *containment vessel* (concrete and steel) provides shielding for the reactor.
- The *fuel rods*, located in the *core*, serve as sources of energy. Uranium-233 and uranium-235 are used as fuels.
- The *moderator*, also located in the core, slows the neutrons so that they will be absorbed by the fuel nuclei. These slow neutrons have kinetic energies that are close to the energies of air molecules at room temperature. For this reason, they are known as *thermal neutrons*. Moderators usually consist of water (containing either hydrogen-1 or hydrogen-2), graphite, or beryllium.
- The *control rods* in the core regulate the rate of fission by absorbing neutrons. Control rods are usually made of cadmium or boron.
- The *coolant* (water or liquid sodium) removes thermal energy from the core.
- The *heat exchanger* receives the thermal energy and produces steam for the generation of electrical energy by the secondary system (turbine) of the reactor.

The most common isotope of uranium, uranium-238, comprises about 99 percent of the naturally occurring element but does *not* undergo fission. Although the isotope uranium-235 does undergo fission, it represents less than 1 percent of the naturally occurring element. For this reason, it is necessary to *enrich* uranium (usually to 3–4%) in order to use it as a fissionable fuel.

When uranium-238 absorbs a neutron, it is converted into uranium-239, which then decays in two steps to form plutonium-239:

$$^{238}_{92}U + ^1_0n \rightarrow ^{238}_{92}U$$

$$^{238}_{92}U \rightarrow ^{239}_{93}Np + ^0_{-1}e$$

$$^{239}_{93}Np \rightarrow ^{239}_{94}Pu + ^0_{-1}e$$

Plutonium-239 *does* fission, and it has been suggested that *breeder reactors*, which would be able to produce their own fuel, be constructed. No breeder reactors have been operated commercially in the United States, however, because of the serious health and environmental hazards attributable to plutonium and also because of the possibility that the plutonium might be used to produce nuclear weapons.

A number of problems are associated with any fission reactor. For example, the heat energy produced by the reactor contributes to *thermal pollution*. There is also the serious problem of *radioactive waste disposal*. Solid and liquid wastes, such as cesium-237 and strontium-90, are placed in corrosion-resistant containers and stored in isolated underground areas. Wastes with low levels of radioactivity may be diluted until they are considered harmless and then released into the environment. Gaseous wastes, such as krypton-85m, nitrogen-16, and radon-222, are stored and allowed to decay until it is considered safe to release them into the atmosphere.

Nuclear reactors are also used to produce radioactive isotopes for a variety of applications. For example, when the stable isotope $^{59}_{27}Co$ is bombarded with neutrons from a reactor, the radioisotope $^{60}_{27}Co$ is produced. As noted in Section 15.5, this radioisotope is used in cancer therapy.

Nuclear Fusion

The joining of light nuclei to form heavier, more stable nuclei is known as **nuclear fusion** and is the process by which stars, including our own Sun, produce their energy. In stars, fusion consists of a series of reactions that depend on the temperature of the particular star. The *net* fusion reaction in our Sun is as follows:

$$4^1_1H \rightarrow ^4_2He + 2^0_{+1}e + 2\gamma + 2v + \text{energy}$$

where γ represents a gamma photon, and v represents another subatomic particle called a *neutrino*.

For a fusion reaction to occur, very high temperatures are needed to give the positively charged nuclei the energy they require to overcome their mutual repulsion. For this reason, all fusion devices are referred to as *thermonuclear devices*.

For a given mass of fuel, nuclear fusion yields more energy than nuclear fission and there are fewer pollution problems with the products of fusion reactions. For this reason, work is being done to develop fusion reactors as a means of producing power. Four fusion reactions under investigation are as follows:

$$_1^2H + {}_1^2H \rightarrow {}_1^3H + {}_1^1H + \text{energy}$$

$$_1^2H + {}_1^2H \rightarrow {}_2^4He + \text{energy}$$

$$_1^2H + {}_1^2H \rightarrow {}_2^3He + {}_0^1n + \text{energy}$$

$$_1^2H + {}_1^3H \rightarrow {}_2^4He + {}_0^1n + \text{energy}$$

At this time, however, no successful fusion reactor has been constructed. The problems to be overcome include the production of a high *ignition* temperature, the packing of nuclei into a space small enough to allow a sufficient number of collisions, the control of the fusion reaction once it has begun, and the development of materials that can withstand the high operating temperatures and radiation levels of the reactor.

END-OF-CHAPTER QUESTIONS

Some questions have the symbol "§2" in front of the question number. This symbol means that the question is based on Section II material.

1. Which particle is electrically neutral?
(1) proton   (2) positron   (3) neutron   (4) electron

2. Isotopes of the same element have the same number of
(1) protons, neutrons, and electrons
(2) protons, but different numbers of neutrons
(3) neutrons, but different numbers of protons
(4) electrons, but different numbers of protons

3. Which particle has the same mass as an electron, but a positive electric charge?
(1) alpha particle   (2) gamma photon
(3) proton   (4) positron

4. The total number of protons and neutrons in the nuclide ^{37}Cl is
(1) 54   (2) 37   (3) 20   (4) 17

5. In the reaction

$$^6_3\text{Li} + ^1_0\text{n} \rightarrow ^4_2\text{He} + X$$

the species represented by X is

(1) ^2H (2) ^3H (3) ^3He (4) ^4He

6. In the reaction

$$^9_4\text{Be} + ^1_1\text{H} \rightarrow ^4_2\text{He} + X$$

which species is represented by X?

(1) ^8Li (2) ^6Li (3) ^8B (4) ^{10}B

7. Given the reaction

$$^{27}_{13}\text{Al} + ^4_2\text{He} \rightarrow X + ^1_0\text{n}$$

When the equation is correctly balanced, the nucleus represented by X is

(1) ^{30}Al (2) ^{30}Si (3) ^{30}P (4) ^{30}S

8. In the reaction

$$^{99\text{m}}\text{Tc} \rightarrow ^{99}\text{Tc} + X$$

X represents

(1) an alpha particle   (2) a beta particle
(3) a positron   (4) gamma radiation

9. In the equation

$$^{39}_{19}\text{K} \rightarrow ^{39}_{20}\text{Ca} + X$$

which particle is represented by X?

(1) a proton   (2) an electron   (3) a positron   (4) a deuteron

10. The nuclear reaction

$$^{14}_6\text{C} \rightarrow ^{14}_7\text{N} + ^0_{-1}\text{e}$$

is an example of

(1) nuclear fusion   (2) nuclear fission
(3) natural transmutation   (4) artificial transmutation

11. Which element has no known stable isotope?

(1) C   (2) K   (3) Po   (4) P

12. When uranium-238 undergoes alpha decay, which daughter nucleus is formed?

(1) ^{234}Th (2) ^{238}Np (3) ^{206}Pb (4) ^{233}U

13. The diagram below represents radiation passing through an electric field.

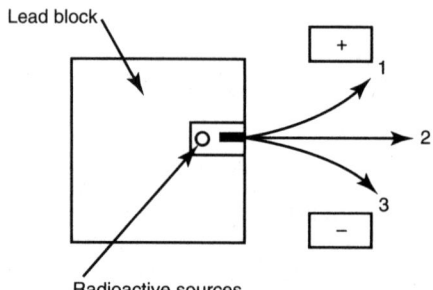

Lead block

+
1

2

3

−

Radioactive sources

The arrow marked 2 most likely represents
(1) an alpha particle   (2) a beta (−) particle
(3) gamma radiation   (4) a proton

14. When a stream of radioactive particles is passed through a pair of oppositely charged parallel plates, which particle would be deflected toward the *negative* plate?
(1) alpha particle   (2) beta (−) particle
(3) gamma photon   (4) neutron

15. After 62.0 hours, 1.0 gram remains unchanged from a sample of ^{42}K. How much ^{42}K was in the original sample?
(1) 8.0 g   (2) 16 g   (3) 32 g   (4) 64 g

16. What mass of a 16-gram sample of cobalt-60 will remain unchanged after 15.78 years?
(1) 1.0 g   (2) 2.0 g   (3) 8.0 g   (4) 4.0 g

17. How many days are required for phosphorus-32 to undergo three half-life periods?
(1) 4.77   (2) 14.3   (3) 28.6   (4) 42.9

18. In 6.20 hours, a 100.-gram sample of an isotope decays to 25.0 grams. What is the half-life, in hours, of this isotope?
(1) 1.60   (2) 3.10   (3) 6.20   (4) 12.4

19. How much of an 8-gram sample of radium-226 will remain unchanged at the end of two half-life periods?
(1) 1 g   (2) 2 g   (3) 3 g   (4) 4 g

20. At the end of 12 days, 1/8 of an original sample of a radioactive element remains. What is the half-life of the element?
(1) 36 days   (2) 48 days   (3) 3 days   (4) 4 days

Base your answers to questions 21–23 on the following graph, which represents the decay curve of a certain radioisotope.

Activity versus Time

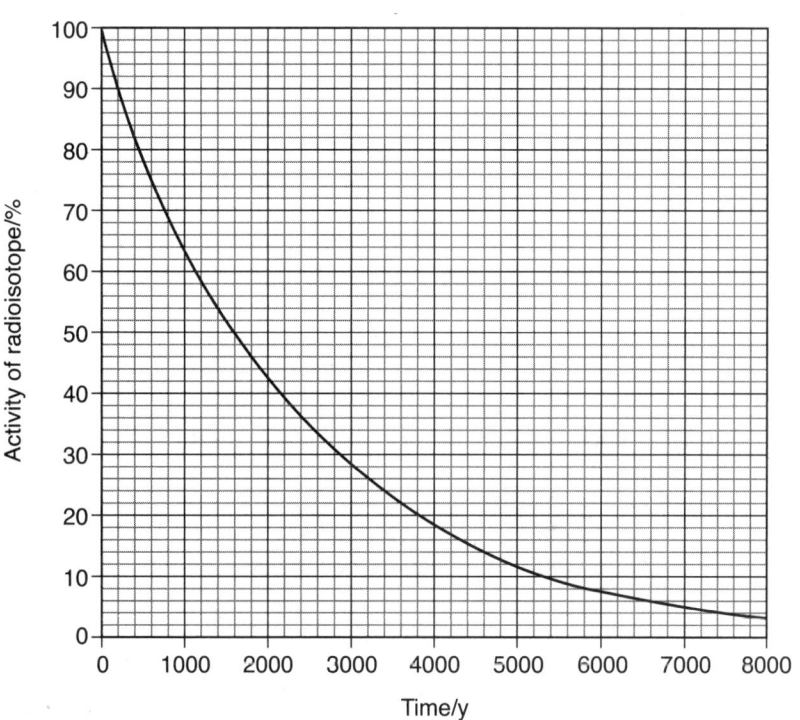

Time/y

21. After approximately how many years will 70 percent of the radioisotope's activity remain?
(1) 500   (2) 800   (3) 2,800   (4) 8,000

22. On the basis of the graph and Reference Table N, this radioisotope is most likely
(1) ^{14}C (2) ^{137}Cs (3) ^{226}Ra (4) ^{232}Th

23. If the temperature at which the radioisotope decayed had been higher, the half-life of the isotope would have been
(1) less (2) greater (3) the same

24. For which application is the ratio of uranium-238 to lead-206 used?
(1) diagnosing thyroid disorders
(2) dating geologic formations
(3) detecting brain tumors
(4) treating cancer patients

25. Which isotopic ratio must be determined when the age of an ancient wooden object is investigated?
(1) ^{235}U to ^{238}U (2) ^{2}H to ^{3}H
(3) ^{16}N to ^{14}N (4) ^{14}C to ^{12}C

26. Radioisotopes used in medical diagnosis should have
(1) short half-lives and be quickly eliminated from the body
(2) short half-lives and be slowly eliminated from the body
(3) long half-lives and be quickly eliminated from the body
(4) long half-lives and be slowly eliminated from the body

27. A radioisotope is called a *tracer* when it is used to
(1) kill bacteria in food
(2) kill cancerous tissue
(3) determine the age of animal skeletal remains
(4) determine the way in which a chemical reaction occurs

Base your answers to questions 28–30 on the nuclear equation given below, which represents an experiment in which aluminum-27 was bombarded by high-energy alpha particles.

$$^{27}_{13}Al + ^{4}_{2}He \rightarrow ^{1}_{0}X + Y$$

28. Particle Y is an isotope of
(1) Al   (2) C   (3) P   (4) Si

29. Particle X is
(1) a proton   (2) an electron (3) a positron (4) a neutron

30. This nuclear reaction may be classified as
(1) artificial transmutation
(2) nuclear fission
(3) nuclear fusion
(4) natural radioactive decay

31. Which particle *cannot* be accelerated by the electric or magnetic fields in a particle accelerator?
(1) neutron
(2) proton
(3) alpha particle
(4) beta (−) particle

32. An isotope of which element may be used as a fuel in a fission reaction?
(1) H
(2) C
(3) Li
(4) U

§2 **33.** Which substance is used in the control rods of a nuclear reactor?
(1) B
(2) He
(3) CO_2
(4) Be

§2 **34.** The main purpose of a moderator in a fission reactor is to
(1) emit neutrons
(2) absorb neutrons
(3) slow down neutrons
(4) speed up neutrons

§2 **35.** The atoms of some elements can be made radioactive by
(1) placing them in a magnetic field
(2) bombarding them with high-energy particles
(3) separating them into their isotopes
(4) heating them to a very high temperature

§2 **36.** Particle accelerators are used primarily to
(1) detect radioactive particles
(2) identify radioactive particles
(3) increase a particle's kinetic energy
(4) increase a particle's potential energy

§2 **37.** Which is a gaseous radioactive waste produced during some fission reactions?
(1) nitrogen-16
(2) thorium-232
(3) uranium-235
(4) plutonium-239

38. In a fusion reaction, a major problem in causing the nuclei to fuse into a single nucleus is the
(1) small mass of the nuclei
(2) large mass of the nuclei
(3) attractions of the nuclei
(4) repulsions of the nuclei

39. In the reaction $^{239}_{93}\text{Np} \rightarrow {}^{239}_{94}\text{Pu} + X$, what does X represent?
(1) a neutron
(2) a proton
(3) an alpha particle
(4) a beta particle

40. Which of these types of nuclear radiation has the greatest penetrating power?
(1) alpha   (2) beta   (3) neutron   (4) gamma

41. What occurs in both fission and fusion reactions?
(1) Small amounts of energy are converted into large amounts of matter.
(2) Small amounts of matter are converted into large amounts of energy.
(3) Heavy nuclei are split into lighter nuclei.
(4) Light nuclei are combined into heavier nuclei.

42. Given the reaction:

$$^{27}_{13}Al + ^{4}_{2}He \rightarrow X + ^{1}_{0}n$$

Which particle is represented by X?

(1) $^{28}_{12}Mg$ (2) $^{28}_{13}Al$ (3) $^{30}_{14}Si$ (4) $^{30}_{15}P$

43. A radioactive isotope has a half-life of 2.5 years. Which fraction of the original mass remains unchanged after 10. years?

(1) $\dfrac{1}{2}$ (2) $\dfrac{1}{4}$ (3) $\dfrac{1}{8}$ (4) $\dfrac{1}{16}$

44. Compared to the half-life and decay mode of the nuclide ^{90}Sr, the nuclide ^{226}Ra has
(1) a longer half-life and the same decay mode
(2) a longer half-life and a different decay mode
(3) a shorter half-life and the same decay mode
(4) a shorter half-life and a different decay mode

Constructed-Response Questions

1. In a laboratory experiment, a student investigated the activity of a radioisotope. The table below represents the data she took.

Time/min	0	100	200	300	400	500	600	700	800	900	1000
Activity Remaining/%	100	64	45	29	19	12	9	6	5	4	1

(a) Plot the data on the graph given below, and draw a *best-fit* radioactive decay curve using the data points.

Activity versus Time

(b) Using the curve you have drawn, determine the half-life of the isotope.

Base your answers to questions 2 through 5 on the reading passage below, the *New York State Regents Reference Tables for Chemistry*, and your knowledge of chemistry.

In the 1920s, paint used to inscribe the numbers on watch dials was composed of a luminescent (glow-in-the-dark) mixture. The powdered paint base was a mixture of radium salts and zinc sulfide. As the paint was mixed, the powdered base became airborne and drifted throughout the workroom. This caused the contents of the workroom, including the painters' clothes and bodies, to glow in the dark.

The paint is luminescent because radiation from the radium salts strikes a scintillator. A scintillator is a material that emits visible light in response to ionizing radiation. In watch dial paint, zinc sulfide acts as the scintillator.

Radium present in the radium salts decomposes spontaneously, emitting alpha particles. These particles can cause damage to the body when they enter human tissue. Alpha particles are especially harmful to the blood, liver, lungs, and spleen because they can alter genetic information in the cells. Radium can be deposited in the bones because it substitutes for calcium.

2. Write the notation for the alpha particles emitted by radium in the radium salts.

3. How can particles emitted from radioactive nuclei damage human tissue?

4. Why does radium substitute for calcium in bones?

5. Explain why zinc sulfide is used in luminescent paint.

Base your answers to questions 6–8 on the information below and on your knowledge of chemistry.

One fission reaction for U-235 is represented by the balanced nuclear equation below:

$$^{235}_{92}U + ^{1}_{0}n \rightarrow ^{140}_{54}Xe + ^{94}_{38}Sr + 2^{1}_{0}n$$

Both radioisotopes produced by this reaction undergo beta decay. The half-life of Xe-140 is 13.6 seconds, and the half-life of Sr-94 is 1.25 minutes.

6. Explain in terms of *both* reactants and products why the reaction represented by the nuclear equation is a fission reaction.

7. In the decay equation of Xe = 140 shown below, what is the missing product?

$$^{140}_{54}Xe \rightarrow ^{0}_{-1}e + \underline{\hspace{1cm}}$$

8. Determine the time required for an original 24.0-gram sample of Sr-94 to decay until only 1.5 grams of the sample remains unchanged.

The answers to these questions are found in Appendix 3.

Chapter Eight

THE ELECTRONIC STRUCTURE OF ATOMS

KEY IDEAS

This chapter continues the development of the atomic model of matter begun in Chapter 2. The Bohr model introduces the idea of quantized electron orbits, and provides an explanation for the emission spectrum of atomic hydrogen. The currently accepted wave-mechanical model extends the placement of electrons within atomic levels, sublevels, and orbitals and is applicable to every element in the Periodic Table.

KEY OBJECTIVES
At the conclusion of this chapter you will be able to:
- Describe the Bohr model and its relationship to atomic spectra.
- Describe the modern wave-mechanical model and its relationship to electron configuration.
- Use the diagonal rule to predict the filling patterns of atoms.
- Define and apply the terms *principal energy level*, *sublevel*, *orbital*, *electron configuration*, *ground state*, *excited state*, *spin state*, *valence electron*, and *Lewis structure* (*electron-dot diagram*).

SECTION I—BASIC
(REGENTS-LEVEL) MATERIAL

NYS REGENTS CONCEPTS AND SKILLS

Note: By the time you have finished Section I, you should have mastered the concepts and skills listed below. The Regents chemistry examination will test your knowledge of these items and your ability to apply them.

Concepts are the *basic ideas* that form the body of the Regents chemistry course (what you need to know!).

Skills are the *activities* that demonstrate your mastery of these concepts (how you show that you know them!).

Following each concept or skill is a page reference (given in parentheses) to this chapter.

8.1 Concept:
Each electron in an atom has its own distinct amount of energy. (Page 220)

8.2 Concept:
When an electron in an atom gains a specific amount of energy, the electron is at a higher energy state (excited state). (Page 221)

Skill:
Distinguish between ground-state configurations (e.g., 2-8-2) and excited-state configurations (e.g., 2-7-3). (Pages 220–221)

8.3 Concept:
When an electron returns from a higher energy state to a lower energy state, a specific amount of energy is emitted. This emitted energy can be used to identify the element. (Page 221)

Skill:
Identify an element by comparing its bright-line spectrum to given spectra. (Page 219)

8.4 Concept:
In the wave mechanical model (electron cloud model), the electrons are in orbitals, which are defined as the regions of most probable electron location (ground state). (Page 222)

8.5 Concept:
The outermost electrons in an atom are called the valence electrons. In general, the number of valence electrons affects the chemical properties of an element. (Pages 222–223)

Skills:
• Draw a Lewis electron-dot structure of an atom. (Page 223)
• Given an electron configuration, such as 2-8-2, distinguish between valence and nonvalence electrons. (Page 223)

8.6 Concept:
Electron dot diagrams (Lewis structures) can represent the valence electron arrangements in elements. (Page 223)

Skill:
Draw the Lewis structures for the elements in Periodic Groups: 1, 2, 13–18. (Page 223)

8.1 INTRODUCTION

In Chapter 2, we examined the development of the atomic models of Dalton, Thomson, and Rutherford. In this chapter, we continue the development of the atomic model into modern times. Our current focus will be on the arrangement of *electrons* within atoms.

8.2 THE BOHR MODEL OF THE ATOM

Rutherford proposed an atomic model in which a positively charged nucleus was surrounded by negative electrons. However, there were serious deficiencies with the Rutherford model's placement of electrons around the nucleus. If the electrons were *stationary*, they should have been drawn into the nucleus, since opposite charges attract one another; if the electrons *orbited* the nucleus (as the planets orbit the Sun), they should have radiated electromagnetic energy continuously. In either case, the atom would collapse! And there was another problem as well....

When an element in the form of a gas at low pressure is excited (by heating or by electricity), it emits visible light. When this light is studied, it is found to be composed of regularly spaced lines. Each element has its own characteristic **visible-line spectrum** by which it can be identified. For example, a portion of the visible-line spectrum of atomic hydrogen is shown in the accompanying diagram. The color of each spectral line is identified by the *wavelength* of the light, which is given in nanometers (nm, 1 nm = 10^{-9} m).

The Rutherford model could not explain why hydrogen (or any other element, for that matter) produced a visible-line spectrum. Did this mean that the Rutherford model had to be abandoned totally? Absolutely not! It meant only that the model needed to be refined.

In 1913, Niels Bohr, a Danish physicist, proposed his own model of the hydrogen atom, a model that answered many of the questions that had confounded Rutherford. In order to do so, however, Bohr had to make a number of assumptions based on the revolutionary ideas first proposed by physicists Max Planck and Albert Einstein.

Bohr proposed a planetary model of the hydrogen atom, as did Rutherford. But in Bohr's model, the electron behaved most unusually:

- The electron could orbit only in certain specified levels, each of which represented a distinct amount of energy. The more energy associated with a level, the farther it was from the nucleus. These **principal energy levels** were said to be *quantized* and were designated by the letters *K, L, M,* and so on. We now designate principal energy levels by numbers—1, 2, 3, 4, and so on.
- The lowest energy level (1 for hydrogen) was called the **ground state**. (Other levels were called **excited states**.) Since the electron could have no lower energy than the ground state, it could not come any closer to the nucleus, and therefore it remained stable.
- In a given orbit, an electron never radiated or absorbed energy. In this case, the electron was said to be in a **stationary state**.
- If an atom absorbed *exactly* the right amount of energy, the electron would rise to a higher energy level. Conversely, if the atom released energy (also in an exact amount), the electron would fall to a lower energy level. The energy that was released appeared as a *photon* of light.

The Periodic Table of the Elements in Appendix 1 lists the ground-state electron configurations of most of the elements, as shown in the accompanying diagram, which is the key to the table:

The electron configuration in the diagram is 2-4, which means that an atom of carbon (C) in the ground state contains two electrons in its first principal energy level and four electrons in its second principal energy level.

TRY IT YOURSELF
The element aluminum has an atomic number of 13. Use the Periodic Table to determine the ground-state electron configuration of an atom of aluminum.

ANSWER

The electron configuration of Al is 2-8-3, that is, two electrons in the first energy level, eight electrons in the second energy level, and three electrons in the third energy level, as shown in the accompanying diagram:

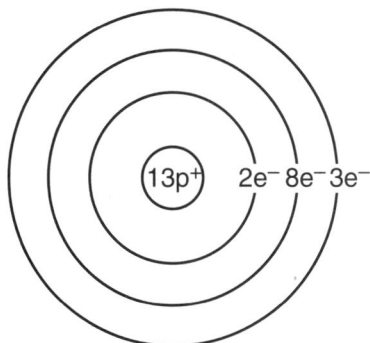

When an atom enters an excited state, one or more electrons are "promoted" to higher levels. For example, the configuration of an excited aluminum atom might be 2-7-4, as shown in the accompanying diagram:

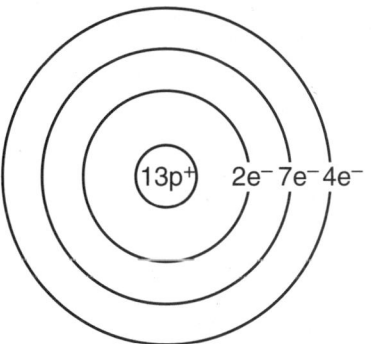

Bohr recognized that *spectral lines are the result of electrons falling from higher energy levels to lower ones.* By applying mathematics to his model, Bohr was able to account precisely for each line in the visible spectrum of hydrogen. This, in itself, provided strong evidence for the correctness of Bohr's model of the hydrogen atom.

Unfortunately, although the Bohr model worked well for hydrogen, it did not work for any other atom—even one with just two electrons (helium). Therefore, once again, the atomic model would have to be modified, hopefully in a way in which the behavior of *all* atoms could be explained. But now, the scientific revolution begun by Planck and Einstein (and continued by Bohr) had attracted a whole new host of players—scientists who would make astounding discoveries about the way nature behaves!

8.3 THE MODERN (WAVE-MECHANICAL) MODEL

The modern model of the atom is based on the premise that an electron (and all other matter) exhibits both particle and wavelike properties, depending on the circumstances. Whereas we can locate a particle with precision, however, we cannot do so in regard to a wave. (For example, we cannot say precisely where a water wave begins or ends.) Therefore, we must find alternative ways of describing an electron. Instead of trying to pinpoint its exact position, we must be content to guess where it is *most probably* located. A very useful analogy is to compare an electron with a cloud. An *electron cloud* is a distribution of probabilities: it is thickest in the regions where we are most likely to find the electron at any given moment and thinnest in the regions where the electron is least likely to be.

This modern model of the atom was developed with a great deal of very high level mathematics well beyond the scope of any high school course. Our goal is to try and understand the model and its applications to the study of chemistry.

In the modern model of the atom, an electron in an atom is described in terms of a concept known as an *atomic orbital*. An **atomic orbital** (**orbital**, for short) may be thought of as the region of space around the nucleus where the probability of locating an electron with a given energy is greatest. Orbitals are associated with characteristic energies, sizes, shapes, and orientations in space.

8.4 VALENCE ELECTRONS

The outermost principal energy level of an atom is important in regard to the way in which simple bonds are formed between atoms. We call this level the *valence level,* and the electrons that occupy the valence level are known as **valence electrons**. If we look through the Periodic Table, or use our diagonal rule, we will discover that valence electrons are found only in the *s* and *p* sublevels. Thus, the number of valence electrons for any element is between one and eight.

PROBLEM
    (a) What is the valence level of sulfur (atomic number 16)?
    (b) How many valence electrons does an atom of sulfur have?

SOLUTIONS
    (a) Referring to the Periodic Table in Appendix 1, we find that the valence level of sulfur is 3.
    (b) An atom of sulfur has six valence electrons.

TRY IT YOURSELF

For an atom of nitrogen (atomic number 7), answer parts (a) and (b) of the problem given on page 222.

ANSWERS

(a) The valence level of nitrogen is 2.

(b) An atom of nitrogen has five valence electrons.

8.5 LEWIS STRUCTURES (ELECTRON-DOT DIAGRAMS)

It would be very useful to be able to see the orbital arrangement of the valence electrons of an element at a glance. Unfortunately, an arrow diagram does not identify the particular element it represents. We can solve this problem by employing a device known as a **Lewis structure** or an **electron-dot diagram**. To draw a Lewis structure, we divide the atom into two parts. The first part, known as the *kernel*, consists of the nucleus plus the inner (nonvalence) electrons of the atom. The second part of the atom is its valence electrons.

(Ground-State) Lewis Electron-Dot Diagrams for Elements 1–20

1. H·	11. Na·
2. He:	12. Mg:
3. Li·	13. A̤l·
4. Be:	14. S̤i·
5. B̈·	15. ·P̈·
6. C̈·	16. :S̈·
7. ·N̈·	17. :C̤l·
8. :Ö·	18. :Är:
9. :F̈·	19. K·
10. :N̈e:	20. Ca:

Note: The positions of the electrons around the symbol of the element is entirely arbitrary. However, the number of paired and unpaired electrons must be correct.

PROBLEM

Describe an atom of carbon (atomic number 6) in terms of the two-part concept explained on the previous page.

SOLUTION

In a carbon atom, the kernel consists of the carbon nucleus and the two electrons in level 1. The valence electrons are the remaining four electrons in level 2.

TRY IT YOURSELF

Answer the problem given above for an atom of sodium (atomic number 11).

ANSWER

The kernel of a sodium atom consists of the sodium nucleus, two electrons in level 1, and eight electrons in level 2. The valence electron is the single electron in level 3.

SECTION II—ADDITIONAL MATERIAL

8.1A ATOMIC ORBITALS AND SUBLEVELS

At this point, it is appropriate to discuss atomic orbitals in more detail. Recall that an atomic **orbital** is the region of space where an electron with a specific energy is most likely to be located. There are many different orbitals, and each orbital is associated with a specific letter: $s, p, d, f, g, \ldots$ Fortunately, we need to be concerned only with s, p, d, and f orbitals. Every atomic orbital, no matter what its designation, can accommodate a maximum of two electrons. An orbital that contains no electron is said to be *empty*; an orbital that contains one electron, to be *half-filled*; an orbital that contains two electrons, to be *filled*.

There is only *one* type of s orbital, and it is present on every principal energy level $(1, 2, 3, \ldots)$. The designation "3s orbital" refers to the s orbital on principal energy level 3. All s orbitals have spherical shapes, as shown in the accompanying diagram:

The p, d, and f orbitals are more complicated. There are *three* types of p orbitals (designated as p_x, p_y, and p_z), and they are located on every energy level *except* level 1. All p orbitals have characteristic "dumbbell" shapes, as shown in the accompanying diagram:

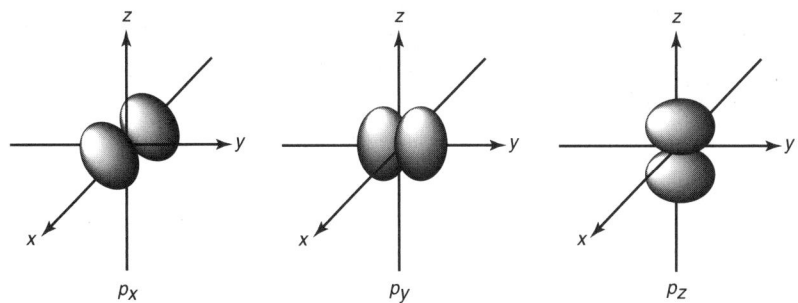

The designation "$4p_y$ orbital" refers to the p_y orbital on energy level 4. There are *five* types of *d* orbitals, and they are located on every energy level *except* levels 1 and 2. Finally, there are *seven* types of *f* orbitals, and they are located on every energy level *except* levels 1, 2, and 3. We shall not be concerned with either the labels or the shapes applied to individual *d* and *f* orbitals because they are too complex.

The three types of *p* orbitals, taken together, constitute a *p* **sublevel** (or **subshell**). Similarly, the five types of *d* orbitals and the seven types of *f* orbitals constitute *d* and *f* sublevels. Therefore, the designation "4*d* sublevel" refers to the five *d* orbitals on energy level 4. The *s* sublevel is peculiar because it contains only the single *s* orbital.

The accompanying table indicates how orbitals and sublevels are arranged on individual principal energy levels.

Principal Energy Level	Type(s) of Sublevel	Number of Orbitals	Maximum Number of Electrons
1	*s*	1	2
Level 1	1 type	1 orbital	2 electrons
2	*s*	1	2
2	*p*	3	6
Level 2	2 types	4 orbitals	8 electrons
3	*s*	1	2
3	*p*	3	6
3	*d*	5	10
Level 3	3 types	9 orbitals	18 electrons
4	*s*	1	2
4	*p*	3	6
4	*d*	5	10
4	*f*	7	14
Level 4	4 types	16 orbitals	32 electrons
SUMMARY			
Level *n* ($n = 1, 2, 3, \ldots$)	*n* types (*s*, *p*, *d*, *f*, others)	n^2 orbitals	$2n^2$ electrons

If we inspect the summary at the bottom of the table, we find that given the number of the level, we can calculate the number of *types* of orbitals, the total number of orbitals, and the maximum number of electrons that are associated with that level.

PROBLEM

Calculate the number of *types* of sublevels, the total number of orbitals, and the maximum number of electrons that are associated with energy level 6.

SOLUTION

On level n, n types of sublevels are present; 6 types of sublevels are present on level 6 ($s, p, d, \ldots$).

On level n, there is a total of n^2 orbitals; on level 6, there is a total of 36 orbitals.

Level n can accommodate a *maximum* of $2n^2$ electrons; level 6 can accommodate a *maximum* of 72 electrons.

8.2A ELECTRON CONFIGURATIONS OF ATOMS

We now build up the atoms for the elements in the Periodic Table.

The basic rule for assigning electrons to atoms is that electrons should occupy the lowest energy states possible. Generally, as the principal energy level number increases, the energy of the electron on that level also increases. Therefore, an electron on level 4 ought to have more energy than an electron on level 3. For atoms with more than one electron (i.e., atoms other than hydrogen), however, this is not always the case, as we will see shortly. Sublevels *within a level* are more predictable. The order of increasing energy is as follows: $s < p < d < f$. Therefore, a $5p$ electron has less energy than a $5d$ electron and more energy than a $5s$ electron.

When we have assigned all of the electrons to a particular atom, the next atom will *repeat* that **electron configuration** and then add one new electron. [This is known as the *Aufbau* ("building-up") principle.]

The Diagonal Rule

To determine the relative energies of sublevels, we will make use of a technique, illustrated below, known as the *diagonal rule*.

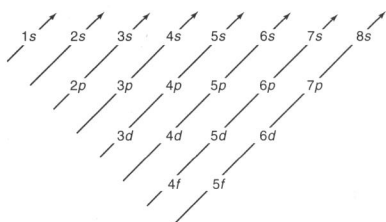

Note in the diagram that each successive line is indented because there are no $1p$, $2d$, or $3f$ sublevels.

The rule is quite simple: We begin at the left and follow each arrow from *tail to head* and then work from *left to right*. The first sublevel we meet is $1s$, and we complete the first arrow. Then we move to the right, where the next sublevel is $2s$. On the third arrow, there are *two* sublevels. Tracing the arrow from tail to head, we meet sublevels $2p$ and $3s$. This series ($1s$, $2s$, $2p$, $3s$) represents the first part of the electron-filling pattern in the Periodic Table.

PROBLEM
Write the electron-filling pattern for the first ten sublevels.

SOLUTION
Using the diagonal rule, we have $1s$, $2s$, $2p$, $3s$, $3p$, $4s$, $3d$, $4p$, $5s$, $4d$.

TRY IT YOURSELF
Complete the electron pattern through the $7s$ sublevel.

ANSWER
$1s$, $2s$, $2p$, $3s$, $3p$, $4s$, $3d$, $4p$, $5s$, $4d$, $5p$, $6s$, $4f$, $5d$, $6p$, $7s$

The answer to this *"Try It Yourself"* exercise provides us with some insights as to how levels and sublevels fill. *Within* a given level, the order of sublevel filling is always $s \rightarrow p \rightarrow d \rightarrow f \rightarrow g$. However, the filling of levels is irregular: we note that level 4 does not *complete* filling ($4f$ sublevel) until level 6 ($6s$ sublevel) has begun!

To apply the diagonal rule, we must remember that the *maximum* numbers of electrons for the sublevels are as follows: $s = 2$; $p = 6$; $d = 10$; $f = 14$; $g = 18$. We simply count electrons until the sublevel is filled, and then we move to the next sublevel in the order given by the diagonal. Each new atom always repeats the pattern of the atom before it.

Let us build up the first 10 elements of the Periodic Table using the diagonal rule. The result is shown in the accompanying table.

Element	Atomic Number	Electron Configuration
H	1	$1s^1$
He	2	$1s^2$
	(Level 1 is complete.)	
Li	3	$1s^2\,2s^1$
Be	4	$1s^2\,2s^2$
B	5	$1s^2\,2s^2\,2p^1$
C	6	$1s^2\,2s^2\,2p^2$
N	7	$1s^2\,2s^2\,2p^3$
O	8	$1s^2\,2s^2\,2p^4$
F	9	$1s^2\,2s^2\,2p^5$
Ne	10	$1s^2\,2s^2\,2p^6$
	(Level 2 is complete.)	

An element that has a total of eight electrons in its *outer* principal energy level, such as neon, is known as a *noble* (or *inert*) gas. Helium is also classified as an inert gas because its only principal energy level is filled (with two electrons). When we reach a noble gas, the *next* element will begin a *new* principal energy level. To conserve space, we usually place the symbol of a noble gas in brackets—for example, [Ne]—rather than write the entire configuration.

PROBLEM

Write the electron configurations of elements 11–20 in the Periodic Table using the diagonal rule and the format given above.

SOLUTION

Element	Atomic Number	Electron Configuration
Na	11	[Ne] $3s^1$
Mg	12	[Ne] $3s^2$
Al	13	[Ne] $3s^2\,3p^1$
Si	14	[Ne] $3s^2\,3p^2$
P	15	[Ne] $3s^2\,3p^3$
S	16	[Ne] $3s^2\,3p^4$
Cl	17	[Ne] $3s^2\,3p^5$
Ar	18	[Ne] $3s^2\,3p^6$
	(Level 3 is *not* complete, but it has 8 valence electrons.)	
K	19	[Ar] $4s^1$
Ca	20	[Ar] $4s^2$

Now suppose we wish to write the electron configuration of element 21 (scandium). Using the diagonal rule, we obtain $1s^2\ 2s^2\ 2p^6\ 3s^2\ 3p^6\ 4s^2\ 3d^1$. The last sublevel of level 3 has begun to fill. We could write this configuration in the "shorthand" notation we used above as [Ar] $4s^2\ 3d^1$, or we could write it as [Ar] $3d^1 4s^2$ in order to keep the principal energy levels in *numerical order*. Reference Table V in Appendix 2 uses the latter method, but the two notations are entirely equivalent.

TRY IT YOURSELF
Write the electron configuration of bromine (Br), element number 35, (a) according to the diagonal rule and (b) according to our "shorthand" notation.

ANSWERS
(a) $1s^2\ 2s^2\ 2p^6\ 3s^2\ 3p^6\ 4s^2\ 3d^{10}\ 4p^5$
(b) [Ar] $4s^2\ 3d^{10}\ 4p^5$ or [Ar] $3d^{10}\ 4s^2\ 4p^5$

The diagonal rule is a *general* rule, and there are exceptions to it. For example, the rule predicts that the electron configuration of an atom of chromium (Cr, atomic number 24) should be [Ar] $3d^4 4s^2$. If we consult Reference Table V in Appendix 2, however, we see that the electron configuration is [Ar] $3d^5 4s^1$. The reason for exceptions such as this one is explored in Chapter 9.

TRY IT YOURSELF
Which element in the Periodic Table lies in the same horizontal row as chromium and also has an irregular filling pattern?

ANSWER
copper (Cu, atomic number 29)

Filling Electrons into Orbitals

At this point, we need to investigate the filling patterns within the *orbitals* of a sublevel. *Any* orbital (regardless of its sublevel) can accommodate zero, one, or two electrons. If the orbital has no electron, we say that it is *empty*; if it has one electron, it is *half-filled*; if it has two electrons, it is *filled*.

Since an s sublevel can accommodate a maximum of two electrons and an orbital can accommodate a maximum of two electrons, it follows that an s sublevel can have only one orbital. Since a p sublevel can accommodate a maximum of six electrons, it follows that a p sublevel has three orbitals. Similarly, a d sublevel has five orbitals and an f sublevel has seven orbitals. It does not matter *where* the sublevel is located; the rule is the same. Thus, the $2p$ sublevel contains three orbitals, as do the $3p$, $4p$, and $5p$ sublevels.

We will indicate the orbitals on a sublevel by using a series of dashes as follows:

——	—— —— ——	—— —— —— —— ——
s orbital (1) on s sublevel	p orbitals (3) on p sublevel	d orbitals (5) on d sublevel

Electrons possess a property known as **spin**. There are only two spin states for an electron, and we indicate these by vertical arrows: ↑ and ↓. These designations are completely arbitrary; we could have used any other pair of matched symbols.

When an orbital is empty, we indicate this by ___.

When an orbital contains only one electron, we indicate this by ↑__.

When an orbital contains two electrons, we indicate this by ↑↓__. This notation shows that the electron "spins" have been "paired," resulting in a more stable condition for the electrons in that orbital.

We are now ready to begin filling the orbitals of each element, using the sublevel chart for elements 1–10 that we developed on page 228.

Electron Configurations

Element	By Sublevels	By Orbitals
H	$1s^1$	↑
He	$1s^2$	↑↓
Li	$1s^2 2s^1$	↑↓ ↑
Be	$1s^2 2s^2$	↑↓ ↑↓
B	$1s^2 2s^2 2p^1$	↑↓ ↑↓ ↑ __ __
C	$1s^2 2s^2 2p^2$	↑↓ ↑↓ ↑ ↑ __
N	$1s^2 2s^2 2p^3$	↑↓ ↑↓ ↑ ↑ ↑
O	$1s^2 2s^2 2p^4$	↑↓ ↑↓ ↑↓ ↑ ↑
F	$1s^2 2s^2 2p^5$	↑↓ ↑↓ ↑↓ ↑↓ ↑
Ne	$1s^2 2s^2 2p^6$	↑↓ ↑↓ ↑↓ ↑↓ ↑↓
		1s 2s $2p_x$ $2p_y$ $2p_z$

Notice, in the accompanying table, what happens when we begin to fill carbon (C): the second $2p$ electron does *not* fill the first orbital; it half-fills the second orbital. This pattern is repeated with nitrogen (N). This half-filling occurs because electrons repel each other and will occupy whichever positions are farthest from each other. When oxygen (O) is filled, the last $2p$ electron has no choice but to enter and fill the first orbital. Meanwhile, we have another rule: *Electrons entering a sublevel will half-fill the orbitals in that sublevel before they fill it completely.*

PROBLEM

Using the preceding tables as a guide, list the orbital filling patterns for elements 11–20.

SOLUTION

Electron Configurations

Element	By Sublevels	By Orbitals
Element 10: Ne	$1s^2\ 2s^2\ 2p^6$ [Ne]	↑↓ ↑↓ ↑↓ ↑↓ ↑↓ 1s 2s $2p_x$ $2p_y$ $2p_z$
Na	[Ne] $3s^1$	[Ne] ↑
Mg	[Ne] $3s^2$	[Ne] ↑↓
Al	[Ne] $3s^2\ 3p^1$	[Ne] ↑↓ ↑ _ _
Si	[Ne] $3s^2\ 3p^2$	[Ne] ↑↓ ↑ ↑ _
P	[Ne] $3s^2\ 3p^3$	[Ne] ↑↓ ↑ ↑ ↑
S	[Ne] $3s^2\ 3p^4$	[Ne] ↑↓ ↑↓ ↑ ↑
Cl	[Ne] $3s^2\ 3p^5$	[Ne] ↑↓ ↑↓ ↑↓ ↑
Ar	[Ne] $3s^2\ 3p^6$ [Ar]	[Ne] ↑↓ ↑↓ ↑↓ ↑↓ 3s $3p_x$ $3p_y$ $3p_z$
K	[Ar] $4s^1$	[Ar] ↑
Ca	[Ar] $4s^2$	[Ar] ↑↓ 4s

In the preceding examples, we listed the electron configurations of the elements that fill their *s* and *p* sublevels and orbitals. In the Periodic Table, these elements are found in Groups 1, 2, and 13–18 (the *s*-block and *p*-block). When we examine these elements in Chapter 9, we will find that their regular properties entitle them to be called *representative elements*.

8.3A LEWIS STRUCTURES AND ATOMIC ORBITALS

In drawing a Lewis structure, we represent the kernel (nucleus plus inner electrons) by the symbol for the element. Thus, C would represent the kernel of an atom of carbon. We then represent the valence electrons as a series of dots surrounding the kernel. However, the dots (electrons) must be placed in a special manner in order to indicate how the valence orbitals are filled. We recall that valence electrons occupy only *s* and *p* sublevels. Since an *s* sublevel has only one orbital and a *p* sublevel has three orbitals (p_x, p_y, and p_z), we have only four positions in which to place dots.

Let us adopt the following scheme for the mythical element X. Imagine that a clock surrounds X. The *s* orbital will be located at 12 o'clock, and the

p_x, p_y, and p_z orbitals at 3, 6, and 9 o'clock, respectively, as indicated in the diagram.

$$p_z \overset{s}{X} p_x$$
$$p_y$$

If an orbital contains no electrons, we will leave the position blank; if it has one electron, we will place a single dot (•) at that position; if it has two electrons, we will place a pair of dots (• •) at that position.

We recall that the electron configuration of oxygen is as follows:

$$1s^2 2s^2 2p^4 \qquad \frac{\uparrow\downarrow}{1s} \quad \frac{\uparrow\downarrow}{2s} \quad \frac{\uparrow\downarrow}{2p_x} \quad \frac{\uparrow}{2p_y} \quad \frac{\uparrow}{2p_z}$$

Applying our rules, we can write the Lewis structure for oxygen as follows:

$$\cdot \overset{\displaystyle ..}{O} :$$
$$\cdot$$

PROBLEM
Draw the Lewis structure for carbon (atomic number 6).

SOLUTION
Carbon has four valence electrons ($2s^2 2p^2$). The Lewis structure is

TRY IT YOURSELF
Draw the Lewis structure for calcium (atomic number 20).

ANSWER
Calcium has two valence electrons ($4s^2$). The Lewis structure is

$$\overset{\displaystyle ..}{Ca}$$

A final note: The "clock" introduced for drawing Lewis structures is *completely arbitrary*. For example, all of the following are correct for the Lewis structure of Ca:

$$\overset{\displaystyle ..}{Ca} \quad Ca: \quad \underset{\displaystyle ..}{Ca} \quad :Ca$$

END-OF-CHAPTER QUESTIONS

Some questions have the symbol "§2" in front of the question number. This symbol means that the question is based on Section II material.

1. The characteristic bright-line spectrum of sodium is produced when its electrons
(1) return to lower energy levels
(2) jump to higher energy levels
(3) are lost by the neutral atoms
(4) are gained by the neutral atoms

2. The following diagram shows the characteristic spectral line patterns of four elements. Also shown are the spectral lines produced by an unknown substance. Which pair of elements is present in the unknown?

(1) Li and Na   (2) Na and H   (3) Li and He   (4) He and H

§2 **3.** The total number of orbitals in an *f* sublevel is
(1) 1   (2) 5   (3) 3   (4) 7

4. Which principal energy level can hold a maximum of 18 electrons?
(1) 5   (2) 2   (3) 3   (4) 4

§2 **5.** The total number of *d* orbitals in the third principal energy level is
(1) 1   (2) 5   (3) 3   (4) 7

§2 **6.** The total number of orbitals in the second principal energy level is
(1) 1   (2) 2   (3) 3   (4) 4

§2 **7.** The maximum number of electrons that a single orbital of the 3*d* sublevel may contain is
(1) 5   (2) 2   (3) 3   (4) 4

§2 **8.** What is the maximum number of sublevels in the third principal energy level?
  (1) 1   (2) 2   (3) 3   (4) 4

§2 **9.** Energy level 2 contains a total of
  (1) 8 sublevels (2) 2 sublevels (3) 3 sublevels (4) 4 sublevels

10. As an electron in an atom moves from the ground state to an excited state, the potential energy of the atom
  (1) decreases   (2) increases   (3) remains the same

11. Which principal energy level change by the electron of a hydrogen atom will cause the greatest amount of energy to be absorbed?
  (1) $n = 2$ to $n = 4$ (2) $n = 2$ to $n = 5$
  (3) $n = 4$ to $n = 2$ (4) $n = 5$ to $n = 2$

§2 **12.** Which principal energy level contains only s and p sublevels?
  (1) 1   (2) 2   (3) 3   (4) 4

§2 **13.** The maximum number of electrons possible in any principal energy level (principal quantum number $= n$) is equal to
  (1) n (2) $2n$ (3) n^2 (4) $2n^2$

14. If 18 electrons completely fill a principal energy level, what is the number of the principal energy level?
  (1) 1   (2) 2   (3) 3   (4) 4

§2 **15.** What is the number of sublevels in an energy level whose number of principal energy levels is n?
  (1) n (2) $2n$ (3) n^2 (4) $2n^2$

§2 **16.** Which principal energy level has no f sublevel?
  (1) 5   (2) 6   (3) 3   (4) 4

17. Which configuration represents an excited atom?
  (1) 2-1   (2) 2-8-1   (3) 2-8-2   (4) 2-7-2

18. An excited atom with the configuration 2-8-6-3 is an atom of?
  (1) Na   (2) K   (3) S   (4) Br

§2 **19.** What is the maximum number of electrons that can occupy a $5f$ sublevel?
  (1) 5   (2) 7   (3) 14   (4) 50

20. An atom contains a total of 25 electrons. When the atom is in the ground state, how many different principal energy levels will contain electrons?
(1) 1   (2) 2   (3) 3   (4) 4

§2 **21.** Energy is released when an electron changes from a sublevel of
(1) $1s$ to $2p$ (2) $2s$ to $3s$ (3) $3s$ to $2s$ (4) $3p$ to $5s$

§2 **22.** What is the total number of occupied sublevels in an atom of chlorine in the ground state?
(1) 1   (2) 5   (3) 3   (4) 9

§2 **23.** An atom in the ground state has 7 valence electrons. Which electron configuration could represent the outermost principal energy level of this atom in the ground state?
(1) $3s^13p^6$ (2) $3s^23p^5$ (3) $3s^13p^43d^2$ (4) $3s^23p^43d^1$

§2 **24.** Which electron configuration represents an atom in an excited state?
(1) $1s^22s^22p^63p^1$ (2) $1s^22s^22p^63s^23p^1$
(3) $1s^22s^22p^63s^23p^2$ (4) $1s^22s^22p^63s^2$

25. What is the total number of principal energy levels that are completely filled in an atom of magnesium in the ground state?
(1) 1   (2) 2   (3) 3   (4) 4

§2 **26.** The total number of completely filled orbitals in an atom of nitrogen in the ground state is
(1) 1   (2) 2   (3) 3   (4) 4

§2 **27.** The neutral atom of an element has an electron configuration of 2-8-1. What is the total number of p electrons in this atom?
(1) 6   (2) 2   (3) 10   (4) 11

28. What is the total number of electrons in the second principal energy level of a calcium atom in the ground state?
(1) 6   (2) 2   (3) 8   (4) 18

§2 **29.** Which is the electron configuration of a noble-gas atom in the excited state?
(1) $1s^1$ (2) $1s^12s^1$ (3) $1s^22s^2$ (4) $1s^22s^22p^2$

§2 **30.** Which is an electron configuration of a fluorine atom in the excited state?

(1) $1s^22s^22p^4$

(2) $1s^22s^22p^5$

(3) $1s^22s^22p^43s^1$

(4) $1s^22s^22p^53s^1$

§2 **31.** Which orbital notation represents the outermost principal energy level of a phosphorus atom in the ground state?

§2 **32.** Which atom in the ground state contains only one orbital that is partially occupied?

(1) Si   (2) Ne   (3) Ca   (4) Na

§2 **33.** An electron in an atom will emit energy when it moves from energy level

(1) $2s$ to $3p$ (2) $2s$ to $2p$ (3) $2p$ to $3s$ (4) $2p$ to $1s$

§2 **34.** The electron configuration of an atom in the ground state is $1s^22s^22p^3$. The total number of occupied atomic orbitals in this atom is

(1) 6   (2) 5   (3) 3   (4) 4

§2 **35.** An atom contains a total of 25 electrons. When the atom is in the ground state, how many different sublevels will contain electrons?

(1) 7   (2) 2   (3) 3   (4) 5

36. Which is the correct Lewis structure of an atom of sulfur in the ground state?

(1) S: (2) •S: (3) •S: (4) :S:

37. If X is the symbol of a noble-gas atom in the ground state, the Lewis structure of the atom could be

(1) X•   (2) X: (3) •X•   (4) :X:

38. Which is the Lewis structure of an atom of boron in the ground state?

(1) •B: (2) B•   (3) •B: (4) B:

39. Which is the correct Lewis structure of an aluminum atom in the ground state?

(1) Al: (2) Ȧl: (3) •Ȧl: (4) •Ȧl:

40. Usually the term *kernel* includes all parts of the atom except
(1) neutrons   (2) protons
(3) valence electrons   (4) orbital electrons

41. What is the total number of valence electrons of an atom of sulfur in the ground state?
(1) 6   (2) 8   (3) 3   (4) 4

42. Which is an electron configuration of chlorine in the excited state?
(1) 2-8-7   (2) 2-8-8
(3) 2-8-6-1   (4) 2-8-7-1

43. The strength of an atom's attraction for electrons in a chemical bond is the atom's
(1) electronegativity
(2) ionization energy
(3) heat of reaction
(4) heat of formation

44. Which ion has the same electron configuration as an atom of He?
(1) H^- (2) O^{2-} (3) Na^+ (4) Ca^{2+}

45. In the ground state, an atom of which element has two valence electrons?
(1) Cr   (2) Cu   (3) Ni   (4) Sc

Constructed-Response Questions

1. Use Reference Table S and the Periodic Table in Appendix 1 to complete the accompanying table.

Element	Symbol	Electron Configuration	Number of Valence Electrons	Lewis Structure
Argon				
Barium				
Carbon				
Chlorine				
Helium				
Nitrogen	N	2-5	5	$\cdot\ddot{N}\cdot$
Oxygen				
Potassium				
Sulfur				
Strontium				

Base your answers to questions 2 and 3 on the electron configuration table shown below.

Element	Electron Configuration
X	2-8-8-2
Y	2-8-7-3
Z	2-8-8-5

2. What is the total number of valence electrons in an atom of electron configuration X?

3. Which electron configuration represents the excited state of a calcium atom?

Base your answers to questions 4 and 5 on the diagram below, which shows bright-line spectra of selected elements.

Bright-Line Spectra

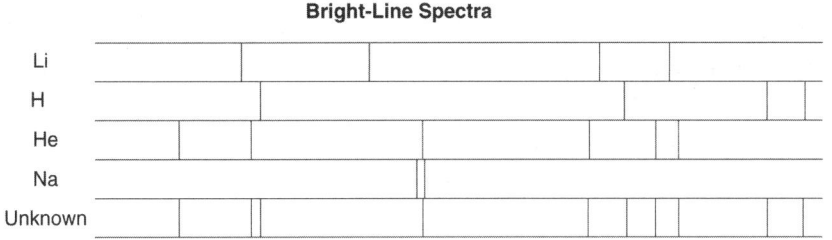

4. Identify the two elements in the unknown spectrum.

5. Explain how a bright-line spectrum is produced in terms of excited state, energy transitions, and ground state.

Base your answers to questions 6–8 on the information below and on your knowledge of chemistry.

The bright-line spectra observed in a spectroscope for three elements and a mixture of two of these elements are represented in the diagram below.

Selected Bright-Line Spectra

6. State evidence from the bright-line spectra that indicates that element *A* is not present in the mixture.

7. Explain why the spectrum produced by a 1-gram sample of Z would have the same spectral lines at the same wavelengths as the spectrum produced by a 2-gram sample of element Z.

8. Describe in terms of *both* electrons and energy states how the light represented by the spectral lines is produced.

The answers to these questions are found in Appendix 3.

Chapter
Nine

CHEMICAL PERIODICITY

KEY IDEAS

This chapter focuses on the periodic properties of the elements. After certain general properties are developed, the elements are studied according to the groups in the Periodic Table in which they are located.

KEY OBJECTIVES

At the conclusion of this chapter you will be able to:

- State the periodic law of Moseley.
- Describe the general arrangement of the elements in the modern Periodic Table with regard to electron configuration.
- Describe how properties of metallic elements differ from those of nonmetallic elements.
- Define the term *metalloid*, and use the Periodic Table to indicate which elements are metalloids.
- Define the term *atomic radius*.
- Define the term *ionic radius*, and indicate how the size of an ion compares with the size of its parent atom.
- Indicate how the following general properties of elements vary within the Periodic Table: metallic character, atomic radius, ionization energies, electron affinity, ionic radius, and electronegativity.
- Compare and contrast the properties of the elements in the various representative groups of the Periodic Table.
- Compare some of the properties of the transition elements with those of the representative elements.
- Describe how the properties of the elements vary across a period.

SECTION I—BASIC (REGENTS-LEVEL) MATERIAL

NYS REGENTS CONCEPTS AND SKILLS

Note: By the time you have finished Section I, you should have mastered the concepts and skills listed on the next page. The Regents chemistry examination will test your knowledge of these items and your ability to apply them.

Concepts are the *basic ideas* that form the body of the Regents chemistry course (what you need to know!).

Skills are the *activities* that demonstrate your mastery of these concepts (how you show that you know them!).

Following each concept or skill is a page reference (given in parentheses) to this chapter.

9.1 Concept:
The placement or location of elements on the Periodic Table gives an indication of the physical and chemical properties of that element. The elements in the Periodic Table are arranged in order of increasing atomic number. (Page 244)

Skill:
Explain the placement of an unknown element in the Periodic Table based on its properties. (Pages 243–245)

9.2 Concept:
Elements can be classified by their properties and located in the Periodic Table as metals, nonmetals, metalloids (B, Si, Ge, As, Sb, Te), or noble gases. (Pages 256–258)

Skill:
Classify elements as metals, nonmetals, metalloids, or noble gases by their properties. (Pages 256–258)

9.3 Concept:
Elements can be differentiated by their physical properties. Physical properties such as density, conductivity, malleability, solubility, and hardness differ among elements. (Page 245)

Skill:
Describe the states (phases) of elements at STP. (Pages 256–258)

9.4 Concept:
Elements can be differentiated by chemical properties. Chemical properties determine how an element behaves during a chemical reaction. (Pages 256–258)

9.5 Concept:
Some elements exist in two or more forms in the same phase. Since these forms differ in their molecular or crystal structures, they also differ in their properties. (Pages 256–258)

9.6 Concept:
For Groups 1, 2, and 13–18 in the Periodic Table, elements within the same group have the same number of valence electrons (helium is an exception) and, therefore, have similar chemical properties. (Page 244)

Skill:
Given the chemical formula of a compound (such as XCl or XCl_2), determine the group of an element (X). (Pages 256–258)

9.7 Concept:
The succession of elements within the same group demonstrates characteristic trends: differences in atomic radius, ionic radius, electronegativity, first ionization energy, and metallic (or nonmetallic) properties. (Pages 248–255)

Skill:
Compare and contrast the properties of elements within a single group (Groups 1, 2, 13–18) in the Periodic Table. (Pages 248–255)

9.8 Concept:
The succession of elements across the same period demonstrates characteristic trends: differences in atomic radius, ionic radius, electronegativity, first ionization energy, and metallic (or nonmetallic) properties. (Pages 248–255)

Skill:
Compare and contrast the properties of elements across a period (for Groups 1, 2, 13–18) in the Periodic Table. (Pages 249–258)

9.9 Concept:
When an atom gains one or more electrons, its radius increases. When an atom loses one or more electrons, its radius decreases. (Page 247)

9.1 INTRODUCTION

The first question we need to ask is this: *Why* do we have a Periodic Table of the Elements? The term *periodic* means that a quantity repeats itself at regular intervals. For example, the motion of the Moon around Earth is periodic because, after a certain length of time (1 month), the motion repeats itself. As early as the Middle Ages, scientists recognized that elements could be differentiated by their physical and chemical properties. These properties are also periodic—*if we list the elements properly*.

If we examine the modern Periodic Table of the Elements in Appendix 1, we find that it is almost, but not quite, rectangular in shape. At the top of the table, two elements (hydrogen, H, and helium, He) are separated from the body of the table, as are two rows of elements at the bottom. This table provides the answer to the question asked above: it allows us to place elements with *similar properties* in close proximity.

9.2   THE PERIODIC TABLE IN HISTORY

Historically, the creation of the modern Periodic Table is credited to chemists Dmitry Mendeleev (Russia) and Lothar Meyer (Germany), who did their work during the nineteenth century. The two men, working independently, produced a table in which elements with similar properties were located in the same vertical column. Using his table, Mendeleev was able to predict the properties of such then-undiscovered elements as germanium (Ge) (which he called *eka-silicon*) and gallium (Ga) (which he called *eka-alominum*). Both Mendeleev and Meyer based their tables on the atomic *masses* of the elements. This arrangement led to some problems, however, because the atomic masses of the elements do not increase regularly. Compare, for example, the atomic masses of tellurium (Te) and iodine (I).

In the early twentieth century, physicist Henry Moseley (England) discovered what we now call our modern periodic law: *The periodic properties of the elements are functions of their atomic numbers.* This means that the Periodic Table should be arranged in order of increasing atomic number—which it is! However, the *basis* of the Periodic Table is the (ground-state) *electron configuration* of each element.

9.3   THE MODERN PERIODIC TABLE

The modern Periodic Table is described in terms of **periods** (the horizontal rows) and **groups** (the vertical columns). Each period is indicated by a number, beginning with 1 and *currently* ending with 7. Elements in the same period have the same number of occupied principal energy levels. For example, the elements sodium (Na) and phosphorus (P) are in the same period—3. Therefore, sodium and phosphorus have *three* occupied principal energy levels. The period number of an element also tells us the *valence* level of the element: the valence level of sodium and of phosphorus is also 3.

The groups begin at the left side of the table (Group 1) and continue through Group 18 on the right side. Elements within a group share similar properties, and this likeness is directly related to the similarities in the electron configurations of their valence levels. For this reason, we also call a group a *chemical family*. For example, the elements in Group 13 constitute a chemical family: Atoms of Group 13 elements each have three valence electrons. Consequently, the properties of the elements in this group are similar to one another. (We note, however, that the word *similar* does *not* mean "identical"!)

9.4 PROPERTIES ASSOCIATED WITH PERIODICITY

Before we describe periodic properties in detail, we need to discuss a number of important factors associated with periodicity.

Metallic Character

Approximately two-thirds of the elements in the Periodic Table are metals. Metallic elements have certain unique properties:

* *Luster*, the mirrorlike shine that reflects light well.
* *Conductivity*, the ability to transfer heat and electrons well.
* *Malleability*, the ability to be rolled or hammered into thin sheets.
* *Ductility*, the ability to be drawn into wire.

Examples of metals are sodium (Na), iron (Fe), and mercury (Hg, a liquid at room temperature).

In contrast, nonmetals tend to be brittle (in the solid phase), to lack luster, and to have poor conductivities. Examples of nonmetals include sulfur (S), helium (He), and bromine (Br_2, a liquid at room temperature).

The bold zigzag line in the Periodic Table separates the metallic elements from the nonmetallic elements. We will refer to it as the *metal-nonmetal line*.

A number of elements, such as boron (B), silicon (Si), germanium (Ge), arsenic (As), antimony (Sb), and tellerium (Te), have properties somewhat between those of metals and those of nonmetals. These *semiconductors* are also known as **metalloids** or *semimetals*.

Atomic Radius

If we think of an atom (whether combined with other atoms or not) as having a roughly *spherical* shape, then its size is best reported as a *radius* (called the **atomic radius**), as shown in the accompanying diagram:

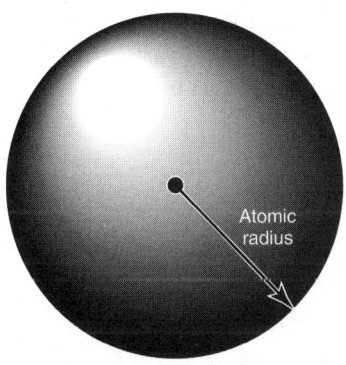

Atomic radius

The radius of a "free" atom cannot be measured *directly*. One way to measure the atomic radius is to calculate a quantity called the **internuclear distance**, which is defined to be the distance between two adjacent atoms when they are closest together. The accompanying diagram illustrates the internuclear distance between two adjacent atoms. We can see from the diagram that the radius of an atom is equal to *one-half* of the internuclear distance.

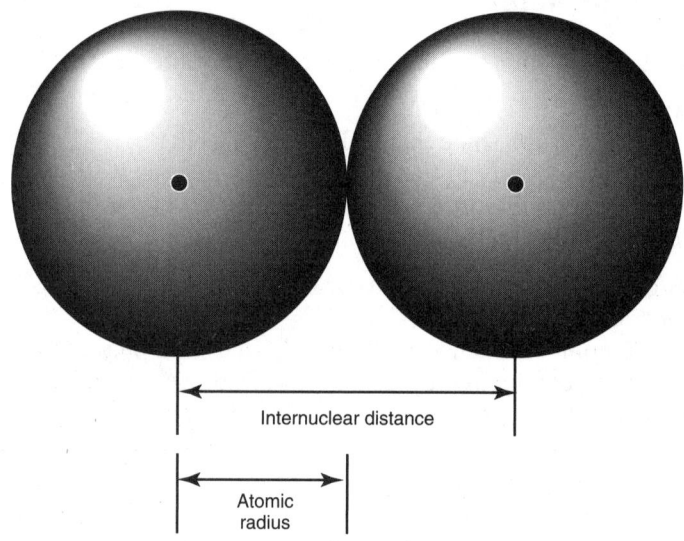

Internuclear distance

Atomic radius

Reference Table S in Appendix 1 lists the atomic radii for most elements. The atomic radius of an atom is reported in *picometers* (pm, 1 pm = 10^{-12} m). For example, according to Table S, the atomic radius of phosphorus (P) is 128 picometers.

First Ionization Energy

An *ion* is an electrically charged atom. Neutral atoms can become charged in either of two ways: they can gain or they can lose electrons. If an atom *loses* one or more electrons, it becomes *positively* charged; if an atom *gains* one or more electrons, it becomes *negatively* charged. The following examples illustrate this principle for an atom of potassium (K) and an atom of sulfur (S):

$$K \rightarrow K^+ + e^-$$
$$S + 2e^- \rightarrow S^{2-}$$

The **first ionization energy** (I_1) measures the ease with which an atom *loses* its *first* electron. To remove an electron from a neutral atom, an amount of energy equal to the first ionization energy must be added:

$$K + I_1 \rightarrow K^+ + e^-$$

The smaller the amount of energy needed, the more easily the electron is lost. Although the first ionization energy may be measured for a single atom, it is generally more useful to measure the amount of energy needed to remove a single electron from each atom in 1 mole of atoms. Therefore, a first ionization energy of 900 kilojoules per mole means that 900 kilojoules is required to remove one electron from each atom in 1 mole of atoms.

First ionization energies of the representative elements range from 393 kilojoules per mole to 2081 kilojoules per mole. Metallic elements generally have low first ionization energies: they tend to lose electrons more easily than nonmetallic elements. This information can be verified by examining Reference Table S in Appendix 1.

PROBLEM
How much energy, in kilojoules, is needed to ionize 0.400 mole of sodium (Na) according to the equation below?

$$Na \rightarrow Na^+ + e^-$$

SOLUTION
According to Reference Table S, the first ionization energy of sodium is 496 kJ/mol, and the amount of energy needed is found by using the factor-label method:

$$0.400 \ \cancel{mol} \cdot 496 \frac{kJ}{\cancel{mol}} = 198 \ kJ$$

TRY IT YOURSELF
The first ionization energy of the element boron (B) is 801 kilojoules per mole. How many moles of boron atoms can be ionized to boron ions (B^+) by the addition of 2000. kilojoules of energy?

ANSWER
2.50 mol

Ionic Radius

As we will see in Chapter 10, atoms can combine with other atoms by forming positive and negative *ions*. Therefore, it is frequently useful to measure the sizes of the corresponding ions these atoms produce. This quantity is known as the **ionic radius**. When an atom forms a *positive ion* by losing one or more electrons, the ionic radius is *smaller* than the atomic radius of the parent atom. Conversely, when an atom forms a *negative ion* by gaining one or more electrons, the ionic radius is *larger* than the atomic radius of the parent atom. The following two examples illustrate this principle:

1. The radius of a lithium atom (Li) is 155 picometers; the radius of a lithium ion (Li$^+$) is 60 picometers.
2. The radius of a chlorine atom (Cl) is 97 picometers; the radius of a chloride ion (Cl$^-$) is 181 picometers.

Electronegativity

Another way in which atoms can combine is by sharing one or more *pairs* of electrons. The **electronegativity** of an atom measures its ability to attract a pair of electrons when bonded to another atom. The higher the electronegativity, the more the atom attracts the shared pair of electrons.

Electronegativities range from a high of 4.0 for fluorine (F) to a low of 0.7 for francium (Fr). Electronegativity values are given in Reference Table S in Appendix 1 of this book.

9.5 VARIATION OF PERIODIC PROPERTIES AMONG THE ELEMENTS

We now examine how the periodic properties introduced in Section 8.4 vary within the Periodic Table. We will restrict our examination to the *representative* elements (Groups 1, 2, 13–18) because these elements are the simplest to study. In order to draw our conclusions, we will consider how the properties of these elements vary *across a period* (i.e., from left to right across the table) and *down a group* (i.e., from top to bottom down the table). When we require quantitative data, we will focus our attention on Period 3 and Group 16 of the table.

The accompanying table is an abbreviated version of the Periodic Table of the Elements in Appendix 1. It shows the periodic properties of the *representative elements* that we shall study in this section.

Metallic Character

As we proceed down a group, metallic character *increases*. As we proceed across a period, metallic character *decreases*.

Elements display metallic properties when they have *low ionization energies* and *relatively few electrons in their valence levels*. Therefore, we would expect that the greatest metallic character would be displayed by elements located in the *lower left* part of the Periodic Table, and the least metallic character by elements located in the *upper right* part of the table. This is, in fact, correct: the lower elements in Groups 1 and 2 are the most metallic elements; the upper elements in Groups 16 and 17, the most nonmetallic.

GROUP

1	2	13	14	15	16	17	18

Row 1:

H							He
37							32
1312							2372
2.1							(–)

	1	2	13	14	15	16	17	18
2	Li 155 520 1.0	Be 112 900 1.6	B 98 801 2.0	C 91 1086 2.6	N 92 1402 3.0	O 65 1314 3.5	F 57 1681 4.0	Ne 51 2081 (–)
3	Na 190 496 0.9	Mg 160 736 1.3	Al 143 578 1.6	Si 132 787 1.9	P 128 1012 2.2	S 127 1000 2.6	Cl 97 1251 3.2	Ar 88 1521 (–)
4	K 235 419 0.8	Ca 197 590 1.0	Ga 141 579 1.8	Ge 137 762 2.0	As 139 944 2.2	Se 140 941 2.6	Br 112 1140 3.0	Kr 103 1351 (–)
5	Rb 248 403 0.8	Sr 215 549 1.0	In 166 558 1.8	Sn 162 709 2.0	Sb 159 831 2.1	Te 142 869 2.1	I 132 1008 2.7	Xe 124 1170 2.6
6	Cs 267 376 0.8	Ba 222 503 0.9	Tl 171 589 2.0	Pb 175 716 2.3	Bi 170 703 2.0	Po 167 812 2.0	At 145 (–) 2.2	Rn 134 1037 (–)
7	Fr 270 393 0.7	Ra 233 (–) 0.9	Nh (–) (–) (–)	Fl (–) (–) (–)	Mc (–) (–) (–)	Lv (–) (–) (–)	Ts (–) (–) (–)	Og (–) (–) (–)

metal   nonmetal

metalloid   no data

SYMBOL →	C
ATOMIC RADIUS / pm →	91
FIRST IONIZATION ENERGY / (kJ/mol) →	1086
ELECTRONEGATIVITY →	2.6

• A dash (–) indicates that data are not available.

249

These variations are borne out by the heavy metal-nonmetal line in the Periodic Table. Its zigzag shape is shifted further to the right as we move down the table, indicating that metallic properties appear in elements that lie further down in the table. Even the Group 16 element polonium (Po) exhibits metallic character because it is far enough down. We must conclude that metallic character does indeed *increase* down a group.

Atomic Radius

As we proceed down a group, we find that the radii of successive atoms *increase*.

The explanation of this observation lies in the fact that, as the period number increases, *new* energy levels are being occupied. Therefore, the outer electrons of successive elements are *farther* away from the nucleus. In addition, the inner electrons serve to *shield* the outer electrons from the nucleus. These two circumstances lead to an *increase* in radius. The accompanying table and graph illustrate this principle for Group 16 of the Periodic Table.

Elements down Group 16	Atomic Radius / pm
O	65
S	127
Se	140
Te	142

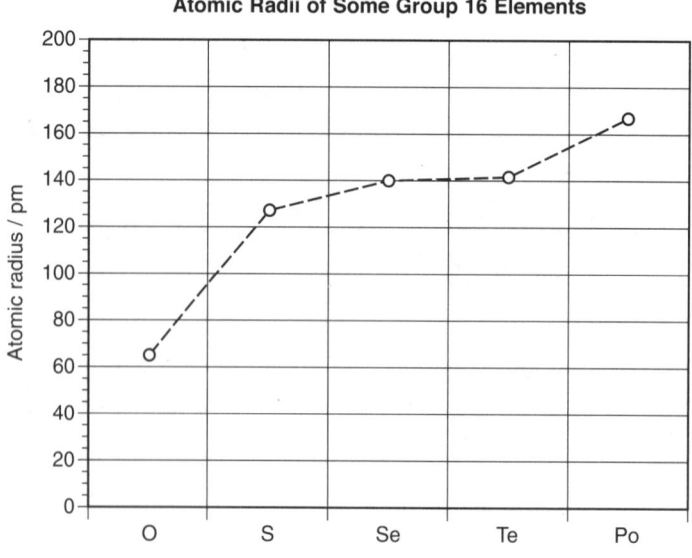

Atomic Radii of Some Group 16 Elements

As we proceed across a period, the radii of successive atoms *decrease*. The explanation of this observation lies in the fact that with each successive element the positive nuclear charge increases, as does the negative electron charge. Since the successive electrons are being added in the *same* principal energy level, the force of attraction between the nucleus and the electrons *increases*, leading to a *decrease* in radius. The accompanying table and graph illustrate this principle for Period 3 of the Periodic Table.

Elements across Period 3	Atomic Radius / pm
Na	190
Mg	160
Al	143
Si	132
P	128
S	127
Cl	97
Ar	88

Atomic Radii of the Period 3 Elements

First Ionization Energy

As we proceed down a group, the first ionization energy of each successive element *decreases*.

Recall that the first ionization energy of an atom is the energy needed to remove a single electron from the atom. If the electron is loosely held, the ionization energy is low; if the electron is strongly held, the ionization energy is high. There is an inverse relationship between the radius of an atom and its first ionization energy: the larger the radius, the more weakly its outer electron is held and, consequently, the more easily it is removed.

If we examine the first ionization energies of the elements down Group 16, as shown in the accompanying table and graph, we find a regular decrease with increasing atomic radius, as we predicted.

Elements down Group 16	First Ionization Energy / (kJ/mol)
O	1314
S	1000
Se	941
Te	869
Po	812

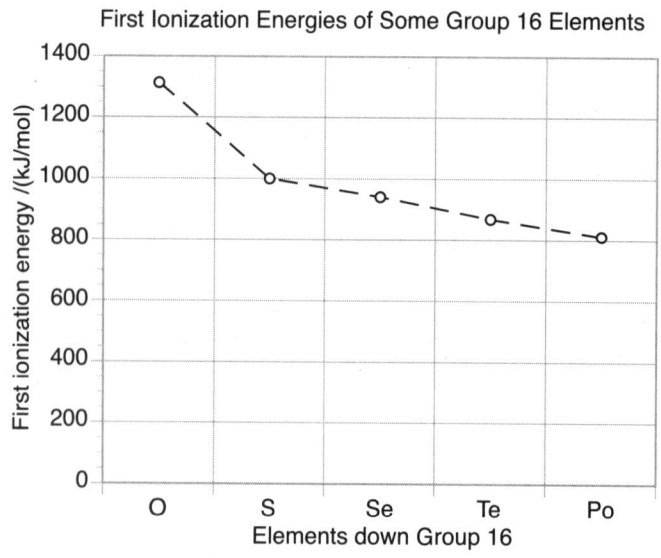

First Ionization Energies of Some Group 16 Elements

As we proceed across a period, the first ionization energy of successive elements *generally increases*. The accompanying table and graph illustrate this trend for Period 3.

We find, however, two surprises: the first ionization energies of aluminum (Al) and sulfur (S) are *lower* than we would expect.

Elements across Period 3	First Ionization Energy / (kJ/mol)
Na	496
Mg	738
Al	578
Si	787
P	1012
S	1000
Cl	1251
Ar	1521

First Ionization Energies of the Period 3 Elements

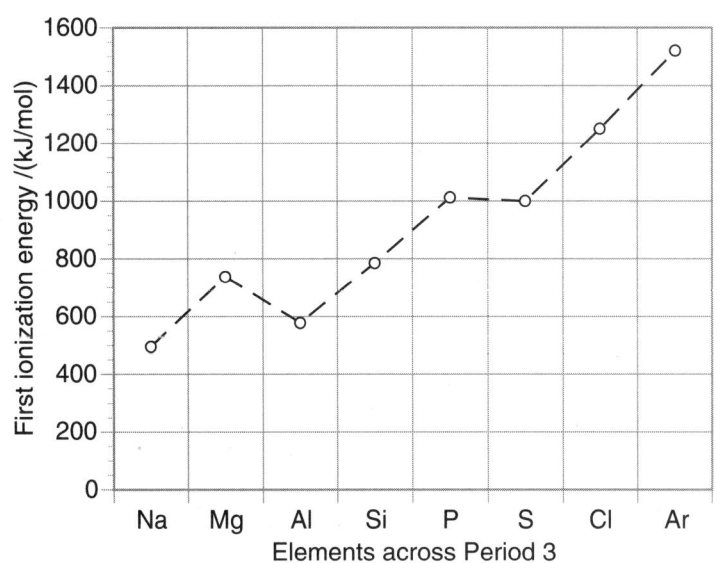

As a result of these two trends in first ionization energies—a *decrease* down a group and an *increase* across a period—we can infer that *elements located at the left side of the Periodic Table and elements located toward the bottom of the table have the greatest tendency to lose electrons.*

Electronegativity

As we proceed down a group, we observe a *decrease* in electronegativity. The variation of electronegativity down a group, as shown in the accompanying table and graph, can be predicted from the variation in the properties discussed on pages 250–253: atomic radius and ionization energy. Elements located successively further down a group have a decreased ability to attract electrons.

Elements down Group 16	Electronegativity
O	3.4
S	2.6
Se	2.6
Te	2.1
Po	2.0

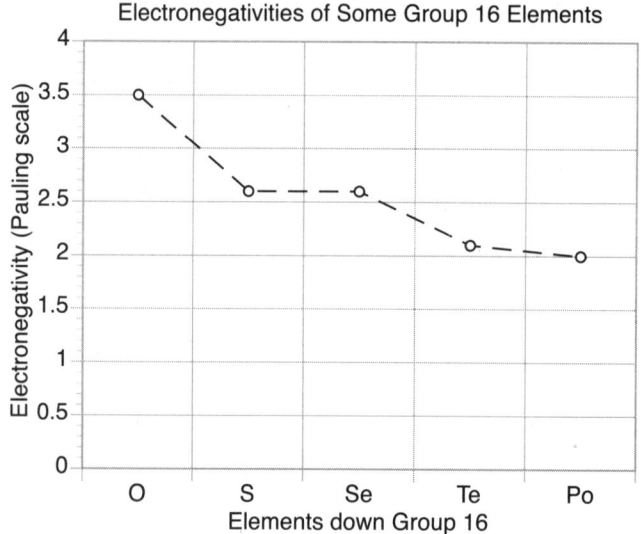

Electronegativities of Some Group 16 Elements

As we proceed across a period, we observe an *increase* in electronegativity. The variation in electronegativity across a period, as shown in the accompanying table and graph, can also be predicted from the variation in the properties discussed on pages 250–253: atomic radius and ionization energy. Elements located successively to the right across a period have an increased ability to attract electrons.

Elements across Period 3	Electronegativity
Na	0.9
Mg	1.3
Al	1.6
Si	1.9
P	2.2
S	2.6
Cl	3.2

Electronegativities of the Period 3 Elements

Elements across Period 3

If we examine the Selected Periodic Properties of the Representative Elements on page 249, we can verify these relationships among the representative elements.

The accompanying table provides a general summary of the variation in three periodic properties of the representative elements within the Periodic Table. (Remember: *There are exceptions to each of these generalizations!*)

Periodic Property	Variation down a Group	Variation across a Period
Atomic radius	Increases	Decreases
First ionization energy	Decreases	Increases
Electronegativity	Decreases	Increases

9.6 THE CHEMISTRY OF THE REPRESENTATIVE GROUPS

We now describe the general properties of the representative elements within a single group. Our purpose is *not* to provide an exhaustive description, but rather to point out the similarities and differences among the elements in that group.

Group 1

The elements in Group 1 are called the *alkali metals*. (Each element has one valence electron.) Elements in this group have the most metallic character, have low ionization energies and electronegativities, and have the largest atoms in their period. Group 1 elements form 1+ ions. Because they are highly reactive, they do not occur as free elements in nature and must be prepared by passing an electric current through their molten compounds. We will discuss this technique in Chapter 15.

Hydrogen (H) is *not* considered an alkali metal because its properties are not similar to those of the other elements in Group 1. This exception is due to the small size of the hydrogen atom and the large attraction of the nucleus for the single electron. Hydrogen is associated with Group 1 because it has one valence electron.

Group 2

The elements in Group 2 are known as the *alkaline earth metals*. (Each element has two valence electrons.) An alkaline earth metal has less metallic character than the alkali metal that precedes it. The elements of Group 2 form 2+ ions, although beryllium (Be), which is relatively small, is known to form covalent bonds (see Chapter 10) with elements such as chlorine (Cl, in $BeCl_2$). The alkaline earth metals are reactive and do not occur as free elements in nature.

Group 13

Each element in Group 13 has three valence electrons. The first element of this group, boron (B), is a metalloid. Because of its small size, it does not form a 3+ ion readily, as do the other elements in this group.

Group 14

Each element in Group 14 has four valence electrons. Carbon (C) is a nonmetal that occurs in two distinct crystalline forms: graphite and diamond. (These forms are known as *allotropes*.) Silicon (Si) and germanium (Ge) are metalloids that are used widely in the computer industry. Tin (Sn) and lead (Pb) are typical metallic substances.

Group 15

Each element in Group 15 has five valence electrons. Nitrogen (N), the first member of the group, exists in nature as N_2, a diatomic molecule with a triple bond (see Chapter 10). The relative inactivity of nitrogen gas in the atmosphere is due to the fact that considerable energy is needed to break the triple bond. Nitrogen is nonmetallic. Phosphorus (P), also a nonmetal, exists in two allotropic forms: red and white.

Group 16

Each element in Group 16 has six valence electrons. Oxygen (O), a nonmetal, exists in nature in two allotropic forms, O_2 and O_3 (ozone). It readily forms a $2-$ ion. Sulfur (S) is also a nonmetal; it exists in several allotropic forms and can form a $2-$ ion although it is less reactive than oxygen. Selenium (Se) and tellurium (Te) are classified as metalloids, and polonium (Po) is considered a metal.

Group 17

The elements in Group 17 are known as the *halogens*. Each element has seven valence electrons. All of these elements are nonmetals that form $1-$ ions. They are the most reactive of the nonmetallic elements, with fluorine (F) being the most reactive element in this group. Fluorine and chlorine (Cl) are gases at room temperature; bromine (Br) is a liquid, and iodine (I) is a solid. Because of their reactivity, the halogens do not occur as free elements in nature.

Group 18

The elements in Group 18 are known as the *noble* (or *inert*) *gases*. Each element other than helium has eight valence electrons. Although helium (He) has only two valence electrons, it is associated with this group because its properties match those of the other elements. The elements of Group 18 are extremely stable and occur as monatomic gases in nature. Although they do not combine readily with other elements, compounds of krypton (Kr) and xenon (Xe) have been prepared.

The Transition Elements

The *transition elements* begin in Period 4 and include Groups 3–11. (Group 12 elements are *not* considered transition elements.) The transition elements exhibit multiple oxidation states because the *two* outermost levels may be involved in chemical combinations. The ions of transition elements are usually colored, both in aqueous solution and as solids.

9.7   THE CHEMISTRY OF A PERIOD

The properties of the elements across a period (left to right) vary much more than the properties of the elements within a single group. Generally, a decrease in atomic radius and an increase in ionization energy and electronegativity occur. The metallic character of successive elements decreases, going from metal to metalloid to nonmetal to noble gas. These changes are accompanied by a gradual change from positive to negative to zero oxidation states.

SECTION II—ADDITIONAL MATERIAL

9.1A   SUBLEVELS AND THE PERIODIC TABLE

Groups 1 and 2 constitute the **s-block** because within these two adjacent groups the *s* sublevels are filled. Groups 13–18 constitute the **p-block** because within these six adjacent groups the *p* sublevels are filled. The *d* sublevels fill in Groups 3–12, and the *f* sublevels fill in the two separated rows at the bottom of the table. These collections of elements are named the **d-block** and the **f-block**, respectively.

The *s*- and *p*-blocks are called the **representative elements**; the *d*-block (through Group 11), the **transition elements**; and the *f*-block, the **lanthanoid** and **actinoid** (or **lanthanide** and **actinide**) *series*.

9.2A SUCCESSIVE IONIZATION ENERGIES

An atom can have as many ionization energies as it has electrons. For example, potassium (K, atomic number 19) has *nineteen* ionization energies. The equations corresponding to the first three ionization energies are as follows:

$$K \quad + \quad I_1 \rightarrow K^+ \quad + \quad e^-$$
$$K^+ \quad + \quad I_2 \rightarrow K^{2+} \quad + \quad e^-$$
$$K^{2+} \quad + \quad I_3 \rightarrow K^{3+} \quad + \quad e^-$$

Successive ionization energies can become very large because of the considerable difficulty in separating negative electrons from increasingly positive ions.

9.3A ELECTRON AFFINITY

Electron affinity measures the energy change that occurs when an atom *gains* a single electron and forms a negative ion. The more energy *released* by the atom, the more stable it becomes when it gains an electron. In this case, the electron affinity is reported as a *negative* number. In some atoms, however, the tendency to gain an electron is negligible, and the electron affinity is reported as zero. In general, nonmetals gain electrons more easily than metals, as the accompanying table shows.

Element	Classification	Electron Affinity (kJ/mol)
Li	Metal	−60
Na	Metal	−53
Mg	Metal	0
C	Nonmetal	−122
S	Nonmetal	−200
Cl	Nonmetal	−349

9.4A ADDITIONAL ASPECTS OF FIRST IONIZATION ENERGY

Explaining the Irregularities of Aluminum and Sulfur

The explanation for the irregularity of *aluminum* lies in the fact that the outer electron is in the $3p$ sublevel. As a result this electron has more energy, and therefore *less* energy is needed to remove it from the atom. The irregularity of *sulfur* is explained by the fact that its outer electron is the fourth electron in the $3p$ sublevel. Removing this electron will leave a half-filled $3p$ sublevel that has more stability. This removal requires *less* energy than removing a single electron in phosphorus, which already has a half-filled $3p$ sublevel. Consequently, the first ionization energy for sulfur is *less* than that for phosphorus.

An Important Principle of First Ionization Energy

These observations allow us to state an important principle: *Half-filled and fully filled sublevels provide an atom with additional stability.* In Chapter 8 (page 229), we saw that two elements, chromium (Cr) and copper (Cu), have irregular electron-filling patterns. This principle explains the observed patterns: The $3d^5 4s^1$ configuration of chromium allows the atom to have two half-filled sublevels; the $3d^{10} 4s^1$ configuration of copper allows the atom to have one fully filled sublevel and one half-filled sublevel.

9.5A VARIATION OF SUCCESSIVE IONIZATION ENERGIES

When we examine the successive ionization energies of the first seven elements across Period 3, as shown in the accompanying table, we see a progressive increase for each element (i.e., $I_1 < I_2 < I_3 < \ldots$). The increases are not smooth, however. For each of the first six elements, there is a very large increase in ionization energy (indicated in the *shaded* boxes in the table). For aluminum (Al), this increase occurs at I_4.

Elements across Period 3	Electron Configuration	Successive Ionization Energies /(kJ/mol)							
		I_1	I_2	I_3	I_4	I_5	I_6	I_7	I_8
Na	[Ne]$3s^1$	496	4562						
Mg	[Ne]$3s^2$	738	1451	7733					
Al	[Ne]$3s^2 3p^1$	578	1817	2745	11,577				
Si	[Ne]$3s^2 3p^2$	787	1577	3232	4356	16,091			
P	[Ne]$3s^2 3p^3$	1012	1907	2914	4964	6274	21,267		
S	[Ne]$3s^2 3p^4$	1000	2252	3357	4556	7004	8496	27,107	
Cl	[Ne]$3s^2 3p^5$	1251	2298	3822	5159	6542	9362	11,018	33,604

In general, when an atom has had all of its *valence* electrons removed, the removal of the *next* electron requires the absorption of a very large quantity of energy because this electron lies within the *noble-gas core* of the atom. Such electrons are held very tightly by the nucleus of the atom.

TRY IT YOURSELF
Explain why the large energy increase for phosphorus (P) occurs at I_6.

ANSWER
An atom of phosphorus has *five* valence electrons $(3s^2 3p^3)$. The sixth electron lies within the noble-gas core of the phosphorus atom.

Because of the great difficulty in removing these inner-core electrons, we can infer that they are unlikely to be removed or (as we shall see in Chapter 9) to participate in any type of chemical combination between atoms. Consequently, we can expect that, according to the table, sodium (Na), magnesium (Mg), and aluminum (Al) will be able to lose one, two, and three electrons, respectively, without too much difficulty.

TRY IT YOURSELF
According to the table, what is the maximum number of electrons you would expect phosphorus (P) and sulfur (S) to be able to lose?

ANSWER
Five and six, respectively.

In reality, phosphorus and sulfur do *not* lose electrons; they tend to *gain* three and two electrons, respectively.

PROBLEM
What happens when an atom of sulfur gains two electrons?

SOLUTION
The sulfur atom becomes an ion with a charge of $2-$ (S^{2-}). The electron configuration of this ion is $[Ne]3s^2 3p^6$, which is the configuration of argon [Ar].

9.6A SYNTHETIC ELEMENTS

A number of elements do *not* occur in nature. These *synthetic elements* include technetium (Tc, atomic number 43) and all of the elements with atomic numbers larger than 92 (the *transuranium* elements), all of which are radioactive. Technetium was the first element to be produced artificially by bombardment; it is also found as a by-product of fission reactors. Transuranium elements are also produced by bombardment reactions, the largest element discovered to date is Oganesson (Og), with an atomic number of 118.

The names and symbols of elements 104–118 are given in the accompanying table.

Atomic Number	Name	Symbol
104	Rutherfordium	Rf
105	Dubnium	Db
106	Seaborgium	Sg
107	Bohrium	Bh
108	Hassium	Hs
109	Neitnerium	Mt
110	Darmstadtium	Ds
111	Roentgenium	Rg
112	Copernicium	Cn
113	Nihonium	Nh
114	Flerovium	Fl
115	Moscovium	Mc
116	Livermorium	Lv
117	Tennessine	Ts
118	Organesson	Og

Naming Newly Discovered Elements

When a new element is discovered, it is given a temporary (Latin) name and a three-letter symbol that is related to its atomic number. For example, when flerovium (Fl, 114) was discovered, it was assigned the temporary name *ununquadium* (symbol, Uuq). Each letter of the symbol corresponds to a digit in the element's atomic number: (U = 1, u = 1, q = 4). Each letter has a Latin root and these roots are joined to form the element's temporary name (*un*, *un*, *quad*, [*ium*]). The accompanying table lists Latin roots and the symbols corresponding to each digit from 0 through 9. As usual the first letter in the three-letter symbol is capitalized.

Digit	Root	Symbol
0	nil	n
1	un	u
2	bi	b
3	tri	t
4	quad	q
5	pent	p
6	hex	h
7	sept	s
8	oct	o
9	enn	e

TRY IT YOURSELF

In the far, far future element 258 may be discovered. Provide its temporary name and symbol by referring to the table given on the bottom of page 262.

ANSWER

bipentoctium, Bpo

END-OF-CHAPTER QUESTIONS

Some questions have the symbol "§2" in front of the question number. This symbol means that the question is based on Section II material.

1. The pair of elements with the most similar chemical properties are
(1) Mg and S   (2) Ca and Br
(3) Mg and Ca   (4) S and Ar

2. More than two-thirds of the elements in the Periodic Table of the Elements are
(1) metalloids   (2) metals
(3) nonmetals   (4) noble gases

3. In the Periodic Tables, all the elements in Group 16 have the same number of
(1) valence electrons   (2) energy levels
(3) protons   (4) neutrons

§2 **4.** In the ground state, how many electrons are in the outermost s sublevel of each element in Group 17?
(1) 5   (2) 2   (3) 7   (4) 8

§2 **5.** The lanthanide and actinide series of elements are found, respectively, in Periods
(1) 6 and 7   (2) 2 and 3
(3) 3 and 4   (4) 4 and 5

6. The elements in the present Periodic Table are arranged according to their
(1) atomic numbers   (2) atomic masses
(3) oxidation states   (4) mass numbers

§2 **7.** The element whose electron configuration is $1s^2 2s^2 2p^6 3s^2$ is a
(1) metalloid   (2) metal
(3) noble gas   (4) nonmetal

8. Which is most likely the first ionization energy, in kilojoules per mole, of a metal?
(1) 520 (2) 1012 (3) 1314 (4) 2081

9. Which element will form an ion whose ionic radius is larger than its atomic radius?
(1) K (2) F (3) Li (4) Mg

10. Atoms of metallic elements tend to
(1) gain electrons and form negative ions
(2) gain electrons and form positive ions
(3) lose electrons and form negative ions
(4) lose electrons and form positive ions

11. Which are two properties of most nonmetals?
(1) low ionization energy and good electrical conductivity
(2) high ionization energy and poor electrical conductivity
(3) low ionization energy and poor electrical conductivity
(4) high ionization energy and good electrical conductivity

12. Which element is considered malleable?
(1) Au (2) H (3) S (4) Rn

13. Element M has an electronegativity of less than 1.2 and reacts with bromine to form the compound MBr_2. Element M could be
(1) Al (2) Na (3) Ca (4) K

14. Which of the following particles has the *smallest* radius?
(1) Na (2) K (3) Na^+ (4) K^+

15. Which element in Period 3 has the least tendency to lose an electron?
(1) Ar (2) Na (3) P (4) Al

16. Which ion is smaller than its corresponding neutral atom?
(1) Al^{3+} (2) Br^- (3) O^{2-} (4) N^{3-}

17. Nonmetals in the solid state are poor conductors of heat and tend to
(1) be brittle
(2) be malleable
(3) have luster
(4) have good electrical conductivity

18. The atomic radius of an oxygen atom is
(1) 167 pm (2) 64 pm (3) 31 pm (4) 598 pm

19. As the atoms of the elements from atomic number 3 to atomic number 9 are considered in sequence from left to right across the Periodic Table, the atomic radius of each successive atom is
(1) smaller, and the nuclear charge is less
(2) smaller, and the nuclear charge is greater
(3) larger, and the nuclear charge is less
(4) larger, and the nuclear charge is greater

20. As the elements in Group 1 are considered in order of increasing atomic number, the atomic radius of each successive element increases. This is primarily due to an increase in the number of
(1) neutrons in the nucleus
(2) electrons in the outermost shell
(3) unpaired electrons
(4) principal energy levels

21. In Period 2, as the elements are considered from left to right, there is a decrease in
(1) ionization energy   (2) atomic mass
(3) metallic character   (4) nonmetallic character

22. Which ion has the largest radius?
(1) I^- (2) Cl^- (3) Br^- (4) F^-

23. Which element in Group 15 has the most metallic character?
(1) N   (2) P   (3) As   (4) Bi

24. Which ion has the smallest radius?
(1) Na^+ (2) K^+ (3) Mg^{2+} (4) Ca^{2+}

25. Which ion has the largest radius?
(1) Na^+ (2) K^+ (3) Mg^{2+} (4) Ca^{2+}

26. Which aqueous solution has a color?
(1) $BaSO_4$(aq)   (2) $CuSO_4$(aq)
(3) $SrSO_4$(aq)   (4) $MgSO_4$(aq)

27. Which element is so active chemically that it occurs naturally only in compounds?
(1) K   (2) Ag   (3) Cu   (4) S

28. Which element is a solid at room temperature and standard pressure?
(1) Br   (2) I   (3) Hg   (4) Ne

29. Ozone is an allotropic form of the element
    (1) O   (2) P   (3) S   (4) C

30. In which group do the elements usually form chlorides that have the general formula MCl_2?
    (1) 1   (2) 2   (3) 17   (4) 18

31. Which of the following elements is most likely to form a compound with radon?
    (1) I   (2) F   (3) Na   (4) Ca

32. An element whose atoms have the electron configuration 2-8-18-1 is
    (1) a transition element   (2) a noble gas
    (3) an alkali metal   (4) an alkaline earth

33. Which is the most active nonmetal in the Periodic Table of the Elements?
    (1) Na   (2) F   (3) I   (4) Cl

34. Element X is in Group 1, and element Y is in Group 16. A compound formed from these two elements is most likely to have the formula
    (1) X_2Y (2) XY_2 (3) X_2Y_7 (4) X_7Y_2

35. As the elements are considered from the top to the bottom of Group 15, which sequence in properties occurs?
    (1) metal → metalloid → nonmetal
    (2) metal → nonmetal → metalloid
    (3) metalloid → metal → nonmetal
    (4) nonmetal → metalloid → metal

36. A chloride dissolves in water to form a colored solution. The chloride could be
    (1) HCl   (2) KCl   (3) $CaCl_2$ (4) $NiCl_2$

37. Which metal atoms can form ionic bonds by losing electrons from both the outermost and next-to-outermost principal energy levels?
    (1) Fe   (2) Pb   (3) Mg   (4) Ca

38. Which of the following periods contains the greatest number of metallic elements?
    (1) 1   (2) 2   (3) 3   (4) 4

39. Which sequence of properties occurs as the atomic numbers increase across Period 3?
(1) metal → metalloid → nonmetal → noble gas
(2) metal → nonmetal → noble gas → metalloid
(3) nonmetal → metalloid → metal → noble gas
(4) nonmetal → metal → noble gas → metalloid

§2 **40.** A newly discovered element has been given the provisional symbol Ubb. According to the chart on the bottom of page 262, the atomic number of this element is
(1) 112   (2) 122   (3) 211   (4) 221

41. Which list of elements consists of a metal, a metalloid, and a noble gas?
(1) aluminum, sulfur, argon
(2) magnesium, selenium, sulfur
(3) sodium, silicon, argon
(4) silicon, phosphorus, chlorine

42. Element X reacts with copper to form the compounds CuX and CuX_2. In which group of the Periodic Table is element X found?
(1) Group 1   (2) Group 2   (3) Group 13   (4) Group 17

43. All elements on the modern Periodic Table are arranged in order of increasing
(1) atomic mass
(2) nuclear mass
(3) number of neutrons per atom
(4) number of protons per atom

44. The valence electron of which atom in the ground state has the greatest amount of energy?
(1) cesium   (2) lithium   (3) rubidium   (4) sodium

Constructed-Response Questions

Base your answers to questions 1 and 2 on the information below.

You are given samples of Na, Ar, As, and Rb.

1. Which two of the given elements have the most similar chemical properties?

2. Explain your answer in terms of the Periodic Table of the Elements.

Base your answers to questions 3–6 on the information below.

Potassium ions are essential to human health. The movement of dissolved potassium ions, $K^+(aq)$, into and out of a nerve cell allows that cell to transmit an electrical impulse.

3. What is the total number of electrons in a potassium ion?

4. Explain, in terms of atomic structure, why a potassium ion is smaller than a potassium atom.

5. What property of potassium ions allows them to transmit an electrical impulse?

6. Based on the Periodic Table, explain why the elements Na and K have similar chemical properties.

Base your answers to questions 7–9 on the information below and on your knowledge of chemistry.

There are six elements in Group 14 of the Periodic Table. One of these elements had the temporary name ununquadium (symbol Uuq). It is now known as flerovium (symbol Fl).

7. Identify an element in Group 14 that is a metalloid.

8. Explain in terms of electron shells why each successive element in Group 14 has a larger atomic radius as you consider the elements in order of increasing atomic number.

9. State the expected number of valence electrons in a ground-state atom of flerovium.

Base your answers to questions 10–12 on your knowledge of chemistry.

10. Compare the radius of a fluoride ion with the radius of a fluorine atom.

11. Explain in terms of subatomic particles why the radius of a lithium ion is smaller than the radius of a lithium atom.

12. Describe the general trend in atomic radius as each element in Period 2 is considered in order from left to right.

The answers to these questions are found in Appendix 3.

Chapter Ten	# CHEMICAL BONDING AND MOLECULAR SHAPE

KEY IDEAS

This chapter focuses on the ways in which atoms, ions, and molecules link together. The properties of substances are discussed in terms of their chemical bonds.

KEY OBJECTIVES
At the conclusion of this chapter you will be able to:
- Explain how energy and stability are related to chemical bond formation.
- Describe how ionic bonding occurs.
- Draw Lewis structures for simple ions.
- Describe how covalent bonding occurs.
- Predict whether a bond between atoms is ionic or covalent.
- Distinguish between polar and nonpolar bonds.
- Draw Lewis structures for covalently bonded molecules.
- Describe coordinate covalent bonding.
- Describe network and metallic bonding.
- Define the terms *polar*, *nonpolar*, and *dipole* as they apply to molecules.
- Describe how the polarity of a molecule is related to its symmetry.
- Define the term *intermolecular attraction (force)*.
- Define and apply the following terms as they apply to inter-molecular attractions: *London dispersion forces, dipole-dipole attraction, hydrogen bonding,* and *van der Waals forces.*
- Describe how molecule-ion attractions occur.
- Relate chemical bond types to the properties of substances.
- Define the term *resonance structure*, and explain how this alternative feature plays a part in covalent bonding.
- Describe how the octet rule is *not* obeyed in certain molecules.
- Use the VSEPR model to predict the shape of a covalently bonded molecule or ion.
- Describe how orbital overlap leads to the formation of a covalent bond.
- Define the terms *sigma bond* and *pi bond*, and indicate how each type formed from the overlapping of atomic orbitals.

- Define the terms *bond energy* and *bond length*, and describe how these quantities are related.
- Define the term *hybrid orbital*, and indicate how *sp*, *sp²*, and *sp³* orbitals are formed.

SECTION I—BASIC (REGENTS-LEVEL) MATERIAL

NYS REGENTS CONCEPTS AND SKILLS

Note: By the time you have finished Section I, you should have mastered the concepts and skills listed below. The Regents chemistry examination will test your knowledge of these items and your ability to apply them.

Concepts are the *basic ideas* that form the body of the Regents chemistry course (what you need to know!).

Skills are the *activities* that demonstrate your mastery of these concepts (how you show that you know them!).

Following each concept or skill is a page reference (given in parentheses) to this chapter.

10.1 Concept:
When a bond is broken, energy is absorbed. When a bond is formed, energy is released. (Pages 272, 297)

10.2 Concepts:
- Atoms attain a stable valence electron configuration by bonding with other atoms. (Page 274)
- Noble gases have stable valence configurations and tend not to bond. (Page 272)

Skill:
Determine the noble-gas configuration an atom will have as a result of bonding. (Pages 273–274)

10.3 Concept:
Two major categories of compounds are ionic and molecular (covalent) compounds. (Pages 273–276)

10.4 Concept:
Chemical bonds are formed when valence electrons are:
(a) transferred from one atom to another (ionic) (Page 273);
(b) shared between atoms (covalent) (Page 275);
(c) mobile within a metal (metallic) (Pages 280–281).

Skill:
Demonstrate bonding using Lewis (electron-dot) structures representing:
(a) the transferring of valence electrons in ionic bonding (Page 274);
(b) the sharing of valence electrons in covalent bonding (Page 275);
(c) the formation of a stable octet as a result of bonding (Page 275).

10.5 Concept:
Electron-dot diagrams (Lewis structures) can represent the valence electron arrangement in compounds and ions. (Pages 273–276)

10.6 Concept:
In a multiple covalent bond, more than one pair of electrons are shared between two atoms. (Page 275)

10.7 Concept:
Electronegativity indicates how strongly an atom of an element attracts electrons in a chemical bond. Electronegativity values are assigned according to arbitrary scales. (Pages 276–277)

10.8 Concept:
The electronegativity difference between two bonded atoms is used to determine the degree of polarity in the bond. (Pages 276–277)

Skill:
Distinguish between nonpolar covalent and polar covalent bonds. (Pages 276–277)

10.9 Concepts:
• Metals tend to react with nonmetals to form ionic compounds. (Page 273)
• Nonmetals tend to react with other nonmetals to form molecular (covalent) compounds. (Page 275)
• Ionic compounds containing polyatomic ions have both ionic and covalent bonding. (Page 279)

10.10 Concept
The polarity of a molecule can be determined by the shape of the molecule and the distribution of charge around it. Symmetrical (nonpolar) molecules include CO_2, CH_4, and diatomic elements. Asymmetrical (polar) molecules include HCl, NH_3, and H_2O. (Page 282)

10.11 Concepts:
• Intermolecular forces created by an unequal distribution of charge result in varying degrees of attraction between molecules. (Pages 283–287)

271

- Hydrogen bonding is an example of a strong intermolecular force. (Pages 285–286)

Skill:
Explain each of the following in terms of intermolecular forces:
(a) vapor pressure;
(b) evaporation rate;
(c) phase changes (Pages 283–287).

10.12 Concept:
Compounds can be differentiated by their chemical and physical properties. (Chart on page 289)

10.13 Concept:
Physical properties of substances can be explained in terms of chemical bonds and intermolecular forces. (These properties include conductivity, malleability, solubility, hardness, melting point, and boiling point.) (Chart on page 289)

Skill:
Compare the physical properties of substances based on chemical bonds and intermolecular forces. (Chart on page 289)

10.1 BONDING AND STABILITY

Now that we have studied the structure of atoms, let us learn how atoms and molecules bond to one another. The first question we need to ask is this: Why do chemical bonds exist at all?

An atom becomes more stable when it acquires the electron configuration of a noble gas. If the valence level of a "stabilized" atom is the first principal energy level, it will contain *two* electrons; otherwise, it will contain *eight* electrons.

Normally, however, the valence levels of atoms are not isoelectronic with the valence levels of noble gases. In order to achieve this state, atoms *bond* with one another in a variety of ways. In most cases, bonded atoms acquire eight electrons in their valence levels. In this case, we say that the bonding atoms have obeyed the **octet rule**.

As any system becomes more stable, its energy *decreases*. This decrease occurs when atoms form chemical bonds with one another. For example, if the energy of two isolated hydrogen (H) atoms is compared with the energy of an H_2 molecule (in which two atoms of hydrogen are *bonded* together), we find that the energy of the molecule is less than the energy of the two isolated atoms. Therefore, the H_2 molecule is more stable than the two isolated hydrogen atoms, as shown in the accompanying diagram.

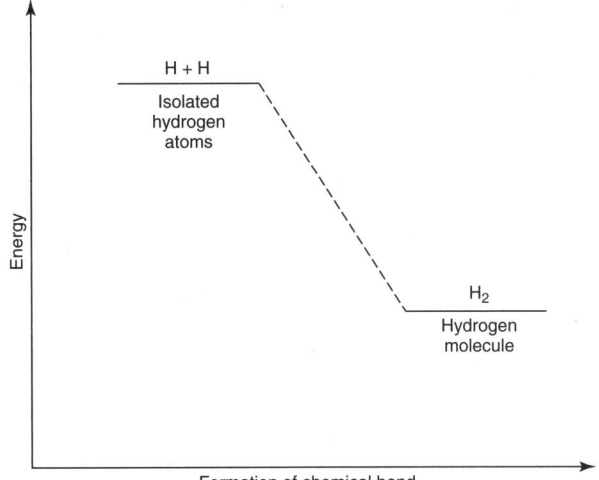

Formation of chemical bond

10.2 IONIC BONDING

Electron Transfer

To achieve noble-gas configurations, a pair of atoms may fill their valence levels by *transferring electrons* between them to form ions with opposite charges. The following examples illustrate how an electron is lost by an atom of potassium and is transferred to an atom of fluorine:

$$K \rightarrow K^+ + e^-$$

$$F + e^- \rightarrow F^-$$

When the pair of ions have been created, their opposite charges cause them to attract strongly. We call this attraction an **ionic bond**. When an ionic bond is formed in a compound such as potassium fluoride (KF), the attraction is not limited to *pairs* of ions. Oppositely charged ions arrange themselves in a regular array (known as a crystal lattice), as shown in the diagram on page 274.

From the diagram it can be seen that KF does *not* exist as individual units or *molecules*. This is true of ionic compounds in general. The formula of an ionic compound simply tells us the *ratio* of positive ions to negative ions. We call this ratio a **formula unit**. In the case of KF, the ratio is 1 to 1.

For electron transfer to occur, one atom must be able to lose electrons readily (i.e., have a low ionization energy) and the other atom must be able to gain electrons readily (i.e., have a sufficiently negative electron affinity). In Section 10.4, we will learn to predict whether two atoms will be able to transfer electrons between them and to form an ionic bond.

Lewis Structures of Simple Ions

The Lewis structures of simple ions such as K^+ and F^- are easy to draw. When the positive ions of the representative elements of Groups 1, 2, and 13 are formed, *all of the valence electrons are lost* and the Lewis structures of ions such as Na^+, Mg^{2+}, and Al^{3+} are written, respectively as:

$$[Na]^+ \quad [Mg]^{2+} \quad [Al]^{3+}$$

When the negative ions of Groups 17, 16, and 15 are formed, *all of the valence levels have been completed and contain eight electrons.* The Lewis structures of ions such as F^-, O^{2-}, and N^{3-} are written, respectively, as:

$$\left[: \ddot{F} :\right]^- \quad \left[: \ddot{O} :\right]^{2-} \quad \left[: \ddot{N} :\right]^{3-}$$

SECTION OF POTASSIUM FLUORIDE (KF) CRYSTAL LATTICE
(TWO-DIMENSIONAL)

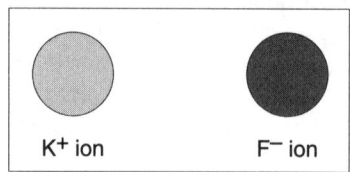

K⁺ ion   F⁻ ion

10.3 COVALENT BONDING

Certain pairs of atoms do not have a strong tendency to transfer electrons between them. Then how do they complete their valence levels? These atoms *share* electrons between them. Electron sharing always occurs in *pairs,* and two atoms may normally share one, two, or three pairs of electrons. The atoms bond together because the shared pair or pairs belong to *both* atoms. This type of bonding is called **covalent bonding**. The following examples show how it is accomplished:

$$\overset{..}{\underset{..}{:}F\overset{xx}{\underset{xx}{}}\overset{x}{\underset{x}{F}}x \qquad H\overset{xx}{\underset{xx}{:}}\overset{x}{\underset{x}{Cl}}x$$

In each molecule a single pair of electrons is shared between the atoms. This type of arrangement is called a *single covalent* bond. Using chemical shorthand, we can let a line represent a pair of electrons. Then, the molecules can be represented as F—F and H—Cl.

However, the sharing of the electron pair is not the same in both molecules. In the F_2 molecule, the pair of electrons is shared *equally* by both atoms. This type of bonding is known as **nonpolar covalent** bonding. In the HCl molecule, however, the electron pair is not shared equally; the chlorine atom has a greater attraction for the shared electrons than the hydrogen atom. This type of bonding is known as **polar covalent** bonding.

Multiple Covalent Bonds

Let us construct a molecule of O_2. An oxygen atom has six valence electrons; and in order to place eight electrons in the valence level of each oxygen atom, it is necessary that the atoms share *two* pairs of electrons, as shown below:

$$:\overset{..}{O}::\overset{..}{O}:$$

This is known as a *double* covalent bond and may be represented as $:\overset{..}{O} = \overset{..}{O}:$. (If you take more advanced courses in chemistry, you will learn why this diagram does not adequately represent the structure of O_2. For the present, however, we will accept it as written.)

PROBLEM
How is a molecule of N_2 constructed?

SOLUTION
A nitrogen atom has five valence electrons. Therefore, it is necessary that the two atoms share *three* pairs of electrons (known as a triple bond). This is represented as follows:

$$: N::: N: \quad or \quad : N \equiv N:$$

Coordinate Covalent Bonding

We can draw the Lewis structure for the ammonium (NH_4^+) ion by adding an H^+ ion to the NH_3 molecule:

$$H \overset{\cdot\cdot}{\underset{\bullet x}{\text{N}}} H \xleftarrow{\quad H^+ \quad} \qquad \longrightarrow \qquad \left[H \overset{\overset{\text{H}}{\cdot\cdot}}{\underset{\underset{\text{H}}{\bullet x}}{\text{N}}} H \right]^+$$

Note that in the NH_3 molecule the hydrogen atoms are bonded to the nitrogen atom with covalent bonds in which each atom contributes one electron to the shared pair. The hydrogen *ion,* however, lacks electrons; and when it bonds, the nitrogen atom must contribute *both* electrons to the shared pair. This type of bond, known as a *coordinate covalent* bond, is found in many molecules and polyatomic ions.

Once the coordinate covalent bond is formed, *it is exactly the same as any other covalent bond.* If we could "view" the nitrogen-hydrogen bonds in an ammonium ion, for example, each bond would be seen to be entirely equivalent to the others.

PROBLEM
Draw the Lewis structure for the H_3O^+ ion, starting with H_2O and H^+.

SOLUTION

$$H \overset{\cdot\cdot}{\underset{\underset{\text{H}}{\bullet x}}{\text{O}}}: \xleftarrow{\quad H^+ \quad} \qquad \longrightarrow \qquad \left[H \overset{\overset{\text{H}}{}}{\underset{\underset{\text{H}}{\bullet x}}{\text{O}}}: \right]^+$$

This polyatomic ion, known as the *hydronium* ion, plays a key role in acid–base chemistry.

10.4 ELECTRONEGATIVITY AND BONDING

How can we tell whether two atoms form an ionic, polar covalent, or nonpolar covalent bond? We compare the *electronegativities* of the atoms. In Chapter 4 we learned that *electronegativity* measures the relative attraction of an atom for shared electrons. The higher the electronegativity, the more the atom attracts the shared pair of electrons. Electronegativities (given in Reference Table S in Appendix 1 of this book) range from a high of 3.98 for fluorine (F) to a low of 0.70 for cesium (Cs).

It is the electronegativity *difference* (Δ) that enables us to predict the type of bond between two atoms. We calculate Δ by subtracting the smaller electronegativity value from the larger one.

First, however, we must recognize that no bond is *completely* ionic or covalent: it is a mixture of both types. We can relate the nature of the bond to the difference in the electronegativities of the bonded atoms. If the difference in the electronegativities of the bonded atoms (Δ) is:

- 0.0–0.4, we classify the bond as *nonpolar covalent*.
- 0.5–1.6, we classify the bond as *polar covalent*.
- 1.7 or above, we classify the bond as *ionic*.

This information is summarized in the accompanying table.

Range of Δ*	Bond Type	Example
0.0−0.4	Nonpolar covalent	BrCl (Δ = 0.2)
0.5−1.6	Polar covalent	HBr (Δ = 0.9)
≥1.7	Ionic	KF (Δ = 3.2)

*Other books may use different ranges to determine bond types. At what precise value of Δ a bond is nonpolar covalent, polar covalent, or ionic is really a judgment call.

TRY IT YOURSELF
Using the electronegativity values in Reference Table S in Appendix 1, classify the type of bond that each of the following pairs of atoms is likely to produce:
(a) Na, Br   (b) H, O   (c) N, N

ANSWERS
(a) ionic (Δ = 2.1) (b) polar covalent (Δ = 1.3) (c) nonpolar covalent (Δ = 0)

10.5 DRAWING THE LEWIS STRUCTURES OF COVALENT MOLECULES AND POLYATOMIC IONS

It is one thing to know which atoms are present in a covalent molecule or polyatomic ion; it is quite another to be able to say *how* the atoms are joined. Fortunately, there is a set of five rules that we can use in order to "build" molecules and polyatomic ions. When we follow these rules, we end up with a representation of a molecule known as a **Lewis structure**, a term familiar to us from Chapter 8 and illustrated for simple ions on page 274 of this chapter. There are (as always!) exceptions to these rules, but we will not worry about them for the moment.

Let us apply the set of rules in order to "build" a molecule of water (H_2O):

1. Count the total number of valence electrons in the molecule or ion.

   In this example, the water molecule has eight valence electrons: six from the oxygen atom and one from each hydrogen atom.

2. If the particle is an ion, add one electron for each negative charge present; subtract an electron for each positive charge present.

   In this example, the water molecule is neutral. No electrons are added or subtracted.

3. Draw a skeleton in which all of the atoms are connected by single bonds. It is always advisable to try a *symmetrical* structure first. Remember: A single bond represents a pair of electrons that belongs to both of the atoms it connects.

   Our skeleton looks like this:

$$H-O$$
$$|$$
$$H$$

   (In Section 10.9 we will learn why the skeletons of H_2O and certain other molecules are written at an angle, rather than in a straight line.)

4. Complete the valence level of each atom in the molecule by adding *unshared pairs* of electrons as necessary. The valence level of a hydrogen atom is complete with two electrons; assume that all other atoms are complete with eight electrons.

   In this example, the valence levels of both hydrogen atoms are already complete. The oxygen atom needs two unshared pairs of electrons:

$$H-\overset{..}{O}:$$
$$|$$
$$H$$

5. Compare the total number of electrons in the skeleton with the number you arrived at in steps 1 and 2. If they agree, you are finished.

   In this example, the numbers of electrons agree and our molecule is "built."

PROBLEM
Draw the Lewis structure of the ClO_2^- ion.

SOLUTION

Each numbered step corresponds to the same one in the rules given on the previous page.

1. The number of valence electrons is 19 (6 from each oxygen and 7 from chlorine).
2. We *add* 1 electron since the ion has a charge of $1-$. This brings the total to 20.
3. The proposed (symmetrical) skeleton is:

$$\begin{array}{c} O-Cl \\ | \\ O \end{array}$$

4. We complete the octet of each atom by adding pairs of *unshared* electrons:

$$\left[\begin{array}{c} :\ddot{O}-\ddot{Cl}: \\ | \\ :\ddot{O}: \end{array}\right]^{-}$$

5. The number of electrons in our skeleton (20) agrees with our count from steps 1 and 2, and we propose the result of step 4 as our Lewis structure.

TRY IT YOURSELF

Draw the Lewis structure of the NH_4^+ ion.

ANSWER

$$\left[\begin{array}{c} H \\ | \\ H-N-H \\ | \\ H \end{array}\right]^{+}$$

What happens if the number of electrons in the skeleton does *not* agree with the number of electrons supplied by the atoms? The next example illustrates how we handle this situation.

Suppose we wish to draw the Lewis structure of the C_2H_4 molecule. Using our set of rules, we arrive at the skeleton:

$$\begin{array}{cc} H & H \\ | & | \\ H-\underset{..}{C}-\underset{..}{C}-H \end{array}$$

When we count the electrons in the skeleton, we find that we have 14. We should have 12 (four from each carbon and one from each hydrogen)! We can correct this situation by introducing *double* or *triple* bonds.

When we form a double bond, we remove *two* unshared pairs of electrons from adjacent atoms and replace them with the double bond. This has the effect of *lowering* the electron count by 2.

Triple bonds are formed by removing *four* unshared pairs of electrons and replacing them with the triple bond. This has the effect of lowering the electron count by 4.

Using this technique, we arrive at the Lewis structure:

$$\begin{array}{ccc} H & & H \\ | & & | \\ H-C & = & C-H \end{array}$$

TRY IT YOURSELF
Draw the Lewis structures of the following molecules and ion:
(a) C_2H_2 (b) CO_2 (c) NO_2^-

ANSWERS

(a) $H-C\equiv C-H$ (b) $\ddot{O}=C=\ddot{O}$ (c) $\left[:\ddot{O}^{\diagup}\overset{\displaystyle \overset{..}{N}}{}\diagdown \ddot{O}: \right]^-$

10.6 NETWORK SOLIDS

Certain atoms, such as carbon, are able to form covalent bonds with other like atoms in a three-dimensional arrangement called a **network solid**. The three-dimensional network solid that consists of carbon is known as the familiar substance *diamond*. Another allotrope of carbon, *graphite,* also exists as a modified network solid, but it consists of a series of loosely associated *two-dimensional sheets*.

Other network solids include silicon carbide (SiC) and silicon dioxide (SiO_2). Network solids, by their very nature, do *not* form individual molecules. As we shall see at the end of Section 10.11, network solids have unique properties because of their extended covalent bonding.

10.7 METALLIC SUBSTANCES

As we learned in Chapter 8, metals have four specific properties (luster, conductivity, malleability, ductility) that arise from the way in which metal atoms bond to one another. Atoms that form metallic substances have low ionization energies and relatively unoccupied valence levels. These conditions allow the valence electrons the freedom and the room in which to move. In a metallic substance, the valence orbitals of the atoms merge with one another, and the valence electrons are *delocalized* over the entire metallic crystal. As a result the valence electrons do not belong to a single atom; they belong to the entire crystal. The remaining part of each atom, which consists of the *kernel* (a positive ion), forms the backbone of the crystal. The accompanying diagram represents a section of a metallic crystal.

Section of Metallic Crystal

Kernel of metal atom (+)

Mobile valence electrons (−)

10.8 DIPOLES AND POLAR MOLECULES

When a bond is classified as polar covalent (such as the H-O bond), the atom with the higher electronegativity (O in this example) has a greater attraction for the shared electrons than the atom with the lower electronegativity (H in this example). As a result, the electron pair is drawn closer to the high-electronegativity atom and the charge around the two atoms becomes unbalanced. The "negative" end of the bond is centered around the atom with the higher electronegativity, and the "positive" end around the atom with the lower electronegativity.

We call this charge imbalance a **dipole.** The positive and negative ends of the dipole are represented by the symbols $\delta+$ and $\delta-$, respectively. The H-O bond may then be represented as follows:

$$\delta+ \ H-O \ \delta-$$

The positive and negative ends of a dipole are not real charges (such as positive and negative *ions*) because no electrons have actually been transferred between the atoms. The dipole represents only an *unbalanced* charge distribution along the bond.

In certain instances, the polar covalent bonds in a molecule will cause the *entire molecule* to have an unbalanced charge distribution. Such molecules are affected by electric fields and are also known as dipoles or as **polar** *molecules*. The HCl molecule has an *unbalanced* charge distribution and is a *polar* molecule. In contrast, the Ne (monatomic) and F_2 (diatomic) molecules have *balanced* charge distributions and are **nonpolar** molecules.

10.9   POLARITY AND MOLECULAR SYMMETRY

We might suppose that *all* molecules with polar covalent bonding would automatically be polar, but this is *not* the case. Whether or not a *molecule* is polar depends on the *symmetry* of its (electron) charge distribution.

For example, H_2O is a polar molecule because its polar covalent bonding *and* its bent shape give it an *asymmetrical* (unbalanced) charge distribution, as illustrated in the diagram below.

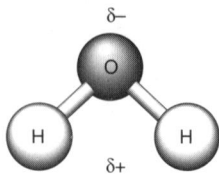

H₂O: A Polar Molecule

In contrast, CO_2 is *not* a polar molecule. While both C=O bonds are polar covalent, the linear shape of the molecule gives it a *symmetrical* (balanced) charge distribution. As a result any dipole arising from one of the C=O bonds is canceled by the other, as shown below.

CO₂: A Nonpolar Molecule

In summary:

- Molecules containing only nonpolar covalent bonds and all monatomic molecules are nonpolar.
  Examples include He, Ar, N_2, P_4, and C (diamond).
- Molecules containing polar covalent bonds with symmetrical charge distributions are nonpolar.
  Examples include CO_2, BF_3, CCl_4, and $BeCl_2$.
- Molecules containing polar covalent bonds with asymmetrical charge distributions are polar.
  Examples include HF, H_2O, NH_3, SO_2, and N_2O.

10.10 INTERMOLECULAR FORCES

As its name implies, **intermolecular forces** are those attractions that exist between molecules (or between molecules and ions). We know that molecules do attract one another because gaseous substances condense to form liquids and these, in turn, condense to form solids. Intermolecular forces are also responsible for the solubilities of substances. For example, molecular substances such as table sugar (sucrose) and table salt (sodium chloride) dissolve in water because of the association of water molecules with the molecules of sucrose and the ions present in sodium chloride. All of these attractions are known collectively as **van der Waals forces**. These intermolecular forces include **dipole forces**, **London dispersion forces**, **hydrogen bonding**, and **ion-dipole forces**.

One important way to compare the strengths of intermolecular forces is to examine the boiling points of various substances. Substances with stronger intermolecular attractions have *higher* boiling points than those with weaker intermolecular attractions.

Dipole Forces

Dipole forces exist among polar molecules such as HCl. Each of the molecules of this substance has both a positive and a negative end. Seeing how dipole attractions can occur is easy. The positive end of the dipole is located near the hydrogen, and the negative end is near the chlorine. Since opposite charges attract, the molecules line up so oppositely charged dipoles are close to one another. As the temperature of HCl gas is reduced, these dipole attractions are responsible for the formation of the liquid phase. This is illustrated below:

$$\delta+ H—Cl\ \delta-\cdots\delta+ H—Cl\ \delta-\cdots\delta+ H—Cl\ \delta-\cdots\delta+ H—Cl\ \delta-$$

(The solid lines represent *covalent bonds*. The dotted lines represent *dipole forces*.)

Hydrogen chloride dissolves readily in water because of the dipole forces that exist between HCl and H_2O molecules. The diagram below is a simplified representation of this association.

$$\cdots H—Cl\cdots \begin{array}{c} H \\ \diagdown \\ O \\ \diagup \\ H \end{array} \cdots H—Cl\cdots \begin{array}{c} H \\ \diagdown \\ O \\ \diagup \\ H \end{array} \cdots$$

Generally, dipole forces depend on the *polarities* of the molecules involved. The more polar the molecule is, the stronger the dipole forces and the higher the boiling point. In order to compare the strengths of dipole forces, we must examine substances that have molar masses ($\mathcal{M}$) close in value. The table below compares three pairs of substances and their respective boiling points. Each pair consists of nonpolar and polar substances with similar molar masses.

Boiling Points of Nonpolar and Polar Substances

Nonpolar Substance	$\mathcal{M}$ (g/mol)	BP (°C)	Polar Substance	$\mathcal{M}$ (g/mol)	BP (°C)
SiH_4	32	−112	PH_3	34	−87
GeH_4	77	−90	AsH_3	78	−55
Br_2	160	+59	ICl	162	+97

In each pair, the polar substance has a higher boiling point than the corresponding nonpolar substance with a similar molar mass due to the presence of dipole forces in each polar substance.

London Dispersion Forces

Wait! There's more! Let us examine the preceding table once again. If we compare the nonpolar substances (SiH_4, GeH_4, and Br_2) and the polar substances (PH_3, AsH_3, and ICl) separately, we observe another trend. As the molar masses of the substances increase, the corresponding boiling points also increase. This trend is shown in the accompanying graph.

Molar Mass and Boiling Point

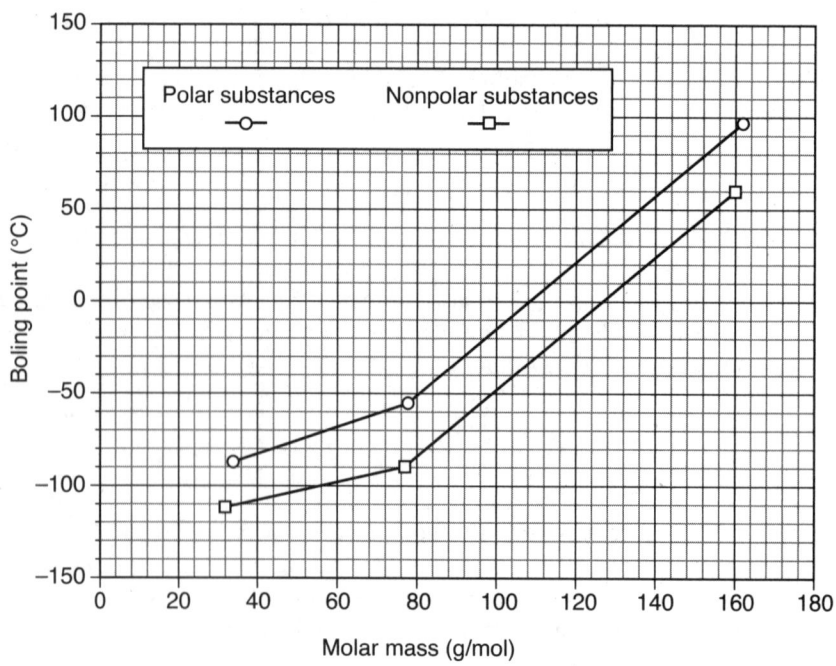

London dispersion forces arise because all molecules, both polar and non-polar, can create temporary dipoles for short periods of time. The resulting forces increase with increasing molecular mass and decrease with increasing distance between adjacent molecules.

Hydrogen Bonding

According to the previous table, the molar mass has a direct influence on the attractions between molecules, as reflected in the boiling point. Nevertheless, some substances seem to defy this general rule. Let's examine the boiling points of the hydrogen compounds of the elements found in Groups 15, 16, and 17. The following table provides these data.

Boiling Points of Various Hydrogen Compounds

Periodic Group	Period	Substance	$\mathcal{M}$ (g/mol)	Boiling Point (°C)
15	2	NH_3	17	−33
	3	PH_3	34	−87
	4	AsH_3	78	−55
	5	SbH_3	125	−17
16	2	H_2O	18	+100
	3	H_2S	34	−60
	4	H_2Se	81	−41
	5	H_2Te	130	−2
17	2	HF	20	+19
	3	HCl	36	−85
	4	HBr	81	−67
	5	HI	128	−35

When we graph these data, we observe some unusual results. We can readily see that each compound formed from a Period 2 element (nitrogen, oxygen, or fluorine) has an unusually high boiling point when compared with those of the other compounds within its group. How can we explain this observation?

First, we note that all of the compounds listed contain *hydrogen*. Second, the unusual behavior occurs only when hydrogen is bonded to a *small, highly electronegative element*, namely nitrogen, oxygen, or fluorine. We attribute this unusual behavior to a special case of dipole attraction called **hydrogen bonding**. This combination of elements allows the hydrogen atom of one molecule to approach the nitrogen, oxygen, or fluorine atom of a *neighboring* molecule very closely, creating an unusually strong attraction between the two polar molecules.

It must be emphasized that hydrogen bonding is *not* covalent bonding. It is a stronger type of dipole force and occurs *between neighboring molecules*. The following diagram provides examples of various types of hydrogen bonds.

$$H—\ddot{O}:\cdots H—\ddot{O}:$$
$$\phantom{H—\ddot{O}:\cdots}H\phantom{—\ddot{O}:}H$$

$$H—\ddot{N}:\cdots H—\ddot{O}:$$

$$H—\ddot{N}:\cdots H—\ddot{N}:$$

$$H—\ddot{O}:\cdots H—\ddot{N}:$$

$$H—\ddot{F}:\cdots H—\ddot{F}:$$

The solid lines represent covalent bonds; the dotted lines, hydrogen bonds.

Ion-Dipole Forces

It is well known that ionic compounds such as NaCl (table salt) dissolve readily in water. This solubility is due to the attraction of the Na^+ and Cl^- ions for the negative and positive ends of polar water molecules. As NaCl is added to water, water molecules surround and separate the ions, destroying the crystal structure in a process known as *hydration*. The accompanying diagram illustrates how molecules and ions attract.

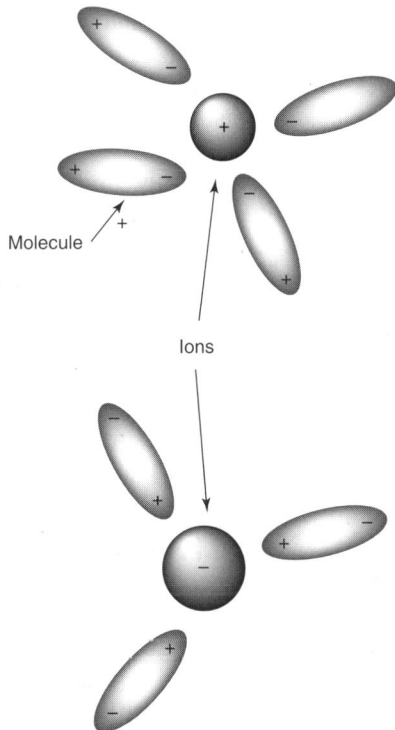

Molecule

Ions

10.11 PHYSICAL AND CHEMICAL PROPERTIES OF BONDED SUBSTANCES: A SUMMARY

The properties of a substance depend on the type of bonding present within the substance. The accompanying chart summarizes the properties of bonded substances.

- *Ionic substances* have high melting and boiling points because of the strong electrical attraction of the ions for one another. These substances usually dissolve in a polar solvent such as water, and their crystals are hard and brittle.

Ionic substances conduct electricity in solution and in the liquid phase, but not in the solid phase because of the rigidity of the ionic crystal.

- *Polar covalent substances* have intermediate melting and boiling points. They also dissolve in polar solvents, but they do not conduct electricity well unless they ionize in solution. Polar covalent substances do not conduct well in either the liquid or the solid phase.
- *Nonpolar substances* have the lowest melting and boiling points because of the weak attraction due to dispersion forces. Nonpolar substances dissolve in nonpolar solvents and do not conduct electricity well in any phase or in solution. Nonpolar crystals tend to be soft and waxy.
- *Metallic substances* have melting and boiling points that vary from low to high, as do the hardnesses of metallic crystals. Metals do not dissolve well in either polar or nonpolar solvents. Metals conduct electricity in the solid and liquid phases.
- *Network substances* have extraordinarily high melting points and form very hard, brittle crystals. They do not generally dissolve in polar or nonpolar solvents, and they are poor conductors of electricity in the solid and liquid phases.

The chart on page 289 summarizes the properties of substances that are consequences of their bond types.

SECTION II—ADDITIONAL MATERIAL

10.1A   RESONANCE STRUCTURES

In the structure below, *either* oxygen atom can be associated with the double bond. Therefore, we could write *two* Lewis structures for the NO_2^- ion. When two or more plausible Lewis structures can be written for a molecule or an ion, each structure is called a **contributing resonance structure**. The *actual* structure of the molecule or ion is a *composite* of its contributing resonance structures and is called a **resonance hybrid**. In the case of the NO_2^- ion, the resonance hybrid is viewed as a structure in which the double bond is actually extended across *both* oxygen atoms, as shown below:

Crystal Type	Particles in Crystal	Principal Attractive Forces Between Particles	Melting Point	Electrical Conductivity of Liquid	Characteristics of Crystal	Conditions for Formation	Examples
Ionic crystals	Positive and negative ions	Electrostatic attractions between ions Very strong	High	High They also conduct electricity in solution.	Hard, brittle. Most dissolve in polar solvents.	Formed between atoms with a large difference (≥ 1.70) in electronegativity.	All salts All metal hydrides KCl $\quad$ CaF_2 CsBr $\quad$ MgO $BaCl_2$ $\quad$ NaCl
Covalent network crystals	Atoms	Covalent bonds Very strong	Very high	Poor	Very hard. Insoluble in most ordinary liquids. Covalent bonds extend from one atom to another in a continuous pattern.	Most formed by two elements of Group 14 or by elements whose average periodic group number is 14.	Diamond $\quad$ AlN SiC $\quad$ $CuCl_2$ BeO $\quad$ Mg_2Sn Mg_2Si
Metallic crystals	Positive metal atom kernels and mobile valence electrons	Metallic bonds Strong	Most are high.	Very high	Most are hard, malleable, ductile. High thermal conductivity. Generally insoluble in liquids. Usually soluble in molten metals.	Formed by elements with low first ionization energies and vacant valence levels.	Cu $\quad$ Bi $\quad$ Zn Fe $\quad$ Li $\quad$ Pt Ca $\quad$ Pb $\quad$ La V $\quad$ Cr $CuZn_3$
Molecular crystals (a) Polar	Polar molecules	Dipole forces Intermediate strength Can be strengthened by hydrogen bonds.	Intermediate	Very low	More fragile than ionic crystals. Most are soluble in polar solvents. Usually liquids or solids at room temperature.	Formed from molecules with asymmetrical charge distributions. Polar covalent bonds are formed between atoms having a moderate difference (0.41–1.69) in electronegativity.	All acids Many organic compounds PF_3 $\quad$ H_2O NH_3 $\quad$ SO_3 $CHCl_3$
(b) Nonpolar	Atoms or nonpolar molecules	Dispersion forces Weak	Low	Extremely low	Very soft. Most are soluble in nonpolar or slightly polar solvents. Usually gases at room temperature.	Formed from atoms or from molecules containing symmetrical charge distributions. Nonpolar covalent bonds are formed between like atoms or atoms having a small difference (≤ 0.40) in electronegativity.	All diatomic molecules Ar $\quad$ Cl_2 H_2 $\quad$ S_8 CH_4 $\quad$ I_2 Ne $\quad$ CCl_4 CO_2 $\quad$ N_2 SF_6 $\quad$ BF_3

10.2A ADDITIONAL TOPICS IN BONDING

Exceptions to the Octet Rule

All of the molecules and ions we have constructed so far have been based on the assumption that all atoms (other than hydrogen) need to have *eight* electrons in their valence levels as a result of bond formation. In reality, this is not always the case. Certain atoms may have fewer—or more—than eight electrons in their valence levels when they bond. The accompanying table lists five molecules that are exceptions to the octet rule.

Molecule	Lewis Structure	Condition of Central Atom's Valence Level
$BeCl_2$	:C̈l — Be — C̈l:	Be has *4* electrons in its valence level.
BF_3	:F̈ F̈: B :F:	B has *6* electrons in its valence level.
NO_2	:O̿ N Ö:	N has *7* electrons in its valence level. One of these electrons is *unpaired*.
PCl_5	:C̈l: Cl: :C̈l—P Cl: :C̈l:	P has *10* electrons in its valence level.
SF_6	:F̈: :F̈ F: S :F̈ F: :F̈:	S has *12* electrons in its valence level.

The NO_2 molecule is unusual because one of its valence electrons is *unpaired*. Molecules and atoms with unpaired electrons are known as *free radicals* and are very reactive. In living tissues, free radicals are known to be destructive agents and are believed to be associated with certain serious health problems.

Molecular Shape and VSEPR

Now that we have learned to construct molecules and ions, we need to know how to predict the three-dimensional shapes they assume. Molecules and ions with just two atoms are easy: they are linear. We can predict the shapes of other molecules and ions by using the **Valence Shell Electron Pair Repulsion** model (**VSEPR**).

VSEPR is based on four premises:

1. Shared and unshared pairs of electrons repel each other.
2. An unshared pair of electrons repels more strongly than a shared pair.
3. *For the purposes of this model,* a single, a double, and a triple bond are all considered equivalent.
4. The shape of a molecule or ion is the result of the shared and unshared pairs of electrons being placed as far from each other as possible.

To apply VSEPR, we look at the *central atom* of the molecule or ion and count the number of shared and unshared pairs of electrons that are associated with it. (Remember: A double or triple bond counts as *one* shared pair of electrons.)

The simplest molecules and ions have *no* unshared pairs of electrons. For example, let us consider the shape of the CO_2 molecule. The carbon atom has two double bonds but no unshared pairs of electrons. The *two* shared pairs of electrons are farthest apart when the angle between them is 180°. Therefore, the CO_2 molecule is *linear:* $O=C=O$.

The accompanying tables and diagrams summarize the shapes of molecules and ions with no unshared pairs of electrons on the central atom. We designate the central atom as A and each atom bonded to it as B.

Number of Shared Pairs	Formula	Shape of Molecule	Bond Angle(s)	Example(s)
2	AB_2	Linear	180°	CO_2, $BeCl_2$
3	AB_3	Trigonal planar	120°	BF_3, NO_3^-
4	AB_4	Tetrahedral	109.5°	CH_4, NH_4^+
5	AB_5	Trigonal bipyramidal	120°, 90°	PCl_5
6	AB_6	Octahedral	90°	SF_6

Arrangement of Electron Pairs about a Central Atom (A) in a Molecule, and Geometry of Some Simple Molecules and Ions in Which the Central Atom Has No Lone Pairs

Number of Electron Pairs	Arrangement of Electron Pairs*	Molecular Geometry*	Examples
2	Linear	B—A—B Linear	$BeCl_2$, $HgCl_2$
3	Trigonal planar	Trigonal planar	BF_3
4	Tetrahedral	Tetrahedral	CH_4, NH_4^+
5	Trigonal bipyramidal	Trigonal bipyramidal	PCl_5
6	Octahedral	Octahedral	SF_6

*Only solid lines represent bonds; the dashed lines show the overall shapes.

TRY IT YOURSELF

Draw the Lewis structure of the SO_3 molecule, and use the VSEPR model to determine its shape.

ANSWER

The SO_3 molecule is a resonance hybrid with the following structure:

Since it has "three" shared and no unshared pairs of electrons, its shape is *trigonal planar*.

Now let us consider molecules and ions whose central atoms have unshared pairs of electrons in addition to their shared pairs. In the H_2O molecule, the oxygen atom has two single bonds and two unshared pairs of electrons, for a total of *four* pairs of electrons. If *all* of these pairs were shared, we would expect, from the preceding tables and diagrams, that the shape would be *tetrahedral*.

Here is the surprise: The four electron pairs *do* form a tetrahedron! However, *the overall shape of the molecule is determined only by the atoms that are present,* not by the unshared pairs of electrons. For this reason, the H_2O molecule is *bent* (or *V-shaped*) as shown below:

Since the unshared pairs of electrons repel more strongly than the bonded pairs, the bonded pairs are forced closer together and the angle between the oxygen and the two hydrogen atoms is less than the expected tetrahedral angle of 109.5°. (The angle is 104.5°.)

The accompanying table and diagrams summarize some of the shapes of molecules and ions with both shared and unshared pairs of electrons on the central atom. The symbol *A* represents the *central* atom. The symbol *B* represents the *peripheral* atom(s). The symbol *E* represents an *unshared* pair or pairs of electrons.

Number of Shared Pairs	Number of Unshared Pairs	Formula	Shape	Example(s)
2	1	AB_2E	Bent	NO_2^-
2	2	AB_2E_2	Bent	H_2O
3	1	AB_3E	Trigonal pyramidal	NH_3, H_3O^+

The "wedge" ($\blacktriangleleft$) in the second and third diagrams indicates that the molecule is *three-dimensional*. The bond with the wedge projects *out of the paper toward you*.

TRY IT YOURSELF
Draw the Lewis structure of the PF_3 molecule, and use the VSEPR model to determine its shape.

ANSWER
The PF_3 molecule has one unshared and three shared pairs of electrons. Its shape is trigonal pyramidal, as shown below.

$$\overset{\overset{\displaystyle\cdot\cdot}{P}}{\underset{\displaystyle F}{F\diagdown\diagup F}}$$

How Covalent Bonding Occurs: Orbital Overlap

We already know that electrons and electron pairs reside in atomic orbitals. How, then, are electrons shared by two atoms? The answer is that the individual orbitals *overlap* to form the covalent bond. The following examples show how *orbital overlap* occurs.

When a hydrogen molecule is formed from two isolated hydrogen atoms, the $1s$ orbitals, each containing one electron, overlap to produce the single bond, as illustrated in the accompanying diagram:

Formation of a Hydrogen Molecule

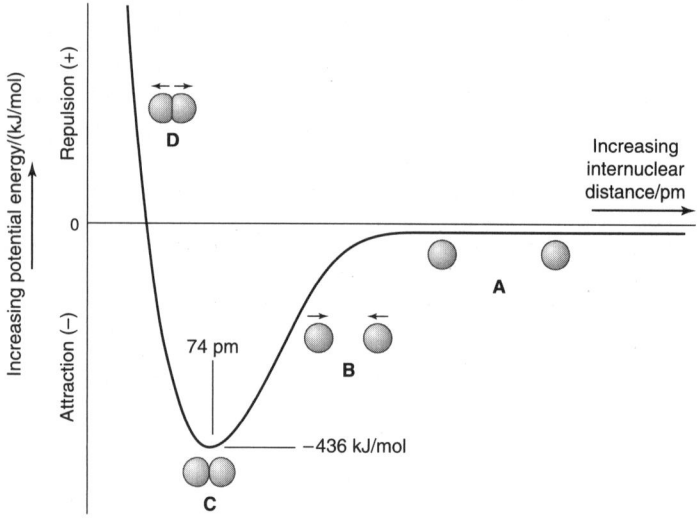

The energy diagram is read from *right to left*.

- At point **A**, the two hydrogen atoms are *isolated*, that is, they do not attract or repel each other. By convention, the potential energy of a system consisting of isolated atoms is set (very close) to 0 kilojoule per mole.
- At point **B**, the hydrogen atoms have moved closer together and a force of attraction exists between them. This configuration is more stable than the one in **A** because the potential energy of the system has been *lowered*. By convention, a force of attraction is represented by a *negative* potential energy value.
- At point **C**, the $1s$ orbitals of the hydrogen atoms have overlapped sufficiently to produce the *maximum* force of attraction and the *minimum* potential energy (-436 kilojoule per mole). At this point, the internuclear distance between the atoms is 74 picometers, and a *stable* H_2 molecule has formed.

At internuclear distances less than 74 picometers, the potential energy of the system rises rapidly and the force between the atoms changes from one of attraction to one of repulsion. Point **D** represents one such *unstable* configuration.

Sigma and Pi Bonds

When a chlorine molecule is formed from two chlorine atoms, the two $3p$ orbitals, each containing one electron, overlap to produce the single bond, as illustrated below.

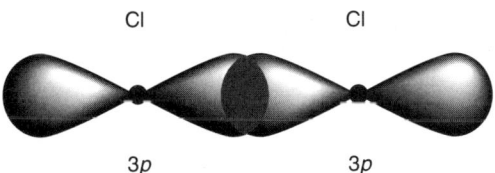

When a hydrogen chloride molecule is formed from a hydrogen and a chlorine atom, the $1s$ orbital of the hydrogen and the $3p$ orbital of the chlorine, each containing one electron, overlap to produce the single bond, as illustrated below.

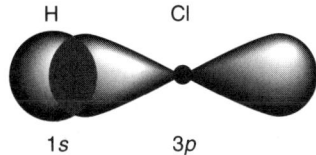

Each of these single bonds lies along a line that joins the nuclei of the two bonding atoms. Such bonds are called **sigma (σ) bonds**. *Almost all single bonds are sigma bonds*.

There is another way in which *p* orbitals can overlap. Since they have a "dumbbell" shape, they are able to overlap sideways (i.e., *perpendicular* to the line joining the nucleii of the bonding atoms). Such bonds are called **pi (π) bonds** and are involved in the formation of double and triple bonds. The formation of a pi bond is illustrated below.

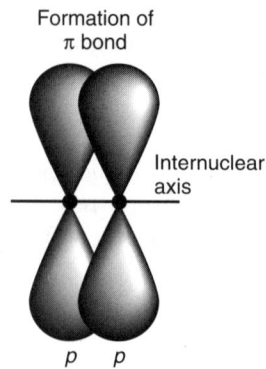

Formation of
π bond

Internuclear
axis

p   *p*

A double bond consists of one sigma bond and one pi bond. *A triple bond* consists of one sigma bond and two pi bonds.

Therefore, a double or triple bond is more complicated than the simple "sharing" of two or three pairs of electrons. In fact, pi bonds are more reactive than sigma bonds, as we will learn in Chapter 11: Organic Chemistry.

Bond Energy and Bond Length

How can we measure the "strength" of a chemical bond? One way is to measure the energy needed to break the bond. This quantity is known as the **bond dissociation energy** or, simply, the **bond energy**. In practice, bond energies are measured for a large variety of compounds and average values are compiled for the bond energy between two atoms. The accompanying table lists a number of average bond energies.

The larger the average bond energy, the stronger the bond. If we examine carbon-to-carbon bonds, for example, we see that the C≡C bond is stronger than the C=C bond, which is, in turn, stronger than the C−C bond. As the table shows, this relationship holds in general. The order of bond energies between two bonded atoms is as follows: triple bond > double bond > single bond.

Bond	Bond Energy / (kJ/mol)	Bond	Bond Energy / (kJ/mol)	Bond	Bond Energy / (kJ/mol)
H−H	436	H−F	565	H−Cl	427
O−H	467	N−H	391	C−H	413
C−C	347	C=C	614	C≡C	839
C−O	358	C=O	745*	C≡O	1072
C−N	305	C=N	615	C≡N	891
N−N	160	N=N	418	N≡N	941
O−O	146	O=O	495		

*The average bond energy for C=O in CO_2 is 799 kJ/mol.

Another quantity directly related to energy is the **bond length**, which is the distance between the nuclei of two bonded atoms. The bond length is measured in picometers (1 pm $= 10^{-12}$ m). In general, the shorter the bond length, the larger the bond energy. The accompanying table lists the bond lengths for carbon-to-carbon and carbon-to-nitrogen bonds.

Bond	Bond Length / pm
C−C	154
C=C	134
C≡C	120
C−N	143
C=N	138
C≡N	116

As we can see from the table, the order of bond lengths is as follows: triple bond < double bond < single bond.

Calculation of Heats of Reaction from Bond Energies

We can test the idea that a chemical reaction involves both the making and breaking of chemical bonds by examining the energies needed to make and break the bonds. The table at the top of the page lists the energies needed to *break* selected bonds. Since bond-breaking is an endothermic process, the values of bond energies are always reported as *positive* numbers. However, this same table also represents bond *formation* if we change the values to *negative* numbers, since bond formation is exothermic.

Let us examine this reaction:

$$CH_4(g) + 2O_2(g) \rightarrow CO_2(g) + 2H_2O(g)$$

| 1 mole | 2 moles | 1 mole | 2 moles |

from the point of view that specific bonds are formed and broken as the reaction occurs. We begin by rewriting the reaction to show the bonds present in each substance:

$$
\begin{array}{c}
\text{H} \\
| \\
\text{H}-\text{C}-\text{H} \\
| \\
\text{H}
\end{array}
+ 2\text{O}{=}\text{O} \rightarrow \text{O}{=}\text{C}{=}\text{O} + 2\ \text{O}\big\langle{}^{\text{H}}_{\text{H}}
$$

| 1 mole | 2 moles | 1 mole | 2 moles |

When we inspect the diagram, we see that the reaction involves the *breaking* of *4* moles of C—H bonds and *2* moles of O=O bonds. We also see that *2* moles of C=O bonds and *4* moles of O—H bonds have been *formed*. Using the values given in the table on page 297, we can calculate the net energy change over the course of this reaction:

4 mol of C—H bonds broken	= 4 mol · (+413 kJ/mol) =	+1652 kJ
2 mol of O=O bonds broken	= 2 mol · (+495 kJ/mol) =	+ 990 kJ
2 mol of C=O bonds formed	= 2 mol · (−799 kJ/mol) =	−1598 kJ
4 mol of O—H bonds formed	= 4 mol · (−467 kJ/mol) =	−1868 kJ
	Net energy change =	− 824 kJ

The value –824 kilojoule represents an estimate of the *heat of reaction* (ΔH) *as calculated from bond energies*. It lies within 3 percent of the value of this heat of reaction (–803 kJ) calculated from the data given in Reference Table W-4, Standard Heats of Formation for Substances, in Appendix 2. The close agreement between the estimated and the measured values of ΔH provides additional evidence that a chemical reaction involves the making and breaking of bonds.

Hybridized Atomic Orbitals

The VSEPR model that we studied on pages 291–294 is a valuable tool for predicting molecular shape. However, it does not really explain *why* molecules such as $BeCl_2$, BF_3, and CH_4 assume their particular shapes (linear, trigonal planar, and tetrahedral, respectively). In particular, the ground-state electron configurations of Be, B, and C seem to suggest that they should *not* form molecules with these shapes.

One explanation for the shapes of $BeCl_2$, BF_3, and CH_4 is provided by the concept of **orbital hybridization**, in which s and p atomic orbitals are "blended" together in various ways.

$BeCl_2$: Before the hybridization process, the ground-state configuration of the Be atom looks like this:

$$\underset{1s}{\uparrow\downarrow}\ \underset{2s}{\uparrow\downarrow}\ \underset{2p}{\rule{0.4cm}{0.4pt}\ \rule{0.4cm}{0.4pt}\ \rule{0.4cm}{0.4pt}}$$

The first step in the hybridization process involves the "promotion" of one of the 2s electrons into an empty 2p orbital, as shown below:

$$\underset{1s}{\uparrow\downarrow}\ \underset{2s}{\uparrow}\ \underset{2p}{\uparrow\ \ \rule{0.4cm}{0.4pt}\ \rule{0.4cm}{0.4pt}}$$

The final step involves the creation of *two hybrid orbitals*, known as *sp* orbitals, as shown below:

$$\underset{1s}{\uparrow\downarrow}\ \ \underset{2(sp)}{\uparrow\ \uparrow}\ \ \underset{2p}{\rule{0.4cm}{0.4pt}\ \rule{0.4cm}{0.4pt}}$$

Each of the orbitals lies 180° apart and allows $BeCl_2$ to be linear.

BF_3: Before the hybridization process, the ground-state configuration of the B atom looks like this:

$$\underset{1s}{\uparrow\downarrow}\ \underset{2s}{\uparrow\downarrow}\ \ \underset{2p}{\uparrow\ \ \rule{0.4cm}{0.4pt}\ \rule{0.4cm}{0.4pt}}$$

The first step in the hybridization process involves the "promotion" of one of the 2s electrons into an empty 2p orbital, as shown below:

$$\underset{1s}{\uparrow\downarrow}\ \underset{2s}{\uparrow}\ \ \underset{2p}{\uparrow\ \uparrow\ \rule{0.4cm}{0.4pt}}$$

The final step involves the creation of *three hybrid orbitals*, known as sp^2 orbitals, as shown below:

$$\underset{1s}{\uparrow\downarrow}\ \ \underset{2(sp^2)}{\uparrow\ \uparrow\ \uparrow}\ \ \underset{2p}{\rule{0.4cm}{0.4pt}}$$

Each of the orbitals lies 120° apart and allows BF_3 to be trigonal planar.

CH_4: Before the hybridization process, the ground-state configuration of the C atom looks like this:

$$\underset{1s}{\uparrow\downarrow}\ \underset{2s}{\uparrow\downarrow}\ \ \underset{2p}{\uparrow\ \uparrow\ \rule{0.4cm}{0.4pt}}$$

The first step in the hybridization process involves the "promotion" of one of the $2s$ electrons into an empty $2p$ orbital, as shown below:

$$\underline{\uparrow\downarrow} \quad \underline{\uparrow} \qquad \underline{\uparrow} \quad \underline{\uparrow} \quad \underline{\uparrow}$$

$$1s \qquad 2s \qquad\qquad 2p$$

The final step involves the creation of *four hybrid orbitals*, known as sp^3 orbitals, as shown below:

$$\underline{\uparrow\downarrow} \qquad \underline{\uparrow} \quad \underline{\uparrow} \quad \underline{\uparrow} \quad \underline{\uparrow}$$

$$1s \qquad\qquad 2(sp^3)$$

Each of the orbitals lies 109.5° apart and allows CH_4 to be tetrahedral.

END-OF-CHAPTER QUESTIONS

Some questions have the symbol "§2" in front of the question number. This symbol means that the question is based on Section II material.

1. Energy is released when the atoms of two elements bond to form a compound. Compared to the total potential energy of the atoms before bonding, the total potential energy of the atoms after bonding is
   (1) lower, and the compound formed is stable
   (2) lower, and the compound formed is unstable
   (3) higher, and the compound formed is stable
   (4) higher, and the compound formed is unstable

2. Which statement is true concerning the following reaction?
   $$N(g) + N(g) \rightarrow N_2(g) + \text{energy}$$
   (1) A bond is broken and energy is absorbed.
   (2) A bond is broken and energy is released.
   (3) A bond is formed and energy is absorbed.
   (4) A bond is formed and energy is released.

3. As a chemical bond forms between hydrogen and chlorine atoms, the potential energy of the atoms
   (1) decreases   (2) increases   (3) remains the same

4. A Ca^{2+} ion differs from a Ca atom in that the Ca^{2+} ion has
   (1) more protons   (2) fewer protons
   (3) more electrons   (4) fewer electrons

5. In which pair do the members have identical electron configurations?
   (1) S^{2-} and Cl^- (2) S and Ar
   (3) K and Na^+ (4) Cl^- and K

6. Which compound is ionic?
   (1) HCl   (2) $CaCl_2$ (3) SO_2 (4) N_2O

7. What is the energy-level electron configuration for a Be^{2+} ion?
   (1) 1   (2) 2   (3) 2-1   (4) 2-2

8. Which formula represents a molecular solid?
   (1) NaCl(s)   (2) $C_6H_{12}O_6$(s)
   (3) Cu(s)   (4) CsF(s)

9. What kind of bond is formed in the reaction below?

$$H : \overset{\displaystyle ..}{\underset{\displaystyle H}{O}} : \; + \; H^+ \longrightarrow \left[H : \overset{\displaystyle \overset{H}{..}}{\underset{\displaystyle H}{O}} : \right]^+$$

   (1) metallic   (2) hydrogen
   (3) network   (4) coordinate covalent

10. Which element at STP exists as monatomic molecules?
    (1) N   (2) O   (3) Cl   (4) Ne

11. Two atoms of element A unite to form a molecule with the formula A_2. The bond between the atoms in the molecule is
    (1) electrovalent   (2) ionic
    (3) nonpolar covalent   (4) polar covalent

12. Which formula represents a molecular solid?
    (1) NaCl(s)   (2) $C_6H_{12}O_6$(s)   (3) Cu(s)   (4) KF(s)

13. Which compound contains both covalent and ionic bonds?
    (1) KCl   (2) NH_4Cl (3) $MgCl_2$ (4) CCl_4

14. In which compound does the bond between the atoms have the *least* ionic character?
    (1) HF   (2) HCl   (3) HBr   (4) HI

15. When a reaction occurs between atoms with ground-state electron configurations 2-1 and 2-7, the predominant type of bond formed is
    (1) polar covalent   (2) nonpolar covalent
    (3) ionic   (4) metallic

§2 **16.** Which of the following molecules has a central atom with fewer than eight electrons in its valence level?
(1) NH_3 (2) BF_3 (3) PCl_5 (4) H_2O

§2 **17.** Which molecule has an atom with an unpaired electron in its valence level?
(1) H_2O (2) NO_2 (3) BF_3 (4) Cl_2

§2 **18.** The shape of a CCl_4 molecule is
(1) linear   (2) trigonal planar
(3) tetrahedral   (4) bent (V-shaped)

§2 **19.** A double bond, such as N=O, consists of
(1) one sigma (σ) bond, only
(2) one pi (π) bond, only
(3) one sigma bond and one pi bond
(4) two pi bonds, only

§2 **20.** Compared to the length of a C—C bond, the length of a C=C bond is
(1) less   (2) greater   (3) equal

21. Which substance is classified as a network solid?
(1) SiO_2 (2) Li_2O (3) H_2O (4) CO_2

22. What type of bonding is present within a network solid?
(1) hydrogen   (2) covalent   (3) ionic   (4) metallic

23. Which element has mobile valence electrons?
(1) S   (2) N   (3) Ca   (4) Cl

24. Which molecule is nonpolar and has a symmetrical shape?
(1) HCl   (2) CH_4 (3) H_2O (4) NH_3

25. Which formula represents a polar molecule?
(1) CH_4 (2) Cl_2 (3) NH_3 (4) N_2

26. Which structural formula represents a nonpolar symmetrical molecule?

(1)
$$H{-}O{-}H \text{ (bent)}$$

(2)
$$\begin{array}{c} H \\ | \\ H{-}C{-}H \\ | \\ H \end{array}$$

(3) H—F

(4)
$$\begin{array}{c} {}_{H}\diagup{}^{N}\diagdown{}_{H} \\ | \\ H \end{array}$$

27. Which substance is made up of molecules that are dipoles?
    (1) N_2 (2) H_2O (3) CH_4 (4) CO_2

28. The weakest intermolecular forces exist between molecules of
    (1) $C_2H_6(\ell)$ (2) $C_3H_8(\ell)$ (3) $C_4H_{10}(\ell)$ (4) $C_5H_{12}(\ell)$

29. The forces of attraction that exist between *nonpolar* molecules are called
    (1) London dispersion   (2) ionic
    (3) covalent   (4) electrovalent

30. The table below shows boiling points for the elements listed.

Element	Normal Boiling Point (°C)
Fluorine	−188.1
Chlorine	−34.6
Bromine	+58.8
Iodine	+184.4

    Which statement best explains the pattern of boiling points relative to molecular size?
    (1) Stronger London dispersion forces occur in larger molecules.
    (2) Weaker London dispersion forces occur in larger molecules.
    (3) Stronger hydrogen bonds occur in larger molecules.
    (4) Weaker hydrogen bonds occur in larger molecules.

31. London dispersion forces of attraction between molecules always decrease with
    (1) increasing molecular size and increasing distance between the molecules
    (2) increasing molecular size and decreasing distance between the molecules
    (3) decreasing molecular size and increasing distance between the molecules
    (4) decreasing molecular size and decreasing distance between the molecules

32. Which of the following liquids has the *weakest* London dispersion forces of attraction between its molecules?
    (1) $Xe(\ell)$ (2) $Kr(\ell)$ (3) $Ne(\ell)$ (4) $He(\ell)$

33. In which of the following liquids are hydrogen bonds between molecules the strongest?
    (1) $HI(\ell)$ (2) $HBr(\ell)$ (3) $HCl(\ell)$ (4) $HF(\ell)$

34. Hydrogen bonds are formed between molecules when hydrogen is covalently bonded to an element that has a
(1) small atomic radius and low electronegativity
(2) large atomic radius and low electronegativity
(3) small atomic radius and high electronegativity
(4) large atomic radius and high electronegativity

35. The kind of attraction that results in the dissolving of sodium chloride in water is
(1) ion-ion   (2) molecule-ion
(3) atom-atom   (4) molecule-atom

36. The attraction between water molecules and an Na^+ ion or a Cl^- ion occurs because water molecules are
(1) linear
(2) symmetrical
(3) polar
(4) nonpolar

37. A characteristic of ionic solids is that they
(1) have high melting points
(2) have low boiling points
(3) conduct electricity
(4) are noncrystalline

38. A pure substance melts at 38°C and does not conduct electricity in either the solid or liquid phase. The substance is classified as
(1) ionic
(2) metallic
(3) electrovalent
(4) molecular

39. Which is a property of most nonmetallic solids?
(1) high thermal conductivity
(2) high electrical conductivity
(3) brittleness
(4) malleability

40. Which of these compounds contains the most polar bond?
(1) H–Br   (2) H–Cl   (3) H–F   (4) H–I

41. Which property is defined as the ability of a substance to be hammered into thin sheets?
(1) conductivity (2) malleability (3) melting point (4) solubility

42. At STP, graphite and diamond are two solid forms of carbon. Which statement explains why these two forms of carbon differ in hardness?
(1) Graphite and diamond have different ionic radii.
(2) Graphite and diamond have different molecular structures.
(3) Graphite is a metal, but diamond is a nonmetal.
(4) Graphite is a good conductor of electricity, but diamond is a poor conductor of electricity.

43. Based on Reference Table S in Appendix 1, an atom of which of the following elements has the *weakest* attraction for electrons in a chemical bond?
(1) polonium   (2) sulfur   (3) selenium   (4) tellurium

44. Which substance in the table below has the strongest intermolecular forces?

Substance	Molar Mass (g/mol)	Boiling Point (kelvins)
HF	20.01	293
HCl	36.46	188
HBr	80.91	207
HI	127.91	237

(1) HF   (2) HCl   (3) HBr   (4) HI

45. Which compound contains both ionic and covalent bonds?
(1) KI   (2) $CaCl_2$ (3) CH_2Br_2 (4) NaCN

46. Which statement explains why a CO_2 molecule is nonpolar?
(1) Carbon and oxygen are both nonmetals.
(2) Carbon and oxygen have different electronegativities.
(3) The molecule has a symmetrical distribution of charge.
(4) The molecule has an asymmetrical distribution of charge.

Constructed-Response Questions

1. Label each of the following compounds as ionic or covalent:
(a) MgO   (b) SO_2 (c) H_2S
(d) BaF_2 (e) CCl_4

2. Write the Lewis structure for the hydrogen peroxide molecule (H_2O_2).

Base your answers to questions 3–6 on the information below. Each molecule listed is formed by sharing electrons between atoms when the atoms within the molecule are bonded together.

Molecule *A*: Cl_2

Molecule *B*: CCl_4

Molecule *C*: NH_3

3. Draw the electron-dot (Lewis) structure for the NH_3 molecule.

4. Explain why CCl_4 is classified as a nonpolar molecule.

5. Explain why NH_3 has stronger intermolecular forces of attraction than Cl_2.

6. Explain how the bonding in KCl is different from the bonding in molecules *A*, *B*, and *C*.

Base your answers to questions 7–11 on the information below and on your knowledge of chemistry.

The Lewis electron-dot diagrams for three substances are shown below.

Diagram 1   Diagram 2   Diagram 3

7. Describe in terms of valence electrons how the chemical bonds form in the substance represented in Diagram 1.

8. Determine the total number of electrons in the bonds between the nitrogen atom and the three hydrogen atoms represented in Diagram 2.

9. Explain in terms of distribution of charge why a molecule of the substance represented in Diagram 3 is nonpolar.

10. Draw a Lewis electron-dot diagram for a molecule of Br_2.

11. Identify the noble gas with the same electron configuration as the positive ion represented in Diagram 1.

The answers to these questions are found in Appendix 3.

Chapter Eleven ORGANIC CHEMISTRY

KEY IDEAS

This chapter focuses on the structure and reactions of the element carbon in organic compounds. The main series of the hydrocarbons and organic compounds containing oxygen and nitrogen are studied in detail.

KEY OBJECTIVES

At the conclusion of this chapter you will be able to:

- Define the term *organic chemistry*, and list the sources of organic materials.
- Compare general properties of organic compounds with those of inorganic compounds.
- Describe the bonding of carbon in simple organic compounds, and write structural formulas for such compounds.
- Define the term *hydrocarbon*, and describe how hydrocarbons are obtained from petroleum.
- Define the term *homologous series*.
- Describe the alkane, alkene, alkyne, and benzene series of hydrocarbons in terms of their general formulas, structural formulas, isomers, and IUPAC names.
- Define the term *functional group*.
- Describe, in terms of their structures and IUPAC names, the following organic oxygen compounds: alcohols, aldehydes, ketones, ethers, and organic acids.
- Describe the following organic nitrogen compounds: amines, amino acids, and amides.
- Define the term *polymer*.
- Describe the following organic reactions: substitution, addition, fermentation, esterification, saponification, combustion, condensation polymerization, and addition polymerization, and provide examples of these reactions.

SECTION I—BASIC (REGENTS-LEVEL) MATERIAL

NYS REGENTS CONCEPTS AND SKILLS

Note: By the time you have finished Section I, you should have mastered the concepts and skills listed below. The Regents chemistry examination will test your knowledge of these items and your ability to apply them.

Concepts are the *basic ideas* that form the body of the Regents chemistry course (what you need to know!).

Skills are the *activities* that demonstrate your mastery of these concepts (how you show that you know them!).

Following each concept or skill is a page reference (given in parentheses) to this chapter.

11.1 Concept:
 • Organic compounds contain carbon atoms that bond to one another in chains, rings, and networks to form a variety of structures. (Pages 309–310)
 • Organic compounds can be named using the IUPAC system. (Pages 314–316)

11.2 Concepts:
 • Hydrocarbons are compounds that contain only carbon and hydrogen. (Page 311)
 • Saturated hydrocarbons contain only single carbon-carbon bonds. (Pages 311–312)
 • Unsaturated hydrocarbons contain at least one multiple (double or triple) carbon-carbon bond. (Pages 316–318)

Skill:
Draw structural formulas for
  (a) alkanes,
  (b) alkenes,
  (c) alkynes,
containing a maximum of ten carbon atoms. (Pages 312–318)

11.3 Concepts:
 • Various categories of organic compounds include:
  (a) organic acids (Page 321);
  (b) alcohols (Pages 319–320);
  (c) esters (Pages 321–322);
  (d) aldehydes (Page 320);
  (e) ketones (Pages 320–321);

(f) ethers (Page 321);

(g) halides (Pages 323–324);

(h) amines (Page 322);

(i) amides (Page 323);

(j) amino acids (Page 322).

- Each of these categories contains molecules that differ in their structures. (Page 318)
- Functional groups impart distinctive physical and chemical properties to organic compounds. (Page 318)

Skills:

- Classify an organic compound based on its structural or condensed structural formula. (Pages 318–323)
- Given the IUPAC name for an organic compound, draw a structural formula for that compound that includes the appropriate functional group or groups (on a straight-chain hydrocarbon backbone). (Chart on page 319)

11.4 Concept:

Isomers of organic compounds have the same molecular formula but different structures and properties. (Pages 313–314)

11.5 Concept:

Types of organic reactions include:

(a) addition (Page 325);

(b) substitution (Page 324);

(c) polymerization (Page 327);

(d) esterification (Pages 325–326);

(e) fermentation (Page 325);

(f) saponification (Page 326);

(g) combustion (Page 326).

Skills:

- Identify types of organic reactions. (Pages 324–327)
- Given a balanced equation for an organic reaction, determine a missing reactant or product. (Pages 324–327)

11.1 ORGANIC CHEMISTRY IS . . . ?

Organic chemistry is essentially the study of carbon and its compounds. Why do we devote an entire chapter to just one element? The answer lies in the fact that the properties of organic compounds and the reactions that they undergo are generally quite different from those of their inorganic counterparts. Organic chemistry is responsible for many of the modern materials we use, and it is the basis of the chemistry of life.

In organic chemistry, the *structure* of a molecule is important in determining its properties and reactivity. While the number of inorganic compounds is in the order of tens of thousands, the number of organic compounds is in the order of millions. The main reason for the existence of so many organic compounds is the ability of carbon atoms to bond with *other carbon atoms* in an almost endless variety of chains, rings, and networks. The raw materials for making organic compounds come from petroleum, plant, and animal sources.

11.2 COMPARISON OF ORGANIC AND INORGANIC COMPOUNDS

To make the difference between organic and inorganic compounds more apparent, let us compare the general properties of each class.

Solubility

Inorganic compounds are generally soluble in polar solvents such as water, while organic compounds are generally soluble in nonpolar solvents. This fact explains why the oil in a salad dressing (organic) does not "mix" with the vinegar (a polar solvent).

Electrolytic Behavior in Solution

When dissolved in solution, many inorganic compounds conduct an electric current because of ionization or dissociation of the solute. This is generally *not* the case with organic compounds.

Melting and Boiling Points

Generally, inorganic compounds have higher melting and boiling points than organic compounds. While it is not unusual for some inorganic compounds to melt at temperatures above 500°C, this is rarely true of organic compounds.

Rate of Reaction

In general, inorganic compounds react more rapidly than organic compounds because the bond rearrangements in inorganic reactions are usually simpler and fewer in number than the bond rearrangements in organic reactions.

Bonding

The properties detailed on the previous page support the view that the existence of ions or polar covalent bonds is commonly associated with inorganic compounds. In organic compounds, however, the bonding is usually covalent with little or no polarity. Carbon has four valence electrons, and in organic compounds it forms a total of four covalent bonds. We will see that carbon atoms may form single, double, or triple bonds by sharing one, two, or three pairs of electrons.

11.3 HYDROCARBONS AND HOMOLOGOUS SERIES

The simplest organic compounds contain only carbon and hydrogen and are called **hydrocarbons**. Hydrocarbons are obtained from the refining of petroleum, a complex mixture of natural hydrocarbons. Petroleum is the starting point for the production of plastics, textiles, rubber, and detergents. The petroleum is separated into components with different *boiling points* by the process of **fractional distillation**. Another process, called **cracking**, is used to break the larger hydrocarbon molecules into smaller ones. This procedure increases the yield of the more important fractions with lower boiling points (such as gasoline). The simplest hydrocarbon is *methane* (CH_4), which is the chief component of natural gas.

Hydrocarbons with related structures and properties are usually separated into "families" known as **homologous series**. Each member of a series differs from the next member by a single carbon atom. With each successive member, the properties of the compounds, such as melting and boiling points, change in a regular way. This classification into scrics greatly simplifies the study of organic chemistry.

The Alkane Series

Alkanes are hydrocarbons that contain only single bonds between carbon atoms. The first member of the alkane series is *methane* (CH_4). In Chapter 9, we discovered that methane is a *tetrahedral* molecule because carbon forms four sp^3 hybrid orbitals—as it does in *every* alkane molecule. The accompanying table lists a number of alkanes together with their molecular formulas and phases at STP.

From the table we see that the name of each member of the alkane series begins with a prefix [e.g., *prop-* ("three") or *pent-* ("five")] that indicates the number of *carbon* atoms in the formula. All alkane names end with the suffix *-ane*. Reference Table P in Appendix 1 also summarizes the names of the organic prefixes for 1–10 carbon atoms.

Name of Alkane	Molecular Formula	Phase at STP
*meth*ane	CH_4	Gas
*eth*ane	C_2H_6	Gas
*prop*ane	C_3H_8	Gas
*but*ane	C_4H_{10}	Gas
*pent*ane	C_5H_{12}	Liquid
*hex*ane	C_6H_{14}	Liquid
*hept*ane	C_7H_{16}	Liquid
*oct*ane	C_8H_{18}	Liquid
*non*ane	C_9H_{20}	Liquid
*dec*ane	$C_{10}H_{22}$	Liquid
...	...	...
octadecane	$C_{18}H_{38}$	Solid

In the table we note also that each alkane contains one more carbon and two more hydrogens than the preceding alkane. The *general* formula for an alkane with *n* carbon atoms is $C_n H_{2n+2}$. We also note that the phase of the substance at STP changes from gas to liquid to solid as the masses of successive molecules increase. This phase change is due to the increased strength of *London dispersion forces,* which we studied in Section 10.10, in more massive molecules.

Chemists write the formulas of organic compounds in several ways:

- A **molecular formula**, such as C_3H_8, indicates how many atoms of each element are present, but it fails to provide information about how the atoms are *connected.*
- A **dashed structural formula** uses straight lines to represent chemical bonds. For example, propane (C_3H_8) is written as follows:

$$
\begin{array}{ccccc}
 & H & H & H & \\
 & | & | & | & \\
H- & C- & C- & C- & H \\
 & | & | & | & \\
 & H & H & H &
\end{array}
$$

For convenience, the angles between bonds are usually drawn as either 90° or 180°. The advantage of a dashed structural formula lies in its ability to display the connections among the atoms. However, more complex formulas can be cumbersome to write in this style.

- A **condensed structural formula** represents a compromise between the simplicity of a molecular formula and the information provided by a dashed structural formula. In this style, atoms are usually written *without* dashes, and hydrogen atoms are written to the *right* of the atom to which they are connected. For example, the condensed structural formula of propane is

$$CH_3CH_2CH_3$$

In this book, dashes occasionally *will* be used with more complex formulas, when the dashes make the structure of the molecule more apparent. For example, the molecule whose dashed structural formula is

$$
\begin{array}{c}
\text{H} \\
| \\
\text{H}-\text{C}-\text{H} \\
\\
\text{H} \quad | \quad \text{H} \\
| \quad | \quad | \\
\text{H}-\text{C}-\text{C}-\text{C}-\text{H} \\
| \quad | \quad | \\
\text{H} \quad \text{H} \quad \text{H}
\end{array}
$$

has its condensed structural formula written as follows:

$$
\begin{array}{c}
\text{CH}_3 \\
| \\
\text{CH}_3\text{CHCH}_3
\end{array}
$$

There are other ways of writing formulas for organic compounds, but they are beyond the scope of this book.

Isomers

The condensed structural formula of butane is of particular interest because *two* molecules have the molecular formula C_4H_{10}. Their structures and common names are as follows:

$$
\text{CH}_3\text{CH}_2\text{CH}_2\text{CH}_3 \qquad
\begin{array}{c}
\text{CH}_3 \\
| \\
\text{CH}_3\text{CHCH}_3
\end{array}
$$

butane   **isobutane**

Substances that have the same molecular formula but are *different* compounds are called **isomers**. Isomers differ from one another in their physical and chemical properties.

Two or more compounds are **constitutional isomers** if they have the same molecular formula but their atoms are connected in a different order. The two butane molecules just shown are constitutional isomers.

PROBLEM
Pentane (C_5H_{12}) consists of three constitutional isomers. Write the condensed structural formula for each isomer.

SOLUTION

$$CH_3CH_2CH_2CH_2CH_3 \qquad \text{[pentane]}$$

$$\begin{array}{c} CH_3 \\ | \\ CH_3CHCH_2CH_3 \end{array} \qquad \text{[isopentane]}$$

$$\begin{array}{c} CH_3 \\ | \\ CH_3 - C - CH_3 \\ | \\ CH_3 \end{array} \qquad \text{[neopentane]}$$

TRY IT YOURSELF

Hexane consists of five constitutional structural isomers. Draw the dashed structural formulas for any *three* of them.

ANSWER

The five dashed structural formulas shown below are presented as hydrocarbon *skeletons*; that is, the hydrogen atoms have been omitted.

(a) C—C—C—C—C—C

(b)
$$\begin{array}{c} C \\ | \\ C - C - C - C - C \end{array}$$

(c)
$$\begin{array}{c} C \\ | \\ C - C - C - C - C \end{array}$$

(d)
$$\begin{array}{c} C \quad C \\ | \quad | \\ C - C - C - C \end{array}$$

(e)
$$\begin{array}{c} C \\ | \\ C - C - C - C \\ | \\ C \end{array}$$

The IUPAC Naming System

As the number of carbon atoms in an organic compound increases, the number of constitutional isomers rises dramatically. Providing a different name for each and every isomer is an impossible task unless a simple and logical way to name organic compounds is available. The *International Union of Pure and Applied Chemistry* (IUPAC) has developed such a system, in which organic compounds are named in much the way people are: each has a *family* name and one or more *given* names. The procedure for naming an organic compound is as follows:

- If a hydrocarbon has *no* branches and has only single carbon-carbon bonds, it is named according to the table given on page 312. For example,

$CH_3CH_2CH_2CH_3$ is named *butane*. (Some texts place the symbol *n-* before the name, as in *n*-butane.)

- First the *parent structure* is determined. The parent structure of an organic compound is found by counting the number of atoms in the *longest, unbroken chain of carbon atoms*. For example, in the skeleton shown below:

$$C - C - C - C - C - C$$
$$\quad\quad | \quad\quad\quad\quad | \quad\quad$$
$$\quad\quad C \quad\quad\quad\quad C \quad\quad$$

there are *six* carbon atoms in the longest, unbroken chain. Since only single bonds are present, the name of the parent structure is *hexane*.

In this structure:

$$C - C - C - C - C - C - C - C$$
$$\quad\quad | \quad\quad\quad\quad\quad\quad | \quad\quad$$
$$\quad\quad C \quad\quad\quad\quad\quad\quad C \quad\quad$$
$$\quad\quad | \quad\quad$$
$$\quad\quad C \quad\quad$$

the longest, unbroken chain contains *nine* carbon atoms, again all with single bonds. No one has said the atoms have to be in a straight line—just an *unbroken* line! The arrows show how the longest chain is formed:

$$C - C \rightarrow C \rightarrow C \rightarrow C \rightarrow C \rightarrow C \rightarrow C$$
$$\quad\quad \uparrow \quad\quad\quad\quad\quad\quad\quad |$$
$$\quad\quad C \quad\quad\quad\quad\quad\quad\quad C$$
$$\quad\quad \uparrow$$
$$\quad\quad C$$

The parent structure of this compound is named *nonane*.

- The full name of an organic compound is determined by modifying the name of the parent structure according to the *functional groups* that are attached to the parent.
- If *no* groups are attached, the parent name is the name of the hydrocarbon. For example, the hydrocarbon shown below is named *butane*.

$$\begin{array}{cccc} H & H & H & H \\ | & | & | & | \\ H - C - C - C - C - H \\ | & | & | & | \\ H & H & H & H \end{array} \quad \text{or} \quad CH_3CH_2CH_2CH_3$$

butane

- When a carbon group is attached to the parent, the group is named by attaching the suffix *-yl* to the prefix that indicates how many carbon atoms the group has. The three simplest carbon groups (written in *condensed* form) are these:

$$CH_3-\qquad\qquad\qquad methyl$$
$$CH_3CH_2-\ or\ C_2H_5-\qquad\quad ethyl$$
$$CH_3CH_2CH_2-\ or\ C_3H_7-\quad propyl$$

The hydrocarbon shown below is named 2-*methylpropane* (written as one word) because the parent (propane) has three carbon atoms and a one-carbon group is attached to it. The number 2 is added to show precisely *where* the group is located on the parent. The carbon atoms of the parent are numbered in order so that the group is assigned the *lowest* number possible—2 in this example.

2-methylpropane

PROBLEM

Name, according to the IUPAC system, the three constitutional isomers of pentane whose condensed structural formulas are shown on page 314.

SOLUTION

The first isomer is named *pentane,* as shown. The second isomer (isopentane) is named 2-*methylbutane.* The third isomer (neopentane) is named 2,2-*dimethylpropane* because the parent structure has three carbon atoms, and each of two *one-carbon groups* is located on the middle carbon of the parent. Note that each group is entitled to its own number.

TRY IT YOURSELF

Name the five constitutional isomers of hexane illustrated on page 315.

ANSWERS

(a) hexane   (b) 2-methylpentane   (c) 3-methylpentane
(d) 2,3-dimethylbutane   (e) 2,2-dimethylbutane

The Alkene Series

Alkenes are hydrocarbons that contain one carbon-carbon *double* bond. The first member of the alkene series, shown on the following page, is *ethene* (C_2H_4, commonly called *ethylene*).

$$\text{H}\diagdown \quad \diagup \text{H}$$
$$\text{C}=\text{C}$$
$$\text{H}\diagup \quad \diagdown \text{H}$$

ethene

The name of each member of the alkene series begins with a prefix that indicates the number of *carbon* atoms in the formula. *All* alkene names end with the suffix *-ene*. The *general* formula for an alkene with n carbon atoms is C_nH_{2n}.

PROBLEM
What is the formula of *propene*?

SOLUTION
Propene is an alkene with three carbon atoms. According to the general formula for alkenes (C_nH_{2n}), it also has six hydrogen atoms. Its molecular formula is C_3H_6, and its structural formula is as follows:

$$\text{H}\diagdown \quad \diagup \text{H}$$
$$\text{C}=\text{C}$$
$$\text{H}\diagup \quad \diagdown \text{CH}_3$$

propene

Note that each carbon atom has a total of four bonds attached to it.

TRY IT YOURSELF
What are the formulas of *butene*?

ANSWER
There are three constitutional isomers of butene. All have the molecular formula C_4H_8. Two of the constitutional isomers differ in the position of the double bond, as shown below.

$$\text{H}\diagdown \quad \diagup \text{H}$$
$$\text{C}=\text{C}$$
$$\text{H}\diagup \quad \diagdown \text{CH}_2\text{CH}_3$$

1-butene

$$\text{CH}_3\diagdown \quad \diagup \text{CH}_3$$
$$\text{C}=\text{C}$$
$$\text{H}\diagup \quad \diagdown \text{H}$$

2-butene

The third isomer has the following structure:

$$\text{H}\diagdown \quad \diagup \text{CH}_3$$
$$\text{C}=\text{C}$$
$$\text{H}\diagup \quad \diagdown \text{CH}_3$$

2-methylpropene

317

The Alkyne Series

Alkynes are hydrocarbons that contain one carbon-carbon *triple* bond. The first member of the alkyne series is *ethyne* (C_2H_2, commonly called *acetylene*):

$$H-C\equiv C-H$$

ethyne

The name of each member of the alkyne series begins with a prefix that indicates the number of *carbon* atoms in the formula. *All* alkyne names end with the suffix *-yne*. The *general* formula for an alkyne with n carbon atoms is C_nH_{2n-2}.

Reference Table Q in Appendix 1 contains a summary of the various types of hydrocarbons.

11.4 FUNCTIONAL GROUPS

Elements *other* than carbon and hydrogen are found in most organic compounds. The elements most commonly present are oxygen, nitrogen, and the halogens (fluorine, chlorine, bromine, and iodine). An organic molecule containing another element is classified according to the manner in which the atoms of this element are *arranged* within the molecule. Such an arrangement of atoms is known as a **functional group**. The presence of functional groups imparts distinctive chemical and physical properties to organic compounds.

Alkyl Groups

The hydrocarbon groups methyl (CH_3-), ethyl (CH_3CH_2-), propyl ($CH_3CH_2CH_2-$), and so on, are known collectively as **alkyl groups**. An alkyl group is formed when a hydrogen atom is removed from an *alkane*. When we want to signify the presence of an alkyl group *without identifying it specifically*, we use the general symbol R. For example, we could write the general formula of an alkane as $R-H$, and the general formula of an alkyne as $R_1-C\equiv C-R_2$. (We use *subscripts* when we want to indicate the possibility that the alkyl groups may be different.) Alkyl groups are particularly useful in writing the general formulas of compounds containing functional groups.

The accompanying table describes the functional groups that we will study in this chapter.

Functional Group	Name of Group	General Formula	Name of Compound
—OH	Hydroxyl	R—OH	Alcohol
—O—	—	R_1—O—R_2	Ether
$\overset{\displaystyle O}{\overset{\|}{-C-}}$	Carbonyl	$R-\overset{\displaystyle O}{\overset{\|}{C}}-H$	Aldehyde
		$R_1-\overset{\displaystyle O}{\overset{\|}{C}}-R_2$	Ketone
$-\overset{\displaystyle O}{\overset{\|}{C}}-OH$	Carboxyl	$R-\overset{\displaystyle O}{\overset{\|}{C}}-OH$	Organic acid
$-\overset{\displaystyle O}{\overset{\|}{C}}-O-$	—	$R_1-\overset{\displaystyle O}{\overset{\|}{C}}-O-R_2$	Ester
$-NH_2$	Amino	$R-NH_2$	Amine
$-\overset{\displaystyle O}{\overset{\|}{C}}-NH_2$	Amido	$R-\overset{\displaystyle O}{\overset{\|}{C}}-NH_2$	Amide
—X*	—	R—X	Haloalkane (Halide)

*X is a halogen atom (F, Br, Cl, or I).

This information is also summarized in Reference Table R in Appendix 1.

The Hydroxyl Group: Alcohols

All *alcohols* contain the *hydroxyl* functional group (—OH). Alcohols can be classified in two ways: (1) according to the number of hydroxyl groups present in a molecule, and (2) according to the way in which the carbon atoms are arranged within the alcohol molecule.

An alcohol is named according to the IUPAC system by adding the suffix *-ol* to the name of the parent hydrocarbon. The simplest *monohydroxy* alcohol contains one carbon atom and is named *methanol* (commonly known as *methyl alcohol* or *wood alcohol*). Methanol is a popular solvent. The next member of this series, *ethanol* (commonly known as *ethyl alcohol* or *grain alcohol*), contains two carbon atoms. Ethanol is used as an antiseptic and is also a component of alcoholic beverages. The structural formulas of methanol and ethanol are as follows:

$$H-\overset{\displaystyle H}{\underset{\displaystyle H}{C}}-O-H \qquad H-\overset{\displaystyle H}{\underset{\displaystyle H}{C}}-\overset{\displaystyle H}{\underset{\displaystyle H}{C}}-O-H$$

$$CH_3OH \qquad\qquad CH_3CH_2OH$$

methanol   ethanol

Below the dashed structural formulas for methanol and ethanol are the *condensed structural formulas* for these compounds.

The Carbonyl Group: Aldehydes and Ketones

One of the most important functional groups in organic chemistry is the **carbonyl** group, which consists of a carbon atom connected to an oxygen atom by a double bond:

$$\diagdown C=O$$

Aldehydes

In **aldehydes**, the carbon of the carbonyl group is bonded to at least *one* hydrogen atom. An aldehyde is named according to the IUPAC system by adding the suffix *-al* to the name of the parent hydrocarbon. Here are the general formula and two examples (*methanal* and *ethanal*):

aldehyde   methanal   ethanal

Methanal (commonly known as *formaldehyde*) is used as a biological preservative.

Ketones

A **ketone** has *no* hydrogen atoms directly attached to the carbonyl group. According to the IUPAC system, a ketone is named by attaching the suffix *-one* to the name of the parent hydrocarbon. The general formula and an example (*propanone*) are given below.

ketone   propanone

Propanone (commonly known as *acetone*) is the principal solvent in nail polish remover.

Ethers

Ethers have the *general* formula R_1-O-R_2. The ether, $C_2H_5-O-C_2H_5$ has the IUPAC name *ethoxyethane*, but it is more commonly known as *diethyl ether*. It was once a popular surgical anesthetic. Ethers may be prepared by dehydrating primary alcohols in the presence of H_2SO_4:

$$2CH_3CH_2OH \xrightarrow{H_2SO_4} C_2H_5OC_2H_5 + H_2O$$

The Carboxyl Group: Organic Acids

The name **carboxyl** is a contraction of *carbonyl* and *hydroxyl*. The carboxyl functional group has this formula:

$$\begin{matrix} & O \\ & \| \\ -&C-OH \end{matrix}$$

which is usually abbreviated as COOH. The presence of the carboxyl group confers acidic properties on the organic compound. According to the IUPAC system, an organic acid is named by adding the suffix *-oic* and the word *acid* to the name of the parent hydrocarbon.

The first member of the series is *methanoic acid* (commonly called *formic acid*), whose formula is HCOOH. The second member is *ethanoic acid* (commonly called *acetic acid*). A 5–6 percent aqueous solution of ethanoic acid is popularly known as *vinegar*. The formula for ethanoic acid is CH_3COOH.

Esters

Esters have this general formula:

$$\begin{matrix} & O & \\ & \| & \\ R_1-&C-O&-R_2 \end{matrix}$$

They are prepared by the reaction between an organic acid and an alcohol (see Section 14.5). An ester is named as a derivative of the acid and alcohol that formed it. For example, the ester

$$CH_3-\overset{\overset{\textstyle O}{\|}}{C}-O-CH_2CH_3$$

is formed by the reaction between *ethanoic acid* and *ethanol*. Its IUPAC name, *ethyl ethanoate*, is derived from the combination of the *alkyl group* found in the alcohol (*ethyl*) and the name of the acid, with the suffix *-ate* replacing *-ic*.

Esters usually have a pleasant odor. The distinctive aromas of many fruits (e.g., bananas) and plants (e.g., wintergreen) are due to esters. The type of organic acid and the type of alcohol present in the ester determine its aroma.

Organic Nitrogen Compounds

Amines

The **amino group** consists of a nitrogen atom bonded to two hydrogen atoms: $-NH_2$.

A hydrocarbon in which one or more hydrogen atoms is replaced by an amino group is known as an *amine*. The simplest amine is *aminomethane*; its structural formula is shown below.

aminomethane

Amino Acids

An **amino acid** is an organic acid that contains one or more amino groups. The simplest amino acids are *aminoethanoic acid* and *aminopropanoic acid,* known, respectively, as *glycine* and *alanine*. Their structures are as follows:

glycine   alanine

Amino acids are the "building blocks" of proteins and also have other functions in living organisms.

Amides

An **amide** is a derivative of an organic acid in which an amino group ($-NH_2$) replaces the $-OH$ on the carboxyl group. Here is the general formula for an amide:

$$\begin{array}{c} O \\ \parallel \\ R-C-NH_2 \end{array}$$

In the IUPAC system, an amide is named by dropping *-oic* from the name of the acid, and adding *-amide* in its place. For example, the amide whose formula is

$$\begin{array}{c} O \\ \parallel \\ CH_3-C-NH_2 \end{array}$$

is derived from *ethanoic acid*. Its IUPAC name is *ethanamide*.

While a detailed study of amides is well beyond the scope of this book, we will see in Section 11.5 that compounds called *polyamides* or, as they are commonly known, *peptides* can be produced by a reaction between amino acids. Polyamides are of great importance in biology.

Haloalkanes (Halides)

A **haloalkane** is a compound in which one or more *hydrogen* atoms in an alkane is replaced by a corresponding number of halogen atoms. The general formula of a haloalkane is $R-X$, in which X represents a halogen atom.

According to the IUPAC system, a haloalkane is named by placing the combining form of the halogen name (*fluoro-, chloro-, bromo-, iodo-*) before the name of the alkane. If necessary, a prefix (*di-, tri-,* etc.) and numbers are assigned to indicate the number of halogen atoms present and their locations.

The formulas and IUPAC names of two haloalkanes are given below.

$$\begin{array}{cc} \begin{array}{c} H \\ | \\ H-C-Cl \\ | \\ H \end{array} & \begin{array}{c} Br \quad Br \\ | \quad\; | \\ H-C-C-H \\ | \quad\; | \\ H \quad H \end{array} \end{array}$$

chloromethane   1,2-dibromoethane

PROBLEM

Name the following haloalkanes according to the IUPAC system:

$$\begin{array}{cccccc} & Br & H & H & Br & H \\ & | & | & | & | & | \\ H- & C- & C- & C- & C- & C-H \\ & | & | & | & | & | \\ & H & H & H & H & H \end{array} \qquad \begin{array}{ccccc} & H & F & H & F \\ & | & | & | & | \\ H- & C- & C- & C- & C-H \\ & | & | & | & | \\ & H & H & H & H \end{array}$$

SOLUTION

The haloalkane on the left has a parent hydrocarbon with five carbon atoms; its name is pentane. Counting from the left, bromine atoms are located on the first and fourth carbon atoms. The haloalkane's name is 1,4-dibromopentane.

The haloalkane on the right has a parent hydrocarbon with four carbon atoms; its name is butane. Since the name must have the *smallest* numbers possible, we count the carbons from right to left; the fluorine atoms are located on the first and third carbon atoms. The haloalkane's name is 1,3-difluoro-butane.

11.5 ORGANIC REACTIONS

Organic reactions generally occur at a slower rate than inorganic reactions, and they usually involve the functional groups of the reactants.

Substitution

In a **substitution** reaction one type of atom or group is replaced with another type. Halogens such as chlorine and bromine may be substituted for the hydrogen atoms of alkanes. Halogen substitution is a stepwise process, as shown below.

$$\begin{array}{lll} CH_4 + Cl_2 & \rightarrow HCl + CH_3Cl & \text{(chloromethane)} \\ CH_3Cl + Cl_2 & \rightarrow HCl + CH_2Cl_2 & \text{(\textit{di}chloromethane)} \\ CH_2Cl_2 + Cl_2 & \rightarrow HCl + CHCl_3 & \text{(\textit{tri}chloromethane)} \\ CHCl_3 + Cl_2 & \rightarrow HCl + CCl_4 & \text{(\textit{tetra}chloromethane)} \end{array}$$

In each step, one hydrogen atom is replaced with a chlorine atom from Cl_2. The remaining chlorine atom combines with the substituted hydrogen to form hydrogen chloride (HCl).

Addition

Addition is a reaction in which atoms are "added across" a double or triple bond. As a result of addition, a double bond is converted into a single bond, and a triple bond into a double bond.

The addition of bromine to ethene is shown below.

$$\underset{\text{ethene}}{\overset{\displaystyle H}{\underset{\displaystyle H}{>}}C = C\overset{\displaystyle H}{\underset{\displaystyle H}{<}} + Br_2 \rightarrow \underset{\text{1.2-dibromoethane}}{H - \overset{\displaystyle Br}{\underset{\displaystyle H}{C}} - \overset{\displaystyle Br}{\underset{\displaystyle H}{C}} - H}$$

Addition reactions take place more rapidly than substitution reactions because the pi (π) bonds in alkenes and alkynes are weaker and more easily broken than the sigma (σ) bonds. Addition is characteristic of unsaturated compounds. The addition of halogen atoms, such as chlorine and bromine, is a common test for unsaturation in hydrocarbons. The addition of hydrogen to an unsaturated compound is usually called *hydrogenation*; it is used commercially to solidify liquid vegetable oils.

Fermentation

Fermentation is a biological process in which oxidation occurs in the absence of oxygen. *Enzymes* produced by living organisms act as the catalysts for this process. For example, yeast enzymes will ferment glucose (a simple sugar) to produce ethanol and carbon dioxide.

$$\underset{\text{glucose}}{C_6H_{12}O_6} \xrightarrow{\text{yeast enzymes}} \underset{\text{ethanol}}{2C_2H_5OH} + \underset{\text{carbon dioxide}}{2CO_2}$$

Esterification

Esterification is the reaction between an acid and an alcohol. The products of an esterification reaction are an *ester* and water:

Esterification: Acid + Alcohol $\rightarrow$ Ester + Water

Esterification occurs slowly and must be catalyzed by acids or bases. It is an example of a *dehydration* or *condensation* reaction. The reaction is reversible and is then called *hydrolysis* because the ester is broken down with water. Illustrations of esterification and hydrolysis follow.

When long-chain organic acids (called *fatty acids*) are esterified with glycerol, the ester is known as a *lipid* or *fat*. If the fatty acids contain a large number of double bonds within the parent hydrocarbons, the fat is said to be unsaturated and is usually liquid. Such is the case with vegetable oils. Animal fats, which are solid, are known as saturated fats because the parent hydrocarbons contain relatively few double bonds.

Saponification

Saponification is the hydrolysis of a fat with a base such as sodium hydroxide (NaOH). The products of saponification are glycerol and salts of the fatty acids, known collectively as *soap:*

$$\text{Saponification: Fat + NaOH} \rightarrow \text{Glycerol + Soap}$$

Soap is a long molecule that is nonpolar at one end and ionic at the other. The nonpolar end can form attractions with other nonpolar substances such as oil, and the ionic end can attract polar solvents such as water. Therefore, soap can bring about the emulsion of oil and water. This property is known as *detergency.*

Combustion

Saturated hydrocarbons such as methane react with oxygen at suitably high temperatures to produce carbon dioxide and water:

$$CH_4 + 2O_2 \rightarrow CO_2 + 2H_2O$$

This property of **combustion** makes alkanes suitable as fuels. In a *limited* supply of oxygen, carbon monoxide or carbon (soot) is produced.

$$2CH_4 + 3O_2 \rightarrow 2CO + 4H_2O$$
$$CH_4 + O_2 \rightarrow C + 2H_2O$$

Polymers and Polymerization

A **polymer** is a large molecule composed of many repeating units called *monomers*. In *polymerization,* the monomer units are joined to produce the polymer:

$$n \text{ monomer} \rightarrow (\text{monomer})_n$$

n units of monomer   **polymer**

Natural polymers include proteins (monomer: amino acids), cellulose and starch (monomer: glucose). *Synthetic polymers* include the plastic polyethylene (monomer: ethene), nylon, and polyester.

SECTION II—ADDITIONAL MATERIAL

11.1A STEREOISOMERISM

Stereoisomers are isomers whose atoms have different *arrangements in space.* Actually, there are *two* stereoisomers of 2-butene! The way in which carbon-carbon double bonds are formed causes the atoms around each carbon to be held rigidly in position. The two stereoisomers of 2-butene are shown below:

cis-2-butene   *trans*-2-butene

In the first stereoisomer, known as the *cis* form, the methyl groups are closer together than in the second stereoisomer, the *trans* form. Therefore, we name the isomer on the left *cis*-2-butene and the one on the right *trans*-2-butene.

TRY IT YOURSELF
In the compound 1,2-dichloroethene, one hydrogen atom on each carbon atom is replaced by a *chlorine* atom. Write the formulas of the two stereoisomers of 1,2-dichloroethene.

ANSWER

cis-1,2-dichloroethene   *trans*-1,2-dichloroethene

11.2A   THE BENZENE SERIES

The homologous **benzene** series of hydrocarbons — also known as the **aromatic hydrocarbons** — has the general formula C_nH_{2n-6}. The first member of the series is named *benzene* and has the molecular formula C_6H_6. The structure of benzene (and of its family) is quite unique: its first and last carbons join together, and for this reason it is known as a *ring* compound. Its structure may be represented as follows:

$$
\begin{array}{c}
H \\
|\\
H \; C \; C \; H \\
C \;\; C \\
C \;\; C \\
H \; C \; C \; H \\
|\\
H
\end{array}
$$

Bonding in Aromatic Compounds

Benzene (and its derivatives) are flat, hexagonal molecules, and the bond angles around each carbon atom are 120°. Each of the six carbon atoms in benzene forms sigma (σ) bonds with the two carbon atoms adjacent to it. In addition, each carbon atom has one $2p$ orbital containing a single electron. The overlap of these p orbitals creates a pi (π) bond that is *extended over the entire molecule*.

For this reason, the carbon-carbon bonding in benzene consists of neither single nor double bonds; benzene has the properties of *both* a saturated and an unsaturated hydrocarbon. Usually, the benzene ring is drawn as a hexagonal skeleton, with the hydrogen atoms omitted, and the extended pi bond is usually represented by a dashed or a solid circle in the center of the hexagon. In an older, but less accurate, way of depicting the benzene molecule, a hexagon with three alternating double bonds is drawn above. The modern representations of the benzene molecule — hexagons with a dashed circle and a solid circle — are shown in the accompanying diagrams.

11.3A PRIMARY, SECONDARY, AND TERTIARY ALCOHOLS

Alcohols (as well as certain other types of organic compounds) are classified as *primary*, *secondary*, or *tertiary* because of the differences in the chemical behavior among these types of alcohols. The classification is based on the *number of carbon atoms attached to the carbon atom that contains the hydroxyl group*:

- An alcohol in which *zero* or *one* carbon atom is bonded to the hydroxyl-containing carbon atom is known as a **primary alcohol**. The general formula and an example (*ethanol*) are given below.

$$R-\underset{\underset{\displaystyle H}{|}}{\overset{\overset{\displaystyle H}{|}}{C}}-OH \qquad\qquad CH_3-\underset{\underset{\displaystyle H}{|}}{\overset{\overset{\displaystyle H}{|}}{C}}-OH$$

primary alcohol   **ethanol**

- An alcohol in which *two* carbon atoms are bonded to the hydroxyl-containing carbon atom is known as a **secondary alcohol**. The general formula and an example (*2-propanol*, commonly called *isopropyl alcohol* or *rubbing alcohol*) are given below.

$$R_1-\underset{\underset{\displaystyle R_2}{|}}{\overset{\overset{\displaystyle H}{|}}{C}}-OH \qquad\qquad CH_3-\underset{\underset{\displaystyle CH_3}{|}}{\overset{\overset{\displaystyle H}{|}}{C}}-OH$$

secondary alcohol   **2-propanol**

The number 2 in 2-propanol stands for the position of the hydroxyl group.
- An alcohol in which *three* carbon atoms are bonded to the hydroxyl-containing carbon atom is known as a **tertiary alcohol**. The general formula and an example (*2-methyl– 2-propanol*) are given below.

$$R_1-\underset{\underset{\displaystyle R_2}{|}}{\overset{\overset{\displaystyle R_3}{|}}{C}}-OH \qquad\qquad CH_3-\underset{\underset{\displaystyle CH_3}{|}}{\overset{\overset{\displaystyle CH_3}{|}}{C}}-OH$$

tertiary alcohol   **2-methyl–2-propanol**

11.4A DIHYDROXY AND TRIHYDROXY ALCOHOLS

As their names imply, *dihydroxy* and *trihydroxy alcohols* contain two and three hydroxyl groups, respectively. Dihydroxy alcohols are also known as *glycols*. The most important dihydroxy alcohol is *1,2-ethanediol* (commonly called *ethylene glycol*), which is a component of automobile antifreeze. The most important trihydroxy alcohol is *1,2,3-propanetriol* (commonly called *glycerol,* which plays a part in biological processes). The structures of these compounds are given below.

$$\begin{array}{cc} \text{OH} \quad \text{OH} & \text{OH} \quad \text{OH} \quad \text{OH} \\ | \quad\quad | & | \quad\quad | \quad\quad | \\ \text{H}-\text{C}-\text{C}-\text{H} & \text{H}-\text{C}-\text{C}-\text{C}-\text{H} \\ | \quad\quad | & | \quad\quad | \quad\quad | \\ \text{H} \quad \text{H} & \text{H} \quad \text{H} \quad \text{H} \end{array}$$

1,2,-ethanediol   **1,2,3-propanetriol**
(ethylene glycol)   **(glycerol)**

11.5A TYPES OF POLYMERIZATION

Condensation Polymerization

In **condensation polymerization**, the monomers are joined by a *dehydration* reaction. The first step in this process is illustrated below, as two monomer units (in this case, dihydroxy alcohols) condense to form a *dimer* and water.

$$\begin{array}{ccc} \text{H} \quad \text{H} & \text{H} \quad \text{H} & \text{H} \quad \text{H} \quad\quad \text{H} \quad \text{H} \\ | \quad\quad | & | \quad\quad | & | \quad\quad | \quad\quad | \quad\quad | \\ \text{HO}-\text{C}-\text{C}-\text{O}(\text{H}+\text{HO})-\text{C}-\text{C}-\text{OH} & \rightarrow & \text{HO}-\text{C}-\text{C}-\text{O}-\text{C}-\text{C}-\text{OH}+\text{H}_2\text{O} \\ | \quad\quad | & | \quad\quad | & | \quad\quad | \quad\quad | \quad\quad | \\ \text{H} \quad \text{H} & \text{H} \quad \text{H} & \text{H} \quad \text{H} \quad\quad \text{H} \quad \text{H} \end{array}$$

monomer   monomer   dimer   water

As additional units are joined to the dimer, the polymer is gradually built up. For condensation polymerization to occur, the monomer must have at least two functional groups that can undergo dehydration.

Examples of synthetic condensation polymers include silicones, nylons, and polyesters.

In living systems, two amino acids can undergo a condensation reaction in which the $-$OH group on the first molecule and the $-$NH$_2$ group on the second molecule interact to eliminate a molecule of H$_2$O. The two molecules then join to produce a special type of amide known as a *dipeptide*. The bond joining the two amino acids is known as a *peptide link*. The accompanying diagram illustrates how this occurs.

Amino acid 1 **Amino acid 2**

$$H-\overset{\overset{\displaystyle H}{|}}{N}-\overset{\overset{\displaystyle H}{|}}{\underset{\underset{\displaystyle R_1}{|}}{C}}-\overset{\overset{\displaystyle O}{\|}}{C}-\left(OH \quad H\right)-\overset{\overset{\displaystyle H}{|}}{N}-\overset{\overset{\displaystyle H}{|}}{\underset{\underset{\displaystyle R_2}{|}}{C}}-\overset{\overset{\displaystyle O}{\|}}{C}-OH$$

$[-H_2O]$ — **Peptide link**

$$H-\overset{\overset{\displaystyle H}{|}}{N}-\overset{\overset{\displaystyle H}{|}}{\underset{\underset{\displaystyle R_1}{|}}{C}}-\overset{\overset{\displaystyle O}{\|}}{C}-\overset{\overset{\displaystyle H}{|}}{N}-\overset{\overset{\displaystyle H}{|}}{\underset{\underset{\displaystyle R_2}{|}}{C}}-\overset{\overset{\displaystyle O}{\|}}{C}-OH$$

Dipeptide

Since each amino acid has *both* amino and hydroxyl groups, a *polyamide* chain containing many amino acid units can be formed. This *natural polymer* is known more commonly as a *polypeptide*. A *protein* is a molecule that contains one or more polypeptide chains. Because of its large size, the synthesis of a protein (and other biological polymers) would be impossibly slow if the process were not catalyzed by other proteins known as *enzymes*.

Addition Polymerization

Monomers that are unsaturated may undergo polymerization by undergoing addition reactions with each other. In this process, called **addition polymerization**, the double or triple bonds are reduced to single or double bonds, as illustrated below, where the "~" symbols indicate the positions where the repeating ethene units are added.

Polyethylene is an example of an addition polymer.

$$n\left(\overset{\displaystyle H}{\underset{\displaystyle H}{}}C=C\overset{\displaystyle H}{\underset{\displaystyle H}{}}\right) \xrightarrow{\text{addition}} \left(\sim\overset{\overset{\displaystyle H}{|}}{\underset{\underset{\displaystyle H}{|}}{C}}-\overset{\overset{\displaystyle H}{|}}{\underset{\underset{\displaystyle H}{|}}{C}}\sim\right)_n$$

n **units of ethene**   **polyethylene**

END-OF-CHAPTER QUESTIONS

Some questions have the symbol "§2" in front of the question number. This symbol means that the question is based on Section II material.

1. All organic compounds must contain the element
   (1) sulfur   (2) nitrogen   (3) carbon   (4) oxygen

2. A characteristic of organic compounds is that they generally
   (1) react rapidly
   (2) dissolve in nonpolar solvents
   (3) are strong electrolytes
   (4) melt at relatively high temperatures

3. Compared with the rate of an inorganic reaction, the rate of an organic reaction is generally
   (1) faster, because organic compounds are usually ionic
   (2) faster, because organic compounds are usually molecular
   (3) slower, because organic compounds are usually ionic
   (4) slower, because organic compounds are usually molecular

4. Organic compounds that are essentially nonpolar and exhibit weak intermolecular forces have
   (1) low melting points
   (2) low vapor pressures
   (3) high conductivities in solution
   (4) high boiling points

5. The members of the alkane series of hydrocarbons are similar in that each member has the same
   (1) empirical formula   (2) general formula
   (3) structural formula   (4) molecular formula

6. Which hydrocarbon has more than one possible structural formula?
   (1) CH_4 (2) C_2H_6 (3) C_3H_8 (4) C_4H_{10}

7. Which is the structural formula of propene?
   (1) $H-C=C-H$ (2) $H-C\equiv C-H$

§2 **8.** Which represents benzene?

9. A compound with the molecular formula C_7H_{16} could be a member of the hydrocarbon series that has the general formula
(1) C_nH_{2n} (2) C_nH_{2n-2} (3) C_nH_{2n-6} (4) C_nH_{2n+2}

10. Which compound has the greatest possible number of isomers?
(1) butane   (2) ethane   (3) pentane   (4) propane

11. Which compound contains a triple bond?
(1) CH_4 (2) C_2H_2 (3) C_3H_6 (4) C_4H_{10}

12. Each member of the alkane series differs from the preceding member by having 1 additional carbon atom and
(1) 1 hydrogen atom   (2) 2 hydrogen atoms
(3) 3 hydrogen atoms   (4) 4 hydrogen atoms

13. Which hydrocarbon ought to have the highest normal boiling point?
(1) butene   (2) ethene   (3) pentene   (4) propene

14. Alkanes differ from alkenes in that alkanes
(1) are hydrocarbons
(2) are saturated compounds
(3) have the general formula C_nH_{2n}
(4) undergo addition reactions

15. Natural gas is composed mostly of
(1) butane   (2) gasoline
(3) methane   (4) propane

§2 **16.** Which method is commonly used to separate the components of petroleum into simpler substances by using their differences in boiling points?

    (1) fractional crystallization   (2) fractional distillation
    (3) esterification   (4) saponification

§2 **17.** Which equation represents a simple example of cracking?

    (1) $N_2 + 3H_2 \xrightarrow{600°C} 2NH_3$
    (2) $S + O_2 \longrightarrow SO_2$
    (3) $C_3H_8 + 5O_2 \longrightarrow 3CO_2 + 4H_2O$
    (4) $C_{14}H_{30} \xrightarrow{600°C} C_7H_{16} + C_7H_{14}$

18. How many hydrogen atoms are there in an alkene containing 20 carbon atoms?

    (1) 34   (2) 20   (3) 42   (4) 40

19. Which compound is an isomer of the substance shown below?

20. The functional group

is always found in an organic

    (1) acid   (2) ester   (3) ether   (4) aldehyde

21. Molecules of 1-propanol and 2-propanol have different
    (1) percentage compositions
    (2) molecular masses
    (3) molecular formulas
    (4) structural formulas

22. Which compound is an organic acid?
    (1) CH_3CH_2OH
    (2) CH_3OCH_3
    (3) CH_3COOH
    (4) CH_3COOCH_3

§2 23. Which compound is a trihydroxy alcohol?
    (1) glycerol   (2) butanol   (3) ethanol   (4) methanol

24. What is the IUPAC name for a compound with the structural formula shown below?

$$
\begin{array}{cccc}
Cl & H & Cl & H \\
| & | & | & | \\
H-C-&C-&C-&C-H \\
| & | & | & | \\
H & H & H & H
\end{array}
$$

    (1) 1,3-dichloropentane
    (2) 2,4-dichloropentane
    (3) 1,3-dichlorobutane
    (4) 2,4-dichlorobutane

25. Which structural formula represents an aldehyde?

    (1)
$$
\begin{array}{cc}
H & O \\
| & || \\
H-C-&C-OH \\
| & \\
H &
\end{array}
$$

    (2)
$$
\begin{array}{ccc}
H & H & H \\
| & | & | \\
H-C-&C-&C-H \\
| & | & | \\
H & H & H
\end{array}
$$

    (3)
$$
\begin{array}{cc}
H & O \\
| & || \\
H-C-&C-H \\
| & \\
H &
\end{array}
$$

    (4)
$$
\begin{array}{cc}
H & H \\
| & | \\
H-C-&C-OH \\
| & | \\
H & H
\end{array}
$$

26. Which is the general formula for ketones?

    (1) $R-C{\overset{O}{\Big\Vert}}{\diagdown}_H$

    (2) $R-C{\overset{O}{\Big\Vert}}{\diagdown}_{OH}$

    (3) $\begin{array}{c} R_1 \\ {\diagdown} \\ C=O \\ {\diagup} \\ R_2 \end{array}$

    (4) R_1-O-R_2

27. Which is the structural formula for diethyl ether?

$$(1) \quad H-\underset{\underset{H}{|}}{\overset{\overset{H}{|}}{C}}-\underset{\underset{H}{|}}{\overset{\overset{H}{|}}{C}}-O-\underset{\underset{H}{|}}{\overset{\overset{H}{|}}{C}}-\underset{\underset{H}{|}}{\overset{\overset{H}{|}}{C}}-H$$

(2) shown above

$$(3) \quad H-\underset{\underset{H}{|}}{\overset{\overset{H}{|}}{C}}-O-\underset{\underset{H}{|}}{\overset{\overset{H}{|}}{C}}-H$$

(4) shown above

§2 **28.** Which structural formula represents a tertiary alcohol?

$$(1) \quad H-\underset{\underset{H}{|}}{\overset{\overset{H}{|}}{C}}-\underset{\underset{H}{|}}{\overset{\overset{H}{|}}{C}}-\underset{\underset{H}{|}}{\overset{\overset{H}{|}}{C}}-\underset{\underset{H}{|}}{\overset{\overset{H}{|}}{C}}-OH$$

$$(2) \quad H-\underset{\underset{OH}{|}}{\overset{\overset{H}{|}}{C}}-\underset{\underset{H}{|}}{\overset{\overset{H}{|}}{C}}-\underset{\underset{H}{|}}{\overset{\overset{H}{|}}{C}}-\underset{\underset{H}{|}}{\overset{\overset{H}{|}}{C}}-H$$

$$(3) \quad H-\underset{\underset{H}{|}}{\overset{\overset{H}{|}}{C}}-\underset{\underset{OH}{|}}{\overset{\overset{H-\overset{H}{\overset{|}{C}}-H}{|}}{C}}-\underset{\underset{H}{|}}{\overset{\overset{H}{|}}{C}}-H$$

$$(4) \quad H-\underset{\underset{H}{|}}{\overset{\overset{H}{|}}{C}}-\underset{\underset{H}{|}}{\overset{\overset{H-\overset{H}{\overset{|}{C}}-H}{|}}{C}}-\underset{\underset{H}{|}}{\overset{\overset{H}{|}}{C}}-OH$$

§2 **29.** What is the total number of hydroxyl groups contained in one molecule of 1,2-ethanediol?

(1) 1   (2) 2   (3) 3   (4) 4

30. Which class of organic compounds contains nitrogen?

(1) amines   (2) aldehydes   (3) ethers   (4) ketones

31. Which is the product of the reaction between ethene and chlorine?

$$(1) \quad H-\underset{\underset{H}{|}}{\overset{\overset{H}{|}}{C}}-\underset{\underset{H}{|}}{\overset{\overset{H}{|}}{C}}-Cl$$

$$(2) \quad H-\underset{\underset{H}{|}}{\overset{\overset{H}{|}}{C}}-Cl$$

$$(3) \quad Cl-\underset{\underset{H}{|}}{\overset{\overset{H}{|}}{C}}-\underset{\underset{H}{|}}{\overset{\overset{H}{|}}{C}}-Cl$$

$$(4) \quad Cl-\underset{\underset{H}{|}}{\overset{\overset{H}{|}}{C}}-Cl$$

32. Which alcohol reacts with C_2H_5COOH to produce the ester $C_2H_5COOC_2H_5$?

(1) CH_3OH (2) C_2H_5OH (3) C_3H_7OH (4) C_4H_9OH

33. Given this equation:

$$C_6H_{12}O_6 \xrightarrow{\text{yeast enzymes}} 2C_2H_5OH + 2CO_2$$

The reaction represented by this equation is called
(1) esterification   (2) saponification
(3) fermentation   (4) polymerization

34. The reaction represented by the equation

$$nC_2H_4 \to (-C_2H_4-)_n$$

is called
(1) saponification   (2) fermentation
(3) esterification   (4) polymerization

35. The reaction

$$C_4H_{10} + Br_2 \to C_4H_9Br + HBr$$

is an example of
(1) substitution   (2) addition
(3) fermentation   (4) polymerization

36. Which physical property makes it possible to separate the components of crude oil (petroleum) by means of distillation?
(1) melting point   (2) conductivity
(3) solubility   (4) boiling point

37. In saturated hydrocarbons, carbon atoms are bonded to each other by
(1) single covalent bonds, only
(2) double covalent bonds, only
(3) alternating single and double bonds
(4) alternating double and triple bonds

38. Which formula correctly represents the product of an addition reaction between ethane and chlorine?
(1) CH_2Cl_2 (2) CH_3Cl
(3) $C_2H_4Cl_2$ (4) C_2H_3Cl

39. Which hydrocarbon is saturated?
(1) propene   (2) ethyne
(3) butene   (4) heptane

40. Which compound is an isomer of pentane?
(1) butane
(2) propane
(3) methylbutane
(4) methylpropane

41. Given the formula representing a compound:

$$H-\underset{\underset{H}{|}}{\overset{\overset{H}{|}}{C}}-\underset{\underset{H}{|}}{\overset{\overset{H}{|}}{C}}-\underset{\underset{H}{|}}{\overset{\overset{H}{|}}{C}}-\underset{}{\overset{\overset{O}{\|}}{C}}-\underset{\underset{H}{|}}{\overset{\overset{H}{|}}{C}}-\underset{\underset{H}{|}}{\overset{\overset{H}{|}}{C}}-H$$

What is an IUPAC name for this compound?
(1) ethyl propanoate
(3) 3-hexanone
(2) propyl ethanoate
(4) 4-hexanone

42. All isomers of octane have the same
(1) molecular formula
(3) physical properties
(2) structural formula
(4) IUPAC name

Constructed-Response Questions

Base your answers to questions 1 through 3 on the information below.

Many artificial flavorings are prepared using the type of organic reaction shown below.

$$H-\underset{\underset{H}{|}}{\overset{\overset{H}{|}}{C}}-\overset{\overset{O}{\|}}{C}-OH \;+\; HO-\underset{\underset{H}{|}}{\overset{\overset{H}{|}}{C}}-\underset{\underset{H}{|}}{\overset{\overset{H}{|}}{C}}-\underset{\underset{H}{|}}{\overset{\overset{H}{|}}{C}}-H \longrightarrow H-\underset{\underset{H}{|}}{\overset{\overset{H}{|}}{C}}-\overset{\overset{O}{\|}}{C}-O-\underset{\underset{H}{|}}{\overset{\overset{H}{|}}{C}}-\underset{\underset{H}{|}}{\overset{\overset{H}{|}}{C}}-\underset{\underset{H}{|}}{\overset{\overset{H}{|}}{C}}-H \;+\; HOH$$

Reactant 1   Reactant 2

1. What type of organic reaction is this?

2. To what class of organic compounds does reactant 2 belong?

3. Draw the structural formula of an isomer of reactant 2.

Base your answers to questions 4–6 on the information below and on your knowledge of chemistry.

The equation below represents a reaction between propene and hydrogen bromide.

$$
\underset{\substack{|\\H}}{\overset{\substack{H\\|}}{H-C}}-\underset{\substack{|\\H}}{\overset{\substack{H\\|}}{C}}=\underset{\substack{}}{\overset{}{C}}-H \ + \ H-Br \ \longrightarrow \ \underset{\substack{|\\H}}{\overset{\substack{H\\|}}{H-C}}-\underset{\substack{|\\Br}}{\overset{\substack{H\\|}}{C}}-\underset{\substack{|\\H}}{\overset{\substack{H\\|}}{C}}-H
$$

Cyclopropane, an isomer of propene, has a boiling point of –33°C at standard pressure and is represented by the formula below.

4. Explain why the reaction shown above can be classified as a synthesis reaction.

5. Identify the class of organic compounds to which the product of the reaction belongs.

6. Explain, in terms of molecular formulas and structural formulas, why cyclopropane is an isomer of propene.

Base your answers to questions 7–9 on the information below and on your knowledge of chemistry.

The diagrams below represent ball-and-stick models of two molecules. (Each ball represents an atom, and each stick represents a chemical bond.)

Key	
●	= an atom of hydrogen
○	= an atom of carbon

Diagram A   Diagram B

7. Draw a Lewis electron-dot diagram for an atom of the element present in all organic compounds.

8. Explain in terms of carbon-to-carbon bonds why the hydrocarbon represented in Diagram B is saturated.

9. Explain why the molecules in Diagrams A and B are isomers of each other.

The answers to these questions are found in Appendix 3.

Chapter Twelve

SOLUTIONS AND THEIR PROPERTIES

KEY IDEAS

This chapter focuses on the properties of the homogeneous mixtures we call solutions. The concepts of solubility, concentration, and colligative properties are explored. The chapter concludes with a discus sion of suspensions and colloidal dispersions.

KEY OBJECTIVES

At the conclusion of this chapter you will be able to:

- Define the terms *solution*, *solute*, and *solvent*.
- Provide examples of various types of solutions.
- Define the terms *miscible*, *saturated*, *unsaturated*, *solubility*, and *supersaturation*.
- Describe the factors that affect the solubility of a substance.
- Use the "solubility rules" to predict the solubility of an ionic compound in water.
- Interpret a solubility curve, and solve problems involving solubility curves.
- Describe how the concentration of a solution can be expressed with respect to the following terms: *percent*, *parts per million*, *mole fraction*, *molarity*, and *molality*, and solve problems involving these measurements of concentration.
- Describe how the solute affects the boiling point and the freezing point of a solution.
- Solve problems involving freezing point depression and boiling point elevation (colligative properties).
- Solve problems involving solutions and chemical equations.
- Define the term *electrolyte*, and indicate why solutions of electrolytes exhibit abnormal behavior.
- Distinguish among suspensions and types of colloidal dispersions.

SECTION I—BASIC (REGENTS-LEVEL) MATERIAL

NYS REGENTS CONCEPTS AND SKILLS

Note: By the time you have finished Section I, you should have mastered the concepts and skills listed below. The Regents chemistry examination will test your knowledge of these items and your ability to apply them.

Concepts are the *basic ideas* that form the body of the Regents chemistry course (what you need to know!).

Skills are the *activities* that demonstrate your mastery of these concepts (how you show that you know them!).

Following each concept or skill is a page reference (given in parentheses) to this chapter.

12.1 Concept:
- A solution is a homogeneous mixture of a solute dissolved in a solvent. (Page 343)
- The solubility of a solute in a given amount of solvent is dependent on the temperature, the pressure, and the chemical natures of the solute and solvent. (Pages 344–345)

Skills:
- Interpret and construct solubility curves. (Pages 345–347)
- Use solubility curves to distinguish among saturated, supersaturated, and unsaturated solutions. (Pages 345–347)
- Apply the "like dissolves like" rule to real-world situations. (Pages 344–345)

12.2 Concept:
The concentration of a solution may be expressed as:
(a) molarity (M) (Pages 349–352);
(b) percent by mass (Pages 347–348);
(c) parts per million (ppm) (Pages 348–349).

Skills:
- Describe the preparation of a solution, given the molarity. (Pages 349–350)
- Calculate solution concentrations in molarity, percent by mass, and parts per million. (Pages 348–349)

12.3 Concepts:
  - The addition of a nonvolatile solute to a solvent causes the boiling point of the solvent to increase and the freezing point of the solvent to decrease. (Page 352)
  - The greater the concentration of particles within the solution, the greater the effect. (Page 352)

12.4 Concepts:
  - An electrolyte is a substance that releases mobile ions into an aqueous solution. As a result, the solution is capable of conducting an electric current. (Pages 352–353)
  - The degree of conductivity of a solution depends on the concentration of ions within the solution. (Pages 352–353)

12.1 A SOLUTION IS . . . ?

A solution is one example of a homogeneous mixture of substances. When we think of the term *solution*, we usually picture a substance, such as table salt, well mixed or *dispersed* throughout another substance, such as water. The result is a water-clear mixture that we call a **solution**. The substance that is dissolved (the salt) is termed the **solute**, while the substance that does the dissolving (the water) is the **solvent**.

Although the solute and the solvent may be in different phases (in our example salt is solid and water is liquid), the solution assumes the phase of the *solvent*. If the solute and the solvent are in the same phase (as is the case with alcohol and water), the solvent is the substance that is present in greater quantity. Solutions in which water is the solvent are termed *aqueous*. There are times when we may be more interested in knowing the *components* of the solution than in identifying which component is the solute and which is the solvent.

Why do substances dissolve in one another? During the dissolving process, the particles of each component must be separated by—and mix with—the particles of every other component in the solution. The accompanying diagram illustrates how this may happen in a two-component solution.

Component *A*

Component *B*

Solution of *A* and *B*

In the formation of a solution, (1) some of the attractive forces between the particles of each component must be overcome, and (2) the particles of all of the components must be able to attract each other with forces of similar strength.

In gases, there are essentially no attractive forces between the particles and therefore all mixtures of gases form solutions. A substance such as sugar forms a solution with water because the dipole attractions among sugar and water molecules in the solution are similar to the intermolecular attractions present in pure sugar and pure water. In contrast, oil and water do *not* form a solution because the weak London dispersion forces among the nonpolar oil molecules are not strong enough to separate the polar water molecules, which are attracted to each other by much stronger dipole forces.

Solutions exist among all three phases of matter, as the accompanying table illustrates.

Phase of Solute	Phase of Solvent	Example
Solid	Solid	Metallic alloy
Solid	Liquid	Table salt–water
Liquid	Liquid	Ethyl alcohol–water
Gas	Liquid	CO_2–water (soda)
Gas	Gas	Air

12.2   SATURATED AND UNSATURATED SOLUTIONS

When a solution of ethyl alcohol and water is prepared, the alcohol and water may be mixed in all proportions. A liquid–liquid solution of this type is said to be **miscible**. Mixtures of *gases* also form infinitely soluble solutions.

Most solutions, however, are not infinitely soluble. An example is the dissolving of table salt in water. After a certain amount of salt has been dissolved, the water will accept no more solute and any remaining salt will drop to the bottom of the vessel. This solution is said to be **saturated**. A solution that has not reached its saturation point is **unsaturated**.

12.3   SOLUBILITY

The quantity of solute needed to *just saturate* a given amount of solvent is known as the **solubility** of the solute-solvent pair. A convenient set of units for specifying the solubilities of most solids is *grams of solute per 100 grams of solvent*. For example, at 50°C, 85 grams of potassium nitrate, KNO_3, will just saturate 100 grams of H_2O. Therefore, the solubility of KNO_3 at 50°C is reported as *85 grams KNO_3 per 100 grams H_2O.*

Solubility depends on factors such as the nature of the solute and of the solvent and the temperature. If the solute is a gas, such as SO_2, its solubility is also affected by the atmospheric pressure.

Chemists frequently use the expression "like dissolves like" to indicate that the solute and the solvent need to have similar bond types in order for a solution to be formed. Substances that are ionic or consist of polar molecules will dissolve in a polar solvent such as water, and nonpolar substances (e.g., oils) will dissolve in nonpolar solvents such as benzene and toluene.

A substance such as potassium nitrate, KNO_3, that can form fairly concentrated solutions (in water) is said to be *soluble*. On the other hand, a substance such as silver iodide, AgI, that can form only very dilute solutions (in water) is termed *nearly insoluble*.

The solubilities of ionic compounds in water follow a number of general rules that are outlined in Reference Table F in Appendix 1.

TRY IT YOURSELF
Use the solubility rules in Reference Table F to classify each of the five compounds given below as *soluble*, *slightly soluble*, or *insoluble*.
(a) $NaNO_3$ (b) $BaSO_4$ (c) $(NH_4)_2S$ (d) AgI (e) $Sr(OH)_2$

ANSWERS
(a) soluble (b) insoluble (c) soluble (d) insoluble (e) slightly soluble

Solubility Curves

A *solubility curve* traces the solubility of a substance with increasing temperature. The accompanying graph displays the solubilities of $CuSO_4$ (a solid) and SO_2 (a gas) in water, as a function of temperature.

Each point on the curves represents the composition of a saturated solution at a different temperature. For example, the solubility of $CuSO_4$ is approximately 57 grams per 100 grams H_2O at 80°C, and the solubility of SO_2 is approximately 9 grams per 100 grams H_2O at 24°C.

As the temperature *increases* between 0°C and 100°C, the solubility of the solid increases, meaning that more solid is needed for a saturated solution. Most (but not all) solids increase their solubilities with increasing temperature. If the temperature of the solution is *decreased*, the solubility of the solid will decrease and any *excess* solute will drop out of the solution. Reference Table G in Appendix 1 contains a number of solubility curves for ionic solids and for gases dissolved in water.

Temperature has the opposite effect on a dissolved gas: when the temperature is *raised*, the solubility of the gas *decreases* and the some of the gas will escape from the solution. If the temperature is lowered, the solvent will be able to dissolve more gas.

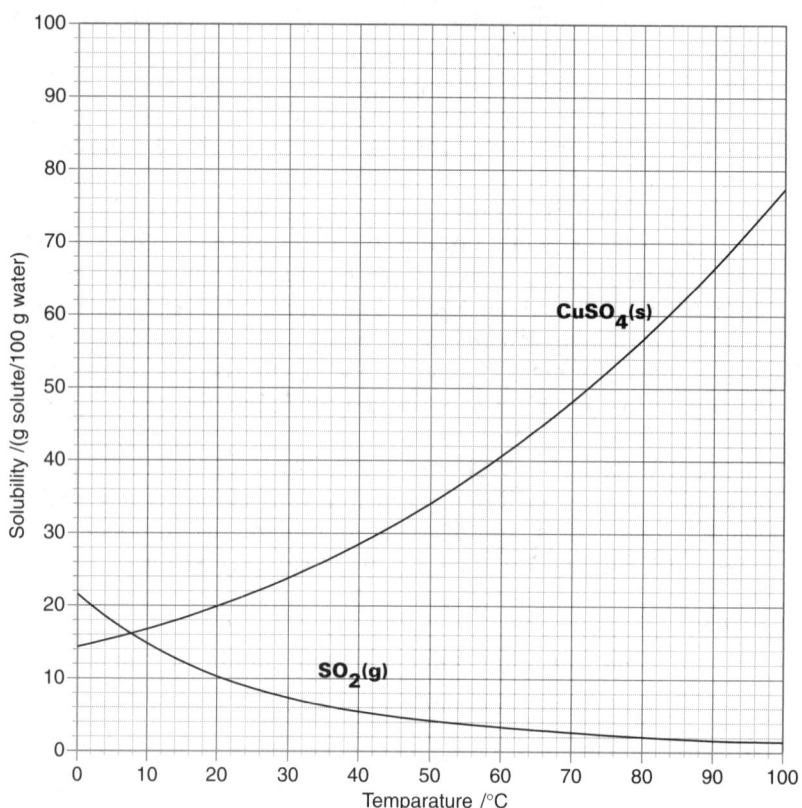

Now let us solve some problems based on the solubility curves in Reference Table G.

PROBLEM

(a) At what temperature will the solubility of KNO_3 be 60 grams per 100 grams H_2O?

(b) At what temperature will the solubility be 120 grams per 100 grams H_2O?

SOLUTIONS

Refer to the solubility curve for KNO_3. (a) At 38°C the solubility of this compound is 60 g/100 g H_2O. (b) At 65°C the solubility of this compound is 120 g/100 g H_2O.

PROBLEM

How much $NaNO_3$ is needed to saturate 50 grams of water at 40°C?

SOLUTION

Refer to the solubility curve for $NaNO_3$. At 40°C, the solubility of this compound is $105 g/100 g\ H_2O$. To saturate *half* the amount of water ($50 g\ H_2O$), half the amount of solute, or 53 g, will be needed.

TRY IT YOURSELF

(a) Which gas is most soluble at 0°C? At 60°C?

(b) In an experiment, 100 grams of water is saturated with NH_3 at 30°C. If the temperature is then lowered to 0°C, how much *additional* NH_3 will be needed to resaturate the solution?

(c) At what temperature do NH_4Cl and HCl have the same solubility?

(d) Which solid shows the smallest *increase* in solubility over the temperature range 0°C–100°C?

ANSWERS

(a) NH_3 (0°C); HCl (60°C)   (b) 45 g (90 g − 45 g)   (c) 58°C   (d) NaCl

Supersaturation

Occasionally, when a solution prepared at a higher temperature is cooled, the solute fails to drop out of the solution. This unstable condition is known as **supersaturation**. Eventually, the excess solute will leave the solution as it returns to its saturation point. Honey is an example of a supersaturated solution. After a period of time, the excess sugar crystallizes at the bottom of the honey jar. [The honey can be restored to its original (supersaturated) state by heating it and then allowing it to cool slowly.]

12.4 CONCENTRATIONS OF SOLUTIONS

When we prepare a solution, we are usually interested in its strength or concentration. **Concentration** is a measure of the quantity of solute dissolved in a given amount of solvent or solution. (This distinction will be made clear below.) For example, we may pick up a bottle that is labeled as *concentrated* (con) or *dilute* (dil) HCl. These terms tell us that the HCl solution is, respectively, relatively strong or relatively weak. However, the terms *concentrated* and *dilute* are not precise enough for most laboratory work.

Percent Concentration

A more precise way of expressing concentration is to specify the quantity of solute that is dissolved in one hundred parts of the solution. This type of expression is known as **percent concentration**. The percent concentration of a solution may be expressed in two ways:

• The **mass percent concentration** is defined as:

$$\text{Mass Percent Concentration} = \frac{\text{Mass of Solute}}{\text{Mass of Solution}} \times 100$$

It is essential that the quantities of the solute and solution be expressed in the same mass units.

PROBLEM

A student dissolves 25.0 grams of glucose in 475 grams of water. What is the mass percent concentration of the glucose solution?

SOLUTIONS

$$\text{Mass percent concentration} = \frac{\text{mass of solute}}{\text{mass of solution}} \times 100$$

$$= \frac{25.0\,\text{g}}{500.\,\text{g}} \times 100 = 5.00\%$$

Parts per Million (PPM)

The concentrations of very dilute solutions are often expressed in **parts per million** (ppm). In fact, scientists usually express the maximum allowable concentrations of toxic or cancer-causing substances in parts per million. For example, the maximum allowable concentration of arsenic in drinking water is 0.05 part per million, or 5 grams of arsenic per 100,000,000 grams of drinking water. We can calculate the concentration of a solution in parts per million by using the following equation:

$$\text{Parts per Million of Solute} = \frac{\text{Mass of Solute}}{\text{Mass of Solution}} \times 10^6$$

It is important to note that the numerator and denominator of this fraction must be expressed in the *same units of mass*.

PROBLEM

A certain gas has a concentration in water of 2 milligrams per 100 grams of water. What is the concentration of the gas in parts per million?

SOLUTION

There are two considerations in solving this problem. First, we must make the mass units of the solute and the solution the same. Second, we need to recognize that this solution is so dilute that the mass of the solution is essentially the mass of the water. With these considerations in mind, let us apply the equation given on page 348 and solve the problem:

$$\text{ppm of solute} = \frac{\text{mass of solute}}{\text{mass of solution}} \times 10^6$$

$$= \frac{2 \text{ mg} \cdot \dfrac{1 \text{ g}}{1000 \text{ mg}}}{100 \text{ g}} \times 10^6$$

$$= 0.00002 \times 10^6 \text{ ppm}$$

$$= 20 \text{ ppm}$$

TRY IT YOURSELF

A 2.5-gram sample of groundwater is found to contain 5.4×10^{-6} gram of Cu^{2+} ion. What is the ion concentration in parts per million?

ANSWER

2.2 ppm

In chemistry, where we are interested in the *number of particles* reacting, we measure the concentration of a solution by stating the number of *moles of solute* contained in the solution. There are three common ways to do this: molarity, molality, and mole fraction. (Mole fractions and molarity will be covered in Section II—Additional Material.)

Molarity

Molarity (*M*) is defined as the number of moles of solute dissolved in 1 liter of solution. In other words, molarity is a ratio between the number of moles of solute and the number of liters of solution.

$$M = \frac{n_{\text{solute}}}{V_{\text{solution (L)}}}$$

For example, if we prepare an aqueous solution by dissolving 4 moles of solute in 8 liters of solution, the concentration of this solution is 4/8 = 0.5 mole per liter. We write this as 0.5 M (read as 0.5 *molar*).

349

We never specify how much *solvent* we actually use. In the example given on the previous page, 4 moles of solute is dissolved in *enough water* to make 8 liters of solution. Usually, we do this by preparing the solution in a *volumetric flask*, that is, a flask in which only the *final volume* (8 liters, in this case) is marked. The solute is added to the flask, and the solvent is poured in gradually (with constant stirring) until the volume of the solution reaches the mark. The accompanying diagram illustrates a volumetric flask.

Final volume
of solution

We use molarity to prepare solutions when we are more concerned about the concentrations of the solutes than about the relationship between solute and solvent.

We can solve problems involving molarity in two ways. The first method involves using the preceding molarity equation.

PROBLEM

How many moles of a solute are needed to prepare 2.5 liters of a 0.60 M solution?

SOLUTION

We rearrange the molarity equation and use the units of molarity (mol/L) to solve for the moles of solute needed:

$$n_{\text{solute}} = M \cdot V_{\text{solution (L)}}$$

$$= 0.60 \frac{\text{mol}}{\cancel{L}} \cdot 2.5 \cancel{L} = 1.5 \text{ mol}$$

We can also use the factor-label method (FLM) to solve molarity problems. The units of molarity (mol/L) can be used as a conversion factor. For example, if our problem involves a 0.60 molar solution, we can use either $\dfrac{0.60 \text{ mol}}{1 \text{ L}}$ or $\dfrac{1 \text{ L}}{0.60 \text{ mol}}$ to solve the problem. Let us solve the problem given above by FLM:

$$\text{mol solute} = 2.5 \ \cancel{L}_{\text{solution}} \cdot \frac{0.60 \text{ mol}_{\text{solute}}}{1 \ \cancel{L}_{\text{solution}}} = 1.5 \text{ mol}_{\text{solute}}$$

It is no accident that the two methods look essentially the same: they are entirely equivalent! Which one should you use? In solving simple problems such as the one given above, use the method that is easier for you. For more complex problems, however, FLM may prove to be your better bet.

TRY IT YOURSELF
A person wishes to prepare a 2.00 molar solution using 2.50 moles of solute. Calculate the final volume of the solution.

ANSWER
1.25 L

The next problem deals with a more complex situation involving molarity, molar mass, and volume. Problems such as this one can be solved by using the factor-label method.

PROBLEM
How many grams of urea (CH_4N_2O; molar mass = 60.1 grams per mole) are needed to make 200. milliliters of a 1.50 M solution?

SOLUTION
Let us write a solution map for this problem:

$$V_{\text{solution (mL)}} \xrightarrow[\text{conversion}]{\text{unit}} V_{\text{solution (L)}} \xrightarrow{\text{molarity}} \text{mol}_{\text{solute}} \xrightarrow{\text{molar mass}} g_{\text{solute}}$$

We can solve the problem using FLM:

$$\text{g urea} = 200. \ \cancel{mL}_{\text{solution}} \cdot \frac{1 \ \cancel{L}_{\text{solution}}}{1000 \ \cancel{mL}_{\text{solution}}} \cdot \frac{1.50 \ \cancel{\text{mol urea}}}{1 \ \cancel{L}_{\text{solution}}} \cdot \frac{60.1 \text{ g urea}}{1 \ \cancel{\text{mol urea}}}$$

$$= 18 \text{ g urea}$$

TRY IT YOURSELF
A 3.00-molar solution of glucose ($C_6H_{12}O_6$; molar mass = 180. grams per mole) is prepared using 60.0 grams of solute. What is the final volume, in liters, of the solution?

ANSWER
0.111 L

12.5 EFFECT OF THE SOLUTE ON THE SOLVENT

When a solute is dissolved in a solvent, the particles of the solute bond with the particles of the solvent, causing the vapor pressure of the solution to be lower than the vapor pressure of the pure solvent at the same temperature. The effect of this vapor pressure lowering is to *raise* or *elevate* the boiling point of the solution and to *lower* or *depress* the freezing point of the solution in comparison to the boiling and freezing points of the pure solvent. For example, when 1 mole of glucose is dissolved in 1 kilogram of water, the boiling point of the solution is 100.52°C and the freezing point is −1.86°C. In contrast, the boiling and freezing points of pure water are 100°C and 0°C, respectively.

Boiling point elevation and freezing point depression are *colligative properties*; that is, they depend on the *number* of solute particles present and the *nature* of the solvent. Colligative properties do *not* depend on the actual *identity* of the solute. For example, if 1 mole of urea were substituted for the 1 mole of glucose in the preceding paragraph, the results would be exactly the same!

12.6 BEHAVIOR OF ELECTROLYTES IN SOLUTION

Certain substances called **electrolytes** produce ions when they dissolve in solution. Because these ions are free to move in solution, the solution conducts electricity: the greater the concentration of ions, the greater the conductivity of the solution.

Ions can be produced in solution in either of two ways.

1. When an ionic substance such as NaCl dissolves in water, the water *separates* the ions present in the NaCl crystal lattice. This process, known as *dissociation*, is represented below.

$$Na^+Cl^-(s) \xrightarrow{H_2O} Na^+(aq) + Cl^-(aq)$$

2. When a polar covalent substance such as HCl dissolves in water, ions are *created* by the interaction between HCl and H_2O molecules. This process, known as *ionization,* is represented below.

$$HCl(g) + H_2O(\ell) \rightarrow H_3O^+(aq) + Cl^-(aq)$$

When the boiling and freezing points of solutions of electrolytes are examined, it is found that they do *not* obey the simple relationship $\Delta t = k \cdot m$. The boiling points are *higher*, and the freezing points *lower*, than expected.

The reason for this abnormal behavior lies in the fact that each mole of dissolved electrolyte produces *more than 1 mole* of particles. For example, in *very dilute* aqueous solutions, 1 mole of $NaNO_3$ produces 2 moles of particles (1 mole of Na^+ and 1 mole of NO_3^-). Therefore, the boiling point elevation and freezing point depression are *twice* those for a molecular substance. Similarly, the effects of dissolving $Ca(NO_3)_2$ (1 mole produces 3 moles of particles) and $Al(NO_3)_3$ (1 mole produces 4 moles of particles) are even greater.

We need to specify that a solution of an electrolyte must be dilute because a phenomenon known as *ion-pairing* lowers the *number* of dissolved particles present in more concentrated solutions. As a result of ion-pairing, the effect that an electrolyte has on raising the boiling point and lowering the freezing point of a more concentrated solution is *less* than we would predict for a dilute solution.

SECTION II—ADDITIONAL MATERIAL

12.1A MOLE FRACTION

Suppose a solution contains two components, A and B. The *mole fraction* (X) of component A is denoted as X_A and is defined as follows:

$$X_A = \frac{\text{moles of A}}{\text{total number of moles}} = \frac{n_A}{n_A + n_B}$$

where n represents the number of moles of each component present. We do not need to specify which component is the solute or the solvent. Note that the mole fraction, X, has no units. Why?

PROBLEM

A solution is prepared by mixing 3.0 moles of ethanol and 9.0 moles of water. Calculate the mole fraction of ethanol in this solution.

SOLUTION

$$X_{ethanol} = \frac{n_{ethanol}}{n_{ethanol} + n_{water}} = \frac{3.0 \text{ mol}}{3.0 \text{ mol} + 9.0 \text{ mol}} = \frac{3.0 \text{ mol}}{12 \text{ mol}} = 0.25$$

This result means that 25 of every 100 particles in the solution are ethanol molecules.

TRY IT YOURSELF
Calculate the mole fraction of water in this solution.

ANSWER
$X_{water} = 0.75$

12.2A   MOLALITY

There are certain situations in which we must know how much solvent is present in a solution. These situations arise when we examine how a solute affects the properties of the solvent. For these instances, we use a method of measuring concentration known as *molality*.

Molality (*m*) is defined as the number of moles of solute dissolved in 1 kilogram of solvent. In other words, molality is a ratio between the number of moles of solute and the mass of the solvent expressed in kilograms:

$$m = \frac{n_{solute}}{kg_{solvent}}$$

For example, if we prepare an aqueous solution by dissolving 0.4 mole of solute in 0.5 kilogram of solvent, the concentration of this solution is 0.4/0.5 = 0.8 mole per kilogram. We write this as 0.8 m (read as 0.8 *molal*).

We never specify the volume of the solution we actually prepare because the volume of the solvent is not directly known to us. (We could obtain this information by measuring the *density of the solution*, but the required technique is beyond the scope of this book.)

PROBLEM
Sucrose (table sugar) has a molar mass of 342 grams per mole. What is the molality of a solution prepared by dissolving 34.2 grams of sucrose in 200. grams of water?

SOLUTION
We solve the problem by using the definition of molality and FLM:

$$m = \frac{n_{\text{solute}}}{kg_{\text{solvent}}} = \frac{34.2 \ \cancel{\text{g sucrose}} \cdot \dfrac{1 \ \text{mol sucrose}}{342 \ \cancel{\text{g sucrose}}}}{200. \ \cancel{\text{g H}_2\text{O}} \cdot \dfrac{1 \ \text{kg H}_2\text{O}}{1000 \ \cancel{\text{g H}_2\text{O}}}}$$

$$= \frac{0.100 \ \text{mol sucrose}}{0.200 \ \text{kg H}_2\text{O}} = 0.500 \ m$$

TRY IT YOURSELF

What is the molality of a solution prepared by dissolving 51 grams of NH_3 in 2.0 kilograms of H_2O?

ANSWER

1.5 m

12.3A DILUTION OF STOCK SOLUTIONS

In many laboratories solutions are frequently prepared as *concentrated stock solutions*. In order to use a *working* solution in an experiment, it is necessary to know how to dilute the stock correctly. The following problem illustrates this point.

PROBLEM

How should a 12 M stock solution be diluted in order to prepare 0.80 liter of a 3.0 M working solution?

SOLUTION

When we dilute a solution, we do *not* change the number of moles of solute present: we simply add more solvent. Therefore, since $n_{\text{solute}} = M \cdot V_{\text{solution (L)}}$, the following relationship must hold true:

$$M_{\substack{\text{stock} \\ \text{solution}}} \cdot V_{\substack{\text{stock} \\ \text{solution}}} = M_{\substack{\text{working} \\ \text{solution}}} \cdot V_{\substack{\text{working} \\ \text{solution}}}$$

We can now solve the problem:

$$12M \cdot V = 3.0 \ \text{M} \cdot 0.80 \ \text{L}$$
$$V = 0.20 \ \text{L}$$

This result means that we must dilute 0.20 L of the 12 M stock solution to 0.80 L with solvent. The resulting solution will have a 3.0 M concentration. Since the volume term appears on *both* sides of the equation, units *other than liters* can be used for volume provided that the *same* units appear on each side.

TRY IT YOURSELF
If 300. milliliters of a 6.00 M stock solution is diluted to produce a 2.00 M working solution, what is the volume of the working solution?

ANSWER
900. mL

12.4A SOLUTIONS AND CHEMICAL EQUATIONS

Since the molarity of a solution depends on the number of moles of solute present, we can solve chemical equation problems that include molarity.
    The following example illustrates how we solve such problems.

PROBLEM
Consider this equation:

$$Cu(s) + 4HNO_3(aq) \rightarrow Cu(NO_3)_2(aq) + 2NO_2(g) + 2H_2O(\ell)$$

How many grams of $NO_2(g)$ will be produced by the complete reaction of 400. milliliters of 2.50 M HNO_3?

SOLUTION
Our solution map for this problem is as follows:

$$\text{mL HNO}_3 \xrightarrow[\text{conversion}]{\text{unit}} \text{L HNO}_3 \xrightarrow{\text{molarity}} \text{mol HNO}_3 \xrightarrow[\text{coefficients}]{\text{equation}} \text{mol NO}_2 \xrightarrow[\text{mass}]{\text{molar}} \text{g NO}_2$$

We now use our trusty friend, FLM, to solve the problem.

$$\text{g NO}_2 = 400. \; \cancel{\text{mL HNO}_3} \cdot \frac{1 \; \cancel{\text{L HNO}_3}}{1000 \; \cancel{\text{mL HNO}_3}} \cdot \frac{2.50 \; \cancel{\text{mol HNO}_3}}{1 \; \cancel{\text{L HNO}_3}}$$

$$\cdot \frac{2 \; \cancel{\text{mol NO}_2}}{4 \; \cancel{\text{mol HNO}_3}} \cdot \frac{46.0 \; \text{g NO}_2}{1 \; \cancel{\text{mol NO}_2}}$$

$$= 23.0 \; \text{g NO}_2$$

TRY IT YOURSELF
In the equation given above, how many grams of Cu(s) will react completely with the quantity of HNO_3 given?

ANSWER
15.9 g

12.5A CALCULATING THE FREEZING AND BOILING POINTS OF SOLUTIONS

If a solute is *molecular* (i.e., not ionic) and is *nonvolatile* (i.e., does not evaporate with the solvent), then the changes in the boiling and freezing points of a solution bear a simple relationship to the *molality* of the solution:

$$\Delta T_b = +k_b \cdot m$$
$$\Delta T_f = -k_f \cdot m$$

where ΔT_b and ΔT_f refer to the *changes* in the boiling and freezing points, respectively. The "+" sign indicates that the boiling point of a solution is elevated; the "−" sign, that the freezing point of a solution is lowered.

In these relationships, k_b and k_f are known as the *boiling point* and *freezing point constants* and they depend on the solvent that is used. (The units of these constants are C°/m.) *The constants k_b and k_f represent the number of degrees that the boiling point of a solvent is raised (or its freezing point is lowered) when 1 mole of a molecular, nonvolatile solute is dissolved in 1 kilogram of the solvent.*

Reference Table W-3 in Appendix 2 lists freezing and boiling point data for selected solvents.

If we inspect this table, we can see that the freezing and boiling points of some solvents (e.g., cyclohexane) are very sensitive to the addition of a solute.

PROBLEM
Calculate the boiling and freezing points of a 1.50 m aqueous solution of a molecular, nonvolatile solute.

SOLUTION

$$\Delta T_b = +k_b \cdot m = +0.513 \text{ C°/m} \cdot 1.50 \text{ m} = +0.77 \text{ C°}$$
$$\Delta T_f = -k_f \cdot m = -1.86 \text{ C°/m} \cdot 1.50 \text{ m} = -2.79 \text{ C°}$$

Since the boiling point of the solution is elevated, we *add* 0.77 C° to the boiling point of pure H_2O (100°C); the solution's boiling point is 100.77°C.

Since the freezing point of the solution is lowered, we *subtract* 2.79 C° from the freezing point of pure H_2O (0°C); the solution's freezing point is −2.79°C.

TRY IT YOURSELF

A 0.80 m solution of a molecular, nonvolatile solute is prepared using benzene as a solvent.

(a) By how many degrees is the freezing point of the solvent lowered and the boiling point raised?

(b) What are the actual boiling and freezing points of the solution?

ANSWERS

(a) $\Delta T_f = -4.1\ C°$; $\Delta T_b = +2.1\ C°$

(b) Freezing point $= 1.4°C$; boiling point $= 82.2°C$

PROBLEM

A solution of a molecular, nonvolatile solute is prepared as follows: 90.0 grams of solute is dissolved in 250. grams of water. The freezing point of the solution is $-3.72°C$.

(a) How many grams of solute would be dissolved in a solution having the same concentration but containing 1 kilogram of water?

(b) What is the freezing point lowering of the solution?

(c) Based on the freezing point lowering, what is the molality of the solution?

(d) What is the molar mass of the solute in grams per mole?

SOLUTIONS

The methods employed in this problem have actually been used to calculate the molar masses of unknown compounds.

(a) Since 250. g of water represents 0.250 kg, a solution containing 1.00 kg of water—four times the amount—would need *four times* the mass of solute in order to have the same concentration. Therefore, the amount of solute needed would be $4 \cdot 90.0\ g = 360.\ g$.

(b) The freezing point of pure water is 0°C. Since the freezing point of the solution is $-3.72°C$, the freezing point has been lowered by the same amount: $-3.72\ C°$.

(c) Since $k_f = -1.86\ C°/m$ for H_2O, the molality of the solution is 2.00.

(d) From part (a) we know that the solution contains 360. g of solute/ 1.00 kg of water. From part (c) we know that the solution contains 2.00 mol of solute/1.00 kg of water. Therefore, the mass of 2.00 mol of solute is 360. g. Since the molar mass is the number of grams in 1 mol, we conclude that the molar mass of the solute is 180. g/mol.

12.6A   SUSPENSIONS AND COLLOIDAL DISPERSIONS

Other examples of homogeneous mixtures include *suspensions* and *colloidal dispersions*. Whether a homogeneous mixture is a solution, a suspension, or a colloidal dispersion depends on the *size* of the particles being dispersed.

The largest particles form **suspensions**. A suspension, such as sand in water, will settle in time.

The smallest particles form *solutions*. In a solution, the solvent particles do not settle; and if the solvent is clear, a light beam passing through the solution will not be visible.

A **colloidal dispersion** is formed when particles that are larger than those in solutions but smaller than those in suspensions are dispersed in a medium. Usually, the dispersed particles range in size from 1 to 1000 nanometers (nm) and consist of very large molecules or clusters of smaller atoms or ions. In reference to colloids, the terms *dispersed substance* and *dispersing medium* are analogous to the terms *solute* and *solvent*. A colloidal dispersion will not settle, and a light beam passing through it will be clearly visible, a phenomenon known as the **Tyndall effect**. An example of the Tyndall effect is the visibility of an automobile headlight beam as it passes through fog.

The accompanying diagram shows how a solution and a colloidal dispersion affect a beam of light.

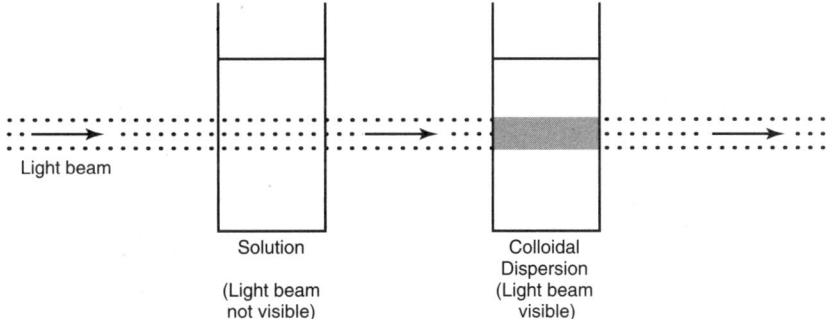

Light beam

Solution
(Light beam
not visible)

Colloidal
Dispersion
(Light beam
visible)

Many familiar mixtures of substances are actually colloidal dispersions. Some of these are listed in the accompanying table.

Colloid Type	Phase of Dispersed Substance	Phase of Dispersing Medium	Examples
Aerosol	Liquid	Gas	Fog, aerosol sprays
Aerosol	Solid	Gas	Smoke
Foam	Gas	Liquid	Whipped cream
Emulsion	Liquid	Liquid	Milk, mayonnaise
Sol	Solid	Liquid	Paint, gelatin
Solid foam	Gas	Solid	Marshmallow
Solid emulsion	Liquid	Solid	Butter, cheese

END-OF-CHAPTER QUESTIONS

Some questions have the symbol "§2" in front of the question number. This symbol means that the question is based on Section II material.

1. A solution containing 95 grams of KNO_3 dissolved in 100 grams of water at 55°C is classified as being
   (1) saturated   (2) supersaturated
   (3) unsaturated   (4) ultrasaturated

2. What is the maximum number of grams of NH_4Cl that will dissolve in 200 grams of water at 70°C?
   (1) 62   (2) 70   (3) 124   (4) 226

3. According to Reference Table G in Appendix 1, approximately how many grams of $KClO_3$ are needed to saturate 100 grams of H_2O at 40°C?
   (1) 6   (2) 14   (3) 37   (4) 47

4. According to Reference Table G, which of the following substances is most soluble at 60°C?
   (1) NH_4Cl (2) $NaNO_3$ (3) $NaCl$ (4) NH_3

5. Which saturated solution is the most concentrated at 20°C?
   (1) $NaCl$ (2) $KClO_3$ (3) KI (4) KNO_3

6. Which compound shows the *least* increase in solubility in water from 50°C to 60°C?
   (1) $KClO_3$ (2) $NaCl$ (3) KNO_3 (4) KI

7. A solution in which an equilibrium exists between dissolved and undissolved solute must be
   (1) saturated   (2) unsaturated
   (3) dilute   (4) concentrated

8. According to Reference Table F in Appendix 1, which compound is most soluble in water?
   (1) $BaCO_3$ (2) $BaSO_4$ (3) $ZnCO_3$ (4) $ZnSO_4$

9. A solution contains 70 grams of KNO_3 in 100 grams of water at 60°C. Approximately how many additional grams of KNO_3 are required to saturate this solution?
   (1) 36   (2) 18   (3) 54   (4) 72

10. At 10.°C, 80 grams of a substance saturates 100. grams of water. The substance could be
    (1) NaCl   (2) NaNO$_3$ (3) NH$_4$Cl   (4) KClO$_3$

11. A solution is prepared by adding 20 grams of an ionic solid to 100 grams of water at 40°C. If the mixture contains undissolved solute, the solid could be
    (1) KNO$_3$ (2) NH$_4$Cl   (3) KClO$_3$ (4) NaNO$_3$

12. According to Reference Table F, which saturated solution is most concentrated?
    (1) AgC$_2$H$_3$O$_2$(aq)   (2) Ag$_2$CO$_3$(aq)
    (3) Ag$_2$S(aq)   (4) AgCl(aq)

§2 13. At which pressure would carbon dioxide gas be most soluble in 100 grams of water at a temperature of 25°C?
    (1) 1 atm   (2) 2 atm   (3) 3 atm   (4) 4 atm

14. As additional KNO$_3$(s) is added to a saturated solution of KNO$_3$ at constant temperature, the concentration of the solution
    (1) decreases   (2) increases   (3) remains the same

15. The maximum quantity of KCl that can dissolve in 150 grams of water at 40°C is closest to
    (1) 17 g   (2) 39 g   (3) 59 g   (4) 210 g

16. A crystal of solute is added to a solution, and the crystal remains undissolved. The solution is best classified as
    (1) dilute   (2) unsaturated
    (3) saturated   (4) supersaturated

§2 17. If 0.50 liter of a 12-molar solution is diluted to 1.0 liter, the molarity of the new solution is
    (1) 2.4   (2) 6.0   (3) 12   (4) 24

18. How many grams of KOH are needed to prepare 250. milliliters of a 2.00 M solution of KOH (molar mass = 56.0 g/mol)?
    (1) 1.00   (2) 2.00   (3) 28.0   (4) 112

19. How many grams of ammonium chloride (molar mass = 53.5 g/mol) are contained in 0.500 liter of a 2.00 M solution?
    (1) 10.0   (2) 26.3   (3) 53.5   (4) 107

20. What is the molarity of a solution of KNO_3 (molar mass = 101 g/mol) that contains 404 grams of KNO_3 in 2.00 liters of solution?
(1) 1.00 M   (2) 2.00 M   (3) 0.500 M   (4) 4.00 M

21. What is the molarity of a solution that contains 20. grams of $CaBr_2$ in 0.50 liter of solution?
(1) 0.50 M   (2) 0.20 M   (3) 5.0 M   (4) 10. M

22. What is the total number of grams of KCl (molar mass = 74.6 g/mol) in 1.00 liter of 0.200-molar solution?
(1) 7.46   (2) 14.9   (3) 22.4   (4) 29.8

23. What is the total number of grams of solute in 500. milliliters of 1 M CH_3COOH (molar mass = 60 g/mol)?
(1) 30.   (2) 60.   (3) 90.   (4) 120

24. Which contains the greatest number of moles of solute?
(1) 0.5 L of 0.5 M solution   (2) 0.5 L of 2 M solution
(3) 2 L of 0.5 M solution   (4) 2 L of 2 M solution

25. When a solution is diluted by the addition of solvent, which quantity does *not* change?
(1) the molarity of the solution
(2) the number of moles of solvent
(3) the mass percent of the solution
(4) the number of moles of the solute

§2 **26.** What is the molality of a solution containing 8.0 moles of solute dissolved in 2.4 kilograms of solvent?
(1) 0.33 m   (2) 3.3 m   (3) 1.9 m   (4) 19 m

§2 **27.** How many moles of solute are dissolved in 5.0 kilograms of solvent if the concentration is 2.0 molal?
(1) 0.40   (2) 2.5   (3) 10.   (4) 25

§2 **28.** What is the mole fraction of ethanol in a solution containing 3.00 moles of ethanol and 5.00 moles of water?
(1) 0.375   (2) 0.600   (3) 0.625   (4) 1.67

29. What is the molar concentration of a 2.00-liter solution containing 200. grams of glucose $(C_6H_{12}O_6)$?
(1) 0.555 M   (2) 1.80 M   (3) 100. M   (4) 180. M

30. 1000 grams of water contains 0.1 gram of a dissolved substance. The concentration of this solution is
(1) 1 ppm   (2) 10 ppm   (3) 100 ppm   (4) 1000 ppm

31. When ethylene glycol (an antifreeze) is added to water, the boiling point of the water
(1) decreases, and the freezing point decreases
(2) decreases, and the freezing point increases
(3) increases, and the freezing point decreases
(4) increases, and the freezing point increases

32. Which solution will freeze at the *lowest* temperature?
(1) 1 mol of sugar in 500 g of water
(2) 1 mol of sugar in 1000 g of water
(3) 2 mol of sugar in 500 g of water
(4) 2 mol of sugar in 1000 g of water

33. A 1-kilogram sample of water will have the highest freezing point when it contains
(1) 1×10^{17} dissolved particles
(2) 1×10^{19} dissolved particles
(3) 1×10^{21} dissolved particles
(4) 1×10^{23} dissolved particles

34. At standard pressure, an aqueous solution of sugar has a boiling point
(1) greater than 100°C and a freezing point greater than 0°C
(2) greater than 100°C and a freezing point less than 0°C
(3) less than 100°C and a freezing point greater than 0°C
(4) less than 100°C and a freezing point less than 0°C

35. Which solution containing 1 mole of solute dissolved in 1000 grams of water has the *lowest* freezing point?
(1) KOH(aq)   (2) $C_6H_{12}O_6$(aq)
(3) C_2H_5OH(aq)   (4) $C_{12}H_{22}O_{11}$(aq)

36. At room temperature, the solubility of which solute in water would be affected *most* by a change in pressure?
(1) methanol   (2) sugar
(3) carbon dioxide   (4) sodium nitrate

37. A solution that is at equilibrium must be
(1) concentrated   (2) saturated
(3) dilute   (4) unsaturated

38. What is the molarity of a solution containing 20 grams of NaOH in 500 milliliters of solution?
(1) 1 M   (2) 2 M
(3) 0.04 M   (4) 0.5 M

39. Based on Reference Table F, in Appendix 1, which of these salts is the best electrolyte?
(1) sodium nitrate   (2) magnesium carbonate
(3) silver chloride   (4) barium sulfate

40. At standard pressure when NaCl is added to water, the solution will have a
(1) higher freezing point and a lower boiling point than water
(2) higher freezing point and a higher boiling point than water
(3) lower freezing point and a higher boiling point than water
(4) lower freezing point and a lower boiling point than water

41. Which expression could represent the concentration of a solution?
(1) 3.5 g   (2) 3.5 M   (3) 3.5 mL   (4) 3.5 mol

42. Which sample, when dissolved in 1.0 liter of water, produces a solution with the highest boiling point?
(1) 0.1 mol KI   (2) 0.2 mol KI
(3) 0.1 mol $MgCl_2$ (4) 0.2 mol $MgCl_2$

43. An aqueous solution has a mass of 400 grams containing 8.5×10^{-3} gram of calcium ions. The concentration of calcium ions in this solution is
(1) 4.3 ppm   (2) 8.5 ppm   (3) 21.25 ppm   (4) 34 ppm

Constructed-Response Questions

1. A solution contains 0.161 mole of solute dissolved in 310. milliliters of solution. What is the molarity of the solution?

2. How many moles of solute are present in 35.1 milliliters of a 0.879 M solution?

3. How many grams of $Ba(OH)_2$ are needed to make 300. milliliters of a 0.692 M solution?

§2 4. A 250.-milliliter solution of 2.11 M H_2SO_4 is diluted to 725 milliliters. Calculate the molarity of the diluted solution.

§2 5. How would you prepare 325 milliliters of a 0.333 M H_2SO_4 solution, using a stock solution of 6.00 M H_2SO_4?

§2 6. What mass of water must be added to 3.75 moles of solute to prepare a 1.66 molal solution?

§2 7. How many grams of water must be added to 11.5 grams of C_2H_5OH to prepare a 0.336 molal solution?

§2 8. A solution is prepared by mixing 115 grams of H_2O, 31.0 grams of C_2H_5OH, and 11.2 grams of $C_6H_{12}O_6$. Calculate the mole fraction (X) of C_2H_5OH in the solution.

§2 9. Given this reaction:

$$2NaOH\ (aq) + H_2SO_4\ (aq) \rightarrow Na_2SO_4\ (aq) + H_2O\ (\ell)$$

What volume of a 0.750 M H_2SO_4 solution is needed to react completely with 15.0 grams of NaOH?

10. The grid below is based on the molarity concept developed in this chapter and is partially filled in. Complete the grid by using the data given in the grid.

Solute	Molar Mass of Solute ($\mathcal{M}$) (g/mol)	Moles of Solute Present (n) (mol)	Mass of Solute Present (m) (g)	Volume of Solution (V)	Molarity (M)
NH_3		2.00		250. mL	
HNO_3			63.0		0.500
H_2SO_4			49.0	5.00 dm^3	
$MgCl_2$		1.00			2.00
$AlPO_4$			244	0.500 L	
$C_6H_{12}O_6$			45.0		2.00
$HC_2H_3O_2$		2.00		4000. cm^3	
KI			498		1.00
$Na_2Cr_2O_7$		3.00			9.00
$B(OH)_3$		4.00		12.0 L	
***			294	2.00 dm^3	1.50

***One of the ten compounds given above.

§2 **11.** A solution is prepared by dissolving 3.86 grams of C_2H_5OH in 325 grams of nitrobenzene (freezing point = 5.76°C; k_f = 6.87 C°/m). Calculate the freezing point of this solution.

§2 **12.** A solvent has a normal boiling point of 90.00°C. When 5.00 grams of C_2H_5OH is dissolved in 455 grams of the solvent, the boiling point of the solution is 93.33°C. Calculate the boiling point elevation constant (k_b) of the solvent.

Base your answers to questions 13–16 on the information below and on your knowledge of chemistry.

In a laboratory investigation, ammonium chloride was dissolved in water. Laboratory procedures and corresponding observations made by a student during the investigation are shown in the table below.

Dissolving NH₄Cl(s) in H₂O(ℓ)

Procedure	Observation
1. Measure the temperature of 10.0 milliliters (10.0 grams) of $H_2O(\ell)$ in a tube.	1. The temperature of the $H_2O(\ell)$ was 25.8°C
2. Add 5.0 grams of the solute, $NH_4Cl(s)$, to the $H_2O(\ell)$.	2. The $NH_4Cl(s)$ settled to the bottom of the test tube.
3. Stir the contents of the test tube for 4 minutes.	3. A small amount of $NH_4Cl(s)$ remained at the bottom of the test tube.
4. Measure the temperature of the $NH_4Cl(aq)$ solution.	4. The temperature of the solution was 11.2°C.

13. Identify *two* types of bonds in the solute.

14. State evidence from the investigation that indicates the $NH_4Cl(aq)$ solution is saturated.

15. State evidence from the investigation that indicates the process of dissolving $NH_4Cl(s)$ in water is endothermic.

16. State the observation that would be made if procedure 3 is repeated with the original temperature of the $H_2O(\ell)$ at 98°C.

The answers to these questions are found in Appendix 3.

Chapter Thirteen

KINETICS AND EQUILIBRIUM

KEY IDEAS

This chapter focuses on the factors that affect the speed of a reaction, the reversibility of reactions, and the way in which reversibility applies to various chemical processes.

KEY OBJECTIVES

At the conclusion of this chapter you will be able to:

- Define the term *chemical kinetics*.
- List and describe the factors that affect the rate of a reaction.
- Define the term *dynamic equilibrium*.
- Provide examples of phase and solution equilibria.
- Define the term *chemical equilibrium*.
- State Le Châtelier's principle.
- Use Le Châtelier's principle to determine the effects of changes in concentration, the addition of a common ion, volume and temperature changes, the addition of an inert gas, and the presence of a catalyst on systems in equilibrium.
- Write equilibrium-constant expressions for homogeneous and heterogeneous chemical reactions.
- Describe the effect of temperature on the equilibrium constant.
- Solve simple problems involving the equilibrium constant.

SECTION I—BASIC (REGENTS-LEVEL) MATERIAL

NYS REGENTS CONCEPTS AND SKILLS

Note: By the time you have finished Section I, you should have mastered the concepts and skills listed below. The Regents chemistry examination will test your knowledge of these items and your ability to apply them.

Concepts are the *basic ideas* that form the body of the Regents chemistry course (what you need to know!).

Skills are the *activities* that demonstrate your mastery of these concepts (how you show that you know them!).

Following each concept or skill is a page reference (given in parentheses) to this chapter.

13.1 Concept:
The rate of a chemical reaction depends on several factors: temperature, concentration, nature of reactants, surface area, and the presence of a catalyst. (Pages 371–373)

Skill:
Read and interpret potential energy diagrams with regard to:
(a) potential energies of reactants and products (Pages 372–373);
(b) activation energy (without and with a catalyst) (Pages 372–373);
(c) heat of reaction (ΔH) (Pages 372–373).

13.2 Concept
A catalyst provides an alternative reaction pathway that has a lower activation energy than an uncatalyzed reaction. (Pages 372–373)

13.3 Concept:
Collision theory states that a reaction is most likely to occur if reactant particles collide with the proper energy and orientation. (Pages 371–372)

Skill:
Use collision theory to explain how each of the following factors affects the rate of a reaction:
(a) temperature (Page 372);
(b) surface area (Page 372);
(c) concentration (Page 372).

13.4 Concept:
Some chemical and physical changes can reach equilibrium. (Pages 373–374)

Skill:
Identify examples of the following phenomena:
(a) solution equilibrium (a saturated solution) (Page 375);
(b) phase equilibrium (Page 374).

13.5 Concepts:
• At equilibrium the rate of the forward reaction equals the rate of the reverse reaction. (Pages 373–374)
• The measurable quantities of reactants and products remain constant at equilibrium. (Pages 373–374)

Skill:
Describe the concentrations of particles and the rates of opposing reactions in an equilibrium system. (Pages 373–374)

13.6   Concept:
Le Châtelier's principle can be used to predict the effects of stress (change in pressure, volume, concentration, and temperature) on a system at equilibrium. (Pages 375–376)

Skill:
Use Le Châtelier's principle to describe the effects of stress on a system in equilibrium. (Pages 375–379)

13.1   CHEMICAL KINETICS

Chemical kinetics is the branch of chemistry that concerns itself with the speed of chemical reactions (known as *reaction rates*) and the way in which these reactions occur (known as *reaction mechanisms*).

Let us consider this reaction:

$$4NH_3(g) + 5O_2(g) \rightarrow 4NO(g) + 6H_2O(g)$$

The study of chemical kinetics has led chemists to the conclusion that a chemical reaction such as this one occurs, not in one step, but in a series of simpler steps involving the collision of one or two particles at a time. These intermediate steps (called *elementary steps*) are usually not observed; nevertheless, they contribute to the overall reaction. Suppose that the reaction $A \rightarrow B$ consists of the three elementary steps shown below.

$$(1) \ A \xrightarrow{\text{fast}} I_1$$
$$(2) \ I_1 \xrightarrow{\text{slow}} I_2$$
$$(3) \ I_2 \xrightarrow{\text{fast}} B$$
....
$$(\text{rxn}) \ A \rightarrow B \ (\text{rate} = ???)$$

Here I_1 and I_2 are intermediates that are not observed when reactant A is converted into product B, and the relative rate of each step is given above the arrow. Of the three elementary steps, (2) has the *slowest* rate and it determines the rate of the *overall* reaction ($A \rightarrow B$). In technical terms, (2) is called a *rate-limiting step*; it serves as a "bottleneck" for the other steps, much as a narrow strip of road serves as a bottleneck for traffic flow on a major highway.

Factors That Affect the Rate of a Reaction

Chemical reactions occur because the reacting particles collide with each other. The speed of a reaction depends on the number of collisions and the fraction of those collisions that are *effective*. For a collision to be effective, the particles must have enough energy and they must be *oriented* properly. Any factor that increases the number of effective collisions will serve to increase the rate of the reaction.

Nature of the Reactants

Since bonds are broken and formed in chemical reactions, the bond types of the reacting substances are an important factor. In solution, ionic compounds tend to react more quickly than covalent compounds because fewer bonds need to be rearranged. For example, the reaction between Fe(s) and O_2(g) (the rusting of iron) is rather slow at room temperature, whereas the ionic reaction between $AgNO_3$(aq) and NaCl(aq) is extremely rapid.

Concentrations of the Reactants

As the concentration of a reactant increases, the number of collisions increases and the rate of the reaction generally increases. For example, in the laboratory production of hydrogen gas, Zn(s) reacts more rapidly with 1 M HCl(aq) than with 0.5 M HCl(aq).

In a system in which gases react, the pressure is directly related to the concentrations of the gases. Therefore, an increase in pressure leads to an increase in reaction rate.

Temperature

An increase in temperature increases the average kinetic energy of the particles, thereby ensuring that the number of effective collisions will increase. Therefore, an increase in temperature always increases the rate of a reaction.

Surface Area

As the surface area of the reacting particles is increased, the number of particles that collide is also increased. The result is an increase in the reaction rate. For example, when the size of Zn(s) particles is reduced, the particles react more rapidly with HCl(aq) to produce hydrogen gas.

Catalysts

If we refer to the potential energy diagrams on pages 115 and 116, we observe that the activation energy is like a hill that must be climbed before the reaction will conclude.

Catalysts *lower the activation energy* of a reaction by providing an alternative reaction pathway and, therefore, increase the speed of the reaction. In effect, the "hill" becomes easier to climb.

The effect of a catalyst is illustrated in the accompanying potential energy diagram. Note that *only* the activation energy is changed by the catalyst. The overall reaction and the value of ΔH, the heat of reaction, remain unchanged by its presence.

Reaction Profile With and Without Catalyst

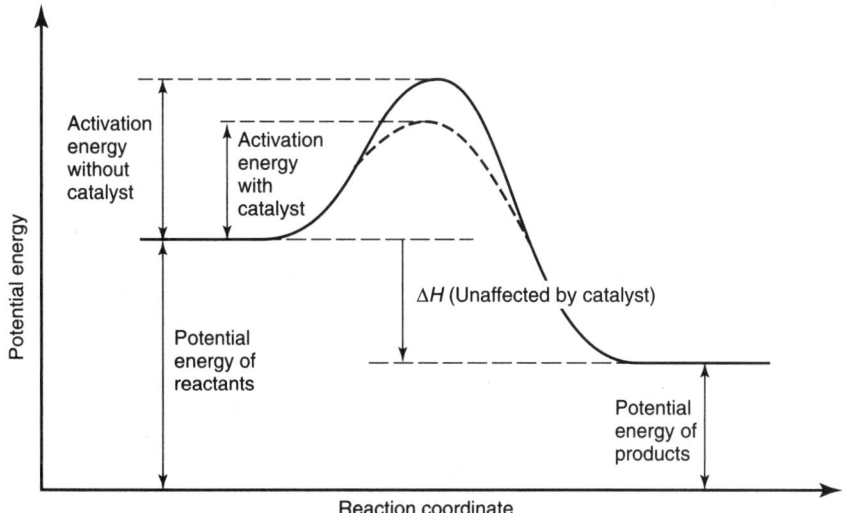

13.2 REVERSIBLE REACTIONS AND DYNAMIC EQUILIBRIUM

When we think of the term *equilibrium*, the first word that usually comes to mind is *balance*. However, balance may be achieved in a variety of ways. For example, a building remains standing, rather than falling down, because all of its stresses and strains are (hopefully) balanced. This is an example of *static equilibrium*.

In chemistry, however, *equilibrium* has a somewhat different meaning. Chemical reactions are usually *reversible*, meaning that they can proceed in *both* directions. We write a reversible reaction with a "split arrow" to indicate this reversibility:

$$A(g) \rightleftharpoons B(g)$$

The equation indicates that A can form B and that B can form A. Let us see how equilibrium can be achieved in this system by referring to the accompanying diagrams of reaction rate and concentration as functions of time.

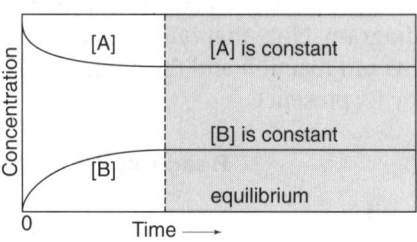

If we refer to the graph on the left, we note that, as time goes on, the rate of the forward reaction (rate$_{A \to B}$) *decreases* because, as we see from the graph on the right, the concentration of A is *decreasing*. Meanwhile, as we can also see from the graphs, the concentration of B is *increasing* and the rate of the reverse reaction (rate$_{A \leftarrow B}$) is *increasing*.

After a time, the rates of the forward and reverse reactions will be equal. At this point, the concentrations of the reactants and the products will no longer change.

We call this condition **dynamic equilibrium**, and we usually indicate it by using split arrows:

$$A(g) \rightleftharpoons B(g)$$

It must be stressed that equilibrium means that the opposing *rates* are equal. At equilibrium, the *concentrations* of the reactants and products need *not* be equal.

In Chapter 5 we studied spontaneous reactions. When a reaction is spontaneous, the change of reactants to products is observable. At equilibrium, however, we observe *neither* the forward nor the reverse reaction, and therefore neither is spontaneous.

13.3 PHASE EQUILIBRIUM

We know that phase changes are reversible, and in closed containers two or more phases may exist in dynamic equilibrium at a given temperature and pressure. For example, if water is placed in a closed container at 25°C, it will reach dynamic equilibrium with its vapor. The pressure created by the gaseous water at equilibrium is the *vapor pressure of water at 25°C*. Similarly, boiling-condensation, melting-freezing, and sublimation-deposition all represent phase equilibria.

13.4 SOLUTION EQUILIBRIUM

Solids Dissolved in Liquids

When a solid dissolves in a liquid and produces a *saturated* solution, the rate of dissolving equals the rate at which the solid leaves the solution (crystallizes). This situation represents dynamic equilibrium in the solution.

For glucose ($C_6H_{12}O_6$) and water, the equilibrium equation is

$$C_6H_{12}O_6(s) \rightleftharpoons C_6H_{12}O_6(aq).$$

For an ionic compound such as NaCl, the equilibrium equation is

$$NaCl(s) \rightleftharpoons Na^+(aq) + Cl^-(aq).$$

Solution equilibria are affected by changes in temperature.

Gases Dissolved in Liquids

In a closed system, such as a bottle of soda, equilibrium may exist between the dissolved gas (CO_2) and the gas present on top of the liquid.

For such a system the equilibrium equation is $CO_2(g) \rightleftharpoons CO_2(aq)$.

Gas-liquid equilibria are affected by changes in temperature and pressure.

13.5 CHEMICAL EQUILIBRIUM

We can now define the term *chemical equilibrium*. **Chemical equilibrium** exists when the forward and reverse rates of a chemical reaction are equal, and, as a result, the concentrations of the reactants and the products remain *constant*. For example, the equation

$$N_2(g) + 3H_2(g) \rightleftharpoons 2NH_3(g)$$

represents a chemical system in a state of equilibrium.

Le Châtelier's Principle

It is possible to disturb any system at equilibrium by applying a *stress* to it. The important question is: How will the system respond to the stress?

This question was answered by the French chemist Henri Le Châtelier in 1888. **Le Châtelier's principle** states:

"When a system at equilibrium is subjected to a stress, the system shifts in order to relieve the effects of the stress and restore the equilibrium conditions as closely as possible."

This principle goes beyond chemistry. For example, our homes are usually programmed to remain at a constant temperature (the equilibrium condition). If the temperature drops suddenly because a window is opened (the stress), the heating system responds (the shift) by putting heat energy back into the system in order to try to restore the original temperature.

We note that, when a system returns to equilibrium, there is a *new equilibrium point* because the original conditions have been changed.

Let us examine how a chemical system at equilibrium reacts to a number of stresses placed upon it. The equilibrium system we will use involves the production of ammonia by the *Haber process:*

$$N_2(g) + 3H_2(g) \rightleftharpoons 2NH_3(g)$$

At equilibrium, the rate of ammonia production equals the rate at which the ammonia is decomposed into its elements, and the concentrations of nitrogen, hydrogen, and ammonia are constant.

First, however, we must define what we mean by a *chemical shift*. Any chemical reaction can proceed in only two directions: forward and reverse. When a reaction responds to a stress, either the forward or the reverse reaction will take over temporarily, and we will observe an increase in the production of products or of reactants. A forward shift is called a shift to the *right*; a reverse shift, a shift to the *left*.

Effect of Concentration

Concentrations are usually expressed in moles per liter. We express the molar concentration of a substance by placing the symbol for the substance in square brackets: for example, $[NH_3]$ means the molar concentration of NH_3.

When the concentration of a reactant or a product is changed, the system will shift in order to restore the original concentration as closely as possible.

Suppose $[N_2]$ is increased in the example of the Haber process given above. Since N_2 has been added to the system, the system will shift to the *right* in order to remove *some* of the additional N_2. As a result, $[NH_3]$ will increase, and $[H_2]$ will decrease because it is needed to react with the N_2. Eventually, a *new* equilibrium point will be established.

Suppose $[NH_3]$ is increased. The system will shift to the *left* in order to remove some of the additional NH_3. As a result, $[N_2]$ and $[H_2]$ will increase.

Suppose $[NH_3]$ is decreased. The system will shift to the *right* in order to replace some of the NH_3. As a result, both $[N_2]$ and $[H_2]$ will decrease.

A reaction in which one of the products is *continuously* removed will *never* achieve equilibrium because the reactants will be used up in the effort to replace the disappearing product. A familiar example is the laboratory preparation of hydrogen gas by reacting zinc with hydrochloric acid:

$$Zn(s) + HCl(aq) \rightarrow H_2(g) + ZnCl_2(aq)$$

Since $H_2(g)$ continuously escapes from the reaction, equilibrium is never attained. This reaction is said to go to *completion*, and we write the equation with the familiar single arrow.

Effect of a Change in Volume

The effect of a change in volume is observed in many systems containing gases. In general, *a change in volume is accomplished by changing the pressure on a gaseous system*: an increase in pressure leads to a reduction in volume, and vice versa.

Reducing the volume (by increasing the pressure) brings all of the particles into closer contact. This stress is relieved by a shift that produces a *smaller* number of particles. In contrast, an increase in volume (i.e., a decrease in pressure) leads to a shift that increases the number of particles present.

Let us again consider the Haber process:

$$N_2(g) + 3H_2(g) \rightleftharpoons 2NH_3(g)$$

The left side of the equation contains 4 moles of particles (1 mole of N_2 and 3 moles of H_2); the right side, 2 moles of particles (NH_3). A shift to the right reduces the number of particles, while a shift to the left increases the number.

Suppose the volume of this system were decreased causing the pressure to increase. The system would shift to the *right* in order to decrease the number of particles present. In this case, $[NH_3]$ would increase; $[N_2]$ and $[H_2]$ would decrease.

Let us now consider this system:

$$H_2(g) + Cl_2(g) \rightleftharpoons 2HCl(g)$$

Changing the volume would *not* cause this system to shift because each side contains the same number of particles (2 moles). A shift in either direction would not relieve the stress.

Changing the Pressure by Adding an Inert Gas

The pressure of an equilibrium system can also be changed by the addition of an inert gas. Suppose helium gas was added to the N_2-H_2-NH_3 system we have been considering. Since helium does *not* react with any of these substances, it would have *no* effect on the equilibrium of the system.

Effect of a Change in Temperature

When we raise the temperature of a system, we do so by adding heat to the system; when we lower the temperature, we remove heat from the system.

Suppose we lower the temperature of an equilibrium system. What is the effect on the system? Lowering the temperature will decrease the rates of both the forward and reverse reactions.

Le Châtelier's principle predicts, however, that the system will shift in order to relieve the stress; that is, the system will try to *replace* some of the lost heat. Since the *exothermic* reaction releases heat, it will be favored when the temperature is lowered. Why is this true? The rate of the exothermic reaction is decreased *less* than the rate of the endothermic reaction.

If, on the other hand, we raise the temperature of an equilibrium system, the rates of both reactions will increase, but the system will shift in order to *absorb* some of the added heat. Since the *endothermic* reaction absorbs heat, it is favored when the temperature is raised. In this case the rate of the endothermic reaction is increased *more* than the rate of the exothermic reaction.

Once again, we consider the Haber process (now rewritten to include the value of ΔH as part of the reaction):

$$N_2(g) + 3H_2(g) \rightleftharpoons 2NH_3(g) + 91.8 \text{ kJ}$$

We see that the forward reaction is *exothermic* and, therefore, the reverse reaction must be *endothermic*.

If the temperature of this system is raised, the reaction will shift toward the left—the endothermic reaction—and $[NH_3]$ will decrease, while $[N_2]$ and $[H_2]$ will increase.

If the temperature of this system is lowered, the reaction will shift toward the right—the exothermic reaction—and $[NH_3]$ will increase, while $[N_2]$ and $[H_2]$ will decrease.

Effect of a Catalyst

For a system in equilibrium, the presence of a catalyst increases the forward and reverse rates *equally*. Therefore, *no* shift in the system will result. If the system is *not* at equilibrium, however, a catalyst will shorten the time needed to reach equilibrium.

Summary of Effects

Direction of Stress	Result of Shift
Concentration decreases.	Some substance is replaced.
Concentration increases.	Some substance is removed.
Volume decreases (pressure increases).	Fewer particles are produced.
Volume increases (pressure decreases).	More particles are produced.
Temperature decreases.	Exothermic reaction is favored.
Temperature increases.	Endothermic reaction is favored.
An inert gas is added.	No shift occurs.
A catalyst is added.	No shift occurs.

TRY IT YOURSELF

For this equilibrium system:

$$904 \text{ kJ} + 6H_2O(g) + 4NO(g) \rightleftharpoons 4NH_3(g) + 5O_2(g)$$

what is the effect of each of the following changes?

(a) decreasing the temperature   (b) decreasing the volume
(c) increasing [NO]   (d) decreasing [H_2O]

ANSWERS

(a) a shift toward the *left*   (b) a shift toward the *right*
(c) a shift toward the *right*   (d) a shift toward the *left*

SECTION II—ADDITIONAL MATERIAL

13.1A   THE COMMON-ION EFFECT

In a saturated solution of an ionic compound such as $AgCl$, the $Ag^+(aq)$ and $Cl^-(aq)$ ions are in equilibrium with the solid $AgCl$:

$$AgCl(s) \rightleftharpoons Ag^+(aq) + Cl^-(aq)$$

If additional $Cl^-(aq)$ were added to this solution (perhaps by adding some NaCl), what would occur? Le Châtelier's principle predicts that the system would shift to the *left* in order to remove some of the added $Cl^-(aq)$ ion. As a result more $AgCl(s)$ would be produced. This shift is known as the *common-ion effect*. In this example, the addition of the common ion $Cl^-(aq)$ *decreased* the solubility of $AgCl$ in water because fewer Ag^+ ions were present in the saturated solution.

13.2A   HETEROGENEOUS EQUILIBRIUM

An equilibrium system is *homogeneous* if all reactants and products are in a single phase (as gases or substances in aqueous solution). In a *heterogeneous* system pure solids or liquids are present; a saturated solution is an example of a heterogeneous equilibrium system. Another example of a heterogeneous equilibrium system is given by the equation

$$C(s) + O_2(g) \rightleftharpoons CO_2(g)$$

and its accompanying diagram, in which solid carbon is in equilibrium with gaseous oxygen and carbon dioxide in a sealed container.

Having examined the conditions that affect systems at equilibrium, we now ask: If more *solid* carbon were added to the container shown in the diagram, how would this system be affected?

The answer is: *Not at all!* Equilibrium is affected by changes in concentration. Adding more solid does *not* change the concentration of the carbon because the solid does not occupy the entire container, as the gases do. The concentration of the solid, which is really its density, remains constant. This would be true also if a pure liquid were part of an equilibrium system.

13.3A   THE EQUILIBRIUM CONSTANT (K_{eq})

When a system reaches equilibrium, it is useful to know the relative amounts of products and reactants present. Certain reactions produce hardly any products, while other reactions leave hardly any reactants remaining. The *equilibrium constant* (represented as K_{eq} or K) provides us with this information.

Defining K_{eq} for a Simple System

For a simple homogeneous system at equilibrium: $A(g) \rightleftharpoons B(g)$, the equilibrium constant expression is defined as follows:

$$K_{eq} = \frac{[B]}{[A]}$$

In defining K_{eq}, we divide the equilibrium concentration of the product, B, by the equilibrium concentration of the reactant, A. If there is more product than reactant at equilibrium, K_{eq} is greater than 1; if there is less product than reactant at equilibrium, K_{eq} is less than 1; if the amounts of product and reactant at equilibrium are approximately equal, K_{eq} is approximately 1.

PROBLEM
For the system discussed above, $[A] = 0.2$ mole per liter and $[B] = 0.02$ mole per liter at equilibrium. Calculate K_{eq} for these conditions.

SOLUTION

$$K_{eq} = \frac{[B]}{[A]} = \frac{0.02 \ \text{mol/L}}{0.2 \ \text{mol/L}} = 0.1$$

The small value of the equilibrium constant (0.1) confirms that, at equilibrium, there is less product than reactant.

TRY IT YOURSELF
In the system given above, if the equilibrium concentration of B is 0.15 mole per liter, what is the equilibrium concentration of A?

ANSWER
1.5 mol/L

Defining K_{eq} for a More Complicated System

For more complicated reactions, the equilibrium constant expression is itself more complicated. For the equilibrium system:

$$aM(g) + bN(g) \rightleftharpoons cP(g) + dQ(g)$$

the equilibrium constant expression is defined as

$$K_{eq} = \frac{[P]^c[Q]^d}{[M]^a[N]^b}$$

We note that, in general, we still divide the products by the reactants, but in a more complicated way: we raise each concentration to the *power of its coefficient*. A small equilibrium constant indicates more reactants at equilibrium, and a large equilibrium constant indicates more products at equilibrium.

PROBLEM
(a)   Write the K_{eq} expression for this reaction: $N_2(g) + 3H_2(g) \rightleftharpoons 2NH_3(g)$.
(b)   Evaluate K_{eq} if $[NH_3] = 0.2$ mole per liter, $[N_2] = 0.04$ mole per liter, $[H_2] = 0.01$ mole per liter.

SOLUTIONS

(a) $K_{eq} = \dfrac{[NH_3]^2}{[N_2][H_2]^3}$ (b) $K_{eq} = \dfrac{(0.2 \text{ mol/L})^2}{(0.04 \text{ mol/L})(0.01 \text{ mol/L})^3} = 1 \times 10^6$

The large value of the equilibrium constant confirms that, at equilibrium, there are more products than reactants. (In more complicated expressions, the units may *not* cancel. In practice, we *ignore* any unit that may be associated with an equilibrium constant.)

PROBLEM
Write the equilibrium expression for this *heterogeneous* system:

$$S(s) + O_2(g) \rightleftharpoons SO_2(g)$$

SOLUTION
The equilibrium constant expression is

$$K_{eq} = \frac{[SO_2]}{[S][O_2]}$$

Since sulfur is present as a solid, its concentration is constant and *is included* in the constant part of the equilibrium expression:

$$K_{eq} \cdot [S] = K = \frac{[SO_2]}{[O_2]}$$

When we write an equilibrium constant expression for a heterogeneous system, *we omit the concentration expressions for solids and liquids*.

TRY IT YOURSELF
Write the equilibrium expression for the system $C(s) + \dfrac{1}{2} O_2(g) \rightleftharpoons CO(g)$.

ANSWER

$$K = \frac{[CO]}{[O_2]^{\frac{1}{2}}}$$

At *constant temperature*, the equilibrium constant is truly *constant*. If we were to disturb a system at equilibrium, the point of equilibrium would shift to relieve the stress. When a new equilibrium was established, however, the value of K_{eq} would not have changed. Only a change in temperature can change K_{eq}.

The Solubility-Product Constant (K_{sp})

The *solubility-product constant* (K_{sp}) is a special type of equilibrium constant that measures the concentrations of ionic compounds in water. A compound with a small K_{sp} value is only slightly soluble in water. For example, $CaSO_4$ ($K_{sp} = 9.1 \times 10^{-6}$) is only very slightly soluble in water, and $BaSO_4$ ($K_{sp} = 9.1 \times 10^{-10}$) is even *less* soluble! In fact, the solubility of $BaSO_4$ is so low that we would classify it as *nearly insoluble* in water.

The K_{sp} expression may be developed by considering the solution equilibrium of an ionic compound AB:

$$AB(s) \rightleftharpoons A^+(aq) + B^-(aq)$$

Since AB is present as a solid, it is absorbed into the equilibrium expression:

$$K_{eq} \cdot [AB] = K_{sp} = [A^+] \cdot [B^-]$$

This expression indicates the reason for calling K_{sp} a solubility *product*; it is equal to the product of the ion concentrations in water.

PROBLEM
(a) Write the solubility equation and K_{sp} expression for $BaSO_4$.
(b) Write the solubility equation and K_{sp} expression for PbI_2.

SOLUTIONS
(a) $BaSO_4(s) \rightleftharpoons Ba^{2+}(aq) + SO_4^{2-}(aq)$; $K_{sp} = [Ba^{2+}] [SO_4^{2-}]$
(b) $PbI_2(s) \rightleftharpoons Pb^{2+}(aq) + 2I^-(aq)$; $K_{sp} = [Pb^{2+}] [I^-]^2$

In Appendix 2 of this book, Reference Table Y-1 lists the solubility-product constants for a number of compounds at 298.15 K.

13.4A PROBLEMS INVOLVING THE EQUILIBRIUM CONSTANT

Up to this point, the only problems we have solved in this chapter involved calculating the value of an equilibrium constant, given the equilibrium concentrations of the reactants and products. It is possible to approach problem solving from the other direction, that is, to use the equilibrium constant to calculate equilibrium concentrations. This method can become quite a complicated process that is far beyond the scope of this book. We will provide two relatively simple examples, however, to show how such problems are approached.

PROBLEM

In an experiment, hydrogen gas (H_2) and iodine gas (I_2) were placed in a container, which was then sealed and allowed to come to equilibrium with hydrogen iodide gas (HI) according to this equation:

$$H_2(g) + I_2(g) \rightleftharpoons 2HI(g)$$

At the start of the experiment, concentrations were $[H_2] = 1.00 \times 10^{-3}$ mole per liter, $[I_2] = 2.00 \times 10^{-3}$ mole per liter, $[HI] = 0.00$ mole per liter.

When equilibrium had been attained, the concentration of HI gas was 1.86×10^{-3} mole per liter.

(a) Calculate the *equilibrium concentrations* of the hydrogen and iodine gases in moles per liter.
(b) Calculate the *equilibrium constant* (K_{eq}) for this system.

SOLUTIONS

(a) We begin by constructing a table, in which all of the given information is summarized:

Reaction: $H_2(g) + I_2(g) \rightleftharpoons 2HI(g)$			
Substance	$[H_2]$ / (mol/L)	$[I_2]$ / (mol/L)	[HI] / (mol/L)
Initial concentration	1.00×10^{-3}	2.00×10^{-3}	0.00
Change in concentration			
Equilibrium concentration			1.86×10^{-3}

Next, we inspect the *coefficients* of the equation. They tell us that, for every 1 mol of $H_2(g)$ that reacts, 1 mol of $I_2(g)$ will react, and 2 mol of HI(g) will be formed. *In other words, the concentrations of $H_2(g)$ and $I_2(g)$ that react must be exactly one-half of the concentration of the HI(g) that is formed.* Since *all* of the HI(g) present at equilibrium (1.86×10^{-3} mol/L) was formed in the reaction (why?), we can conclude that one-half of this concentration, 9.3×10^{-4} mol/L, of $H_2(g)$ and $I_2(g)$ reacted.

We return to our table and add this information in the appropriate spaces. We use a "−" sign to indicate *reacted* and a "+" sign to indicate *formed*.

Reaction: $H_2(g) + I_2(g) \rightleftharpoons 2HI(g)$			
Substance	[H_2] / (mol/L)	[I_2] / (mol/L)	[HI] / (mol/L)
Initial concentration	1.00×10^{-3}	2.00×10^{-3}	0.00
Change in concentration	-9.3×10^{-4}	-9.3×10^{-4}	$+1.86 \times 10^{-3}$
Equilibrium concentration			1.86×10^{-3}

Finally, we calculate each equilibrium concentration by *adding* the *change* in concentration to the *initial* concentration. Our table is now complete.

Reaction: $H_2(g) + I_2(g) \rightleftharpoons 2HI(g)$			
Substance	[H_2] / (mol/L)	[I_2] / (mol/L)	[HI] / (mol/L)
Initial concentration	1.00×10^{-3}	2.00×10^{-3}	0.00
Change in concentration	-9.3×10^{-4}	-9.3×10^{-4}	$+1.86 \times 10^{-3}$
Equilibrium concentration	7.00×10^{-5}	1.07×10^{-3}	1.86×10^{-3}

At equilibrium, the concentration of $H_2(g)$ is 7.00×10^{-5} mol/L, and the concentration of $I_2(g)$ is 1.07×10^{-3} mol/L.

(b) To calculate the equilibrium constant for this system, we write the K_{eq} expression, substitute the values that appear in the table, and perform the appropriate calculations:

$$K_{eq} = \frac{[HI]^2}{[H_2] \cdot [I_2]} = \frac{(1.86 \times 10^{-3})^2}{(7.00 \times 10^{-5}) \cdot (1.07 \times 10^{-3})} = 46.2$$

In the next problem, we are given the equilibrium constant for the system and the initial concentrations, and are required to calculate the equilibrium concentrations of the reactant and product.

PROBLEM

In an experiment, dinitrogen tetroxide gas (N_2O_4) was placed in a 1-liter container, which was then sealed and allowed to come to equilibrium with nitrogen dioxide gas (NO_2) according to this equation:

$$N_2O_4(g) \rightleftharpoons 2NO_2(g) \qquad K_{eq} = 0.211$$

At the start of the experiment, the concentrations were [N_2O_4] = 1.00 mole per liter, [NO_2] = 0.00 mole per liter. Equilibrium is established over a period of time. What are the *equilibrium concentrations* of $NO_2(g)$ and $N_2O_4(g)$?

SOLUTION

We begin by inspecting the *coefficients* of the equation. They tell us that for every *1* mol of $N_2O_4(g)$ that react 2 mol of $NO_2(g)$ will be formed.

We know, however, that all 1.00 mol of the $N_2O_4(g)$ will *not* react because the system will reach equilibrium before that can happen. Then how many moles of $N_2O_4(g)$ *will* react? *We don't know!* Let us simply say that x mol of $N_2O_4(g)$ have reacted when equilibrium is attained. This means that $2x$ mol of $NO_2(g)$ will be formed. (Why?)

At this point we need to stress that all equilibrium calculations involve *concentrations* (in mol/L). In this example, however, the reaction container is a *1*-L vessel; therefore, the number of moles is equivalent to the number of moles per liter.

We now construct a table that summarizes our problem.

Reaction: $N_2O_4(g) \rightleftharpoons 2NO_2(g)$ $K_{eq} = 0.211$		
Substance	$[N_2O_4]/(mol/L)$	$[NO_2]/(mol/L)$
Initial concentration	1.00	0.00
Change in concentration	$-x$	$+2x$
Equilibrium concentration	$1.00 - x$	$0.00 + 2x = 2x$

We use a "−" sign to indicate *reacted* and a "+" sign to indicate *formed*.

The next step is to write the equilibrium expression for the system, substitute the equilibrium concentration values given in the table, and solve for x, $1.0 - x$, and $2x$:

$$K_{eq} = \frac{[NO_2]^2}{[N_2O_4]} = 0.211 = \frac{(2x)^2}{1.0 - x}$$

To arrive at the answer, we need to solve the quadratic equation that the equilibrium expression produces:

$$4x^2 + 0.211x - 0.211 = 0$$

When the equation is solved, we find that $x = 0.205$ mol/L.

Therefore, at equilibrium:

$[NO_2] = 2x = 0.410$ mol/L   and $[N_2O_4] = 1 - x = 0.795$ mol/L

By the way, quadratic equations have *two* solutions. We *rejected* the second solution because it leads to a nonsensical answer. Can you explain why?

13.5A APPLICATIONS OF CHEMICAL EQUILIBRIUM

The Haber Process

We have learned that ammonia is produced commercially by the reaction

$$N_2(g) + 3H_2(g) \rightleftharpoons 2NH_3(g) + \text{heat}$$

and that the formation of ammonia is favored by low temperatures and high pressures. The problem with using low temperatures is that the *rate* of ammonia formation is slow. If the temperature is raised, however, the reaction is shifted toward the *left* (the endothermic reaction), thereby decreasing the *yield* of ammonia.

In practice, an intermediate temperature (about 500°C) is used as a compromise. With a pressure of 200–1000 atmospheres and the addition of a catalyst, a reasonable rate of ammonia formation is maintained.

The Contact Process

Sulfuric acid (H_2SO_4) is one of the world's most important substances. It is prepared by a four-step method known as the *contact process*:

1. $S(s) + O_2(g) \rightleftharpoons SO_s(g) + \text{heat energy}$
Sulfur is burned to form sulfur dioxide gas.
2. $2SO_2(g) + O_2(g) \rightleftharpoons 2SO_3(g) + \text{heat}$
The sulfur dioxide is oxidized to sulfur trioxide gas.
3. $SO_3(g) + H_2SO_4(\ell) \rightarrow H_2S_2O_7(\ell)$
Sulfur trioxide dissolves readily in concentrated sufluric acid to form *fuming sulfuric acid*.
4. $H_2S_2O_7(\ell) + H_2O(\ell) \rightarrow 2H_2SO_4(\ell)$
The fuming sulfuric acid is dissolved in water to produce more sufluric acid.

The net reaction is as follows:

$$SO_3(g) + H_2O(\ell) \rightarrow H_sSO_4(\ell)$$

The sulfur trioxide must be dissolved in concentrated sulfuric acid (step 3) because it does not dissolve well in water.

The formation of SO_3 (step 2), like the formation of ammonia, is favored by high pressure and low temperature. As with the Haber process, a compromise temperature (450–575°C) is used, along with a catalyst, to maintain reasonable formation rates.

END-OF-CHAPTER QUESTIONS

Some questions have the symbol "§2" in front of the question number. This symbol means that the question is based on Section II material.

1. A 1-cubic-centimeter cube of sodium reacts more rapidly in water at 25°C than does a 1-cubic-centimeter cube of calcium at 25°C. This difference in rate of reaction is most closely associated with the different
   (1) surface areas of the metal cubes
   (2) natures of the metals
   (3) densities of the metals
   (4) concentrations of the metals

2. When a catalyst is added to a chemical reaction, there is a change in the
   (1) heat of reaction
   (2) rate of reaction
   (3) potential energy of the reactants
   (4) potential energy of the products

3. Raising the temperature speeds up the rate of a chemical reaction by increasing
   (1) the effectiveness of the collisions, only
   (2) the frequency of the collisions, only
   (3) both the effectiveness and the frequency of the collisions
   (4) neither the effectiveness nor the frequency of the collisions

4. Given the reaction

$$A + B \rightarrow C + D$$

The reaction will most likely occur at the greatest rate if A and B represent
   (1) nonpolar molecular compounds in the solid phase
   (2) ionic compounds in the solid phase
   (3) solutions of nonpolar molecular compounds
   (4) solutions of ionic compounds

5. A student adds two 50-milligram pieces of Ca(s) to water. A reaction takes place according to the following equation:

$$Ca(s) + 2H_2O(\ell) \rightarrow Ca(OH)_2(aq) + H_2(g)$$

Which change could the student have made that would most likely have increased the rate of the reaction?
(1) used ten 10-mg pieces of Ca(s)
(2) used one 100-mg piece of Ca(s)
(3) decreased the amount of the water
(4) decreased the temperature of the water

6. In order for a chemical reaction to occur, there must always be
(1) an effective collision between reacting particles
(2) a bond that breaks in a reactant particle
(3) reacting particles with a high charge
(4) reacting particles with a high kinetic energy

7. In a gaseous system, temperature remaining constant, an increase in pressure by decreasing the volume will lead to
(1) an increase in the activation energy
(2) a decrease in the activation energy
(3) fewer particles being formed
(4) more particles being formed

8. At room temperature, which reaction would be expected to have the fastest reaction rate?
(1) $Pb^{2+}(aq) + S^{2-}(aq) \rightarrow PbS(s)$
(2) $2H_2(g) + O_2(g) \rightarrow 2H_2O(\ell)$
(3) $N_2(g) + 2O_2(g) \rightarrow 2NO_2(g)$
(4) $2KClO_3(s) \rightarrow 2KCl(s) + 3O_2(g)$

9. Given this reaction at equilibrium:

$$2CO(g) + O_2(g) \rightleftharpoons 2CO_2(g)$$

Which statement regarding this reaction is *always* true?
(1) The rates of the forward and reverse reactions are equal.
(2) The reaction occurs in an open system.
(3) The masses of the reactants and the products are equal.
(4) The concentrations of the reactants and the products are equal.

10. Which factors must be equal in a reversible chemical reaction at equilibrium?
(1) the concentrations of the reactants and products
(2) the potential energies of the reactants and products
(3) the activation energies of the forward and reverse reactions
(4) the rates of reaction of the forward and reverse reactions

11. The diagram that follows shows a bottle containing $NH_3(g)$ dissolved in water. How can the equilibrium

$$NH_3(g) \rightleftharpoons NH_3(aq)$$

be reached?

(1) Add more water. (2) Add more $NH_3(g)$.
(3) Cool the contents. (4) Stopper the bottle.

12. Given the reaction

$$2Na(s) + 2H_2O(\ell) \rightarrow 2Na^+(aq) + 2OH^-(aq) + H_2(g)$$

This reaction goes to completion because one of the products formed is
(1) an insoluble base (2) a soluble base
(3) a precipitate (4) a gas

13. In all reversible chemical reactions, equilibrium has been reached when
(1) the opposing reactions cease
(2) the molar concentrations of the reactants and products are equal
(3) any one substance leaves the field of reaction
(4) the rates of the opposing reactions are equal

14. Given the reaction

$$HC_2H_3O_2(aq) + H_2O(\ell) \rightleftharpoons H_3O^+(aq) + C_2H_3O_2^-(aq)$$

When the reaction reaches a state of equilibrium, the concentrations of the reactants
(1) are less than the concentration of the products
(2) are equal to the concentrations of the products
(3) begin decreasing
(4) become constant

15. A flask at 25°C is partially filled with water and stoppered. After a period of time the water level remains constant. Which relationship best explains this observation?
(1) The rate of condensation exceeds the rate of evaporation.
(2) The rates of condensation and evaporation are both zero.
(3) The rate of evaporation exceeds the rate of condensation.
(4) The rate of evaporation equals the rate of condensation.

16. When $AgNO_3(aq)$ is mixed with $NaCl(aq)$, a reaction occurs that tends to go to completion because
(1) a gas is formed   (2) water is formed
(3) a weak acid is formed   (4) a precipitate is formed

17. A solution in which equilibrium exists between undissolved and dissolved solute is always
(1) saturated   (2) unsaturated
(3) dilute   (4) concentrated

18. Given this system at equilibrium:

$$X_2(g) + 2Y_2(g) \rightleftharpoons 2XY_2(g) + 80 \text{ kcal}$$

The equilibrium point will shift to the right if the volume is
(1) increased and the temperature is increased
(2) increased and the temperature is decreased
(3) decreased and the temperature is increased
(4) decreased and the temperature is decreased

§2 **19.** Given this system at equilibrium:

$$PbCl_2(s) \rightleftharpoons Pb^{2+}(aq) + 2Cl^-(aq)$$

When $KCl(s)$ is added to the system, the equilibrium shifts to the
(1) right, and the concentration of $Pb^{2+}(aq)$ ions decreases
(2) right, and the concentration of $Pb^{2+}(aq)$ ions increases
(3) left, and the concentration of $Pb^{2+}(aq)$ ions decreases
(4) left, and the concentration of $Pb^{2+}(aq)$ ions increases

20. Given this system at equilibrium:

$$H_2(g) + I_2(g) + heat \rightleftharpoons 2HI(g)$$

The equilibrium will shift to the right if there is an increase in
(1) temperature
(2) pressure
(3) concentration of $HI(g)$
(4) volume of the reaction container

21. Which system at equilibrium will shift to the right when the pressure is increased?

(1) $NaCl(s) \overset{H_2O}{\rightleftharpoons} Na^+(aq) + Cl^-(aq)$

(2) $C_2H_5OH(\ell) \overset{H_2O}{\rightleftharpoons} C_2H_5OH(aq)$

(3) $NH_3(g) \overset{H_2O}{\rightleftharpoons} NH_3(aq)$

(4) $C_6H_{12}O_6(s) \overset{H_2O}{\rightleftharpoons} C_6H_{12}O_6(aq)$

§2 **22.** Given this system at equilibrium:

$$C(s) + O_2(g) \rightleftharpoons CO_2(g)$$

As $C(s)$ is added to the system, the amount of $CO_2(g)$ will
(1) decrease   (2) increase   (3) remain the same

23. Given this system at equilibrium:

$$A(g) + B(g) \rightleftharpoons C(g)$$

If the concentration of A is increased at constant temperature, which will also increase?
(1) the rate of the forward reaction
(2) the value of the equilibrium constant
(3) the activation energy of the forward reaction
(4) the concentration of B

24. Given this system at equilibrium:

$$2SO_2(g) + O_2(g) \rightleftharpoons 2SO_3(g) + 197 \text{ kJ}$$

The amount of SO_3 will increase if there is
(1) an increase in temperature
(2) a decrease in pressure (an increase in volume)
(3) an increase in the concentration of SO_2
(4) a decrease in the concentration of O_2

25. Given this system at equilibrium:

$$2H_2(g) + O_2(g) \rightleftharpoons 2H_2O(g) + heat$$

Which concentration changes occur when the temperature of the system is increased?
(1) The $[H_2]$ decreases and the $[O_2]$ decreases.
(2) The $[H_2]$ increases and the $[O_2]$ decreases.
(3) The $[H_2]$ decreases and the $[O_2]$ increases.
(4) The $[H_2]$ increases and the $[O_2]$ increases.

26. Given this equilibrium system:

$$N_2(g) + 3H_2(g) \rightleftharpoons 2NH_3(g)$$

If the volume of the reaction vessel is decreased at constant temperature, there will be an increase in the number of moles of
(1) N_2, only
(2) H_2, only
(3) NH_3, only
(4) N_2, H_2, and NH_3

§2 **27.** For the system

$$4NH_3(g) + 5O_2(g) \rightleftharpoons 4NO(g) + 6H_2O(g)$$

the correct equilibrium expression is

(1) $K_{eq} = \dfrac{[NO]^4[H_2O]^6}{[NH_3]^4[O_2]^5}$
(2) $K_{eq} = \dfrac{[NO]^4+[H_2O]^6}{[NH_3]^4+[O_2]^5}$

(3) $K_{eq} = \dfrac{[4NO][6H_2O]}{[4NH_3][5O_2]}$
(4) $K_{eq} = \dfrac{[4NO]+[6H_2O]}{[4NH_3]+[5O_2]}$

§2 **28.** Given this solution at equilibrium:

$$CaF_2(s) \rightleftharpoons Ca^{2+}(aq) + 2F^-(aq)$$

What is the solubility product (K_{sp}) expression?
(1) $K_{sp} = [Ca^{2+}][F^-]$
(2) $K_{sp} = [Ca^{2+}][2F^-]$
(3) $K_{sp} = [Ca^{2+}][F^-]^2$
(4) $K_{sp} = [Ca^{2+}]2[F^-]$

29. Given this system at equilibrium:

$$2CO(g) + O_2(g) \rightleftharpoons 2CO_2(g)$$

The correct equilibrium expression for this system is

(1) $K_{eq} = \dfrac{[2CO][O_2]}{[CO_2]}$
(2) $K_{eq} = \dfrac{[CO]^2[O_2]}{[CO_2]^2}$

(3) $K_{eq} = \dfrac{[2CO_2]}{[2CO][O_2]}$
(4) $K_{eq} = \dfrac{[CO_2]^2}{[CO]^2[O_2]}$

§2 **30.** What is the equilibrium expression for the following reaction?

$$4Al(s) + 3O_2(g) \rightleftharpoons 2Al_2O_3(s)$$

(1) $K_{eq} = [O_2]^3$

(2) $K_{eq} = [3O_2]$

(3) $K_{eq} = \dfrac{1}{[O_2]^3}$

(4) $K_{eq} = \dfrac{1}{[3O_2]}$

§2 **31.** According to Reference Table Y-1, which compound is more soluble than $BaSO_4$?
(1) AgBr   (2) CuCl   (3) AgI   (4) $ZnCO_3$

§2 **32.** Given this system at equilibrium:

$$H_2(g) + \frac{1}{2}O_2(g) \rightleftharpoons H_2O(g) + heat$$

The value of the equilibrium constant for this reaction can be changed by
(1) changing the pressure
(2) changing the temperature
(3) adding more O_2
(4) adding a catalyst

§2 **33.** According to Reference Table Y-1, which compound is least soluble in water?
(1) AgBr   (2) $BaSO_4$ (3) CuCl   (4) $ZnCO_3$

§2 **34.** Which system has the following equilibrium expression?

$$K_{eq} = \frac{[A][B]^2}{[AB_2]}$$

(1) $AB_2(g) \rightleftharpoons A(g) + 2B(g)$
(2) $2AB(g) \rightleftharpoons A(g) + B_2(g)$
(3) $A(g) + 2B(g) \rightleftharpoons AB_2(g)$
(4) $A(g) + B_2(g) \rightleftharpoons 2AB(g)$

35. Given this system at equilibrium:

$$N_2(g) + 3H_2(g) \rightleftharpoons 2NH_3(g) + heat$$

Which stress on the system at equilibrium favors the production of $NH_3(g)$?
(1) decreasing the concentration of $N_2(g)$
(2) decreasing the concentration of $H_2(g)$
(3) reducing the volume of the system
(4) increasing the temperature of the system

36. Which statement best explains the role of a catalyst in a chemical reaction?
(1) A catalyst is consumed in the reaction but not regenerated.
(2) A catalyst limits the amount of reactants used.
(3) A catalyst changes the kinds of products produced.
(4) A catalyst provides an alternate reaction pathway that requires less activation energy.

37. In most aqueous reactions, as temperature increases, the effectiveness of collisions between reacting particles
(1) decreases
(2) increases
(3) remains the same
(4) reverses

38. Based on the nature of the reactants in each of the equations below, which reaction will occur at the *fastest* rate?
(1) $C(s) + O_2(g) \rightarrow CO_2(g)$
(2) $NaOH(aq) + HCl(aq) \rightarrow NaCl(aq) + H_2O(\ell)$
(3) $CH_3OH(\ell) + CH_3COOH(\ell) \rightarrow CH_3COOCH_3(aq) + H_2O(\ell)$
(4) $CaCO_3(s) \rightarrow CaO(s) + CO_2(g)$

39. Given the reaction at equilibrium:

$$A(g) + B(g) \rightleftharpoons AB(g) + \text{heat}$$

The concentration of $A(g)$ can be increased by
(1) lowering the temperature
(2) adding a catalyst
(3) increasing the concentration of $AB(g)$
(4) increasing the concentration of $B(g)$

40. Given the reaction:

$$AgCl(s) \overset{H_2O}{\rightleftharpoons} Ag^+(aq) + Cl^-(aq)$$

Once equilibrium is reached, which statement is accurate?
(1) The concentration of $Ag^+(aq)$ is greater than the concentration of $Cl^-(aq)$.
(2) The $AgCl(s)$ will be completely consumed.
(3) The rates of the forward and reverse reactions are equal.
(4) The entropy of the forward reaction will continue to increase.

41. As the temperature of a reaction increases, it is expected that the reacting particles collide
(1) more often and with greater force
(2) more often and with less force
(3) less often and with greater force
(4) less often and with less force

42. Given the equation for a system at equilibrium:

$$N_2(g) + 3H_2(g) \rightleftharpoons 2NH_3 + energy$$

If only the concentration of $N_2(g)$ is increased, the concentration of
(1) $NH_3(g)$ increases
(2) $NH_3(g)$ remains the same
(3) $H_2(g)$ increases
(4) $H_2(g)$ remains the same

Constructed-Response Questions

1. Consider the following system at equilibrium:

$$2NO_2(g) \rightleftharpoons N_2(g) + 2O_2(g) + heat$$

A vessel contains $NO_2(g)$, $N_2(g)$, and $O_2(g)$ at equilibrium. How will each of the following stresses affect the equilibrium position?
(a) More $NO_2(g)$ is added to the vessel.
(b) Some $N_2(g)$ is removed from the vessel.
(c) The volume of the vessel is reduced to one-half its original value.
(d) Some $He(g)$ is added to the vessel.
(e) The temperature of the system is increased.

§2 **2.** Write the equilibrium constant expression for each of the following systems:
(a) $4NH_3(g) + 7O_2(g) \rightleftharpoons 4NO_2(g) + 6H_2O(g)$
(b) $Fe_2O_3(s) + 3H_2(g) \rightleftharpoons 2Fe(s) + 3H_2O(g)$

§2 **3.** Consider the following system:

$$A(g) + 2B(g) \rightleftharpoons 3C(g) + 2D(g)$$

(a) Write the equilibrium constant expression for this system.

(b) The table below contains a set of equilibrium concentrations for the reactants and products when they reach equilibrium at 298 K. Use these data to calculate the value of K_{eq} for the system given on the previous page at 298 K.

Substance	Equilibrium Concentration (mol/L)
A	0.35
B	0.20
C	1.0
D	0.50

§2 **4.** In a certain experiment, 1.000 mole of $N_2(g)$ and 1.000 mole of H_2 were allowed to react at 500°C in a 1.000-liter flask. The relevant reaction was as follows:

$$N_2(g) + 3H_2(g) \rightleftharpoons 2NH_3(g)$$

After the system reached equilibrium, the flask was found to contain 0.921 mole of $N_2(g)$. Calculate the equilibrium concentrations of $H_2(g)$ and $NH_3(g)$.

For questions 5 and 6, use the given equilibrium equation at 298 K:

$$KNO_3(s) + 34.89 \text{ kJ} \overset{H_2O}{\rightleftharpoons} K^+(aq) + NO_3^-(aq)$$

5. Describe, in terms of Le Châtelier's principle, why an increase in temperature increases the solubility of KNO_3.

6. The equation indicates that KNO_3 has formed a saturated solution. Explain, in terms of equilibrium, why the solution is saturated.

Human blood contains dissolved carbonic acid, H_2CO_3, in equilibrium with carbon dioxide and water. The equilibrium system is shown below:

$$H_2CO_3(aq) \rightleftharpoons CO_2(aq) + H_2O(\ell)$$

7. Explain, using Le Châtelier's principle, why decreasing the concentration of CO_2 decreases the concentration of H_2CO_3.

The answers to these questions are found in Appendix 3.

<table>
<tr><td>Chapter
Fourteen</td><td colspan="2"># ACIDS AND BASES</td></tr>
</table>

Chapter
Fourteen

ACIDS AND BASES

KEY IDEAS

This chapter focuses on the ways in which acids and bases are defined and the reactions that they undergo. The strengths of acids and bases are discussed, and the important laboratory procedure of acid–base titration is introduced.

KEY OBJECTIVES

At the conclusion of this chapter you will be able to:
- Define acids and bases operationally.
- State and apply the Arrhenius definitions of acids and bases.
- Solve acid–base titration problems.
- State and apply the Brønsted–Lowry definitions of acids and bases.
- Define the term *amphiprotic* (*amphoteric*), and apply the concept.
- Define the term *conjugate acid–base pairs*, and recognize these pairs in an acid–base reaction.
- Compare the relative strengths of conjugate acid–base pairs.
- Relate ionization constants K_a and K_b to acid–base strength.
- Solve K_w problems.
- Define the terms *pH* and *pOH*, and apply these definitions.
- Describe the properties of acid–base indicators.
- Indicate which salts are likely to produce acidic, basic, or neutral solutions when dissolved in water.
- Describe the acid–base properties of metallic and nonmetallic oxides in water.
- Define the terms *Lewis acid* and *Lewis base*, and apply these concepts.

SECTION I—BASIC
(REGENTS-LEVEL) MATERIAL

NYS REGENTS CONCEPTS AND SKILLS

Note: By the time you have finished Section I, you should have mastered the concepts and skills listed on the following pages. The Regents chemistry examination will test your knowledge of these items and your ability to apply them.

Concepts are the *basic ideas* that form the body of the Regents chemistry course (what you need to know!).

Skills are the *activities* that demonstrate your mastery of these concepts (how you show that you know them!).

Following each concept or skill is a page reference (given in parentheses) to this chapter.

14.1 Concept:
 • An electrolyte is a substance that, when dissolved in water, forms a solution capable of conducting an electric current. (Page 401)
 • The ability of a solution to conduct an electric current depends on the concentration of ions. (Page 401)

14.2 Concepts
 • The behaviors of many acids and bases can be explained by the Arrhenius theory. (Page 403)
 • Arrhenius acids and bases are electrolytes. (Page 403)

Skill:
Given the properties of a substance, identify it as an Arrhenius acid or base. (Page 403)

14.3 Concept:
Arrhenius acids yield hydrogen ions [H+ (aq)] as the only positive ions in an aqueous solution. The hydrogen ion can also be written as H_3O^+ (aq) (hydronium ion). (Page 403)

14.4 Concept:
Arrhenius bases yield hydroxide ions [OH– (aq)] as the only negative ions in an aqueous solution. (Page 403)

14.5 Concept:
In the process of neutralization, an Arrhenius acid and an Arrhenius base react to form a salt and water. (Page 403)

Skill:
Write simple neutralization reactions, given the reactants (acid and base). (Page 403)

14.6 Concept:
Titration is a laboratory process in which a volume of solution of known concentration is used to determine the concentration of another solution. (Pages 403–406)

Skill:
Calculate the concentration or the volume of a solution, given titration data. (Pages 403–406)

14.7 Concept:
There are alternative acid base theories. One theory states that an acid is a proton (H+) donor and a base is a proton (H+) acceptor. (Pages 406–408)

14.8 Concepts:
- The acidity or alkalinity of a solution can be measured by its pH value. (Pages 408–409)
- The relative level of acidity or alkalinity of a solution can be shown by using indicators. (Pages 409–410)

Skills:
- Identify a solution as acidic, basic, or neutral, based on the pH value of the solution. (Pages 408–409)
- Interpret changes in acid-base indicator color. (Pages 409–410)

14.9 Concept:
On the pH scale, each decrease of one unit of pH represents a tenfold increase in hydronium ion concentration. (Page 408)

14.1 OPERATIONAL DEFINITIONS OF ACIDS AND BASES

Defining something operationally means describing it by what it does. When we define acids and bases operationally, we are interested in observing their characteristic properties; we are not immediately concerned with why they exhibit these properties.

- Acids and bases are *electrolytes:* they conduct electricity when dissolved in water. (Conductivity depends on the ion concentration.)
- Dilute solutions of acids taste sour; dilute solutions of bases taste bitter.
- Aqueous solutions of bases feel slippery.
- Acids and bases cause color changes in solutions of substances known as acid–base indicators. (We will have more to say about acid–base indicators in Section 14.6.)
- Acids react with certain metals to liberate hydrogen gas:

$$Zn(s) + 2HCl(aq) \rightarrow H_2(g) + ZnCl_2(aq)$$

$$\underset{\text{metal}}{} \qquad \underset{\text{acid}}{} \qquad \underset{\text{hydrogen gas}}{}$$

Not all metals, however, liberate hydrogen gas when treated with acid. For example, copper does not react with HCl.
- Acids and bases *neutralize* each other. If an acid reacts with a hydroxide base, a salt and water are formed:

$$HCl(aq) + NaOH(aq) \rightarrow NaCl(aq) + H_2O(\ell)$$

$$\underset{\text{acid}}{} \qquad \underset{\text{hydroxide base}}{} \qquad \underset{\text{salt}}{} \qquad \underset{\text{water}}{}$$

The salt formed in this reaction, sodium chloride, is an ionic compound whose positive ion (Na^+) comes from the base and whose negative ion (Cl^-) comes from the acid. Common hydroxide bases include the metals of Groups 1 and 2: $NaOH$, KOH, $LiOH$, $Ca(OH)_2$, $Mg(OH)_2$.

Some common acids and the negative ions they generate in neutralization are listed in the accompanying table.

Acid	Name	Ion	Name of Salt
HCl	Hydrochloric	Cl^-	Chloride
HNO_3	Nitric	NO_3^-	Nitrate
H_2SO_4	Sulfuric	HSO_4^-	Hydrogen sulfate
		SO_4^{2-}	Sulfate
H_2CO_3	Carbonic	HCO_3^-	Hydrogen carbonate
		CO_3^{2-}	Carbonate
HBr	Hydrobromic	Br^-	Bromide
HF	Hydrofluoric	F^-	Fluoride
HNO_2	Nitrous	NO_2^-	Nitrite
H_3PO_4	Phosphoric	$H_2PO_4^-$	Dihydrogen phosphate
		HPO_4^{2-}	Hydrogen phosphate
		PO_4^{3-}	Phosphate
$HOCl$	Hypochlorous	ClO^-	Hypochlorite
$HClO_2$	Chlorous	ClO_2^-	Chlorite
$HClO_3$	Chloric	ClO_3^-	Chlorate
$HClO_4$	Perchloric	ClO_4^-	Perchlorate

A shorter listing of common acids and bases appears in Reference Tables K and L in Appendix 1.

PROBLEM
(a) Write the equation for the neutralization of chlorous acid by potassium hydroxide. Name the salt produced.
(b) An acid–base neutralization produces calcium nitrate as the salt. What are the names of the acid and base?

SOLUTIONS
(a) $HClO_2 + KOH \rightarrow KClO_2 + H_2O$. The salt is potassium chlorite.
(b) The acid is nitric acid; the base is calcium hydroxide.

TRY IT YOURSELF
What are the name and the formula of the salt produced by the reaction of magnesium hydroxide and chloric acid?

ANSWER
magnesium chlorate; $Mg(ClO_3)_2$

14.2 ARRHENIUS DEFINITIONS OF ACIDS AND BASES

The Swedish chemist Svante Arrhenius was the first person to propose a theory as to how solutions of electrolytes are able to conduct electric currents. He recognized that a substance such as NaCl has to dissociate in water and produce Na^+ and Cl^- ions. Arrhenius extended this idea to acids and bases, giving us the following Arrhenius definitions:

Arrhenius acid: *A substance that releases H^+ ions in aqueous solution.*
Arrhenius base: *A substance that releases OH^- ions in aqueous solution.*

Therefore, when HCl dissolves in water, its acidic properties are due to the H^+ ions it releases:

$$HCl(aq) \xrightarrow{H_2O} H^+(aq) + Cl^-(aq)$$

Similarly, when NaOH dissolves in water, its basic properties are due to the OH^- ions it releases:

$$NaOH(aq) \xrightarrow{H_2O} Na^+(aq) + OH^-(aq)$$

We are now able to understand what happens when an acid such as HCl neutralizes a base such as NaOH:

$$H^+(aq) + Cl^-(aq) + Na^+(aq) + OH^-(aq) \rightarrow H_2O(\ell) + Na^+(aq) + Cl^-(aq)$$
$$\text{Net reaction: } H^+(aq) + OH^-(aq) \rightarrow H_2O(\ell)$$

Neutralization is the reaction between H^+ and OH^- ions to form H_2O. The Na^+ and Cl^- ions are unchanged by the reaction and are called *spectator ions*. When the water is removed, the salt NaCl remains behind.

14.3 ACID–BASE TITRATION

Suppose we find a bottle labeled "dilute HCl" in the laboratory. A logical question would be: How dilute is the HCl? We are really asking what the molar concentration of the acid is.

The molar concentration of an unknown acid or base can readily be determined by a procedure known as *acid–base titration*. In this procedure, we perform an acid–base neutralization, proceeding as follows:

1. The volume of the unknown (acid or base) is measured carefully, and an *acid–base indicator* is added.
2. A *standard solution* (base or acid) of known concentration is slowly and accurately added to the unknown.
   (Usually, a long, narrow tube known as a *burette,* shown at the right, is used to deliver the standard solution.)
3. When the indicator changes color, this signal indicates that neutralization has occurred. This visual signal is known as the *end point* of the titration.
4. The volume of the unknown and the volume and molarity of the standard solution are recorded.
5. The molarity of the unknown is calculated.

When an acid is neutralized by a base in water, the net equation for the neutralization is

$$H^+(aq) + OH^-(aq) \rightarrow H_2O(\ell)$$

This equation indicates that *1 mole* of H^+ (acid) is needed to neutralize *1 mole* of OH^- (base).

It follows that the number of moles of H^+ and OH^- must always be equal when neutralization occurs:

$$n_{acid} = n_{base}$$

We know from Chapter 12 on solutions that, in expressions where M is the molarity, n is the number of moles, and V is the volume of the solution in liters. If we solve the equation $M = \dfrac{n}{V}$ for n, we obtain

$$n = M \cdot V$$

Therefore, when acid–base neutralization occurs, we can write

$$\boxed{M_A \cdot V_A = M_B \cdot V_B}$$

This equation, known as the titration equation, appears also in Reference Table T in Appendix 1. The symbols M_A and M_B represent the molarities of the H^+ and OH^- ions, respectively. The volumes of the acid and base (V_A and V_B) can be measured in any units (L, mL, etc.) as long as the same units appear on both sides of the equation.

PROBLEM

In an acid–base titration, standard 0.10 M KOH is used to determine the concentration of an unknown HNO_3 solution.

(a) Write the equation for the neutralization.

(b) If 50. milliliters of KOH is needed to neutralize 200. milliliters of HNO_3, calculate the concentration of the acid.

SOLUTIONS

(a) $HNO_3(aq) + KOH(aq) \rightarrow H_2O(\ell) + KNO_3(aq)$

(b) $(M \cdot V)_{acid} = (M \cdot V)_{base}$

$M_{acid} \cdot 200. \text{ mL} = 0.10 \text{ M} \cdot 50. \text{ mL}$

$M_{acid} = 0.025 \text{ M}$

TRY IT YOURSELF

An acid–base titration reaches its end point when 100 milliliters of 0.2 M HCl is added to 50 milliliters of NaOH. What is the concentration of the NaOH?

ANSWER

0.4 M

For an acid such as H_2SO_4 or a base such as $Ba(OH)_2$, we must be careful in using the titration equation. In a 1-molar solution of H_2SO_4, for example, M_A is *2 molar* because both hydrogen atoms can ionize. Similarly, in a 0.2-molar solution of $Ba(OH)_2$, M_B is *0.4 molar* because of the presence of two OH^- ions in the formula.

PROBLEM

In an acid–base titration, standard 0.1 M H_2SO_4 is used to determine the concentration of an unknown NaOH solution.

(a) Write the equation for the neutralization.

(b) If 40 milliliters of H_2SO_4 is needed to neutralize 100 milliliters of NaOH, calculate the concentration of the base.

SOLUTIONS

(a) $H_2SO_4(aq) + 2NaOH(aq) \rightarrow 2H_2O(\ell) + Na_2SO_4(aq)$

(b) $M_A \cdot V_A = M_B \cdot V_B$

$0.2 \text{ M} \cdot 40 \text{ mL} = M_B \cdot 100 \text{ mL}$

$M_B = 0.08 \text{ M}$

TRY IT YOURSELF

What volume of 0.200 M H_2SO_4 is needed to neutralize 150. milliliters of 0.600 M NaOH?

ANSWER
225 mL

Lest we forget, this "Try It Yourself" problem can also be solved by the factor-label method used in Chapter 12. First we need to write the balanced equation for the neutralization reaction:

$$2NaOH(aq) + H_2SO_4(aq) \rightarrow 2H_2O(\ell) + Na_2SO_4(aq)$$

Then we construct a solution map of the problem:

$$\text{mL NaOH} \xrightarrow[\text{conversion}]{\text{volume}} \text{L NaOH} \xrightarrow{\text{molarity}} \text{mol NaOH} \xrightarrow[\text{coefficient}]{\text{equation}}$$

$$\text{mol } H_2SO_4 \xrightarrow{\text{molarity}} \text{L } H_2SO_4 \xrightarrow[\text{conversion}]{\text{volume}} \text{mL } H_2SO_4$$

The solution of this problem is given below.

$$V = 150. \text{ mL NaOH} \cdot \left(\frac{1 \text{ L NaOH}}{1000 \text{ mL NaOH}}\right) \cdot \left(\frac{0.600 \text{ mol NaOH}}{1 \text{ L NaOH}}\right)$$

$$\cdot \left(\frac{1 \text{ mol } H_2SO_4}{2 \text{ mol NaOH}}\right) \cdot \left(\frac{1 \text{ L } H_2SO_4}{0.200 \text{ mol } H_2SO_4}\right) \cdot \left(\frac{1000 \text{ mL } H_2SO_4}{1 \text{ L } H_2SO_4}\right)$$

$$= 225 \text{ mL } H_2SO_4$$

As you can see, in such cases it is simpler to rely on the general titration equation given on page 404.

14.4 BRØNSTED–LOWRY DEFINITIONS OF ACIDS AND BASES

In order to explain acid–base reactions more fully, Johannes Brønsted and T. M. Lowry extended the definitions of acids and bases as follows:

> **Brønsted–Lowry acid:**
> *Any species that can donate a proton (H⁺ ion) to another species, that is, a proton donor.*
> **Brønsted–Lowry base:**
> *Any species that can accept a proton (H⁺ ion) from another species, that is, a proton acceptor.*

(The term *species* indicates any particle: an atom, a molecule, or an ion.)

The Brønsted–Lowry definitions do not replace the Arrhenius theory; they extend it. Any substance that is an Arrhenius acid is also a Brønsted–Lowry acid; any substance that is an Arrhenius base is also a Brønsted–Lowry base.

The Brønsted–Lowry definitions explain why substances such as HCl behave as acids in water:

$$HCl(aq) + H_2O(\ell) \rightarrow H_3O^+(aq) + Cl^-(aq)$$

We see from the equation that the HCl donates a proton to the H_2O molecule, forming the H_3O^+ (hydronium) ion. (The proton forms a coordinate covalent bond with the oxygen atom.) It is the hydronium ion that is responsible for the acidic properties of acids in water solution.

If we review the equation for neutralization given in Section 14.2, we see why the OH^- ion is a Brønsted–Lowry base: it accepts a proton from the acid and forms water.

According to Arrhenius, only metallic hydroxides are bases. The Brønsted–Lowry definitions explain why a substance such as NH_3, which has no hydroxide ion, behaves as a base in water:

$$NH_3(aq) + H_2O(\ell) \rightarrow NH_4^+(aq) + OH^-(aq)$$

We see from the equation that the NH_3 molecule accepts a proton from the H_2O molecule. (The proton forms a coordinate covalent bond with the nitrogen atom.)

According to the Brønsted–Lowry definitions, every equation that has an acid must also have a base! A substance cannot donate a proton unless there is another substance that will accept it. In this sense, all such reactions are properly called *acid–base reactions*. In the reaction between HCl and H_2O, H_2O is the Brønsted–Lowry base; in the reaction between NH_3 and H_2O, H_2O is the Brønsted–Lowry acid.

PROBLEM

In the reaction $NH_3(aq) + HBr(aq) \rightarrow NH_4^+(aq) + Br^-(aq)$, identify the Brønsted–Lowry acid and the Brønsted–Lowry base.

SOLUTION

To solve this problem, we need to examine both sides of the equation. We see that NH_3 becomes NH_4^+ and HBr becomes Br^-. This could have happened only if HBr donated the proton to NH_3.

Therefore, HBr is the Brønsted–Lowry acid and NH_3 is the Brønsted–Lowry base.

TRY IT YOURSELF

In the reaction $H_2O(\ell) + NH_2^-(aq) \rightarrow NH_3(aq) + OH^-(aq)$, identify the Brønsted–Lowry acid and the Brønsted–Lowry base.

ANSWER

H_2O is the Brønsted–Lowry acid; NH_2^- is the Brønsted–Lowry base.

14.5 THE pH SCALE OF ACIDITY AND BASICITY

To measure the degree of acidity or basicity of an aqueous solution, chemists have come to rely on the **pH scale**, a scale that depends on the concentration of H^+ ions in solution. The pH scale generally runs from 0 to 14, and a change of *one* pH unit represents a *tenfold* change in the H^+ ion concentration of the solution. For example, a solution with a pH of 2 has 10 times the H^+ ion concentration of a solution with a pH of 3. Similarly, a solution with a pH of 4 has 1/100 the H^+ ion concentration of a solution with a pH of 2. The table that follows illustrates how pH is related to the degree of acidity and basicity.

pH Values and Acid-Base Character of an Aqueous Solution

pH	Solution Is:	pH	Solution Is:
0	Strongly acidic	8	Weakly basic
1	Strongly acidic	9	Moderately basic
2	Strongly acidic	10	Moderately basic
3	Moderately acidic	11	Moderately basic
4	Moderately acidic	12	Strongly basic
5	Moderately acidic	13	Strongly basic
6	Weakly acidic	14	Strongly basic
7	Neutral		

To give you an idea of how the concept of pH relates to the "real" world, examine the next table: it lists the pH values of some common fluids.

The pH Values of Some Common Fluids

Fluid	pH Value
Gastric juice (stomach)	1.0-2.0
Lemon juice	2.4
Vinegar	3.0
Grapefruit juice	3.2
Orange juice	3.5
Urine	4.8-7.5
Water (exposed to air)	5.5
Saliva	6.4-6.9
Milk	6.5
Water (pure)	7.0
Blood	7.35-7.45
Tears	7.4
Milk of magnesia	10.6
Ammonia (household)	11.5

14.6 ACID–BASE INDICATORS

Acid–base indicators are substances (usually of plant origin) that change color with changing pH. Every indicator can act either as a Brønsted–Lowry acid or base, and each form has its own specific color. The color change of an indicator takes place over a specific range of pH. Outside its pH range, an indicator cannot be used effectively since it exists only in its acidic or basic form.

Reference Table M in Appendix 1 lists some common indicators, the pH ranges over which their color changes occur, and their colors at the low and high ends of the pH range.

PROBLEM

The following three indicators appear in Reference Table M in Appendix 1: methyl orange, bromthymol blue, thymol blue. Which of these indicators can distinguish between a solution whose pH is 4.6 and a solution whose pH is 7.9?

SOLUTION

The following table lists the indicators and their colors at the two pH values:

Indicator	Color at pH 4.6	Color at pH 7.9
Methyl orange	Yellow	Yellow
Bromthymol blue	Yellow	Blue
Thymol blue	Yellow	Yellow

As we can see, only bromthymol blue will distinguish between the two solutions.

TRY IT YOURSELF

Which indicator in Reference Table M can be used to distinguish between the solutions in each of the following pairs?
(a) Solution *A*: pH = 2.0; solution *B*: pH = 5.0
(b) Solution *C*: pH = 5.0; solution *D*: pH = 9.0
(c) Solution *E*: pH = 7.0; solution *F*: pH = 10.0

ANSWERS

(a) methyl orange
(b) bromthymol blue
(c) thymol blue

SECTION II—ADDITIONAL MATERIAL

14.1A   AMPHIPROTIC (AMPHOTERIC) SUBSTANCES

We have seen that H_2O is an example of a substance that can behave as an acid in one reaction and as a base in another. Such substances are called **amphiprotic** or **amphoteric**.

14.2A   ACID–BASE EQUILIBRIA

Like other chemical reactions, acid–base reactions are equilibrium systems and therefore have both forward and reverse reactions. (Note: From now on, we will omit the phases of the reactants and products unless they have special significance.) Consider the acid HA ionizing in water:

$$HA + H_2O \rightleftharpoons H_3O^+ + A^-$$

In the forward reaction: $HA + H_2O \rightarrow H_3O^+ + A^-$, the acid, HA, donates a proton to the base, H_2O.

In the reverse reaction: $HA + H_2O \leftarrow H_3O^+ + A^-$, the H_3O^+ ion acts as an acid and donates a proton to the A^- ion, which acts as a base.

Every acid–base reaction has two acids and two bases. To summarize:

$$HA + H_2O \rightleftharpoons H_3O^+ + A^-$$
$$\text{acid} \qquad \text{base} \qquad \text{acid} \qquad \text{base}$$

Similarly, when a base, B, ionizes in water, it behaves as in any other acid–base reaction:

$$B + H_2O \rightleftharpoons BH^+ + OH^-$$
$$\text{base} \quad \text{acid} \quad \text{acid} \quad \text{base}$$

PROBLEM

In the acid–base reaction $HNO_2 + H_2O \rightleftharpoons H_3O^+ + NO_2^-$, identify the acids and bases.

SOLUTION

In the forward reaction, we see that HNO_2 becomes the NO_2^- ion on the right side of the equation. Therefore, we can conclude that HNO_2 donates the proton and is the acid. It follows that the H_2O, which accepts the proton, is the base.

In the reverse reaction, we see that the H_3O^+ ion becomes H_2O on the left side of the equation. Therefore, we can conclude that the H_3O^+ ion donates the proton and is the acid. It follows that the NO_2^- ion, which accepts the proton, is the base. To summarize:

$$HNO_2 + H_2O \rightleftharpoons H_3O^+ + NO_2^-$$
$$\text{acid} \qquad \text{base} \qquad \text{acid} \qquad \text{base}$$

TRY IT YOURSELF

Identify the acids and bases in the following reaction:

$$NH_3 + HCO_3^- \rightleftharpoons NH_4^+ + CO_3^{2-}$$

ANSWER

$$NH_3 + HCO_3^- \rightleftharpoons NH_4^+ + CO_3^{2-}$$
$$\text{base} \qquad \text{acid} \qquad \text{acid} \qquad \text{base}$$

14.3A CONJUGATE ACID–BASE PAIRS

If we examine all of the acid–base equilibria shown on page 411, we observe that the acid on the left side of the equation becomes the base on the right side. Similarly, the base on the left side of the equation becomes the acid on the right side. These are known as **conjugate acid–base pairs**. In the reaction

$$HNO_2 + H_2O \rightleftharpoons H_3O^+ + NO_2^-$$

$$\text{acid} \qquad \text{base} \qquad \text{acid} \qquad \text{base}$$

HNO_2 and NO_2^- are a conjugate acid–base pair, as are also H_2O and H_3O^+.

For example, we may say that HNO_2 is the conjugate acid of NO_2^-, or that NO_2^- is the conjugate base of HNO_2.

Note that a conjugate acid and its conjugate base differ by an H^+:

> **Conjugate Acid = H$^+$ + Conjugate Base**

PROBLEM
(a) What is the conjugate base of HF?
(b) What is the conjugate base of H_2O?
(c) What is the conjugate base of $H_2PO_4^-$?
(d) What is the conjugate base of NO_3^-?

SOLUTIONS
A conjugate base has one less H^+ than its conjugate acid. Therefore, we will "subtract" an H^+ from each of the four compounds to determine its conjugate base:

(a) $HF - H^+ = F^-$ (b) $H_2O - H^+ = OH^-$ (c) $H_2PO_4^- - H^+ = HPO_4^{2-}$
(d) There is no conjugate base! NO_3^- has no H^+ to donate.

TRY IT YOURSELF
(a) What is the conjugate acid of Br^-?
(b) What is the conjugate acid of H_2O?
(c) What is the conjugate acid of S^{2-}?

ANSWERS
(a) $Br^- + H^+ = HBr$ (b) $H_2O + H^+ = H_3O^+$ (c) $S^{2-} + H^+ = HS^-$

The accompanying table lists a number of conjugate acid–base pairs. If we scan the list, we see that certain species (e.g., NH_3, HSO_4^-) appear as both conjugate acids and conjugate bases. These species are exhibiting amphoteric behavior. Can you find other amphoteric species in this table?

RELATIVE STRENGTHS OF CONJUGATE ACID–BASE PAIRS

Formula	Conjugate Acid Name	Formula	Conjugate Base Name
$HClO_4$	perchloric acid	ClO_4^-	perchlorate ion
HI	hydriodic acid	I^-	iodide ion
HBr	hydrobromic acid	Br^-	bromide ion
HCl	hydrochloric acid	Cl^-	chloride ion
H_2SO_4	sulfuric acid	HSO_4^-	hydrogen sulfate ion
HNO_3	nitric acid	NO_3^-	nitrate ion
H_3O^+	hydronium ion	H_2O	water
HSO_4^-	hydrogen sulfate ion	SO_4^{2-}	sulfate ion
H_3PO_4	phosphoric acid	H_2PO4^-	dihydrogen phosphate ion
HF	hydrofluoric acid	F^-	fluoride ion
HNO_2	nitrous acid	NO_2^-	nitrite ion
HCO_2H	formic (methanoic) acid	CO_2H^-	formate (methanoate) ion
$HC_2H_3O_2$	acetic (ethanoic) acid	$C_2H_3O_2^-$	acetate (ethanoate) ion
H_2CO_3	carbonic acid	HCO_3^-	hydrogen carbonate ion
$H_2PO_4^-$	dihydrogen phosphate ion	HPO_4^{2-}	hydrogen phosphate ion
NH_4^+	ammonium ion	NH_3	ammonia
HCN	hydrocyanic acid	CN^-	cyanide ion
HCO_3^-	hydrogen carbonate ion	CO_3^{2-}	carbonate ion
HPO_4^{2-}	hydrogen phosphate ion	PO_4^{3-}	phosphate ion
H_2O	water	OH^-	hydroxide ion
NH_3	ammonia	NH_2^-	amide ion

Acid strength increases — *Strong acids* / *Weak acids*

Base strength increases

14.4A NEUTRALIZATION (REVISITED)

At the beginning of this chapter, we said that acids and bases neutralized each other. However, we were talking about Arrhenius acids and bases, in which H^+ ions and OH^- ions combine to produce water. In the Brønsted–Lowry sense, not every acid–base reaction is a neutralization. *For acid–base neutralization to occur, the solvent of the system must be produced from its conjugate acid and base.*

We have been studying reactions in water. The conjugate acid of water is the H_3O^+ ion; the conjugate base is the OH^- ion. Therefore, neutralization involves the reaction

$$H_3O^+(aq) + OH^-(aq) \rightarrow 2H_2O(\ell)$$

Since water is a covalently bonded molecule, the reaction goes nearly to completion and removes the acidic and basic ions from the aqueous solution.

14.5A STRENGTHS OF CONJUGATE ACID–BASE PAIRS

If a conjugate acid is strong, then its conjugate base is weak; if a conjugate acid is weak, then its conjugate base is strong. This relationship follows from the Brønsted–Lowry definitions of acids and bases. If a conjugate acid is strong, it is able to donate its proton readily. Therefore, its conjugate base is not able to hold onto the proton and is weak. Similarly, a weak conjugate acid is not able to donate a proton readily because its conjugate base holds onto it tightly and is strong.

In the preceding subsection, the table of conjugate acid-base pairs lists the strengths of conjugate acids in descending order, and conjugate bases in ascending order.

PROBLEM
The following four conjugate acids are listed in descending order of strength:

$$HI > HF > H_2CO_3 > NH_3$$

List the conjugate bases of these acids, also in descending order of strength.

SOLUTION
First, we find the conjugate bases of the acids by "subtracting" H^+ from each acid on the list:

$$HI - H^+ = I^-, \qquad HF - H^+ = F^-, \qquad H_2CO_3 - H^+ = HCO_3^-,$$
$$NH_3 - H^+ = NH_2^-$$

We know that there is an inverse relationship between the strength of a conjugate acid and the strength of its conjugate base. Since HI is the strongest acid in the list, its conjugate base, I^-, must be the weakest; since NH_3 is the weakest conjugate acid, its conjugate base, NH_2^-, must be the strongest.

With this information, we can arrange the conjugate bases in descending order:

$$NH_2^- > HCO_3^- > F^- > I^-$$

TRY IT YOURSELF
If OH^- is a stronger conjugate base than HSO_4^-, what can we conclude about their conjugate acids?

ANSWER
H_2SO_4 is a stronger acid than H_2O.

14.6A IONIZATION CONSTANTS OF ACIDS AND BASES (K_a and K_b)

Since acid–base reactions are equilibrium systems, we can write equilibrium constants for them. These constants give us information about the relative strengths of the acid and base.

In water solution, an acid ionizes according to the equation

$$HA(aq) + H_2O(\ell) \rightleftharpoons H_3O^+(aq) + A^-(aq)$$

The stronger an acid is in water, the more H_3O^+ will be produced. Therefore, the concentration of hydronium ion is a measure of the strength of the acid in water solution.

The equilibrium constant for this reaction is as follows:

$$K_{eq} = \frac{[H_3O^+][A^-]}{[HA][H_2O]}$$

In dilute solutions (which is all we concern ourselves with) the concentration of water is constant. Therefore it can be absorbed into K_{eq}, yielding the constant K_a:

$$K_a = \frac{[H_3O^+][A^-]}{[HA]}$$

The constant K_a is known as the ionization constant for the acid HA. If we examine the K_a expression given above, we see that the hydronium-ion concentration $[H_3O^+]$ appears in the numerator of the fraction. It follows that the stronger the acid is in water, the larger is its K_a value.

Reference Table Y-2 in Appendix 2 lists the K_a values of a number of acids in descending order. For example, nitrous acid (HNO_2), whose K_a is 7.2×10^{-4}, is a stronger acid than hypochlorous acid (HOCl), whose K_a is 2.9×10^{-8}. A **monoprotic acid** is a molecule or ion that is capable of donating *only one proton* to a Brønsted base.

If we examine Reference Table Y-2, we note that almost all of the K_a values are relatively small numbers. These low K_a values mean that these acids are relatively weak in water and the hydronium-ion concentration at equilibrium is small.

Acids that are capable of donating two or three protons are called, respectively, **diprotic** and **triprotic** acids. Collectively, they are known as **polyprotic** acids. Such acids undergo a process known as *stepwise ionization*, in which one proton is released in each step. For example, the diprotic acid carbonic acid (H_2CO_3) ionizes in *two* steps:

$$(1)\ H_2CO_3 + H_2O \rightleftharpoons H_3O^+ + HCO_3^-$$

$$(2)\ HCO_3^- + H_2O \rightleftharpoons H_3O^+ + CO_3^{2-}$$

Each step has its own distinct K_a value, written as $K_{a1}, K_{a2}, \ldots$.

Reference Table Y-2 in Appendix 2 *also* lists the ionization constants for a number of diprotic and triprotic acids. Note that the K_a value for each step is *much larger* than the value for the *succeeding* step (that is, $K_{a1} \gg K_{a2} \gg \ldots$). This means that, as stepwise ionization occurs, it becomes increasingly difficult to remove a proton.

A strong acid ionizes almost completely in water:

$$HA(aq) + H_2O(\ell) \xrightarrow{\sim 100\%} H_3O^+(aq) + A^-(aq)$$

In this case, the concentration of un-ionized HA(aq) at equilibrium is nearly 0, and the value of K_a is so large that it cannot be evaluated accurately. This is the case for the first six acids ($HClO_4$, HI, HBr, HCl, H_2SO_4, and HNO_3) in the table on page 413. This also means that $H_3O^+(aq)$ is the strongest acid that can be present in water solution. As a result, it is impossible to compare the relative strengths of the first five acids in water solution. To make such a comparison, other solvents must be used.

In water solution, a base ionizes according to the equation

$$B(aq) + H_2O(\ell) \rightleftharpoons BH^+(aq) + OH^-(aq)$$

The stronger a base is in water, the more OH^- will be produced. Therefore, the concentration of hydroxide ion is a measure of the strength of the base in water solution.

The ionization constant for a base, K_b, measures the relative strength of the base in water solution. The larger the value of K_b, the stronger the base. The expression for this constant is as follows:

$$K_b = \frac{[BH^+][OH^-]}{[B]}$$

Reference Table Y-2 in Appendix 2 lists the ionization constants for a number of weak bases. Note that all of these substances contain nitrogen, which can accept a proton from a Brønsted acid. In general, aqueous solutions of nitrogen-containing compounds are weakly basic in water. The characteristically bitter taste of coffee is due to the presence of the weak nitrogen base, caffeine ($C_8H_{10}N_4O_2$).

A strong base such as NaOH or KOH dissociates almost completely in aqueous solution:

$$NaOH \xrightarrow[H_2O]{\sim 100\%} Na^+(aq) + OH^-(aq)$$

Therefore, K_b values for these bases cannot be calculated accurately. This also means that OH⁻(aq) is the strongest base that can be present in water solution. As a result, it is impossible to compare the relative strengths of bases that are stronger than OH⁻ (such as NH_2^-) in water solution. To make such a comparison, other solvents must be used.

14.7A IONIZATION CONSTANT OF WATER (K_w)

If we examine the electrical conductivity of water, we find that water conducts poorly. This poor conductivity indicates the presence of a small amount of ions, which arise from the self-ionization of the water:

$$H_2O(\ell) + H_2O(\ell) \rightleftharpoons H_3O^+(aq) + OH^-(aq)$$

In writing an equilibrium constant for this reaction, we recall that the concentration of water is a constant and can be absorbed into K_{eq}. The ionization constant for water, K_w, can be expressed as follows:

$$K_w = [H_3O^+] \cdot [OH^-]$$

At 298 K, the value of K_w is 1.0×10^{-14}. In pure water only H_3O^+ and OH⁻ ions are produced; therefore, they are produced in equal amounts. We can use this fact to calculate $[H_3O^+]$ and $[OH^-]$ in pure water. If x represents $[H_3O^+]$, it must also represent $[OH^-]$:

$$1 \times 10^{-14} = K_w = [H_3O^+] \cdot [OH^-] = x \cdot x = x^2$$
$$1 \times 10^{-14} = x^2$$
$$x = 1 \times 10^{-7} \text{ M} = [H_3O^+] = [OH^-]$$

In pure water at 298 K, therefore, the concentrations of hydronium and hydroxide ions are each 1×10^{-7} M.

Suppose another substance is dissolved in the water. The fact that K_w is a constant tells us that the product of the hydronium and hydroxide ion concentrations in the water will *always* be 1.0×10^{-14} at 298 K.

If the substance is an acid, it will increase the hydronium ion concentration in the water. Therefore, the hydroxide ion concentration must *decrease* if the product is to remain at 1.0×10^{-14}. Similarly, if the substance is a base, it will increase the hydroxide ion concentration in the water. It follows that the hydronium ion concentration will *decrease*.

PROBLEM
What are $[H_3O^+]$ and $[OH^-]$ in 0.10 M HCl, a strong acid?

SOLUTION

A strong acid such as HCl ionizes completely in water:

$$HCl(aq) + H_2O(\ell) \xrightarrow{\sim 100\%} H_3O^+(aq) + Cl^-(aq)$$

At equilibrium all of the HCl is converted to $H_3O^+(aq)$. It follows that

$$[H_3O^+] = 0.10 \text{ M} = 1.0 \times 10^{-1} \text{ M}$$

Since

$$K_w = [H_3O^+] \cdot [OH^-] = 1.0 \times 10^{-14}$$

we have

$$(1.0 \times 10^{-1} \text{ M}) \cdot [OH^-] = 1.0 \times 10^{-14}$$
$$[OH^-] = 1.0 \times 10^{-13} \text{ M}$$

TRY IT YOURSELF

What are $[H_3O^+]$ and $[OH^-]$ in 0.010 M KOH, a strong base?

ANSWER

$$[OH^-] = 0.010 \text{ M} = 1.0 \times 10^{-2} \text{ M}$$
$$[H_3O^+] = 1.0 \times 10^{-12} \text{ M}$$

Let us summarize what we mean when we say that an aqueous solution is acidic, basic, or neutral (at 298 K):

Acidic: $[H_3O^+] > [OH^-]$; $[H_3O^+] > 1 \times 10^{-7}$ M and $[OH^-] < 1 \times 10^{-7}$ M

Basic: $[H_3O^+] < [OH^-]$; $[H_3O^+] < 1 \times 10^{-7}$ M and $[OH^-] > 1 \times 10^{-7}$ M

Neutral: $[H_3O^+] = [OH^-]$; $[H_3O^+] = 1 \times 10^{-7}$ M and $[OH^-] = 1 \times 10^{-7}$ M

In *every* case: $[H_3O^+] \cdot [OH^-] = K_w = 1 \times 10^{-14}$.

14.8A   A MORE DETAILED LOOK AT pH AND pOH

Now let us construct a table that actually relates the pH of an aqueous solution to $[H_3O^+]$ and $[OH^-]$ at 298 K.

pH (pOH) and $[H_3O^+]$ ($[OH^-]$) in an Aqueous Solution

pH	$[H_3O^+]$	$[OH^-]$	pOH
0	1×10^0 M	1×10^{-14} M	14
1	1×10^{-1} M	1×10^{-13} M	13
2	1×10^{-2} M	1×10^{-12} M	12
3	1×10^{-3} M	1×10^{-11} M	11
4	1×10^{-4} M	1×10^{-10} M	10
5	1×10^{-5} M	1×10^{-9} M	9
6	1×10^{-6} M	1×10^{-8} M	8
7	1×10^{-7} M	1×10^{-7} M	7
8	1×10^{-8} M	1×10^{-6} M	6
9	1×10^{-9} M	1×10^{-5} M	5
10	1×10^{-10} M	1×10^{-4} M	4
11	1×10^{-11} M	1×10^{-3} M	3
12	1×10^{-12} M	1×10^{-2} M	2
13	1×10^{-13} M	1×10^{-1} M	1
14	1×10^{-14} M	1×10^0 M	0

We note that in every case

$$[H_3O^+] \cdot [OH^-] = 1 \times 10^{-14} = K_w$$

We also note that we really don't need to know both $[H_3O^+]$ and $[OH^-]$ in order to determine how acidic or basic a solution is: either one will do. In practice, we use $[H_3O^+]$. To simplify matters even further, we can use the exponent of the hydronium ion concentration, which we call **pH** (the term stands for "power of the hydronium ion"). In our table pH values are given in the left column.

We can also use the exponent of $[OH^-]$ to develop a **pOH** scale that runs from 14 to 0, as shown in the table. A pOH less than 7 indicates a basic solution; the lower the pOH below 7, the more basic the solution. A pOH greater than 7 indicates an acidic solution; the higher the pOH above 7, the more acidic the solution. A pOH of 7 indicates a neutral solution. Since the pOH scale is based on the exponents of $[OH^-]$, a change of one pOH unit equals a tenfold change in $[OH^-]$. In our table pOH values are given in the fourth column.

Just as there is a relationship between $[H_3O^+]$ and $[OH^-]$, there is also a relationship between pH and pOH. If we inspect our table, we see that

$$\boxed{\text{pH} + \text{pOH} = 14}$$

PROBLEM

What is the pH of a solution whose $[OH^-]$ is 1×10^{-5} M? (Don't use the table!)

419

SOLUTION

We must find $[H_3O^+]$ in order to solve the problem.

$$[H_3O^+] \cdot (1.0 \times 10^{-5} \text{ M}) = 1.0 \times 10^{-14}$$
$$[H_3O^+] = 1.0 \times 10^{-9} \text{ M}$$
$$pH = 9$$

TRY IT YOURSELF

What is the pOH of the solution described above?

ANSWER

5

General Definitions of pH and pOH

Since the table on page 419 lists only concentrations of the form 1.0×10^{-x} M, we cannot use it to calculate the pH or pOH of *every* aqueous solution. For example, suppose $[H_3O^+]$ in an aqueous solution is 2.0×10^{-3} M. How can we calculate the pH of this solution?

The general definitions of pH and pOH, which can be used for *any* aqueous solution, are as follows:

$$pH = -\log_{10} [H_3O^+]$$
$$pOH = -\log_{10} [OH^-]$$

In our example, if $[H_3O^+]$ in an aqueous solution is 2.0×10^{-3} M, then

$$pH = -\log_{10} (2.0 \times 10^{-3}) = 2.7$$

PROBLEM

Calculate the pOH and pH of a 0.0003 M aqueous solution of KOH, a strong base.

SOLUTION

Since KOH is a strong base, it dissociates nearly 100% in aqueous solution and $[OH^-] = 0.0003$ M.

$$pOH = -\log_{10} [OH^-] = -\log_{10} (0.0003 \text{ M}) = 3.5$$
$$pH = 14 - pOH = 10.5$$

TRY IT YOURSELF

Calculate the pH and pOH of 0.04 M HCl, a strong acid.

ANSWER
pH = 1.4; pOH = 12.6

14.9A   HYDROLYSIS OF SALTS IN AQUEOUS SOLUTIONS

If we test an aqueous NaCl solution, we find that its pH is 7. Since NaCl is the product of the neutralization of HCl by NaOH, we are not surprised that the solution is neutral.

However, not all salts produce neutral solutions when dissolved in water. For example, the salt NH_4Cl produces an acidic solution; that is, its pH is less than 7. The salt K_2CO_3 produces a basic solution; its pH is greater than 7. To learn why this occurs, we must look at the acids and bases that produce the salt.

If the acid and base are both strong, the neutralization is complete and the salt will produce a neutral solution. We recall that HI, HBr, HCl, HNO_3, and H_2SO_4 are considered strong acids. The Group 1 hydroxides (LiOH, NaOH, KOH, RbOH, and CsOH) are considered strong bases. Any of these combinations (strong acid and strong base) will produce a salt whose aqueous solution is neutral.

Let us analyze why the salt produced by the interaction of a *strong acid and a weak base* produces an *acidic* solution when dissolved in water. The salt NH_4Cl is produced by the reaction of HCl (a strong acid) and NH_3 (a weak base). When this salt is dissolved in water, NH_4^+ and Cl^- ions are present. If we examine the conjugate acid-base pairs table on page 413, we find that:

- The Cl^- ion is a *weaker base* than a molecule of H_2O. Therefore, Cl^- ions have no tendency to accept protons from water molecules, and no additional OH^- ions will be produced in the solution.
- The NH_4^+ ion is a *stronger acid* than a molecule of H_2O. Therefore, NH_4^+ ions can donate protons to water molecules, and additional H_3O^+ ions will be produced in the solution.

As a result, the solution contains a slight excess of H_3O^+ ions and is acidic. We call this process, in which NH_4^+ ions interact with H_2O molecules, **hydrolysis**, and we say that NH_4^+ ions *hydrolyze* in water, but Cl^- ions do not.

Using this analysis, we can conclude that *any combination of a strong acid and a weak base will produce a salt whose aqueous solution is acidic.*

We can perform a similar analysis for salts that form from the interaction of a *weak acid and a strong base*. The salt $NaC_2H_3O_2$ (sodium acetate) is formed from the interaction of $HC_2H_3O_2$ (acetic acid, a weak acid) and

NaOH (sodium hydroxide, a strong base). In aqueous solution, Na^+ and $C_2H_3O_2^-$ ions are present. A Na^+ ion will not hydrolyze because it is neither a conjugate acid nor a conjugate base. However, if we examine the table on p. 399, we find that the $C_2H_3O_2^-$ ion is a stronger base than a molecule of H_2O. Therefore, $C_2H_3O_2^-$ ions will hydrolyze in water by accepting protons from water molecules, and additional OH^- ions will be produced in the solution.

As a result, the solution contains an excess of OH^- ions and is basic. Using this analysis, we can conclude that *any combination of a weak acid and a strong base will produce a salt whose aqueous solution is basic*.

PROBLEM

Using the analyses described above, explain why an aqueous solution of KCl (derived from the strong acid HCl and the strong base KOH) will be *neutral*.

SOLUTION

The aqueous solution contains K^+ and Cl^- ions. A K^+ ion does not hydrolyze in water because it is neither a conjugate acid nor a conjugate base. A Cl^- ion does not hydrolyze in water because it is a weaker base than a molecule of H_2O. As a result, no additional H_3O^+ or OH^- ions will be produced and the solution will be neutral.

What will happen if a weak acid reacts with a weak base? We cannot easily predict the outcome; the salt may produce an acidic, basic, or neutral solution in water. To answer the question, we would need to know more about the relative strengths of the acid and base that react. This is the stuff of more advanced chemistry courses!

PROBLEM

For each of the following salts, identify its parent acid and base. Then indicate the kind of aqueous solution—acidic, basic, or neutral—that the salt ought to produce.
(a) LiF   (b) RbI   (c) $Al(NO_3)_3$ (d) $MgCl_2$

SOLUTIONS

	Parent Acid	Parent Base	Result
(a)	HF (weak)	LiOH (strong)	Basic solution
(b)	HI (strong)	RbOH (strong)	Neutral solution
(c)	HNO_3 (strong)	$Al(OH)_3$ (weak)	Acidic solution
(d)	HCl (strong)	$Mg(OH)_2$ (weak)	Acidic solution

TRY IT YOURSELF

For each salt, tell whether it will dissolve in water to produce an acidic, basic, or neutral solution.
(a) NaBr   (b) $CuSO_4$ (c) K_3PO_4

ANSWERS
(a) NaBr(aq) is neutral. (b) $CuSO_4$(aq) is acidic. (c) K_3PO_4(aq) is basic.

14.10A ACID–BASE PROPERTIES OF OXIDES

Certain oxides can combine *directly* with water to produce an acid or a base.
The metallic oxides of Groups 1 and 2 form bases:

$$K_2O(s) + H_2O\ (\ell) \rightarrow 2KOH(aq)$$
$$CaO(s) + H_2O\ (\ell) \rightarrow Ca(OH)_2(aq)$$

Such substances are called *basic oxides*.
The nonmetallic oxides of Groups 14, 15, 16, and 17 form acids:

$$CO_2(g) + H_2O\ (\ell) \rightarrow H_2CO_3(aq)$$
$$P_2O_5(g) + 3H_2O\ (\ell) \rightarrow 2H_3PO_4(aq)$$
$$SO_3(g) + H_2O\ (\ell) \rightarrow H_2SO_4(aq)$$
$$Cl_2O_7(g) + H_2O\ (\ell) \rightarrow 2HClO_4(aq)$$

Such substances are called *acidic oxides*. Other oxides may be acidic, basic, or even amphoteric.

14.11A LEWIS DEFINITIONS OF ACIDS AND BASES

The Brønsted–Lowry definition of acid–base behavior, which extended the Arrhenius definition, was in turn extended by the Lewis definition of acid–base behavior (formulated by G.N. Lewis in the early 1920s). The Brønsted–Lowry theory classifies acids and bases in terms of how they interact with protons; the Lewis theory, in terms of how they interact with electron pairs. The Lewis definitions are as follows:

Lewis acid: *Any species that can accept a pair of electrons from another species, that is, an* electron-pair acceptor.

Lewis base: *Any species that can donate a pair of electrons to another species, that is, an* electron-pair donor.

In the Lewis system, H^+ is an acid because it can accept a pair of electrons from a Lewis base such as OH^-, which can donate the pair of electrons, as shown on the next page.

$$H^+ \quad + \quad :\!\overset{\cdot\cdot}{\underset{\cdot\cdot}{O}}\!-\!H^- \quad \longrightarrow \quad :\!\overset{\overset{\textstyle H}{\mid}}{\underset{\cdot\cdot}{O}}\!-\!H$$

Lewis Acid	Lewis Base	Lewis Acid–Base
(electron-pair acceptor)	(electron-pair donor)	Product

We might be tempted to ask what all the shouting is about since H^+ and OH^- are very nicely defined in terms of the Brønsted–Lowry model. A second example, however, will make the distinction more apparent.

In Chapter 9 we discovered that the compound BF_3 was stable even though boron contains only six electrons in its outer principal energy level. When BF_3 combines with NH_3, boron forms a coordinate covalent bond with nitrogen. In terms of the Lewis definitions, BF_3 is a Lewis acid and NH_3 is a Lewis base, as shown below.

$$\overset{\textstyle F}{\underset{\textstyle F}{F\!-\!B}} \quad + \quad \overset{\textstyle H}{\underset{\textstyle H}{:\!N\!-\!H}} \quad \longrightarrow \quad \overset{\textstyle F\ \ H}{\underset{\textstyle F\ \ H}{F\!-\!B\!:\!N\!-\!H}}$$

Lewis Acid	Lewis Base	Lewis Acid–Base
(electron-pair acceptor)	(electron-pair donor)	Product

What is remarkable is that BF_3 cannot be considered a Brønsted–Lowry acid because it has no proton to donate! The Lewis definitions of acids and bases play important roles in understanding reactions that occur by means of shifting electron-pairs. This statement is particularly true in organic chemistry.

END-OF-CHAPTER QUESTIONS

Some questions have the symbol "§2" in front of the question number. This symbol means that the question is based on Section II material.

1. Which substance is classified as a salt?
   (1) $Ca(OH)_2$
   (2) $C_2H_4(OH)_2$
   (3) CCl_4
   (4) $CaCl_2$

2. Water containing dissolved electrolyte conducts electricity because the solution contains mobile
   (1) electrons   (2) molecules   (3) atoms   (4) ions

3. What is the name for the sodium salt of the acid $HClO_2$? (*Hint:* See the table on page 402.)
   (1) sodium chlorite   (2) sodium chloride
   (3) sodium chlorate   (4) sodium perchlorate

4. According to the Arrhenius theory, when an acid substance is dissolved in water it will produce a solution containing only one kind of positive ion. To which ion does the theory refer?
   (1) acetate   (2) hydrogen   (3) chloride   (4) sodium

5. Which equation represents a neutralization reaction?
   (1) $H^+(aq) + OH^-(aq) \rightarrow H_2O\ (\ell)$
   (2) $Ag^+(aq) + I^-(aq) \rightarrow AgI(s)$
   (3) $Zn(s) + 2HCl(aq) \rightarrow ZnCl_2(aq) + H_2(g)$
   (4) $NaCl(aq) + AgNO_3(aq) \rightarrow NaNO_3(aq) + AgCl(s)$

6. When substance X is dissolved in water, the only positive ions in the solution are hydronium ions. Substance X could be
   (1) NaOH   (2) NaH   (3) HNO_3 (4) NH_3

7. Which substance can act as an Arrhenius acid in aqueous solution?
   (1) NaI   (2) HI   (3) LiH   (4) NH_3

8. Which species is classified as an Arrhenius base?
   (1) CH_3OH (2) LiOH   (3) NH_3 (4) BF_3

9. A water solution contains 0.50 mole of HCl. How much KOH should be added to the HCl solution to exactly neutralize it?
   (1) 1.0 molc   (2) 2.0 moles   (3) 0.25 mole   (4) 0.50 mole

10. A 30-milliliter sample of HCl is completely neutralized by 10. milliliters of a 1.5 M NaOH solution. What is the molarity of the HCl solution?
    (1) 0.25   (2) 0.50   (3) 1.5   (4) 4.5

11. How much water is formed when 1.0 mole of HCl reacts completely with 1.0 mole of NaOH?
    (1) 1.0 mole   (2) 2.0 moles   (3) 0.50 mole   (4) 0.25 mole

12. What volume of a 0.200 M NaOH solution is needed to exactly neutralize 40.0 milliliters of a 0.100 M H_2SO_4 solution?
    (1) 10.0 mL   (2) 20.0 mL   (3) 40.0 mL   (4) 80.0 mL

§2 **13.** Which ion is the conjugate base of H_2SO_4?
(1) SO_3^{2-} (2) S^{2-} (3) HSO_3^- (4) HSO_4^-

§2 **14.** According to the table on page 413, ammonia can act as
(1) a Brønsted acid, only
(2) a Brønsted base, only
(3) either a Brønsted acid or a Brønsted base
(4) neither a Brønsted acid nor a Brønsted base

§2 **15.** Which of the following Brønsted acids has the strongest conjugate base?
(1) HF   (2) HCl   (3) HBr   (4) HI

§2 **16.** Which of the following is the *weakest* Brønsted acid?
(1) HBr   (2) H_2S (3) H_2O (4) NH_3

17. According to the Brønsted–Lowry theory, an acid is any species that can
(1) donate a proton   (2) accept a proton
(3) donate an electron   (4) accept an electron

18. In the reaction

$$HSO_4^- + H_2O \rightarrow H_3O^+ + SO_4^{2-}$$

the HSO_4^- ion is
(1) a proton donor   (2) an amphiprotic substance
(3) a proton acceptor   (4) a base

§2 **19.** According to the Brønsted–Lowry theory, neutralization occurs in aqueous solution when there is a reaction between
(1) H_3O^+ and H_2O (2) OH^- and H_2O
(3) H_3O^+ and OH^- (4) H_2O and H_2O

20. In the reaction

$$NH_2^- + HOH \rightleftharpoons NH_3 + OH^-$$

the two Brønsted–Lowry bases are
(1) OH^- and NH_2^- (2) HOH and NH_3
(3) OH^- and HOH   (4) OH^- and NH_3

§2 **21.** The ionization constants (K_a values) of four acids are shown below. Which K_a represents the *weakest* of these acids?
(1) $K_a = 1.0 \times 10^{-5}$ (2) $K_a = 1.0 \times 10^{-4}$
(3) $K_a = 7.1 \times 10^{-3}$ (4) $K_a = 1.7 \times 10^{-2}$

22. According to the table on page 413, which 0.1 M acid solution contains the highest concentration of H_3O^+ ions?
(1) H_2S (2) H_3PO_4 (3) HNO_2 (4) HF

23. The diagram below illustrates an apparatus used to test the conductivities of various solutions.

When the switch is closed, which of the following 1–molar solutions will cause the bulb to glow most brightly?
(1) ammonia (2) acetic acid
(3) carbonic acid (4) sulfuric acid

§2 **24.** A solution of hydrochloric acid contains
(1) fewer hydronium ions than chloride ions
(2) fewer hydroxide ions than hydronium ions
(3) more hydronium ions than chloride ions
(4) more hydroxide ions than hydronium ions

§2 **25.** As the H_3O^+ ion concentration of a solution increases at 298 K, K_w
(1) decreases (2) increases (3) remains the same

§2 **26.** The OH^- ion concentration is greater than the H_3O^+ ion concentration in a water solution of
(1) CH_3CH_2OH (2) $Ba(OH)_2$
(3) HCl (4) H_2SO_4

§2 **27.** The K_w value for a sample of water at 1 atmosphere and 298 K will be most likely to change when there is an increase in the
(1) concentration of H^+ ions
(2) concentration of OH^- ions
(3) pressure
(4) temperature

§2 **28.** Given:

$$K_w = [H^+][OH^-] = 1 \times 10^{-14} \text{ at 298 K.}$$

What is the concentration of H^+ in pure water at 298 K?
(1) 1×10^{-7} mol/L   (2) 1×10^{7} mol/L
(3) 1×10^{-14} mol/L   (4) 1×10^{14} mol/L

§2 **29.** Which statement best describes a solution with a pH of 3?
(1) It has an H_3O^+ ion concentration of 1×10^3 mol/L and is acidic.
(2) It has an H_3O^+ ion concentration of 1×10^{-3} mol/L and is acidic.
(3) It has an H_3O^+ ion concentration of 1×10^3 mol/L and is basic.
(4) It has an H_3O^+ ion concentration of 1×10^{-3} mol/L and is basic.

§2 **30.** What is the OH^- ion concentration of an aqueous solution that has a pH of 11?
(1) 1.0×10^{-11} mol/L   (2) 1.0×10^{-3} mol/L
(3) 3.0×10^{-1} mol/L   (4) 11×10^{-1} mol/L

§2 **31.** What is the pOH of a solution whose OH^- ion concentration is 1×10^{-3} M?
(1) 14   (2) 11   (3) 3   (4) 7

§2 **32.** The H_3O^+ ion concentration of an aqueous solution is 1×10^{-4} mole per liter. This solution has a pH of
(1) 4 and a pOH of 4   (2) 4 and a pOH of 10
(3) 10 and a pOH of 4   (4) 10 and a pOH of 10

§2 **33.** An aqueous solution has a hydronium ion concentration of 6.0×10^{-9} mole per liter at 25°C. The pH of this solution is closest to
(1) 5.8   (2) 6.0   (3) 8.2   (4) 9.0

34. If an aqueous solution turns phenolphthalein pink, the solution will turn litmus
(1) red   (2) purple   (3) blue   (4) violet

35. An indicator was used to test a water solution with a pH of 12. Which indicator color would be observed?
(1) colorless with litmus
(2) red with litmus
(3) colorless with phenolphthalein
(4) pink with phenolphthalein

36. According to Reference Table M, at what pH would the indicators bromthymol blue, thymol blue, and methyl orange all appear as yellow?
(1) 1.9   (2) 2.9   (3) 4.7   (4) 9.8

§2 **37.** When K_2CO_3 is dissolved in water, the resulting solution turns litmus paper
(1) red, and is acidic   (2) blue, and is acidic
(3) red, and is basic   (4) blue, and is basic

§2 **38.** Which salt will hydrolyze in water to produce a basic solution?
(1) BaI_2 (2) $NaNO_2$ (3) $CaCl_2$ (4) K_2SO_4

§2 **39.** Oxides of nonmetals react with water to form
(1) acids   (2) bases   (3) salts   (4) anhydrides

§2 **40.** Which of the following is the best classification for the compound BF_3 in this reaction: $BF_3 + NH_3 \rightarrow BF_3NH_3$?
(1) Lewis acid   (2) Lewis base
(3) Arrhenius acid   (4) Arrhenius base

41. Which chemical equation represents the reaction of an Arrhenius acid with an Arrhenius base?
(1) $HC_2H_3O_2 + NaOH(aq) \rightarrow NaC_2H_3O_2(aq) + H_2O(\ell)$
(2) $C_3H_8(g) + 5\ O_2(g) \rightarrow 3\ CO_2(g) + 4\ H_2O(\ell)$
(3) $Zn(s) + 2\ HCl(aq) \rightarrow ZnCl_2(aq) + H_2(g)$
(4) $BaCl_2(aq) + Na_2SO_4(aq) \rightarrow BaSO_4(s) + 2\ NaCl(aq)$

42. When the pH of a solution changes from a pH of 5 to a pH of 3, the hydronium ion concentration is
(1) 0.01 of the original content
(2) 0.1 of the original content
(3) 10 times the original content
(4) 100 times the original content

43. A sample of $Ca(OH)_2$ is considered to be an Arrhenius base because it dissolves in water to produce
(1) Ca^{2+} ions as the only positive ion in solution
(2) H_3O^+ ions as the only positive ions in solution
(3) OH^- ions as the only negative ions in solution
(4) H^- ions as the only negative ions in solution

44. Which reaction occurs when hydrogen ions react with hydroxide ions to form water?
(1) substitution   (2) saponification
(3) ionixation   (4) neutralization

45. Which of these 1 M solutions has the highest pH?
(1) CH_3OH (2) NaOH   (3) HCl   (4) NaCl

46. Phenolphthalein is pink in an aqueous solution having a pH of
(1) 5   (2) 2   (3) 7   (4) 12

47. The pH of a solution is 7. When acid is added to the solution, the hydronium ion concentration becomes 100 times greater. What is the pH of the new solution?
(1) 1   (2) 5   (3) 9   (4) 14

Constructed-Response Questions

§2   **1.** The substance methylamine (CH_3NH_2) can act as a Brønsted–Lowry base in water.
    (a) Write the reaction between methylamine and water. [*Hint*: Note the structural similarity of methylamine to ammonia (NH_3).]
    (b) Write the K_b expression for methylamine.

§2   **2.** (a) Calculate the pH of a solution in which $[OH^-] = 2.00 \times 10^{-4}$ M.
    (b) Calculate the pOH of a solution in which $[H_3O^+] = 5.00 \times 10^{-3}$ M.

Base your answers to questions 3 through 6 on the information and data table below.

A titration setup was used to determine the unknown molar concentration of a solution of NaOH. A 1.2 M HCl solution was used as the titration standard. The following data were collected.

	Trial 1	Trial 2	Trial 3	Trial 4
Amount of HCl Standard Used	10.0 mL	10.0 mL	10.0 mL	10.0 mL
Initial NaOH Buret Reading	0.0 mL	12.2 mL	23.2 mL	35.2 mL
Final NaOH Buret Reading	12.2 mL	23.2 mL	35.2 mL	47.7 mL

3. Calculate the volume of NaOH solution used to neutralize 10.0 mL of the HCl standard solution in trial 3.

4. According to Reference Table M in Appendix 1, which indicator would be most appropriate in determining the end point of this titration? Give one reason for choosing this indicator.

5. Calculate the average molarity of the unknown NaOH solution for all four trials.

6. Explain why it is better to use the average data from multiple trials rather than the data from a single trial to calculate the results of the titration.

Base your answers to questions 7–10 on the information below and on your knowledge of chemistry.

A NaOH(aq) solution and an acid–base indicator are used to determine the molarity of an HCl(aq) solution. A 25.0-milliliter sample of HCl(aq) is exactly neutralized by 15.0 milliliters of 0.20 M NaOH(aq).

7. Identify the laboratory process mentioned in this passage.

8. Write the complete equation for the neutralization reaction that occurs.

9. Based on the data given, the calculated molarity of the HCl(aq) solution should be expressed to what number of significant figures?

10. Using the data, determine the concentration of the HCl(aq).

Base your answers to questions 11–13 on the information below and on your knowledge of chemistry.

A company produces a colorless vinegar that is 5.0% $HC_2H_3O_2$ in water. Using thymol blue as an indicator, a student titrates a 15.0-milliliter sample of the vinegar with 43.1 milliliters of 0.30 M NaOH(aq) solution until the acid is neutralized.

11. Based on Reference Table M in Appendix 1, what is the color of the indicator in the vinegar solution before any base is added?

12. Identify the negative ion in the NaOH(aq) used in this titration.

13. Using the titration data, determine the molarity of the $HC_2H_3O_2$ in the vinegar sample.

The answers to these questions are found in Appendix 3.

Chapter
Fifteen

REDUCTION-OXIDATION (REDOX) AND ELECTROCHEMISTRY

KEY IDEAS

This chapter focuses on the roles of oxidation and reduction in chemistry. In particular, the production of electrical energy by electrochemical cells is examined, and a number of important commercial applications are presented.

KEY OBJECTIVES
At the conclusion of this chapter you will be able to:
- Formally define the terms *oxidation* and *reduction*.
- Define the term *redox reaction*, and identify these reactions.
- Write oxidation and reduction half-reactions for redox equations.
- Balance redox equations by the half-reaction and ion-electron methods.
- Predict when a redox reaction will occur spontaneously.
- Draw and label a simple voltaic cell with a porous barrier and with a salt bridge.
- Describe electron and ion movements in a simple voltaic cell.
- Define the terms *standard electrode potential, standard reduction potential*, and *standard oxidation potential*.
- Use standard potentials to calculate the potential difference of a voltaic cell operating under standard conditions.
- Define *electrolysis*, and draw and label a simple electrolytic cell.
- Describe the electrolysis of fused salts, water, and brine.
- Describe the principles of electroplating.
- Apply redox and electrochemistry to real-world applications.

SECTION I—BASIC (REGENTS-LEVEL) MATERIAL

NYS REGENTS CONCEPTS AND SKILLS

Note: By the time you have finished Section I, you should have mastered the concepts and skills listed on the following pages. The Regents chemistry examination will test your knowledge of these items and your ability to apply them.

Concepts are the *basic ideas* that form the body of the Regents chemistry course (what you need to know!).

Skills are the *activities* that demonstrate your mastery of these concepts (how you show that you know them!).

Following each concept or skill is a page reference (given in parentheses) to this chapter.

15.1 Concept:
An oxidation-reduction (redox) reaction involves the transfer of electrons (e^-). (Page 437)

Skill:
Determine a missing product in a balanced redox equation. (Pages 437–438)

15.2 Concept:
Reduction is the gain of electrons. (Page 436)

15.3 Concept:
A half-reaction can be written to represent reduction. (Page 436)

Skill:
Write and balance reduction half-reactions for free elements and their monatomic ions. (Page 436)

15.4 Concept:
Oxidation is the loss of electrons. (Page 436)

15.5 Concept:
A half reaction can be written to represent oxidation. (Page 436)

Skill:
Write and balance oxidation half-reactions for free elements and their monatomic ions. (Page 436)

15.6 Concept:
In a redox reaction the number of electrons lost is equal to the number of electrons gained. (Pages 437–440)

15.7 Concepts:
- Oxidation numbers (states) can be assigned to atoms and ions. (Page 436)
- Changes in oxidation numbers indicate that oxidation or reduction has occurred. (Page 437)

15.8 Concepts:
- An electrochemical cell can be either voltaic or electrolytic. (Pages 441–444)
- In an electrochemical cell, oxidation occurs at the anode and reduction at the cathode. (Page 442)

Skill:
Compare and contrast voltaic and electrolytic cells.
(Pages 442, 443–444)

15.9   Concept:
A voltaic cell spontaneously converts chemical energy to electrical energy. (Pages 441–442)

Skills:
• Given a redox equation, identify the parts of a voltaic cell:
    (a) anode (Pages 441–442);
    (b) cathode (Pages 441–442);
    (c) salt bridge (Pages 441–444).
• Identify the direction of electron flow in a voltaic cell.
  (Pages 441–442)
• Use an activity series to determine whether a redox reaction is spontaneous. (Pages 438–440)

15.10   Concept:
An electrolytic cell requires electrical energy to produce chemical change. This process is known as electrolysis. (Pages 438–444)

Skills:
• Given a redox equation, identify the parts of an electrolytic cell:
    (a) anode (Page 444);
    (b) cathode (Page 444).
• Identify the direction of electron flow in an electrolytic cell.
  (Page 444)

15.1   WHAT ARE OXIDATION AND REDUCTION?

Most of us are already acquainted with the term *oxidation*. Historically, *oxidation* means the combination of a substance with oxygen, as in the following example:

$$2Mg + O_2 \rightarrow 2MgO$$

Oxygen is called the *oxidizing agent*, and magnesium is the substance that is oxidized.

Historically, *reduction* is associated with the removal of oxygen from a substance, as in the following example:

$$CuO + H_2 \rightarrow Cu + H_2O$$

Hydrogen is called the *reducing agent*, and copper(II) oxide is the substance that is reduced.

15.2 FORMAL DEFINITIONS OF OXIDATION AND REDUCTION

At this point, you should review the section on assigning oxidation numbers (Chapter 3, pages 62–63). Remember that charges are written so that the sign is placed *after* the number, as in 2+, and oxidation numbers are written so that the sign of the number is placed *before* the number itself, as in +2. Oxidation and reduction arise out of the competition for electrons in a chemical reaction.

Let us examine the oxidation of magnesium by oxygen, this time including the oxidation number for each atom in the equation:

$$2Mg^0 + O^0_2 \rightarrow 2Mg^{2+}O^{2-}$$

We see that magnesium has increased its oxidation number from 0 to 2+. We know from Chapter 9 on chemical bonding that charges arise from the loss or gain of electrons. We can conclude that each magnesium atom lost two electrons when it became oxidized, and we can write this oxidation as a half-reaction:

$$Mg^0 \rightarrow Mg^{2+} + 2e^- \quad \text{(oxidation half-reaction)}$$

The formal definition of **oxidation** is as follows:

> **Oxidation:** *a loss of electrons*

Now let us examine the reduction of copper(II) oxide by hydrogen and also include the oxidation numbers for each element in the equation:

$$Cu^{2+}O^{2-} + H^0_2 \rightarrow Cu^0 + H^{+1}_2O^{-2}$$

We note that the reduction of copper(II) oxide does not involve the oxygen because the oxidation number of oxygen remains unchanged. The oxidation number of copper, however, has changed: it has been decreased from +2 to 0. We conclude that the copper gained two electrons when it was reduced, and we can also write this reduction as a half-reaction:

$$Cu^{2+} + 2e^- \rightarrow Cu^0 \quad \text{(reduction half-reaction)}$$

The formal definition of **reduction** is as follows:

> **Reduction:** *a gain of electrons*

As a result of these formal definitions, we have expanded what we mean by oxidation and reduction. For example, the presence of oxygen is not even required. Any atom that loses one or more electrons is oxidized, and any atom that gains electrons is reduced.

Not only are oxidation and reduction opposite processes, but also one cannot occur without the other! If an atom loses electrons, there must be another atom that will gain them. We call all reactions involving oxidation and reduction **redox reactions**.

In Section 15.1, in our example involving the oxidation of magnesium, we called oxygen the *oxidizing agent*. We see that its oxidation number decreases; therefore, it has been reduced. This is a general rule: *In a redox reaction, the oxidizing agent is reduced.*

In our example involving copper(II) oxide and hydrogen, we called hydrogen the *reducing agent*. We see that its oxidation number has been increased; therefore, it has been oxidized. This is also a general rule: *In a redox reaction, the reducing agent is oxidized.*

15.3 REDOX EQUATIONS

Is this (unbalanced) equation:

$$NH_3 + O_2 \rightarrow NO + H_2O$$

a redox equation? To answer this question, we rewrite the equation and include the oxidation number for each element. Then, we look for changes that occurred in the oxidation numbers as a result of the reaction: an increase means oxidation, and a decrease means reduction.

$$N^{-3}H^{+1}_3 + O^0_2 \rightarrow N^{+2}O^{2-} + H^+_2O^{-2}$$

In this reaction, nitrogen changes its oxidation number from -3 to $+2$; its oxidation number increases, and therefore nitrogen has been oxidized. Oxygen changes its oxidation number from 0 to -2; its oxidation number decreases, and therefore oxygen has been reduced. Hydrogen, whose oxidation number is not changed, is neither oxidized nor reduced. This is indeed a redox equation. As with all other chemical reactions, mass, energy, and electric charge are *conserved* within redox reactions.

PROBLEM
Indicate whether each of the following equations is a redox equation. If it is, write the half-reactions representing oxidation and reduction.
(a) $HNO_3 + I_2 \rightarrow HIO_3 + NO_2 + H_2O$ (unbalanced)
(b) $HCl + NaOH \rightarrow H_2O + NaCl$

SOLUTIONS
(a) $H^{+1}N^{+5}O_3^{-2} + I^0_2 \rightarrow H^{+1}I^{+5}O_3^{-2} + N^{+4}O_2^{-2} + H^{+1}_2O^{-2}$
N changes its oxidation number from $+5$ to $+4$; it is reduced.
$N^{+5} + e^- \rightarrow N^{+4}$ (reduction half-reaction)
I changes its oxidation number from 0 to $+5$; it is oxidized.

$$I^0 \rightarrow I^{+5} + 5e^- \quad \text{(oxidation half-reaction)}$$

This equation represents a redox equation.

(b) $H^{+1}Cl^{-1} + Na^+O^{-2}H^{+1} \rightarrow H^{+1}{}_2O^{-2} + Na^+Cl^-$

No atom changes its oxidation number.

This equation does *not* represent a redox equation.

TRY IT YOURSELF

Indicate whether each of the following unbalanced equations is a redox equation.

(a) $Cu + HNO_3 \rightarrow Cu(NO_3)_2 + NO_2 + H_2O$

(b) $CH_4 + O_2 \rightarrow CO_2 + H_2O$

ANSWERS

Both (a) and (b) are redox equations.

15.4 SPONTANEOUS REDOX REACTIONS

The accompanying diagram represents a simple laboratory experiment. A strip of silver-gray zinc metal is placed in a beaker containing a (blue) solution of aqueous Cu^{2+} ions, and a strip of red copper metal is inserted into a second beaker containing a (colorless) solution of aqueous Zn^{2+} ions. In the beaker at the left, we will observe that the solution loses its blue color and the zinc strip becomes coated with copper metal. We observe no reaction in the beaker at the right.

The loss of the blue color and the appearance of the metallic deposit indicate that Cu^{2+} ion is being converted to $Cu(s)$ by reduction. The half-reaction that occurs is

$$Cu^{2+}(aq) + 2e^- \rightarrow Cu(s)$$

The two electrons are supplied by the Zn(s), which is oxidized to Zn^{2+}. The half-reaction that occurs is

$$Zn(s) \rightarrow Zn^{2+}(aq) + 2e^-$$

The net equation for this reaction is

$$Zn(s) + Cu^{2+}(aq) \rightarrow Zn^{2+}(aq) + Cu(s)$$

This redox reaction is clearly spontaneous under laboratory conditions; therefore, the reverse reaction—between Cu(s) and Zn^{2+}(aq)—cannot be spontaneous. This is why we see no reaction in the beaker on the right side of the diagram.

Like all redox reactions, this reaction involves a competition for electrons. Here, Zn(s) loses electrons more easily than Cu(s), and Cu^{2+}(aq) gains electrons more easily than Zn^{2+}(aq), as we observe.

How can we predict whether another, similar experiment will yield a spontaneous redox reaction? The *Activity Series* in Reference Table J in Appendix 1 lists a number of *metals* in (descending) order of their abilities to be *oxidized*. The table also lists a number of *nonmetals* in (descending) order of their abilities to be *reduced*. Both series are based on H_2, known as the *hydrogen standard*.

For example, in Reference Table J, manganese (Mn) is listed higher than nickel (Ni). This means that the oxidation half-reaction

$$Mn(s) \rightarrow Mn^{2+}(aq) + 2e^-$$

is more likely to occur than the oxidation half-reaction

$$Ni(s) \rightarrow Ni^{2+}(aq) + 2e^-$$

The spontaneous reaction that will occur between Mn/Mn^{2+} and Ni/Ni^{2+} is

$$Mn(s) + Ni^{2+}(aq) \rightarrow Mn^{2+}(aq) + Ni(s)$$

Notice that the oxidation of Mn to Mn^{2+} has *forced the reduction* of Ni^{2+} to Ni. We can always use Reference Table J and a simple rule to determine how a spontaneous redox reaction between two metals (including H_2) will occur: *The metal that is listed higher in the table will be oxidized, and the metal that is listed lower in the table will be reduced.*

PROBLEM

Use Reference Table J to write the balanced spontaneous redox reaction between the following pairs of metals and their ions:

$$Au(s)/Au^{3+}(aq)$$

$$Sn(s)/Sn^{2+}(aq)$$

SOLUTION

We see that Sn appears higher on Reference Table J than Au. Therefore, Sn(s) will be oxidized and $Au^{3+}(aq)$ will be reduced. The half-reactions are

$$Sn(s) \rightarrow Sn^{2+}(aq) + 2e^-$$

$$Au^{3+}(aq) + 3e^- \rightarrow Au(s)$$

In order to combine these two half-reactions, we must be certain that the number of electrons lost by Sn(s) is equal to the number of electrons gained by $Au^{3+}(aq)$. This is accomplished by multiplying the first half-reaction by 3 and the second half-reaction by 2. As a result, 6 electrons will be transferred during the reaction. The final *balanced* redox equation is

$$3Sn(s) + 2Au^{3+}(aq) \rightarrow 3Sn^{2+}(aq) + 2Au(s)$$

TRY IT YOURSELF

Use Reference Table J to write the balanced spontaneous redox reaction between the following pairs of metals and their ions:

$$Fe(s)/Fe^{2+}(aq)$$

$$Pb(s)/Pb^{2+}(aq)$$

ANSWER

$$Fe(s) + Pb^{2+}(aq) \rightarrow Fe^{2+}(aq) + Pb(s)$$

Pairs of nonmetals listed in Reference Table J (F_2, Cl_2, Br_2, and I_2) are handled similarly, *except* that the higher nonmetal is *reduced* and lower one *oxidized*.

PROBLEM

Use Reference Table J to write the balanced spontaneous redox reaction between the following pairs of metals and their ions:

$$Br_2(\ell)/Br^-(aq)$$

$$Cl_2(g)/Cl^-(aq)$$

SOLUTION

Since Cl_2 is higher than Br_2, $Cl_2(g)$ will be reduced to $Cl^-(aq)$, and $Br^-(aq)$ will be oxidized to $Br_2(\ell)$ The relevant half reactions are

$$Cl_2(g) + 2e^- \rightarrow 2Cl^-(aq)$$

$$2Br^-(aq) \rightarrow Br_2(\ell) + 2e^-$$

The balanced redox equation is

$$Cl_2(g) + 2Br^-(aq) \rightarrow 2Cl^-(aq) + Br_2(\ell)$$

15.5 ELECTROCHEMICAL CELLS

An **electrochemical cell** is a device that relates electricity to oxidation-reduction reactions. There are two types of electrochemical cells: *voltaic cells* and *electrolytic cells*.

Voltaic Cells

A **voltaic cell** (also known as a *Galvanic cell*) uses a spontaneous redox reaction to provide a source of electrical energy. It is designed so that the oxidation and reduction half-reactions occur in separate half-cells that are connected to one another.

Under standard laboratory conditions, electrochemical cells are constructed to operate at 298 K and 100 kilopascals (nearly 1 atmosphere). All solutions have a 1 M concentration, and any gases present have a partial pressure of 100 kilopascals.

A simple design for a voltaic cell involving the zinc-copper reaction described at the beginning of Section 13.6 is shown in the accompanying diagram.

$Zn(s) \rightarrow Zn^{2+}(aq) + 2e^-$
(oxidation half-reaction)

$Cu^{2+}(aq) + 2e^- \rightarrow Cu(s)$
(reduction half-reaction)

Net Reaction: $Zn(s) + Cu^{2+}(aq) \rightarrow Cu(s) + Zn^{2+}(aq)$

Oxidation occurs in the left half-cell. The electrons that are released travel through the external wire and enter the Cu(s). Reduction occurs in the right half-cell. The metal strips are called *electrodes*. The electrode at which oxidation occurs is the *anode*; the electrode at which reduction occurs, the *cathode*. *The direction of the electron flow in a voltaic cell is always from the anode (where oxidation occurs) to the cathode (where reduction occurs).*

As the reaction progresses, an electrical imbalance in the half-cells will stop the voltaic cell from functioning. However, the *porous barrier* allows the migration of positive and negative ions between the half-cells, keeping them electrically neutral. In the diagram, the directions of the positive and negative ion flows are shown.

If we connect the external wire to an external circuit, we will have a usable source of electrical energy; if it were feasible to use this voltaic cell commercially, we would call it a *battery*! *The anode of a voltaic cell is the negative terminal* because electrons are flowing out of it; *the cathode is the positive terminal* because electrons are flowing into it.

After a period of time, the redox reaction of the cell reaches equilibrium and the cell no longer operates. When this occurs in a battery, we say that the battery is dead!

Voltaic Cells with Salt Bridges

We can also build a voltaic cell whose half-cells are completely separated, but we must provide a device known as a *salt bridge* to allow for the flow of ions between the half-cells. As shown in the accompanying diagram, the salt bridge contains an electrolyte, such as potassium chloride (KCl), that is dispersed throughout a gel, such as agar. The gel provides firmness but allows ions to flow through it.

Since Zn^{2+}(aq) is produced in the left half-cell, negative ions are required to keep the solution electrically neutral. Therefore, Cl^- ions flow into this half-cell. At the same time, Cu^{2+}(aq) is being used up in the right half-cell, which requires positive ions to keep the solution electrically neutral. Therefore, K^+ ions flow into this half-cell.

Electrolytic Cells and Electrolysis

Electrolytic cells are electrochemical cells that are used to force nonspontaneous redox reactions to occur. Unlike voltaic cells, they do not generate electrical energy—they use it. In practice, an electric current is passed through the substance inside the cell. The process is called *electrolysis* (meaning "to break apart using electricity").

Operation of an Electrolytic Cell

The redox reaction

$$2NaCl \rightarrow 2Na + Cl_2$$

is *not spontaneous*.

To make this reaction occur, we must perform the electrolysis on the fused or molten salt, $NaCl(\ell)$. In the liquid phase, NaCl is a "soup" of freely moving Na^+ and Cl^- ions. The accompanying diagram illustrates how an electrolytic cell operates.

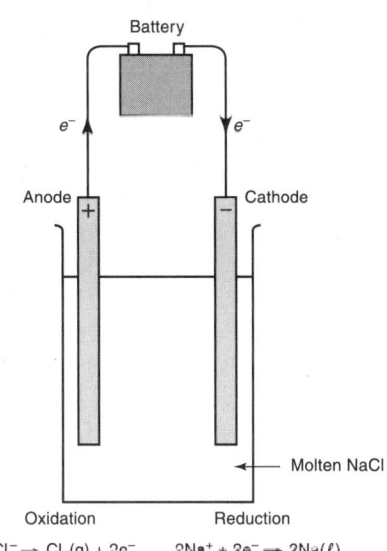

Oxidation · · · · · · · · · · Reduction

$2Cl^- \rightarrow Cl_2(g) + 2e^-$ · · · $2Na^+ + 2e^- \rightarrow 2Na(\ell)$

The source of electricity establishes positive and negative electrodes. Ions of opposite charge migrate to these electrodes, and the half-reaction occurs. Because oxidation occurs at the positive electrode, this is the anode of the electrolytic cell. Similarly, the negative electrode, at which reduction occurs, is the cathode.

SECTION II—ADDITIONAL MATERIAL

15.1A BALANCING REDOX EQUATIONS BY THE HALF-REACTION METHOD

Redox equations may be very difficult to balance simply by inspection. A good case in point is this redox equation:

$$Cu + HNO_3 \rightarrow Cu(NO_3)_2 + NO + H_2O$$

Redox equations can be balanced, however, by identifying the elements that are oxidized and reduced and then writing their half-reactions. The seven steps in this **half-reaction method**, with the equation above used as an example, are as follows:

1. Rewrite the equation with its oxidation numbers.
   $$Cu^0 + H^{1+}N^{5+}O_3^{2-} \rightarrow Cu^{2+}(N^{5+}O_3^{2-})_2 + N^{2+}O^{2-} + H_2^{+1}O^{2-}$$
2. Identify the elements that are oxidized and reduced.
   Cu is oxidized; its oxidation number increases from 0 to +2.
   N is reduced; its oxidation number decreases from +5 to +2.
3. Write the oxidation and reduction half-reactions.
   $Cu^0 \rightarrow Cu^{2+} + 2e^-$ (oxidation half-reaction)
   $N^{5+} + 3e^- \rightarrow N^{2+}$ (reduction half-reaction)
4. Multiply each half-reaction by an appropriate number so that the number of electrons lost by the oxidized element is equal to the number of electrons gained by the reduced element. (This "balancing" step is necessary because *electric charge must be conserved*.)
   $3 \cdot (Cu^0 \rightarrow Cu^{2+} + 2e^-) = 3Cu^0 \rightarrow 3Cu^{2+} + 6e^-$
   $2 \cdot (N^{5+} + 3e^- \rightarrow N^{2+}) = 2N^{5+} + 6e^- \rightarrow 2N^{2+}$
5. Add the two "balanced" half-reactions, eliminating the electrons. We call the result the balanced *skeleton* redox equation.
   $3Cu^0 \rightarrow 3Cu^{2+} + 6e^-$
   $2N^{5+} + 6e^- \rightarrow 2N^{2+}$

   $3Cu^0 + 2N^{5+} \rightarrow 3Cu^{2+} + 2N^{2+}$
6. Insert the coefficients from the skeleton into the original equation by matching each element and its oxidation number. There is one exception: Do *not* insert the coefficient of any item that appears in more than *one* place in the equation.

$$3Cu^0 + \_H^{1+}N^{+5}O_3^{2-} \rightarrow 3Cu^{2+}(H^{+1}N^{+5}O_3^{-2})_2 + 2N^{2+}O^{2-} + \_H_2^{1+}O^{2-}$$

Since N^{+5} appears twice, we do *not* insert the coefficient for it.

$$3Cu + \_HNO_3 \rightarrow 3Cu(NO_3)_2 + 2NO + \_H_2O$$

7. Balance the rest of the equation (the nonredox part) by *inspection*.

$$3Cu + 8HNO_3 \rightarrow 3Cu(NO_3)_2 + 2NO + 4H_2O$$

TRY IT YOURSELF

Balance each of the following redox equations by the half-reaction method:

(a) $Cu + HNO_3 \rightarrow Cu(NO_3)_2 + NO_2 + H_2O$

(b) $HNO_3 + I_2 \rightarrow HIO_3 + NO_2 + H_2O$

ANSWERS

(a) $Cu + 4HNO_3 \rightarrow Cu(NO_3)_2 + 2NO_2 + 2H_2O$

(b) $10HNO_3 + I_2 \rightarrow 2HIO_3 + 10NO_2 + 4H_2O$

15.2A BALANCING REDOX EQUATIONS BY THE ION–ELECTRON METHOD

Another technique, known as the **ion–electron method**, is useful for balancing *ionic* redox equations. The redox equation between copper and nitric acid:

$$Cu + HNO_3 \rightarrow Cu(NO_3)_2 + NO + H_2O$$

can be written as an ionic equation, stripped of all the items that do not actually participate in oxidation-reduction:

$$Cu + NO_3^- \rightarrow Cu^{2+} + NO \quad \text{(acidic solution)}$$

We need to indicate that the solution is *acidic*, because HNO_3 is not written out.

The ion–electron method used to balance this ionic redox equation is simpler than balancing by the half-reaction method because we do not have to assign oxidation numbers for individual elements. We balance for both mass and charge in a step-by-step fashion:

1. Separate the equation into half-reactions by selecting similar items on both sides of the arrow.

$$Cu \rightarrow Cu^{2+}$$
$$NO_3^- \rightarrow NO$$

2. Balance the half-reactions for all atoms that are neither hydrogen nor oxygen. The Cu and N atoms are already balanced.

3. Balance the half-reactions for oxygen by adding one H_2O molecule for each O atom needed. Add the H_2O molecules to the appropriate side of the half-reaction.

$$Cu \rightarrow Cu^{2+}$$
$$NO_3^- \rightarrow NO + 2H_2O$$

4. Balance the half-reactions for hydrogen by adding one H^+ ion for each H atom needed. Add the H^+ ions to the appropriate side of the half-reaction. (Remember: Acid is present.)

$$Cu \rightarrow Cu^{2+}$$
$$4H^+ + NO_3^- \rightarrow NO + 2H_2O$$

5. Balance the half-reactions for electric charge by adding electrons (e^-) to the appropriate side of the half-reaction.

$$Cu \rightarrow Cu^{2+} + 2e^- \quad \text{(oxidation half-reaction)}$$
$$3e^- + 4H^+ + NO_3^- \rightarrow NO + 2H_2O \quad \text{(reduction half-reaction)}$$

6. Multiply the half-reactions in order to conserve electric charge. (The number of electrons lost must equal the number of electrons gained.)

$$3 \cdot (Cu \rightarrow Cu^{2+} + 2e^-) = 3Cu \rightarrow 3Cu^{2+} + 6e^-$$
$$2 \cdot (3e^- + 4H^+ + NO_3^- \rightarrow NO + 2H_2O) =$$
$$6e^- + 8H^+ + 2NO_3^- \rightarrow 2NO + 4H_2O$$

7. Add the half-reactions, but eliminate the electrons. If H_2O and/or H^+ appears on *both* sides of the equation, subtract the smaller amount from each side of the equation.
The resulting equation is now balanced.

$$3Cu + 8H^+ + 2NO_3^- \rightarrow 3Cu^{2+} + 2NO + 4H_2O$$

PROBLEM
Balance the following ionic redox equation by the ion-electron method:

$$NO_3^- + I_2 \rightarrow HIO_3 + NO_2 \quad \text{(acidic solution)}$$

SOLUTION
We will balance the equation step by step according to the method given above.
1. $NO_3^- \rightarrow NO_2$
   $I_2 \rightarrow HIO_3$
2. $NO_3^- \rightarrow NO_2$
   $I_2 \rightarrow 2HIO_3$
3. $NO_3^- \rightarrow NO_2 + H_2O$
   $6H_2O + I_2 \rightarrow 2HIO_3$

446

4. $2H^+ + NO_3^- \rightarrow NO_2 + H_2O$

$6H_2O + I_2 \rightarrow 2HIO_3 + 10H^+$

5. $e^- + 2H^+ + NO_3^- \rightarrow NO_2 + H_2O$

$6H_2O + I_2 \rightarrow 2HIO_3 + 10H^+ + 10e^-$

6. $10 \cdot (e^- + 2H^+ + NO_3^- \rightarrow NO_2 + H_2O)$

$1 \cdot (6H_2O + I_2 \rightarrow 2HIO_3 + 10H^+ + 10e^-)$

7. $20H^+ + 10NO_3^- + 6H_2O + I_2 \rightarrow 2HIO_3 + 10H^+ + 10NO_2 + 10H_2O$

This reduces to

$10H^+ + 10NO_3^- + I_2 \rightarrow 2HIO_3 + 10NO_2 + 4H_2O$

If the reaction should occur in *basic* solution, we balance it as though it were in acid solution. Then, at the last step, we add enough OH^- ions (to both sides of the equation) to remove all of the H^+ present. Each $H^+ - OH^-$ combination creates one H_2O molecule.

PROBLEM

Balance the following ionic redox equation, which occurs in basic solution:

$$NO_2^- + Al \rightarrow AlO_2^- + NH_3$$

SOLUTION

If this reaction were balanced in *acid* solution, it would look like this:

$$2H_2O + NO_2^- + 2Al \rightarrow 2AlO_2^- + NH_3 + H^+$$

To remove one H^+, we add one OH^- to each side of the equation:

$$OH^- + 2H_2O + NO_2^- + 2Al \rightarrow 2AlO_2^- + NH_3 + H^+ + OH^-$$

The equation becomes

$$OH^- + 2H_2O + NO_2^- + 2Al \rightarrow 2AlO_2^- + NH_3 + H_2O$$

which reduces to

$$OH^- + H_2O + NO_2^- + 2Al \rightarrow 2AlO_2^- + NH_3$$

TRY IT YOURSELF

Balance each of the following redox equations by the ion-electron method:

(a) $Fe^{2+} + MnO_4^- \rightarrow Mn^{2+} + Fe^{3+}$ (acid solution)

(b) $IO_3^- + H_2S \rightarrow I_2 + SO_3^{2-}$ (basic solution)

ANSWERS

(a) $5Fe^{2+} + MnO_4^- + 8H^+ \rightarrow Mn^{2+} + 5Fe^{3+} + 4H_2O$

(b) $6IO_3^- + 5H_2S + 4OH^- \rightarrow 3I_2 + 5SO_3^{2-} + 7H_2O$

15.3A HALF-CELL POTENTIALS AND CELL VOLTAGE

Now that we have "constructed" a voltaic cell from zinc and copper, our next question is: How good is it? We measure the energy output of electrical devices in relation to the amount of charge they transfer. This quantity is known as *potential difference* or *voltage*, and its SI unit is the volt (V).

Under standard conditions, the zinc-copper cell would have a maximum voltage of 1.10 volts. As the cell continued to operate, however, its voltage would drop and finally become 0 when the cell reached equilibrium. Table Z, *Standard Electrode Potentials*, in Appendix 2 can be used to calculate the *maximum* standard voltage of a voltaic cell. Before we perform these calculations, however, we need to know more about the numerical values given in Table Z.

For each reduction half-reaction in the table, a **standard reduction potential** (which is one type of *standard electrode potential*) has been assigned. Its symbol is either $\mathscr{E}°$ or $E°$, and it is also measured in volts. (We will use the script form, $\mathscr{E}°$, in this book.) Note that some of the values in Table Z are positive, some are negative, and one is equal to 0. *The more positive a reduction potential is, the more easily the reduction half-reaction associated with it can occur.*

We could just as well have built a similar table based on oxidation half-reactions and assigned another type of standard electrode potential, known as a **standard oxidation potential**, to each half-reaction. How would these two tables compare? Since an oxidation half-reaction is the reverse of a reduction half-reaction, *the standard oxidation potential is simply the standard reduction potential with its sign reversed*. The following example illustrates this principle:

Half-Reaction	$\mathscr{E}°$	
$Cu^{2+} + 2e^- \rightarrow Cu(s)$	+0.34 V	(reduction potential)
$Cu(s) \rightarrow Cu(aq) + 2e^-$	−0.34 V	(oxidation potential)

In this book, we will use standard *reduction* potentials exclusively. The maximum standard cell voltage ($\mathscr{E}°_{cell}$) is calculated by subtracting the standard reduction potential for the half-cell that is reduced from the standard reduction potential for the half-cell that is oxidized:

$$\mathscr{E}°_{cell} = \mathscr{E}°_{reduced} - \mathscr{E}°_{oxidized}$$

$$\mathscr{E}°_{cell} = \mathscr{E}°_{cathode} - \mathscr{E}°_{anode}$$

In Table Z, we find the following listings for the copper-zinc electrochemical cell that we constructed above:

Half-Reaction	$\mathscr{E}°$
$Cu^{2+}(aq) + 2e^- \rightarrow Cu(s)$	$+0.34$ V
$Zn^{2+}(aq) + 2e^- \rightarrow Zn(s)$	-0.76 V

In this case, Cu^{2+} is reduced since its reduction potential is more positive than that of Zn^{2+}. Therefore, we need to reverse the sign of the reduction potential for Zn (in order to convert it into an oxidation potential) and to add the two values:

$$\mathscr{E}°_{cell} = \mathscr{E}°_{Cu^{2+}|Cu} - \mathscr{E}°_{Zn|Zn^{2+}} = (+0.34 \text{ V}) - (-0.76 \text{ V}) = +1.10 \text{ V}$$

Note that we indicate a reduction or oxidation potential by writing a "shorthand" version of the half-reaction as a subscript of $\mathscr{E}°$. For example, the notation "$Cu^{2+}|Cu$" is an abbreviated way of writing the reduction half-reaction $Cu^{2+}(aq) + 2e^- \rightarrow Cu(s)$, and $Zn|Zn^{2+}$ is an abbreviated way of writing the oxidation half-reaction $Zn(s) \rightarrow Zn^{2+}(aq) + 2e^-$.

The positive sign of $\mathscr{E}°_{cell}$ is not simply a result of the arithmetic: it tells us that the cell's redox reaction is spontaneous as written.

PROBLEM
Calculate $\mathscr{E}°_{cell}$ for the following:

Half-Reaction	$\mathscr{E}°$
$Sn^{2+}(aq) + 2e^- \rightarrow Sn(s)$	-0.14 V
$Ag^+(aq) + e^- \rightarrow Ag(s)$	$+0.80$ V

SOLUTION
The half-cell that will be reduced will have the more positive $\mathscr{E}°$. This is the $Ag^+|Ag$ half-cell. Therefore:

$$\mathscr{E}°_{cell} = \mathscr{E}°_{Ag^+|Ag} - \mathscr{E}°_{Sn|Sn^{2+}} = (0.80 \text{ V}) - (-0.14 \text{ V}) = 0.94 \text{ V}$$

The fact that Sn(s) loses two electrons, whereas Ag^+ gains only one electron, is ignored when $\mathscr{E}°_{cell}$ is calculated.

TRY IT YOURSELF
Using the values given in Reference Table Z, calculate $\mathscr{E}°_{cell}$ for an electrochemical cell that has this redox reaction:

$$Fe(s) + Pb^{2+}(aq) \rightarrow Fe^{2+}(aq) + Pb(s)$$

ANSWER
$\mathscr{E}°_{cell} = +0.31$ V

15.4A   THE STANDARD HYDROGEN HALF-CELL

The $\mathcal{E}°$ value assigned to each half-reaction (and its accompanying half-cell) in Reference Table Z is determined by connecting that half-cell to a standard hydrogen half-cell and measuring the voltage of this special electrochemical cell.

The *standard hydrogen half-cell* consists of hydrogen gas (at a partial pressure of 100 kPa) bubbled into a solution of a strong acid, such as HCl, that yields a $[H_3O^+]$ equal to 1 M. This half-cell operates at a temperature of 298 K and has the following reversible half-reactions:

$$2H^+ \text{ (aq, 1 M)} + 2e^- \rightleftharpoons H_2 \text{ (g, 100 kPa)}$$

By convention, $\mathcal{E}°$ for either half-reaction of the standard hydrogen half-cell is set at exactly 0.00 volt. An illustration of the special voltaic cell used to measure $\mathcal{E}°$ for a half-cell follows.

If the half-cell is *reduced* (that is, it causes the standard hydrogen half-cell to be *oxidized*), its $\mathcal{E}°$ is reported as a *positive* value.

If the half-cell is *oxidized* (that is, it causes the standard hydrogen half-cell to be *reduced*), it is reported as a *negative* value.

PROBLEM
A half-cell consisting of Zn(s) and Zn^{2+}(aq, 1 M) is connected to a standard hydrogen half-cell.
(a) Using Reference Table Z, determine each of the half-reactions that occurs in the cell.
(b) Write the overall redox reaction for this electrochemical cell.
(c) Calculate $\mathcal{E}°_{cell}$.

SOLUTION

(a) Since $\mathcal{E}°$ for Zn has a *negative* value, its half-cell will undergo *oxidation* when the cell operates. Therefore, the two half-reactions are

$$Zn(s) \rightarrow Zn^{2+}(aq) + 2e^-$$
$$2H^+(aq) + 2e^- \rightarrow H_2(g)$$

(b) Overall redox reaction: $Zn(s) + 2H^+(aq) \rightarrow Zn^{2+}(aq) + H_2(g)$

(c) $\mathcal{E}°_{cell} = \mathcal{E}°_{H^+|H_2} - \mathcal{E}°_{Zn|Zn^{2+}} = (+0.00 \text{ V}) - (-0.76 \text{ V}) = +0.76 \text{ V}$

TRY IT YOURSELF

Repeat the problem on the previous page for a half-cell consisting of Cu(s) and Cu^{2+}(aq, 1 M).

ANSWERS

(a) $H_2(g) \rightarrow 2H^+(aq) + 2e^-$
    $Cu^{2+}(aq) + 2e^- \rightarrow Cu(s)$

(b) Overall redox reaction: $H_2(g) + Cu^{2+}(aq) \rightarrow 2H^+(aq) + Cu(s)$

(c) $\mathcal{E}°_{cell} = \mathcal{E}°_{Cu^{2+}|Cu} - \mathcal{E}°_{H_2|H^+} = (+0.34 \text{ V}) - (+0.00 \text{ V}) = +0.34 \text{ V}$

15.5A ELECTROLYSIS OF WATER AND AQUEOUS NaCl (BRINE)

Pure water is a very poor conductor of electricity. We can electrolyze water, however, by adding an electrolyte such as H_2SO_4. The reactions that occur at the electrodes are as follows:

Anode (oxidation): $2H_2O(\ell) \rightarrow O_2(g) + 4H^+(aq) + 4e^-$
Cathode (reduction): $4H_2O(\ell) + 4e^- \rightarrow 2H_2(g) + 4OH^-(aq)$

Net reaction: $6H_2O(\ell) \rightarrow 2H_2(g) + O_2(g) + 4H^+(aq) + 4OH^-(aq)$

When the H^+ and OH^- ions come into contact, 4 moles of H_2O are formed $[4H^+(aq) + 4OH^-(aq) \rightarrow 4H_2O(\ell)]$, and the net reaction reduces to

$$2H_2O(\ell) \rightarrow 2H_2(g) + O_2(g)$$

If an aqueous solution of NaCl (known as brine) is electrolyzed, the reactions that occur at the electrodes are as follows:

Anode (oxidation): $2Cl^-(aq) \rightarrow Cl_2(g) + 2e^-$
Cathode (reduction): $2H_2O(\ell) + 2e^- \rightarrow H_2(g) + 2OH^-(aq)$

Net reaction: $2H_2O(\ell) + 2Cl^-(aq) \rightarrow H_2(g) + Cl_2(g) + 2OH^-(aq)$

As we can see, chlorine gas is produced at the positive electrode but sodium metal is *not* produced at the negative electrode; water reacts and produces OH^- ions and H_2 gas instead. The explanation of why brine electrolyzes as it does is quite complicated and lies beyond the scope of this book.

15.6A ELECTROPLATING

Electroplating is a process in which an electric current is used to deposit a layer of metal, such as silver, on the object to be plated. An electroplating cell is a special adaptation of the electrolytic cell. The accompanying diagram illustrates how electroplating is accomplished.

$$Ag^+ + e^- \rightarrow Ag$$
(A layer of silver is deposited on the material to be plated.)

$$Ag \rightarrow Ag^+ + e^-$$

The plating solution is a salt of the metal. A bar of metal serves as the anode (positive electrode) and provides the positive ions needed for plating. The material to be plated acts as the cathode (negative electrode) and receives the positive ions that are reduced to the metallic layer of electroplate.

15.7A ADDITIONAL APPLICATIONS OF REDOX AND ELECTROCHEMISTRY

Reduction of Metals

Most metals do not occur freely in nature; they occur in the oxidized state (i.e., as positive ions). A metal that belongs to Group 1 or 2 is recovered by electrolysis of its fused salt. Other metals are recovered by reduction of their

452

ores; the method used depends on the activity of the metal and the nature of the ore.

In the production of chromium metal, chromium(III) oxide is fairly stable and is treated with aluminum metal, which is a relatively strong reducing agent:

$$2Al + Cr_2O_3 \rightarrow Al_2O_3 + 2Cr$$

Metals such as zinc and iron are extracted by the reduction of their oxides. Carbon, in the form of coke, or carbon monoxide is used as the reducing agent:

$$ZnO + C + heat \rightarrow Zn + CO$$
$$Fe_2O_3 + 3CO + heat \rightarrow 2Fe + 3CO_2$$

Electrolytic Purification of Metals

Electrolytic methods are useful in purifying metals. For example, a cell similar to the electroplating cell shown in Section 15.6A can be used to purify copper from its ores. The anode is made from impure copper, and purified copper (more than 99.5 percent purified) is collected at the cathode. The reactions are as follows:

$$Cathode: Cu^{2+}(aq) + 2e^- \rightarrow Cu(s)$$
$$Anode: Cu(s) \rightarrow Cu^{2+}(aq) + 2e^-$$

The impurities, iron and zinc, are also oxidized at the anode and enter the solution as $Fe^{2+}(aq)$ and $Zn^{2+}(aq)$. Their electrode potentials are too negative, however, to allow their reduction at the cathode. Other impurities, such as silver and gold, remain in their metallic states and are valuable by-products of this process.

Preventing the Corrosion of Metals

Corrosion occurs when a metal is attacked slowly by elements in its environment. In many cases, the metal ceases to be useful. Corrosion is a redox reaction, and agents such as moisture may contribute to the process.

Metals may be protected from corrosion in a variety of ways. Aluminum oxidizes rapidly, but the oxide adheres tightly to the metal and forms a self-protective coating that shields the metal underneath from further oxidation. Zinc also produces a self-protective oxide coating.

When a metal such as iron oxidizes, however, the oxide (rust) flakes off and allows the oxidation of fresh metal. Iron can be protected by plating it with a corrosion-resistant metal such as chromium or nickel or by coating it with paint, oil, or porcelain.

Another protection technique is to coat the iron with zinc—a process known as *galvanizing*. If the coating is broken, exposing fresh iron and zinc, the more active zinc will be oxidized first and will produce a protective oxide coating.

END-OF-CHAPTER QUESTIONS

Some questions have the symbol "§2" in front of the question number. This symbol means that the question is based on Section II material.

1. How many moles of electrons are needed to reduce 1 mole of Cu^{2+} to $Cu(s)$?
   - (1) 1
   - (2) 2
   - (3) 3
   - (4) 4

2. A reducing agent is a substance that
   - (1) gains protons
   - (2) loses protons
   - (3) gains electrons
   - (4) loses electrons

3. Metals located near the top of Reference Table J tend to have
   - (1) high ionization energies and high electronegativities
   - (2) high ionization energies and low electronegativities
   - (3) low ionization energies and high electronegativities
   - (4) low ionization energies and low electronegativities

4. Which half-reaction correctly represents oxidation?
   - (1) $Mg + 2e^- \rightarrow Mg^{2+}$
   - (2) $Mg^{2+} + 2e^- \rightarrow Mg$
   - (3) $Mg^{2+} \rightarrow Mg + 2e^-$
   - (4) $Mg \rightarrow Mg^{2+} + 2e^-$

5. In the reaction

$$Pb + 2Ag^+ \rightarrow Pb^{2+} + 2Ag$$

the Ag^+ is
   - (1) reduced, and the oxidation number changes from +1 to 0
   - (2) reduced, and the oxidation number changes from +2 to 0
   - (3) oxidized, and the oxidation number changes from 0 to +1
   - (4) oxidized, and the oxidation number changes from +1 to 0

6. In the reaction

$$2Al(s) + 3Cu^{2+}(aq) \rightarrow 2Al^{3+}(aq) + 3Cu(s)$$

Al(s)
(1) gains protons   (2) loses protons
(3) gains electrons   (4) loses electrons

7. In the reaction

$$4Zn + 10HNO_3 \rightarrow 4Zn(NO_3)_2 + NH_4NO_3 + 3H_2O$$

the zinc is
(1) reduced, and the oxidation number changes from 0 to +2
(2) oxidized, and the oxidation number changes from 0 to +2
(3) reduced, and the oxidation number changes from +2 to 0
(4) oxidized, and the oxidation number changes from +2 to 0

8. All redox reactions involve
(1) the gain of electrons, only
(2) the loss of electrons, only
(3) both the gain and the loss of electrons
(4) neither the gain nor the loss of electrons

§2 **9.** In the reaction

$$2Fe^{2+} + Cl_2 \rightarrow 2Fe^{3+} + 2\,Cl^-$$

which is the oxidizing agent?
(1) Fe^{2+} (2) Cl_2 (3) Fe^{3+} (4) Cl^-

§2 **10.** In the reaction

$$3Cu + 8HNO_3 \rightarrow 3Cu(NO_3)_2 + 2NO + 4H_2O$$

the reducing agent is
(1) Cu   (2) N^{5+} (3) Cu^{2+} (4) N^{2+}

11. Which is a redox reaction?
(1) $CaCO_3 \rightarrow CaO + CO_2$
(2) $NaOH + HCl \rightarrow NaCl + H_2O$
(3) $2NH_4Cl + Ca(OH)_2 \rightarrow 2NH_3 + 2H_2O + CaCl_2$
(4) $2H_2O \rightarrow 2H_2 + O_2$

12. Given the reaction

$$3Ag + Au^{3+} \rightarrow 3Ag^+ + Au$$

Which equation correctly represents the oxidation half-reaction?
(1) $3Ag + 3e^- \rightarrow 3Ag^+$ (2) $3Ag \rightarrow 3Ag^+ + 3e^-$
(3) $Au^{3+} + 3e^- \rightarrow Au$ (4) $Au^{3+} \rightarrow Au + 3e^+$

13. When the redox equation

$$\_Cr^{3+}(aq) + \_Mn(s) \rightarrow \_Mn^{2+}(aq) + \_Cr(s)$$

is completely balanced, the coefficient of $Cr^{3+}(aq)$ will be
(1) 1   (2) 2   (3) 3   (4) 4

§2 **14.** When the reaction

$$\_Cu(s) + \_HNO_3(aq) \rightarrow \_Cu(NO_3)_2(aq) + \_NO_2(g) + \_H_2O(\ell)$$

is completely balanced using smallest whole numbers, the coefficient of $HNO_3(aq)$ will be
(1) 1   (2) 2   (3) 3   (4) 4

§2 **15.** When the reaction

$$\_K_2Cr_2O_7 + \_HCl \rightarrow \_KCl + \_CrCl_3 + \_Cl_2 + \_H_2O$$

is completely balanced using smallest whole numbers, the coefficient of Cl_2 will be
(1) 1   (2) 2   (3) 3   (4) 4

§2 **16.** The redox reaction below occurs in *acid* solution:

$$\_Cu(s) + \_NO_3^- \rightarrow \_Cu^{2+} + \_NO_2$$

When the equation in correctly balanced, the coefficient of NO_3^- is
(1) 1   (2) 2   (3) 3   (4) 4

17. According to Reference Table J, which will reduce Mg^{2+} to $Mg(s)$?
(1) Fe(s)   (2) Ba(s)   (3) Pb(s)   (4) Ag(s)

18. According to Reference Table J, which of the following nonmetals is most easily reduced?
(1) F_2 (2) Cl_2 (3) Br_2 (4) I_2

19. According to Reference Table J, which element will react spontaneously with Al^{3+}?
(1) Cu   (2) Au   (3) Li   (4) Ni

§2 **20.** According to Reference Table Z in Appendix 2, which ion can be both an oxidizing agent and a reducing agent?
(1) Fe^{2+} (2) Cu^{2+} (3) Al^{3+} (4) Cr^{3+}

21. Which metal will react spontaneously with H_2SO_4?
(1) Au   (2) Ag   (3) Cu   (4) Mg

22. Which will oxidize $Zn(s)$ to Zn^{2+}, but will *not* oxidize $Pb(s)$ to Pb^{2+}?
(1) Al^{3+} (2) Au^{3+} (3) Co^{2+} (4) Mg^{2+}

§2 **23.** Which half-reaction is the arbitrary standard used in the measurement of the standard electrode potentials in Reference Table Z in Appendix 2?
(1) $2H^+ + 2e^- \rightarrow H_2(g)$
(2) $2H_2O + 2e^- \rightarrow 2OH^- + H_2(g)$
(3) $F_2(g) + 2e^- \rightarrow 2F^-$
(4) $Li^+ + e^- \rightarrow Li(s)$

§2 **24.** Given the reaction

$$2Fe^{3+} + 2I^- \rightarrow 2Fe^{2+} + I_2$$

The net potential difference $(\mathcal{E}°)$ for the overall reaction is
(1) 1.00 V   (2) 1.31 V   (3) 2.08 V   (4) 0.23 V

25. In a voltaic cell composed of two half-cells, *ions* are allowed to flow from one half-cell to another by means of
(1) electrodes   (2) an external conductor
(3) a voltmeter   (4) a salt bridge

§2 **26.** What is the voltage of a voltaic cell that has reached equilibrium?
(1) 1   (2) greater than 1
(3) between 0 and 1   (4) 0

Base your answers to questions 27 and 28 on the diagram below, which represents a voltaic cell at 298 K. The equation that accompanies the diagram represents the net cell reaction.

$$2Al(s) + 3Cu^{2+}(aq) \rightarrow 2Al^{3+}(aq) + 3Cu(s)$$

§2 **27.** When switch S is closed, the maximum potential difference ($\mathscr{E}°$) for the cell will be
(1) 1.32 V   (2) 2.00V   (3) −1.32 V   (4) −2.00 V

28. When switch S is closed, electrons in the external circuit will flow from
(1) Al to Al^{3+} (2) Al to Cu
(3) Cu to Al   (4) Cu to Cu^{2+}

§2 **29.** In order for a redox reaction to be spontaneous, the potential ($\mathscr{E}°$) for the overall reaction must be
(1) greater than 0   (2) 0
(3) between 0 and −1   (4) less than −1

Base your answers to questions 30 and 31 on the diagram below, which represents the electroplating of a metal fork with Ag(s). (An electroplating cell is a special kind of *electrolytic* cell.)

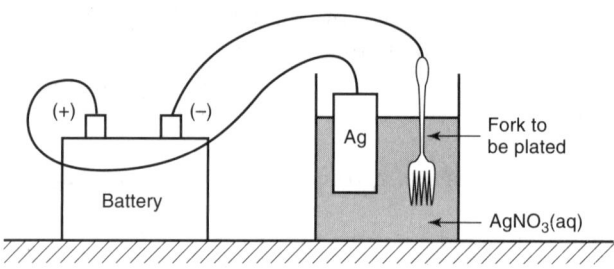

§2 **30.** Which part of the electroplating system is provided by the fork?
(1) the anode, which is the negative electrode
(2) the cathode, which is the negative electrode
(3) the anode, which is the positive electrode
(4) the cathode, which is the positive electrode

§2 **31.** Which equation represents the half-reaction that takes place at the fork?
(1) $Ag^+ + NO_3^- \rightarrow AgNO_3$
(2) $AgNO_3 \rightarrow Ag^+ + NO_3^-$
(3) $Ag^+ + e^- \rightarrow Ag(s)$
(4) $Ag(s) \rightarrow Ag^+ + e^-$

32. If redox reactions are forced to occur by use of an externally applied electric current, the procedure is called
(1) neutralization   (2) esterification
(3) electrolysis   (4) hydrolysis

33. In an electrolytic cell, a negative ion will be attracted to the
(1) positive electrode and oxidized
(2) positive electrode and reduced
(3) negative electrode and oxidized
(4) negative electrode and reduced

34. Which half-reaction occurs at the cathode in an electrolytic cell in which an object is being plated with copper?
(1) $Cu(s) \rightarrow Cu^{2+} + 2e^-$
(2) $Cu(s) + 2e^- \rightarrow Cu^{2+}$
(3) $Cu^{2+} \rightarrow Cu(s) + 2e^-$
(4) $Cu^{2+} + 2e^- \rightarrow Cu(s)$

§2 **35.** Which metal is *not* obtained from its ore by electrolytic reduction?
(1) Na   (2) Li   (3) Au   (4) K

36. In which substance does chlorine have an oxidation number of +1?
(1) Cl_2 (2) HCl   (3) HClO   (4) $HClO_2$

37. Which statement is true for any electrochemical cell?
(1) Oxidation occurs at the anode, only.
(2) Reduction occurs at the anode, only.
(3) Oxidation occurs at both the anode and the cathode.
(4) Reduction occurs at both the anode and the cathode.

38. A diagram of a chemical cell and an equation are shown below.

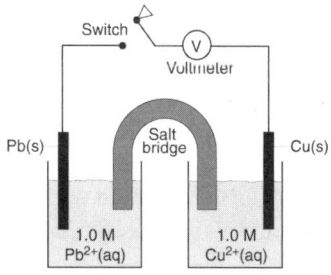

When the switch is closed, electrons will flow from
(1) the Pb(s) to the Cu(s)
(2) the Cu(s) to the Pb(s)
(3) the Pb^{2+}(aq) to the Pb(s)
(4) the Cu^{2+}(aq) to the Cu(s)

39. When a neutral atom undergoes oxidation, the atom's oxidation state
   (1) decreases as it gains electrons
   (2) decreases as it loses electrons
   (3) increases as it gains electrons
   (4) increases as it loses electrons

40. Given the equation:

$$C(s) + H_2O(g) \rightarrow CO(g) + H_2(g)$$

Which species undergoes reduction?
   (1) $C^0(s)$ (2) H^{1+} (3) C^{2+} (4) $H_2^0(g)$

41. Where do reduction and oxidation occur in an electrolytic cell?
   (1) Both occur at the anode.
   (2) Both occur at the cathode.
   (3) Reduction occurs at the anode, and oxidation occurs at the cathode.
   (4) Reduction occurs at the cathode, and oxidation occurs at the anode.

Constructed-Response Questions

1. Zinc metal can be reacted with sulfuric acid to produce hydrogen gas according to the balanced equation

$$Zn(s) + H_2SO_4(aq) \rightarrow ZnSO_4(aq) + H_2(g)$$

The balanced *redox* equation for this reaction is

$$Zn(s) + 2H^+(aq) \rightarrow Zn^{2+}(aq) + H_2(g)$$

(a) Write the oxidation and reduction half-reactions for this equation.

When copper metal [Cu(s)] is mixed with sulfuric acid, *no* generation of hydrogen gas occurs. Using Reference Table J in Appendix 1:

(b) Explain why zinc reacts, but copper will not.
(c) Write the balanced redox equation that *can* occur between the pairs:

$$Cu(s)/Cu^{2+}(aq) \text{ and } H_2(g)/H^+(aq)$$

§2 **2.** Balance each of the following redox equations:
   (a) $KOH + Zn + H_2O + KNO_3 \rightarrow K_2Zn(OH)_4 + NH_3$
   (b) $H_2SO_4 + FeSO_4 + KClO_3 \rightarrow Fe_2(SO_4)_3 + KCl + H_2O$
   (c) $H_3AsO_4 + Zn + HNO_3 \rightarrow AsH_3 + Zn(NO_3)_2 + H_2O$
   (d) $H_2O + P_4 + LiOH \rightarrow PH_3 + LiH_2PO_2$
   (e) $Zn(s) + NO_3^-(aq) \rightarrow Zn^{2+}(aq) + NH_4^+(aq)$ (acidic solution)
   (f) $MnO_4^-(aq) + CN^-(aq) \rightarrow MnO_2(s) + CNO^-(aq)$ (basic solution)

§2 **3.** (a) Write the half-reaction for the reduction of 1.000 mole of $Cu^{2+}(aq)$ to $Cu(s)$.
   (b) How many moles of electrons are needed in this half-reaction?
   (c) If the electric charge on 1.000 mole of electrons is 96,470 coulombs, calculate the total charge that is needed to reduce all of the $Cu^{2+}(aq)$.
   (d) One ampere of electric current is equal to an electric charge flow of one coulomb per second. If the electric charge you calculated in part was supplied over a period of 5.000 hours, how much electric current was needed for the reduction?

Base your answers to questions 4–6 on the diagram of a voltaic cell provided below and on your knowledge of chemistry.

4. On the diagram, indicate with one or more arrows the direction of electron flow through the wire.

5. Write an equation for the half-reaction that occurs at the zinc electrode.

6. Explain the function of the salt bridge.

Base your answers to questions 7–9 on the information below and on your knowledge of chemistry.

One type of voltaic cell, called a mercury battery, uses zinc and mercury(II) oxide to generate an electric current. Mercury batteries were used because of their miniature size even though mercury is toxic. The overall reaction for a mercury battery is given in the equation below.

$$Zn(s) + HgO(s) \rightarrow ZnO(s) + Hg(\ell)$$

7. Determine the change in oxidation number of the zinc while the battery operates.

8. Compare the number of moles of electrons lost to the number of moles of electrons gained during the reaction.

9. Using the information in the passage, state *one* risk and *one* benefit of using a mercury battery.

Base your answers to questions 10–12 on the information below and on your knowledge of chemistry.

Early scientists defined oxidation as a chemical reaction in which oxygen is combined with another element to produce the oxide of the element. An example of oxidation based on this definition is the combustion of methane represented by the balanced equation (Equation 1) below.

$$\text{Equation 1: } CH_4(g) + 2O_2(g) \rightarrow CO_2(g) + 2H_2O(g)$$

The definition of oxidation has since been expanded to include many reactions that do not involve oxygen. One example of oxidation based on the expanded definition is the reaction between magnesium ribbon and powdered sulfur represented by the balanced equation (Equation 2) below.

$$\text{Equation 2: } Mg(s) + S(s) \rightarrow MgS(s)$$

10. Based on the information given in the passage, state why early scientists classified the reaction represented by Equation 1 as oxidation.

11. Determine the change in the oxidation number of carbon in the reaction represented by Equation 1.

12. Write a balanced half-reaction equation for the oxidation that occurs in the reaction represented by Equation 2.

The answers to these questions are found in Appendix 3.

Chapter Sixteen

THE CHEMISTRY LABORATORY

KEY IDEAS

This chapter is a review of the laboratory skills you have acquired and the laboratory activities you have performed during your chemistry course. Included in the chapter are listings of safety procedures and of basic skills and activities, illustrations of common laboratory equipment, and a guide for preparing laboratory reports.

KEY OBJECTIVES

At the conclusion of this chapter you will have reviewed the following information:

- Which safety procedures are basic in the laboratory.
- Which measuring devices are commonly used in the laboratory.
- Which basic laboratory skills you should be familiar with.
- Which equipment is common to most laboratories.
- Which laboratory activities are commonly employed in a chemistry laboratory.
- How colors are used to identify various substances.
- What general guidelines should be observed in constructing a laboratory report.

16.1 INTRODUCTION

This chapter is quite different from the 15 chapters that precede it. You cannot learn laboratory skills by reading a book any more than you can learn to drive a car by reading a book! Learning about the laboratory can take place only in the laboratory.

In this chapter, we will briefly review the activities that you should have performed and the skills that you should have acquired during the course of the year. You should supplement this chapter by referring to your laboratory manual and/or notebook.

16.2 SAFETY PROCEDURES

While each chemistry laboratory has its own procedures to ensure safety, a number of rules for students are common to every laboratory. Included are the following:

- Never be in a laboratory without a teacher.
- Do not bring any food into the laboratory.
- If your hair is long, tie it back to avoid catching fire.
- Always wear safety goggles in the laboratory.
- Do not handle equipment until you receive instructions.
- Always be alert when you handle chemicals or work with open flames.
- Learn and follow the correct procedures for diluting acids, heating substances, mixing chemicals, and working with glass.
- Never get "creative" and mix chemicals "to see what will happen."

You can undoubtedly add to this list. The point is that you must practice safety procedures at all times!

16.3 USING MEASURING DEVICES

Most of the work in a chemistry laboratory involves measurement. Devices that you may use include a balance, ruler, graduated cylinder, thermometer, burette, and eudiometer tube.

When you measure the level of a liquid, as with a burette or graduated cylinder, your eye should be level with the *bottom* of the meniscus of the liquid, and the reading should be taken at that point, as illustrated in the accompanying diagram.

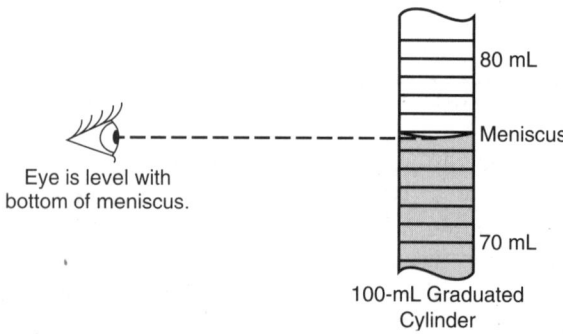

Eye is level with
bottom of meniscus.

80 mL

Meniscus

70 mL

100-mL Graduated
Cylinder

In general, when you use any measuring device that has a ruled scale, such as a ruler, balance, or graduated cylinder, you may estimate to $\frac{1}{10}$ of the smallest division present. For example, in the preceding diagram, the smallest divisions on the cylinder are 1 milliliter each. Therefore, you read this measurement as 75.7 milliliters.

16.4 BASIC LABORATORY SKILLS

During your chemistry laboratory course, you acquired a number of basic skills. You should be familiar with the following skills:

- How to adjust a (Bunsen) burner.
- How to cut, bend, and fire-polish glass tubing.
- How to separate a mixture with a funnel and filter paper.
- How to dilute acids and other liquids safely.
- How to remove solids from containers.
- How to pour liquids safely.
- How to heat materials in test tubes, beakers, and flasks safely.

16.5 IDENTIFICATION OF COMMON LABORATORY APPARATUS

The illustrations on pages 466 and 467 show the laboratory equipment that you use most often during your study of chemistry. You should review how each item is used in the laboratory.

The Chemistry Laboratory

Test tube rack

Test tubes

Burette/test tube clamp

Clamp holder

3-prong jaw clamp

Forceps

Double burette clamp

Stirring rod

Condenser

Rubber policeman

Graduated pipette

Burette

Graduated cylinder

Pinch clamp

Pipette bulb

Ring clamp

Gas collecting tube

Volumetric pipette

Thermometer

Screw clamp

Platform balance (triple beam)

Beaker

Mortar and pestle

Crucible and cover

Watch glass

Evaporating dish

Pneumatic trough

Florence flask

Wide-mouth collecting bottle

Plastic wash bottle

Dropper pipette

Funnel

Erlenmeyer flask

Safety goggles

Rubber stoppers

Glass rod with nichrome wires (for flame testing)

Test tube holder

Crucible tongs

Metal spatula

Test tube brush

Triangular file

Clay triangle

Ring stand

Ceramic square

Wire gauze

Scoopula

Rubber tubing

Burner

Tripod

16.6   BASIC LABORATORY ACTIVITIES

The following is a list of other laboratory activities that are common to most courses in chemistry. You should use your laboratory manual and notebook to review these activities.

- How to construct a heating or cooling curve and to interpret points on the curve.
- How to measure the heat of a simple chemical reaction.
- How to identify exothermic and endothermic processes using temperature measurements.
- How to determine the solubility of a substance at various temperatures and to construct a solubility curve for that substance.

- How to distinguish between inorganic and organic substances by comparing their solubilities, melting points, and electrical conductivities.
- How to identify the ions of such metals as lithium, sodium, potassium, strontium, and copper by means of a flame test.
- How to find the percent (by mass) of water in a crystal.
- How to determine the molar volume of a gas generated in an experiment.
- How to perform an acid–base titration.
- How to prepare and collect a gas such as hydrogen or carbon dioxide.

16.7   THE ROLE OF COLORS IN IDENTIFYING SUBSTANCES

Colors play an important role in the chemistry laboratory, particularly in identifying elements, ions, and compounds.

Flame Tests

All of us are aware of the beautiful colors that fireworks produce. The ions of certain elements produce the characteristic colors of various types of fireworks. In the laboratory, we generally prepare an aqueous solution of certain specific elements, dip a platinum or Nichrome wire loop into the solution, and place the loop into the flame of a laboratory burner. The resulting color of the flame is used to identify the element that produced the color.

Flame Test Colors of Elemental Ions

Elemental Ion	Flame Test Color
Li^+	Deep red (crimson)
Na^+	Yellow
K^+	Pale violet
Ca^{2+}	Orange-red
Sr^{2+}	Red
Ba^{2+}	Yellow-green
Cu^{2+}	Blue-green

Sometimes the flame test of a mixture of Na^+ ions and K^+ ions in aqueous solution will show the yellow color of Na^+, but not the pale violet color of K^+ because of the difference in the intensities of the two colors. In this case, using a special type of glass, known as cobalt blue glass, absorbs the yellow color and allows the pale violet color to be observed in the flame.

Colors of Ions in a Solid and/or in Aqueous Solution

Many students notice that solid copper(II) sulfate is blue in color, as is its aqueous solution. In certain cases, a (positive) metallic ion or a polyatomic ion containing a transition metal will yield a solid or aqueous solution that is colored.

Colors of Solids and Aqueous Solutions of Ions

Ion	Associated Color
Cu^+	Green
Cu^{2+}	Blue
Fe^{2+}	Yellow-green (depending on the anion)
Fe^{3+}	Orange-red (depending on the anion)
Co^{2+}	Pink
Cr^{3+}	Violet ($Cr(NO_3)_3$) to Green ($CrCl_3$)
Ni^{2+}	Green
Mn^{2+}	Pink
MnO_4^-	Purple
CrO_4^{2-}	Yellow
$Cr_2O_7^{2-}$	Orange
$FeSCN^{2+}$	Deep red
$CoCl_4^{2-}$	Blue

Note that various ions such as Al^{3+}, Li^+, Na^+, K^+, Mg^{2+}, Ca^{2+}, Ba^{2+}, Pb^{2+}, Pb^{4+}, Sr^{2+}, and Zn^{2+} are colorless in aqueous solution but most (though not all) of their solid compounds are white. A notable exception is PbI_2, which is an intensely yellow solid.

Colors of Elements and Certain Compounds

Several elements and compounds can be readily identified by their characteristic colors.

Element or Compound and Associated Color

Elemental or Compound	Associated Color
F_2	Pale yellow gas
Cl_2	Green-yellow gas
Br_2	Deep red liquid
I_2	Metallic gray solid; violet vapor
S_8	Yellow solid
Cu	Red metallic solid
Au	Yellow metallic solid
NO_2	Brown gas

In addition, the sulfides of transition metals tend to be black solids. Metals that consist of very small particles appear black because they reflect light poorly.

16.8 GUIDELINES FOR LABORATORY REPORTS

Although each teacher has his or her own rules for keeping a laboratory notebook, there are a number of general guidelines for laboratory reports that you should bear in mind.

- Make your reports concise. A wordy report tends to obscure important details.
- Arrange the information in your reports logically. There should be an introduction, a materials and methods section, a section for observations and results, and a section for discussion and conclusions.
- Never erase data that you consider unacceptable. Draw a single line through questionable material. (It is entirely possible that "bad" data may turn out to be correct!)
- Whenever possible, present your data and results in the form of tables and graphs.
- List your observations clearly.
- Be sure that your conclusions follow directly and logically from your observations.

END-OF-CHAPTER QUESTIONS

1. Which procedure represents the safest technique to use for diluting a concentrated acid?
   (1) Add the acid to the water quickly.
   (2) Add the water to the acid quickly.
   (3) Add the acid slowly to the water with steady stirring.
   (4) Add the water slowly to the acid with steady stirring.

2. Which procedure is the safest to follow when using an open flame to heat the contents of a test tube that contains a flammable mixture?
   (1) Cork the test tube and then heat it gently near the bottom only.
   (2) Heat the open test tube gently near the bottom only.
   (3) Cork the test tube and place it in a beaker of water, then heat the water in the beaker.
   (4) Place the open test tube in a beaker of water, then heat the water in the beaker.

3. The diagram below represents a portion of a triple beam balance. If the beams are in balance, with the riders in the position shown, what is the total mass, in grams, of the object being massed?

(1) 460.62   (2) 466.20   (3) 466.62   (4) 460.20

4. A student has to measure the diameter of a test tube (as shown in the diagram below) in order to calculate the tube's volume.

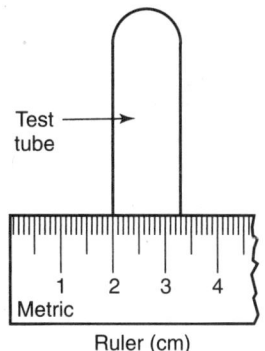

Based on the diagram, the tube's diameter is closest to
(1) 1.25 cm   (2) 2.32 cm   (3) 3.25 cm   (4) 12.5 cm

5. The diagram below represents a portion of a burette.

What is the reading of the meniscus?
(1) 39.2 mL   (2) 39.5 mL   (3) 40.7 mL   (4) 40.9 mL

6. The diagram below shows a section of a 100-milliliter graduated cylinder:

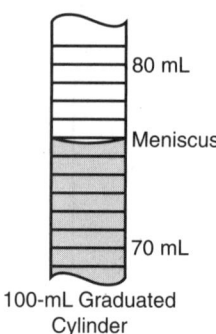

100-mL Graduated
Cylinder

In order to read the volume of the liquid most accurately, the eye of the student should be
(1) level with the meniscus   (2) above the meniscus
(3) at the bottom of the cylinder   (4) at the top of the cylinder

7. As a result of dissolving a salt in water, a student found that the temperature of the water increased. From this observation alone, the student should conclude that the dissolving of the salt
(1) produced an acid solution   (2) produced a basic solution
(3) was endothermic   (4) was exothermic

8. A Bunsen burner flame is sooty black and mixed with an orange-yellow color. Which is the probable reason for this condition?
(1) No oxygen is mixing with the gas.
(2) No gas is mixing with the oxygen.
(3) Insufficient oxygen is mixing with the gas.
(4) Insufficient gas is mixing with the oxygen.

9. A student obtained the following data in determining the solubility of $NaNO_3$:

Temperature (°C)	Solubility (g $NaNO_3$/100 g H_2O)
0	73
10	80
20	88
30	97
40	105
50	115
60	124
70	134
80	145

Which set of coordinates graphically present the data in the table most clearly?

(1)

(2)

(3)

(4)

10. The volume of an acid required to neutralize exactly 15.00 milliliters of a base could be measured most precisely if it were added to the base solution from a
(1) 100-mL graduate
(2) 125-mL Erlenmeyer flask
(3) 50-mL burette
(4) 50-mL beaker

11. Which diagram represents a graduated cylinder?

(1)   (2)   (3)   (4)

12. Refer to the experimental data below.

Mass of empty graduated cylinder = 141 g
Mass of graduated cylinder and distilled water = 163 g
Volume of distilled water = 25.3 mL

Based on the experimental data collected, what is the density of the distilled water?
(1) 1.0 g/mL
(2) 0.253 g/mL
(3) 0.87 g/mL
(4) 1.15 g/mL

13. Which diagram below represents a pipette?

(1)   (2)   (3)   (4)

14. Which piece of laboratory apparatus would most likely be used to evaporate a 1-milliliter sample of a solution to dryness?
(1) a volumetric flask   (2) a burette
(3) a pipette   (4) a watch glass

15. Which diagram below represents an Erlenmeyer flask?

(1)   (2)   (3)   (4)

16. In a laboratory experiment, a student collects a gas by water displacement. Which piece of apparatus is best used to measure the volume of the gas as it is generated?
(1) a pipette   (2) a volumetric flask
(3) a eudiometer tube   (4) a beaker

17. The heat energy given off by an acid solution as it is neutralized by a basic solution is best measured by using a
(1) pipette   (2) pH meter
(3) calorimeter   (4) test tube

18. In order to measure the solubility of a compound in water at 20°C, a student prepared a saturated solution of the compound at that temperature and then evaporated it to dryness. The student's experimental data are given below.

Mass of the saturated solution before evaporation   =   18.8 g
Mass of the solid remaining after evaporation   =   8.8 g

Based on the data and on Reference Table G in Appendix 1, the compound could have been
(1) $NaNO_3$ (2) $KClO_3$ (3) NaCl   (4) NH_4Cl

19. The process of filtration is performed in the laboratory to
(1) form precipitates
(2) remove water from solutions
(3) separate dissolved particles from the solvent
(4) separate insoluble substances in an aqueous mixture

20. Which graph shown below could represent the uniform cooling of a substance, starting with the gaseous phase and ending with the solid phase?

(1)

(2)

(3)

(4)

21. A 1.20-gram sample of a hydrated salt is heated to a constant mass of 0.80 gram. What was the percent by mass of water contained in the original sample?
(1) 20 (2) 33 (3) 50 (4) 67

22. A student obtained the following data while cooling a substance. The substance was originally in the liquid phase at a temperature below its boiling point.

Time (min)	0.5	1.0	1.5	2.0	2.5	3.0	3.5	4.0	4.5	5.0	5.5	6.0
Temperature (°C)	70.	63	57	54	53	53	53	53	53	52	51	48

What is the freezing point of the substance?
(1) 70.°C (2) 59°C (3) 53°C (4) 48°C

Base your answers to questions 23 and 24 on the table below, which shows the data collected during the heating of a 5.0-gram sample of a hydrated salt.

Mass of Salt (g)	Heating Time (min)
5.0	0.0
4.1	5.0
3.1	10.
3.0	15.
3.0	30.
3.0	60.

23. After 60. minutes, how many grams of water appear to remain in the salt?
   (1) 0.00　　　　(2) 2.0　　　　(3) 1.9　　　　(4) 0.90

24. What is the percent of water in the original sample?
   (1) 82.　　　　(2) 60.　　　　(3) 30.　　　　(4) 40.

25. Which piece of glassware is best for transferring water to an Erlenmeyer flask?
   (1) beaker
   (2) test tube
   (3) volumetric flask
   (4) condenser

26. In an experiment, a student determined the normal boiling points of four unknown liquids. The collected data were organized into the table below.

Unknown	Normal Boiling Point (°C)
A	10
B	33
C	78
D	100

Which unknown liquid has the weakest attractive forces between its molecules?
   (1) A　　　　(2) B　　　　(3) C　　　　(4) D

27. A student investigated samples of four different substances in the solid state. The table below is a record of the properties observed when each solid was tested.

Property	Sample I	Sample II	Sample III	Sample IV
Melting point	High	Low	High	Low
Solubility in water	Soluble	Insoluble	Insoluble	Soluble
Stability at high temperature	Stable	Decomposes	Stable	Stable
Electrical conductivity	Good	Poor	Poor	Good

According to the tabulated results, which of the solids investigated had the characteristics most closely associated with those of an organic compound?
(1) Sample I   (2) Sample II
(3) Sample III   (4) Sample IV

28. A 9.90-gram sample of a hydrated salt is heated to a constant mass of 6.60 grams. What was the percent, by mass, of water contained in the sample?
(1) 66.7   (2) 50.0   (3) 33.3   (4) 16.5

29. The data below were obtained during the determination of the percent of water in $BaCl_2 \cdot 2H_2O$.

	Mass
Empty crucible	45.12 g
Crucible + $BaCl_2 \cdot 2H_2O$	75.82 g
Crucible + contents after first heating	72.60 g
Crucible + contents after second heating	71.34 g
Crucible + contents after third heating	71.34 g

The percent of water in $BaCl_2 \cdot 2H_2O$ may be expressed as

(1) $\dfrac{30.70}{75.82} \times 100\%$

(2) $\dfrac{30.70}{45.12} \times 100\%$

(3) $\dfrac{71.34}{75.82} \times 100\%$

(4) $\dfrac{4.48}{30.70} \times 100\%$

30. The following procedures are carried out during a laboratory activity to determine the mass, in grams, of $CuSO_4$ in a sample of $CuSO_4 \cdot 5H_2O$:

Step 1. Determine the mass, in grams, of the crucible and
$CuSO_4 \cdot 5H_2O$.
Step 2. Determine the mass, in grams, of the crucible and $CuSO_4$.
Step 3. Determine the mass, in grams, of $CuSO_4 \cdot 5H_2O$.
Step 4. Determine the mass, in grams, of the empty crucible.
Step 5. Determine the mass, in grams, of $CuSO_4$.

Which sequence of steps would produce the desired result using only *three* weighings?
(1) 1, 3, 4, 2, 5   (2) 2, 4, 1, 5, 3
(3) 3, 4, 1, 2, 5   (4) 4, 1, 3, 2, 5

31. All of the following are good laboratory safety practices EXECPT
(1) wearing eye goggles
(2) placing unused chemicals back into their original containers
(3) adding acid to water when diluting an acid
(4) allowing heated glass to cool before touching it

32. A student observes that an aqueous solution of a compound is blue. Which ion could the solution contain?
(1) Zn^{2+}
(2) Na^+
(3) Al^{3+}
(4) Cu^{2+}

33. In order to identify the metallic element in an ionic compound, a student places a small amount of the compound onto a wire loop and places the loop into a flame. If the flame is pale violet, the compound most likely contains
(1) K^+
(2) Na^+
(3) Li^+
(4) Ca^{2+}

34. A student wishes to prepare one liter of a 0.2-molar aqueous solution. Which piece of laboratory apparatus will yield the most precise final volume?
(1) beaker
(2) graduated cylinder
(3) volumetric flask
(4) buretette

35. When a flame test on a solid was performed, a student observed a bright orange-yellow flame. When the student used cobalt blue glass, a pale violet color flame was observed. The solid most likely contained
(1) sodium and potassium
(2) lithium and strontium
(3) copper and barium
(4) calcium and zinc

Constructed-Response Questions

Five pieces of laboratory equipment are listed below.

A. graduated cylinder
B. triple-beam balance
C. beaker
D. filter paper
E. Nichrome loop

Match laboratory activities 1–4 with the equipment in the list above. A choice may be used once, more than once, or not at all

1. used to measure 100 mL of NaCl(aq)

2. used in flame tests to insert a substance into the flame

3. used to separate sand from water

4. used to measure the mass of a solid

The answers to these questions are found in Appendix 3.

GLOSSARY

absolute zero The lowest possible temperature, written as 0 K or $-273.15°C$.

accuracy The closeness of a measurement to an accepted value; see also **precision**.

acid See **Arrhenius acid**; **Brønsted–Lowry acid**; **Lewis acid**.

acid ionization constant (K_a) A constant whose value indicates the relative strength of an acid in aqueous solution.

activated complex The intermediate state between reactants and products in a chemical reaction; the peak of the potential energy diagram.

activation energy The minimum energy needed to initiate a reaction.

addition polymerization The joining of unsaturated monomers by a series of addition reactions.

addition reaction The process in which a substance reacts across a double or triple bond in an organic compound.

alcohol An organic compound containing one or more hydroxyl ($-OH$) groups.

aldehyde An organic compound containing a carbonyl group with at least one hydrogen atom attached to the carbonyl carbon.

alkali metal Any Group 1 element, excluding hydrogen.

alkaline earth element Any Group 2 element.

alkane A hydrocarbon containing only single bonds between adjacent carbon atoms.

alkene A hydrocarbon containing one double bond between two adjacent carbon atoms.

alkyl group An open-chained hydrocarbon less one hydrogen atom; for example, CH_3 = methyl group, C_2H_5 = ethyl group. Unspecified alkyl groups are designated by the letter R.

alkyne A hydrocarbon containing one triple bond between two adjacent carbon atoms.

allotrope A specific form of an element that can exist in more than one form; graphite and diamond are allotropes of the element carbon.

alloy A solid metallic solution.

alpha decay The radioactive process in which an alpha particle is emitted.

alpha particle (α) A helium-4 nucleus.

amide A functional group that combines an amino and a carbonyl group; $-CONH_2$.

amine A hydrocarbon derivative containing an amino group.

amino acid An organic compound containing at least one amino group and one carboxyl group.

amino group An ammonia molecule less one hydrogen atom; $-NH_2$.

ampere (A) The SI unit of electric current.

amphiprotic Pertaining to a substance that can act as a Brønsted–Lowry acid or base by donating or accepting H^+ ions.

amphoteric See **amphiprotic**.

anhydrous Pertaining to a compound from which the water of crystallization has been removed.

anode The electrode at which oxidation occurs.

aqueous Pertaining to a solution in which water is the solvent.

aromatic hydrocarbon Any ring hydrocarbon whose electronic structure is related to that of benzene.

Arrhenius acid Any substance that releases H^+ ions in water.

Arrhenius base Any substance that releases OH^- ions in water.

atmosphere (atm) A unit of pressure equal to 101.3 kilopascals.

atom The basic unit of an element.

atomic mass The weighted average of the masses of the naturally occurring isotopes of an element.

atomic mass unit (u) One-twelfth the mass of a carbon-12 atom.

atomic number The number of protons in the nucleus of an atom; the atomic number defines the element.

atomic radius An estimate of the size of an atom.

Avogadro's hypothesis Equal volumes of gases, measured at the same temperature and pressure, contain equal numbers of particles.

Avogadro's law At constant temperature and pressure, the volume of an ideal gas is directly proportional to the number of gas particles present; $\dfrac{V_1}{n_1} = \dfrac{V_2}{n_2}$.

Avogadro's number (N_A) The number of particles in 1 mole; 6.02×10^{23}.

base See **Arrhenius base**; **Brønsted–Lowry base**; **Lewis base**.

base ionization constant (K_b) A constant whose value indicates the relative strength of a base in aqueous solution.

battery A commercial voltaic cell.

benzene C_6H_6; the parent hydrocarbon of all aromatic compounds.

beta decay The radioactive process in which a beta particle is emitted.

beta (−) particle (β^-) An electron.

beta (+) particle (β^+) A positron.

binary compound A compound containing two elements.

binding energy The energy released when a nucleus is assembled from its nucleons.

boiling The transition of liquid to gas; boiling occurs when the vapor pressure of a liquid equals the atmospheric pressure above the liquid.

boiling point The temperature at which boiling occurs; the temperature at which the liquid and vapor phases of a substance are in equilibrium.

boiling point elevation The increase in the boiling point of a solvent due to the presence of solute particles.

bond energy The energy needed to break a chemical bond.

Boyle's law At constant temperature and mass, the pressure of an ideal gas is inversely proportional to its volume; $P_1V_1 = P_2V_2$.

breeder reactor A fission reactor that generates its own nuclear fuel.

bright-line spectrum The lines of visible light emitted by elements as electrons fall to lower energy levels.

Brønsted–Lowry acid A substance that can donate H^+ ions.

Brønsted–Lowry base A substance that can accept H^+ ions.

burette A long, graduated tube used to deliver a controlled volume of liquid precisely.

calorie A quantity of energy; 1 calorie is approximately equal to 4.2 joules.

carbonyl group The functional group characteristic of aldehydes and ketones; $>C=O$.

carboxyl group The functional group characteristic of organic acids; $-COOH$.

catalyst A substance that speeds a chemical reaction by lowering the activation energy of the reaction.

cathode The electrode at which reduction occurs.

cathode rays A stream of electrons emitted at the negative electrode when electricity is passed through a gas at very low pressure.

Celsius (C) scale The temperature scale on which the freezing and boiling points of water (at 1 atm) were originally set at 0 and 100, respectively. The Celsius scale is now defined in terms of the Kelvin scale: $C = K − 273.15$.

chain reaction A chemical or nuclear reaction in which one step supplies energy or reactants for the next step.

Charles's law At constant pressure and mass, the volume of an ideal gas is directly proportional to the Kelvin temperature; $\dfrac{V_1}{T_1} = \dfrac{V_2}{T_2}$.

chemical bond The stabilizing of two atoms by sharing or transferring electrons.

chemical change See **chemical reaction**.

chemical energy The part of internal energy that is associated with the bonds and intermolecular attractions of substances.

chemical equation A shorthand listing of reactants, products, and molar quantities in a chemical reaction.

chemical equilibrium The state in which the rates of the forward and reverse reactions are equal.

chemical family See **group**.

chemical property A property that describes the composition and reactivity of a substance.

chemical reaction A process in which one or more substances are converted into other substances.

chromatography The process in which the components of certain mixtures are separated because of their differences in solubility.

cis-trans isomerism A form of stereoisomerism across the double bond of an alkene. In the *cis* isomer, two identical atoms lie on the *same side* of the molecule across the double bond. In the *trans* isomer, the two atoms lie on *opposite sides* of the molecule across the double bond.

coefficient A number in a chemical equation that indicates how many particles of a reactant or product are required or formed in the reaction.

colligative property A property that depends on the number of particles present rather than the type of particle; see also **boiling point elevation**; **freezing point depression**.

combined gas law At constant mass, the product of the pressure and volume divided by the Kelvin temperature is a constant; $\dfrac{P_1 V_1}{T_1} = \dfrac{P_2 V_2}{T_2}$.

common ion effect An equilibrium shift caused by the addition of an ion present in a reaction.

compound A combination of two or more elements with a fixed composition by mass.

concentrated Pertaining to a solution that contains a relatively large quantity of solute.

concentration The "strength" of a solution; the quantity of solute relative to the quantity of solvent.

condensation The change of gas to liquid.

condensation polymerization The joining of monomers by a series of dehydration reactions.

conjugate acid-base pair Two particles that differ by a single H^+ ion; HCl and Cl^- are a conjugate acid-base pair.

constitutional isomers Isomers that differ because their atoms are connected in different orders.

control rod The part of a fission reactor that controls the rate of fission by absorbing neutrons.

coordinate covalent bond A single covalent bond in which the pair of electrons is supplied by one atom.

coulomb (C) The SI unit of electric charge.

covalent bond A chemical bond formed by the sharing of electrons.

cracking The process of breaking large hydrocarbon molecules into smaller ones in order to increase the yield of compounds such as gasoline.

critical temperature The temperature above which a gas cannot be liquefied, regardless of the applied pressure.

crystal A solid whose particles are arranged in a regularly repeating pattern.

crystallization The process in which a solute separates from its solution; also called *precipitation*.

Dalton's law of partial pressures The sum of the partial pressures of the gases in a container is equal to the total pressure of all of the gases in the container; $P_i = X_i \cdot P_t$. See also **mole fraction**.

decomposition A reaction in which a compound forms two or more simpler substances.

density Mass per unit volume; $d = \dfrac{m}{v}$.

deposition The direct transition from gas to solid.

deuterium The isotope of hydrogen with a mass number of 2.

diatomic molecule A neutral particle consisting of two atoms; Br_2 and CO are diatomic molecules.

diffusion The movement of one substance through another.

dihydroxy alcohol An organic compound with two hydroxyl groups.

dilute (adjective) Pertaining to a solution that contains a relatively small quantity of solute; (verb) to reduce the concentration of a solution by adding solvent.

dipole An unsymmetrical charge distribution in a neutral molecule.

dipole-dipole attraction The attractive force between two oppositely charged dipoles of neighboring polar molecules.

dispersion forces The attractive forces between neighboring nonpolar molecules; also called *London dispersion forces*.

dissociation The separation of an ionic compound in solution into positive and negative ions.

distillation The simultaneous boiling of a liquid and condensation of its vapor.

double bond A covalent bond in which two pairs of electrons are shared by two adjacent atoms.

ductility The property of a substance that allows it to be drawn into a wire; metallic substances possess ductility.

dynamic equilibrium The state in which the rates of opposing processes are equal; see also **chemical equilibrium**; **phase equilibrium**; **solution equilibrium**.

effusion The escape of a gas from a small porous opening; see also **Graham's law of effusion**.

Einstein's mass-energy relationship The equivalence of mass and energy; $E = mc^2$.

electrochemical cell A voltaic cell or an electrolytic cell.

electrode A conductor in an electrochemical or electrolytic cell that serves as the site of oxidation or reduction.

electrode potential A measurement, expressed in volts (V), that indicates the relative ease with which an oxidation or reduction half-reaction can occur. Electrode potentials are determined with respect to a standard hydrogen half-cell; see also **standard hydrogen half-cell**.

electrolysis A nonspontaneous redox reaction driven by an external source of electricity.

electrolyte A substance whose aqueous solution conducts electricity.

electrolytic cell A device for carrying out electrolysis.

electron The elementary unit of negative charge.

electron affinity The energy change that occurs when an atom or ion gains an electron.

electron-dot diagram See **Lewis structure**.

electronegativity The measure of an atom's attraction for a bonded pair of electrons.

electroplating The use of an electric current to deposit a layer of metal on a negatively charged object.

element A substance all of whose atoms have the same atomic number.

elementary reactions A series of simpler reactions that are the "building blocks" of a more complex reaction.

empirical formula A formula in which the elements are present in the smallest whole-number ratio; NO_2 is an empirical formula, but C_2H_4 is not.

endothermic reaction A reaction that absorbs energy; ΔH is positive for an endothermic reaction.

end point The point in a titration that signals that equivalent quantities of reactants have been added.

energy A quantity related to an object's capacity to do work.

enthalpy change (ΔH) The heat energy absorbed or released by a system at constant pressure.

entropy (S) The measure of the randomness or disorder of a system.

entropy change (ΔS) An increase or decrease in the randomness of a system.

equilibrium See **dynamic equilibrium**.

ester The organic product of esterification.

esterification The reaction of an acid with an alcohol to produce an ester and water.

ethanoic acid CH_3COOH; acetic acid.

ethanol CH_3CH_2OH; ethyl (grain) alcohol.

ethene C_2H_4; ethylene; the parent of the alkene family of hydrocarbons.

ether An organic compound containing the arrangement $R-O-R$.

ethyne C_2H_2; acetylene; the parent of the alkyne family of hydrocarbons.

evaporation The surface transition of liquid to gas.

excited state A condition in which one or more electrons in an atom are no longer in the lowest possible energy state.

exothermic reaction A reaction that releases energy; ΔH is negative for an exothermic reaction.

factor-label method (FLM) A technique for solving problems by using relationships among units.

fermentation The (anaerobic) oxidation of a sugar such as glucose to produce ethanol and carbon dioxide; the reaction is catalyzed by enzymes.

filtrate The part of a mixture that remains in solution after any residue is separated from it.

filtration The process in which a precipitate is recovered from a mixture.

first ionization energy The quantity of energy needed to remove the most loosely held electron from an isolated neutral atom.

first law of thermodynamics Energy is conserved in any process; $\Delta E = q + w$.

fission A nuclear reaction in which a heavy nuclide splits to form lighter nuclides and energy.

fission reactor A device for producing electrical energy by means of a controlled fission reaction.

formula mass The sum of the masses of the atoms in a formula; units are atomic mass units (u) or grams per mole (g/mol).

formula unit The notation that expresses the relative numbers of positive and negative ions in an ionic compound; the formula unit $CaCl_2$ indicates that calcium chloride contains two Cl^- ions for every Na^+ ion.

fractional distillation The separation of organic substances based on differences in their boiling points.

free-energy change (ΔG) A quantity that determines whether a reaction is spontaneous; $\Delta G = \Delta H - T\,\Delta S$; a negative value of ΔG indicates a spontaneous reaction.

freezing The transition from liquid to solid.

freezing point The temperature at which freezing occurs.

freezing point depression (lowering) The decrease in the freezing point of a solvent due to the presence of solute particles.

fuel rod The part of a nuclear reactor that contains the fissionable material.

functional group An atom or group of atoms that confers specific properties on an organic molecule.

fundamental forces The four basic forces in the universe: gravitational, electromagnetic, strong, weak; also called *fundamental interactions*.

fundamental interactions See **fundamental forces**.

fusion A synonym for *melting*; also, a nuclear process in which light nuclides join to form heavier nuclides and produce radiant energy.

fusion reactor An experimental device for producing a controlled fusion reaction and generating electrical energy from it.

gas The phase in which matter has neither definite shape nor definite volume.

glycol See **dihydroxy alcohol**.

Graham's law of effusion At constant temperature and pressure, the rate of effusion of a gas is inversely proportional to the square root of its molar mass; $\dfrac{r_1}{r_2} = \sqrt{\dfrac{M_2}{M_1}}$.

gram (g) A metric unit of mass.

gram-atomic mass The molar mass of an element expressed in grams per mole (g/mol).

gram-formula mass See **molar mass**.

gram-molecular mass The molar mass of a molecule.

ground state The electron configuration of an atom in the lowest energy state.

group The elements within a single vertical column of the Periodic Table.

half-cell The part of an electrochemical cell in which oxidation or reduction occurs.

half-cell potential See **electrode potential**.

half-life The time needed for a substance to decay to one-half its initial mass.

half-reaction The oxidation or reduction portion of a redox reaction.

halogen An element in Group 17 of the Periodic Table; F, Cl, Br, I, At.

heat The energy transferred between two objects when they are at different temperatures.

heat of formation (ΔH_{f°) The heat absorbed or released when 1 mole of a compound is formed from its elements under standard conditions.

heat of fusion (H_f) The heat absorbed when a unit mass of solid changes to liquid at its melting point; $H_{f(\text{ice})} = 333.6$ joules per gram.

heat of reaction (ΔH) The heat absorbed or released as a result of a chemical reaction.

heat of vaporization (H_v) The heat absorbed when a unit mass of liquid changes to gas at its boiling point; $H_{v(\text{water})} = 2259$ joules per gram.

heavy water A molecule of water in which the hydrogen atoms have a mass number of 2; deuterium oxide.

Hess's law The heat absorbed or released in a given reaction can be expressed as the sum of the heats associated with other reactions if these reactions can be "added" to produce the original reaction.

heterogeneous mixture A nonuniform mixture.

homogeneous mixture A mixture with a uniform distribution of particles; a solution is one example of a homogeneous mixture.

homologous series A group of organic compounds with related structures and properties; each successive member of the series differs from the one before it by a specific number of carbon and hydrogen atoms (usually CH_2).

hybrid orbital An orbital formed by the "mixing" of individual atomic orbitals.

hydrate A crystalline compound that has water molecules incorporated into its crystal structure; common examples include $CuSO_4 \cdot 5H_2O$ and $Na_2SO_4 \cdot 10H_2O$ [also written as $CuSO_4(H_2O)_5$ and $Na_2SO_4(H_2O)_{10}$].

hydration The association of water molecules with an ion or another molecule.

hydride A binary compound of an active metal and hydrogen; the H^- ion.

hydrocarbon An organic compound composed of carbon and hydrogen.

hydrogen bond An unusually strong intermolecular attraction that results when hydrogen is bonded to a small, highly electronegative atom such as F, O, or N.

hydrogen half-cell See **standard hydrogen half-cell**.

hydrolysis A reaction in which a water molecule breaks a chemical bond; the reaction between certain salts and water to produce an excess of hydronium or hydroxide ions.

hydronium ion H_3O^+; the conjugate acid of H_2O; responsible for acidic properties in water solutions.

hydroxide ion OH^-; the conjugate base of H_2O; responsible for basic properties in water solutions.

ideal gas A model of a gas in which the particles have no volume, do not attract or repel each other, and collide without loss of energy; real gases approximate ideal gas behavior under conditions of low pressure and high temperature.

ideal gas law The relationship obeyed by an ideal gas; $PV = nRT$, where n is the number of moles of gas and R is the universal gas constant.

indicator A substance that undergoes a color change to signal a change in chemical conditions; acid-base indicators change color over specified pH ranges.

inert gas An element in Group 18 of the Periodic Table; Ne, Ar, Kr, Xe, Rn. (He is also associated with Group 18.)

inorganic compound A compound that is not a hydrocarbon derivative.

internal energy (E) The total energy possessed by a system.

ion A particle in which the numbers of protons and electrons are not equal.

ion-dipole attraction The attractive force between an ion and the oppositely charged dipole of a neighboring polar molecule.

ionic bond The electrostatic attraction of positive and negative ions in an ionic compound; an electronegativity difference of 1.7 or greater indicates the presence of an ionic bond.

ionic compound A substance whose particles consist of positive and negative ions.

ionic radius The size of a monatomic ion; positive ions are generally smaller than their neutral atoms, while negative ions are generally larger.

ionization constant An equilibrium constant expression that describes an ionization reaction; a relatively large value indicates a more ionized substance; K_a, K_b, and K_w are examples of ionization constants.

ionization energy The quantity of energy needed to remove an electron from an atom or ion; see also **first ionization energy**.

isomers Different compounds that have the same molecular formula; see also **constitutional isomers; enantiomers; stereoisomers**.

isotopes Atoms having the same atomic number but different mass numbers; atoms of the same element with differing numbers of neutrons.

IUPAC International Union of Pure and Applied Chemistry; the scientific group responsible for all major policies in chemistry, including the naming of elements and compounds.

joule (J) The SI unit of work and energy.

Kelvin (K) The SI unit of temperature. A temperature *difference* of 1 K is equal to a temperature *difference* of 1 C°.

kernel The nucleus and inner (nonvalence) electrons of an atom.

ketone An organic compound containing a carbonyl group with no hydrogen atoms directly attached to the carbonyl carbon.

kilo- The metric prefix signifying 1000.

kilojoule (kJ) 1000 joules.

kilopascal (kPa) 1000 pascals.

kinetic energy The energy associated with the motion of an object.

kinetic-molecular theory (KMT) The theory that explains the structures and behaviors of idealized models of gases, liquids, and solids.

Le Châtelier's principle When a system at equilibrium is subjected to a stress, the system will shift in order to lessen the effects of the stress. Eventually, a new equilibrium point is established.

Lewis acid Any substance that can accept a pair of electrons.

Lewis base Any substance that can donate a pair of electrons.

Lewis structure A shorthand notation for illustrating the ground-state valence electron configuration of an atom.

limiting reactant Any reactant that is entirely consumed in a chemical reaction.

liquid The phase in which matter has a definite volume but an indefinite shape; a liquid takes the shape of its container.

liter (L) A unit of volume in the metric system; 1 liter = 1000 cubic centimeters; 1 liter = 1 cubic decimeter; 1 liter is approximately equal to 1 quart.

litmus An acid-base indicator that is red in acidic solutions and blue in basic solutions.

London dispersion forces See **dispersion forces**.

macromolecule A giant molecule formed by network bonding or by polymerization.

malleability The property by which a substance is able to be formed into various shapes; metallic substances possess malleability.

manometer A curved tube used to determine gas pressure by measuring the difference in the heights of the liquid that partially fills the tube. One end of the tube can be in contact with the atmosphere (*open-tube manometer*) or be isolated from it (*closed-tube manometer*).

mass defect The difference in mass between a nucleus and the nucleons that compose it; the energy equivalent of mass defect is known as *binding energy*.

mass number The number of nucleons in a nuclide.

melting The transition from solid to liquid.

melting point The temperature at which the vapor pressure of a solid equals the vapor pressure of the liquid; the temperature at which the solid and liquid phases of a substance are in equilibrium; see also **freezing point**.

meniscus The curved surface of a liquid, caused by the attraction of the particles of the liquid and the container holding the liquid (e.g., water in a graduated cylinder), or by the mutual attraction of the particles of the liquid (e.g., mercury).

metal A substance composed of atoms with low ionization energies and relatively vacant valence levels; Na, Fe, Ag, and Ba are metallic substances.

metallic bond The delocalization of the valence electrons among the kernels of the metal atoms; "mobile valence electrons immersed in a sea of positive ions."

metalloid An element that has both metallic and nonmetallic properties; examples of metalloids include B, Ge, Si, and Te.

methanal HCHO; formaldehyde; the simplest aldehyde.

methane CH_4; the parent of the alkane family of hydrocarbons.

methanoic acid HCOOH; formic acid; the simplest organic acid.

methanol CH_3OH; methyl (wood) alcohol; the simplest alcohol.

milli- The metric prefix signifying 1/1000.

milliliter (mL) 1/1000 liter; 1 liter = 1000 milliliters.

miscible Able to be mixed in a solution of liquids that is soluble in all proportions; ethanol and water are a miscible pair of liquids.

mixture A material consisting of two or more components and having a variable composition.

moderator A substance used to produce slow neutrons and promote nuclear fission; graphite and heavy water are used as moderators in fission reactors.

molal boiling point constant (k_b) The increase in the boiling point of a solvent when 1 mole of particles is dissolved in 1 kilogram of solvent; k_b for H_2O is 0.52 C° per molal.

molal freezing point constant (k_f) The decrease in the freezing point of a solvent when 1 mole of particles is dissolved in 1 kilogram of solvent; k_f for H_2O is 1.86 C° per molal.

molality The concentration of a solution, measured as the number of moles of solute per kilogram of solvent.

molar mass The mass of any atom, element, ion, or compound expressed in grams per mole (g/mol).

molarity The concentration of a solution, measured as the number of moles of solute per liter of solution.

molar volume The volume occupied by 1 mole of an ideal gas; 22.4 liters at STP.

mole The number of atoms contained in 12 grams of carbon-12; see also **Avogadro's number**.

molecular formula A chemical formula that lists the number of atoms present but does not show the arrangement of the atoms in space.

molecular mass The sum of the masses of the atoms in a molecule; units are atomic mass units (u) or grams per mole (g/mol).

molecule The smallest unit of a nonionic substance; Ar, Cl_2, and NH_3 are molecules.

mole fraction (X) A measure of concentration that expresses the ratio of the number of moles of a given substance to the total number of moles present;

$$X_i = \frac{n_i}{n_{\text{total}}}.$$

monatomic ion An ion containing only one atom; Ca^{2+} is a monatomic ion; see also **simple ion**.

monatomic molecule A molecule consisting of one atom; Xe and He are monatomic molecules.

monomer The basic unit of a polymer; the monomer of a protein is an amino acid; the monomer of starch is glucose.

monoprotic acid A Brønsted–Lowry acid that contains one ionizable hydrogen atom.

network solid A substance formed by a two- or three-dimensional web of covalent bonds to produce a macromolecule; diamond and SiO_2 are network solids.

neutralization The reaction of equivalent amounts of hydronium and hydroxide ions in aqueous solution; the principal product is water. When the water is evaporated, the spectator ions form a salt; see also **spectator ion**.

neutron A neutral nuclear particle with a mass comparable to that of a proton.

noble gas See **inert gas**.

nonelectrolyte A substance whose aqueous solution does not conduct electricity; glucose is a nonelectrolyte.

nonmetal A substance that does not have characteristic metallic properties; C and S are nonmetallic elements.

nonpolar bond A covalent bond in which the electron pair or pairs are shared equally by both atoms.

nonpolar molecule A molecule containing only nonpolar bonds, such as N_2, or a molecule with a symmetrical charge distribution, such as CCl_4 and CO_2.

normal boiling point The boiling temperature of a substance at a pressure of 1 atmosphere.

nuclear equation A shorthand listing of reactant and product nuclides in a nuclear reaction.

nucleon A constituent of an atomic nucleus; a proton or a neutron.

nucleus The portion of the atom that contains more than 99.9 percent of the atom's mass; the nucleus is small, dense, and positively charged.

orbital A designation that describes the orientation of the electron cloud in space; p_x, p_y, and p_z are orbital designations.

order of magnitude The power of 10 closest to the value of a measurement.

ore A native mineral from which a metal or metals can be extracted.

organic chemistry The study of the hydrocarbons and their derivatives; the chemistry of carbon.

organic compound A compound that is a hydrocarbon or a hydrocarbon derivative.

oxidation The loss or apparent loss of electrons in a chemical reaction.

oxidation number (state) The charge that an atom has or appears to have when certain arbitrary rules are applied; oxidation numbers are useful for identifying the atoms that are oxidized and reduced in a redox reaction.

oxidation potential An electrode potential for an oxidation half-reaction.

oxidizing agent The particle in a redox reaction that causes another particle to be oxidized; as a result, an oxidizing agent is reduced.

paraffin A common name for a mixture of solid alkanes; another name for paraffin is wax.

partial pressure The individual pressure due to each gas in a mixture of gases.

particle accelerator A device that uses electric and/or magnetic fields to accelerate charged subatomic particles in order to study them further.

parts per million (ppm) A unit of concentration that expresses the mass of a solute dissolved in 1 million parts of a very dilute solution;

$$\text{ppm} = \frac{\text{mass}_{solute}}{\text{mass}_{solution}} \times 10^6.$$

pascal (Pa) The SI unit of pressure.

percent composition by mass The number of grams of an element (or group of elements) present in 100 grams of an ion or compound.

percent concentration A unit of concentration that expresses the mass of a solute dissolved in 100 parts of a solution;

$$\text{percent } (\%) = \frac{\text{mass}_{solute}}{\text{mass}_{solution}} \times 10^2.$$

percent yield The ratio of the actual yield of a product of a chemical reaction to the predicted yield, multiplied by 100.

period One of the horizontal rows of the Periodic Table; the period number indicates the valence level of an element.

periodic law The properties of elements recur at regular intervals and depend on their nuclear charges; "The properties of elements are a periodic function of their atomic numbers."

peroxide The O_2^{2-} ion; H_2O_2 and BaO_2 are peroxides.

petroleum Crude oil containing a mixture of hydrocarbons.

pH A scale of acidity and basicity based on the hydronium ion concentration in aqueous solution; $pH = -\log [H_3O^+]$. At 298 K, a pH less than 7 indicates an acidic solution; a pH greater than 7, a basic solution; a pH of 7, a neutral solution.

phase equilibrium The state in which the rates of opposing phase changes (freezing–melting, boiling–condensation, sublimation–deposition) are equal.

phenolphthalein An acid–base indicator that is colorless in acidic solutions and pink in basic solutions.

photon A fundamental unit of radiant energy; a quantum of radiation.

physical change A change that occurs with no change in the composition of a substance. Phase changes are physical changes.

physical property A property that can be measured without changing the composition of a substance. Density is a physical property.

pOH A scale of acidity and basicity based on the hydroxide ion concentration in aqueous solution; $pOH = -\log [OH^-]$. At 298 K, a pOH less than 7 indicates a basic solution; a pOH greater than 7, an acidic solution; a pOH of 7, a neutral solution; $pOH = 14 - pH$.

polar bond A covalent bond in which the electron pair or pairs are shared unequally by two atoms; the atom with the larger electronegativity has more of the electron density surrounding it.

polar molecule A molecule with an unsymmetrical charge distribution, such as H_2O; a dipole.

polyatomic ion An ion composed of more than one atom; SO_4^{2-} is a polyatomic ion.

polymer A macromolecule consisting of a chain of simpler units; polyethylene is a polymer of ethene.

polymerization See **polymer**; **addition polymerization**; **condensation polymerization**.

polyprotic acid A Brønsted–Lowry acid that contains more than one ionizable hydrogen atom.

positron A positively charged electron; a particle of antimatter.

potential energy The energy associated with the position of an object; a "stored" form of energy.

precipitate A deposit formed by the appearance of an excess of solid solute in a saturated solution.

precipitation See **crystallization**.

precision The closeness of a series of measurements to one another; see also **accuracy**.

pressure The force exerted on an object divided by the surface area of the object; $P = \dfrac{F}{A}$.

primary alcohol An alcohol in which two or three hydrogen atoms are bonded to the carbon atom containing the hydroxyl group.

principal energy level An integer, beginning with 1, that describes the approximate distance of an electron from the nucleus of an atom.

product(s) The substance or substances that are formed in a chemical process; products are on the right side of a chemical equation.

propanone $(CH_3)_2CO$; acetone; the simplest ketone.

proton A nuclear particle with a positive charge equal to the negative charge on the electron; a nucleon.

pure substance An element or a compound.

quark One of the fundamental particles of matter, always found in combination with each other. Nuclear particles, such as protons and neutrons, are comprised of groups of three quarks each.

491

radiant energy Electromagnetic energy; visible light and X rays are examples of radiant energy.

radioactive dating The use of radioactive isotopes to measure the age of an object.

radioactive tracer A radioisotope used to indicate the path of an atom in a chemical reaction.

radioactivity The spontaneous breakdown of a radioactive nuclide.

radioisotope A radioactive isotope.

rate-limiting step The slowest of the elementary reactions in a complex reaction.

reactant(s) The substance or substances that react in a chemical process; reactants are on the left side of a chemical equation.

redox reaction A chemical reaction in which oxidation-reduction takes place.

reducing agent The particle in a redox reaction that causes another particle to be reduced; as a result, a reducing agent is oxidized.

reduction The gain or apparent gain of electrons in a chemical reaction.

reduction potential An electrode potential for a reduction half-reaction.

roasting The reaction of a metallic ore with oxygen in order to produce the metallic oxide from which the metal may be obtained.

salt The spectator-ion product of a neutralization reaction; see also **spectator ion**.

salt bridge A device for allowing the flow of ions in a voltaic cell.

saponification The reaction of an ester with a base to produce an alcohol and the sodium salt of an organic acid; soap is produced by saponifying fats with NaOH.

saturated hydrocarbon A hydrocarbon containing only single carbon-carbon bonds.

saturated solution A solution in which the pure solute is in equilibrium with the dissolved solute; a solution that contains the maximum amount of dissolved solute.

secondary alcohol An alcohol in which only one hydrogen atom is bonded to the carbon atom containing the hydroxyl group.

second law of thermodynamics A statement that relates spontaneous processes to the entropy (disorder) of the universe.

significant digit(s) [figure(s)] The number or numbers that are part of a measurement. If there are two or more, all but the last figure is known; the last figure is the experimenter's best estimate.

simple ion An ion containing only one atom; a monatomic ion; Na^+ is a simple ion.

single bond A covalent bond in which one pair of electrons is shared by two adjacent atoms.

solid The phase in which matter has both definite shape and definite volume.

solubility The amount of solute needed to produce a saturated solution with a given amount of solvent.

solubility product constant (K_{sp}) The equilibrium constant expression that provides a measure of the solubility of an ionic compound; a large K_{sp} value generally indicates a more soluble salt.

solute(s) The substance or substances dissolved in a solution.

solution A homogeneous mixture whose particles are extremely small.

solution equilibrium The state in which the undissolved and dissolved solutes are in dynamic equilibrium; in a solid-liquid solution, the rate of dissolving equals the rate of crystallization.

solvent The substance in which the solute is dissolved.

spectator ion An ion that does not take part in a chemical reaction; in the (acid-base) reaction between NaOH(aq) and HCl(aq), Na^+(aq) and Cl^-(aq) are spectator ions; in the (redox) reaction between Zn(s) and $Cu(NO_3)_2$(aq), NO_3^-(aq) is the spectator ion.

spontaneous reaction A reaction that can occur under a specified set of conditions without the application of external work.

standard atmosphere (atm) A unit of pressure; 1 standard atmosphere = 101.325 kilopascals (exactly); 760 millimeters Hg.

standard cell potential ($\mathscr{E}_{cell}°$) The maximum potential difference of an electrochemical cell, measured in volts, under standard conditions.

standard electrode (reduction) potential ($\mathscr{E}°$) An oxidation potential or a reduction potential measured under standard conditions.

standard hydrogen half-cell A hydrogen gas-hydrogen ion electrode whose electrode potential under standard conditions is set at 0 volt.

standard solution A solution whose concentration is accurately known; a standard solution is used for analyzing other substances.

standard state A set of conditions imposed on a pure substance that includes a pressure of 100.00 kilopascals and a specified temperature (usually 298.15 K).

standard-state pressure A pressure of 100.00 kilopascals (exactly).

standard temperature and pressure (STP) 273.15 Kelvin and 1 atmosphere.

stereoisomers Isomers in which the atoms are connected in the same order but are arranged differently in space. *Cis-trans* isomerism is an example of stereoisomerism.

Stock system A systematic method of naming chemical compounds in which the *positive* oxidation number is written as a Roman numeral in parentheses after the element. For example, the Stock name of the compound Fe_2O_3 is iron(III) oxide.

stoichiometry The study of quantitative relationships in substances and reactions; chemical mathematics.

strong acid In aqueous solution, a substance that ionizes almost completely to hydronium ion.

strong base In aqueous solution, a substance that ionizes or dissociates almost completely to hydroxide ion.

structural formula A chemical formula that illustrates the spatial arrangement of each atom.

sublevel The property of an electron that describes the shape of its cloud; the letter designations for sublevels are *s, p, d, f, ...*

sublimation The direct transition from solid to gas; CO_2(s) and I_2(s) sublime at atmospheric pressure.

supersaturated solution A solution that contains more dissolved solute than a saturated solution at the same temperature; supersaturation is an unstable condition.

surroundings In thermodynamics, the part of the universe that is *not* included in the system being studied.

system In thermodynamics, the part of the universe singled out for study.

temperature A measure of the average kinetic energy of the particles of a substance.

tertiary alcohol An alcohol in which no hydrogen atoms are bonded to the carbon atom containing the hydroxyl group.

thermal energy The part of internal energy that is associated with random molecular motions.

titration The addition of a known volume of a standard solution in order to determine the concentration of an unknown solution.

torr A unit of pressure equivalent to 1 millimeter of mercury (mm Hg).

transition element An element in Groups 3–11 of the Periodic Table.

transmutation The conversion of one element to another by a nuclear process.

trihydroxy alcohol An alcohol with three hydroxyl groups; glycerol (1,2,3-propanetriol) is a trihydroxy alcohol.

triple bond A covalent bond in which three pairs of electrons are shared by two adjacent atoms.

triple point The unique temperature and pressure at which the solid, liquid, and gaseous phases can coexist in equilibrium. The triple point of water is used to set the fixed point on the Kelvin temperature scale (273.16 K).

tritium The radioactive isotope of hydrogen with a mass number of 3.

unsaturated hydrocarbon A hydrocarbon containing double and/or triple carbon-carbon bonds.

unsaturated solution A solution that contains less dissolved solute than a saturated solution at the same temperature.

valence electron An electron in the outermost principle energy level.

valence shell electron-pair repulsion method (VSEPR) A technique for determining the shape of a molecule or polyatomic ion by considering the number of shared and unshared pairs of electrons around the central atom of the molecule or ion.

van der Waals forces All forces involving attractions of polar molecules and nonpolar molecules; see also **dipole-dipole attraction**; **dispersion forces**; **ion-dipole attraction**.

vapor pressure The pressure produced by a solid or a liquid when it is in equilibrium with its gas phase.

volt (V) The unit of electric potential; it measures the work output per unit of electric charge transferred.

voltaic cell A device that uses a spontaneous redox reaction to generate electrical energy.

water of hydration The water molecules that are part of the crystalline structure of certain compounds.

weak acid In aqueous solution, a substance that is poorly ionized and produces only a small concentration of hydronium ion.

weak base In aqueous solution, a substance that is poorly ionized or dissociated and produces only a small concentration of hydroxide ion.

NEW YORK STATE REGENTS REFERENCE TABLES FOR CHEMISTRY

TABLE A
Standard Temperature and Pressure

Name	Value	Unit
Standard Pressure	101.3 kPa 1 atm	kilopascal atmosphere
Standard Temperature	273 K 0°C	kelvin degree Celsius

TABLE B
Physical Constants for Water

Heat of Fusion	334 J/g
Heat of Vaporization	2260 J/g
Specific Heat Capacity of $H_2O(\ell)$	4.18 J/g•K

TABLE C
Selected Prefixes

Factor	Prefix	Symbol
10^3	kilo-	k
10^{-1}	deci-	d
10^{-2}	centi-	c
10^{-3}	milli-	m
10^{-6}	micro-	μ
10^{-9}	nano-	n
10^{-12}	pico-	p

TABLE D
Selected Units

Symbol	Name	Quantity
m	meter	length
g	gram	mass
Pa	pascal	pressure
K	kelvin	temperature
mol	mole	amount of substance
J	joule	energy, work, quantity of heat
s	second	time
min	minute	time
h	hour	time
d	day	time
y	year	time
L	liter	volume
ppm	parts per million	concentration
M	molarity	solution concentration
u	atomic mass unit	atomic mass

TABLE E
Selected Polyatomic Ions

Formula	Name	Formula	Name
H_3O^+	hydronium	CrO_4^{2-}	chromate
Hg_2^{2+}	mercury(I)	$Cr_2O_7^{2-}$	dichromate
NH_4^+	ammonium	MnO_4^-	permanganate
$C_2H_3O_2^-$ CH_3COO^- } acetate		NO_2^-	nitrite
		NO_3^-	nitrate
CN^-	cyanide	O_2^{2-}	peroxide
CO_3^{2-}	carbonate		
HCO_3^-	hydrogen carbonate	OH^-	hydroxide
		PO_4^{3-}	phosphate
$C_2O_4^{2-}$	oxalate	SCN^-	thiocyanate
ClO^-	hypochlorite	SO_3^{2-}	sulfite
ClO_2^-	chlorite	SO_4^{2-}	sulfate
ClO_3^-	chlorate	HSO_4^-	hydrogen sulfate
ClO_4^-	perchlorate	$S_2O_3^{2-}$	thiosulfate

TABLE F
Solubility Guidelines for Aqueous Solutions

Ions That Form *Soluble* Compounds	Exceptions
Group 1 ions (Li^+, Na^+, etc.)	
ammonium (NH_4^+)	
nitrate (NO_3^-)	
acetate ($C_2H_3O_2^-$ or CH_3COO^-)	
hydrogen carbonate (HCO_3^-)	
chlorate (ClO_3^-)	
halides (Cl^-, Br^-, I^-)	when combined with Ag^+, Pb^{2+}, or Hg_2^{2+}
sulfates (SO_4^{2-})	when combined with Ag^+, Ca^{2+}, Sr^{2+}, Ba^{2+}, or Pb^{2+}

Ions That Form *Insoluble* Compounds*	Exceptions
carbonate (CO_3^{2-})	when combined with Group 1 ions or ammonium (NH_4^+)
chromate (CrO_4^{2-})	when combined with Group 1 ions, Ca^{2+}, Mg^{2+}, or ammonium (NH_4^+)
phosphate (PO_4^{3-})	when combined with Group 1 ions or ammonium (NH_4^+)
sulfide (S^{2-})	when combined with Group 1 ions or ammonium (NH_4^+)
hydroxide (OH^-)	when combined with Group 1 ions, Ca^{2+}, Ba^{2+}, Sr^{2+}, or ammonium (NH_4^+)

*compounds having very low solubility in H_2O

TABLE G
Solubility Curves at Standard Pressure

TABLE H
Vapor Pressure of Four Liquids

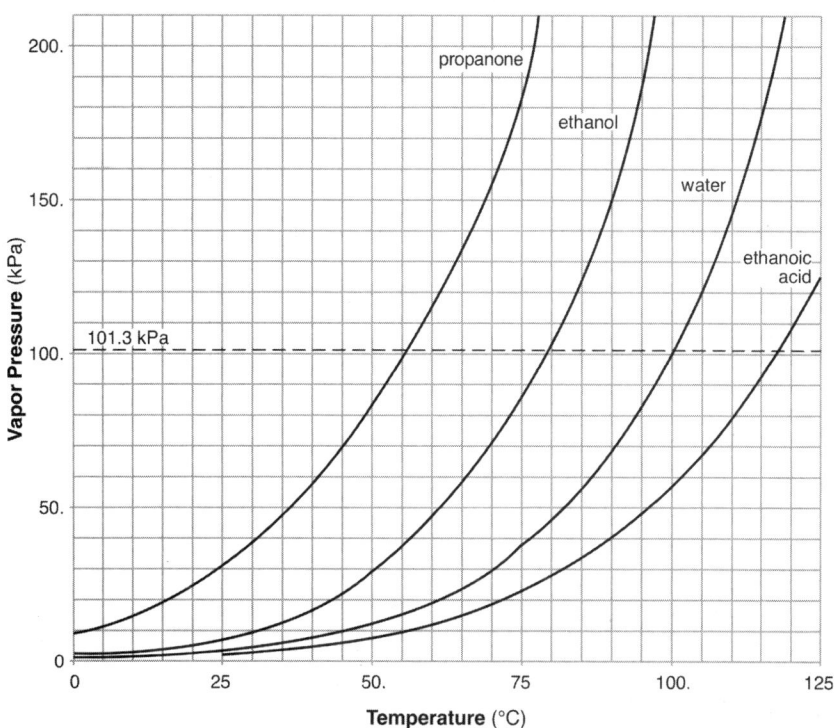

TABLE I
Heats of Reaction at 101.3 kPa and 298 K

Reaction	ΔH (kJ)*
$CH_4(g) + 2O_2(g) \longrightarrow CO_2(g) + 2H_2O(\ell)$	−890.4
$C_3H_8(g) + 5O_2(g) \longrightarrow 3CO_2(g) + 4H_2O(\ell)$	−2219.2
$2C_8H_{18}(\ell) + 25O_2(g) \longrightarrow 16CO_2(g) + 18H_2O(\ell)$	−10943
$2CH_3OH(\ell) + 3O_2(g) \longrightarrow 2CO_2(g) + 4H_2O(\ell)$	−1452
$C_2H_5OH(\ell) + 3O_2(g) \longrightarrow 2CO_2(g) + 3H_2O(\ell)$	−1367
$C_6H_{12}O_6(s) + 6O_2(g) \longrightarrow 6CO_2(g) + 6H_2O(\ell)$	−2804
$2CO(g) + O_2(g) \longrightarrow 2CO_2(g)$	−566.0
$C(s) + O_2(g) \longrightarrow CO_2(g)$	−393.5
$4Al(s) + 3O_2(g) \longrightarrow 2Al_2O_3(s)$	−3351
$N_2(g) + O_2(g) \longrightarrow 2NO(g)$	+182.6
$N_2(g) + 2O_2(g) \longrightarrow 2NO_2(g)$	+66.4
$2H_2(g) + O_2(g) \longrightarrow 2H_2O(g)$	−483.6
$2H_2(g) + O_2(g) \longrightarrow 2H_2O(\ell)$	−571.6
$N_2(g) + 3H_2(g) \longrightarrow 2NH_3(g)$	−91.8
$2C(s) + 3H_2(g) \longrightarrow C_2H_6(g)$	−84.0
$2C(s) + 2H_2(g) \longrightarrow C_2H_4(g)$	+52.4
$2C(s) + H_2(g) \longrightarrow C_2H_2(g)$	+227.4
$H_2(g) + I_2(g) \longrightarrow 2HI(g)$	+53.0
$KNO_3(s) \xrightarrow{H_2O} K^+(aq) + NO_3^-(aq)$	+34.89
$NaOH(s) \xrightarrow{H_2O} Na^+(aq) + OH^-(aq)$	−44.51
$NH_4Cl(s) \xrightarrow{H_2O} NH_4^+(aq) + Cl^-(aq)$	+14.78
$NH_4NO_3(s) \xrightarrow{H_2O} NH_4^+(aq) + NO_3^-(aq)$	+25.69
$NaCl(s) \xrightarrow{H_2O} Na^+(aq) + Cl^-(aq)$	+3.88
$LiBr(s) \xrightarrow{H_2O} Li^+(aq) + Br^-(aq)$	−48.83
$H^+(aq) + OH^-(aq) \longrightarrow H_2O(\ell)$	−55.8

*The ΔH values are based on molar quantities represented in the equations. A minus sign indicates an exothermic reaction.

TABLE J
Activity Series**

Most Active	Metals	Nonmetals	Most Active
	Li	F_2	
	Rb	Cl_2	
	K	Br_2	
	Cs	I_2	
	Ba		
	Sr		
	Ca		
	Na		
	Mg		
	Al		
	Ti		
	Mn		
	Zn		
	Cr		
	Fe		
	Co		
	Ni		
	Sn		
	Pb		
	H_2		
	Cu		
	Ag		
Least Active	Au		Least Active

**Activity Series is based on the hydrogen standard. H_2 is *not* a metal.

TABLE K
Common Acids

Formula	Name
$HCl(aq)$	hydrochloric acid
$HNO_2(aq)$	nitrous acid
$HNO_3(aq)$	nitric acid
$H_2SO_3(aq)$	sulfurous acid
$H_2SO_4(aq)$	sulfuric acid
$H_3PO_4(aq)$	phosphoric acid
$H_2CO_3(aq)$ or $CO_2(aq)$	carbonic acid
$CH_3COOH(aq)$ or $HC_2H_3O_2(aq)$	ethanoic acid (acetic acid)

TABLE L
Common Bases

Formula	Name
$NaOH(aq)$	sodium hydroxide
$KOH(aq)$	potassium hydroxide
$Ca(OH)_2(aq)$	calcium hydroxide
$NH_3(aq)$	aqueous ammonia

TABLE M
Common Acid–Base Indicators

Indicator	Approximate pH Range for Color Change	Color Change
methyl orange	3.1–4.4	red to yellow
bromthymol blue	6.0–7.6	yellow to blue
phenolphthalein	8–9	colorless to pink
litmus	4.5–8.3	red to blue
bromcresol green	3.8–5.4	yellow to blue
thymol blue	8.0–9.6	yellow to blue

Source: *The Merck Index*, 14th ed., 2006, Merck Publishing Group

TABLE N
Selected Radioisotopes

Nuclide	Half-Life	Decay Mode	Nuclide Name
^{198}Au	2.695 d	β^-	gold-198
^{14}C	5715 y	β^-	carbon-14
^{37}Ca	182 ms	β^+	calcium-37
^{60}Co	5.271 y	β^-	cobalt-60
^{137}Cs	30.2 y	β^-	cesium-137
^{53}Fe	8.51 min	β^+	iron-53
^{220}Fr	27.4 s	α	francium-220
^{3}H	12.31 y	β^-	hydrogen-3
^{131}I	8.021 d	β^-	iodine-131
^{37}K	1.23 s	β^+	potassium-37
^{42}K	12.36 h	β^-	potassium-42
^{85}Kr	10.73 y	β^-	krypton-85
^{16}N	7.13 s	β^-	nitrogen-16
^{19}Ne	17.22 s	β^+	neon-19
^{32}P	14.28 d	β^-	phosphorus-32
^{239}Pu	2.410×10^4 y	α	plutonium-239
^{226}Ra	1599 y	α	radium-226
^{222}Rn	3.823 d	α	radon-222
^{90}Sr	29.1 y	β^-	strontium-90
^{99}Tc	2.13×10^5 y	β^-	technetium-99
^{232}Th	1.40×10^{10} y	α	thorium-232
^{233}U	1.592×10^5 y	α	uranium-233
^{235}U	7.04×10^8 y	α	uranium-235
^{238}U	4.47×10^9 y	α	uranium-238

Source: *CRC Handbook of Chemistry and Physics*, 91st ed., 2010–2011, CRC Press

TABLE O
Symbols Used in Nuclear Chemistry

Name	Notation	Symbol
alpha particle	^4_2He or $^4_2\alpha$	α
beta particle	$^0_{-1}\text{e}$ or $^0_{-1}\beta$	β^-
gamma radiation	$^0_0\gamma$	γ
neutron	^1_0n	n
proton	^1_1H or ^1_1p	p
positron	$^0_{+1}\text{e}$ or $^0_{+1}\beta$	β^+

TABLE P
Organic Prefixes

Prefix	Number of Carbon Atoms
meth-	1
eth-	2
prop-	3
but-	4
pent-	5
hex-	6
hept-	7
oct-	8
non-	9
dec-	10

TABLE Q
Homologous Series of Hydrocarbons

Name	General Formula	Examples	
		Name	Structural Formula
alkanes	C_nH_{2n+2}	ethane	H—C—C—H (with H atoms above and below each C)
alkenes	C_nH_{2n}	ethene	C=C (with H atoms attached)
alkynes	C_nH_{2n-2}	ethyne	H—C≡C—H

Note: n = number of carbon atoms

504

TABLE R
Organic Functional Groups

Class of Compound	Functional Group	General Formula	Example
halide (halocarbon)	$-F$ (fluoro-) $-Cl$ (chloro-) $-Br$ (bromo-) $-I$ (iodo-)	$R-X$ (X represents any halogen)	$CH_3CHClCH_3$ 2-chloropropane
alcohol	$-OH$	$R-OH$	$CH_3CH_2CH_2OH$ 1-propanol
ether	$-O-$	$R-O-R'$	$CH_3OCH_2CH_3$ methyl ethyl ether
aldehyde	$\overset{\displaystyle O}{\overset{\|}{-C-H}}$	$\overset{\displaystyle O}{\overset{\|}{R-C-H}}$	$\overset{\displaystyle O}{\overset{\|}{CH_3CH_2C-H}}$ propanal
ketone	$\overset{\displaystyle O}{\overset{\|}{-C-}}$	$\overset{\displaystyle O}{\overset{\|}{R-C-R'}}$	$\overset{\displaystyle O}{\overset{\|}{CH_3CCH_2CH_2CH_3}}$ 2-pentanone
organic acid	$\overset{\displaystyle O}{\overset{\|}{-C-OH}}$	$\overset{\displaystyle O}{\overset{\|}{R-C-OH}}$	$\overset{\displaystyle O}{\overset{\|}{CH_3CH_2C-OH}}$ propanoic acid
ester	$\overset{\displaystyle O}{\overset{\|}{-C-O-}}$	$\overset{\displaystyle O}{\overset{\|}{R-C-O-R'}}$	$\overset{\displaystyle O}{\overset{\|}{CH_3CH_2COCH_3}}$ methyl propanoate
amine	$\overset{\displaystyle }{\underset{}{-N-}}$	$\overset{\displaystyle R'}{\overset{\|}{R-N-R''}}$	$CH_3CH_2CH_2NH_2$ 1-propanamine
amide	$\overset{\displaystyle O}{\overset{\|}{-C-NH}}$	$\overset{\displaystyle O \quad R'}{\overset{\|\quad\|}{R-C-NH}}$	$\overset{\displaystyle O}{\overset{\|}{CH_3CH_2C-NH_2}}$ propanamide

Note: R represents a bonded atom or group of atoms.

Periodic Table of the Elements

Period

1

| 1 | H 1.00794 +1 −1 1 1 |

Group

1   2

| 2 | Li 6.941 +1 3 2-1 | Be 9.01218 +2 4 2-2 |
| 3 | Na 22.98977 +1 11 2-8-1 | Mg 24.305 +2 12 2-8-2 |

Group

	1	2	3	4	5	6	7	8	9
4	K 39.0983 +1 19 2-8-8-1	Ca 40.08 +2 20 2-8-8-2	Sc 44.9559 +3 21 2-8-9-2	Ti 47.867 +2 +3 +4 22 2-8-10-2	V 50.9415 +2 +3 +4 +5 23 2-8-11-2	Cr 51.996 +2 +3 +6 24 2-8-13-1	Mn 54.9380 +2 +3 +4 +7 25 2-8-13-2	Fe 55.845 +2 +3 26 2-8-14-2	Co 58.9332 +2 +3 27 2-8-15-2
5	Rb 85.4678 +1 37 2-8-18-8-1	Sr 87.62 +2 38 2-8-18-8-2	Y 88.9059 +3 39 2-8-18-9-2	Zr 91.224 +4 40 2-8-18-10-2	Nb 92.9064 +3 +5 41 2-8-18-12-1	Mo 95.94 +6 42 2-8-18-13-1	Tc (98) +4 +6 +7 43 2-8-18-13-2	Ru 101.07 +3 44 2-8-18-15-1	Rh 102.906 +3 45 2-8-18-16-1
6	Cs 132.905 +1 55 2-8-18-18-8-1	Ba 137.33 +2 56 2-8-18-18-8-2	La 138.9055 +3 57 2-8-18-18-9-2	Hf 178.49 +4 72 *18-32-10-2	Ta 180.948 +5 73 -18-32-11-2	W 183.84 +6 74 -18-32-12-2	Re 186.207 +6 +7 75 -18-32-13-2	Os 190.23 +3 +4 76 -18-32-14-2	Ir 192.217 +3 +4 77 -18-32-15-2
7	Fr (223) +1 87 -18-32-18-8-1	Ra (226) +2 88 -18-32-18-8-2	Ac (227) +3 89 -18-32-18-9-2	Rf (261) +4 104	Db (262) 105	Sg (266) 106	Bh (272) 107	Hs (277) 108	Mt (276) 109

| Ce 140.116 +3 +4 58 | Pr 140.908 +3 59 | Nd 144.24 +3 60 | Pm (145) +3 61 | Sm 150.36 +2 +3 62 |
| Th 232.038 +4 90 | Pa 231.036 +4 +5 91 | U 238.029 +3 +4 +5 +6 92 | Np (237) +3 +4 +5 +6 93 | Pu (244) +3 +4 +5 +6 94 |

*denotes the presence of (2-8-) for elements 72 and above

**The systematic names and symbols for elements of atomic numbers 113 and above will be used until the approval of trivial names by IUPAC.

Source: *CRC Handbook of Chemistry and Physics*, 91st ed., 2010–2011, CRC Press

KEY

Atomic Mass → | 12.011 | −4 ← Selected Oxidation States
                                +2
Symbol → **C** | +4
                              Relative atomic masses are based
                              on ¹²C = 12 (exact)
Atomic Number → | 6
Electron Configuration → | 2-4

Note: Numbers in parentheses are mass numbers of the most stable or common isotope.

| 18 |
| 4.00260   0 |
| **He** |
| 2 / 2 |

Group

| 10 | 11 | 12 | 13 | 14 | 15 | 16 | 17 | 18 |

58.693 +2 +3	63.546 +1 +2	65.409 +2	69.723 +3	72.64 +2 +4	74.9216 −3 +3 +5	78.96 −2 +4 +6	79.904 −1 +1 +5	83.798 0 +2
Ni 28 2-8-16-2	**Cu** 29 2-8-18-1	**Zn** 30 2-8-18-2	**Ga** 31 2-8-18-3	**Ge** 32 2-8-18-4	**As** 33 2-8-18-5	**Se** 34 2-8-18-6	**Br** 35 2-8-18-7	**Kr** 36 2-8-18-8
106.42 +2 +4	107.868 +1	112.41 +2	114.818 +3	118.71 +2 +4	121.760 −3 +3 +5	127.60 −2 +4 +6	126.904 −1 +1 +5 +7	131.29 0 +2 +4 +6
Pd 46 2-8-18-18	**Ag** 47 2-8-18-18-1	**Cd** 48 2-8-18-18-2	**In** 49 2-8-18-18-3	**Sn** 50 2-8-18-18-4	**Sb** 51 2-8-18-18-5	**Te** 52 2-8-18-18-6	**I** 53 2-8-18-18-7	**Xe** 54 2-8-18-18-8
195.08 +2 +4	196.967 +1 +3	200.59 +1 +2	204.383 +1 +3	207.2 +2 +4	208.980 +3 +5	(209) +2 +4	(210)	(222) 0
Pt 78 -18-32-17-1	**Au** 79 -18-32-18-1	**Hg** 80 -18-32-18-2	**Tl** 81 -18-32-18-3	**Pb** 82 -18-32-18-4	**Bi** 83 -18-32-18-5	**Po** 84 -18-32-18-6	**At** 85 -18-32-18-7	**Rn** 86 -18-32-18-8
(281)	(280)	(285)	(284)	(289)	(288)	(292)	(?)	(294)
Ds 110	**Rg** 111	**Cn** 112	**Uut** 113**	**Uuq** 114	**Uup** 115	**Uuh** 116	**Uus** 117	**Uuo** 118

151.964 +2 +3	157.25 +3	158.925 +3	162.500 +3	164.930 +3	167.259 +3	168.934 +3	173.04 +2 +3	174.9668 +3
Eu 63	**Gd** 64	**Tb** 65	**Dy** 66	**Ho** 67	**Er** 68	**Tm** 69	**Yb** 70	**Lu** 71
(243) +3 +4 +5 +6	(247) +3	(247) +3 +4	(251) +3	(252) +3	(257) +3	(258) +2 +3	(259) +2 +3	(262) +3
Am 95	**Cm** 96	**Bk** 97	**Cf** 98	**Es** 99	**Fm** 100	**Md** 101	**No** 102	**Lr** 103

TABLE S
Properties of Selected Elements

Atomic Number	Symbol	Name	First Ionization Energy (kJ/mol)	Electro-negativity	Melting Point (K)	Boiling* Point (K)	Density** (g/cm³)	Atomic Radius (pm)
1	H	hydrogen	1312	2.2	14	20.	0.000082	32
2	He	helium	2372	—	—	4	0.000164	37
3	Li	lithium	520.	1.0	454	1615	0.534	130.
4	Be	beryllium	900.	1.6	1560.	2744	1.85	99
5	B	boron	801	2.0	2348	4273	2.34	84
6	C	carbon	1086	2.6	—	—	—	75
7	N	nitrogen	1402	3.0	63	77	0.001145	71
8	O	oxygen	1314	3.4	54	90.	0.001308	64
9	F	fluorine	1681	4.0	53	85	0.001553	60.
10	Ne	neon	2081	—	24	27	0.000825	62
11	Na	sodium	496	0.9	371	1156	0.97	160.
12	Mg	magnesium	738	1.3	923	1363	1.74	140.
13	Al	aluminum	578	1.6	933	2792	2.70	124
14	Si	silicon	787	1.9	1687	3538	2.3296	114
15	P	phosphorus (white)	1012	2.2	317	554	1.823	109
16	S	sulfur (monoclinic)	1000.	2.6	388	718	2.00	104
17	Cl	chlorine	1251	3.2	172	239	0.002898	100.
18	Ar	argon	1521	—	84	87	0.001633	101
19	K	potassium	419	0.8	337	1032	0.89	200.
20	Ca	calcium	590.	1.0	1115	1757	1.54	174
21	Sc	scandium	633	1.4	1814	3109	2.99	159
22	Ti	titanium	659	1.5	1941	3560.	4.506	148
23	V	vanadium	651	1.6	2183	3680.	6.0	144
24	Cr	chromium	653	1.7	2180.	2944	7.15	130.
25	Mn	manganese	717	1.6	1519	2334	7.3	129
26	Fe	iron	762	1.8	1811	3134	7.87	124
27	Co	cobalt	760.	1.9	1768	3200.	8.86	118
28	Ni	nickel	737	1.9	1728	3186	8.90	117
29	Cu	copper	745	1.9	1358	2835	8.96	122
30	Zn	zinc	906	1.7	693	1180.	7.134	120.
31	Ga	gallium	579	1.8	303	2477	5.91	123
32	Ge	germanium	762	2.0	1211	3106	5.3234	120.
33	As	arsenic (gray)	944	2.2	1090.	—	5.75	120.
34	Se	selenium (gray)	941	2.6	494	958	4.809	118
35	Br	bromine	1140.	3.0	266	332	3.1028	117
36	Kr	krypton	1351	—	116	120.	0.003425	116
37	Rb	rubidium	403	0.8	312	961	1.53	215
38	Sr	strontium	549	1.0	1050.	1655	2.64	190.
39	Y	yttrium	600.	1.2	1795	3618	4.47	176
40	Zr	zirconium	640.	1.3	2128	4682	6.52	164

TABLE S
Properties of Selected Elements *(Continued)*

Atomic Number	Symbol	Name	First Ionization Energy (kJ/mol)	Electro-negativity	Melting Point (K)	Boiling* Point (K)	Density** (g/cm³)	Atomic Radius (pm)
41	Nb	niobium	652	1.6	2750.	5017	8.57	156
42	Mo	molybdenum	684	2.2	2896	4912	10.2	146
43	Tc	technetium	702	2.1	2430.	4538	11	138
44	Ru	ruthenium	710.	2.2	2606	4423	12.1	136
45	Rh	rhodium	720.	2.3	2237	3968	12.4	134
46	Pd	palladium	804	2.2	1828	3236	12.0	130.
47	Ag	silver	731	1.9	1235	2435	10.5	136
48	Cd	cadmium	868	1.7	594	1040.	8.69	140.
49	In	indium	558	1.8	430.	2345	7.31	142
50	Sn	tin (white)	709	2.0	505	2875	7.287	140.
51	Sb	antimony (gray)	831	2.1	904	1860.	6.68	140.
52	Te	tellurium	869	2.1	723	1261	6.232	137
53	I	iodine	1008	2.7	387	457	4.933	136
54	Xe	xenon	1170.	2.6	161	165	0.005366	136
55	Cs	cesium	376	0.8	302	944	1.873	238
56	Ba	barium	503	0.9	1000.	2170.	3.62	206
57	La	lanthanum	538	1.1	1193	3737	6.15	194
Elements 58–71 have been omitted.								
72	Hf	hafnium	659	1.3	2506	4876	13.3	164
73	Ta	tantalum	728	1.5	3290.	5731	16.4	158
74	W	tungsten	759	1.7	3695	5828	19.3	150.
75	Re	rhenium	756	1.9	3458	5869	20.8	141
76	Os	osmium	814	2.2	3306	5285	22.587	136
77	Ir	iridium	865	2.2	2719	4701	22.562	132
78	Pt	platinum	864	2.2	2041	4098	21.5	130.
79	Au	gold	890.	2.4	1337	3129	19.3	130.
80	Hg	mercury	1007	1.9	234	630.	13.5336	132
81	Tl	thallium	589	1.8	577	1746	11.8	144
82	Pb	lead	716	1.8	600.	2022	11.3	145
83	Bi	bismuth	703	1.9	544	1837	9.79	150.
84	Po	polonium	812	2.0	527	1235	9.20	142
85	At	astatine	—	2.2	575	—	—	148
86	Rn	radon	1037	—	202	211	0.009074	146
87	Fr	francium	393	0.7	300.	—	—	242
88	Ra	radium	509	0.9	969	—	5	211
89	Ac	actinium	499	1.1	1323	3471	10.	201
Elements 90 and above have been omitted.								

* boiling point at standard pressure
** density of solids and liquids at room temperature and density of gases at 298 K and 101.3 kPa
— no data available
Source: *CRC Handbook for Chemistry and Physics*, 91st ed., 2010–2011, CRC Press

TABLE T
Important Formulas and Equations

Density	$d = \dfrac{m}{V}$	d = density m = mass V = volume
Mole Calculations	number of moles = $\dfrac{\text{given mass}}{\text{gram-formula mass}}$	
Percent Error	% error = $\dfrac{\text{measured value} - \text{accepted value}}{\text{accepted value}} \times 100$	
Percent Composition	% composition by mass = $\dfrac{\text{mass of part}}{\text{mass of whole}} \times 100$	
Concentration	parts per million = $\dfrac{\text{mass of solute}}{\text{mass of solution}} \times 1\,000\,000$	
	molarity = $\dfrac{\text{moles of solute}}{\text{liter of solution}}$	
Combined Gas Law	$\dfrac{P_1V_1}{T_1} = \dfrac{P_2V_2}{T_2}$	P = pressure V = volume T = temperature
Titration	$M_AV_A = M_BV_B$	M_A = molarity of H^+ M_B = molarity of OH^- V_A = volume of acid V_B = volume of base
Heat	$q = mC\Delta T$ $q = mH_f$ $q = mH_v$	q = heat H_f = heat of fusion m = mass H_v = heat of vaporization C = specific heat capacity ΔT = change in temperature
Temperature	$K = °C + 273$	K = kelvin $°C$ = degree Celsius

ADDITIONAL REFERENCE TABLES FOR CHEMISTRY

TABLE U
Selected Physical Constants

Name	Symbol or Abbreviation	Value
Avogadro constant	N_A	6.022×10^{23}/mol
Boltzmann constant	k	1.381×10^{-23} J/K
Elementary charge	e	1.602×10^{-19} C
Speed of light	c	2.998×10^8 m/s
Planck constant	h	6.626×10^{-34} J·s
Molar gas constant	R	8.315 Pa·m³/mol·K 0.08206 L·atm/mol·K
Atomic mass unit	u (amu)	1.661×10^{-27} kg
Proton mass	m_p	1.673×10^{-27} kg 1.007 u
Neutron mass	m_n	1.675×10^{-27} kg 1.009 u
Electron mass	m_e	9.109×10^{-31} kg 0.0005486 u
Standard-state pressure	—	100.0 kPa
Standard atmosphere	atm	101.3 kPa 760.0 mm Hg 760.0 torr
Triple point of water	T_{tr}	273.16 K
Standard temperature	—	273.15 K
Standard temperature and pressure	STP	273.15 K and 1 atm
Molar volume of ideal gas at STP	V_m	22.41 L/mol

TABLE V
Elements of the Periodic Table
(in Alphabetical Order)

Name	Atomic Number	Symbol	Location in Periodic Table*	Electron Configuration†
Actinium	89	Ac	P7, G3	$[Rn]6d^17s^2$
Aluminum	13	Al	P3, G13	$[Ne]3s^23p^1$
Americium	95	Am	AS	$[Rn]5f^77s^2$
Antimony	51	Sb	P5, G15	$[Kr]4d^{10}5s^25p^3$
Argon	18	Ar	P3, G18	$[Ne]3s^23p^6$
Arsenic	33	As	P4, G15	$[Ar]3d^{10}4s^24p^3$
Astatine	85	At	P6, G17	$[Xe]4f^{14}5d^{10}6s^26p^5$
Barium	56	Ba	P6, G2	$[Xe]6s^2$
Berkelium	97	Bk	AS	$[Rn]5f^97s^2$
Beryllium	4	Be	P2, G2	$[He]2s^2$
Bismuth	83	Bi	P6, G14	$[Xe]4f^{14}5d^{10}6s^26p^3$
Bohrium	107	Bh	P7, G7	$[Rn]5f^{14}6d^57s^2$
Boron	5	B	P2, G13	$[He]2s^22p^1$
Bromine	35	Br	P4, G17	$[Ar]3d^{10}4s^24p^5$
Cadmium	48	Cd	P5, G12	$[Kr]4d^{10}5s^2$
Calcium	20	Ca	P4, G2	$[Ar]4s^2$
Californium	98	Cf	AS	$[Rn]5f^{10}7s^2$
Carbon	6	C	P2, G14	$[He]2s^22p^2$
Cerium	58	Ce	LS	$[Xe]4f^15d^16s^2$
Cesium	55	Cs	P6, G1	$[Xe]6s^1$
Chlorine	17	Cl	P3, G17	$[Ne]3s^23p^5$
Chromium	24	Cr	P4, G6	$[Ar]3d^54s^1$
Cobalt	27	Co	P4, G9	$[Ar]3d^74s^2$
Copernicium	112	Cn	P7, G12	$[Rn]5f^{14}6d^{10}7s^2$
Copper	29	Cu	P4, G11	$[Ar]3d^{10}4s^1$
Curium	96	Cm	AS	$[Rn]5f^76d^17s^2$
Darmstadtium	110	Ds	P7, G10	$[Rn]5f^{14}6d^97s^1$
Dubnium	105	Db	P7, G5	$[Rn]5f^{14}6d^37s^2$
Dysprosium	66	Dy	LS	$[Xe]4f^{10}6s^2$
Einsteinium	99	Es	AS	$[Rn]5f^{11}7s^2$
Erbium	68	Er	LS	$[Xe]4f^{12}6s^2$
Europium	63	Eu	LS	$[Xe]4f^76s^2$
Fermium	100	Fm	AS	$[Rn]5f^{12}7s^2$

Name	Atomic Number	Symbol	Location in Periodic Table*	Electron Configuration†
Flerovium	114	Fl	P7, G14	$[Rn]5f^{14}6d^{10}7s^27p^2$
Fluorine	9	F	P2, G17	$[He]2s^22p^5$
Francium	87	Fr	P7, G1	$[Rn]7s^1$
Gadolinium	64	Gd	LS	$[Xe]4f^75d^16s^2$
Gallium	31	Ga	P4, G13	$[Ar]3d^{10}4s^24p^1$
Germanium	32	Ge	P4, G14	$[Ar]3d^{10}4s^24p^2$
Gold	79	Au	P6, G11	$[Xe]4f^{14}5d^{10}6s^1$
Hafnium	72	Hf	P6, G4	$[Xe]4f^{14}5d^26s^2$
Hassium	108	Hs	P7, G8	$[Rn]5f^{14}6d^67s^2$
Helium	2	He	P1, G18	$1s^2$
Holmium	67	Ho	LS	$[Xe]4f^{11}6s^2$
Hydrogen	1	H	P1, G1	$1s^1$
Indium	49	In	P5, G13	$[Kr]4d^{10}5s^25p^1$
Iodine	53	I	P5, G17	$[Kr]4d^{10}5s^25p^5$
Iridium	77	Ir	P6, G9	$[Xe]4f^{14}5d^76s^2$
Iron	26	Fe	P4, G8	$[Ar]3d^64s^2$
Krypton	36	Kr	P4, G18	$[Ar]3d^{10}4s^24p^6$
Lanthanum	57	La	P6, G3	$[Xe]5d^16s^2$
Lawrencium	103	Lr	AS	$[Rn]5f^{14}6d^17s^2$
Lead	82	Pb	P6, G10	$[Xe]4f^{14}5d^{10}6s^26p^2$
Lithium	3	Li	P2, G1	$[He]2s^1$
Livermorium	116	Lv	P7, G16	$[Rn]5f^{14}6d^{10}7s^27p^4$
Lutetium	71	Lu	LS	$[Xe]4f^{14}5d^16s^2$
Magnesium	12	Mg	P3, G2	$[Ne]3s^2$
Manganese	25	Mn	P4, G7	$[Ar]3d^54s^2$
Meitnerium	109	Mt	P7, G9	$[Rn]5f^{14}6d^77s^2$
Mendelevium	101	Md	AS	$[Rn]5f^{13}7s^2$
Mercury	80	Hg	P6, G12	$[Xe]4f^{14}5d^{10}6s^2$
Molybdenum	42	Mo	P5, G6	$[Kr]4d^55s^1$
Moscovium	115	Mc	P7, G15	$[Rn]5f^{14}6d^{10}7s^27p^3$
Neodymium	60	Nd	LS	$[Xe]4f^46s^2$
Neon	10	Ne	P2, G18	$[He]2s^22p^6$
Neptunium	93	Np	AS	$[Rn]5f^46d^17s^2$
Nickel	28	Ni	P4, G10	$[Ar]3d^84s^2$
Nihonium	113	Nh	P7, G13	$[Rn]5f^{14}6d^{10}7s^27p^1$
Niobium	41	Nb	P5, G5	$[Kr]4d^45s^1$

Name	Atomic Number	Symbol	Location in Periodic Table*	Electron Configuration†
Nitrogen	7	N	P2, G15	$[He]2s^22p^3$
Nobelium	102	No	AS	$[Rn]5f^{14}7s^2$
Oganesson	118	Og	P7, G18	$[Rn]5f^{14}6d^{10}7s^27p^6$
Osmium	76	Os	P6, G8	$[Xe]4f^{14}5d^66s^2$
Oxygen	8	O	P2, G16	$[He]2s^22p^4$
Palladium	46	Pd	P5, G10	$[Kr]4d^{10}$
Phosphorus	15	P	P3, G15	$[Ne]3s^23p^3$
Platinum	78	Pt	P6, G10	$[Xe]4f^{14}5d^96s^1$
Plutonium	94	Pu	AS	$[Rn]5f^67s^2$
Polonium	84	Po	P6, G16	$[Xe]4f^{14}5d^{10}6s^26p^4$
Potassium	19	K	P4, G1	$[Ar]4s^1$
Praseodymium	59	Pr	LS	$[Xe]4f^36s^2$
Promethium	61	Pm	LS	$[Xe]4f^56s^2$
Protactinium	91	Pa	AS	$[Rn]5f^26d^17s^2$
Radium	88	Ra	P7, G2	$[Rn]7s^2$
Radon	86	Rn	P6, G18	$[Xe]4f^{14}5d^{10}6s^26p^6$
Rhenium	75	Re	P6, G7	$[Xe]4f^{14}5d^56s^2$
Rhodium	45	Rh	P5, G9	$[Kr]4d^85s^1$
Roentgenium	111	Rg	P7, G11	$[Rn]5f^{14}6d^{10}7s^1$
Rubidium	37	Rb	P5, G1	$[Kr]5s^1$
Ruthenium	44	Ru	P5, G8	$[Kr]4d^75s^1$
Rutherfordium	104	Rf	P7, G4	$[Rn]5f^{14}6d^27s^2$
Samarium	62	Sm	LS	$[Xe]4f^66s^2$
Scandium	21	Sc	P4, G3	$[Ar]3d^14s^2$
Seaborgium	106	Sg	P7, G6	$[Rn]5f^{14}6d^47s^2$
Selenium	34	Se	P4, G16	$[Ar]3d^{10}4s^24p^4$
Silicon	14	Si	P3, G14	$[Ne]3s^23p^2$
Silver	47	Ag	P5, G11	$[Kr]4d^{10}5s^1$
Sodium	11	Na	P3, G1	$[Ne]3s^1$
Strontium	38	Sr	P5, G2	$[Kr]5s^2$
Sulfur	16	S	P3, G16	$[Ne]3s^23p^4$
Tantalum	73	Ta	P6, G5	$[Xe]4f^{14}5d^36s^2$
Technitium	43	Tc	P5, G7	$[Kr]4d^55s^2$
Tellurium	52	Te	P5, G16	$[Kr]4d^{10}5s^25p^4$
Tennessine	117	Ts	P7, G17	$[Rn]5f^{14}6d^{10}7s^27p^5$
Terbium	65	Tb	LS	$[Xe]4f^96s^2$

Name	Atomic Number	Symbol	Location in Periodic Table*	Electron Configuration†
Thallium	81	Tl	P6, G13	$[Xe]4f^{14}5d^{10}6s^26p^1$
Thorium	90	Th	AS	$[Rn]6d^27s^2$
Thulium	69	Tm	LS	$[Xe]4f^{13}6s^2$
Tin	50	Sn	P5, G14	$[Kr]4d^{10}5s^25p^2$
Titanium	22	Ti	P4, G4	$[Ar]3d^24s^2$
Tungsten	74	W	P6, G6	$[Xe]4f^{14}5d^46s^2$
Uranium	92	U	AS	$[Rn]5f^36d^17s^2$
Vanadium	23	V	P4, G5	$[Ar]3d^34s^2$
Xenon	54	Xe	P5, G18	$[Kr]4d^{10}5s^25p^6$
Ytterbium	70	Yb	LS	$[Xe]4f^{14}6s^2$
Yttrium	39	Y	P5, G3	$[Kr]4d^15s^2$
Zinc	30	Zn	P4, G12	$[Ar]3d^{10}4s^2$
Zirconium	40	Zr	P5, G4	$[Kr]4d^25s^2$

*Refers to the Periodic Table in Appendix 1: P = period; G = group; LS = lanthanide series (elements 58–71); AS = actinide series (elements 90–103).

†Ground-state configuration

TABLE W-1
Specific Heats of Selected Substances and Mixtures*

Substance	C/(J/g · K)	Substance	C/(J/g · K)	Substance	C/(J/g · K)
Ag (s)	0.235	Cu (s)	0.385	Mg (s)	1.02
Al (s)	0.897	Fe (s)	0.449	Mn (s)	0.479
AlF_3 (s)	0.895	Glass (ℓ)	0.753	Na (s)	1.23
As (s)	0.329	He (g)	5.19	NaCl (s)	0.858
Au (s)	0.129	Hg (ℓ)	0.140	Pb (s)	0.129
BeO (s)	1.02	H_2O (s)	2.06	Si (s)	0.714
Ca (s)	0.647	H_2O (ℓ)	4.19	SiO_2 (s)	0.740
$CaCO_3$ (s)	0.920	H_2O (g)	2.02	Sn (s)	0.228
$CaSO_4$ (s)	0.732	In (s)	0.233	Steel (s)	0.449
C_6H_6 (ℓ)	1.74	In (ℓ)	0.216	Steel (ℓ)	0.719
$C_6H_5CH_3$ (ℓ)	1.80	K (s)	0.757	Ti (s)	0.523
CH_3CH_2OH (ℓ)	2.42	Kerosene (ℓ)	2.09	W (s)	0.132
CH_3COOH	2.05	Li (s)	3.58	Zn (s)	0.388

* (s) = solid phase, (ℓ) = liquid phase, (g) = gas phase. Note: 1 K = 1 C°.

TABLE W-2
Heats (Enthalpies) of Fusion and Vaporization
(Including Melting Points and Boiling Points)

Substance	M.P./°C	DH_f/(J/g)	B.P./°C	DH_v/(J/g)
Aluminum [Al]	660	397	2450	11400
Ammonia [NH_3]	−77.8	33.0	−33.4	137
Copper [Cu]	1083	134	1187	5060
Ethanol [C_2H_5OH]	−114	104	78	854
Gold [Au]	1063.00	64.4	2660	1580
Helium [He]	−269.65	5.23	−268.93	20.9
Iron [Fe]	1808	289	3023	6340
Lead [Pb]	327.3	24.5	1750	870
Nitrogen [N_2]	−209.97	25.5	−195.81	201
Oxygen [O_2]	−218.79	13.8	−182.97	213
Silver [Ag]	960.80	88.2	2193	2330
Sulfur [S]	119	38.1	444.60	326
Tungsten [W]	3410	184	5900	4800
Water [H_2O]	0.00	333	100.00	2260

TABLE W-3
Molal Freezing and Boiling Point Data*

Substance	F.P./°C	k_f/(C°·kg/mol)	B.P./°C	k_b/(C°·kg/mol)
Acetic acid [$HC_2H_3O_2$]	16.7	3.63	118	3.22
Benzene [C_6H_6]	5.53	5.07	80.1	2.64
Camphor [$C_{10}H_{16}O$]	178	37.8	207	—
Cyclohexane [C_6H_{12}]	6.54	20.8	80.7	2.92
Nitrobenzene [$C_6H_5NO_2$]	5.76	6.87	211	5.20
Phenol [C_6H_5OH]	40.9	6.84	182	3.54
Water [H_2O]	0.00	1.86	100.	0.513

* A dash (−) indicates that data are not available.

TABLE W-4
Standard Heats of Formation for Substances
(at 298.15 K and 100 kPa)

Substance	$DH°_f$/(kJ/mol)	Substance	$DH°_f$/(kJ/mol)
Aluminum		**Copper**	
Al(s)	0	Cu(s)	0
$AlCl_3$(s)	−704.2	CuCl(s)	−137.2
Al_2O_3(s)	−1675.7	$CuCl_2$(s)	−220.1
Barium		CuO(s)	−157.3
Ba(s)	0	$CuSO_4$(s)	−771.4
$BaCO_3$(s)	−1213.0	**Fluorine**	
$BaCl_2$(s)	−855.0	F_2(g)	0
$Ba(NO_3)_2$(s)	−988.0	F(g)	79.4
$BaSO_4$(s)	−1473.2	**Hydrogen**	
Bromine		H_2(g)	0
$Br_2(\ell)$	0	HBr(g)	−36.3
Br_2(g)	30.9	HCl(g)	−92.3
Br(g)	111.9	HCN(g)	135.1
Calcium		HF(g)	−273.3
Ca(s)	0	HI(g)	26.5
$CaCO_3$(s) [calcite]	−1207.6	$HNO_3(\ell)$	−174.1
$CaCl_2$(s)	−795.4	$H_2O(\ell)$	−285.8
CaF_2(s)	−1228.0	H_2O(g)	−241.8
$Ca(NO_3)_2$(s)	−938.2	$H_2O_2(\ell)$	−187.8
CaO(s)	−634.9	H_2O_2(g)	−136.3
$Ca(OH)_2$(s)	−985.2	H_2S(g)	−20.6
$Ca_3(PO_4)_2$(s)	−4120.8	$H_2SO_4(\ell)$	−814.0
$CaSO_4$(s)	−1434.5	**Iodine**	
Carbon		I_2(s)	0
C(s) [graphite]	0	I_2(g)	62.4
C(s) [diamond]	1.9	I(g)	106.8
CH_4(g)	−74.6	$ICl(\ell)$	−23.9
C_2H_2(g)	227.4	ICl(g)	17.8
C_2H_4(g)	52.4	**Iron**	
C_2H_6(g)	−84.0	Fe(s)	0
C_3H_8(g)	−103.8	$FeCl_2$(s)	−341.8
$C_6H_6(\ell)$	49.1	$FeCl_3$(s)	−399.5
$CH_3OH(\ell)$	−239.2	Fe_2O_3(s)	−824.2
$C_2H_5OH(\ell)$	−277.6	Fe_3O_4(s)	−1118.4
CO(g)	−110.5	**Lead**	
CO_2(g)	−393.5	Pb(s)	0
Chlorine		$PbCl_2$(s)	−359.4
Cl_2(g)	0	PbS(s)	−100.4
Cl(g)	121.3	$PbSO_4$(s)	−920.0
ClO_2(g)	102.5	**Lithium**	
Cl_2O(g)	80.3	Li(s)	0
Chromium		LiCl(s)	−408.6
Cr(s)	0	LiF(s)	−616.0
$CrCl_2$(s)	−395.4	$LiNO_3$(s)	−483.1
$CrCl_3$(s)	−556.5	LiOH(s)	−484.9

TABLE W-4: *(Continued)*

Substance	$DH°_f/(\text{kJ/mol})$	Substance	$DH°_f/(\text{kJ/mol})$
Magnesium		**Silver**	
Mg(s)	0	Ag(s)	0
$MgCl_2(s)$	−641.3	AgBr(s)	−100.4
$MgF_2(s)$	−1124.2	AgCl(s)	−127.0
MgO(s)	−601.6	AgI(s)	−61.8
$Mg(OH)_2(s)$	−924.5	$AgNO_3(s)$	−124
Manganese		**Sodium**	
Mn(s)	0	Na(s)	0
$MnO_2(s)$	−520.0	Na(g)	107.5
Nitrogen		NaBr(s)	−361.1
$N_2(g)$	0	NaCl(s)	−411.2
N(g)	472.7	$Na_2CO_3(s)$	−1130.7
$NH_3(g)$	−45.9	NaF(s)	−576.6
$N_2H_4(\ell)$	50.6	NaH(s)	−56.3
$NH_4Cl(s)$	−314.4	NaOH(s)	−425.6
$NH_4NO_3(s)$	−365.6	$Na_2SO_4(s)$	−1387.1
NO(g)	91.3	**Sulfur**	
$NO_2(g)$	33.2	S(s) [rhombic]	0
$N_2O(g)$	81.6	$SF_6(g)$	−1220.5
$N_2O_4(g)$	11.1	$SO_2(g)$	−296.8
Oxygen		$SO_3(g)$	−395.7
$O_2(g)$	0	**Tin**	
$O_3(g)$	142.7	Sn(s) [white]	0
Potassium		Sn(s) [gray]	−2.1
K(s)	0	$SnCl_4(\ell)$	−511.3
KBr(s)	−393.8	**Zinc**	
KCl(s)	−436.5	Zn(s)	0
$KClO_3(s)$	−397.7	$ZnCl_2(s)$	−415.1
$KClO_4(s)$	−432.8	ZnO(s)	−350.5
KF(s)	−567.3		
$KMnO_4(s)$	−837.2		
KOH(s)	−424.6		
$K_2SO_4(s)$	−1437.8		

TABLE X
Vapor Pressures of Water at Various Temperatures

t_c/°C	V.P. /mmHg	V.P. /kPa	t_c/°C	V.P. /mmHg	V.P. /kPa
0	4.585	0.6113	21	18.66	2.488
1	4.929	0.6572	22	19.84	2.645
2	5.296	0.7061	23	21.08	2.810
3	5.686	0.7581	24	22.39	2.985
4	6.102	0.8136	25	23.77	3.169
5	6.545	0.8726	26	25.22	3.363
6	7.016	0.9354	27	26.75	3.567
7	7.516	1.002	28	28.37	3.782
8	8.048	1.073	29	30.06	4.008
9	8.612	1.148	30	31.84	4.246
10	9.212	1.228	40	55.37	7.381
11	9.848	1.313	50	92.59	12.34
12	10.52	1.403	60	149.5	19.93
13	11.24	1.498	70	233.8	31.18
14	11.99	1.599	80	355.3	47.37
15	12.79	1.706	90	525.9	70.12
16	13.64	1.819	100	760.0	101.3
17	14.54	1.938	105	906.0	120.8
18	15.48	2.064	110	1074	143.2
19	16.48	2.198	115	1268	169.0
20	17.54	2.339	120	1489	198.5

TABLE Y-1
Solubility-Product Constants (at 298.15 K and 100 kPa)

Formula	K_{sp}	Formula	K_{sp}
Bromides		**Iodides**	
AgBr	5.35×10^{-13}	AgI	8.52×10^{-17}
CuBr	6.27×10^{-9}	CuI	1.27×10^{-12}
$PbBr_2$	6.60×10^{-6}	PbI_2	9.80×10^{-28}
Carbonates		**Phosphates**	
Ag_2CO_3	8.46×10^{-12}	Ag_3PO_4	8.89×10^{-17}
$BaCO_3$	2.58×10^{-9}	$Ca_3(PO_4)_2$	2.07×10^{-33}
$CaCO_3$	3.36×10^{-9}	$Cd_3(PO_4)_2$	2.53×10^{-33}
$MgCO_3$	6.82×10^{-6}	$Cu_3(PO_4)_2$	1.40×10^{-37}
$SrCO_3$	5.60×10^{-10}	$Mg_3(PO_4)_2$	1.04×10^{-24}
$ZnCO_3$	8×10^{-28}	$Ni_3(PO_4)_2$	4.74×10^{-32}
Chlorides		**Sulfates**	
AgCl	1.77×10^{-10}	Ag_2SO_4	1.20×10^{-5}
CuCl	1.72×10^{-7}	$BaSO_4$	1.08×10^{-10}
$PbCl_2$	1.70×10^{-5}	$CaSO_4$	4.93×10^{-5}
Chromate		$PbSO_4$	2.53×10^{-8}
Ag_2CrO_4	1.12×10^{-12}	**Sulfides**	
Fluorides		CdS	1.40×10^{-29}
BaF_2	1.84×10^{-6}	CuS	1.27×10^{-36}
CaF_2	3.45×10^{-11}	HgS	1.55×10^{-52}
MgF_2	5.16×10^{-11}	MnS	4.65×10^{-14}
PbF_2	3.30×10^{-8}	PbS	9.05×10^{-29}
Hydroxides		ZnS	2.93×10^{-25}
$Ca(OH)_2$	5.02×10^{-6}	**Sulfites**	
$Cd(OH)_2$	7.20×10^{-15}	Ag_2SO_3	1.50×10^{-14}
$Fe(OH)_2$	4.87×10^{-17}	$BaSO_3$	5.00×10^{-10}
$Fe(OH)_3$	2.79×10^{-39}		
$Mg(OH)_2$	5.61×10^{-12}		
$Ni(OH)_2$	5.48×10^{-16}		
$Zn(OH)_2$	3.00×10^{-17}		

TABLE Y-2
Ionization Constants of Weak Acids and Bases
(at 298.15 K and 100 kPa)

$$K_w = [H_3O^+] \cdot [OH^-] = 1.0 \times 10^{-14}$$

Monoprotic Acids (acids with one ionizable hydrogen atom)

Formula	Name	K_a
HIO_3	iodic acid	1.6×10^{-1}
HNO_2	nitrous acid	7.2×10^{-4}
HF	hydrofluoric acid	6.6×10^{-4}
$HCHO_2$	formic acid	1.8×10^{-4}
$HC_3H_5O_3$	lactic acid	1.4×10^{-4}
$HC_7H_5O_2$	benzoic acid	6.3×10^{-5}
$HC_4H_7O_2$	butanoic acid	1.5×10^{-5}
HN_3	hydrazoic acid	1.9×10^{-5}
$HC_2H_3O_2$	acetic acid	1.8×10^{-5}
$HC_3H_5O_2$	propanoic acid	1.3×10^{-5}
HOCl	hypochlorous acid	2.9×10^{-8}
HCN	hydrocyanic acid	6.2×10^{-10}
HC_6H_5O	phenol	1.0×10^{-10}
H_2O_2	hydrogen peroxide	2.2×10^{-12}

Diprotic Acids (acids with two ionizable hydrogen atoms)

Formula	Name	K_{a1}	K_{a2}	K_{a3}
H_2SO_4	sulfuric acid	Large	1.1×10^{-2}	
H_2CrO_4	chromic acid	5.0	1.5×10^{-6}	
$H_2C_2O_4$	oxalic acid	5.4×10^{-2}	5.3×10^{-5}	
H_3PO_3	phosphorous acid	3.7×10^{-2}	2.1×10^{-7}	
H_2SO_3	sulfurous acid	1.3×10^{-2}	6.2×10^{-8}	
H_2SeO_3	selenous acid	2.3×10^{-3}	5.4×10^{-9}	
$H_2C_3H_2O_4$	malonic acid	1.5×10^{-3}	2.0×10^{-6}	
$H_2C_8H_4O_4$	phthalic acid	1.1×10^{-3}	3.9×10^{-6}	
$H_2C_4H_4O_6$	tartaric acid	9.2×10^{-4}	4.3×10^{-5}	
H_2CO_3	carbonic acid	4.4×10^{-7}	4.7×10^{-11}	

Triprotic Acids (acids with three ionizable hydrogen atoms)

Formula	Name	K_{a1}	K_{a2}	K_{a3}
H_3PO_4	phosphoric acid	7.1×10^{-3}	6.3×10^{-8}	4.2×10^{-13}
H_3AsO_4	arsenic acid	6.0×10^{-3}	1.0×10^{-7}	3.2×10^{-12}
$H_3C_6H_5O_7$	citric acid	7.4×10^{-4}	1.7×10^{-5}	4.0×10^{-7}

Bases

Formula	Name	K_b
$(CH_3)_2NH$	dimethylamine	6.9×10^{-4}
CH_3NH_2	methylamine	4.2×10^{-4}
$CH_3CH_2NH_2$	ethylamine	4.3×10^{-4}
$(CH_3)_3N$	trimethylamine	6.3×10^{-5}
NH_3	ammonia	1.8×10^{-5}
N_2H_4	hydrazine	8.5×10^{-7}
C_5H_5N	pyridine	1.5×10^{-9}
$C_6H_5NH_2$	aniline	7.4×10^{-10}

TABLE Z
Standard Electrode Potentials
(at 298.15 K and 100 kPa)

Half-reaction	$E°/V$	Half-reaction	$E°/V$
$F_2 + 2e^- \rightarrow 2F^-$	2.87	$O_2 + 2H_2O + 4e^- \rightarrow 4OH^-$	0.40
$Ag^{2+} + e^- \rightarrow Ag^+$	1.99	$Cu^{2+} + 2e^- \rightarrow Cu$	0.34
$Co^{3+} + e^- \rightarrow Co^{2+}$	1.82	$Hg_2Cl_2 + 2e^- \rightarrow 2Hg + 2Cl^-$	0.34
$H_2O_2 + 2H^+ + 2e^- \rightarrow 2H_2O$	1.78	$AgCl + e^- \rightarrow Ag + Cl^-$	0.22
$Ce^{4+} + e^- \rightarrow Ce^{3+}$	1.70	$SO_4^{2-} + 4H^+ + 2e^-$	
$PbO_2 + 4H^+ + SO_4^{2-} + 2e^-$		$\rightarrow H_2SO_3 + H_2O$	0.20
$\rightarrow PbSO_4 + 2H_2O$	1.69	$Cu^{2+} + e^- \rightarrow Cu^+$	0.16
$MnO_4^- + 4H^+ + 3e^- \rightarrow MnO_2 + 2H_2O$	1.68	$\mathbf{2H^+ + 2e^- \rightarrow H_2}$	**0.00**
$2e^- + 2H^+ + IO_4^- \rightarrow IO_3^- + H_2O$	1.60	$Fe^{3+} + 3e^- \rightarrow Fe$	−0.036
$MnO_4^- + 8H^+ + 5e^- \rightarrow Mn^{2+} + 4H_2O$	1.51	$Pb^{2+} + 2e^- \rightarrow Pb$	−0.13
$Au^{3+} + 3e^- \rightarrow Au$	1.50	$Sn^{2+} + 2e^- \rightarrow Sn$	−0.14
$PbO_2 + 4H^+ + 2e^- \rightarrow Pb^{2+} + 2H_2O$	1.46	$Ni^{2+} + 2e^- \rightarrow Ni$	−0.23
$Cl_2 + 2e^- \rightarrow 2Cl^-$	1.36	$PbSO_4 + 2e^- \rightarrow Pb + SO_4^{2-}$	−0.35
$Cr_2O_7^{2-} + 14H^+ + 6e^- \rightarrow 2Cr^{3+} + 7H_2O$	1.33	$Cd^{2+} + 2e^- \rightarrow Cd$	−0.40
$O_2 + 4H^+ + 4e^- \rightarrow 2H_2O$	1.23	$Fe^{2+} + 2e^- \rightarrow Fe$	−0.44
$MnO_2 + 4H^+ + 2e^- \rightarrow Mn^{2+} + 2H_2O$	1.21	$Cr^{3+} + e^- \rightarrow Cr^{2+}$	−0.50
$IO_3^- + 6H^+ + 5e^- \rightarrow \frac{1}{2}I_2 + 3H_2O$	1.20	$Cr^{3+} + 3e^- \rightarrow Cr$	−0.73
$Br_2 + 2e^- \rightarrow 2Br^-$	1.09	$Zn^{2+} + 2e^- \rightarrow Zn$	−0.76
$VO_2^+ + 2H^+ + e^- \rightarrow VO^{2+} + H_2O$	1.00	$2H_2O + 2e^- \rightarrow H_2 + 2OH^-$	−0.83
$AuCl_4^- + 3e^- \rightarrow Au + 4Cl^-$	0.99	$Mn^{2+} + 2e^- \rightarrow Mn$	−1.18
$NO_3^- + 4H^+ + 3e^- \rightarrow NO + 2H_2O$	0.96	$Al^{3+} + 3e^- \rightarrow Al$	−1.66
$ClO_2 + e^- \rightarrow ClO_2^-$	0.95	$H_2 + 2e^- \rightarrow 2H^-$	−2.23
$2Hg^{2+} + 2e^- \rightarrow Hg_2^{2+}$	0.91	$Mg^{2+} + 2e^- \rightarrow Mg$	−2.37
$Ag^+ + e^- \rightarrow Ag$	0.80	$La^{3+} + 3e^- \rightarrow La$	−2.37
$Hg_2^{2+} + 2e^- \rightarrow 2Hg$	0.80	$Na^+ + e^- \rightarrow Na$	−2.71
$Fe^{3+} + e^- \rightarrow Fe^{2+}$	0.77	$Ca^{2+} + 2e^- \rightarrow Ca$	−2.76
$O_2 + 2H^+ + 2e^- \rightarrow H_2O_2$	0.68	$Ba^{2+} + 2e^- \rightarrow Ba$	−2.90
$MnO_4^- + e^- \rightarrow MnO_4^{2-}$	0.56	$K^+ + e^- \rightarrow K$	−2.92
$I_2 + 2e^- \rightarrow 2I^-$	0.54	$Li^+ + e^- \rightarrow Li$	−3.05
$Cu^+ + e^- \rightarrow Cu$	0.52		

ANSWERS TO END-OF-CHAPTER QUESTIONS

Chapter 1

1. (4)	**7.** (4)	**13.** (2)	**19.** (4)	**25.** (2)
2. (4)	**8.** (3)	**14.** (3)	**20.** (2)	**26.** (4)
3. (1)	**9.** (1)	**15.** (2)	**21.** (4)	**27.** (3)
4. (3)	**10.** (4)	**16.** (1)	**22.** (2)	**28.** (3)
5. (4)	**11.** (4)	**17.** (3)	**23.** (2)	
6. (1)	**12.** (2)	**18.** (1)	**24.** (3)	

Constructed-Response Questions

1. (a) 6.3477×10^4 (b) 2.30×10^{-4}

2. (a) 654,000 (b) 0.00555

3. $7.00 \ \cancel{\text{daddles}} \cdot \left(\dfrac{1 \ \text{piddle}}{3.33 \ \cancel{\text{daddles}}} \right) = 2.10 \ \text{piddles}$

4. 21 g

5. $f = \dfrac{cd}{abe}$

6. (a)

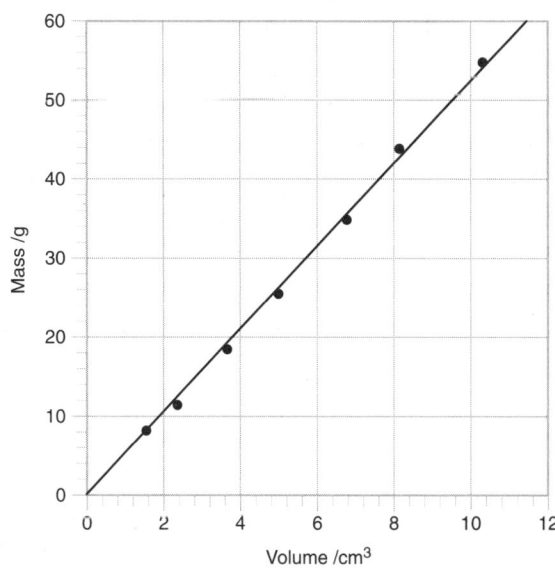

Mass versus Volume

(b) density (from slope) = 5.26 g/cm³

7. (a) 4   (b) 5   (c) 2   (d) 3   (e) 4

8. (a) 343.2 m   (b) 0.0165 kJ   (c) 2.2 cm^2 (d) 5.4 g/L

9. Sample 3

10. 2 molecules of compound y_2z

11. A compound must consist of two or more elements.

Chapter 2

1. (4)	8. (1)	15. (3)	22. (4)	29. (3)
2. (1)	9. (4)	16. (3)	23. (4)	30. (4)
3. (1)	10. (1)	17. (3)	24. (3)	31. (4)
4. (3)	11. (1)	18. (3)	25. (4)	32. (2)
5. (2)	12. (1)	19. (3)	26. (2)	33. (3)
6. (2)	13. (3)	20. (2)	27. (3)	34. (1)
7. (2)	14. (3)	21. (1)	28. (1)	

Constructed-Response Questions

1.

Symbol	^{39}K	^{51}V	^{197}Au	^{222}Rn	$^{31}P^{3-}$	$^{206}Pb^{2+}$	$^{79}Se^{2-}$	$^{59}Ni^{2+}$
Protons	19	23	79	86	15	82	34	28
Neutrons	20	28	118	136	16	124	45	31
Electrons	19	23	79	86	18	80	36	26
Atomic no.	19	23	79	86	15	82	34	28
Mass no.	39	51	197	222	31	206	79	59
Net charge	0	0	0	0	3−	2+	2−	2+

2.

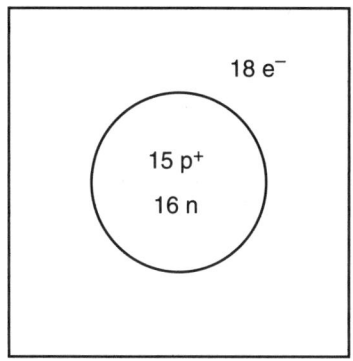

18 e⁻

15 p⁺

16 n

Model of $^{31}P^{3-}$ Ion

Chapter 3

1. (1)	**10.** (2)	**19.** (2)	**28.** (1)	**37.** (2)
2. (1)	**11.** (3)	**20.** (1)	**29.** (4)	**38.** (3)
3. (3)	**12.** (1)	**21.** (2)	**30.** (3)	**39.** (4)
4. (3)	**13.** (2)	**22.** (3)	**31.** (2)	**40.** (3)
5. (2)	**14.** (2)	**23.** (2)	**32.** (2)	**41.** (4)
6. (4)	**15.** (1)	**24.** (2)	**33.** (1)	
7. (4)	**16.** (2)	**25.** (1)	**34.** (1)	
8. (3)	**17.** (3)	**26.** (2)	**35.** (3)	
9. (2)	**18.** (1)	**27.** (2)	**36.** (4)	

Constructed-Response Questions

1.

		bromide Br^-	carbonate CO_3^{2-}	chlorate ClO_3^-	chloride Cl^-	chromate CrO_4^{2-}	nitrate NO_3^-	phosphate PO_4^{3-}	sulfate SO_4^{2-}	sulfide S^{2-}
aluminum	Al^{3+}	$AlBr_3$	$Al_2(CO_3)_3$	$Al(ClO_3)_3$	$AlCl_3$	$Al_2(CrO_4)_3$	$Al(NO_3)_3$	$AlPO_4$	$Al_2(SO_4)_3$	Al_2S_3
ammonium	NH_4^+	NH_4Br	$(NH_4)_2CO_3$	NH_4ClO_3	NH_4Cl	$(NH_4)_2CrO_4$	NH_4NO_3	$(NH_4)_3PO_4$	$(NH_4)_2SO_4$	$(NH_4)_2S$
barium	Ba^{2+}	$BaBr_2$	$BaCO_3$	$Ba(ClO_3)_2$	$BaCl_2$	$BaCrO_4$	$Ba(NO_3)_2$	$Ba_3(PO_4)_2$	$BaSO_4$	BaS
calcium	Ca^{2+}	$CaBr_2$	$CaCO_3$	$Ca(ClO_3)_2$	$CaCl_2$	$CaCrO_4$	$Ca(NO_3)_2$	$Ca_3(PO_4)_2$	$CaSO_4$	CaS
copper (II)	Cu^{2+}	$CuBr_2$	$CuCO_3$	$Cu(ClO_3)_2$	$CuCl_2$	$CuCrO_4$	$Cu(NO_3)_2$	$Cu_3(PO_4)_2$	$CuSO_4$	CuS
iron (II)	Fe^{2+}	$FeBr_2$	$FeCO_3$	$Fe(ClO_3)_2$	$FeCl_2$	$FeCrO_4$	$Fe(NO_3)_2$	$Fe_3(PO_4)_2$	$FeSO_4$	FeS
iron (III)	Fe^{3+}	$FeBr_3$	$Fe_2(CO_3)_3$	$Fe(ClO_3)_3$	$FeCl_3$	$Fe_2(CrO_4)_3$	$Fe(NO_3)_3$	$FePO_4$	$Fe_2(SO_4)_3$	Fe_2S_3
lead (II)	Pb^{2+}	$PbBr_2$	$PbCO_3$	$Pb(ClO_3)_2$	$PbCl_2$	$PbCrO_4$	$Pb(NO_3)_2$	$Pb_3(PO_4)_2$	$PbSO_4$	PbS
lead (IV)	Pb^{4+}	$PbBr_4$	$Pb(CO_3)_2$	$Pb(ClO_3)_4$	$PbCl_4$	$Pb(CrO_4)_2$	$Pb(NO_3)_4$	$Pb_3(PO_4)_4$	$Pb(SO_4)_2$	PbS_2
magnesium	Mg^{2+}	$MgBr_2$	$MgCO_3$	$Mg(ClO_3)_2$	$MgCl_2$	$MgCrO_4$	$Mg(NO_3)_2$	$Mg_3(PO_4)_2$	$MgSO_4$	MgS
mercury (I)	Hg_2^{2+}	Hg_2Br_2	Hg_2CO_3	$Hg_2(ClO_3)_2$	Hg_2Cl_2	Hg_2CrO_4	$Hg_2(NO_3)_2$	$(Hg_2)_3(PO_4)_2$	Hg_2SO_4	Hg_2S
mercury (II)	Hg^{2+}	$HgBr_2$	$HgCO_3$	$Hg(ClO_3)_2$	$HgCl_2$	$HgCrO_4$	$Hg(NO_3)_2$	$Hg_3(PO_4)_2$	$HgSO_4$	HgS
potassium	K^+	KBr	K_2CO_3	$KClO_3$	KCl	K_2CrO_4	KNO_3	K_3PO_4	K_2SO_4	K_2S
silver	Ag^+	$AgBr$	Ag_2CO_3	$AgClO_3$	$AgCl$	Ag_2CrO_4	$AgNO_3$	Ag_3PO_4	Ag_2SO_4	Ag_2S
sodium	Na^+	$NaBr$	Na_2CO_3	$NaClO_3$	$NaCl$	Na_2CrO_4	$NaNO_3$	Na_3PO_4	Na_2SO_4	Na_2S
zinc	Zn^{2+}	$ZnBr_2$	$ZnCO_3$	$Zn(ClO_3)_2$	$ZnCl_2$	$ZnCrO_4$	$Zn(NO_3)_2$	$Zn_3(PO_4)_2$	$ZnSO_4$	ZnS

2. $S = 4+, O = 2-$

3. $K = 1+, Mn = 7+, O = 2-$

4. (a) $3Zn(OH)_2 + 2H_3PO_4 \rightarrow 6H_2O + Zn_3(PO_4)_2$
   (b) $6CO_2 + 6H_2O \rightarrow C_6H_{12}O_6 + 6O_2$

5. (a) sodium iodide   (b) carbon(II) sulfide
   (c) iron(III) chloride   (d) phosphorus(V) oxide
   (e) chlorine(VII) oxide

6. $Mg + Br_2 \rightarrow MgBr_2$

7. $2FeO \rightarrow 2Fe + O_2$

8. $2Al + 3Zn(NO_3)_2 \rightarrow 3Zn + 2Al(NO_3)_3$

9. $2Na_3PO_4 + 3Pb(NO_3)_2 \rightarrow 6NaNO_3 + Pb_3(PO_4)_2$

   According to Reference Table F, $Pb_3(PO_4)_2$ is the precipitate.

10. decomposition

11. The total number of atoms (4 H and 2 O) on both sides of the equation are equal.

12. 4 molecules of O_2 are produced.

Chapter 4

1. (3)	11. (1)	21. (3)	31. (3)	41. (1)
2. (2)	12. (4)	22. (3)	32. (2)	42. (3)
3. (1)	13. (2)	23. (1)	33. (2)	43. (3)
4. (2)	14. (3)	24. (1)	34. (3)	44. (1)
5. (3)	15. (1)	25. (3)	35. (2)	45. (3)
6. (1)	16. (2)	26. (4)	36. (2)	46. (4)
7. (2)	17. (2)	27. (2)	37. (2)	47. (1)
8. (4)	18. (3)	28. (1)	38. (3)	48. (1)
9. (4)	19. (2)	29. (4)	39. (2)	49. (2)
10. (4)	20. (4)	30. (1)	40. (3)	

Constructed-Response Questions

Substance	Molar Mass ($\mathcal{M}$) (g/mol)	Mass of Substance (m) (g)	Moles of Substance (n) (mol)	Number of Particles of Substance (N)
O_3	48	24	0.50	3.0×10^{23}
NH_3	17	170	10.	6.0×10^{24}
F_2	38	38	1.0	6.0×10^{23}
CO_2	44	4.4	0.10	6.0×10^{22}
NO_2	46	9.2	0.20	1.2×10^{23}
Ne	20	5.0	0.25	1.5×10^{23}
N_2O	44	88	2.0	1.2×10^{24}
NH_3	17	8.5	0.50	3.0×10^{23}

2. (a) 82 g/mol (b) 314 g/mol

3. (a) 1.0 mol (b) 6.0×10^{23} molecules

4. 154 g

5. $Na_2B_4O_7$

6. S_2Cl_2

7. (a) 3.93 g (b) 47.2 g

8. 87.4 g

9. 26.6%

10. (a) 58.3 g/mol (b) 0.144 mol

11. $(10.01)(0.1991) + (11.01)(0.8009) = 10.81$

12. $CaSO_4 = (40 \text{ u}) + (32 \text{ u}) + (64 \text{ u}) = 136 \text{ u}$

$2H_2O = (4 \text{ u}) + (32 \text{ u}) = 36 \text{ u}$

$GFM = 136 \text{ u} + 36 \text{ u} = 172 \text{ u}$

13. $\%H_2O = \left(\dfrac{36 \text{ u}}{172 \text{ u}} \right) \times 100\% = 21\%$

14. 71.6%

15. $(0.57)(12.90 \text{ u}) + (0.43)(12.90 \text{ u})$

16. $2Sb_2S_3(s) + 9O_2(g) \rightarrow 2Sb_2O_3(s) + 6SO_2(g)$

Chapter 5

1. (1)	**11.** (4)	**21.** (1)	**31.** (1)	**41.** (2)
2. (3)	**12.** (4)	**22.** (3)	**32.** (4)	**42.** (2)
3. (2)	**13.** (4)	**23.** (4)	**33.** (1)	**43.** (2)
4. (4)	**14.** (3)	**24.** (4)	**34.** (2)	**44.** (2)
5. (4)	**15.** (1)	**25.** (2)	**35.** (1)	
6. (4)	**16.** (2)	**26.** (2)	**36.** (3)	
7. (2)	**17.** (1)	**27.** (1)	**37.** (2)	
8. (1)	**18.** (2)	**28.** (2)	**38.** (1)	
9. (4)	**19.** (2)	**29.** (4)	**39.** (2)	
10. (1)	**20.** (4)	**30.** (1)	**40.** (1)	

Constructed-Response Questions

1. 6.026 kJ released

2. -179.0 kJ

3. -166.6 kJ/mol

4. *D*

5. *C*

6. Any substance that absorbs heat ($+\Delta H$) when it dissolves in water could be used. Examples include $KNO_3(s)$ and $NH_4NO_3(s)$.

7. The bracelet's temperature increased because heat flowed from the student's body to the copper bracelet.

8. 0.474 mol

9. $q = (30.1 \text{ g})(0.385 \text{ J/g} \cdot \text{K})(19°C - 33°C)$

Chapter 6

1. (2)	**12.** (2)	**23.** (4)	**34.** (4)	**45.** (3)
2. (2)	**13.** (3)	**24.** (1)	**35.** (1)	**46.** (4)
3. (3)	**14.** (3)	**25.** (3)	**36.** (3)	**47.** (2)
4. (2)	**15.** (1)	**26.** (1)	**37.** (4)	**48.** (2)
5. (1)	**16.** (1)	**27.** (1)	**38.** (3)	**49.** (3)
6. (4)	**17.** (1)	**28.** (2)	**39.** (2)	**50.** (2)
7. (4)	**18.** (3)	**29.** (4)	**40.** (4)	**51.** (3)
8. (3)	**19.** (1)	**30.** (4)	**41.** (4)	**52.** (2)
9. (4)	**20.** (4)	**31.** (1)	**42.** (2)	
10. (2)	**21.** (4)	**32.** (3)	**43.** (4)	
11. (2)	**22.** (2)	**33.** (2)	**44.** (3)	

Constructed-Response Questions

1. 1093 torr

2. 939°C (1212 K)

3. 98.000 kPa

4. 630. mL at STP

5. $\dfrac{\text{rate of } N_2}{\text{rate of } CO_2} = \dfrac{1.25}{1}$

6. 0.423 g/L

7. (a) 23 L   (b) 250.0 L

8. Either graph is acceptable.

 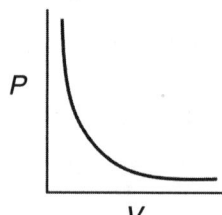

9. Use the combined gas law. Since the temperature is held constant, it is omitted from the equation.

   (6.2 mL) • (1.4 atm) = (3.1 mL) • P_2

   P_2 = 2.8 atm

10. The average kinetic energy decreases.

11. freezing (or solidification or change from liquid to solid)

12. 2.0 atm

13. The frequency of collisions between the gas molecules increases.

14. 710 (±10) mmHg

15. 114 (±2)°C

16. Liquid *A* will evaporate more rapidly. Compared with the vapor pressure of liquid *B*, the vapor pressure of liquid *A* is higher because liquid *A* has weaker intermolecular forces.

Chapter 7

1. (3)	**11.** (3)	**21.** (2)	**31.** (1)	**41.** (2)
2. (2)	**12.** (1)	**22.** (3)	**32.** (4)	**42.** (4)
3. (4)	**13.** (3)	**23.** (3)	**33.** (1)	**43.** (4)
4. (2)	**14.** (1)	**24.** (2)	**34.** (3)	**44.** (2)
5. (2)	**15.** (3)	**25.** (4)	**35.** (2)	
6. (2)	**16.** (2)	**26.** (1)	**36.** (3)	
7. (3)	**17.** (4)	**27.** (4)	**37.** (1)	
8. (4)	**18.** (2)	**28.** (3)	**38.** (4)	
9. (2)	**19.** (2)	**29.** (4)	**39.** (4)	
10. (3)	**20.** (4)	**30.** (1)	**40.** (4)	

Constructed-Response Questions

1. (a)

Activity versus Time

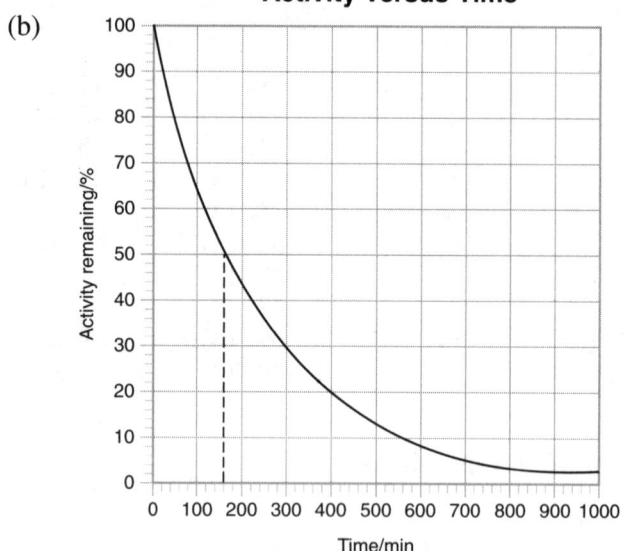

Activity versus Time

(b)

Half-life (as determined from the graph) = 160 min

2. $_2^4He$

3. Alpha particles harm human tissue by altering genetic information in human cells.

4. Radium and calcium are located in the same periodic group and have similar chemical properties.

5. Zinc sulfide emits visible light in response to ionizing radiation.

6. The U-235 nucleus splits into smaller nuclei.

7. Cs-140

8. 5.00 min

Chapter 8

1. (1)	11. (2)	21. (3)	31. (2)	41. (1)
2. (4)	12. (2)	22. (2)	32. (4)	42. (3)
3. (4)	13. (4)	23. (2)	33. (4)	43. (1)
4. (3)	14. (3)	24. (1)	34. (2)	44. (1)
5. (2)	15. (1)	25. (2)	35. (1)	45. (3)
6. (4)	16. (3)	26. (2)	36. (3)	
7. (2)	17. (4)	27. (1)	37. (2)	
8. (3)	18. (2)	28. (3)	38. (4)	
9. (2)	19. (3)	29. (2)	39. (2)	
10. (2)	20. (4)	30. (3)	40. (3)	

Constructed-Response Questions

1.

Element	Symbol	Electron Configuration	Number of Valence Electrons	Lewis Structure
Argon	Ar	2-8-8	8	:Är:
Barium	Ba	2-8-18-18-8-2	2	Bä
Carbon	C	2-4	4	C̈·
Chlorine	Cl	2-8-7	7	·C̈l:
Helium	He	2	2	Hë
Nitrogen	N	2-5	5	·N̈·
Oxygen	O	2-6	6	·Ö:
Potassium	K	2-8-8-1	1	K̇
Strontium	Sr	2-8-18-8-2	2	S̈r
Sulfur	S	2-8-6	6	·S̈:

2. 2 valence electrons

3. element Y (2-8-7-3)

4. H and He

5. As electrons transition from excited states to the ground state, photons are emitted. These emissions produce bright-line spectra.

6. Not all of the wavelengths of element A are shown in the spectrum of the mixture.

7. The wavelengths of the spectral lines for element Z are independent of the mass of the sample.

8. Different colors of light are produced when electrons return from higher energy states to lower energy states.

Chapter 9

1. (3)	**11.** (2)	**21.** (3)	**31.** (2)	**41.** (3)
2. (2)	**12.** (1)	**22.** (1)	**32.** (1)	**42.** (4)
3. (1)	**13.** (3)	**23.** (4)	**33.** (2)	**43.** (2)
4. (2)	**14.** (3)	**24.** (3)	**34.** (1)	**44.** (2)
5. (1)	**15.** (1)	**25.** (2)	**35.** (4)	
6. (1)	**16.** (1)	**26.** (2)	**36.** (4)	
7. (2)	**17.** (1)	**27.** (1)	**37.** (1)	
8. (1)	**18.** (2)	**28.** (2)	**38.** (4)	
9. (2)	**19.** (2)	**29.** (1)	**39.** (1)	
10. (4)	**20.** (4)	**30.** (2)	**40.** (2)	

Constructed-Response Questions

1. Na and Rb

2. They are located in the same group (Group 1) of the Periodic Table.

3. A K^+ ion has a total of 18 electrons.

4. A potassium atom has the electron configuration 2-8-8-1. A K^+ ion has the configuration 2-8-8. When a potassium atom forms an ion, the fourth shell is lost, thereby decreasing the radius.

5. A K^+ ion is electrically charged and is mobile within nerve fibers.

6. Na and K are located in the same group (Group 1) of the Periodic Table.

7. silicon (Si) or germanium (Ge)

8. The atomic radius of these elements increases down the group because each successive element has one more electron shell.

9. 4 valence electrons

10. The radius of a fluoride ion is larger than the radius of a fluorine atom.

11. A lithium atom loses its second-shell valence electron. So the lithium ion has only one shell of electrons.

12. As the elements of Period 2 are considered from left to right, the atomic radius generally decreases (with the exception of Ne).

Chapter 10

1. (1)	**11.** (3)	**21.** (1)	**31.** (3)	**41.** (2)
2. (4)	**12.** (2)	**22.** (2)	**32.** (4)	**42.** (2)
3. (1)	**13.** (2)	**23.** (3)	**33.** (4)	**43.** (1)
4. (4)	**14.** (4)	**24.** (2)	**34.** (3)	**44.** (1)
5. (1)	**15.** (3)	**25.** (3)	**35.** (2)	**45.** (4)
6. (2)	**16.** (2)	**26.** (2)	**36.** (3)	**46.** (3)
7. (2)	**17.** (2)	**27.** (2)	**37.** (1)	
8. (2)	**18.** (3)	**28.** (1)	**38.** (4)	
9. (4)	**19.** (3)	**29.** (1)	**39.** (3)	
10. (4)	**20.** (1)	**30.** (1)	**40.** (3)	

Constructed-Response Questions

1. (a) ionic   (b) covalent   (c) covalent   (d) ionic   (e) covalent

2.

H
:O—O:
H

3.

H:N:H
    H

4. Although a CCl_4 molecule has polar bonds, its symmetry makes the entire molecule nonpolar.

5. The molecules of NH_3 are polar while the molecules of Cl_2 are not.

6. In molecules A, B, and C, electrons are shared. In KCl, the electrons are transferred from K to Cl.

7. Valence electrons are lost by potassium and gained by bromine.

8. 6

9. The charge is symmetrically distributed.

10. Either diagram is acceptable.

:Br:Br: :Br—Br:

11. argon (Ar)

Chapter 11

1. (3)	**11.** (2)	**21.** (4)	**31.** (3)	**41.** (3)
2. (2)	**12.** (2)	**22.** (3)	**32.** (2)	**42.** (1)
3. (4)	**13.** (3)	**23.** (1)	**33.** (3)	
4. (1)	**14.** (2)	**24.** (3)	**34.** (4)	
5. (2)	**15.** (3)	**25.** (3)	**35.** (1)	
6. (4)	**16.** (2)	**26.** (3)	**36.** (4)	
7. (4)	**17.** (4)	**27.** (1)	**37.** (1)	
8. (1)	**18.** (4)	**28.** (3)	**38.** (3)	
9. (4)	**19.** (3)	**29.** (2)	**39.** (4)	
10. (3)	**20.** (1)	**30.** (1)	**40.** (3)	

Constructed-Response Questions

1. esterification

2. alcohol

3. Either diagram is acceptable.

$$
\begin{array}{ccccc}
& H & OH & H & \\
& | & | & | & \\
H- & C- & C- & C & -H \\
& | & | & | & \\
& H & H & H &
\end{array}
$$

$$
\begin{array}{ccccc}
& H & H & & H \\
& | & | & & | \\
H- & C- & C- & O- & C-H \\
& | & | & & | \\
& H & H & & H
\end{array}
$$

4. Two substances react to produce one substance.

5. halide (or alkyl halide or halogenalkane)

6. The compounds have the same molecular formulas but different structural formulas.

7. Any of the following diagrams is acceptable.

8. The molecule in Diagram B has only single carbon-to-carbon bonds.

9. The molecules have the same molecular formulas but different structural formulas.

Chapter 12

1. (1)	**10.** (2)	**19.** (3)	**28.** (1)	**37.** (2)
2. (3)	**11.** (3)	**20.** (2)	**29.** (1)	**38.** (1)
3. (2)	**12.** (1)	**21.** (2)	**30.** (3)	**39.** (1)
4. (2)	**13.** (4)	**22.** (2)	**31.** (3)	**40.** (3)
5. (3)	**14.** (3)	**23.** (1)	**32.** (3)	**41.** (2)
6. (2)	**15.** (3)	**24.** (4)	**33.** (1)	**42.** (4)
7. (1)	**16.** (3)	**25.** (4)	**34.** (2)	**43.** (3)
8. (4)	**17.** (2)	**26.** (2)	**35.** (1)	
9. (1)	**18.** (3)	**27.** (3)	**36.** (3)	

Constructed-Response Questions

1. 0.519 M

2. 0.0309 mol

3. 35.6 g

4. 0.728 M

5. Dilute 18.0 mL of the H_2SO_4 with enough water to make 325 mL of solution.

6. 2.26 kg

7. 744 g

8. $X = 0.0946$

9. 0.251 L

10.

Solute	Molar Mass of Solute ($\mathcal{M}$) (g/mol)	Moles of Solute Present (n) (mol)	Mass of Solute Present (m) (g)	Volume of Solution (V)	Molarity (M)
NH_3	17	2.00	34.0	250. mL	8.00
HNO_3	63	1.00	63.0	2.00 L	0.500
H_2SO_4	98	0.500	49.0	5.00 dm^3	0.100
$MgCl_2$	95	1.00	95.0	0.500 L	2.00
$AlPO_4$	122	2.00	244	0.500 L	4.00
$C_6H_{12}O_6$	180	0.250	45.0	0.125 L	2.00
$HC_2H_3O_2$	60	2.00	120.	4000. cm^3	0.500
KI	166	3.00	498	3.00 L	1.00
$Na_2Cr_2O_7$	262	3.00	786	0.333 L	9.00
$B(OH)_3$	62	4.00	248	12.0 L	0.333
H_2SO_4	98	3.00	294	2.00 dm^3	1.50

11. 3.58°C

12. 14.0 C°/m

13. ionic bonds and polar covalent bonds

14. The solution is saturated because some $NH_4Cl(s)$ remained undissolved at the bottom of the test tube.

15. The dissolving of $NH_4Cl(s)$ is endothermic because the temperature of the solution at the end of the procedure is lower than the initial temperature of the water.

16. No solute remained in the bottom of the test tube.

Chapter 13

1. (2)	11. (4)	21. (3)	31. (2)	41. (1)
2. (2)	12. (4)	22. (3)	32. (2)	42. (1)
3. (3)	13. (4)	23. (1)	33. (4)	
4. (4)	14. (4)	24. (3)	34. (1)	
5. (1)	15. (4)	25. (4)	35. (3)	
6. (1)	16. (4)	26. (3)	36. (4)	
7. (3)	17. (1)	27. (1)	37. (2)	
8. (1)	18. (4)	28. (3)	38. (2)	
9. (1)	19. (3)	29. (4)	39. (3)	
10. (4)	20. (1)	30. (3)	40. (3)	

Constructed-Response Questions

1. (a) System is shifted to the right.
   (b) System is shifted to the right.
   (c) System is shifted to the left.
   (d) System is unaffected.
   (e) System is shifted to the left.

2. (a) $K_{eq} = \dfrac{[NO_2]^4[H_2O]^6}{[NH_3]^4[O_2]^7}$

   (b) $K = \dfrac{[H_2O]^3}{[H_2]^3}$

3. (a) $K_{eq} = \dfrac{[C]^3[D]^2}{[A][B]^2}$

   (b) $K_{eq} = 18$

4. $[H_2]_{eq} = 0.763$ mol/L, $[NH_3]_{eq} = 0.158$ mol/L

5. The forward reaction (the dissolving of KNO_3 in water) is endothermic. Raising the temperature favors the endothermic reaction.

6. The system is in equilibrium because the rate of dissolving is equal to the rate of crystallizing.

7. As the concentration of $CO_2(aq)$ decreases, the reaction shifts to the right, lowering the concentration of $H_2CO_3(aq)$.

Chapter 14

1. (4)	**11.** (1)	**21.** (1)	**31.** (3)	**41.** (1)
2. (4)	**12.** (3)	**22.** (2)	**32.** (2)	**42.** (4)
3. (1)	**13.** (4)	**23.** (4)	**33.** (3)	**43.** (3)
4. (2)	**14.** (3)	**24.** (2)	**34.** (3)	**44.** (4)
5. (1)	**15.** (1)	**25.** (3)	**35.** (4)	**45.** (2)
6. (3)	**16.** (4)	**26.** (2)	**36.** (3)	**46.** (4)
7. (2)	**17.** (1)	**27.** (4)	**37.** (4)	**47.** (2)
8. (2)	**18.** (1)	**28.** (1)	**38.** (2)	
9. (4)	**19.** (3)	**29.** (2)	**39.** (1)	
10. (2)	**20.** (1)	**30.** (2)	**40.** (1)	

Constructed-Response Questions

1. (a) $CH_3NH_2(aq) + H_2O(\ell) \rightleftharpoons CH_3NH_3^+(aq) + OH^-(aq)$

   (b) $K_b = \dfrac{[CH_3NH_3^+][OH^-]}{[CH_3NH_2]}$

2. (a) pH = 10.3   (b) pOH = 11.7

3. Amount of NaOH used: (35.2 mL − 23.2 mL) = 12.0 mL

4. Bromthymol blue. The indicator changes color around the end point of the titration (pH = 7.0).

5. $M_A V_A = M_B V_B$

   $$M_B = \frac{M_A V_A}{V_B} = \frac{(1.2 \text{ M}) \bullet (10.0 \text{ mL})}{(12.0 \text{ mL})} = 1.0 \text{ M}$$

6. Multiple trials tend to cancel out experimental errors found in the individual trials.

7. titration

8. $HCl(aq) + NaOH(aq) \rightarrow H_2O(\ell) + NaCl(aq)$

9. 2

10. 0.12 M
11. yellow
12. hydroxide ion (OH^-)
13. 0.86 M

Chapter 15

1. (2)	10. (1)	19. (3)	28. (2)	37. (1)
2. (4)	11. (4)	20. (1)	29. (1)	38. (1)
3. (4)	12. (2)	21. (4)	30. (2)	39. (4)
4. (4)	13. (2)	22. (3)	31. (3)	40. (2)
5. (1)	14. (4)	23. (1)	32. (3)	41. (4)
6. (4)	15. (3)	24. (4)	33. (1)	
7. (2)	16. (2)	25. (4)	34. (4)	
8. (3)	17. (2)	26. (4)	35. (3)	
9. (2)	18. (1)	27. (2)	36. (3)	

Constructed-Response Questions

1. (a) $Zn(s) \rightarrow Zn^{2+}(aq) + 2e^-$
   $2H^+(aq) + 2e^- \rightarrow H_2(g)$
   (b) Zn is *higher* in Reference Table J than H_2; Cu is *lower* in Reference Table J than H_2.
   (c) $Cu^{2+}(aq) + H_2(g) \rightarrow Cu(s) + 2H^+(aq)$

2. (a) $7KOH + 4Zn + 6H_2O + KNO_3 \rightarrow 4K_2Zn(OH)_4 + NH_3$
   (b) $3H_2SO_4 + 6FeSO_4 + KClO_3 \rightarrow 3Fe_2(SO_4)_3 + KCl + 3H_2O$
   (c) $H_3AsO_4 + 4\ Zn + 8HNO_3 \rightarrow AsH_3 + 4Zn(NO_3)_2 + 4H_2O$
   (d) $3H_2O + 3LiOH + P_4 \rightarrow PH_3 + 3LiH_2PO_2$
   (e) $4Zn + 10H^+ + NO_3^- \rightarrow 4Zn^{2+} + NH_4^+ + 3H_2O$
   (f) $H_2O + 2MnO_4^- + 3CN^- \rightarrow 2MnO_2 + 2OH^- + 3CNO^-$

3. (a) $Cu^{2+}(aq) + 2e^- \rightarrow Cu(s)$

(b) 2.000 moles of electrons are needed

(c) $2.000 \; \cancel{mole} \cdot \left(\dfrac{96,470 \; C}{1 \; \cancel{mole}} \right) = 1.929 \times 10^5 C$

(d) $5.000 \; \cancel{h} \cdot \left(\dfrac{3600. \; s}{1.000 \; \cancel{h}} \right) = 1.800 \times 10^4 s$

$$\text{Current} = \frac{1.929 \times 10^5 \; C}{1.800 \times 10^4 s} = 10.72 \; A$$

4.

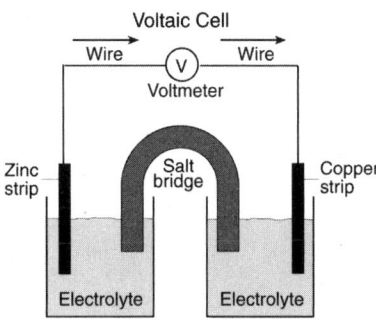

5. $Zn^0 \rightarrow Zn^{2+} + 2e^-$

6. The flow of ions through the salt bridge maintains the electrical neutrality of each half-cell.

7. The oxidation number changes from 0 to +2.

8. The number of moles of electrons lost is equal to the number of moles of electrons gained.

9. Risk: Mercury is toxic.

Benefit: Mercury batteries are small in size.

10. The methane (CH_4) is combined with oxygen (O_2) to form two oxides (H_2O and CO_2).

11. The oxidation number changes from −4 to +4.

12. $Mg \rightarrow Mg^{2+} + 2e^-$

Chapter 16

1. (3)	**8.** (3)	**15.** (4)	**22.** (3)	**29.** (4)
2. (4)	**9.** (2)	**16.** (3)	**23.** (1)	**30.** (4)
3. (2)	**10.** (3)	**17.** (3)	**24.** (4)	**31.** (2)
4. (1)	**11.** (1)	**18.** (1)	**25.** (1)	**32.** (4)
5. (2)	**12.** (3)	**19.** (4)	**26.** (1)	**33.** (1)
6. (1)	**13.** (3)	**20.** (2)	**27.** (2)	**34.** (3)
7. (4)	**14.** (4)	**21.** (2)	**28.** (3)	**35.** (1)

Constructed-Response Questions

1. A
2. E
3. D
4. B

ANSWERING CONSTRUCTED-RESPONSE QUESTIONS

A *constructed-response question* (also known as a *free-response question*) is an examination question that requires the test taker to do more than to choose among several responses or to fill in a blank. You may need to perform numerical calculations, draw or interpret a graph, draw a diagram, provide an extended written response to a question or problem, or write and balance a chemical equation. This appendix is designed to provide you with a number of general guidelines for answering constructed-response questions. *At this time, you should review the material in Chapter 1, pages 23–31.*

A. SOLVING PROBLEMS INVOLVING NUMERICAL CALCULATIONS

If you are presented with a numerical problem, you are advised to:

• Provide the appropriate equation(s).
• Substitute values and units into the equation(s).
• Display the answer, with appropriate units.

Chemistry courses use various types of metric units (SI and non-SI) to describe matter and energy, and you are expected to have some familiarity with them.

You may or may not be penalized if your answer to a calculation has an incorrect number of significant digits. It is always good practice, however, to pay attention to this detail.

You should write your solution in a clear, logical fashion. Many (if not all) teachers become more than a little annoyed if they have to jump visually around your paper in order to find your next calculation. It is also a good idea to identify your answer clearly, either by enclosing it in a box or by writing the word *answer* next to it.

A final caution: If you provide the correct answer but do not show any work, most teachers will award little or no credit for the problem!

Here are a sample problem and a model solution.

PROBLEM
A 5.00-gram object has a density of 4.00 grams per cubic centimeter. Calculate the volume of this object.

SOLUTION

$$d = \frac{m}{V}$$

Rearranging the equation gives

$$V = \frac{m}{d} = \frac{5.00 \text{ g}}{4.00 \text{ g/cm}^3}$$

$$\boxed{V = 1.25 \text{ cm}^3}$$

B. SOLVING PROBLEMS THAT REQUIRE THE GRAPHING OF EXPERIMENTAL DATA

If you are presented with a problem that requires you to graph experimental data, you are advised to:

- Label both axes with the appropriate variables and units.
- Divide the axes so that the given values occupy a significant portion of the graph. This allows a trend in the data to be observed.
- Plot all data points accurately.
- Depending on the instructions given, connect the data points with straight lines or draw a *best-fit* line or curve.
- If a part of the question requires that the slope of a line be calculated, calculate the slope *from the line*, not from individual data points.

Generally a graph should have a title, and the *independent variable* is usually displayed along the *x*-axis.

Here are a sample problem and a step-by-step model solution.

PROBLEM

A student attempts to estimate absolute zero in the following way: He subjects a sample of gas (at constant pressure and mass) to varying temperatures and measures the gas volume at each of the temperatures. The accompanying table contains his experimental data.

Temperature/°C	Volume/mL
−100	128
−60	148
−40	156
0	204
40	222
80	272
160	310

(a) Using axes that are appropriately labeled and scaled, draw a graph that accurately displays the student's data.

(b) Estimate the student's value for absolute zero by extending the graph to the Celsius temperature at which the volume of the gas is 0 milliliter.

SOLUTION

(a) The first step is to construct an appropriate set of axes if one is not provided. We will assume that you must start from scratch. Since temperature is the independent variable, you need to place it along the x-axis. Also, since absolute zero is $-273.15°C$, you must scale the axes properly. Here is one appropriate set of axes:

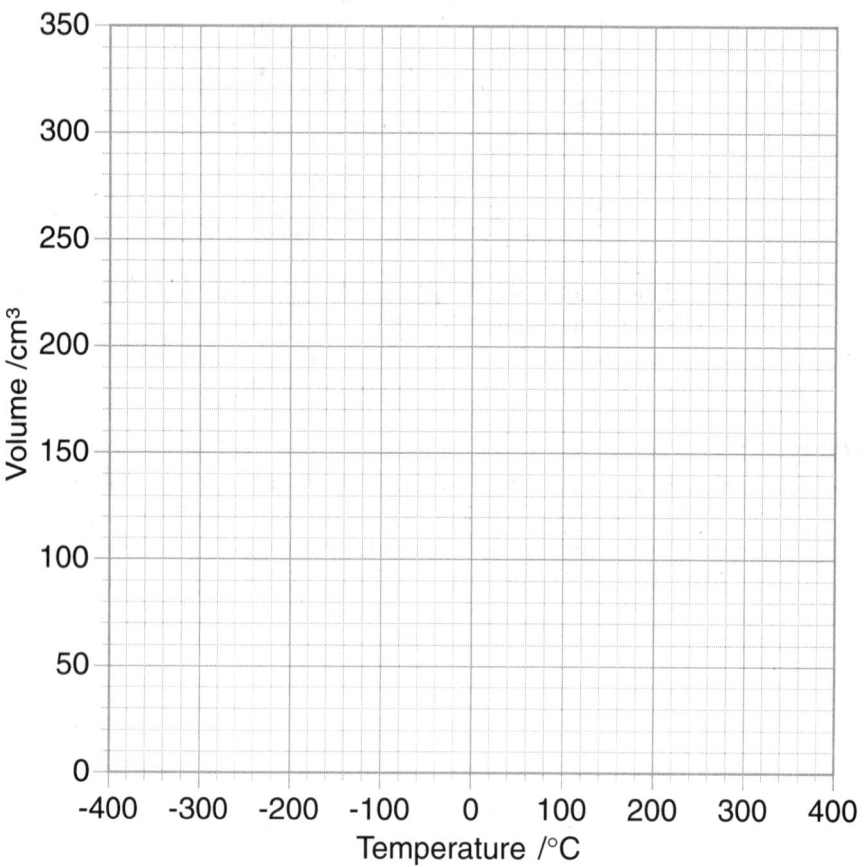

Second, you must plot the data points carefully on the axes as shown below:

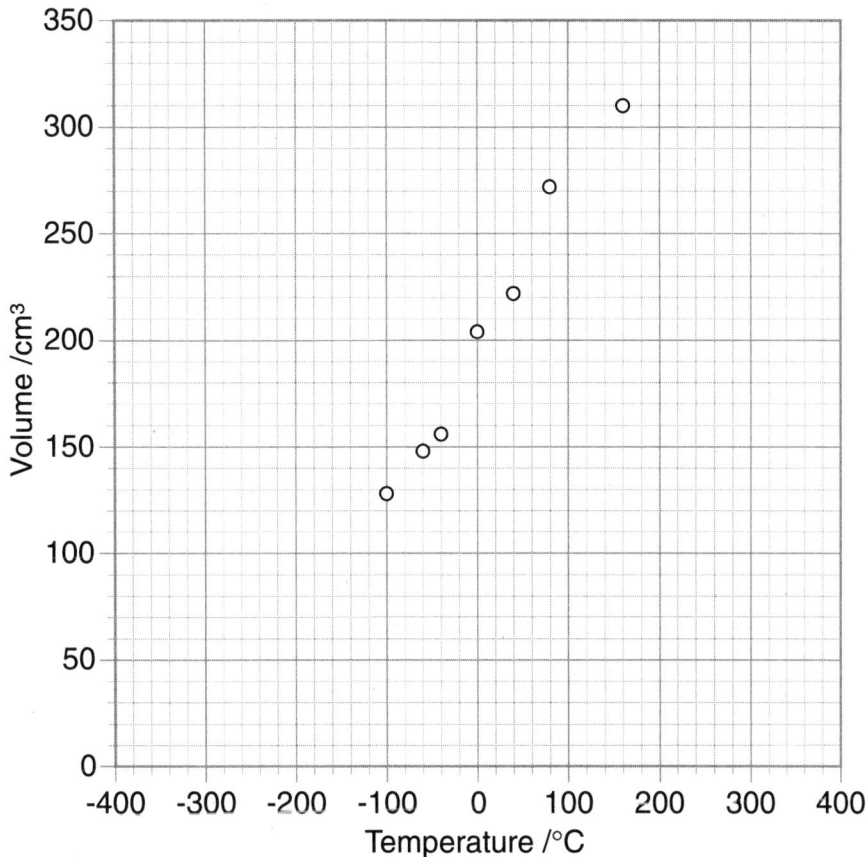

Volume versus Temperature

Your third task is to draw the graph. Examining *all* of the plotted points, you note that they fall approximately on a *straight line*. Therefore, the next step is to draw a *best-fit* straight-line graph. This is a graph in which the data points are most closely distributed on both sides of the line. The accompanying graph shows the best-fit straight line.

Volume versus Temperature

Note that the graph is extended beyond the data points. Such extensions are known as *extrapolated data*, and they are based on the assumption that the gas will continue to behave as it did within the experimental range for which the student collected data.

(b) Your final task is to inspect the graph closely and to estimate absolute zero. The calculated value is –263°C. (This corresponds to an experimental error of 3.7%.)

C. SOLVING PROBLEMS REQUIRING THE DRAWING OF A DIAGRAM

Since you may be presented with a problem requiring you to draw a diagram, you are advised to bring a straightedge with you. Then:

- Use the straightedge as needed.
- Draw the diagram neatly and accurately, and label it clearly.

The accompanying diagram represents a zinc–copper electrochemical cell containing an agar–KCl salt bridge. This is the type of diagram you may be asked to draw as part of an examination question.

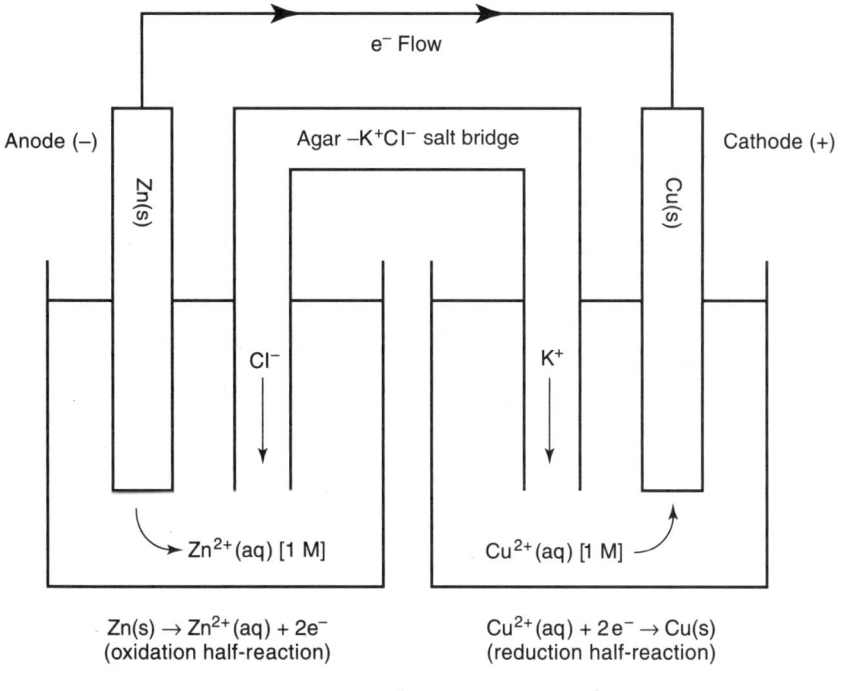

Zn(s) → Zn²⁺(aq) + 2e⁻ (oxidation half-reaction)

Cu²⁺(aq) + 2e⁻ → Cu(s) (reduction half-reaction)

Net reaction: Zn(s) + Cu²⁺(aq) → Cu(s) + Zn²⁺(aq)

Remember: your teacher has every right to expect you to draw an *accurate*, *complete*, and *neat* diagram!

D. WRITING AN EXTENDED RESPONSE TO AN EXAMINATION QUESTION

If the examination question requires an extended response, you are advised to:

- Use complete, clear sentences that make sense to the reader. You may decide to structure your response in outline or in paragraph form.
- Use correct chemistry in your explanations.

A sample question and its model answer are given below.

QUESTION
Describe in detail the technique for determining the concentration of a dilute hydrochloric acid solution using a dilute sodium hydroxide solution of known concentration.

Name any equipment or other chemicals that you would use. *You need not describe any mathematical calculations.*

MODEL ANSWER
Since this question requires an extended answer containing a series of steps, we decide to use an outline form.

The technique is called *titration*, and it is described in the steps given below.

- Place a known volume of the hydrochloric acid solution in a beaker of suitable size.
- Add a drop or two of an appropriatee acid–base indicator such as phenolphthalein.
- Pour the sodium hydroxide solution carefully into a burette, using a small funnel.
- Open the stopcock to allow the trapped air to escape. Then close the stopcock, and wipe the tip of the burette with a tissue to remove any clinging liquid.
- Record the initial volume in the burette.
- Add the base slowly to the acid solution, with continuous stirring, until the phenolphthalein just changes from colorless to faint pink.
- Record the final volume in the burette.
- Repeat the experiment at least once.
- Rinse the apparatus with water to remove all traces of acid and base.
- Calculate the concentration of the hydrochloric acid solution.

E. WRITING AND BALANCING EQUATIONS IN A CHEMISTRY EXAMINATION

For all of the suggestions that follow, the reaction that occurs between aqueous solutions of sodium sulfate and barium nitrate is used as an example and is referred to as the given reaction.

• If you are asked to write a *word equation*, be certain to include the correct names of the reactants and products and their phases in the reaction.

For the given reaction, the word equation is

barium nitrate(aq) + sodium sulfate(aq) → barium sulfate(s) + sodium nitrate(aq)

• If you are asked to write a *balanced* equation, you are usually expected to balance using *smallest whole-number coefficients*.

For the given reaction, the balanced equation is

$$Ba(NO_3)_2(aq) + Na_2SO_4(aq) \rightarrow BaSO_4(s) + 2NaNO_3(aq)$$

• If you are asked to write an *ionic* equation occurring in aqueous solution, you must reduce everything to its component ions except *insoluble* compounds, such as $BaSO_4(s)$, and (of course!) *covalently* bonded substances, such as $H_2O(\ell)$.

For the given reaction, the ionic equation is

$$Ba^{2+}(aq) + 2NO_3^-(aq) + 2Na^+(aq) + SO_4^{2-}(aq) \rightarrow$$
$$BaSO_4(s) + 2Na^+(aq) + 2NO_3^-(aq)$$

• If you are asked to write a *net* ionic equation, you should omit all *spectator ions*, that is, all ions appearing *unchanged on both sides of the equation*.

For the given reaction, the net ionic equation is

$$Ba^{2+}(aq) + SO_4^{2-}(aq) \rightarrow BaSO_4(s)$$

Appendix 5

THE NEW YORK STATE REGENTS EXAMINATION IN CHEMISTRY

The New York State Regents Chemistry—Physical Setting examination consists of questions that are based on twelve subject areas as follows:

- Math Skills
- Reading Skills
- Atomic Concepts
- Periodic Table
- Moles/Stoichiometry
- Chemical Bonding
- Physical Behavior of Matter
- Kinetics/Equilibrium
- Organic Chemistry
- Oxidation-Reduction
- Acids, Bases, and Salts
- Nuclear Chemistry

The Chemistry—Physical Setting examination takes 3 hours and includes four parts: A, B-1, B-2, and C. You should be prepared to answer questions in multiple-choice format as well as questions that require more extended responses. Parts A, B-1, B-2, and C are divided as follows:

Part	Question Type(s)	Description of Questions
A	Multiple-choice	*Content-based* questions that will test your knowledge and understanding of the material contained in the NYS Chemistry Core.
B-1	Multiple-choice	*Content- and skills-based* questions that will test your ability to apply, analyze, and evaluate the NYS Chemistry Core material.
B-2	Constructed-response	*Content- and skills-based* questions that will test your ability to apply, analyze, and evaluate the NYS Chemistry Core material.
C	Constructed-response and/or *extended* constructed-response	*Content-based and application* questions that will test your ability to apply your knowledge of chemistry concepts and skills to "real-world" situations.

You may be required to graph data, complete a data table, label or draw diagrams, design experiments, make calculations, or write short or more extended responses. In addition you may be required to hypothesize, interpret, analyze, evaluate data, or apply your scientific knowledge and skills to real-world situations.

Some of the questions will require use of the New York State Regents Reference Tables for Chemistry in Appendix 1, pages 495–510.

Note that you will be required to answer ALL of the questions on the Regents examination.

Examination
June 2019
Chemistry—Physical Setting

PART A

Answer all questions in this part.

Directions (1–30): For *each* statement or question, write in the answer space the *number* of the word or expression that, of those given, best completes the statement or answers the question. Some questions may require the use of the *2011 Edition Reference Tables for Physical Setting/Chemistry*.

1 Which particles are found in the nucleus of an argon atom?

(1) protons and electrons
(2) positrons and neutrons
(3) protons and neutrons
(4) positrons and electrons

1 _____

2 The diagram below represents a particle traveling through an electric field.

An electric field exists between the two plates.

Which particle remains undeflected when passing through this electric field?

(1) proton
(2) electron
(3) neutron
(4) positron

2 _____

3 The mass of an electron is

(1) equal to the mass of a proton
(2) equal to the mass of a neutron
(3) greater than the mass of a proton
(4) less than the mass of a neutron

3 _____

4 Compared to the energy of an electron in the second shell of an atom of sulfur, the energy of an electron in the

(1) first shell is lower
(2) first shell is the same
(3) third shell is lower
(4) third shell is the same

4 _____

5 In the ground state, an atom of which element has seven valence electrons?

(1) sodium
(2) phosphorus
(3) nitrogen
(4) fluorine

5 _____

6 Which information is sufficient to differentiate a sample of sodium from a sample of silver?

(1) the mass of each sample
(2) the volume of each sample
(3) the reactivity of each sample with water
(4) the phase of each sample at room temperature

6 _____

7 Graphite and diamond are two forms of solid carbon at STP. These forms have

(1) different molecular structures and different properties
(2) different molecular structures and the same properties
(3) the same molecular structures and different properties
(4) the same molecular structures and the same properties   7 _____

8 As the first five elements in Group 14 are considered in order from top to bottom, there are changes in both the

(1) number of valence shell electrons and number of first shell electrons
(2) electronegativity values and number of first shell electrons
(3) number of valence shell electrons and atomic radii
(4) electronegativity values and atomic radii   8 _____

9 Which statement explains why NaBr is classified as a compound?

(1) Na and Br are chemically combined in a fixed proportion.
(2) Na and Br are both nonmetals.
(3) NaBr is a solid at 298 K and standard pressure.
(4) NaBr dissolves in H_2O at 298 K.   9 _____

10 Which two terms represent types of chemical formulas?

(1) fission and fusion
(2) oxidation and reduction
(3) empirical and structural
(4) endothermic and exothermic   10 _____

11 During all chemical reactions, charge, mass and energy are

(1) condensed
(2) conserved
(3) decayed
(4) decomposed

11 _____

12 The degree of polarity of a covalent bond between two atoms is determined by calculating the difference in their

(1) atomic radii
(2) melting points
(3) electronegativities
(4) ionization energies

12 _____

13 Which substance can *not* be broken down by a chemical change?

(1) ammonia
(2) magnesium
(3) methane
(4) water

13 _____

14 Which statement describes the components of a mixture?

(1) Each component gains new properties.
(2) Each component loses its original properties.
(3) The proportions of components can vary.
(4) The proportions of components cannot vary.

14 _____

15 Table sugar can be separated from a mixture of table sugar and sand at STP by adding

(1) sand, stirring, and distilling at 100.°C
(2) sand, stirring, and filtering
(3) water, stirring, and distilling at 100.°C
(4) water, stirring, and filtering

15 _____

16 Which statement describes the particles of an ideal gas, based on the kinetic molecular theory?

(1) The volume of the particles is considered negligible.
(2) The force of attraction between the particles is strong.
(3) The particles are closely packed in a regular, repeating pattern.
(4) The particles are separated by small distances, relative to their size.　　16 _____

17 During which two processes does a substance release energy?

(1) freezing and condensation
(2) freezing and melting
(3) evaporation and condensation
(4) evaporation and melting　　17 _____

18 Based on Table I, which compound dissolves in water by an exothermic process?

(1) NaCl
(2) NaOH
(3) NH_4Cl
(4) NH_4NO_3　　18 _____

19 At STP, which property of a molecular substance is determined by the arrangement of its molecules?

(1) half-life
(2) molar mass
(3) physical state
(4) percent composition　　19 _____

20 Equilibrium can be reached by

(1) physical changes, only
(2) nuclear changes, only
(3) both physical changes and chemical changes
(4) both nuclear changes and chemical changes　　20 _____

21 Which value is defined as the difference between the potential energy of the products and the potential energy of the reactants during a chemical change?

(1) heat of fusion
(2) heat of reaction
(3) heat of deposition
(4) heat of vaporization   21 _____

22 The effect of a catalyst on a chemical reaction is to provide a new reaction pathway that results in a different

(1) potential energy of the products
(2) heat of reaction
(3) potential energy of the reactants
(4) activation energy   22 _____

23 Chemical systems in nature tend to undergo changes toward

(1) lower energy and lower entropy
(2) lower energy and higher entropy
(3) higher energy and lower entropy
(4) higher energy and higher entropy   23 _____

24 The atoms of which element bond to one another in chains, rings, and networks?

(1) barium
(2) carbon
(3) iodine
(4) mercury   24 _____

25 What is the general formula for the homologous series that includes ethene?

(1) C_nH_{2n}
(2) C_nH_{2n-6}
(3) C_nH_{2n-2}
(4) C_nH_{2n+2} 25 _____

26 When an F atom becomes an F⁻ ion, the F atom

    (1) gains a proton
    (2) loses a proton
    (3) gains an electron
    (4) loses an electron   26 _____

27 Which substance is an Arrhenius base?

    (1) HNO_3
    (2) H_2SO_3
    (3) $Ca(OH)_2$
    (4) CH_3COOH 27 _____

28 In which type of nuclear reaction do two light nuclei combine to produce a heavier nucleus?

    (1) positron emission
    (2) gamma emission
    (3) fission
    (4) fusion   28 _____

29 Using equal masses of reactants, which statement describes the relative amounts of energy released during a chemical reaction and a nuclear reaction?

    (1) The chemical and nuclear reactions release equal amounts of energy.
    (2) The nuclear reaction releases half the amount of energy of the chemical reaction.
    (3) The chemical reaction releases more energy than the nuclear reaction.
    (4) The nuclear reaction releases more energy than the chemical reaction.   29 _____

30 The ratio of the mass of U-238 to the mass of Pb-206 can be used to

    (1) diagnose thyroid disorders
    (2) diagnose kidney function
    (3) date geological formations
    (4) date once-living things   30 _____

PART B–1

Answer all questions in this part.

Directions (31–50): For *each* statement or question, write in the answer space the *number* of the word or expression that, of those given, best completes the statement or answers the question. Some questions may require the use of the *2011 Edition Reference Tables for Physical Setting/Chemistry.*

31 The bright-line spectra of four elements, *G, J, L,* and *M,* and a mixture of *at least two* of these elements is given below.

Bright-Line Spectra

Wavelength (nm)

Which elements are present in the mixture?
(1) *G* and *J*   (3) *M, J,* and *G*
(2) *G* and *L*   (4) *M, J,* and *L*   31 _____

32 Which electron configuration represents an atom of chlorine in an excited state?

(1) 2-8-7-2   (3) 2-8-8

(2) 2-8-7   (4) 2-7-8   32 _____

33 A student measures the mass and volume of a sample of aluminum at room temperature, and calculates the density of Al to be 2.85 grams per cubic centimeter. Based on Table S, what is the percent error for the student's calculated density of Al?

(1) 2.7%   (3) 5.6%

(2) 5.3%   (4) 95%   33 _____

34 Magnesium and calcium have similar chemical properties because their atoms in the ground state have

(1) equal numbers of protons and electrons
(2) equal numbers of protons and neutrons
(3) two electrons in the first shell
(4) two electrons in the outermost shell   34 _____

35 As the elements in Period 2 of the Periodic Table are considered in order from left to right, which property generally *decreases*?

(1) atomic radius
(2) electronegativity
(3) ionization energy
(4) nuclear charge   35 _____

36 Given the balanced equation for the reaction of butane and oxygen:

$$2C_4H_{10} + 13O_2 \rightarrow 8CO_2 + 10H_2O + energy$$

How many moles of carbon dioxide are produced when 5.0 moles of butane react completely?

(1) 5.0 mol
(2) 10. mol
(3) 20. mol
(4) 40. mol   36 _____

37 What is the percent composition by mass of nitrogen in the compound N_2H_4 (gram-formula mass = 32 g/mol)?

(1) 13%   (3) 88%

(2) 44%   (4) 93%   37 _____

38 Which ion in the ground state has the same electron configuration as an atom of neon in the ground state?

(1) Ca^{2+} (3) Li^+

(2) Cl^- (4) O^{2-} 38 _____

39 The molar masses and boiling points at standard pressure for four compounds are given in the table below.

Compound	Molar Mass (g/mol)	Boiling Point (K)
HF	20.01	293
HCl	36.46	188
HBr	80.91	207
HI	127.91	237

Which compound has the strongest intermolecular forces?

(1) HF   (3) HBr

(2) HCl   (4) HI   39 _____

40 Which particle model diagram represents xenon at STP?

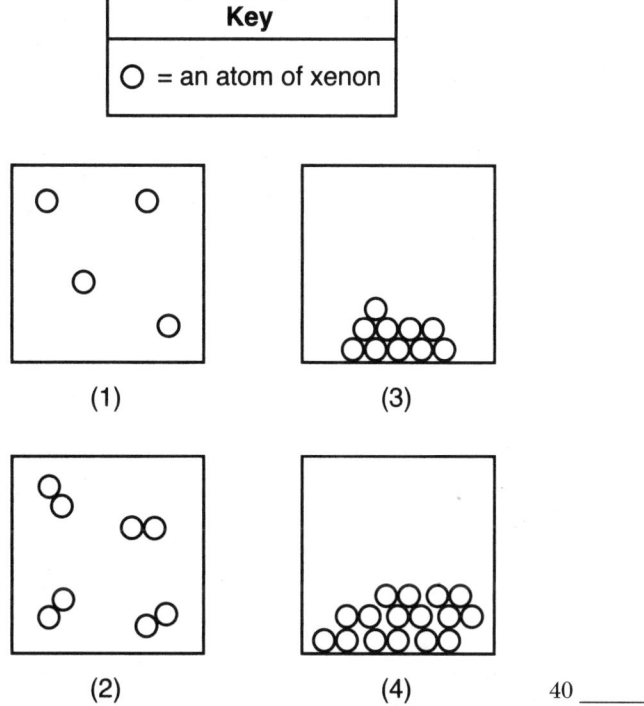

41 What is the amount of heat absorbed when the temperature of 75 grams of water increases from 20.°C to 35°C?

(1) 1100 J   (3) 6300 J
(2) 4700 J   (4) 11 000 J   41 _____

42 Which sample of HCl(aq) reacts at the fastest rate with a 1.0-gram sample of iron filings?

(1) 10. mL of 1 M HCl(aq) at 10.°C
(2) 10. mL of 1 M HCl(aq) at 25°C
(3) 10. mL of 3 M HCl(aq) at 10.°C
(4) 10. mL of 3 M HCl(aq) at 25°C   42 _____

43 Given the equation representing a system at equilibrium:

$$N_2O_4(g) \rightleftharpoons 2NO_2(g)$$

Which statement describes the concentration of the two gases in this system?

(1) The concentration of $N_2O_4(g)$ must be less than the concentration of $NO_2(g)$.
(2) The concentration of $N_2O_4(g)$ must be greater than the concentration of $NO_2(g)$.
(3) The concentration of $N_2O_4(g)$ and the concentration of $NO_2(g)$ must be equal.
(4) The concentration of $N_2O_4(g)$ and the concentration of $NO_2(g)$ must be constant.

43 _____

44 Given the equation representing a system at equilibrium:

$$PCl_5(g) + energy \rightleftharpoons PCl_3(g) + Cl_2(g)$$

Which change will cause the equilibrium to shift to the right?

(1) adding a catalyst
(2) adding more $PCl_3(g)$
(3) increasing the pressure
(4) increasing the temperature

44 _____

45 Given the formula representing a molecule:

$$
\begin{array}{c}
\text{H\ \ H\ \ H\ \ H\ \ H} \\
|\ \ \ |\ \ \ |\ \ \ |\ \ \ | \\
\text{H}-\text{C}-\text{C}-\text{C}-\text{C}-\text{C}-\text{N} \overset{\nearrow \text{H}}{\searrow_{\text{H}}} \\
|\ \ \ |\ \ \ |\ \ \ |\ \ \ | \\
\text{H\ \ H\ \ H\ \ H\ \ H}
\end{array}
$$

A chemical name for this compound is

(1) pentanone
(2) 1-pentanol
(3) 1-pentanamine
(4) pentanamide

45 _____

46 Given the formula of a compound:

$$H-C\equiv C-\overset{\displaystyle H}{\underset{\displaystyle H}{\overset{|}{\underset{|}{C}}}}-\overset{\displaystyle H}{\underset{\displaystyle H}{\overset{|}{\underset{|}{C}}}}-H$$

This compound is classified as an

(1) aldehyde   (3) alkyne

(2) alkene   (4) alcohol   46 _____

47 Which equation represents fermentation?

(1) $C_2H_4 + H_2O \rightarrow CH_3CH_2OH$

(2) $C_2H_4 + HCl \rightarrow CH_3CH_2Cl$

(3) $C_6H_{12}O_6 \rightarrow 2CH_3CH_2OH + 2CO_2$

(4) $2CH_3CHO \rightarrow C_3H_5CHO + H_2O$ 47 _____

48 Given the equation representing a reaction:

$3CuCl_2(aq) + 2Al(s) \rightarrow 3Cu(s) + 2AlCl_3(aq)$

The oxidation number of copper changes from

(1) +1 to 0   (3) +2 to +1

(2) +2 to 0   (4) +6 to +3   48 _____

49 Given the equation representing a reversible reaction:

$CH_3COOH(aq) + H_2O(\ell) \rightleftharpoons CH_3COO^-(aq) + H_3O^+(aq)$

According to one acid-base theory, the two H^+ donors in the equation are

(1) CH_3COOH and H_2O

(2) CH_3COOH and H_3O^+

(3) CH_3COO^- and H_2O

(4) CH_3COO^- and H_3O^+ 49 _____

50 Which nuclear equation represents a spontaneous decay?

(1) $^{222}_{86}\text{Rn} \rightarrow {}^{218}_{84}\text{Po} + {}^{4}_{2}\text{He}$

(2) $^{27}_{13}\text{Al} + {}^{4}_{2}\text{He} \rightarrow {}^{30}_{15}\text{P} + {}^{1}_{0}\text{n}$

(3) $^{235}_{92}\text{U} + {}^{1}_{0}\text{n} \rightarrow {}^{139}_{56}\text{Ba} + {}^{94}_{36}\text{Kr} + 3{}^{1}_{0}\text{n}$

(4) $^{7}_{3}\text{Li} + {}^{1}_{1}\text{H} \rightarrow {}^{4}_{2}\text{He} + {}^{4}_{2}\text{He}$

50 _____

PART B–2

Answer all questions in this part.

Directions (51–65): Record your answers on the answer sheet provided. Some questions may require the use of the *2011 Edition Reference Tables for Physical Setting/Chemistry*.

51 Draw a structural formula for methanal.   [1]

Base your answers to questions 52 through 54 on the information below and on your knowledge of chemistry.

The atomic mass and natural abundance of the naturally occurring isotopes of hydrogen are shown in the table below.

Naturally Occurring Isotopes of Hydrogen

Isotope	Common Name of Isotope	Atomic Mass (u)	Natural Abundance (%)
H-1	protium	1.0078	99.9885
H-2	deuterium	2.0141	0.0115
H-3	tritium	3.0160	negligible

The isotope H-2, also called deuterium, is usually represented by the symbol "D." Heavy water forms when deuterium reacts with oxygen, producing molecules of D_2O.

52 Explain, in terms of subatomic particles, why atoms of H-1, H-2, and H-3 are each electrically neutral.   [1]

53 Determine the formula mass of heavy water, D_2O.   [1]

54 Based on Table *N*, identify the decay mode of tritium.   [1]

Base your answers to questions 55 through 57 on the information below and on your knowledge of chemistry.

At 23°C, 85.0 grams of $NaNO_3(s)$ are dissolved in 100. grams of $H_2O(\ell)$.

55 Convert the temperature of the $NaNO_3(s)$ to kelvins.   [1]

56 Based on Table G, determine the additional mass of $NaNO_3(s)$ that must be dissolved to saturate the solution at 23°C.   [1]

57 State what happens to the boiling point and freezing point of the solution when the solution is diluted with an additional 100. grams of $H_2O(\ell)$.   [1]

Base your answers to questions 58 through 61 on the information below and on your knowledge of chemistry.

A 200.-milliliter sample of $CO_2(g)$ is placed in a sealed, rigid cylinder with a movable piston at 296 K and 101.3 kPa.

58 State a change in temperature and a change in pressure of the $CO_2(g)$ that would cause it to behave more like an ideal gas.   [1]

59 Determine the volume of the sample of $CO_2(g)$ if the temperature and pressure are changed to 336 K and 152.0 kPa.   [1]

60 State, in terms of both the frequency and force of collisions, what would result from decreasing the temperature of the original sample of $CO_2(g)$, at constant volume.   [1]

61 Compare the mass of the original 200.-milliliter sample of $CO_2(g)$ to the mass of the $CO_2(g)$ sample when the cylinder is adjusted to a volume of 100. milliliters. [1]

Base your answers to questions 62 through 65 on the information below and on your knowledge of chemistry.

Cobalt-60 is an artificial isotope of Co-59. The incomplete equation for the decay of cobalt-60, including beta and gamma emissions, is shown below.

$$_{27}^{60}Co \rightarrow X + _{-1}^{0}e + _{0}^{0}\gamma$$

62 Explain, in terms of *both* protons and neutrons, why Co-59 and Co-60 are isotopes of cobalt. [1]

63 Compare the penetrating power of the beta and gamma emissions. [1]

64 Complete the nuclear equation, *on your answer sheet*, for the decay of cobalt-60 by writing a notation for the missing product. [1]

65 Based on Table N, determine the total time required for an 80.00-gram sample of cobalt-60 to decay until only 10.00 grams of the sample remain unchanged. [1]

PART C

Answer all questions in this part.

Directions (66–85): Record your answers on the answer sheet provided. Some questions may require the use of the *2011 Edition Reference Tables for Physical Setting/Chemistry.*

Base your answers to questions 66 through 69 on the information below and on your knowledge of chemistry.

During a laboratory activity, appropriate safety equipment was used and safety procedures were followed. A laboratory technician heated a sample of solid $KClO_3$ in a crucible to determine the percent composition by mass of oxygen in the compound. The unbalanced equation and the data for the decomposition of solid $KClO_3$ are shown below.

$$KClO_3(s) \rightarrow KCl(s) + O_2(g)$$

Lab Data and Calculated Results

Object or Material	Mass (g)
empty crucible and cover	22.14
empty crucible, cover, and $KClO_3$	24.21
$KClO_3$	2.07
crucible, cover, and KCl after heating	23.41
KCl	?
O_2	0.80

66 Write a chemical name for the compound that decomposed. [1]

67 Based on the lab data, show a numerical setup to determine the number of moles of O_2 produced. Use 32 g/mol as the gram-formula mass of O_2. [1]

68 Based on the lab data, determine the mass of KCl produced in the reaction. [1]

69 Balance the equation *on your answer sheet* for the decomposition of $KClO_3$, using the smallest whole-number coefficients. [1]

Base your answers to questions 70 through 73 on the information below and on your knowledge of chemistry.

A bottled water label lists the ions dissolved in the water. The table below lists the mass of some ions dissolved in a 500.-gram sample of the bottled water.

Ions in 500. g of Bottled Water

Ion Formula	Mass (g)
Ca^{2+}	0.040
Mg^{2+}	0.013
Na^+	0.0033
SO_4^{2-}	0.0063
HCO_3^-	0.180

70 State the number of significant figures used to express the mass of hydrogen carbonate ions in the table above. [1]

71 Based on Table F, write the formula of the ion in the bottled water table that would form the *least* soluble compound when combined with the sulfate ion. [1]

72 Show a numerical setup for calculating the parts per million of the Na^+ ions in the 500.-gram sample of the bottled water.   [1]

73 Compare the radius of a Mg^{2+} ion to the radius of a Mg atom.   [1]

Base your answers to questions 74 through 77 on the information below and on your knowledge of chemistry.

Ethyl ethanoate is used as a solvent for varnishes and in the manufacture of artificial leather. The formula below represents a molecule of ethyl ethanoate.

```
    H  O      H  H
    |  ||     |  |
H—C—C—O—C—C—H
    |         |  |
    H         H  H
```

74 Identify the element in ethyl ethanoate that makes it an organic compound.   [1]

75 Write the empirical formula for this compound.   [1]

76 Write the name of the class of organic compounds to which this compound belongs.   [1]

77 Determine the number of electrons shared in the bond between a hydrogen atom and a carbon atom in the molecule.   [1]

Base your answers to questions 78 through 80 on the information below and on your knowledge of chemistry.

An operating voltaic cell has magnesium and silver electrodes. The cell and the ionic equation representing the reaction that occurs in the cell are shown below.

Voltaic Cell

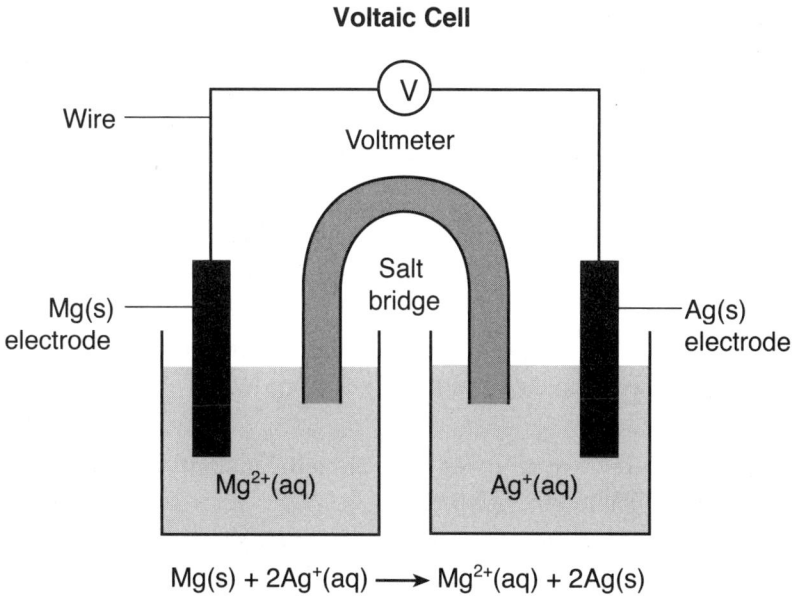

$$Mg(s) + 2Ag^+(aq) \longrightarrow Mg^{2+}(aq) + 2Ag(s)$$

78 State the purpose of the salt bridge in this cell.   [1]

79 Write a balanced equation for the half-reaction that occurs at the magnesium electrode in this cell.   [1]

80 Explain, in terms of electrical energy, how electrolysis reactions differ from voltaic cell reactions.   [1]

Base your answers to questions 81 through 85 on the information below and on your knowledge of chemistry.

In a laboratory investigation, an HCl(aq) solution with a pH value of 2 is used to determine the molarity of a KOH(aq) solution. A 7.5-milliliter sample of the KOH(aq) is exactly neutralized by 15.0 milliliters of the 0.010 M HCl(aq). During this laboratory activity, appropriate safety equipment is used and safety procedures are followed.

81 Determine the pH value of a solution that is ten times *less* acidic than the HCl(aq) solution.   [1]

82 State the color of the indicator bromcresol green if it is added to a sample of the KOH(aq) solution.   [1]

83 Complete the equation *on your answer sheet* by writing the chemical formula for *each* product.   [1]

84 Show a numerical setup for calculating the molarity of the KOH solution.   [1]

85 Explain, in terms of aqueous ions, why 15.0 mL of a 1.0 M HCl(aq) solution is a better conductor of electricity than 15.0 mL of a 0.010 M HCl(aq) solution. [1]

PART B–2

51

52 _____

_____

_____

53 _____ **u**

54 _____

55 _____ **K**

56 _____ **g**

57 Boiling point: _____

Freezing point: _____

58 Temperature: _____

Pressure: _____

59 _____ **mL**

60 _____

_____

_____

61 _____

_____

_____

62 _____

_____

_____

63 _____

_____

_____

64 $_{27}^{60}\text{Co} \rightarrow$ _____ $+ \ _{-1}^{0}\text{e} + \ _{0}^{0}\gamma$

65 _____ **y**

PART C

66 _____

67

68 _____ **g**

69 _____ $KClO_3(s) \rightarrow$ _____ $KCl(s)$ + _____ $O_2(g)$

70 _____

71 _____

72

73 _____

_____

_____

74 _____

75 _____

76 _____

77 _____

78 _____

_____

_____

79 _____

80 _____

_____

_____

81 _____

82 _____

83 HCl(aq) + KOH(aq) → _____ + _____

84

85 _____

_____

_____

Answers
June 2019

Chemistry—Physical Setting

Answer Key

PART A

1. 3	7. 1	13. 2	19. 3	25. 1
2. 3	8. 4	14. 3	20. 3	26. 3
3. 4	9. 1	15. 4	21. 2	27. 3
4. 1	10. 3	16. 1	22. 4	28. 4
5. 4	11. 2	17. 1	23. 2	29. 4
6. 3	12. 3	18. 2	24. 2	30. 3

PART B–1

31. 2	36. 3	41. 2	46. 3
32. 4	37. 3	42. 4	47. 3
33. 3	38. 4	43. 4	48. 2
34. 4	39. 1	44. 4	49. 2
35. 1	40. 1	45. 3	50. 1

INDEX